GEOMETRIC DIMENSIONG AND TOLERANCING TEXTBOOK AND WORKBOOK (per ASME Y14.5-2018)

James D. Meadows

GEOMETRIC DIMENSIONING AND TOLERANCING

Bibliography

ASME Y14.5-2018 Dimensioning and Tolerancing

ASME Y14.5-2009 Dimensioning and Tolerancing

ASME Y14.5M-1994 Dimensioning and Tolerancing

ASME Y14.5.1M-2018 Mathematical Definition of Dimensioning and Tolerancing Principles

ASME Y14.43-2011 Dimensioning and Tolerancing Principles for Gages and Fixtures

ASME Y14.1 Decimal Inch Drawing Sheet Size and Format

ASME Y14.1M Metric Drawing Sheet Size and Format

ASME Y14.2 Line Conventions and Lettering

ASME Y14.3 Orthographic and Pictorial Views

ASME Y14.24 Types and Applications of Engineering Drawings

ASME Y14.34 Associated Lists

ASME Y14.35 Revision of Engineering Drawings and Associated Documents

ASME Y14.36 Surface Texture Symbols

ASME Y14.38 Abbreviations and Acronyms for Use on Drawings and Related Documents

ASME Y14.41 Digital Product Definition Data Practices

ASME Y14.100 Engineering Drawing Practices

ASME Y14.8 Castings and Forgings

All of the Y14 standards above are published by The American Society of Mechanical Engineers, New York, New York.

German Democratic Standard TGL39-097. Methods of Measuring Deviations from Cylindricity, Germany.

ISO 1101 Geometrical Tolerancing-Tolerances of form, orientation, location and run-out. International Organization of Standards.

Geometric Dimensioning and Tolerancing-Applications, Analysis, Gauging & Measurement [per ASME Y14.5-2018] by James D. Meadows co-published by The American Society of Mechanical Engineers, New York

Geometric Dimensioning and Tolerancing-Applications, Analysis & Measurement [per ASME Y14.5-2009] by James D. Meadows published by The American Society of Mechanical Engineers, New York

Workbook and Answerbook for Geometric Dimensioning and Tolerancing [per ASME Y14.5-2009] by James D. Meadows published in 2009 by James D. Meadows & Associates, Inc.

Tolerance Stack-Up Analysis for Plus and Minus Tolerancing and Geometric Dimensioning and Tolerancing by James D. Meadows published in 2010 by James D. Meadows & Associates, Inc.

Geometric Dimensioning and Tolerancing, Applications and Techniques for Use in Design, Manufacturing, and Inspection by James D. Meadows published in 1995 by Marcel Dekker, Inc., New York, New York.

Geometric Dimensioning and Tolerancing Workbook/Answerbook by James D. Meadows published in 1997 by Marcel Dekker, Inc., New York, New York.

Geometric Dimensioning and Tolerancing in 2007 by James D. Meadows published in 2007 by James D. Meadows & Associates, Inc., Hendersonville, TN.

Geometric Dimensioning and Tolerancing in 2007 Workbook and Answerbook by James D. Meadows published in 2007 by James D. Meadows & Associates, Inc., Hendersonville, TN.

Measurement of Geometric Tolerances in Manufacturing by James D. Meadows published in 1998 by Marcel Dekker, Inc., New York, New York.

Differences and Similarities between ASME and ISO Dimensioning and Tolerancing Standards by James D. Meadows published in 2006 by James D. Meadows & Associates, Inc.

DataMyte Handbook a practical guide to computerized data collection for Statistical Process Control by DataMyte Business Allen-Bradley Company, Inc.

Statistical Tolerancing, A Solution to Procurement Problems-US Army Tank Command.

GEOMETRIC DIMENSIONG AND TOLERANCING TEXTBOOK AND WORKBOOK (per ASME Y14.5-2018)

JAMES D. MEADOWS

James D. Meadows & Associates, Inc.

Hendersonville, Tennessee

Published and distributed by:
James D. Meadows & Associates, Inc.
170 E. Main, D-137
Hendersonville, TN 37075
Phone: (615) 924-3600
JDMeadows@geotolmeadows.com
jamesmeadows@comcast.net
www.geotolmeadows.com

ISBN: 978-0-578-68108-5
Printed in the United States of America

1 2 3 4 5 6 7 8 9 10 Printing

Foreword

Sometimes a book must follow how the author teaches a topic. If it's a book about Geometric Dimensioning and Tolerancing, it must tell the rules, show the symbology, teach the logic of how to read the language and how to apply it. But it must then allow the learner to practice those skills. That is what this book does. It combines a textbook and a workbook to allow all facets of learning to reside in one place.

This textbook has been written for anyone whose work requires them to communicate, interpret or manufacture products through the use of engineering drawings and/or CAD models that use Geometric Dimensioning and Tolerancing. Students will learn the new ASME Y14.5-2018 standard on Dimensioning and Tolerancing, as well as the differences between that standard and prior revisions of Y14.5. It shows the new ASME Y14.5-2018 symbology, rules and basic principle revisions. The result of attaining this knowledge is that: product representations are able to be more specific in conveying functional requirements, products can be more easily manufactured, and appropriate inspection techniques are clear. Designers, manufacturing engineers, quality engineers, inspectors, product engineers and process engineers are just a few of the job categories that can benefit from the material contained in this textbook. Even those who have been trained in GD&T will need to be made aware of the changes that have been instituted into this latest Y14.5 revision. This book covers all of the basics on how to interpret, apply and measure GD&T per ASME Y14.5-2018. It also allows the learner to practice those techniques that are explained in great detail by the author.

Books, such as this one, are evolutions of explanations given hundreds of times by teachers and consultants trying to find the optimal way of giving their knowledge to others. The topics covered are refined, and our ways of conveying them to our students and readers get better over time. I've been teaching and consulting on this material for decades. I've written many textbooks, workbooks, articles, newsletters and tests on GD&T, Tolerance Stack-Up Analysis and Measurement, Inspection and Gauging. In the beginning, I thought that knowing a topic was all that was necessary to explain that topic. But, after a more than a few blank stares, and a ton of questions, I realized that just wasn't true. So, I learned how to think like a student again and to anticipate many of the questions they would have, and to answer them before they were asked. I also learned there is a time for the teacher to get out of the way, to stop lecturing and allow the students to practice what they have learned. This book has many problems to solve, questions to answer and chances to apply the techniques to parts and assemblies.

As a teacher, I used to believe that the most difficult topics were the ones that the most time should be devoted to. Then, I realized that the topics that were most beneficial to allowing professionals to do their jobs more efficiently were most worthy of study and explanation. Writers and teachers learn over time, just like students do.

I've learned that short, simple books usually sell better than long, comprehensive books. But I've also learned that the reader of a short, simple book is cheated and has been given a false belief that they know all they need to know. So, I write books that are long enough to allow the student to be immersed in the topic. I want the readers to have all of the information and practice they need to apply this knowledge to their products effectively. I want to give them all of the basic information, but also to provide them with more complex information to use when they realize that a little bit of knowledge is simply not enough to do a good job. I also love to show examples of how to do the many things that Y14.5 states you can do, but doesn't show you how to do, then to allow the student to practice on a variety of real products and assemblies that I've collected from the hundreds of companies I've worked for in over 4 decades of full time GD&T consulting.

This book combines basic and advanced information about the principles and applications of plus and minus tolerancing, geometric dimensioning and tolerancing, tolerance stack-up analysis, statistical tolerancing, inspection, gauge and fixture design and how to read geometric controls as though the symbology was a group sentences trying to describe how a part functions. It is an extremely comprehensive book. What is different about this book is that it is a combined textbook and work book. So, the reader reads, learns, sees how to apply the symbology for achieving product functionality, manufacturability and inspectability. Then the student practices these techniques to hone the skills they have acquired.

I sincerely hope the information contained in this book helps you.

James D. Meadows

Table of Contents

Chapter 1

Geometric Dimensioning and Tolerancing

INTRODUCTION

GEOMETRIC PRODUCT DEFINITION PRINCIPLES

During World War Two the industrialized world realized that the plus and minus tolerancing system for defining parts and assemblies sufficiently for products to function and assemble with interchangeability was not viable. Parts were produced in one facility, while their mating parts were produced in another and all were shipped to another to be assembled.

That's when the grinders came out and the rework began to grind the parts to fit. Very few assemblies had consistently interchangeable parts. Matched sets of parts were created to allow the machines of war to function. So, when parts broke down on the field of battle, it tended to render the entire mechanism useless. Vehicles and weapons of all varieties suffered breakdowns that could not easily be repaired by carrying a satchel of replacement parts.

So, it was determined that the language of plus and minus tolerancing for defining complicated geometric shapes, angles and locations was insufficient and ambiguous to such a degree as to render it almost useless.

When the war ended, all of the industrialized nations still standing began to look for a better system. Committees were formed in each country to take what they had been using to define products to be mass produced and improve on it. It was also quickly realized that an international effort was warranted, due to the fact that designers, engineers and skilled tradesmen were tasked to define and create products that had originated in countries other than their own.

Over many decades and voluminous revisions, standards began to evolve that created a system of geometric characteristic symbols, formulas and rules that allowed products to be defined with more and more specificity. This new language of geometric definition came to be known as Geometric Dimensioning and Tolerancing. The problem with that term was that it implied there was another, just as acceptable language, with its own standard know as Plus and Minus Tolerancing. That just wasn't the case. For example, in the United States, the standards were just called Dimensioning and Tolerancing, not Geometric Dimensioning and Tolerancing. It employed the best of the old plus and minus tolerancing system and coupled it with geometric symbology, rules and principles to define and control the dimensions and tolerances in a way that allowed parts meant to function and mate in assemblies to do so with interchangeability, even if they had been made to the limits of their tolerances. It was, unquestionably, a better system and continues to be improved by the (American Society of Mechanical Engineers) ASME Y14 committees to this day. The Dimensioning and Tolerancing standard was designated as Y14.5. In the beginning, the versions of Y14.5 issued were known as (American Standards Association) ASA Y14.5-1957, (United States of America Standards Institute) USASI Y14.5-1966, (American National Standards Institute) ANSI Y14.5-1973 and ANSI Y14.5M-1982. ANSI is the representative group for the United States in the International Organization of Standards (ISO). Eventually, since Y14.5 was sponsored by ASME (as they sponsor many standards committees and subcommittees) in the 1994 revision and subsequent revisions of the Y14.5 standard in 2009 and 2018, it was decided that instead of using the ANSI designation in the title, the standard would become known as an ANSI approved document, but would be called ASME Y14.5.

FOCUS AND EXCEL
(MAXIMIZING YOUR ABILITIES)

Somewhere, perhaps a long time ago, someone such as a college professor told you that in order to be a good engineer, designer, measurement expert or professional in general, that you had to be able to think about a hundred different things at once and do it well. That person, or persons, whoever they were, lied to you. I've trained upwards of sixty thousand people in my career and only about five of them could think globally and do it well. And they weren't happy people! There is a high suicide rate in this IQ category. Not something I covet, nor could ever achieve. Most people in the professions we (the people likely reading this) occupy are above average. That's what we are; above average. Even if you are a genius, chances are you are only a genius in a couple of things, not in everything. Think of the man so many of us associate with genius; Albert Einstein. Even he wasn't a genius in everything. Would you, for example, want him as your hair stylist?

Most of us would be wise to maximize our intelligence by focusing our attention on one thing at a time, giving that thing all of our attention, accomplishing the task it presents, then moving on and finding something else to do, until all facets of a project are complete. It's like putting together a jigsaw puzzle. We'll keep in mind what the picture on the front of the puzzle box looks like, but we can't forget that we only have one piece of it in our hand at a time, and focus all of our attention on that piece until we either fit it into the proper place or decide to put it down and focus on another piece. Eventually, we will complete the entire puzzle. We won't miss even the smallest detail. By following this tact we will do a better job. It will be more accurate, more error-free, more complete and believe it or not, it will be done faster.

People who look at things globally, most often diffuse their intelligence so much that they become less competent, often incompetent. They panic and freeze up or do things slowly or in a chaotic frenzy. An example is when someone complains about their workload. They look at a desktop buried in projects and proclaim, "My God! I'll never be able to do all this!" You sometimes read about them bringing a gun to work to punish those who have caused their troubles. Even then their approach probably reflects their lack of focus, shooting anyone that happens to be there. More than likely, the source of their problems looks back at them from the mirror each morning. If they had just asked themselves the right questions in the first place, instead of trying to exceed their intellectual abilities, they probably would have been okay. Yes, there is a stack of projects on your desk, but: 1) Isn't one of them due first? and 2) In that file that is due first, isn't there one task that needs to be done before any others? In other words, begin...do something, but not just anything. Take the time to put things in order, then focus on accomplishing the ordered items. Recognize your weaknesses and make them your strengths. No, unless you have infinite experience and a stratospheric IQ, you can't accomplish or even think about everything at once. Quit crying. It's a fact.

Now that you're over that hump, focus on what you are good at, doing one thing at a time and doing it well, maybe even better than anyone else could do it (at least anyone available today). Begin, focus, persevere, be tenacious, finish. Begin again. It doesn't even matter what your job is, this works. Just do it. Quit staring at it. The pile isn't getting smaller. Unless your boss starts giving you less work, and if so, watch your back. The old manager's adage is to give the work to those able to accomplish it. If Joe has a lot of projects and you don't, it isn't time to celebrate. It's time to update your resume'. The boss has discovered your depth. He's standing in it and his socks aren't even wet.

You can be less talented, less brilliant than those around you and still run rings around them. Just focus all your abilities, all your attention and peck away at it. Pretty soon you'll realize you can do almost anything, especially if you're willing to admit you don't know everything in the whole wide world and ask others for advice from their areas of expertise.

Don't worry about showing weakness of knowledge. In the long run you won't look weak. You'll look curious and confident and successful. And if they don't want to help you, and berate you for not knowing something already, the hell with them, ask someone else.

To paraphrase Henry Ford, "I might not know the answer to that, but give me five minutes and I'll hire someone that does", or words to that effect. Take the criticism of those crevice dwellers that would tear down your ego and wave at them as you rise above and watch them sinking slowly in their own self-assigned importance. And if you have a bad day here and there, it goes with the job of pushing air in and out. Take it seriously, but don't agonize over it. Take it from a guy who watched his life's blood flow out of an ulcer the size of death's door. With the right perspective, life and the job can be fun and even the tribulations become amusing.

William Shakespeare portrayed that perspective in Macbeth. He wrote: "Tomorrow, and tomorrow, and tomorrow, creeps in this petty pace from day to day, to the last syllable of recorded time; and all our yesterdays have lighted fools the way to dusty death. Out, out, brief candle! Life's but a walking shadow; a poor player, that struts and frets his hour upon the stage, and then is heard no more: it is a tale told by an idiot, full of sound and fury, signifying nothing."

Philosophy and Logic

To be successful at geometric definition, you must first define perfection, then state how far from perfect a feature, then a part, then an assembly may be from perfection before it fails to function. To do this well, one must understand the logic of the process.

I liken it to a trip, a train ride, sometimes short, sometimes long, but with a definite beginning, middle and an end. You decide to define a product. It must be functional, manufacturable and inspectable. So, plan your trip, you board the train and you make the proper stops. With each stop, you define a facet of the feature, a geometric characteristic it must possess and comply with, in order to complete the trip. It is a piece of the mosaic. Each piece involves one or more of four characteristics. These are size, form, angle and location. These are what we define and what we tolerance.

But, to begin, we plan and then execute. It begins. I was once in front of a group of engineers wherein my task was to use my knowledge of geometric definition and their product knowledge to define their product. It was a large group for such an endeavor, about 60 professionals. As with most engineers, they believed that their product, its geometries, its complexities were unique, something never seen before throughout the history of mankind. It was comprised of curved plastics, wires, computer boards and housings. Just before I begin to speak, one person, in the middle of the room rose and yelled, "This is impossible! It can't be done!" Other members of the group, began to agree. So, after a few moments, I said, "Okay. Then let's give up.", and I began to pack up my books. The gentleman standing looked startled and said, "Well, we can't give up." So, I responded, as I unpacked my books, "Then sit down and let's get started." Because it's one or the other. Either start, figure it out and maybe succeed, or don't even try and, therefore, definitely fail. Perfection is God's domain. We humans never achieve it. But we can do our best, give it our optimal effort, using all of the gifts He has given us. I find that once you've looked over, then studied in great depth and comprehended the standards, the rules, the concepts, the geometric characteristics, the definitions and then get to know all of the assets you have to work with for manufacturing and inspection, the most valuable tools you can employ (and do it better than any computer), are logic and following a known philosophy. Relate every feature to every other part feature. **Calculate the tolerances. Don't guess. Know.**

But, to begin, we plan and then execute. It begins. I was once in front of a group of engineers wherein my task was to use my knowledge of geometric definition and their product knowledge to define their product. It was a large group for such an endeavor, about 60 professionals. As with most engineers, they believed that their product, its geometries, its complexities were unique, something never seen before throughout the history of mankind. It was comprised of curved plastics, wires, computer boards and housings. Just before I begin to speak, one person, in the middle of the room rose and yelled, "This is impossible! It can't be done!" Other members of the group, began to agree. So, after a few moments, I said, "Okay. Then let's give up.", and I began to pack up my books. The gentleman standing looked startled and said, "Well, we can't give up." So, I responded, as I unpacked my books, "Then sit down and let's get started." Because it's one or the other. Either start, figure it out and maybe succeed, or don't even try and, therefore, definitely fail. Perfection is God's domain. We humans never achieve it. But we can do our best, give it our optimal effort, using all of the gifts He has given us. I find that once you've looked over, then studied in great depth and comprehended the standards, the rules, the concepts, the geometric characteristics, the definitions and then get to know all of the assets you have to work with for manufacturing and inspection, the most valuable tools you can employ (and do it better than any computer), are logic and following a known philosophy. Relate every feature to every other part feature. **Calculate the tolerances. Don't guess. Know.**

1. Board the train, assign a primary datum feature and consider it for a **form control to insure stability in function and repeatability in measurements**. You don't want to relate and measure subsequent features to a rocking or wobbling datum feature, so give it a sufficient form control to limit this.
2. Decide whether you want to measure from the exterior features to the interior features, or from the interior features to the exterior, or a bit of both. This will set up the map your train will travel. Relate datum features to each other in an order they will be used. You may change that order later, but use them in that order at least once, or choose a different order.
3. If you don't want to relate all part features to those datums, you can establish new datum features once you reach them in this linear progression of geometric definition. Each time you switch datum features, you accumulate more and more stack-up error to the original set of datums you've established. This can be overdone or not done enough to affect the part's manufacturability and functionality.
4. Calculate the tolerances. Use fixed or floating fastener equations. Know manufacturing's capabilities. Calculate the cost of tolerances.
5. If you want all parts that mate to be accepted by inspection and they don't have to seal or spin and balance use the maximum material condition (MMC) symbol where it is allowed, and the maximum material boundary (MMB) symbol after datum features of size. It lowers the product cost and allows simple receiver gages to be used. If sealing or balance is critical, consider not using the MMB symbol after datum features of size. It allows parts to be bought by inspection that can only be assembled off center. And if it spins really fast, you might even consider not using the MMC symbol after geometric tolerances. This allows features to move out of position more as you depart from MMC. This does not affect assembly of parts on center of the assembly, but it means individual features may move off center more as holes are produced larger and shafts are produced smaller. This may adversely affect balance for parts that rotate at high rates of RPMs.
6. Each set of new datum features may require a new set up and that affects inspection. It may or may not affect manufacturing. Because manufacturing can make the part any

way their equipment, knowledge and skills dictate, as long as the finished product meets the requirements you have spelled out.

7. Remember, you are not creating a process engineering sheet. You are creating the finished product requirements. If you can do both in one definition and the product functions as well as needed, the birds will sing perched on your windowsill.
8. Relate every feature to every other feature to within a tolerance. The tolerance may be miniscule or it may be huge, but a tolerance it shall have. When I was a young man, I saw an engineer give a hole six inches of positional tolerance. I laughed at him and said, "Charlie, six inches of positional tolerance?" He looked me over, as though I had left a slime trail leading to his desk, and said, "Look sonny, this is a drainage hole. You know what happens, if it is produced within that six inches of location tolerance?" I allowed my eyebrows to raise as though in anticipation of his answer. He said, with disgust, "It drains.", then he turned away, as though dismissing me as an idiot. But I learned something from that. The size of the geometric tolerance can convey that a feature is only as important, as critical, as the tolerances applied to it. A huge tolerance states, this isn't very important, but is exactly as important as it's allowed deviation from perfection. Likewise, a tight tolerance can (I've used tolerances that must be held to within millionths of an inch) convey immense criticality.
9. There is no implied tolerance of perpendicularity, coaxiality or symmetry. The limits of size do not control the relationship between individual features. Features shown perpendicular, coaxial or symmetrical to each other, must be controlled with a tolerance to avoid an incomplete part definition. In other words, be thorough.
10. Calculate wall thicknesses to make sure the product's structural integrity is not compromised. If you don't do this, part's can collapse under its own weight or break apart as it rotates.
11. Ensure that seals will seal and that rotating parts will maintain proper balance. Test them as they function, under the conditions that they function. Consider all factors, such as temperature, humidity, speed and anything that will affect the product when in use.
12. Do Tolerance Stack-Up Analysis on your assemblies. Sometimes, even if the individual parts assemble and function well from one part to the next, when all of the parts are assembled, they may not work well as a finished assembly.
13. Make all of the train stops necessary to define the limits of imperfection that can be allowed. Take your time and make good decisions. Enjoy the ride. It is a blending of math, logic and can weave itself into a beautiful, logical work of art. You are the artist. And what could be more beautiful to a technical person than a mechanical, electrical or other type of object that when manifested from an idea into physical reality, works great, looks gorgeous, is affordable to the customer and assures you employment for as long as you want or need it?

Don't answer. The question is rhetorical. All aboard.

Chapter 1

Geometric Dimensioning and Tolerancing

•Symbols, Rules and Formulas, Principles and Concepts

Chapter Objectives

Readers will learn:

1. The 14 geometric characteristic symbols per ASME Y14.5-2009 and the 12 geometric characteristic symbols per ASME Y14.5-2018, the 5 categories they fall within (form, orientation, profile, runout and location), when material condition symbols may be used and when datum references are allowed.
2. A variety of symbols that are used on design drawings to replace words.
3. What a basic dimension is.
4. Rule 1, implied RFS and RMB, formulas for calculating mating part tolerances, how to calculate a virtual condition (MMC and MMB Concept), selection of datum features criteria and what MMC, LMC, MMB and LMB stand for.
5. What geometric characteristic symbols are used on planar surfaces, what each tolerance means and how each compares to the others in power and versatility.
6. What geometric characteristic symbols are used on round surfaces, what each tolerance means and how each compares to the others in power and versatility.
7. The types of geometric controls and what they are used to control.
8. Rules for displaying tolerances in millimeters, inches, angular units, radii, statistically calculated tolerances, measurement compliance temperature, part restraint and insignificant zeros.

GEOMETRIC CHARACTERISTICS

Category and Geometric Characteristic	Symbol	Material Condition Symbols Allowed (Modifiers)	Datum Reference Allowed
FORM			
Flatness	⏥	Ⓜ or Ⓛ for derived median plane; No for surface	No
Straightness	—	Ⓜ or Ⓛ for derived median line; No for surface	No
Circularity (Roundness)	○	No	No
Cylindricity	⌭	No	No
PROFILE			
Profile of a surface	⌓	No for feature; Ⓜ or Ⓛ allowed for datum features of size	Allowed but not required
Profile of a line	⌒	No for feature; Ⓜ or Ⓛ allowed for datum features of size	Allowed but not required
ORIENTATION			
Parallelism	∥	Ⓜ or Ⓛ allowed for feature of size Ⓜ or Ⓛ allowed for datum features of size	Required
Perpendicularity	⊥	Ⓜ or Ⓛ allowed for feature of size Ⓜ or Ⓛ allowed for datum features of size	Required
Angularity	∠	Ⓜ or Ⓛ allowed for feature of size Ⓜ or Ⓛ allowed for datum features of size	Required
RUNOUT			
Total Runout	⌰	No	Required
Circular Runout	↗	No	Required
LOCATION			
Position	⌖	Ⓜ or Ⓛ allowed for all features Ⓢ implied if none specified Ⓜ or Ⓛ allowed for all datum features of size. Ⓢ implied if none specified	Required except for feature-to-feature only situations
Concentricity*	◎	No	Required
Symmetry*	⌯	No	Required

[RFS concept is implied for all geometric tolerances and RMB is implied for all datum features of size where no modifier is specified.] ***Concentricity and Symmetry Symbology has been removed from ASME Y14.5-2018.**

ADDITIONAL SYMBOLOGY IN Y14.5

SYMBOL FOR	ASME Y14.5
Arc Length	⌒100
Maximum Material Condition	Ⓜ
Least Material Condition	Ⓛ
Regardless of Feature Size	Ⓢ implied
Projected Tolerance Zone	Ⓟ
Tangent Plane	Ⓣ
Free State	Ⓕ
Reference Dimension (Parenthesis on each side of any dimension)	(1.25)
Statistical Tolerance	‹ST›
Dimension is Not to Scale	<u>20</u>
Datum Target	A1 OR Ø5/A1 OR A1 Ø5
Basic Dimension (theoretically exact dimension to show size, orientation or location with its tolerance given in feature control frames, established tooling or gaging tolerances or notes.)	[100]

From-To Symbol and Dynamic Profile. Other symbols introduced in the Y14.5-2018 standard are one symbol that indicates that a specification transitions from one location to a second location. Example: **A→B** and another that shows Dynamic Profile: △ . See next page and profile of a surface section of this book.

CHART OF SYMBOLS IN ASME Y14.5

SYMBOL FOR	ASME Y14.5
Counterbore	⌴
Countersink	⌵
Depth or Deep	↧
Times, Places, or By	X
Square Shape	□
All Around (Profile)	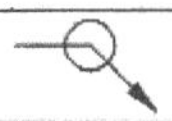
All Over (Profile)	
Radius	R
Spherical Radius	SR
Controlled Radius	CR
Diameter	⌀
Spherical Diameter	S⌀
Between, such as B ◄──► C, commonly used to show extent of control as in profiles.	◄──►
From-to, such as X ──► Y, used to show how a specification transitions from one location to another.	──►
Dimension Origin	
Slope (Flat Taper)	⌳
Conical Taper	⌲
Dynamic Profile	△

Symbols Introduced in Y14.5-2009

▷	Translation Symbol
Ⓤ	Unequally Disposed or Unilateral Tolerance
Ⓘ	Independency (Size Tolerance Does Not Control Form)
⟨CF⟩	Continuous Feature (Treat Multiple Features as One)
(All Over symbol)	All Over (Control Applies in all Views)

Symbols Given Additional Context in Y14.5-2009

Ⓜ	Means: Applies at Maximum Material Condition when it is used after a geometric tolerance. Means: Applies at Maximum Material Boundary when it is used after a datum reference.
Ⓛ	Means: Applies at Least Material Condition when it is used after a geometric tolerance. Means: Applies at Least Material Boundary when it is used after a datum reference.
Ⓢ Implied	When no material condition symbol is specified, it means: a) Applies at Regardless of Feature Size when it is implied after a geometric tolerance. b) Applies at Regardless of Material Boundary when it is implied after a datum reference.

Symbol Introduced in Y14.5-2009

SPOTFACE	⌴SF⌴Ø20

When the symbol for spotface is used, either the remaining thickness of material may be specified, or the depth of the spotface may be given. The spotface depth is the minimum needed to clean up the material of the surface to the diameter of the spotface. The spotface may be noted using the symbol and its diameter without showing it on the design drawing or in the CAD model. The surface to receive the spotface must be clearly indicated. If desired, a filet radius may be given on the drawing to relieve sharp corners for the spotface.

For example:

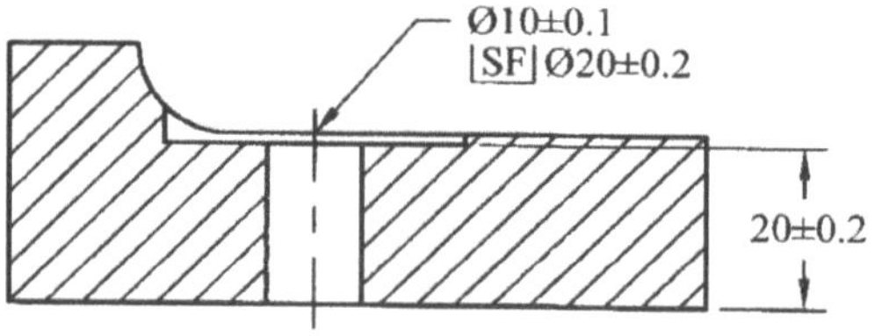

The Dynamic Profile Symbol was introduced in 2018
Example:

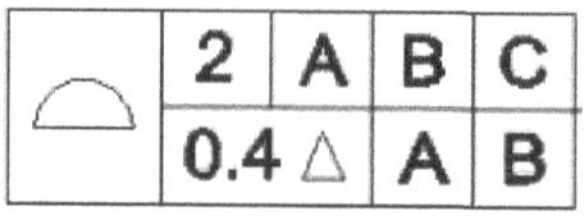

Food Chain of Symbology

if Used on <u>Planar Surfaces</u> as a <u>Surface</u> <u>Control</u>

⌓ = **Profile of the surface.** Tolerance zone consists of two uniform boundaries that follow the basically dimensioned or implied profile. All elements of the controlled surface(s) must reside within the tolerance zone. Datums may be referenced but are not required. It always controls 3D form but can be used to control size, angles and location.

∠ = **Angularity of the surface.** Tolerance zone consists of two parallel planes separated by the angularity tolerance between which all elements of the surface must reside. These parallel planes are at the specified basic angle(s) to the datum(s) referenced. It controls flatness and the specified basic angle(s) to the datum(s) referenced in the feature control frame.

⊥ = **Perpendicularity of the surface.** Tolerance zone consists of two parallel planes separated by the perpendicularity tolerance between which all elements of the surface must reside. These parallel planes are perpendicular (90°) to the primary and perhaps a secondary datum plane. It controls flatness and 90° angles.

// = **Parallelism of the surface.** Tolerance zone consists of two parallel planes separated by the parallelism tolerance between which all elements of the surface must reside. These parallel planes are parallel (0°) to the primary datum plane. It controls flatness and 0°/180° angles.

▱ = **Flatness of the surface.** Tolerance zone consists of two parallel planes separated by the flatness tolerance between which all elements of the surface must reside. It controls pits, bumps and curves, which is straightness in all directions.

— = **Straightness of linear line elements.** Tolerance zone consists of two parallel straight lines separated by the straightness tolerance. Each line element in the plane of the view has its own separately verifiable tolerance zone. It controls pits, bumps and curves on each line under control.

Food Chain of Symbology

Commonly Used on <u>Round Surfaces</u> as a <u>Surface</u> <u>Control</u>

⌰ = **Total Runout of the surface.** If used on a cylindrical surface, the tolerance zone consists of two concentric cylinders separated by the runout tolerance between which all elements of the surface must reside. The tolerance zone is centered on the datum axis. As the part continuously revolves 360° while moving the indicator longitudinally down the entire surface, the full indicator movement may not exceed the runout tolerance. It controls cylindricity and centering to the datum axis. In other words, it controls everything that Circular Runout controls plus taper and surface straightness.

↗ = **Circular Runout of circular cross sections of the surface.** Like controlling a stack of coins, each circular cross section, while centered on a datum axis, has a tolerance zone consisting of two concentric circles between which all points on the circle under test must reside. The full indicator movement for each cross section independently verified may not exceed the runout tolerance. It controls pits, bumps, flats, ovals, centering to the datum axis and feature derived median line straightness.

These two geometric characteristics <u>are not</u> surface controls.

◎*= **Concentricity of the median points.** This coaxiality-type control has a cylindrical tolerance zone which confines all median points created by taking differential measurements on the part surface 180° apart (opposed points). The tolerance zone is centered on the datum axis. Unlike the other symbols on this page, concentricity is not a surface control, but rather tries to balance material on each side of the datum axis. See the note at the bottom of the page.

— | Ø = **Straightness of a derived median line.** This axis control has a cylindrical tolerance zone which confines the central points taken normal to the minimum circumscribed cylinder (if a shaft) or to the maximum inscribed cylinder (if a hole). These central points reflect the cross-sectional axes and together constitute a derived median line. Confines axial bowing. Like concentricity, this is not a surface control. See the note at the bottom of the page.

⌭ = **Cylindricity of a surface.** Tolerance zone consists of two concentric cylinders separated by the cylindricity tolerance between which all elements of the surface must reside. The full indicator movement in continuous 360° revolutions, while moving longitudinally down the surface, may not exceed the cylindricity tolerance. It controls circularity, straightness and taper.

○ = **Circularity of round line elements.** Tolerance zone consists of two concentric circles separated by the circularity tolerance. All points on the circle under consideration must reside between the two concentric circles. Each circular cross-section (like a stack of coins) of the surface has its own separately verifiable tolerance zone. The full indicator movement, in a 360° revolution, may not exceed the circularity tolerance. It controls pits, bumps, flats and ovals.

Note: As mentioned in the definitions above, unlike the other symbols on this page, concentricity and straightness of the derived median line are not surface controls. ***Remember, Concentricity symbology does not appear in ASME Y14.5-2018.**

Maximum Material Condition (MMC). Definition per ASME Y14.5-2018: It is the condition in which a feature of size contains the maximum amount of material within the stated limits of size. For example, the minimum hole diameter or the maximum shaft diameter is its maximum material condition.

Ⓜ = MMC = Smallest hole or largest shaft.

Least Material Condition (LMC). Definition per ASME Y14.5-2018: It is the condition in which a feature of size contains the least contains the least amount of material within the stated limits of size. For example, the maximum hole diameter or the minimum shaft diameter is its least material condition.

Ⓛ = LMC = Largest hole or smallest shaft

Regardless of Feature Size (RFS). Definition per ASME Y14.5-2018: Indicates a geometric tolerance applies at any actual size of the actual mating envelope of the feature of size. The symbol for regardless of feature size (formerly a circled S) is implied for all geometric tolerances unless specified otherwise.

Regular Feature of Size. One cylindrical or spherical surface, a circular element, a set of two opposed elements or opposed parallel surfaces, each of which is associated with a size dimension.

Irregular Feature of Size. The two types of irregular features of size are: a) a feature or collection of features that may contain or be contained by an actual mating envelope which is a sphere, cylinder or pair of parallel planes, and b) a feature or collection of features that may contain or be contained by an actual mating envelope other than a sphere, cylinder or pair of parallel planes.

Rule #1

Size limits control surface form. Unless otherwise specified, for rigid features the LMC is measured for violations at cross-sections and the MMC is measured to verify compliance with an envelope of perfect form at MMC. The new Independency Symbol placed near the size dimension of a feature negates this rule.

Rule #2 (RFS for geometric tolerances and RMB for datum features is implied)

For all geometric characteristic symbols used, where no Ⓜ (maximum material condition symbol) or Ⓛ (least material condition symbol) is specified in the feature control frame after the geometric tolerance, the regardless of feature size (RFS) concept is implied for the geometric tolerance.

For all geometric characteristic symbols used, where no Ⓜ (maximum material boundary symbol-MMB) or Ⓛ (least material boundary symbol-LMB) is specified in the feature control frame **after datum features**, the regardless of material boundary concept (RMB) is implied.

Definition of and Guidance for the Selection of Datum Features.

Datum features are real. They are physical features (surfaces) that generate datums (theoretical axes or planes) from which we measure either angles, location or both.

Datum features should most often be:

1) Functional (serving some purpose in the way part works).
2) Representative of seating features, mating features and alignment features.
3. Accessible (which is influenced by it having sufficient surface area to stabilize the part in the way it fits into the assembly).
4) Repeatable (which is greatly influenced by the feature's precision of form).

Virtual Condition MMC Concept
The virtual condition (MMC Concept) is the worst mating boundary.

The virtual condition of mating features should be compatible. The virtual condition of the mating hole and shaft should be the same if one is using 100% of the available arithmetically calculated tolerance.

Functional gages are sized at the Virtual Condition for features being gaged and at the maximum material boundary (MMB) of the datum features they are to simulate. The MMB is calculated using the virtual condition formulas as follows.
To calculate the virtual condition (MMC Concept):

- for holes
  MMC of the hole minus the geometric tolerance at MMC = virtual condition

- for shafts
  MMC of the shaft plus the geometric tolerance at MMC = virtual condition

Floating Fastener Formula:

  MMC hole
- MMC shaft (or screw)
  Geometric Tolerance for all holes

Fixed Fastener Formulas:

  MMC hole
- MMC shaft (or screw)
  Geo. Tol. to be divided between the two mating features (parts)

  Virtual Condition hole (MMC Concept)
- MMC shaft (or screw)
  Geo. Tol. for shaft

  MMC hole
- Virtual Condition shaft (or screw) (MMC concept)
  Geo. Tol. for hole

Virtual Condition-ASME Y14.5-2018
A constant boundary generated by the collective effects of a considered feature of size's specified MMC or LMC material condition and the geometric tolerance for that material condition.

Resultant Condition-ASME Y14.5-2018
The single worst-case boundary generated by the collective effects of a feature of size's specified MMC or LMC, the geometric tolerance for that material condition, the size tolerance, and the additional geometric tolerance derived from the feature's departure from its specified material condition.

In ASME Y14.5M-1994, resultant condition was defined as "The variable boundary generated by the collective effects of a size feature's specified MMC or LMC material condition, the geometric tolerance for that material condition, the size tolerance, and the additional geometric tolerance derived from the feature's departure from its specified material condition.

Rule Regarding the Use of the Ⓢ Symbol

The old Rule 2a which, prior to Y14.5-2009, stated that a position tolerance was allowed to use a redundant Ⓢ symbol as a clarification of the applicability of the regardless of feature size concept is no longer true. Unless otherwise specified at MMC or LMC, all geometric tolerances are implied at RFS (regardless of feature size). Unless otherwise specified at MMB (maximum material boundary) or LMB (least material boundary) all datum features are implied to apply at regardless of material boundary (RMB).

Y14.5-2009 & Y14.5-2018 Terms for Concepts: MMB, LMB and RMB

Maximum Material Boundary (MMB) is the limit that is defined by a tolerance or combination of tolerances, existing outside of the material of a feature or features. It is calculated by the following formulas:

For holes, slots (or other internal features of size), the maximum material boundary is the MMC of the feature minus the applicable geometric tolerance at MMC. The applicable geometric tolerance is (unless otherwise specified) the smallest geometric tolerance that exists between the datum feature being considered and any datums that precede it in the feature control frame. This boundary is also referred to as the inner boundary.

For shafts, tabs (or other external features of size), the maximum material boundary is the MMC of the feature plus the applicable geometric tolerance at MMC. The applicable geometric tolerance is (unless otherwise specified) the smallest geometric tolerance that exists between the datum feature being considered and any datums that precede it in the feature control frame. This boundary is also referred to as the outer boundary.

Least Material Boundary (LMB) is the limit defined by a tolerance or combination of tolerances, existing inside the material of a feature or features. It is calculated by the following formulas:

For holes, slots (or other internal features of size), the least material boundary is the LMC of the feature plus the applicable geometric tolerance at LMC. The applicable geometric tolerance is (unless otherwise specified) the smallest geometric tolerance that exists between the datum feature being considered and any datums that precede it in the feature control frame. This boundary is also referred to as the outer boundary.

For shafts, tabs (or other external features of size), the least material boundary is the LMC of the feature minus the applicable geometric tolerance at LMC. The applicable geometric tolerance is (unless otherwise specified) the smallest geometric tolerance that exists between the datum feature being considered and any datums that precede it in the feature control frame. This boundary is also referred to as the inner boundary.

Regardless of Material Boundary (RMB) is a similar perfect counterpart of a tolerance limit which grows or shrinks through a tolerance zone from MMB towards LMB until it makes maximum contact with the extremities of a feature. It is calculated by the following formulas:

For holes, slots (or other internal features of size), the regardless of material boundary for datum features controlled with position, perpendicularity, angularity or parallelism is the actual mating size of the datum feature minus the applicable geometric tolerance used by the produced datum feature at that size.

For shafts, tabs (or other external features of size), the regardless of material boundary for datum features controlled with position, perpendicularity, angularity or parallelism is the actual mating size of the feature plus the applicable geometric tolerance used by the produced datum feature at that size.

Actual Minimum Material Envelope. It is an envelope that is within the material as shown in the illustrations that follow this definition.

a) It is a similar perfect feature(s) counterpart of largest size that can be inscribed within an external feature(s) so that it just contacts the surface at the lowest points.
b) It is a similar perfect feature(s) counterpart of smallest size that can be circumscribed about the feature so that it just contacts the surface at its lowest points.

There are two types of actual minimum material envelopes.

1) One type is not related to datums. This one is called the unrelated actual minimum material envelope.
2) The other type is related to datums. This one is called the related actual minimum material envelope.

The following illustration compares the "actual mating envelope" to the "actual minimum material envelope" for internal and external features of size.

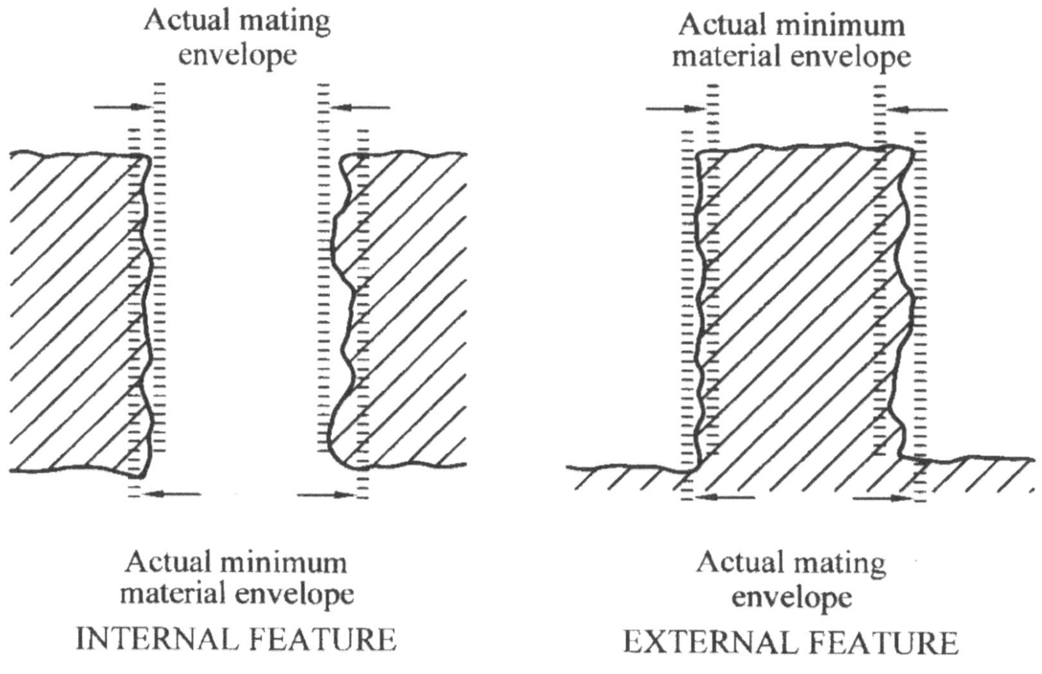

Derived Median Plane and Derived Median Line (Terms per Y14.5-2009 & Y14.5-2018 Symbol for Derived Median Plane Control)

The derived median plane of a feature is an irregular plane created by the median points of all line segments bounded by the feature taken normal to the centerplane of the unrelated actual mating envelope. Previously, in ASME Y14.5M-1994, this control was used with Straightness. In ASME Y14.5-2009 and ASME Y14.5-2018, it is used with flatness instead. Unlike most geometric characteristics, with flatness of the derived median plane, bonus tolerance is calculated locally and could result in a non-uniform tolerance zone size on the part as produced.
Example:

Part Specification

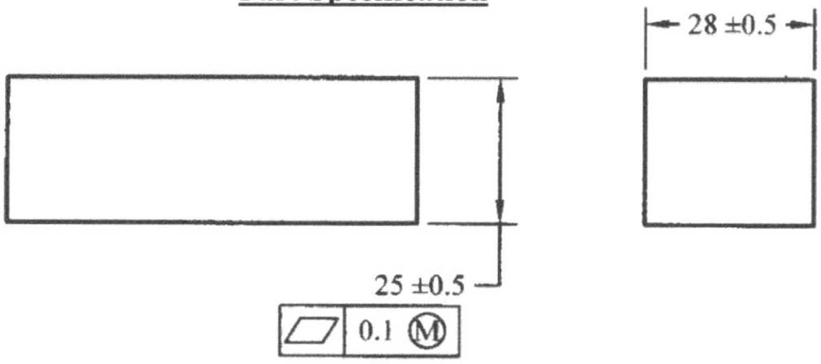

Tolerance Zone

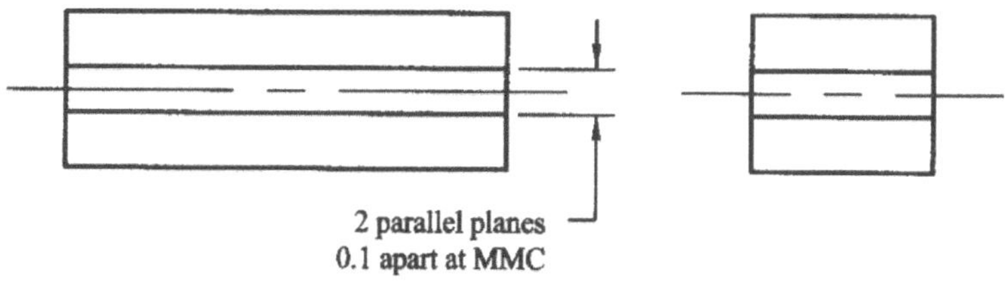

Part as Produced

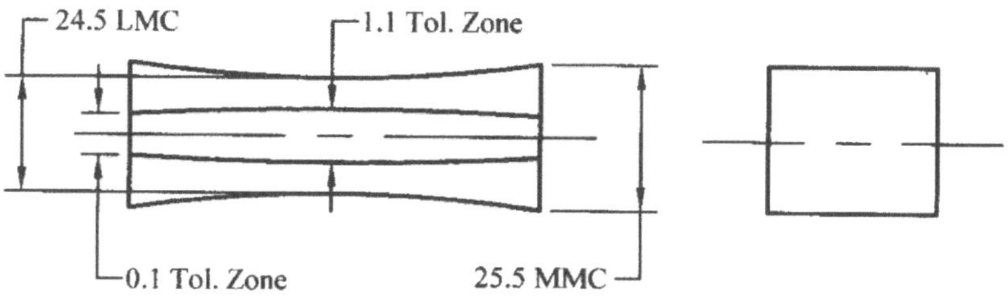

The derived median line of a diameter is an irregular line (axis) created by the median points of the diameter's cross-sections taken normal to the axis of the unrelated actual mating envelope. Since Y14.5-2009, and in the current revision of the Y14.5 standard (2018), this control is used with straightness, just as it was in the previous versions of the Y14.5 standard. As with the above illustration for flatness of a derived median plane, bonus tolerance for straightness of the derived median line is calculated locally.

Geometric Dimensioning and Tolerancing

Geometric dimensioning and tolerancing is a language of symbols based on good, sound logic. It allows a designer to express his or her thoughts and design requirements in a clear, concise manner. Tolerance zones, which in the past have not been allowed to be the same shape as the feature being protected and controlled, can now be defined as those shapes, for example, cylindrical tolerance zones for cylindrical features. This does not dictate that tolerance zones of other shapes stop being used, but does afford us the opportunity to more fully define our parts in more functionally precise ways never before available. It assures us that, if properly defined, parts made per design drawing requirements will also now fulfill design needs.

Surfaces, Features, Features of Size, Datum Features, Datum Features of Size and Datums

A part has surfaces. These surfaces either alone or at times in combination with one another can be called features. These features actually exist on the part. They can be planar surfaces like the individual sides of a part, in which case each of these individual features or surfaces has no size. They can also be features of size like, for example, a shaft or a hole or even two opposing planar surfaces (features). Features of size can be measured for size. They are features that are cylindrical, spherical or consist of two parallel planar opposing surfaces. It is from features (both those of size and those not of size) that datum features are chosen. Surfaces, features, features of size, datum features and datum features of size all exist. They are real and can be touched. Datums are derived from datum features. They are manifested by the perfect geometric counterparts of the imperfect datum features. They may be planes, points, lines or axes (the intersection of two mutually perpendicular planes). They do not exist in reality. They are perfect and, therefore, can only be simulated. Datum feature simulators are commonly used to construct origins of measurement for location, orientation, profile or runout. Some common datum feature simulators are surface plates, angle plates, chucks, mandrels and machine tables. **In Y14.5-2018 standard, the term for a perfect, therefore imaginary, datum feature simulator, which in the Y14.5-2009 standard was called a Theoretical Datum Feature Simulator was replaced by a term from Y14.5-1994. That term is True Geometric Counterpart or TGC. It is used a lot in the Y14.5-2018 standard.**

Datum features are simulated by equipment/tools to act in place of their perfect geometric counterpart datum planes, axes or points from which features are measured and geometrically controlled and on which parts are set-up for manufacturing and inspection purposes. Datum features should be chosen on the basis of their function, repeatability, accessibility and their representation of mating features.

Sometimes datum targets are used and if basic dimensions, rather than directly toleranced dimensions, are specified to show datum target location or area size. In these cases, the tolerance that applies to the target location or size is to be the established tooling or gaging tolerances of that facility. If your facility has not created a tolerance specifically for datum target locations, the Y14.43 standard on Dimensioning and Tolerancing Principles for Gages and Fixtures has a formula you can use to calculate the applicable tolerances. Y14.43 states the following:

Datum Target Basic Dimensions. When datum targets are defined with basic dimensions, ASME Y14.5 states that tooling or gaging tolerances apply. This method benefits greatly from a company's internal process standard (or other general or local tolerance specifications) to provide the applicable tolerances to the basic dimensions. When such documentation exists, it is recommended that it be referenced on the workpiece drawing or in a document referenced thereon. However, when there is no company documentation or workpiece definition, and basic dimensions are shown (or implied) to define the form, size, orientation, or location of datum target simulators, and no tolerance is given, this Standard defines default tolerances for use on the basic dimensions for the datum target simulators as follows: (a) For Form: 10% or less of the largest tolerance that controls form given to the surfaces on which the datum targets reside. (b) For Orientation: 10% or less of the largest tolerance that controls orientation between the surfaces on which the datum targets reside. This control is typically used with target line and area simulators. (c) For Location: 10% or less of the largest tolerance that controls location between the datum targets and the part features located (usually by position or profile) from them. (d) For Size: 10% or less of the largest tolerance that controls size given to the feature on which the datum targets reside. If no size tolerance for the workpiece feature is given, then10% or less of the largest tolerance that controls form shall be used.

NOTE: ASME Y14.5-2018 indicates the use of this Standard for information on datum feature simulator tolerances and toleranced relationships between the simulators. Reference on a workpiece drawing to this Standard is only required when the workpiece drawing is based on using the defaults contained herein. Where it is desired to invoke the defaults of this Standard, the workpiece drawing shall contain a note such as "Default datum target tolerances per ASME Y14.43."

Datum Target Toleranced Dimensions. When datum targets are defined with toleranced dimensions, this Standard recommends that a note be added to the workpiece drawing or in a document referenced thereon indicating if 100% of the tolerances specified with the datum target are for use in the tooling and gaging processes or if some other interpretation is intended. Considerations for the development of datum target tolerances should include the following: (a) Using 10% of the datum feature form, size, orientation or location tolerance to shape or to relate datum targets to each other. (b) In some instances, it is necessary to define the dimension and tolerances of the target simulator, such as the spherical diameter of a simulator. Due to flaws in the part surface, these dimensions and tolerances will determine how effectively the simulator touches the point(s) on the part surface.

Some ASME Y14.5 users have mistakenly assumed the general tolerance notes given in the drawing's title tolerance block apply to the basic dimensions. This is never true, unless a geometric tolerance with a geometric characteristic capable of controlling/tolerancing the basic dimension appears in the general tolerance note. Those two-place and three-place dimensional tolerances and angle tolerances apply to otherwise untoleranced dimensions on the drawing--but not to the basic dimensions. If not shown in the general tolerance note, the tolerance that applies to the location of the features controlled by those basic dimensions is to be found in a feature control frame attached to or pointing at those features.

Basic dimensions don't mean the part or features on the part have to be perfect in location or size. They only display what perfection is, then a tolerance of allowed deviation is stated, usually shown inside a feature control frame. This tolerance can be very small (costly) or extremely large (inexpensive to comply with).

As you can see, there are many more pieces of information needed to accurately define the part. **Some of those might be**:

1) datum features--surfaces from which perfect planes can be established and used as origins of measurement for the hole location and perpendicularity, because position controls location and orientation (in this case, perpendicularity)

2) desired hole size

3) hole size tolerance

4) a feature control frame with

- a geometric characteristic symbol
- a tolerance zone descriptor (for instance a diameter sign indicating an axis control)
- a tolerance of location
- a material condition symbol standing for Maximum Material Condition , Least Material Condition or Regardless of Feature Size (implied)
- primary, secondary and tertiary datum features

- for datum features of size, a material boundary condition, such as Maximum Material Boundary (MMB), Least Material Boundary (LMB) or Regardless of Material Boundary (RMB, which is implied, if no MMB or LMB is specified).

FITS AND ALLOWANCES

The assurance that mating features will indeed assemble is one of, if not the, greatest achievement of the geometric dimensioning and tolerancing system. When, for example, a shaft at MMC is compared to its mating feature (a hole) at MMC, if neither has been given a geometric tolerance of form, orientation, position, etc., and/or a general note that allows either to exceed a boundary (envelope) of perfect form as dictated by their MMC limits, then this situation is considered the worst fit situation (refer to the Taylor Principle).

In a clearance fit, when the MMC of the shaft is subtracted from the MMC of the hole, the difference will always be a positive number or zero (a fit with zero difference between the size of the hole and the size of the shaft is categorized as a line fit). This, theoretically, will result in a non-material interference situation 100% of the time. In other words, in their worst cases, largest (MMC) shaft and smallest (MMC) hole, the features will go together.

Clearance fits are the most often discussed fit situation in this text. Other fits are interference and transition. In an interference fit, the mating features are designed so that no matter what size the features are made to (within their size limits) there will always be material interference between the two. In a transition fit situation, the mating features are designed so that if the features are made within their size limits, material interference at mating might, but does not necessarily, occur. This situation will not be covered in this text.

In ANSI Y14.5-1973, there was a definition for the term allowance. Since this deals directly with the fit situations in this section, I will quote from it. *"Allowance - The intentional difference between the MMC limits of size of mating parts. It is the minimum clearance (positive allowance) or maximum interference (negative allowance) between such parts."*

Clearance fits are those mating situations wherein the external features of size, if produced within size limits, are always smaller than their mating internal features of size. Interference fits are those mating situations wherein the external features of size, if produced within size limits, are always larger than their mating internal features of size. Transition fits are those mating situations wherein the size limits of mating features overlap so that the external features of size may be either smaller or larger than their mating internal features of size. Line fits are those mating situations wherein the largest external feature of size is the same size as the smallest internal mating feature of size.

Allowance is the intentional difference between the maximum material conditions of mating features. It may be either positive (clearance) or negative (interference).

RULES, CONCEPTS, CHARACTERISTICS AND UNTOLERANCED DIMENSIONS

INDIVIDUAL OR RELATED TO DATUMS

There are five categories of geometric characteristic symbology. They are as follows:

1) **Form:** contains flatness, straightness, circularity and cylindricity. The form characteristics are always individual (not related to datums).

2) **Profile:** contains profile of a line and profile of a surface. The profile characteristics may, but are not required to use datums.

3) **Orientation:** contains perpendicularity, parallelism and angularity. The orientation characteristics require the use of datums.

4) **Runout:** contains circular runout and total runout. The runout controls require the use of datums.

5) **Location:** contains position, symmetry and concentricity. The location controls require the use of datums. There are, however, instances where position is used to locate a pattern of features, such as holes, to one another only. In these instances, it is sometimes not desirable to locate those features within the pattern to datums. One example of this would be if the pattern of features was to be used as the origin of measurement and orientation (as datums) for all the other features on the part.

Of the following controls:

Form Controls ⏥ — ○ ⌭

Profile Controls ⌒ ⌓

Orientation Controls ⊥ // ∠

Runout Controls ↗ ⌰

Location Controls ⌖

Concentricity and Symmetry were removed from the ASME Y14.5 standard in the 2018 revision. Prior to that, they were location controls.

some are considered to be associated with the control of a feature's shape without consideration of other features--therefore **INDIVIDUAL**.

Some geometric characteristics are controls of the interrelationship of the feature to which the characteristic symbol has been assigned to another feature's perfect geometric counterpart. This type of control is one which relates a feature or features to a datum or datums. These controls are considered **RELATED**.

Both profile of a line and profile of a surface are allowed to be either related or individual. However, profile of a surface is most often related to a datum or datums and profile of a line is often individual (not related to a datum or datums).

The form controls, which include the geometric characteristic symbols of flatness, straightness, circularity (or roundness), and cylindricity, are individual controls (not related to datums). In other words, features that are flat, round, straight or cylindrical are not compared to other features but are compared only to perfect geometric counterparts of themselves. However, all the other geometric characteristics (besides profile of a surface, profile of a line and sometimes position) are always related to datums. For example, when you say some feature is parallel, the first response you should get is "Parallel to what?" Likewise, for angularity (45° to what?) and perpendicularity (Perpendicular to what?) and runout, concentricity, symmetry and usually position.

Once a geometric characteristic is on a drawing and associated with a feature, it is easy to tell whether it is related or individual. If one or more datum feature symbols are used in the feature control frame, the feature is related to one or more datum features.

Position is one geometric characteristic which is usually thought of as associated with datums. For example: a pattern of holes is often positioned to a datum reference frame consisting of three mutually perpendicular planes. It is true that if only one hole or shaft uses a positional control, it must include a datum or datums to be sensical. However, if more than one hole or shaft (a pattern of holes or shafts) is being positioned, it is quite permissible and sensible to exclude datum features from the positional control. If datums are not included, the control is simply trying to convey that the features within the pattern are being "positioned" only to one another. If datums are also included in the control, they convey that in addition to holding the location of one feature to all others within the pattern, location must also be held to the datums.

MATERIAL CONDITIONS

The maximum material condition size requirement can be reduced to a 2 point opposing points measurement requirement by the use of the LMC material condition symbol (B) for the feature in the feature control frame. This also changes the LMC condition check of the feature to one for an envelope of perfect form at LMC.
For example:

⌖	Ø.010 Ⓛ	A	B	C

Once these size limit requirements (MMC and LMC) have been met, then you must check the geometric control imposed by the feature control frame. These requirements can be easily interpreted once you understand the symbols and the ramifications of each.

The material condition concepts of MMC and RFS are those used most commonly. For clearance situations of mating features, the M or maximum material condition symbol is often the most desirable. It provides for more geometric tolerance (drawn from the size limits) as the feature departs from its MMC size. This makes acceptable parts easier to make and, consequently, more economical. It makes inspection faster and easier if receiver gages, or as they are often referred to--functional gages--are used.

These gages are possible to use when the MMC symbol is given because of the increase in tolerance provided by any feature departure (within size limits) from MMC. The gage size is derived for external features by adding the geometric tolerance in the feature control frame to the MMC size of the feature. For internal features, the functional gage size is derived by subtracting the original geometric tolerance in the feature control frame from the MMC size of the feature. See further explanations of functional gaging in this text. The MMC symbol may be used both for features of size and for datum features of size where permitted by the rules of the geometric characteristic being utilized.

The regardless of feature size material condition concept is usually invoked for features that (in their functional situations) must be tightly or more uniformly controlled. RFS says, in essence, that no matter what size the feature is made to (within its size limits), the tolerance in the feature control frame applies. In the straightness control example in FIG. 1, because RFS is implied, the feature has an allowed out-of-straightness control for its derived median line of Ø0.04 whether the feature is made at Ø16.00 or Ø15.89 or anywhere in between (Ø15 89- 16.00).

FIG. 1

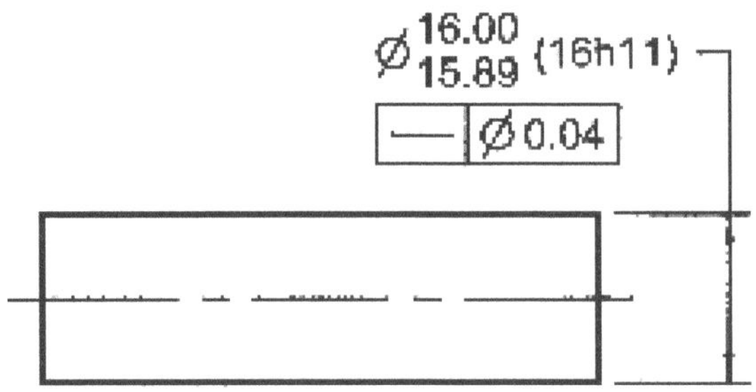

In ASME Y14.5, General Rule #2 states that for all geometric characteristic tolerances, RFS applies with respect to the geometric tolerance, datum reference (or both) where no modifying symbol is specified. MMC or LMC must be specified on the drawing where they are required for a feature of size or datum feature of size.

An alternate practice (referred to as Rule #2a) was allowed in Y14.5-1994 for a position tolerance wherein RFS was actually allowed to be specified by including the circled S material condition symbol inside the feature control frame, with respect to the geometric tolerance and/or datum features of size referenced in the control. This alternate practice could be classified as a clarifying redundancy and a carry-over from ANSI Y14.5M-1982 in that if the regardless of feature size symbol is not specified (nor another modifier), it is implied per ASME Y14.5M-1994. This alternate practice was removed from ASME Y14.5-2009.

Screw threads: Unless otherwise specified, the axis of a threaded feature is dictated by its pitch diameter. Each geometric tolerance of orientation, position and datum reference specified for a threaded feature refers to this axis. Any desired exceptions to this practice (such as major dia. or minor dia.) must be specified beneath the feature control frame or datum feature symbol.

Gears and splines: When gears or splines are given tolerances of orientation or location, or made datum references, it must be specified beneath the feature control frame or datum feature symbol as to which specific feature of the gear or spline each applies. Common used designations include: MAJOR DIA., MINOR DIA., and PITCH DIA.

The maximum material condition symbol allows a tolerance zone to grow when used on size features as the features depart in size from maximum material condition. Functional gaging, a receiver type gage, may be used (but is not required) to inspect geometrically controlled features of size when the features use the maximum material condition symbol in their feature control frame. These functional gages are made to reflect the combined effect of MMC and any applicable geometric tolerance (a boundary defined as virtual condition). In these instances, functional gages are made to the size of the virtual condition boundary of the features being inspected.

The least material condition symbol allows a geometric tolerance growth as the feature of size departs from least material condition. When this symbol is used, the MMC envelope is converted to an opposing points verification, while the LMC opposing points check is converted to an envelope requirement.

The regardless of feature size concept allows no growth in the tolerance zone.

In summary, for tolerances of position where the maximum material or least material condition concept is required, MMC or LMC must be specified on the drawing with respect to the individual tolerance, datum reference, or both, as applicable. But, as with all other geometric characteristics, RFS applies with respect to the individual tolerance, datum reference, or both where no modifying symbol is specified. MMC or LMC must be specified on the drawing as required for other geometric characteristics, just as they must be for position.

UNTOLERANCED DIMENSIONS

There are several dimensions on a drawing not considered associated with tolerances--at least not immediately. The closer we look, however, we come to realize this can be a matter of how one chooses to interpret the situation.

Basic dimensions, for example, are given no tolerance right next to the dimension. They locate, in many instances, the exact true position--the perfect location, if you will--of features. But, as we all know, nothing in the real world is made perfect. Consequently, the tolerance for this perfect location (its allowed deviation from true position) is to be found in a feature control frame. In this case, it is not that the basic dimension has no tolerance, just that the tolerance is not found next to the basic (boxed) dimension. The tolerance is, instead, found in the feature control frame that controls the feature(s) and includes the datums from which the basic dimensions emanate.

In certain situations where basic dimensions are used to describe the location and/or size of a datum target area, the tolerance for these basic dimensions is assumed to reside in established tooling or gaging tolerances for that product or facility. If no tooling or gaging tolerances have been established for this situation (datum targets), toleranced dimensions are legal and should be considered instead of basic dimensions.

Reference dimensions are another instance where tolerances are not stated. The dimension is usually enclosed in parentheses (preferred). In past practices, this was designated with the notation "REF" after it. Reference dimensions are redundant dimensions, something that has been stated elsewhere on the drawing or something that can be calculated given other dimensions or information on the drawing. They are, many times, the sum of several nominal dimensions already stated. Reference dimensions are given for convenience and are not meant to dictate manufacturing or inspection criteria.

Single limit dimensions are considered non-toleranced. Maximum and minimum dimensions are single limit dimensions. Minimum dimensions are used, many times, to preserve minimum wall thicknesses between edges of internal features of size (holes, slots, etc.) and part edges. Maximum dimensions might be used to state a hole's depth to preserve a minimum of material thickness between the hole's bottom and the underside of the part. Both maximum and minimum dimensions are used where intent is obvious and the only important dimension is the one stated. The other unspecified limit is dictated or controlled by other feature dimensions on the drawing. MAX or MIN is put after the dimension to which it applies. Some other uses for single limit dimensions are chamfers, corner radii and thread lengths.

In summary, one of the fundamental rules of tolerancing is that all dimensions shall have a tolerance. Basic dimensions, although having no tolerance directly applied to the dimension, draw their tolerances from feature control frames or, when used to define datum target location and datum target area size, draw their tolerance from established tooling or gaging tolerances.

Reference dimensions, however, are usually untoleranced. This is the case because they are auxiliary pieces of information--redundant, helpful, but not to be inspected.

Single limit dimensions are dimensions that state either MAX or MIN. The other limit not stated is either calculable from other information given on the drawing, or not considered needed. If a MIN dimension is given and no maximum is calculable, the maximum is considered to be infinity. If a MAX dimension is given and no minimum is calculable, the minimum is considered to be zero.

COMPONENTS COMMON TO A FEATURE CONTROL FRAME

A feature control frame consists of components. These components have specific names and each falls into a major category which generally describes the function of the component. All symbols that show the geometric form or relationship being toleranced are referred to as geometric characteristic symbols. The diameter symbol is often referred to as a zone descriptor because it shows the shape of the tolerance zone. Any tolerances shown inside the control are known as geometric tolerances. An MMC, LMC or RFS symbol, if included or implied, is known as either a modifier or material condition symbol.

If datum features are referenced, they are read from left to right as primary, then secondary and then tertiary datum features. Some feature control frames contain only a geometric characteristic symbol and a geometric tolerance, while others use more--or even all--of the components discussed here. Feature control frames will sometimes appear in combination with other needed information, such as other feature control frames often stacked one atop another, or datum feature symbols which designate that the feature will be used to measure other features from, or to orient those other features. These datum features and the relationships held to them will be called out in subsequent feature control frames found on the drawing.

Projected tolerance zone symbols are sometimes included inside the control to more functionally define orientation and position for such features as threaded holes or holes that will have pins tightly pressed into them later in the assembly. For example:

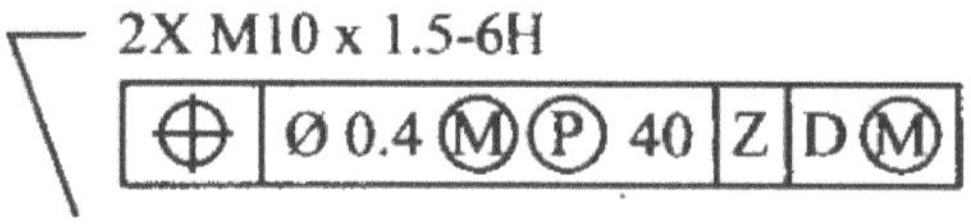

Composite feature control frames contain one geometric characteristic symbol, often position, for two levels of control. When this is used, it shows the location relationship between the features must be held to a tighter tolerance than the pattern of features' location to any datums listed in the upper level of the control. In these situations, the tolerance in the upper datum-to-pattern relating tier of the control is always larger than the tolerance in the lower feature-to-feature relating tier of the control. If any datums are used in the lower feature-to-feature relating control, they merely act to refine the orientation of the feature relating tolerance zones. Any datums used in this lower tier of a composite positional control are not meant to, nor are they capable of refining location to those datums to within the smaller tolerance listed. See FIG. 6 item D.

If location refinement is desired, two or more separate position symbols in two or more separate feature control frames can be used. They may be stacked one atop another or separated.

FIG. 2

The above represents the least one would find in a feature control frame. The amount of data included in any feature control frame will depend on design requirements and individual geometric characteristic limitations. For example, the most components used in an individual form control would be for straightness of a derived median line at MMC. It might appear as follows:

FIG. 3

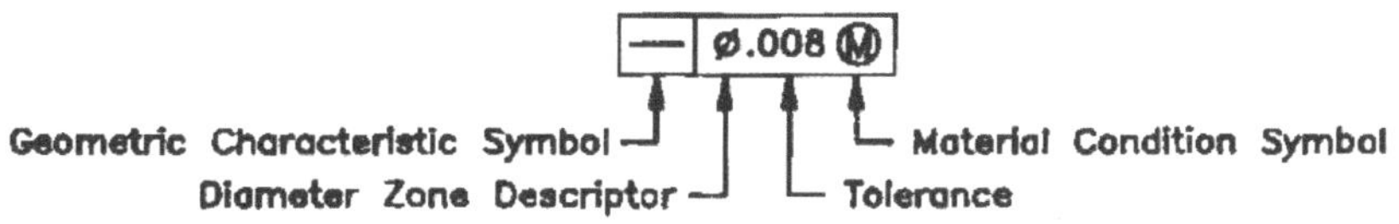

Related geometric characteristics have different requirements, each of which will be discussed in detail in later units of this text. However, a good example of a characteristic which allows the feature control frame to expand its data is shown below using position. For example:

FIG. 4

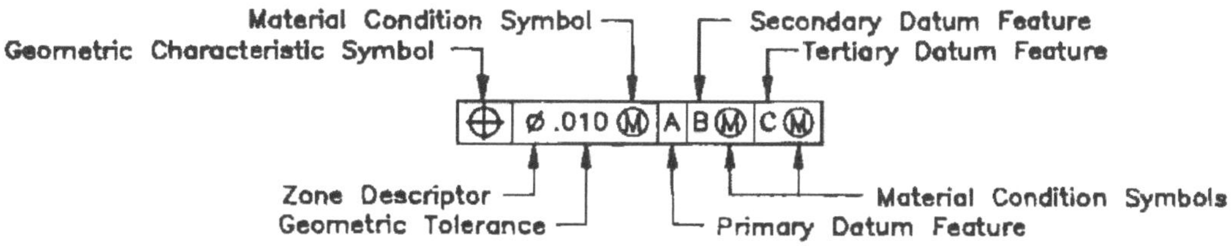

Sometimes one feature is controlled to an axis which is formed by two coaxial datum features simultaneously. This is often referred to as a compound datum feature. For example, see FIG. 5.

In the previous example, we know datum feature C has its own axis and datum feature D has its own axis, but the larger diameter is controlled to the perfect axis formed between the two. Consequently, the control C-D is shown. When this technique is used, some thought should be given to controlling the coaxiality of datum features C and D to one another prior to using them to generate one common compound datum axis. If datum features C and D are grossly out-of-coaxiality to one another, repeatability of measurements taken from compound datum axis C-D may be unreliable.

For the sake of introduction, the datum feature symbols were identified as such on the drawing in FIG. 5-5. This, of course, is not done on regular drawings, nor will it again be done on drawings in this text. Sometimes, additional pieces of information are combined with feature control frames. For example, see FIG. 5-6.

FIG. 5

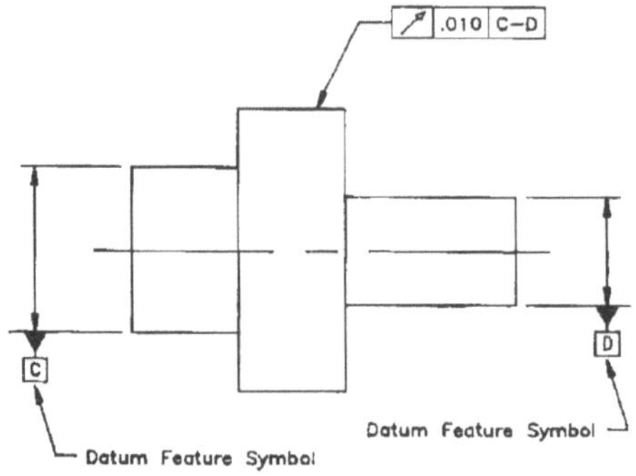

FIG. 6

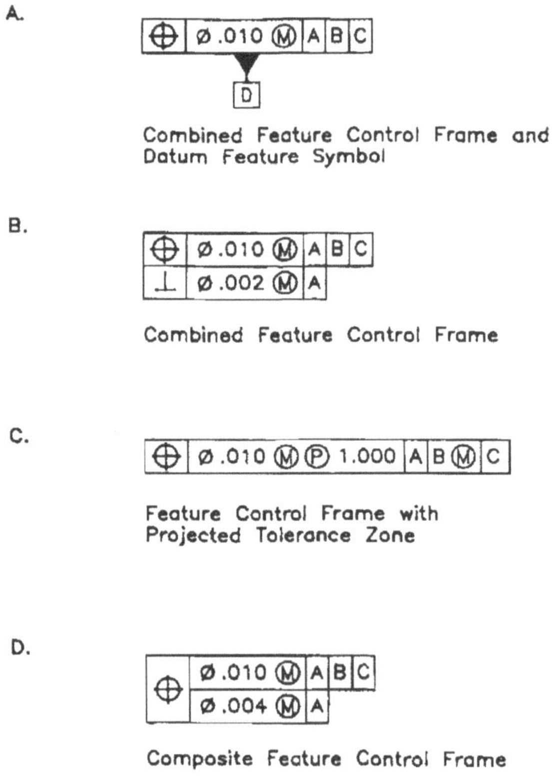

GD&T General Principles--an Introduction

The Four Things We Tolerance

There are only four tolerances we can apply. These are tolerances on: size, form (shape), angle and location.

The Three Types of Controls

There are three types of geometric characteristic controls. They are: surface controls, axis controls and centerplane controls. These are covered by the fourteen geometric characteristic symbols in five categories. The categories are: Form, Orientation, Profile, Runout and Location.

Surface Controls

Surface controls are used to tolerance the shape of a surface. Examples are flatness, straightness, circularity, cylindricity, profile, runout, perpendicularity, angularity and parallelism. Some of these geometric characteristic symbols can be used to control axes or centerplanes instead of surfaces. Some are relationship controls that reference datums and others are not.

Axis Controls

Axis controls are used to tolerance the shape, orientation (angle) and/or location of a feature's axis. These controls are only used on features that are nominally round, such as cylinders and spheres. Often a diameter symbol is required to create a cylindrical tolerance zone to confine the axis. Examples are position, perpendicularity, angularity, parallelism, straightness (of the derived median line) and pre-2018, concentricity (of the median points). **The concentricity geometric characteristic symbol was removed from the ASME Y14.5-2018 standard.**

Centerplane Controls

Centerplane controls are the third type of control. They are used to control and confine the centerplane of a feature's width. Examples of this concept are used in position, perpendicularity, angularity, flatness of the derived median plane [a concept that was new to the Y14.5 2009 standard and carried over to the Y14.5-2018 standard] and **symmetry (of the median points) prior to the issuance of the Y14.5-2018, when the geometric characteristic symbol of symmetry was removed from the standard.**

Tolerances

There are two types of features of size. All are associated with a size dimension.
A "Regular Feature of Size" is:
1) one cylindrical surface
2) one spherical surface

3) a set of opposed parallel surfaces

4) a circular element
5) a set of two opposed elements

An "Irregular Feature of Size" was a concept created for ASME Y14.5-2009 and carried over to the ASME Y14.5 2018 standard.

An "Irregular Feature of Size is":

1) a feature or collection of features that may contain or be contained by an actual mating envelope which is a sphere, a cylinder or a pair of parallel planes.

2) a feature or collection of features that may contain or be contained by an actual mating envelope other than a sphere, cylinder or pair of parallel planes.

Actual Mating Envelope

This envelope is outside of the material. It is a similar perfect feature(s) counterpart of smallest size that can be contracted about an external feature(s) or largest size that can be expanded within an internal feature(s) so that it coincides with the surface(s) at the highest points. There are two types of actual mating envelopes as described below.

Unrelated Actual Mating Envelope vs. a Related Actual Mating Envelope

1) An Unrelated Actual Mating Envelope is a similar perfect feature(s) counterpart expanded within an internal feature(s) or contracted about an external feature(s) and not constrained to any datum reference frame.

2) A Related Actual Mating Envelope is a similar perfect feature(s) counterpart expanded within an internal feature(s) or contracted about an external feature(s) while constrained either in orientation or location or both to the applicable datum(s).

It should be noted that features such as these will often have multiple Related Actual Mating Envelopes depending on the datum or datums to which they are constrained.

What a Feature of Size Needs.

A feature of size often needs:

1) a desired size
2) a tolerance on that size
3) a desired location and/or orientation, and
4) a tolerance on that location and/or orientation.

Often a person confuses these items, thinking a size automatically includes a tolerance and so they do not state one. This happens with location as well. A location and/or orientation will often be shown on the drawing or in the CAD model without saying what tolerance on that location and/or orientation is acceptable. Size, orientation and location show dimensions, and one of the basic rules for dimensioning is that all dimensions need a tolerance--because perfect parts cannot be produced.

Tolerances of (1) size and (2) location, orientation, profile or runout are requirements that often must be considered independently of one another. Size limits can be stated several different ways. Size tolerances control form tolerances unless otherwise specified. See flexible parts, average dimensions and stock in the as-furnished condition for exceptions to this rule.

General Rules for Dimensioning and Tolerancing

- A zero basic dimension applies where axes, center planes or surfaces are shown coincident (in the same place) on a drawing and geometric tolerances establish the relationship between the features. [Although this concept is not new to previous Y14.5 standards, it was explicitly stated for the first time in the Y14.5-2009 standard and carried over into the Y14.5-2018 standard.]

- A general note should appear on all drawings, such as "Unless otherwise specified, all dimensions are in inches." or "Unless otherwise specified, all dimensions are in millimeters.".

- When metric dimensions are shown on a drawing that is in inches, the metric dimensions will be designated with a local note, such as "mm", next to the dimension to show that dimension

- When inch dimensions are shown on a drawing that is in millimeters, the inch dimensions will be designated with the local note "IN." next to the dimension to show that dimension is in inches.

- Where a coordinate system is shown on the drawing, it shall be right-handed unless otherwise specified. Each axis shall be labeled and the positive direction shown. [These rules were new to the Y14.5-2009 standard and restated in Y14.5-2018]. Where a model coordinate system is shown on the drawing, it shall be in compliance with ASME Y14.41M.

Inch Tolerances

- When inch dimensions are used, both the dimension and the tolerance will use an equal number of decimal places.
  For example:

  $$1.000^{+.005}_{-.000}$$

- When the tolerance is zero for either the plus or minus value, the appropriate plus and minus signs are both shown and the values of tolerance will use an equal number of decimal places.
  For example:

  $$1.000^{+.005}_{-.000} \quad \underline{\text{NOT}} \quad 1.000^{+.005}_{0}$$

- When a limit dimension is used, both extremes of the dimensional values will use an equal number of decimal places. For example: .950
  .948

- No zero precedes the decimal point when the value is less than one inch. For example: .954, not 0.954

- In ASME Y14.5-2009 and ASME Y14.5-2018, when basic dimensions are used, the associated feature control frame tolerance no longer has to use an equal number of decimal places as the basic dimensions it tolerances. For example, a basic dimension of [2.000] may now be toleranced by a feature control frame such as [⌖|∅.01Ⓜ|A|B|C]. **This is a change from ASME Y14.5M-1994.**

Millimeter Tolerances

- When expressing a tolerance for a metric dimension and either the plus or minus tolerance is zero, the zero is specified as a single zero without a decimal point. Also, no plus or minus sign is shown for the zero tolerance.
  For example:

  $50\ ^{0}_{-0.01}$ OR $50\ ^{+0.01}_{0}$

- When the tolerance is bilateral, it is expressed with an equal number of decimal places for both the plus and minus tolerances.
  For example:

  $50\ ^{+0.02}_{-0.01}$

- When limit dimensioning is used, both dimensional limit extremes will have an equal number of decimal places. For example: 30.25
  29.50
- As with other metric dimensions, when basic dimensions are used, if the basic dimension is a whole number, neither the decimal point nor a zero is shown. The associated feature control frame tolerance contains the number of decimal places without additional trailing zeros.

- A zero is used to precede the decimal point when a metric dimension is less than one millimeter. For example: 0.25, not .25

Angular Units and Tolerances

Angular dimensions are shown on a drawing in either degrees and decimal portions of a degree or in degrees, minutes and seconds. The universal symbols for degrees, minutes and seconds are specified after their numerical value.

- Where angular tolerances are expressed in degrees, both the plus and minus tolerance values and the angle will use the same number of decimal places. For example: $30.0^{\circ} \pm 0.2^{\circ}$, not $30^{\circ} \pm 0.2^{\circ}$. For Example: $30^{\circ} \pm 0^{\circ}\ 30'$ not $30^{\circ} \pm 30'$.

- A 90° angle is implied on a 2 dimensional orthographic drawing where features are drawn at right angles and on centerlines depicting a 90° relationship. An angle other than 90° would have to be specified. The tolerance on a 90 degree angle must be specified. Some of the ways to show this tolerance are:

1) letting the title block general note cover it--a note such as: "Unless otherwise specified, the tolerance on all angles is plus or minus one degree."
2) specify a perpendicularity tolerance in a feature control frame.
3) specify another type of geometric tolerance, such as position of the feature which includes a datum feature meant to control perpendicularity (along with any other datums which may be needed for location).

- The 90 degree angle becomes basic when surfaces or features of size are depicted at 90 degrees and are located or defined by basic dimensions. Again, these basic 90 degree angles must be toleranced in an appropriate manner.

Radii

The numerical value of a radius is preceded by the letter R for radius.

- When a radius center is unlocated by dimensions, the radius center is not shown. When this is done, it must be clearly indicated that the radius arc location is controlled by other features that are dimensioned on the drawing. An example would be when the unlocated radius is to be blended to two located radii. Another example is the use of a radius on the corner of a 90Δ angle.
- On a 2D orthographic drawing, the local note "TRUE R" precedes the numerical value of a radius when it is shown in a view that does not show the radius' true shape.
- As with diameters, where multiple radii of the same numerical value are to be shown on the same drawing, the number of radii followed by the "X" symbol for "times" or "number of places" followed by the numerical value of the radii may be used. For example: 8X R1.5
- For spherical radii, the designation "SR" is used.
- For fully-rounded ends of features with overall dimensions (such as an elongated hole), the letter "R" is used to indicate a radius, but no numerical value is needed. However, if the controlled feature is only partially rounded on the ends, the letter "R" is followed by the numerical value of the radius.
- The tolerance zone created by the symbol "R" (for radius), its value, and tolerance is defined by two arcs that represent the minimum and maximum value of the radius. The surface of the radius simply must reside between the two arcs to be acceptable.
- When the symbol "CR" is used, the concept of controlled radius is invoked. A tolerance zone defined by two arcs representing the minimum and maximum radii is created tangent to the adjacent surfaces. The controlled surface of the radius must not only reside within the tolerance zone, but the contour must be a fair curve and may not have reversals.

Statistical Tolerancing

Statistical tolerancing was a concept added to the dimensioning and tolerancing standard in 1994. It allows the use of tolerances on an assembly based on statistics.

- A commonly used assembly tolerance formula is: the tolerance likely to be consumed by manufacturing is equal to the square root of the sum of the squares of the individual tolerances. This type of prediction has the advantage of allowing the increase of individual feature tolerances for the reduction of part costs. Methods such as this may be employed when a manufacturing system is brought under control through the use of statistical process control. It should not be used when a manufacturing facility/process is out of statistical control or in statistical chaos, producing parts without appropriate repeatability and/or accuracy. The more restrictive tolerances derived by arithmetic stacking limits should be employed in these situations.

- A note may be added to a drawing such as "Features identified as statistically toleranced using the symbol ST shown in a hexagonal box shall be produced with statistical process controls, or to the more restrictive arithmetic limits", in which case a feature will be listed with two tolerances--one tighter tolerance for use in facilities out of statistical control and one a looser ST (Statistical Tolerance) to be used by manufacturers under statistical control and using statistical methodology in their production procedures. If the tolerance is a statistical geometric tolerance, the symbol is used in the feature control frame after the geometric tolerance and any modifier it uses. For more information, see the unit on Statistical Tolerancing.

When Tolerances Apply

- Any dimensions and tolerances given a feature to be plated or coated shall specify if the dimensions apply before or after plating/coating. A general or local note to this effect is recommended.

- All dimensions and tolerances apply in the free state and at 68 degrees F (20 degrees C) unless specified otherwise. Parts to be restrained while inspected must state so with a drawing note. As a clarification, it is permissible to use the free state symbol (a circled F) inside the feature control frame to indicate the part is measured in the free state. This is done when a note or other indication appears that states, "UNLESS OTHERWISE SPECIFIED ALL GEOMETRIC TOLERANCES ARE INSPECTED WHILE THE PART IS RESTRAINED". However, even without the use of the free state symbol, the part is measured in the free state unless a restrained condition note is specified.

- The same is true of the temperature at which a part is measured. Unless a note is used to specify that a part's toleranced dimensions apply at a temperature other than 68 degrees F (20 degrees C), the part must meet its dimensional requirements at that temperature.

- All dimensions and tolerances apply only at the drawing level at which they are shown. For example, dimensional requirements given at the detail drawing level do not have to be met at the assembly drawing level. Therefore, it is recommended that each level of drawing retain or specify dimensions, tolerances and relationships important at that level that may otherwise not be met.

- Dimensions not shown in model views are defined within the model and are considered basic. (ASME Y14.5-2018)

Absolutes

- All limits of a dimension are absolute, regardless of the number of decimal places. It is as though an infinite number of zeros existed after the last number given.
  For example: 15.3 means 15.30---0
  15.0 15.00---0

- When the part is inspected, the measured value is compared directly with these limits. Any measured dimension outside the limits is in violation of the tolerance, regardless of the number of decimal places shown in the dimensional limits.

Plus and Minus Tolerancing

- Bilateral (tolerance on both the plus and minus sides). Examples are shown in inches.
  a. Example b. Example

  $.500 \pm .010$ $.500^{+.003}_{-.001}$

- Unilateral (tolerance on only the plus or the minus side, but not both at the same time)
  a. Example b. Example

  $.500^{+.005}_{-.000}$ $.500^{+.000}_{-.006}$

Limit Dimensioning

Limit dimensioning is stating the actual boundaries of size in which you must stay.

- Linear, or one next to the other, has the smallest number preceding the larger, separated by a dash. Example: .500 - .510

- When one limit is placed above the other, the larger number is positioned over the smaller.
  Example: .510
  .500

Countersinks and How to Display Them

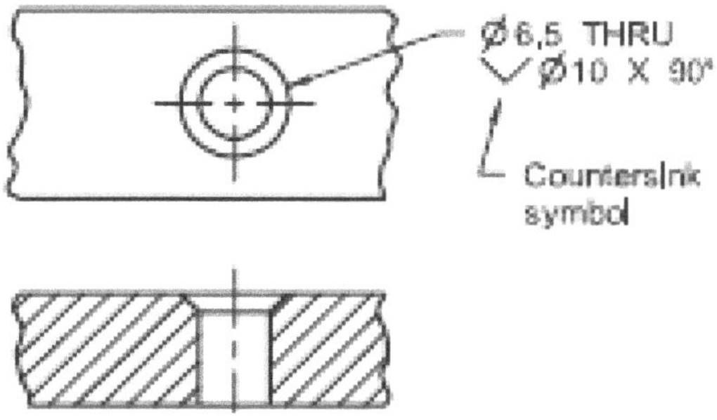

HOW TO READ A FEATURE CONTROL FRAME

OUTLINE

• Symbols and the functional information they convey

ESSENTIAL TERMS

feature control frame; Datum Feature; Diameter; Location; Perpendicularity; Functional Gage

LEARNING OBJECTIVES

After reading and discussing this chapter, you should be able to do the following:

1. Read a feature control frame as though it was a sentence written in symbology to convey functional requirements.

HOW TO READ
A FEATURE CONTROL FRAME

Symbology contained in a feature control frame can convey information vital to the functionality of a part. In order to understand this symbology, one must not only be able to read the individual name of a symbol but also the implications and ramifications of the symbols when strung together in sentence or paragraph form. This section begins to explain some of the vital nuances in reading part functionality from a feature control frame.

Just as we have replaced many of the words on signs with symbols (as "No Smoking" signs now appear as a guy smoking a cigarette with a dagger stuck in his chest and "Don't Walk" signs are now a guy plastered on the front of a Buick), we are replacing many of the notes on our engineering drawings with symbols. These symbols can, indeed, be read as one would read a sentence to describe how a part functions or is to be measured or made. For example:

FIG. 7

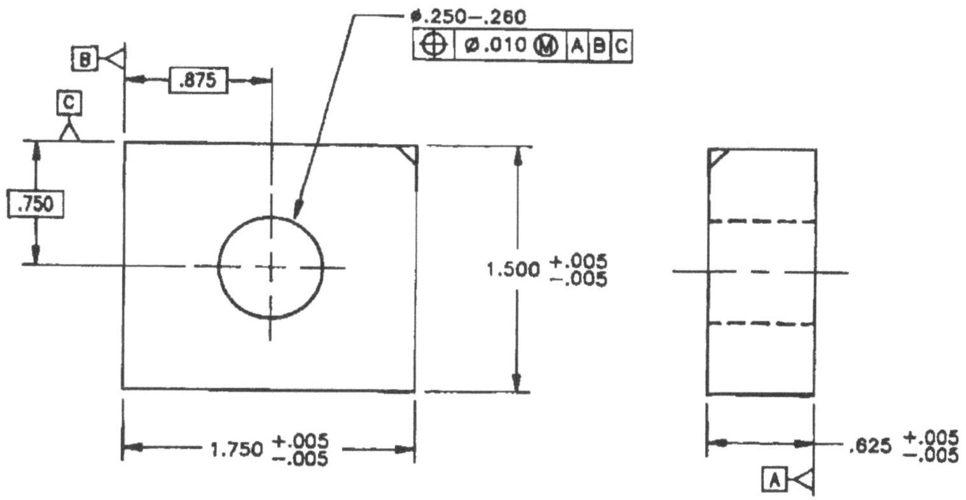

This feature control frame can be read, starting with the diameter sign:

Ø = The axis may be out of

⊕ = Position

Then, reading the diameter sign again:

Ø = a diameter of .010 = .010

M = if produced at maximum material condition (a diameter of .250)

A = to A for perpendicularity

B = and B for location (holding the .875 dimension)

C = and C for location (holding the .750 dimension)

So, as a complete sentence, it may read, "The axis of this hole may be out of position a diameter of .010 if the hole is produced at a diameter of .250 holding perpendicularity to datum A, and location to datums B and C." It may also be read, "Position of the axis of this hole must be held to A for perpendicularity and to B and C for location to within a diameter of .010 if the hole is produced at .250." It also implies some important information. It implies that this hole may be further out of its perfect position if the hole grows. For example, .011 would be allowed for a hole made at a diameter of .251 and .012 for a .252 hole.

This is so because the hole uses the maximum material condition symbol. The maximum material condition symbol allows the geometric tolerance of location to grow as the hole grows and thereby creates a reduction in part cost by the acceptance of more of the parts that have been produced. This increase in geometric positional tolerance is only allowed because the functional requirements of the part are not endangered by this tolerance zone growth. This is based on the information that using the maximum material condition symbol implies. It says that a larger hole has to be less perfectly located than a smaller hole. This is most often true when a hole mates. A shaft will more easily fit into a larger hole; therefore, the location of that larger hole can afford to be less perfect.

In fact, we may choose to read the control in an entirely different manner because of the circled M in use. For example, we might read it as: "Position of a diameter that mates while the part is seated on A and the hole is located from B and C." And, if it does indeed mate, we have to assume we are trying to preserve a mating boundary. That mating boundary is the collective effect of the actual size of the hole and the positional imperfection allowed to it.

For example, when a hole moves or leans out of its perfect location in space, its available area for mating at that location and angle closes up by the amount that it has moved or leaned. Since we cannot allow it to infringe on the boundary in which the mating shaft will reside or interference may occur, we protect that mating boundary (virtual condition-MMC concept).
So, a hole produced at Ø.250 may move or lean only Ø.010, while a hole produced at Ø.255 may move or lean out of its perfect location Ø.015 because we have determined the mating boundary to be protected to be Ø.240 and:

Ø.250 - .010 = Ø.240	Ø.251 - .011 = Ø.240	Ø.252 - .012 = Ø.240
Ø.255 - .015 = Ø.240	Ø.258 - .018 = Ø.240	Ø.260 - .020 = Ø.240

Therefore, we have determined that if the holes produced at these sizes are out of their perfect location by no more than an amount that would allow them to mate with a Ø.240 or smaller shaft, they will function and should be accepted by inspection as being within their functional limits of hole location.

This is only a bare-bones sketch of all the information that can be packed into one of these controls.

Here's another example:

FIG. 8

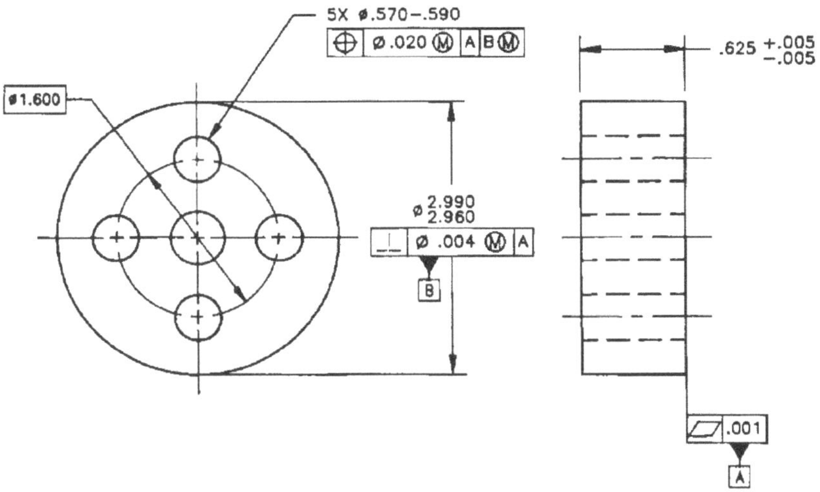

This positional control can be read as "Position of diameters that mate while the part is seated on A and the holes are centered to B while B mates." It tells us much about the function of the part. If round holes mate, what do they mate with? They mate with round shafts. If so, what is the largest shaft that will mate with the holes in a worst-case situation? The shafts could be no larger than Ø.570 - .020 = Ø.550, so the five mating shafts will have worst conditions of Ø.550 or smaller. They will mate while the plate they are mounted on seats on A. They will do so while datum feature B mates.

If B (a round shaft) mates, what does it mate with? It mates with a round hole. If so, what is the smallest hole that will receive B in a worst-case situation? The hole could be no smaller than Ø2.990 + .004 = Ø2.994. So, the part (or perhaps the portion of the part) that mates with this part consists of five pins in a cylindrical cavity. The pins are a maximum of Ø.550 and the cavity is a minimum of Ø2.994.

If drawn correctly, every picture tells a story. The story is one of function; and, although the mating part may not be sized at exactly Ø.550 and Ø2.994, if this part as produced was capable of fitting such a part, it would also be capable of fitting the worst mating part (if it was also dimensioned properly) without interference. This is so because what we have described here is a receiver-type functional gage. This functional gage would be capable of checking both the perpendicularity control shown and the positional control. It would not check feature size limits

.

The gage may or may not be needed for inspection purposes. Certainly, the part could be inspected in many different ways. However, the process we just went through (of visualizing the worst conditions allowed) is important and allows us to understand exactly what the drawing asks for and why.

The untoleranced gage may appear as follows:

FIG. 9

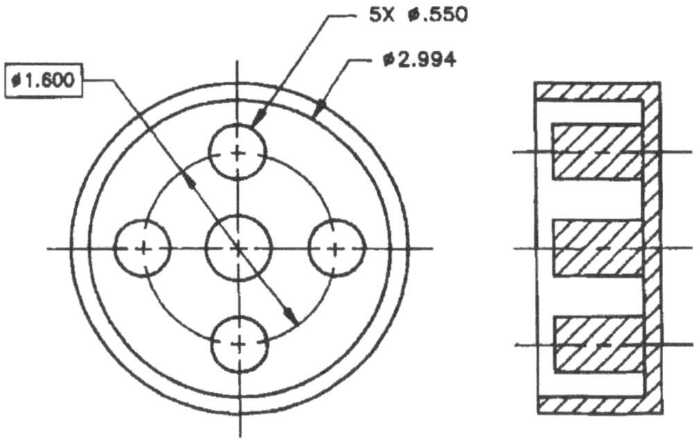

With only some of the dimensions and tolerances, the gage may appear as follows:

FIG. 10

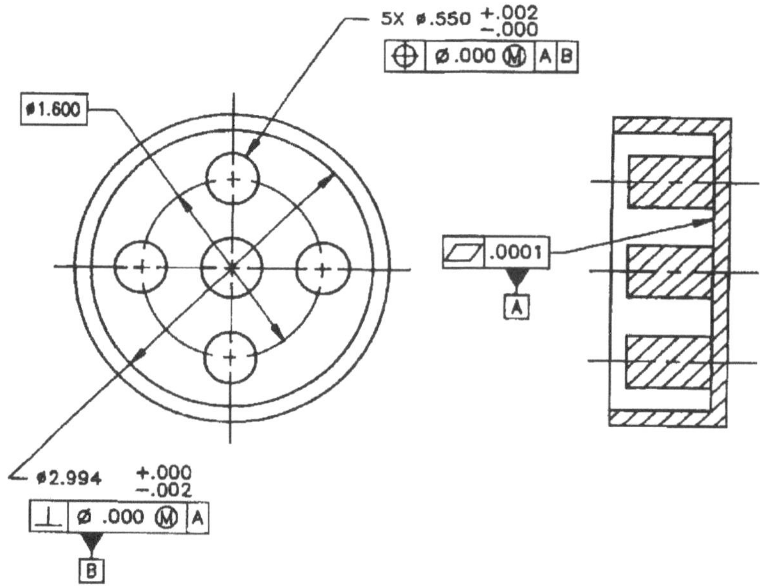

Chapter
2

Selecting a Tolerancing Approach

•Datums & Datum Features
•Steps in a Tolerancing Scheme
•How to Read a Feature Control Frame
• Datum Targets-See Separate Objectives for this Section

Chapter Objectives

Readers will learn:

1. How to label a surface as a datum feature for measurements of orientation and/or location.
2. What a datum and a datum feature are and why they are chosen.
3. How to tolerance datum features and how to relate them to one another.
4. How to calculate and assign geometric tolerances.
5. When a Simultaneous Requirement applies and what it means.
6. How to sequence a common tolerancing scheme.
7. How to read a geometric control as though it was a sentence.

Datums and Datum Features

Datum Feature Symbols

FIGURE 11 [Datum Feature Symbols - Some Typical Ways to Specify a Datum Feature]

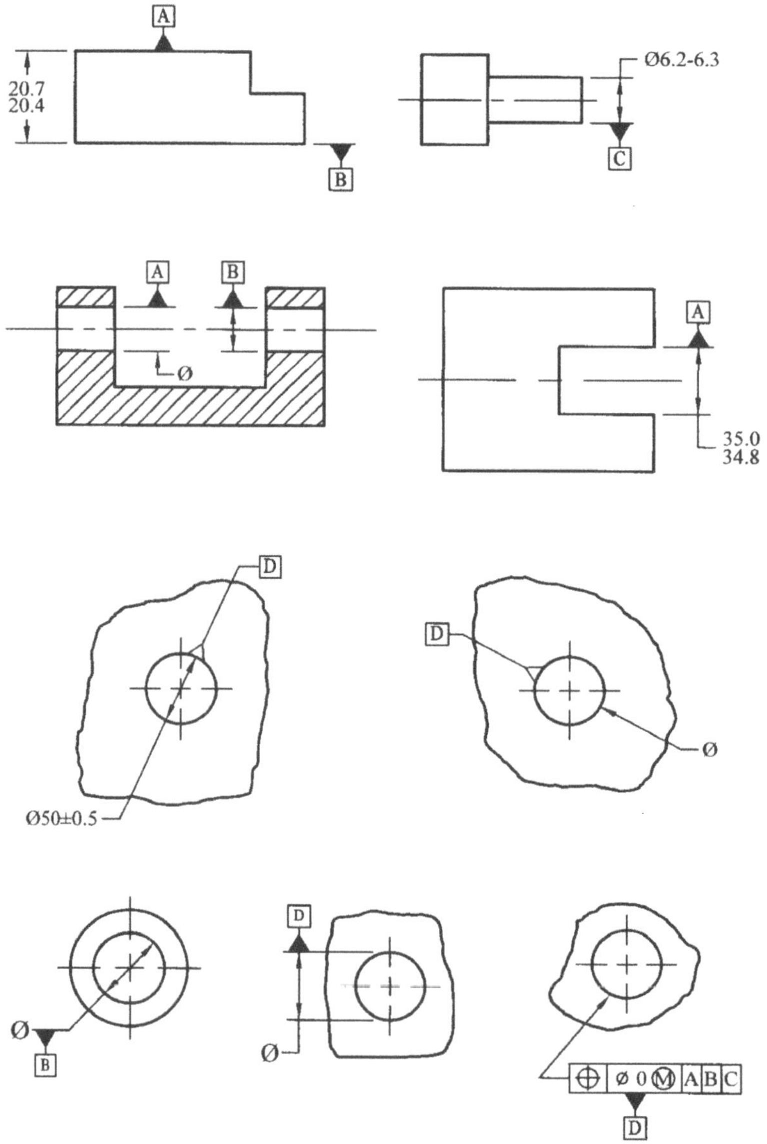

The datum feature symbol consists of a capital letter inside of a square and has a leader line that extends from the frame to the feature, terminating with a triangle. The triangle may be either filled in or not. The letters I, O and Q are not used for datum features. All other letters may be used and are to be assumed as either primary, secondary or tertiary based on their appearance in the feature control frame as read from left to right, not on the letter's alphabetical order. When the alphabet has been exhausted, datum feature designations may begin again by using a double alpha series, such as AA, AB, BA, BB, BZ, etc., as the user deems appropriate. The datum feature symbol may be used in more than one place on the drawing to clarify that a particular feature is, indeed, the datum feature. In these instances, the datum feature symbol need not be identified as reference.

The datum feature symbol is shown applied to the surface outline, dimension line, extension line, phantom line, chain line or feature control frame. For example, it may be placed on the outline of a feature surface or on an extension line of the feature outline. If it is meant to designate the one surface only as the datum feature, and not a datum feature of size, it is important in that instance to clearly separate the datum feature symbol from the dimension line. If meant as a datum feature of size, the datum feature symbol can be placed on an extension of the dimension line of a size feature. This will designate that the datum generated by the datum feature of size is either a datum axis, a datum centerplane or (as is the case with a spherical feature) a datum point. For the designation of a datum feature of size, the dimension line associated with the datum feature symbol may consist of one line with an arrow at both ends or one line with an arrow only at one end and terminating at the other with the datum feature triangle.

Other ways to show a datum feature of size are:

- to place the datum feature symbol on the outline of a cylindrical feature surface or on an extension line of the feature outline, separated from the size dimension. This designates that the datum is the axis of the cylindrical feature.
- to accommodate Computer Aided Design (CAD) systems, the triangle may be tangent to the feature.
- to place the datum feature symbol alone on a dimension leader line to the feature size dimension where no geometric tolerance and feature control frame are used.

For complex or irregular features, or as a clarifying redundancy to re-identify previously established datum axes or planes as reference in other places on the same sheet or on other sheets of a multi-sheet drawing, the datum feature symbol may be placed on the planes or axes established by datum targets.

The datum feature symbol may also be placed below and attached to the feature control frame to show that the controlled feature or features are generating a plane or axis from which we measure.

For a visual display of these techniques, see FIGURE 11 of optimal ways to designate a datum feature.

Datum Features

Datum features are real surfaces on a part used to construct imaginary axes or planes from which to measure angle and/or location. Datum reference frames are coordinate systems that consist of three mutually perpendicular planes used to stabilize parts, to set up on and measure from.

Datum features are selected for a variety of reasons, sometimes to expedite manufacturing processing of the part. Whenever possible, datum features should be selected to be functional part features that seat, mate and/or align the part with the next part in the assembly. It is especially important that the primary datum feature be of sufficient surface area to stabilize the part in the same way it seats in the assembly. Surfaces to which the part bolts, surfaces that dictate the angle at which the part fits into the assembly, and surfaces that will come into the most physical contact with the next part in the assembly make good primary datum features. These surfaces should be considered for a form control (such as flatness for a planar primary datum feature or cylindricity for a cylindrical primary datum feature) to ensure repeatability in measurements taken from the primary datum plane or axis generated by this surface.

Geometric tolerances that are related to a datum reference frame don't take into account variations in form, orientation, or location of the datum features. **Datum features shall be controlled by applying appropriate geometric tolerances and/or by size dimensions. To make it possible to calculate the true geometric counterpart boundaries of each datum feature in a datum reference frame, a relationship between the datum features shall be specified.**

Defining, Tolerancing and Qualifying Datum Features

The application of datum features and geometric tolerances can best be understood in terms of a linear logic. Progressively more restrictive and powerful control is applied as one progresses through the part definition until a point is reached of maximum control.

So, how does one select the geometric characteristic symbols and the datum features?

Primary Datum Feature

The primary datum feature, once selected, must be toleranced for form. ASME Y14.5 states that size tolerance controls form tolerance (Rule #1), but if the size tolerance is too liberal to control form well enough, a form control is used to refine it. A flatness control is commonly considered for a planar surface. A cylindricity control is commonly considered for a cylindrical surface. The tolerance chosen is most often decided by determining manufacturing capability at an affordable cost to the customer.

Secondary Datum Feature

The secondary datum feature is to be selected based on its functional ability to give an origin of measurement that mates and/or aligns the piece part in the assembly. For example, see FIGURE 12 on the next page.

Datum feature B has a tighter tolerance than the two holes subsequently positioned to it. If the mating shafts that fit into these holes fit with less clearance into B than the other holes, B has a better chance of aligning the two parts in the assembly. **Features such as this, with the tightest fit from part to part, doing the alignment, make good datum features.** This secondary datum feature must then be toleranced in its relationship to the primary datum. The relationship between the axis of datum feature B and datum plane A is depicted as one of perpendicularity. So, perpendicularity is implied, but we must specify how far from a perfect 90° angle it may stray. This tolerance could be specified in degrees, such as 90° ± 1 °. But since it is a mating feature, it would be better to calculate the tolerance using a formula that would specify the tolerance in a perpendicularity control – in this case, in the thousandths of an inch.

If the assembly condition is categorized as a fixed fastener assembly (clearance holes on this part, but threaded holes on the next part into which screws are inserted to bolt the parts together), a formula for that situation should be used.

FIGURE 12

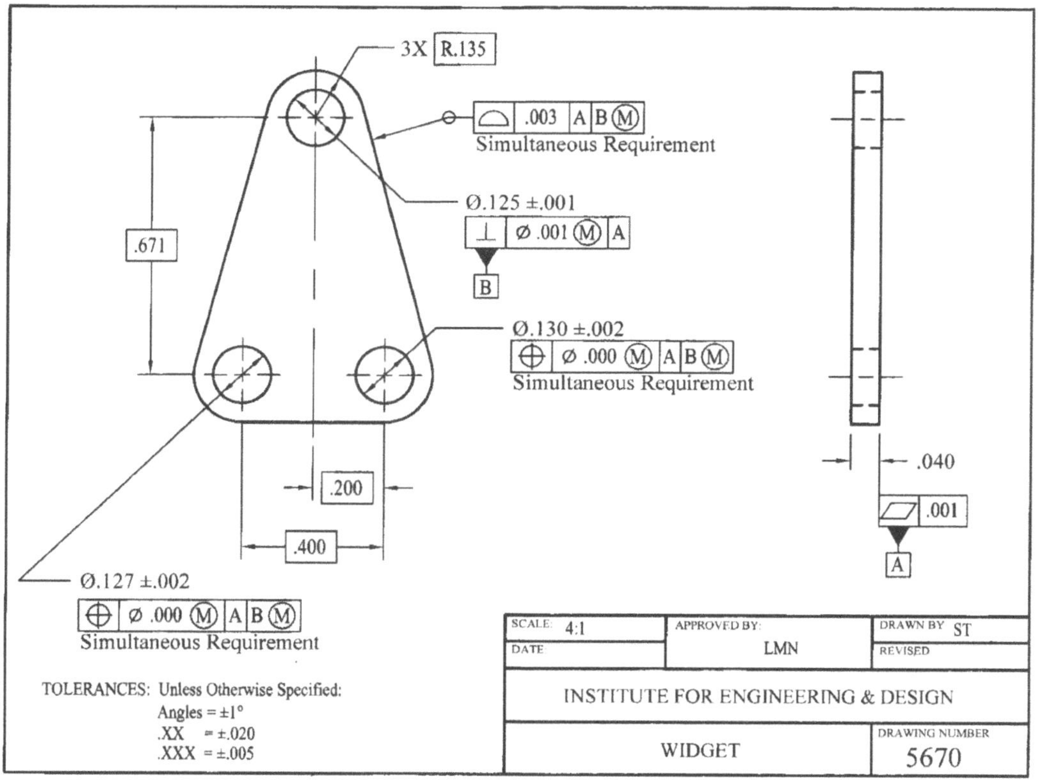

This formula is:

MMC Clearance Holes
- <u>MMC Screws</u>
Geometric tolerance to be divided between the clearance holes and the threaded holes

If the mating part simply consists of a plate with three shafts mounted on it, the same formula would be used, but the MMC of the shafts would replace the MMC of the screws. For example:

.124	=	MMC of Datum Hole B
- .122	=	MMC of Shaft on Mating Part
.002	=	Geometric tolerance to be divided between the shaft and the hole

If the parts were equally hard to manufacture, the tolerance could be divided equally between them as .001 apiece.

So, the perpendicularity tolerance of datum feature B to datum plane A would be:

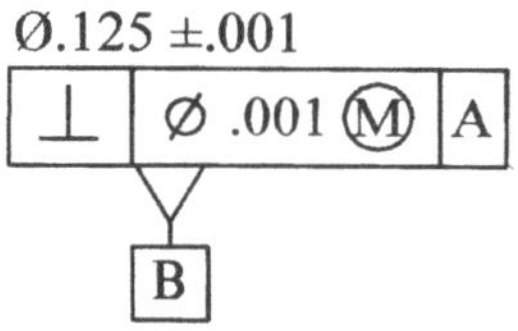

Once B has been established, the coordinate system of three mutually perpendicular planes is complete. Datum A constitutes the primary datum plane and datum axis B is created by two planes intersecting at the center of the hole.

These two planes are perpendicular to each other and to datum plane A. They are the secondary and tertiary (third) datum planes of the three-plane datum reference frame. Often, we would have to create a fourth datum plane to orient the planes crossing at the axis of B, since they are oriented only to each other and A. This fourth auxiliary datum plane would give inspectors an angle of measurement while measuring distance from datum axis B. Instead, on this part, a rule has been taken advantage of.

The rule states that ***all other features located from the same datums (in this case, A and B at MMB) are automatically oriented to each other. They constitute a single pattern of features that must rotate as a unit, if they rotate at all.*** Each still is entitled to its individual tolerance of either position or profile, but these individual tolerance zones are bound together by their basic dimensions.

If they shift or rotate or even rock on datum A, they must do so together as a group. The local note beneath each control of "SIMULTANEOUS REQUIREMENT", which can also be abbreviated as "SIM REQT", is just meant as a clarifying redundancy for those reading the drawing that may not be aware of this useful and powerful rule. ***This rule is also known as the Simultaneous Gaging Requirement Rule*** because the profile and two holes referenced to A and B at MMB, if gaged, would be required to be gaged with one gage for all three controls.

Simultaneous Requirement Rule: Simultaneous requirements apply to position and profile tolerances that are related to the same datum reference frame. The same datum reference frame is defined as using the same datum features referenced in the same order of precedence with same modifiers applied to the datum feature references. The default of simultaneous requirements requires that all applicable geometric tolerance zones are related to the same datum reference frame and are appropriately constrained as a single pattern to a single datum reference frame.

If the part had a different geometry, datums would still have been toleranced to relate them to each other with geometric controls that escalate in power as they progress.

Each subsequent control would reference all of the previously defined datums until they had all been related to one another to within known tolerances. Only then should the datum reference frame be used to relate other features or patterns to. For example:

FIGURE 13

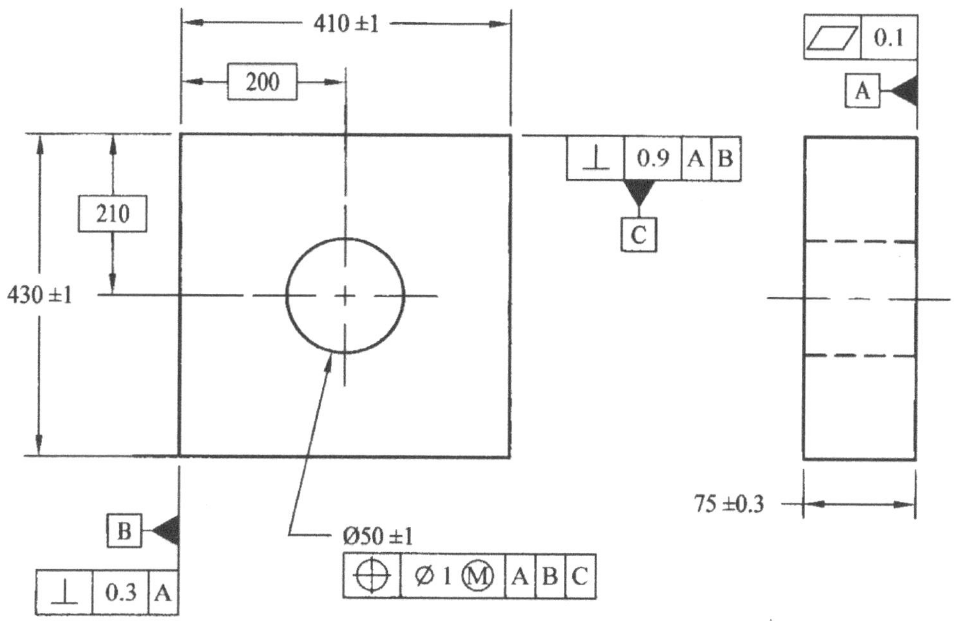

STEP 1:
The primary datum feature is given a three-dimensional form control. A planar surface, such as this, is given a flatness control. It references no datums (flatness disallows datum references) since it is first.

STEP 2:
The secondary datum feature is selected on the basis of how the part is dimensioned. It is a surface from which a dimension originates to locate the hole. There are two of these (looking at the front view):

1) the top surface from which the 210mm dimension originates and
2) the left side from which the 200mm dimension originates.

We will select the one with the most surface area since it will stabilize the part better. The left edge is labeled as a datum feature (B) and related to datum A. Their depicted relationship is one of 90º, so we assign it a perpendicularity control to A. For measurement repeatability, we would prefer perpendicularity tolerance on datum feature B be larger than the tolerance (of flatness) on datum feature A.

If not overridden by mating part formulas or cost considerations, ***a good rule of thumb is the tolerance on B should be 3 times the tolerance on A.*** This acts to limit A from rocking so much when inspecting B as to get significantly different results each time the part is reset and the perpendicularity control measured. This geometric tolerance is shown as:

⊥	0.3	A

Notice that only datum A has been referenced in this feature control frame. That is because at this stage of the part's geometric definition, only datum A exists. Only after completing this control on datum feature B may datum B be referenced in subsequent feature control frames.

<u>STEP 3</u>:

The remaining surface from which a dimension originates (the 210mm dimension) that locates the hole is used as the tertiary datum feature. This will construct a plane that will contact the surface with a minimum of one high point of contact, while it is perpendicular to plane A (three points minimum of high point contact on surface A) and plane B (two points minimum contact on surface B). These form the datum reference frame (coordinate system) that consists of three mutually perpendicular planes from which the hole will be measured.

But, before that happens, we must relate the third plane to the first two. It is perpendicular to both. We must keep the order of reference the same as the planes were constructed in (A primary and B secondary). If we do not reference the datums in the same order, the plane each surface creates will be formed with different points of contact. For example, if B was referenced as a primary datum, it would require a minimum of three points of high point contact from surface B. This plane would be at an entirely different orientation than a secondary datum plane B.

If we keep using the **Rule of Thumb** that says ***tolerances used on the datum features should be three times greater for each subsequently established datum feature in the datum reference frame***, datum feature C will receive a tolerance of 0.9. This control then reads:

⊥	0.9	A	B

<u>STEP 4</u>:

The hole is finally positioned to all three datum planes. The geometric tolerance is derived from the appropriate fixed or floating fastener formula. It is referenced at MMC, since it is a mating feature and the larger the hole is produced, the easier it will mate. The MMC reference allows the position tolerance to grow as the hole is produced larger than its MMC.

The control appears as:

⌖	Ø1Ⓜ	A	B	C

It can be read as:

⌖ = Position

Ø = of the axis of this diameter

1 = must be held to within 1mm

Ⓜ = if the hole is produced at its maximum material condition (Ø49)

A = to A for perpendicularity

B = and to B for a location of 200

C = and to C for a location of 210

DATUMS

DATUM FEATURES

Datum features are chosen from real physical features on the part. They are chosen from which to construct datums (imaginary points, axes and planes). These imaginary datums do not exist in the real world, so we use manufacturing and inspection equipment to simulate their existence. We recognize that this equipment is not perfect, as are the imaginary datums, but is the best we can do to recreate the perfect geometric counterpart of the datum surface(s). This simulation equipment creates a reasonable facsimile of a point, axis or plane from contact with the datum feature surface. We then orient or locate other feature surfaces, axes, centerplanes, or tangent planes from these simulated datum points, axes and planes.

Datum features are either permanent on the part or temporary (to be removed or altered during the manufacturing process). Temporary datums are often used as in-process datums to establish more permanent datums to be used in the final inspection process.

Sometimes, because of rough surfaces--such as forgings or castings--where machining is out of the question because of cost factors, datum target points, lines or areas are used instead of entire surfaces from which to establish datums. This is also sometimes done with weldments and similar parts prone to warpage or other distortions. Datum targets are even occasionally more functional than would be the use of entire surfaces for the establishment of datums for orientation or location. These portions of surfaces will often act to create a datum reference frame to eliminate the six degrees of freedom (three rotational and three linear) a part experiences in free space. Datum features should be:

- **functional** (serving a purpose in preserving how a part operates),

- **representative** of mating or seating features and/or alignment edges (to assure that if inspected and accepted while oriented to, or located from, datums constructed from those features, that the controlled features will, indeed, mate and/or seat),

- **accessible** to those in manufacturing and inspection,

- **repeatable** from manufacturing operation to manufacturing operation and department to department. Datum features are often made more repeatable in the planes, points or axes they generate by the form and orientation controls they are given. For example, the better formed a primary datum feature, the more likely a subsequent set-up will be to recreate the same datum. One should not give up part functionality to gain producibility or inspectability, but producibility and inspectability should occupy places high on the priority list when datum features are selected.

A drawing should be a physical description--a factual statement--of how a part works, and the datum selection is key to making it such. Concessions can and should be made to manufacturing for ease and consistency of set-up when these concessions do not impair the functionality. Some additional controls will occasionally be added to allow surfaces which are not as functional to be used as datum features, but, if over done, this can make parts harder to produce. Sometimes, the judgement of the team will justify these less functional datum features, but this must be done with the overall goal in mind --a low cost, functional product. Generally, when this "Indirect Datum" method is used, geometric tolerances must be tightened to hold the more functional relationships needed, than if features are controlled directly to the functional datums.

Here are just a few of the procedures to be used as "Rules of Thumb" for the selection of symbology, datum features and material condition symbols. Please use these to spur your decision making, not to replace it. Like any tools, they are only as good as the craftsman using them.

- Select those surfaces that seat on one another as primary datum features if you wish them to do so with good (for example, 3-point high point) contact. Control these planar surfaces with flatness, unless size limits do so sufficiently. Remember, we orient to and measure from datum planes and, generally speaking, the better formed the datum features, the closer they will be to these primary planes.

- Watch not only the surfaces that seat on one another, but those that fit inside one another--like functional diameters. They, too, make good datum features. If these diameters are used as primary datum features, consider giving them a cylindricity control so as to make them more repeatable, to taper less and to be able to strike datum axes from which to measure and orient. If a diameter is used as a secondary datum feature, consider either a perpendicularity control back to the planar primary surface (if all other features on the part are to be measured from these) or a positional control (if this datum feature is to be initially measured from others for location).

- Remember, not all functional features can be datum features. Sometimes, it is easier to work with a small number of datum features and derive your functional requirements by controlling other important features back to that same set of datums. This approach allows fewer set-ups and saves time for manufacturing and inspection.

- When several choices seem equally functional, choose the one that has the most surface area and/or is the most accessible on which to set up.

- A datum reference frame is not only to relate to the drawing reader the important interrelationships, it is to stabilize the part and to create an orientation that is a physical description of how the part is oriented in its assembly.

- Switching from one datum reference frame to another on one part or specifying composite positional tolerancing often indicates that it is unimportant for all datum features on mating parts to align. An error of alignment between original and subsequent datum reference frames must be expected as tolerance accumulation is created when switching from one set of datums to another. This "datum alignment" between mating parts is also implied as not a priority when composite positional tolerancing is used.

- Switching datums in mid-stream is often an indication that the part has more than one mating component. In these situations, one mating component may need to relate to one set of datums, while another relates to a different set.

- Just as often, if one mating part with many mating features mates with a single mating part and its corresponding mating features, only one set of datums (one datum reference frame) is used.

- Try not to allow your datum features to conflict with one another. There are, after all, only 6 degrees of freedom. Datum features that try to eliminate the same degrees of freedom as others already in that control often conflict and/or confuse.

- Only use the maximum material condition symbol when an alteration in the size of the feature(s) or datum feature(s) really will have an effect on the allowed geometric imperfection of the controlled feature or pattern of features. One should always ask, "Does it mate?" If so, and all one cares about is that they go together, not with pins in the very center of holes (anywhere in the holes will do) and not with the datum features of size centered between the mating parts, then circled M's all around usually work fine.

When one sees people use the MMC symbols for parts that don't mate or next to datum features of size that don't mate while the controlled features mate, one should automatically question it. Sometimes, this is done only for the convenience of functional gage use and although one can understand that need in some instances, how functional can the gage be if it is checking the part on a false basis? (Saying, in essence, that an alteration in the size of the datum feature has an effect on whether or not the part will mate--when it really has none.)

- Use of the least material condition symbol is usually an indication that mating is not the most important requirement, but that it is the preservation of material on the part. One example of this is when wall thicknesses are the main consideration. Another might be on casting drawings wherein the preservation of enough material to allow clean-up to occur in accordance with a matching requirement found on a subsequent machining drawing is vital.

- The regardless of feature size concept (implied symbol) is of use when describing that the worsening of a geometric shape or geometric relationship based on feature or datum feature size is undesirable. For example, when balance is required, the use of the RFS concept usually makes it easier to balance a part than would use of the MMC concept (although RFS rarely assures balance by itself). It is especially useful after cylindrical datum features of size when they are used for location on spinning parts.

The establishment of datum reference frames has come under close scrutiny recently by members of the ASME, ANSI and measurement communities. The Dimensioning and Tolerancing Standard (ASME Y14.5) makes statements to the effect that:

- When magnified, flat surfaces of manufactured parts are seen to have irregularities; contact is made with a datum plane at a number of surface extremities or high points.

- Therefore, it is necessary to establish a method for simulating the theoretical reference frame from the actual features of the part.

- Planes are simulated in a mutually perpendicular relationship to provide direction as well as the origin for related dimensions and measurements.

- A datum feature is selected on the basis of its geometric relationship to the toleranced feature and the requirements of the design. To ensure proper part interface and assembly, corresponding features of mating parts are also selected as datum features whenever possible.

- If not sufficiently accurate, datum features from which datum planes and datum axes are formed may need to be controlled by specifying appropriate geometric tolerances. Where control of the entire feature becomes impractical, use of datum targets may be considered.

- But, if datum targets are not used, datum planes and axes are established by datum feature surface extremities or high points. For example: a planar primary datum feature relates the part to the datum reference frame by bringing a minimum of 3 high points on the surface into contact with the first datum plane. The part is further related to the frame by bringing at least 2 high points of the secondary planar datum feature into contact with the second datum plane. The relationship is completed by bringing at least 1 high point of the tertiary planar datum feature into contact with the third datum plane.

The problems faced in complying with datum establishment procedures are complicated by imperfect datum surfaces that rock or wobble when contacted at the high points. This creates a candidate set of planes or axes from which to choose and from which to measure. Where a nominally flat surface is specified as a datum feature, the corresponding datum is simulated by a plane contacting the high points of that surface. If irregularities on the surface of a primary or secondary datum feature are such that the part is unstable (that is, it wobbles) when brought into contact with the corresponding surface of a fixture, the part may be adjusted to an optimum position, if necessary, to simulate the datum.

Unfortunately, the statement "may be adjusted to an optimum position" is so ambiguous it is difficult to comply. Should one merely rock the surface one way or another and apply pressure until it stabilizes, or should the sur-face be shimmed up until it no longer wobbles or rocks? If it is to be shimmed, is one to use shims of equal heights on each end, shim up only one end, or shim to create an equal angle (if the rock point is not in the middle of the part)?

The ASME committees have considered this problem and issued clarifying statements in ASME Y14.5.1-1994 (Mathematical Definition of Dimensioning and Tolerancing Principles). These statements issued in 1994 guided us to the effect that a rocking datum feature (primary) or set of datum features (primary and secondary) will generate more than one possible datum plane or axis each, and/or multiple datum reference frames which may be established for a single feature control frame. Another way of looking at it is that the part may move in a datum reference frame. The set of datum reference frames may be viewed as a candidate datum reference frame set from which one must choose to measure. If the controlled feature complies with its geometric tolerance to at least one datum reference frame in the set, the feature is viewed as in compliance with its geometric tolerance.

Because of these insights and clarifying statements (mostly originating in ASME Y14.5.1), compliance with "may be adjusted to an optimum position" is much better defined and, as such, much less ambiguous, and constructing a valid datum plane or axis is easier. One such recommendation is that (in simplified terms) the primary datum feature be divided into thirds, and any candidate datum plane may not have all of the high point contact points residing in either of the outer thirds of the surface. This type of clarification on constructing a valid datum plane makes for a more complete explanation.

In the revised version of Y14.5.1 approved in 2020, a mathematical formula is given to find the single best datum of all the possible datums that could be constructed from a wobbling or rocking datum feature. One must wonder how long before this new formula for establishing the one datum recognized as the correct datum will be instituted in measurement systems throughout the world.

Another problem in establishing repeatable datum planes simulated by a plane contacting the high points of that surface has come into evidence with the widespread use of the computerized Coordinate Measuring Machine. In the past, many operators of these machines have chosen not to fixture their parts, to establish datums from high point contact--for example up against mutually perpendicular flat rails or angle plates--from which probed points can be taken. Since it is faster to probe arbitrary points on the surface instead of taking the time to fixture the part and probe the fixture, many CMM sales people push surface probing to establish datum planes or axes. Even though good fixtures touch the datum surface high points and, if probed, the fixture can simulate quite accurately high point mutually perpendicular datum planes and/or axes, a person trying to make a sale does not always care about truth, justice and the good of mankind. If one salesperson recommends fixturing, it might cost a sale to another salesperson who says fixturing on his or her particular machine is not necessary.

It is a self-perpetuating falsehood, that if one expects to sell a CMM, one must tell, because others will--and there is a lot of competition out there. So many say, with a straight face, "With a CMM it is never necessary to fixture a part." They are wrong, of course, because without fixturing you don't stand much of a chance of forming datum planes or axes from high point contact. You will, instead, get, for example, a plane which is the average plane formed between the arbitrary points you have probed. But it is faster to surface probe instead of doing it correctly and CMM users want to believe it works because it is easier and . . . speed is a great selling point. Datum planes (if formed in this way) are often not repeatable from set-up to set-up; and, therefore, results in measurement data collected from a feature may vary greatly.

Non-repeatability is not the only problem when no fixturing is used and arbitrary points on the feature surface are probed. Constructing datum planes in this way, when high point contact is called for, is simply not in compliance with the Dimensioning and Tolerancing Standard. And, if the primary datum surface is not sufficiently well formed and secondary and tertiary surfaces are not well enough formed and oriented, parts may be accepted that do not meet the geometric tolerances specified (had the feature been inspected properly), or parts that may be technically acceptable may be rejected as being out of tolerance. Parts that will not function or mate are sometimes accepted, then when assembly is attempted, since measurements have not been taken from the datum feature high point planes, the high points of the datum surfaces interfere with other datum surface high points of mating parts.

On the other hand, some of the parts that have been rejected may function and assemble just fine. Clearly, this is a problem wherein the speed of initial measurement must be weighed against the cost of scrapping good parts, reviewing and rechecking parts that have failed the initial improper inspection procedure, and/or accepting low quality, non-functioning parts. Proper fixturing is recommended so that the fixtures, which contact the datum surface high points, can be probed in place of having to probe the actual datum surface at arbitrary points.

DATUMS

(HOW TO SELECT DATUM FEATURES)

Often the functional surfaces are completely ignored for use as datum features, and it isn't until way down line that someone realizes this can cause assembly and other functionality problems. Many times, although the manufacturing procedure has been adequately described on the drawing, it has nothing to do with whether or not the parts will work - and that if inspected per those "manufacturing datums" shown on the drawing, the only information or assurances that can be given by the inspectors are that the parts fit the drawings. But they still will have no idea whether the parts will mate or function. And it gets worse...

As one person told me, "We have a problem at our company that could put us out of business! We have been using datums that have nothing whatsoever to do with the way our parts work."

Sometimes, when this is brought to light, the solution is suggested by someone to 'Add a multitude of geometric interrelationships between those no-function datums and the functional surfaces that have not been used.' Extra parallelism, perpendicularity, angularity and profile controls are added, until the drawing is so full of geometric symbology that the part configuration is hard to make out.

"No problem," they say, "add more views and more pages, a few more drawing notes here and a home phone number there in case of confusion." Of course, these extra controls mean extra manufacturing operations. Machining, forming or molding costs grow geometrically (sorry about that). And the end result is that looking at our drawings gives a visual impact so full of geometric confusion it sends everyone for the Tums and aspirin concession. They shrug and ask, "What can I do?"

The purpose of the "Designing for Manufacturability" wave was never to increase product cost and design out functionality. It was to get the design and manufacturing personnel communicating as part of the same team, realizing their common goals, sharing their knowledge and experience to turn out low cost designs that preserve part function. As the ads say, sometimes a great idea gets lost.

As I am fond of explaining, "If manufacturing wants to use a catapult to fling my parts high into the air and cannon fire to pepper them with holes - and when they fall to the ground they wish to release rats to scurry up and gnaw in the final details, as long as I can hook up a team of horses to drag the remains over to the inspection department and they tell me the parts fit the drawing requirements (which reflect the functional needs), then I will expire with a smile on my face."

A drawing should be a physical description (a factual statement) of how a part works, and the datum selection is key to making it such. Concessions can and should be made to manufacturing when they do not impair the functionality. Some additional controls will occasionally be added to allow surfaces which are not as functional to be used as datums, but for the most part, this is viewed as over design. Sometimes, the judgment of the team will justify these concessions, but they must be made with the overall goal in mind - a low-cost, functional product.

I would like to relate just a few of the procedures I have used as "Rules of Thumb" over the years for the selection of symbology, datum features and material condition symbols. **Please use these to spur your decision making, not to replace it. Like any tools, these are only as good as the craftsman using them.**

- Select those surfaces that seat on one another as primary datum features if you wish them to do so with good (for example 3 point high point) contact. Control these plane surfaces with flatness (unless size limits do so sufficiently). Remember, we orient to and measure from datum planes, and generally speaking the better formed the datum features, the closer they will be to these primary planes.
- Watch not only the surfaces that seat on one another, but those that fit inside one another, like functional diameters. They too, make good datum features. If these diameters are used as primary datum features, consider giving them a cylindricity control, so as to make them more repeatable, to taper less, to be able to strike datum axes from which to measure and orient. If a diameter is used as a secondary datum feature, consider either a perpendicularity control back to the plane primary surface (if all other features on the part are to be measured from these), or a positional control (if this datum feature is to be measured from others for location).
- Remember, not all functional features can be datum features. Sometimes it is easier to work with a small number of datum features and derive your functional requirements by controlling other important features back to that same set of datums.
- When several choices seem equally functional, choose the one that has the most surface area and/or is the most accessible on which to set up.
- A datum reference frame is not only to relate to the drawing reader the important interrelationships, it is to stabilize the part and to create an orientation that is a physical description of how the **part** is oriented in its assembly.
- Switching from one datum reference frame to another on one part, or specifying composite positional tolerancing often indicates that it is unimportant
for all datum features on mating parts to align. An error in alignment between original and subsequent datum reference frames must be expected, as tolerance accumulation is created when switching from one set to datums to another. This "datum alignment" between mating parts is also not a priority when composite positional tolerancing is used.
- Switching datums in mid-stream is just as likely an indication that the part has more than one mating component. In these situations, one mating component may need to relate to one set of datums, while another relates to a different set.
- Just as often, if one mating part with many mating features mates with a single mating part and its corresponding mating features, only one set of datums (one datum reference frame) is used.
- Try not to allow your datum features to conflict with one another. There are, after all, only six degrees of freedom. Datum features that try to eliminate the same degrees of freedom as others already in that control often conflict and/or confuse. Besides, whichever datum gets there first gets the job.
- Only use the Maximum Material Condition Symbol when an alteration in the size of the feature(s) or datum feature(s) really will have an effect on the allowed geometric imperfection of the controlled feature or pattern of features. I always ask "Does it mate?" If so, and all I care about is that they go together, not with pins in the center of holes (anywhere in the holes will do), and not with the datum features of size centered between the mating parts, then circled M's all around usually works fine. When I see people use the MMC symbols for parts that don't mate, or next to datum features of size that don't mate **while** the controlled features mate, I automatically question it. Sometimes, this is done only for the convenience of functional gage use and although

I can understand that need in some instances, how functional can the gage be if it is checking the part on a false basis? (Saying, in essence, that an alteration in the size of the datum feature has an effect on whether or not the part will mate - when it really has none.)

- Use of the Least Material Condition Symbol is usually an indication that mating is not the most important requirement, but that it is the preservation of material on the part. One example of this is when wall thicknesses are the main consideration. Another might be on casting drawings wherein the preservation of enough material to allow clean-up to occur in accordance with a machining requirement found on a subsequent drawing is vital.
- The Regardless of Feature Size implied symbol is of use when describing that the worsening of a geometric shape or geometric relationship based on feature or datum feature size is undesirable. For example, when balance is required, the use of the RFS concept usually makes it easier to balance a part than would use of the MMC concept (although RFS rarely assures balance by itself). It is especially useful after cylindrical datum features of size when they are used for location on spinning parts. The RFS concept is also useful for uniformity of fit, equal airspace between mating features and uniformity of pressure for hydraulics.

Datum Targets

•Flexible Parts
•Finding the Datum Planes
•Fixtures

Section Objectives

Readers will learn:

1. How to specify datum target points, lines and areas on part surfaces.
2. To stabilize an inherently unstable part, such as a forging, casting, weldment, sheet metal or plastic molded part.
3. To fixture a part to datum targets with the least possibility of damage.
4. What a GD&T sheet is.
5. How to geometrically tolerance a sheet metal panel.
6. How to reference a CAD model for needed dimensions not found on a GD&T sheet.
7. How to use notes to specify restraint for allow flexible parts during inspection.
8. How to use a general tolerance note that includes profile of a surface controls.
9. How to design a fixture to simulate datum targets.

Flexible Parts

Datum targets are points, lines or areas on a surface commonly used to stabilize an inherently unstable part such as a rough part like a casting or forging or a warped part such as a weldment. They are also regularly used on parts such as sheet metal or plastic parts with free form curved surfaces.

Datum targets are used to establish datum planes or axes from which to measure. Datum targets have no inherent uses other than what they are referenced for in a feature control frame, such as a position control or a profile control (or sometimes in drawing notes). If they are not referenced, they should not be on the part drawing. One only knows what to use them for when the geometric controls or notes tell us.

The following examples show some simple explanations of how to fixture datum target points, lines and areas. Subsequent illustrations show some common uses for them on a variety of workpieces.

To begin, the datum target symbol is depicted as follows:

FIGURE 14

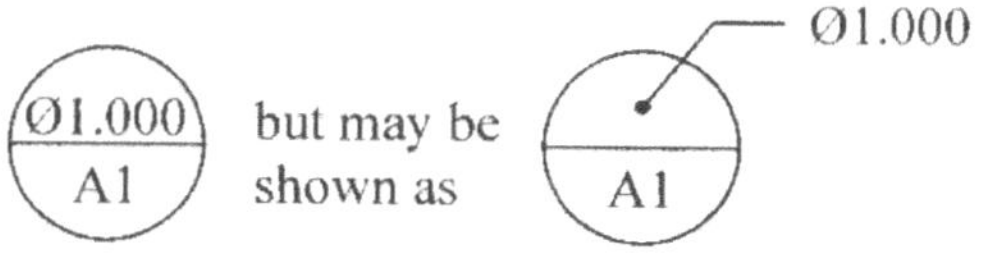

This symbol denotes a datum target area with a diameter of 1.000. If a point or a line of contact was intended, the top half of the datum target symbol would be blank. As has been stated, basic or toleranced dimensions are significant in the designation of target point, line and area location. They are also used to define the diameter of area contact.

Sometimes, it is just as desirable to use partial datum features. These may be shown as datum target areas or as a datum feature whose area is defined with a chain line. An example of the use of a chain line to specify a partial datum is shown below on a rectangular part.

FIGURE 15

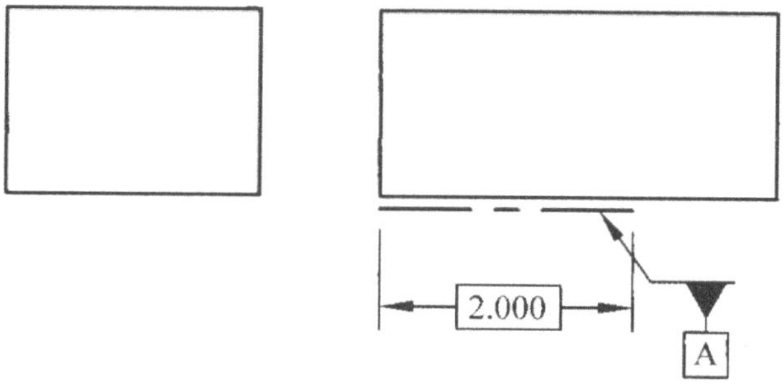

FIGURE 16 [Datum Target Assignment]

Point Contact

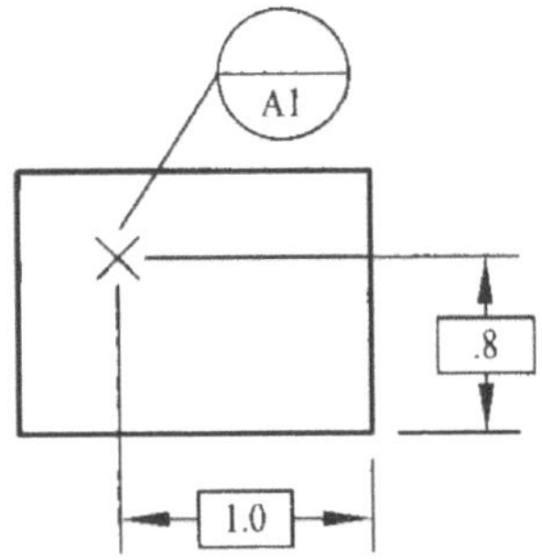

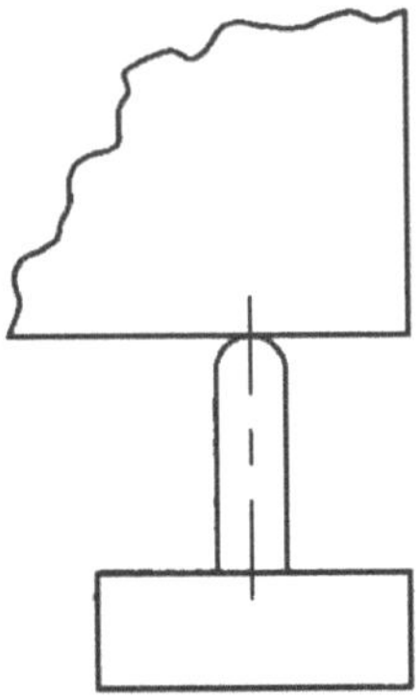

The datum target area is commonly contacted with a spherically tipped pin or the pointed end of a cone.

Line Contact

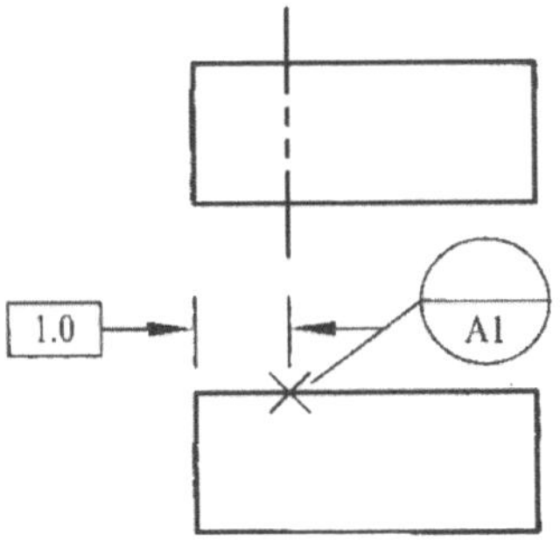

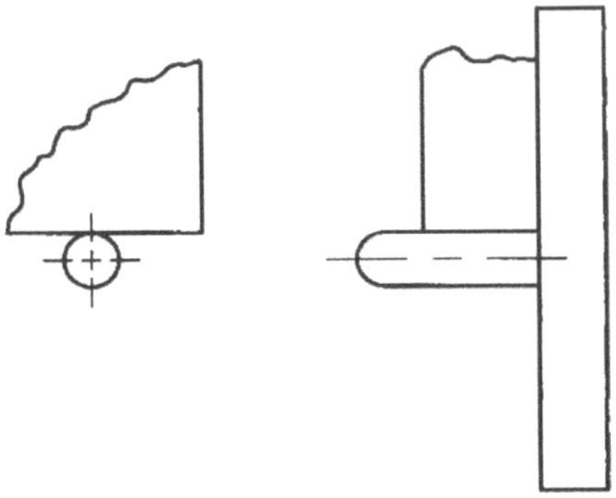

A datum target line is established by contacting the surface with the side of a cylindrical pin as shown.

Area Designations

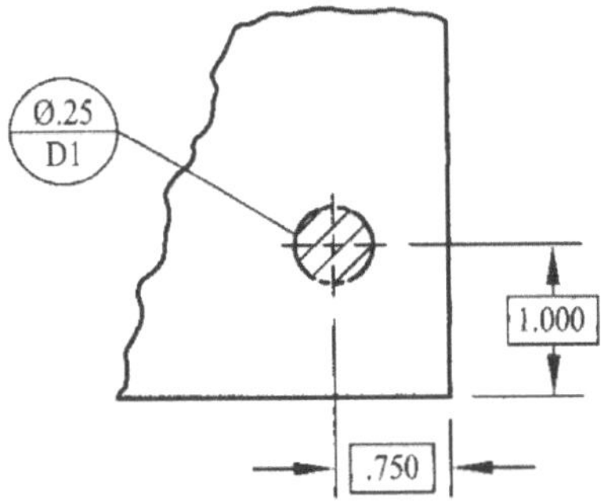

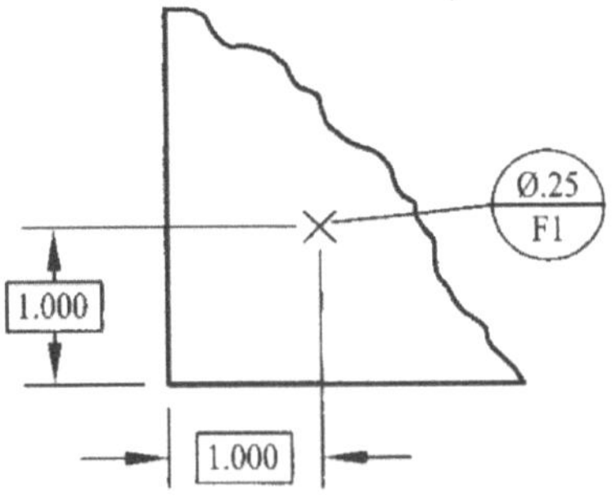

Datum target areas of any configuration may be used to establish a datum. Circular areas of contact, as shown, are common. When flat areas of contact such as these are used, they require an attempt at full flat area contact by the datum target area simulator.

The tolerance applied to the location basic dimensions for datum targets (points, lines and areas of contact), as well as the size tolerance for datum target areas, is the tooling or gaging tolerance established for that situation at your company. However, it is still acceptable per ASME Y14.5-2018 to use plus or minus toleranced dimensions to locate a datum target point, line or area or to size a datum target area. These plus or minus toleranced dimensions would be used instead of basic dimensions. Although legal, this option is not preferred over basic dimensions for datum targets.

FIGURE 17 [Datum Reference Frame for a Part using Datum Target Areas]

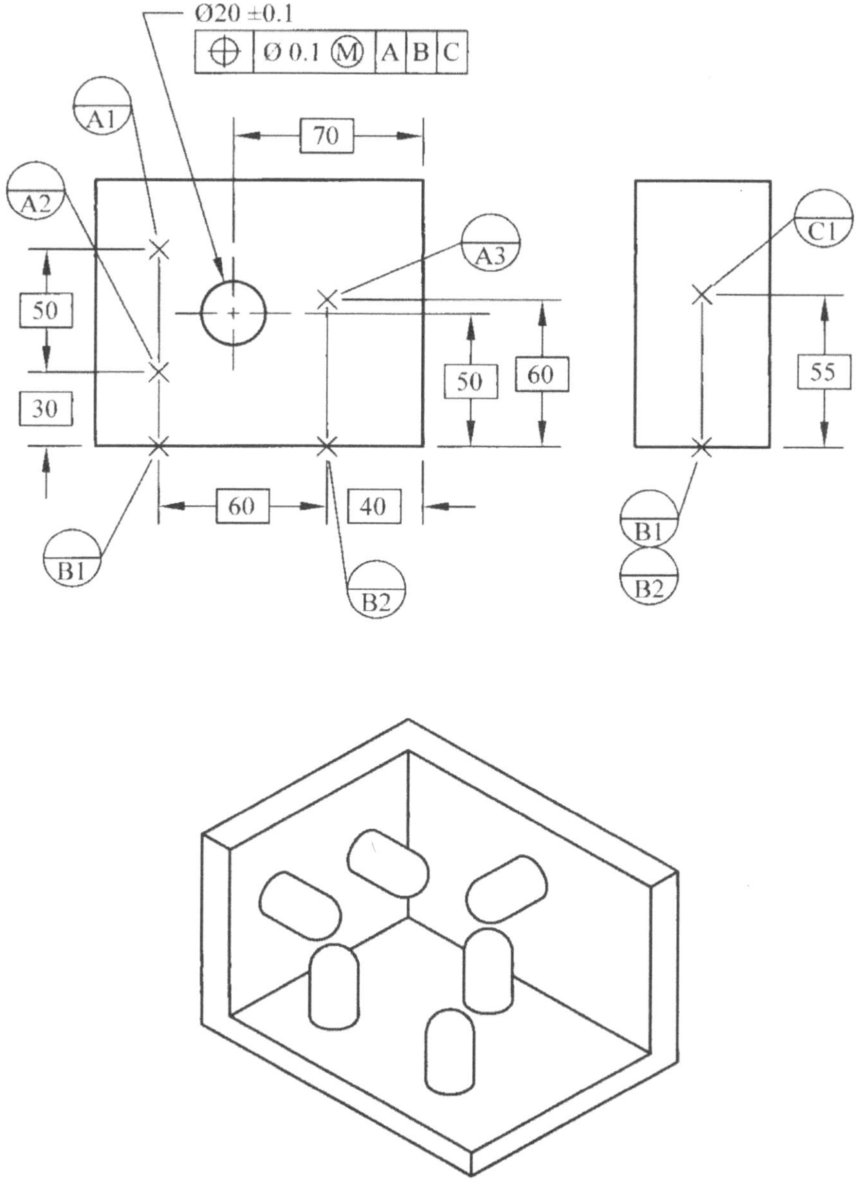

FIGURE 18 [Datum Targets–Area Contact Flat Contact Required to Simulate Datum]

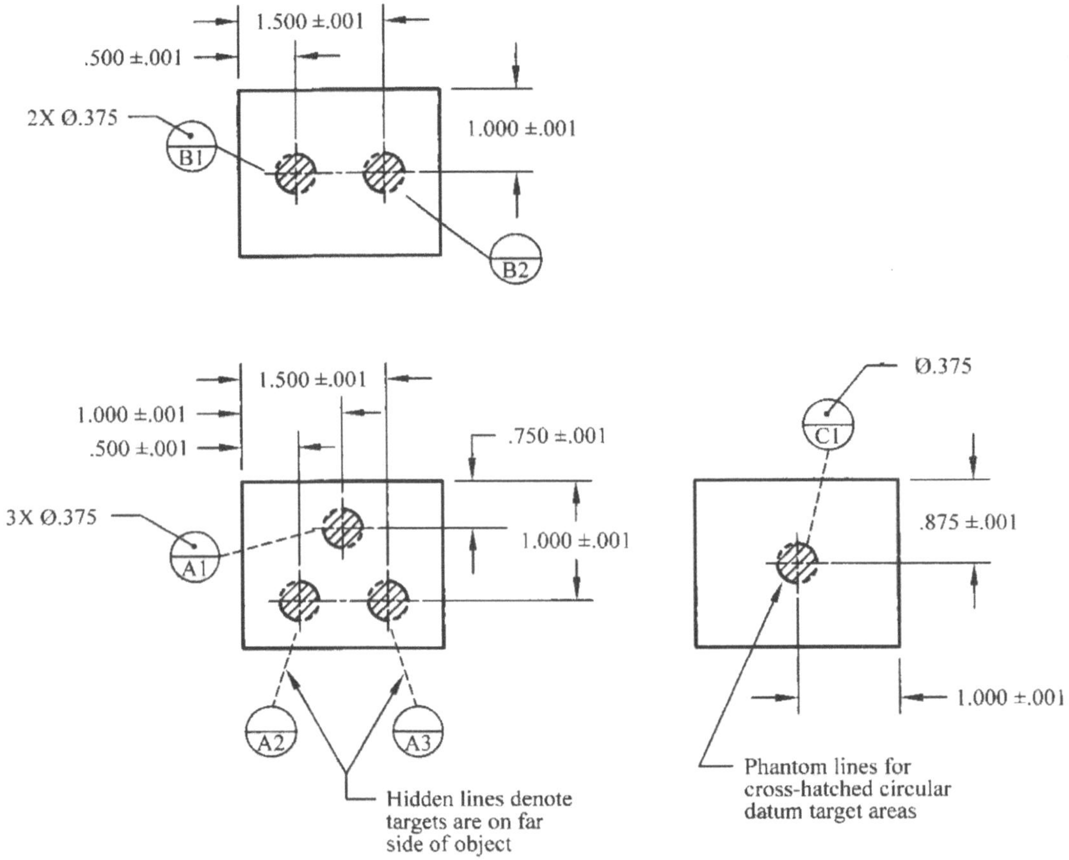

FIGURE 18a

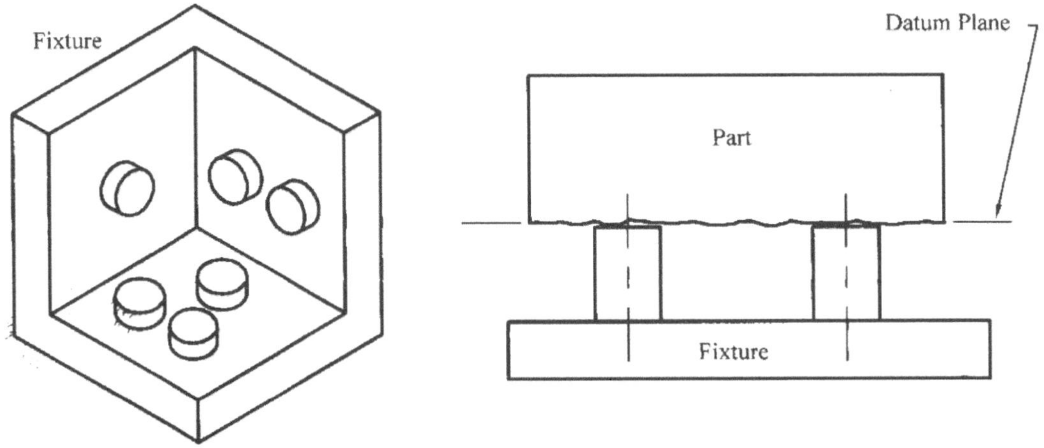

The fixture shown for the part in FIGURE 18 has protruding pins which correspond to the size and location of the datum target areas defined on the part drawing. The part is placed on the fixture in an unrestrained condition, unless restrained contact is specifically spelled out in a drawing note. Full area contact is simulated by the highest points within all areas: minimum of 3 high points for datum A, minimum of 2 high points for datum B and a minimum of 1 high point of contact for datum C.

FIGURE 19

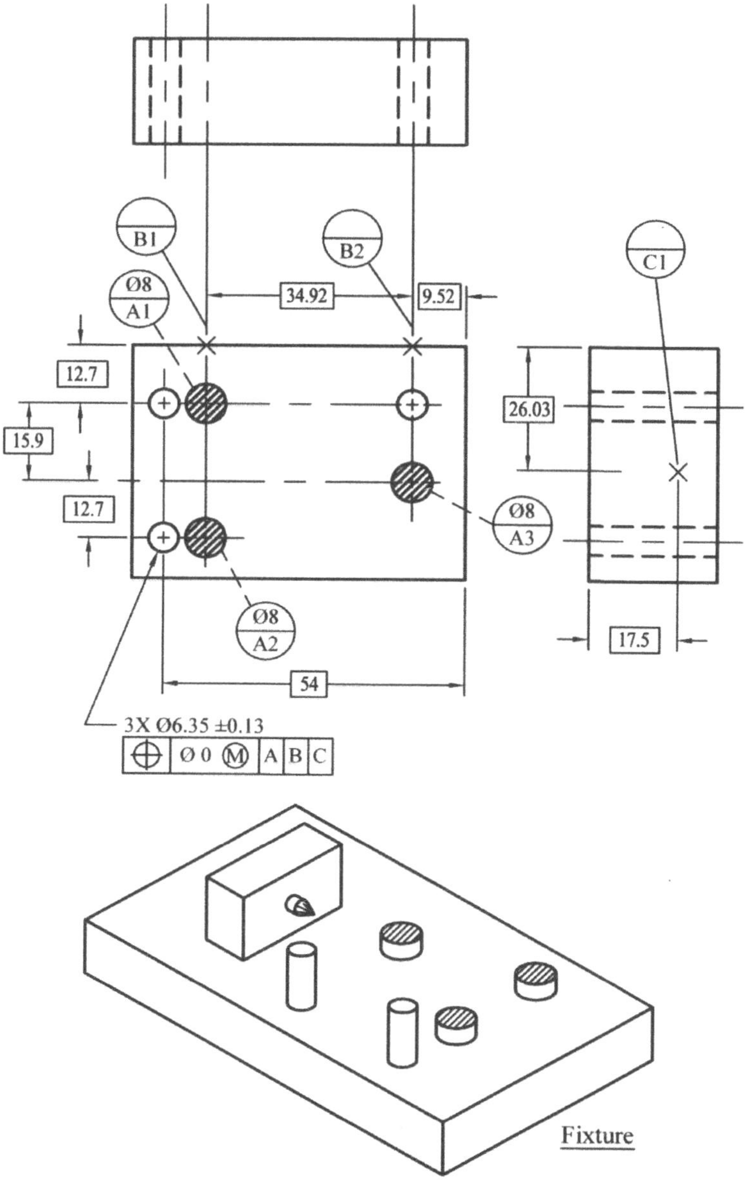

FIGURE 20 [GD&T Sheet]

This is a GD&T sheet. GD&T sheets are commonly used as a vehicle to show the reader all of a part's GD&T requirements. The CAD model or other orthographic drawings would be used to show missing dimensions. Datum targets are often used on sheet metal automobile panels.

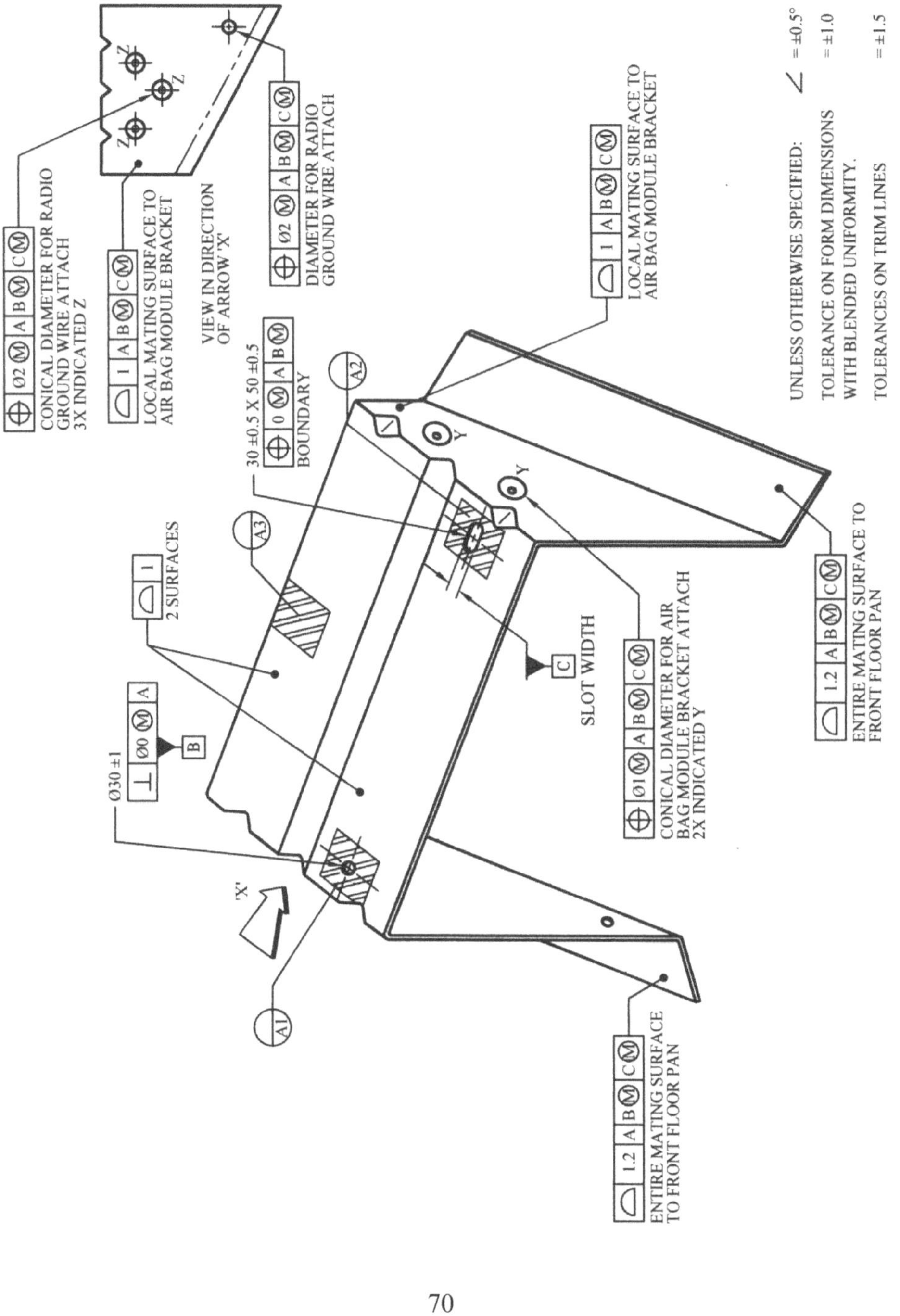

FIGURE 21 [Another GD&T Sheet]

This is another example of a GD&T sheet for a sheet metal part. Notice datum target area C1 is the area around a hole. If the hole receives a bolt and force is applied to the panel in that area, Notes 2 and 3 show the measurement trying to duplicate that functional usage.

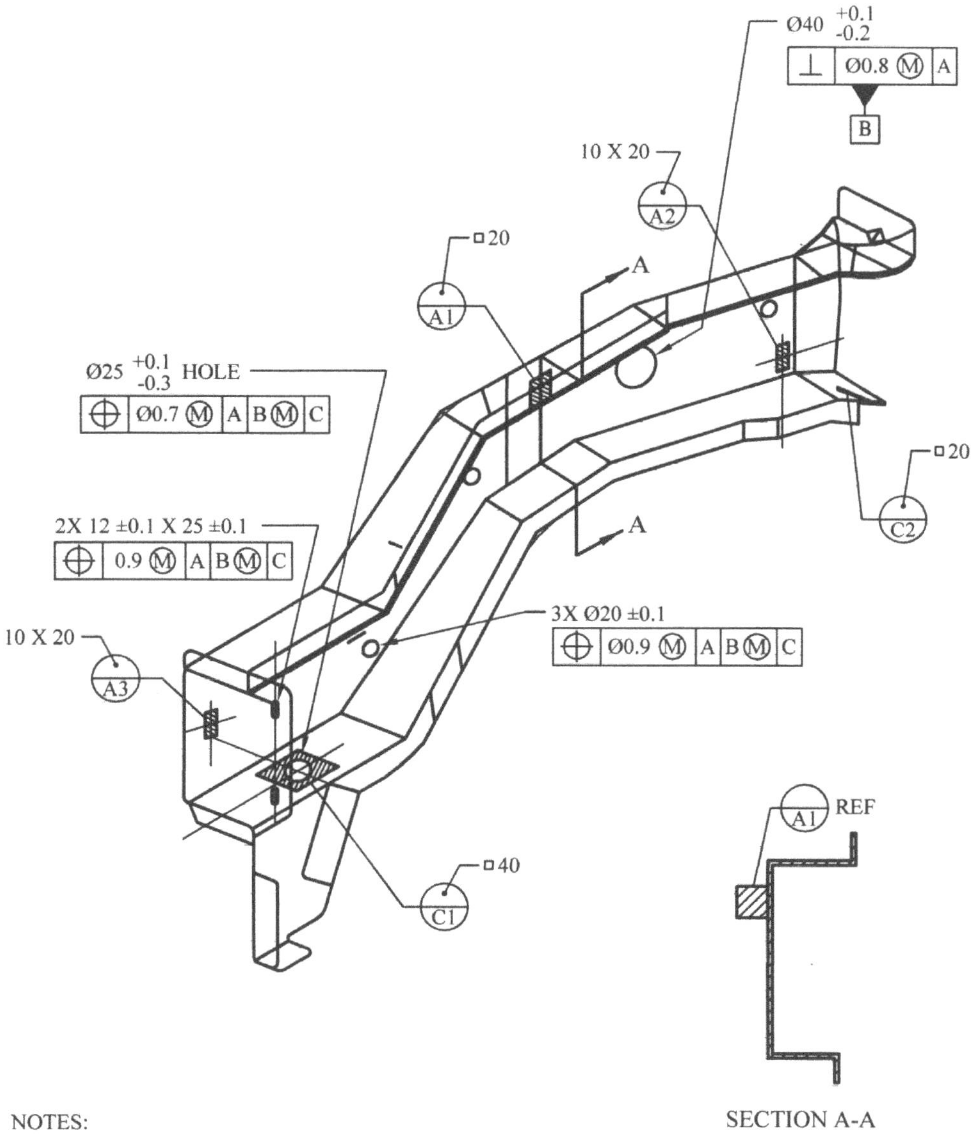

NOTES:

1. TO BE INTERPRETED PER ASME Y14.5M 1994.
2. ALL GEOMETRIC TOLERANCES TO BE INSPECTED WHILE WORKPIECE IS RESTRAINED AGAINST ALL DATUM TARGET AREAS.
3. ALL DATUM TARGET AREAS ARE ON FAR SIDE AS SHOWN FOR A1 IN SECTION A-A.

UNLESS OTHERWISE SPECIFIED:

ALL SURFACES MUST COMPLY WITH

⌓	1	A	B Ⓜ	C

Chapter 3

Bonus Tolerancing Formulas

for Position, Perpendicularity, Angularity and Parallelism Modified at MMC

Objectives: Readers will learn to calculate bonus tolerance and allowed vs. actual deviation from true position.

HOLE

(Step 1)

Actual Size [max. inscribed cylinder]
-MMC
Bonus Geometric Tolerance

(Step 2)

Bonus Tolerance
+Original Geometric Tolerance
Total Geometric Tolerance

SHAFT

(Step 1)

MMC
-Actual Size [min. circumscribed cylinder]
Bonus Geometric Tolerance

(Step 2)

Bonus Tolerance
+Original Geometric Tolerance
Total Geometric Tolerance

The MMC symbol is allowed where it has been determined that function and fit between mating features of size will not be endangered by the addition of a proportional extra deviation from perfect form, orientation or position as the features depart from MMC (staying within their size limits). The examples given in this section deal with a positional control, but, as stated, the formulae for calculating total geometric tolerance are valid for any appropriate application of the MMC symbol to the regular feature of size being controlled (although with straightness and flatness controls at MMC, actual size is actual local size).

A different situation arises when the MMB symbol is applied to datum features of size. For example:

| ⌖ | Ø0.25 | B Ⓜ |

or

| ⊥ | 0.1 | B Ⓜ |

You can see in these two controls that we are not worried about the extra tolerance to be gained as the features being controlled depart from MMC because they are controlled closely by the RFS concept. The features get no extra expansion in the size of their geometric tolerance zones as the features are made larger or smaller (within size limits). However, each has a datum feature of size called out at MMB that it is being controlled to. This means that as the datum feature of size departs from its own MMB, an additional shift in the tolerance zone of the feature or pattern of features being controlled to this datum is allowed. This concept is known as datum feature shift.

Sometimes, the amount of this additional shift is easy to figure. When the part and feature geometry is very simple (perhaps, for example, one feature controlled to one datum regular feature of size--like a plain surface to a shaft, or a clearance hole to a clearance hole, or even a coaxial situation of position with one shaft diameter controlled to another shaft diameter), the additional shift of the feature's tolerance zone is easy to calculate because it is usually a shift directly proportional to the datum feature's departure from its MMB size. But, in some situations, this allowable shift is more difficult to visualize. The more complicated part geometries (the more features and datum features of size involved) make it somewhat more difficult for a floor inspector to determine the effect this allowable shift has on part acceptance without the use of receiver (functional) gaging.

The use of receiver gages makes the calculation of bonus tolerance <u>and</u> allowed shift of tolerances unnecessary for part acceptance or rejection. The gage automatically does these calculations by either accepting or rejecting the part. However, it must be noted that unless the functional receiver type gages are soft gages (computer generated in software), the physical gage gives only attribute data (good vs. bad) and does not give variables data (how good or bad and why). Also, Coordinate Measuring Machines are getting better at determining datum feature shift (a.k.a. pattern shift) because of software improvements.

[**Note:** Paper gaging has been used by some inspection departments as a useful, inexpensive tool to augment their inspection procedures. When used correctly, it can, in many instances, be used in place of receiver gages. Paper gaging is used in conjunction with open set-ups (variables data collectors, such as probes and indicators). Paper gaging is simply a term used to describe that collected measurement data has been graphed out. This gives a visual display of how much and in what direction part features have deviated from perfect.]

The following charts show examples of how to calculate bonus tolerances.

FIGURE 22 [Examples of Calculating Bonus Tolerances]

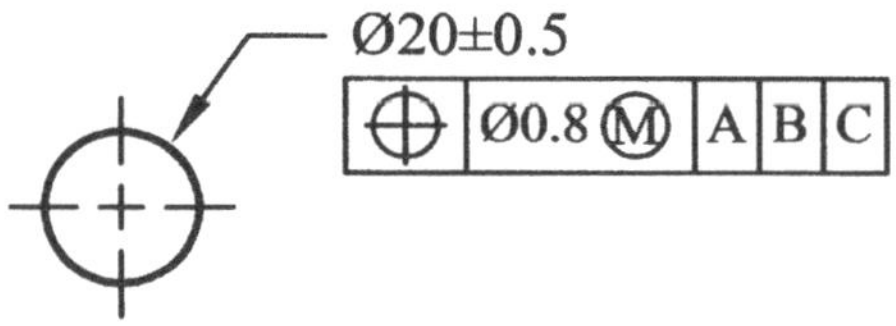

IF A HOLE

	Actual	Less MMC	Bonus Equals	Original Pos. Tol.	Total
A.	20.5	19.5	1	0.8	1.8
B.	20.3	19.5	0.8	0.8	1.6
C.	20.1	19.5	0.6	0.8	1.4
D.	20	19.5	0.5	0.8	1.3
E	19.8	19.5	0.3	0.8	1.1
F.	19.6	19.5	0.1	0.8	0.9
G.	19.5	19.5	0	0.8	0.8

IF A SHAFT

	MMC	Less Actual	Bonus Equals	Original Pos. Tol.	Total
H.	20.5	19.5	1	0.8	1.8
I.	20.5	19.6	0.9	0.8	1.7
J.	20.5	19.8	0.7	0.8	1.5
K.	20.5	20	0.5	0.8	1.3
L.	20.5	20.1	0.4	0.8	1.2
M.	20.5	20.3	0.2	0.8	1
N.	20.5	20.5	0	0.8	0.8

For a **25-26mm diameter shaft**, the completed table below shows the allowable total tolerance using the following formula: **Bonus + Original Tolerance = Total Tolerance**.

FIGURE 23

Actual Feature Size	26	25.8	25.6	25.4	25.1
⌖ | Ø0.5 Ⓜ | A | B | C	A 0.5	B 0.7	C 0.9	D 1.1	E 1.4
// | Ø0.4 Ⓜ | A	F 0.4	G 0.6	H 0.8	I 1	J 1.3
⌖ | Ø0.5 | A | B | C	K 0.5	L 0.5	M 0.5	N 0.5	O 0.5
⊥ | Ø0 Ⓜ | A	P 0	Q 0.2	R 0.4	S 0.6	T 0.9
⌖ | Ø0.2 Ⓜ | L | D | C	U 0.2	V 0.4	W 0.6	X 0.8	Y 1.1
— | Ø0.3 Ⓜ	Z 0.3	AA 0.5	BB 0.7	CC 0.9	DD 1.2
⌭ | 0.05	EE 0.05	FF 0.05	GG 0.05	HH 0.05	II 0.05
⌖ | Ø0.08 Ⓜ | J | C | E	JJ 0.08	KK 0.28	LL 0.48	MM 0.68	NN 0.98

Allowed vs. Actual Deviation from True Position

The allowed deviation from true position is the size of the controlled feature's geometric tolerance zone. In the case of those features with cylindrical tolerance zones, the allowed deviation is given on the basis of a diametrical tolerance zone constructed about the true position axis. Allowed deviation takes into consideration the original tolerance given in the feature control frame plus any additional (bonus) tolerance drawn from the feature size.

This bonus tolerance is allowed for those holes or shafts using the MMC symbol. As they depart from MMC toward LMC, extra tolerance for the allowed deviation from true position is gained. In other words, the tolerance grows. The actual axis of the hole or shaft as produced must lie within this cylindrical tolerance zone which is the allowed deviation from true position (diameter basis) that is acceptable.

The actual deviation from true position may be determined by an open set-up. Coordinate Measuring Machines, optical and video comparators, height gages, gage pins, indicators and other inspection equipment may be used to locate the actual hole or shaft axis. Once this is found, other calculations are required. First, measuring in a straight line from the location datums, it must be determined how far the feature axis has been produced from its true position. This deviation must be found first measuring along the X-axis, then along the Y-axis from the datum planes.

FIGURE 24

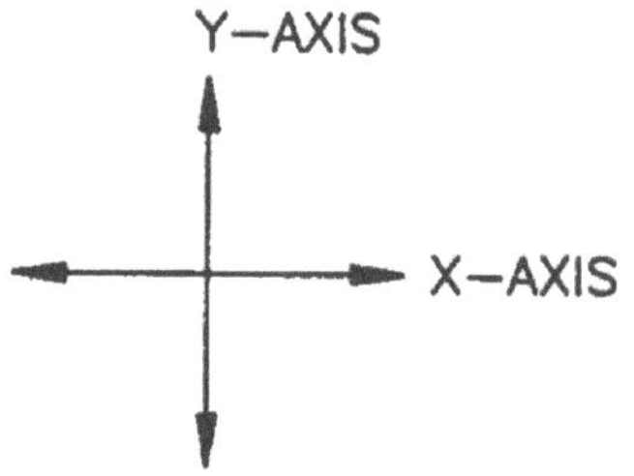

For example, if the feature's true position is drawn as in FIGURE 5-24, we know exactly where true position is.

After the part is produced, we must determine how far the actual hole axis is from datum C. Let's say we've checked and found the maximum deviation point of the axis is 50.12 from datum C. This is a deviation of 0.12 from true position along the X-axis. Checking along the Y-axis from datum B, we find a maximum deviation point of 63.4 or 0.1 from true position. If we use those figures (0.12 and 0.1) in the hole's actual deviation from true position radially, we get the following:

FIGURE 25

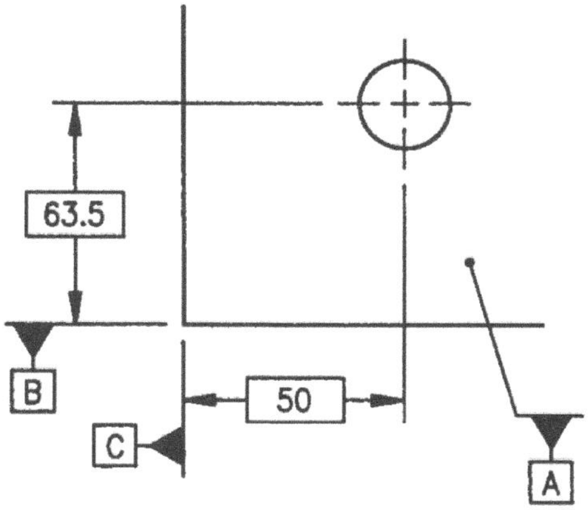

Radial Deviation from ⌖ $= \sqrt{x^2 + y^2}$

$= \sqrt{0.12^2 + 0.1^2}$

$= \sqrt{0.0144 + 0.01}$

$= \sqrt{0.0244}$

$= 0.1562049$

But since we are really interested in how large a diameter would have to be drawn around true position to encompass the actual hole axis, we must multiply this answer by two. The following formula is more appropriate for our needs.

Diametrical Deviation from ⌖ $= 2\sqrt{x^2 + y^2}$

$= 2\sqrt{0.12^2 + 0.1^2}$

$= 0.3124098$

Note: The formula for calculating the axis deviation from true position of a **spherical diameter** is:

$$2\sqrt{x^2 + y^2 + z^2}$$

FIGURE 26 [Inch Conversion Chart: Coordinate Measurement to Calculate the Diameter of the Actual Deviation from True Position (Z)]

INCH CONVERSION CHART

Coordinate Measurement to Calculate the
Diameter of the Actual Deviation from True Position (Z)

.020	.0400	.0402	.0404	.0408	.0412	.0418	.0424	.0431	.0439	.0447	.0456	.0466	.0477	.0488	.0500
.019	.0380	.0382	.0385	.0388	.0393	.0398	.0405	.0412	.0420	.0429	.0439	.0449	.0460	.0472	.0484
.018	.0360	.0362	.0365	.0369	.0374	.0379	.0386	.0394	.0403	.0412	.0422	.0433	.0444	.0456	.0469
.017	.0340	.0342	.0345	.0349	.0354	.0360	.0368	.0376	.0385	.0394	.0405	.0416	.0428	.0440	.0453
.016	.0321	.0322	.0325	.0330	.0335	.0342	.0349	.0358	.0367	.0377	.0388	.0400	.0412	.0425	.0439
.015	.0301	.0303	.0306	.0310	.0316	.0323	.0331	.0340	.0350	.0360	.0372	.0384	.0397	.0410	.0424
.014	.0281	.0283	.0286	.0291	.0297	.0305	.0313	.0322	.0333	.0344	.0356	.0369	.0382	.0396	.0410
.013	.0261	.0263	.0267	.0272	.0278	.0286	.0295	.0305	.0316	.0328	.0340	.0354	.0368	.0382	.0397
.012	.0241	.0243	.0247	.0253	.0260	.0268	.0278	.0288	.0300	.0312	.0325	.0339	.0354	.0369	.0384
.011	.0221	.0224	.0228	.0234	.0242	.0250	.0261	.0272	.0284	.0297	.0311	.0325	.0340	.0356	.0372
.010	.0201	.0204	.0209	.0215	.0224	.0233	.0244	.0256	.0269	.0283	.0297	.0312	.0328	.0344	.0360
.009	.0181	.0184	.0190	.0197	.0206	.0216	.0228	.0241	.0254	.0269	.0284	.0300	.0316	.0333	.0350
.008	.0161	.0165	.0171	.0179	.0189	.0200	.0213	.0226	.0241	.0256	.0272	.0288	.0305	.0322	.0340
.007	.0141	.0146	.0152	.0161	.0172	.0184	.0198	.0213	.0228	.0244	.0261	.0278	.0295	.0313	.0331
.006	.0122	.0126	.0134	.0144	.0156	.0170	.0184	.0200	.0216	.0233	.0250	.0268	.0286	.0305	.0323
.005	.0102	.0108	.0117	.0128	.0141	.0156	.0172	.0189	.0206	.0224	.0242	.0260	.0278	.0297	.0316
.004	.0082	.0089	.0100	.0113	.0128	.0144	.0161	.0179	.0197	.0215	.0234	.0253	.0272	.0291	.0310
.003	.0063	.0072	.0085	.0100	.0117	.0134	.0152	.0171	.0190	.0209	.0228	.0247	.0267	.0286	.0306
.002	.0045	.0056	.0072	.0089	.0108	.0126	.0146	.0165	.0184	.0204	.0224	.0243	.0263	.0283	.0303
.001	.0028	.0045	.0063	.0082	.0102	.0122	.0141	.0161	.0181	.0201	.0221	.0241	.0261	.0281	.0301
	.001	.002	.003	.004	.005	.006	.007	.008	.009	.010	.011	.012	.013	.014	.015

Deviation
in X or Y Axis

Equation: $Z = 2\sqrt{X^2 + Y^2}$

z=the diameter of actual deviation from true position of the feature axis.

x=radial deviation from true position along the axis.

y=radial deviation from true position along y axis.

FIGURE 27

METRIC CONVERSION CHART

Coordinate Measurement to Calculate the Diameter of the Actual Deviation from True Position (Z)

Ø POSITION ZONE (Z)

Y \ X	0.02	0.04	0.06	0.08	0.10	0.12	0.14	0.16	0.18	0.20	0.22	0.24	0.26	0.28	0.30	0.32	0.34	0.36	0.38	0.40	0.42	0.44	0.46	0.48	0.50
0.50	1.001	1.003	1.007	1.013	1.020	1.028	1.038	1.050	1.063	1.077	1.093	1.109	1.127	1.146	1.166	1.187	1.209	1.232	1.256	1.281	1.306	1.332	1.359	1.386	1.414
0.48	0.961	0.963	0.967	0.973	0.981	0.990	1.000	1.012	1.025	1.040	1.056	1.073	1.092	1.111	1.132	1.154	1.176	1.200	1.224	1.250	1.276	1.302	1.330	1.358	1.386
0.46	0.921	0.923	0.928	0.934	0.941	0.951	0.962	0.974	0.988	1.003	1.020	1.038	1.057	1.077	1.098	1.121	1.144	1.168	1.193	1.219	1.246	1.273	1.301	1.330	1.359
0.44	0.881	0.884	0.888	0.894	0.902	0.912	0.923	0.936	0.951	0.967	0.984	1.002	1.022	1.043	1.065	1.088	1.112	1.137	1.163	1.189	1.217	1.245	1.273	1.302	1.332
0.42	0.841	0.844	0.849	0.855	0.863	0.874	0.885	0.899	0.914	0.930	0.948	0.967	0.988	1.010	1.032	1.056	1.081	1.106	1.133	1.160	1.188	1.217	1.246	1.276	1.306
0.40	0.801	0.804	0.809	0.816	0.825	0.835	0.848	0.862	0.877	0.894	0.913	0.933	0.954	0.977	1.000	1.024	1.050	1.076	1.103	1.131	1.160	1.189	1.219	1.250	1.281
0.38	0.761	0.764	0.769	0.777	0.786	0.797	0.810	0.825	0.841	0.859	0.878	0.899	0.921	0.944	0.968	0.994	1.020	1.047	1.075	1.103	1.133	1.163	1.193	1.224	1.256
0.36	0.721	0.724	0.730	0.738	0.747	0.759	0.773	0.788	0.805	0.824	0.844	0.865	0.888	0.912	0.937	0.963	0.990	1.018	1.047	1.076	1.106	1.137	1.168	1.200	1.232
0.34	0.681	0.685	0.691	0.699	0.709	0.721	0.735	0.752	0.769	0.789	0.810	0.832	0.856	0.881	0.907	0.934	0.962	0.990	1.020	1.050	1.081	1.112	1.144	1.176	1.209
0.32	0.641	0.645	0.651	0.660	0.671	0.684	0.699	0.716	0.734	0.755	0.777	0.800	0.825	0.850	0.877	0.905	0.934	0.963	0.994	1.024	1.056	1.088	1.121	1.154	1.187
0.30	0.601	0.605	0.612	0.621	0.632	0.646	0.662	0.680	.0700	0.721	0.744	0.768	0.794	0.821	0.849	0.877	0.907	0.937	0.968	1.000	1.032	1.065	1.098	1.132	1.166
0.28	0.561	0.566	0.573	0.582	0.595	0.609	0.626	0.645	0.666	0.688	0.712	0.738	0.764	0.792	0.821	0.850	0.881	0.912	0.944	0.977	1.010	1.043	1.077	1.111	1.146
0.26	0.522	0.526	0.534	0.544	0.557	0.573	0.591	0.611	0.632	0.656	0.681	0.708	0.735	0.764	0.794	0.825	0.856	0.888	0.921	0.954	0.988	1.022	1.057	1.092	1.127
0.24	0.482	0.487	0.495	0.506	0.520	0.537	0.556	0.577	0.600	0.625	0.651	0.679	0.708	0.738	0.768	0.800	0.832	0.865	0.899	0.933	0.967	1.002	1.036	1.073	1.109
0.22	0.442	0.447	0.456	0.468	0.483	0.501	0.622	0.544	0.569	0.595	0.622	0.651	0.681	0.712	0.744	0.777	0.810	0.844	0.878	0.913	0.948	0.984	1.020	1.056	1.093
0.20	0.402	0.408	0.418	0.431	0.447	0.466	0.488	0.512	0.538	0.566	0.595	0.625	0.656	0.688	0.721	0.755	0.789	0.824	0.859	0.894	0.930	0.967	1.003	1.040	1.077
0.18	0.362	0.369	0.379	0.394	0.412	0.433	0.456	0.482	0.509	0.538	0.569	0.600	0.632	0.666	0.700	0.734	0.769	0.805	0.841	0.877	0.914	0.951	0.988	1.025	1.063
0.16	0.322	0.330	0.342	0.358	0.377	0.400	0.425	0.453	0.482	0.512	0.544	0.577	0.611	0.645	0.680	0.716	0.752	0.788	0.825	0.862	0.899	0.936	0.974	1.012	1.050
0.14	0.283	0.291	0.305	0.322	0.344	0.369	0.396	0.425	0.456	0.488	0.522	0.556	0.591	0.626	0.662	0.699	0.735	0.773	0.810	0.848	0.885	0.923	0.962	1.000	1.038
0.12	0.243	0.253	0.268	0.288	0.312	0.339	0.369	0.400	0.433	0.466	0.501	0.537	0.573	0.609	0.646	0.684	0.721	0.759	0.797	0.835	0.874	0.912	0.951	0.990	1.028
0.10	0.204	0.215	0.233	0.256	0.283	0.312	0.344	0.377	0.412	0.447	0.483	0.520	0.557	0.595	0.632	0.671	0.709	0.747	0.786	0.825	0.863	0.902	0.941	0.981	1.020
0.08	0.165	0.179	0.200	0.226	0.256	0.288	0.322	0.358	0.394	0.431	0.468	0.506	0.544	0.582	0.621	0.660	0.699	0.738	0.777	0.816	0.855	0.894	0.934	0.973	1.013
0.06	0.126	0.144	0.170	0.200	0.233	0.268	0.305	0.342	0.379	0.418	0.456	0.495	0.534	0.573	0.612	0.651	0.691	0.730	0.769	0.809	0.849	0.888	0.928	0.967	1.007
0.04	0.089	0.113	0.144	0.179	0.215	0.253	0.291	0.330	0.369	0.408	0.447	0.487	0.526	0.566	0.605	0.645	0.685	0.724	0.764	0.804	0.844	0.884	0.923	0.963	1.003
0.02	0.057	0.089	0.126	0.165	0.204	0.243	0.283	0.322	0.362	0.402	0.442	0.482	0.522	0.561	0.601	0.641	0.681	0.721	0.761	0.801	0.841	0.881	0.921	0.961	1.001

⇦ X ⇨

FORMULA: $Z = 2\sqrt{X^2 + Y^2}$

Once computed, this actual diametrical deviation from true position must be compared against the allowed deviation. If the allowed deviation (or actual tolerance zone size) is larger than the actual diametrical deviation, the feature position is acceptable. If not, the feature must be reworked or rejected. Reworking is sometimes possible. For example, if a hole is not already made at LMC, it may be opened up (drilled larger). If the hole is modified with a MMC symbol, this procedure will enlarge the allowed deviation from true position (tolerance zone size).

The control given on the figure below can, if examined closely, explain not only how to calculate allowed additional positional tolerance but also why it is allowed. In the inspection of a part, we are often trying to simply discern whether or not the produced part or parts will function. If so, we accept them. If not, rejection or rework of the part is recommended.

FIGURE 28

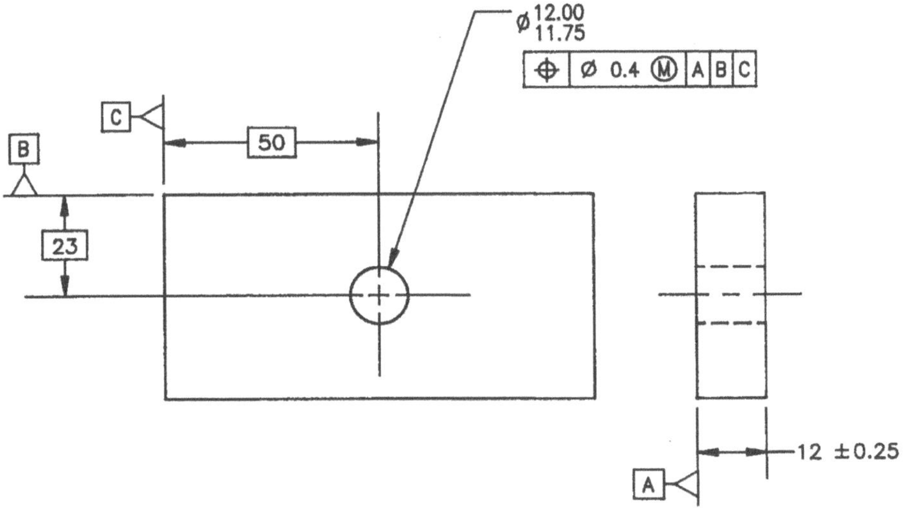

In order to make a good judgment as to the capability of a part to function, one would like to know how the part is to be used. Sometimes, the inspector has only the design drawing from which to work. It is unfortunate when one is not given an intimate knowledge of features to be inspected. But, the reality is that this is often the case. The inspector must read the drawing as though it tells a story about the needed characteristics for conformance in order to function. If the drawing is well done, the datums and feature control frames can, indeed, give the inspector the information necessary to make good judgments about the part and also to make valid recommendations to manufacturing regarding adjustments to part processing to improve these desired characteristics.

If we read the aforementioned drawing and focus on the symbols/components of the positional control, it can tell the story of what is expected of the controlled hole. It says in part, "Position of a diameter that mates while the part is seated on surface A and the hole is located from B and C." Of course, it could be read in the more traditional manner of, "The axis may be out of position a diameter of 0.4 if the hole is produced at maximum material condition holding a relationship of perpendicularity to primary datum plane A, distance from secondary datum plane B and distance from tertiary datum plane C."

But to decode the function of the hole, the first reading and its subsequent logical implications are probably more helpful. If we derive from the positional control that we are positioning a hole that mates while A seats with a three-point high point contact and we locate/measure the hole from planes B and C, we must then ask a series of questions. "If the round hole mates, what does it mate with?" Without a quantum leap in logic, we may speculate that a round hole mates with a round shaft. We might then ask, "What is the largest shaft that, in theory, could enter this hole at the desired angle and location from the datums if the hole was made within its limits of size and also out of perpendicularity and location to the listed datums the maximum amount allowed at that size?"

The answer is to be found by simply subtracting the allowed geometric tolerance from the produced hole size in each instance. If done, one finds that a constant boundary of virtual condition (MMC concept) has been protected on each hole produced. This boundary is perfectly cylindrical, perfectly oriented to datum plane A, and perfectly located from datum planes B and C. The job of the hole is to stay outside of this boundary. The mating pin/shaft is designed, dimensioned and toleranced (for size and position) to reside within this boundary.

If this is discovered to be the case in parts that have been produced, the inspector accepts the parts with a fair amount of confidence that he has proven that the inspected features will work/assemble. In order to determine the boundary size on the parts under discussion, one may go to the columns labeled actual hole size (maximum inscribed cylinder) and column A, which is the allowed diameter of the deviation from true position, and subtract these two numbers. For example:

	Actual Hole Size		A		Virtual Condition
(1)	11.75	-	0.40	=	11.35
(2)	11.80	-	0.45	=	11.35
(3)	11.90	-	0.55	=	11.35
(4)	12.00	-	0.65	=	11.35

The mating boundary remained constant even though the size of the produced hole and geometric positional tolerance allowed changes. The inspector is saying, "If the hole does not violate this boundary, it will mate with the worst case mating shaft if it also does not violate the boundary." If the hole resides outside of the boundary and the shaft resides inside of the boundary (which has its center/axis at true position), no interference of material will occur. Therefore, the parts will assemble and should be passed on by the inspector to assembly.

The following illustration shows a part as specified, then being measured as produced. It is first judged as a good part using the method of verifying that the virtual condition boundary has not been violated. Then, it is measured and judged to be within its positional tolerance zone. These methods are the two most common in verifying positional tolerance.

FIGURE 29 [Tolerance Zone vs. Boundary Verification]

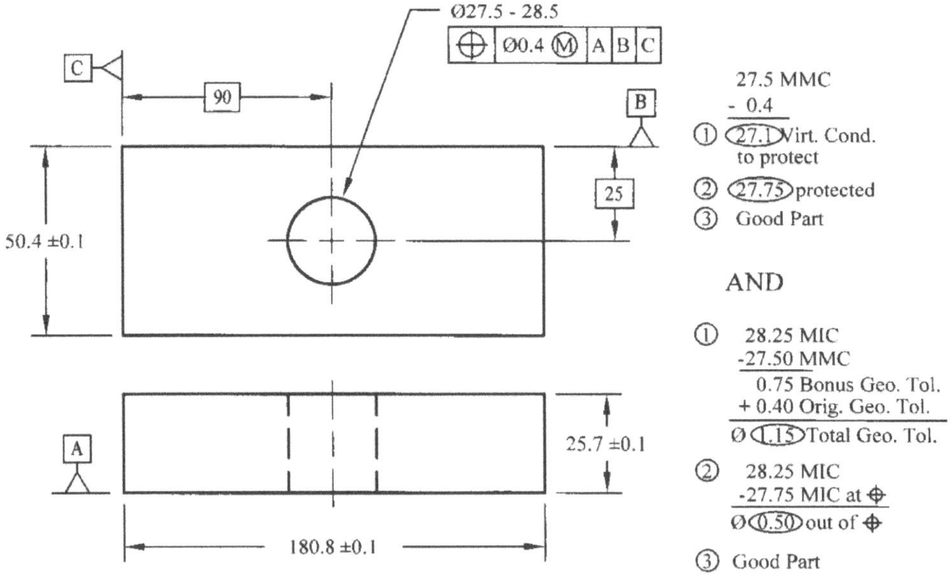

Part 1 as produced being inspected:

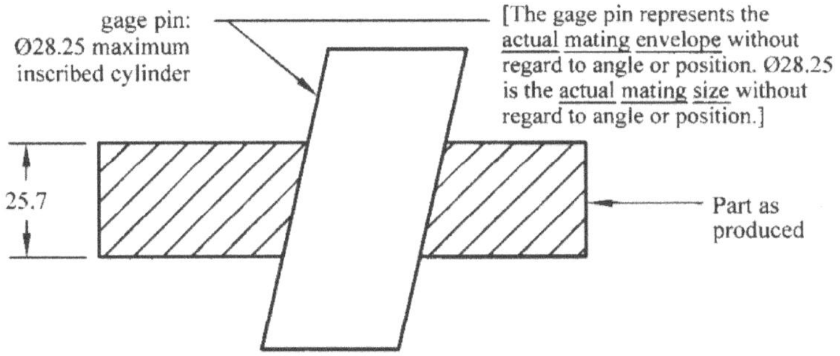

Part 1 as produced being inspected:

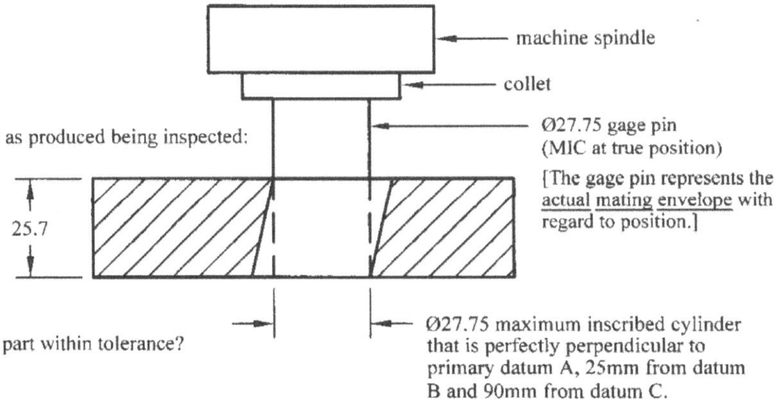

1. Is this part within tolerance?

Chapter 4

MAXIMUM MATERIAL CONDITION, LEAST MATERIAL CONDITION, AND REGARDLESS OF FEATURE SIZE

Objectives: Readers will learn the meaning and uses of maximum material condition, least material condition and regardless of feature size modifiers.

MAXIMUM MATERIAL CONDITION

Meaning and Uses

Maximum material condition is the condition of a feature wherein the feature contains the most material. It is often thought of as the heaviest feature. It is the smallest hole or the largest shaft. In the ASME Y14.5, maximum material condition is described as "The condition in which a feature of size contains the maximum amount of material within the stated limits of size--for example, minimum hole diameter, maximum shaft diameter."

A shaft or other external feature such as a pin, a tab, or a feature made up of opposing external sides of a part are at maximum material condition (MMC) when they are at their largest allowable size. Some like to refer to it as when the feature is the heaviest. For example, a shaft with a size of Ø.125 ± .010 would be at MMC when it measures Ø.135 or its largest size.

A hole or other internal feature such as a keyway, slot or the feature made up of opposing internal sides of a part are at maximum material condition when they are at their smallest allowable size. If a hole was drilled bigger than its smallest allowable size, more material would have to be removed from the hole; therefore, it wouldn't have its maximum amount of material. If a hole with size limits Ø.480 - .500 was produced at Ø.480, it would be at its smallest allowable size--its maximum material condition.

In feature control frames, the MMC concept (circled M) is usually used for mating features. If a shaft is to be inserted into a hole, the shaft's geometry of form, orientation or location need not be as perfect if the shaft is made at a smaller size--just as the hole's geometry can be less perfect if the hole is produced at a larger size. This additional geometric tolerance based on size departure from maximum material condition is often termed "bonus tolerance".

It allows more parts to pass the inspection procedure based on functional requirements. It lets the inspectors buy off parts that will mate.

Many of these parts may not have been allowed to pass the inspection process if they had been controlled on a regardless of feature size basis, even though they may easily have been able to mate with mating features. The MMC concept allows the acceptance of parts that have mating features that may mate with pins more off to one side of the holes rather than directly in the center of the holes. It allows the acceptance of pins that may lean more inside a mating hole as the pin shrinks in size. If the need is simply that all parts that mate are to be accepted, then the MMC principle may be employed without endangering functionality.

If used after a geometric tolerance in the feature control frame, the effect on part balance is not usually a major factor because it does not require a shift of the part off center to be assembled. However, the use of the MMC control, if used after a datum feature of size, allows the acceptance of parts that may require the shifting of the part's weight off the center of the datum axis. This must be used with caution, as it may endanger the balance of a rotating part.

The MMC symbol is usually used to preserve a mating boundary, a constant boundary of virtual condition--the worst mating condition. For example:

Ø.250-.260 hole

⌖	Ø.014 Ⓜ	A	B	C

FORMULA: The maximum inscribed cylinder size minus the allowed geometric tolerance at that size equals the worst mating boundary--virtual condition.

Ø.250 - .014 = Ø.236
Ø.251 - .015 = Ø.236
Ø.252 - .016 = Ø.236
Ø.253 - .017 = Ø.236
Ø.254 - .018 = Ø.236

Ø.255 - .019 = Ø.236
Ø.256 - .020 = Ø.236
Ø.257 - .021 = Ø.236
Ø.258 - .022 = Ø.236
Ø.259 - .023 = Ø.236
Ø.260 - .024 = Ø.236

As the hole moves or leans out of perfect position, it has the effect of closing the hole's mating boundary at true position by the amount that it moves or leans. It may not measure outside of its size limits of Ø.250 - .260, but the effect of having less than perfect position or orientation closes the space available at that location and angle for a feature expected to mate with the hole at that location and angle. It is closed by the amount the hole moves or angles

away from perfect. The result is that the mating shaft must reside within the boundary of Ø.236 and the hole may not violate that boundary if the features are to mate without interference.
If it is also necessary to calculate a minimum wall thickness between this hole and another surface, it would be helpful to know how much larger an area this hole is allowed to sweep (in a worst case situation) than the biggest hole. The addition of the allowed geometric tolerance to the largest hole size allowed (LMC) would calculate this sweep boundary. For example: Ø.260 + .024 = Ø.284. If one wished to calculate a minimum wall thickness of this hole's surface to, say, an edge on the part from which it was given a basic dimension, the Ø.284 could be divided by two and the answer subtracted from the basic dimension given from datum B or C to the center of the hole (to begin the process).

General Rule of Thumb:

- Use the maximum material condition symbol when mating is most important. Use it when you wish the inspector to be able to accept all parts that can be assembled. It allows the acceptance of all parts that assemble, whether they be assembled with the pin in the middle or near the side of the mating hole. For example:

⌖	Ø.010 Ⓜ	A	B	C

- If a circled M is used after the datum feature of size, it is now called a maximum material boundary symbol (MMB) allows some parts to be accepted that can only be assembled off center, imbalanced, with unequal gaps between mating features. MMB is another term for virtual condition. For example:

⌖	Ø.010 Ⓜ	A	B Ⓜ

  If a pattern of bolt holes is positioned to a datum feature of size specified at MMB, as the datum feature of size departs from MMB (or virtual condition if the datum feature of size is secondary or tertiary, instead of primary) the pattern of bolt holes may shift location (as a group) away from the datum feature axis. If shifting occurs and assembly is attempted with the mating datum features of size centered to one another, the bolt holes on one part will probably not align with the bolt holes on the mating part. However, since there will be clearance between the mating datum features (because of the datum feature's departure from MMB), you may now shift the entire part off the datum center to align the bolt holes on the mating parts. This will allow you to assemble the part, but the entire weight of the part will be off to one side. If the part is a spinning part, it will be more difficult to balance in this assembled condition. If the parts are panels to be assembled, such as vehicle panels, gaps may occur on opposite sides that are not equal, making the panels appear ill fitting (either a cosmetic or functional defect).

It's not that the fits will be looser than if they had been controlled at RFB (Regardless of Material Boundary-which allows no datum feature shift/pattern shift), just that the fits may not be as uniform.

Use the MMC (maximum material condition) symbol after the positional tolerance to protect a constant mating boundary. This boundary is the collective effect of the actual feature size and its allowed geometric tolerance. For Example:

Ø.250-.260 hole

⌖	Ø.010 Ⓜ	A	B	C

Hole Size	Minus	Allowed Positional Tolerance		Equals Mating Boundary
Ø.250	-	.010	=	Ø.240
Ø.251	-	.011	=	Ø.240
Ø.255	-	.015	=	Ø.240
Ø.260	-	.020	=	Ø.240

This Ø.240 boundary is a constant boundary referred to as a virtual condition boundary. Any mating pin to be assembled at this location, at this angle, must be compatible. In other words, with a worst mating condition of its own that is of a Ø.240 or smaller (unless a transition or an interference fit is desired).

LEAST MATERIAL CONDITION
Meaning and Uses

Least material condition is the condition of a feature wherein the feature con-tains the least material. It is often thought of as the lightest feature. It is the largest hole or the smallest shaft. The definition for least material condition is in many ways an opposite of maximum material condition. ASME Y14.5 lists it as "The condition in which a feature of size contains the least amount of material within the stated limits of size--for example, maximum hole diameter, minimum shaft diameter."

So, instead of a shaft or tab or other external feature being at its largest size as with maximum material condition, least material condition (LMC) is when that external feature is at its smallest size (having the least material) within its size limits. For example, a shaft with size limits of

Ø.500 - .510 is at its LMC when it measures Ø.500. An internal feature, however, like a hole or keyway, is at its least material condition when it is at its largest size. For example, a hole with size limits of Ø.750 - .760 will be at its least material condition when produced at Ø.760.
In feature control frames, the LMC concept (circled L) is usually used for features when the preservation of material is of great importance and cost is a significant factor. It is used when wall thickness is thought to be endangered and the holes stand a chance of approaching a breakout condition. It is also used on casting drawings to assure that in subsequent machining operations enough material is available to allow a machine cut to clean up the part surface. It is capable of alleviating the machinist's frustration of starting out machining material and ending up machining nothing but air. Since wall thickness, instead of mating, is of major concern, if a hole is made smaller and a shaft made bigger, a greater positional tolerance can be allowed. This additional geometric tolerance, based on size departure from least material condition, is often termed "bonus tolerance". It allows more parts to pass the inspection procedure based on functional requirements than if the feature had been controlled on a regardless of feature size basis.

It could, for example, for a positional control, let the inspector buy off parts that have enough preserved material at that location to assure breakout will not occur. Many of these parts may not have been allowed to pass the inspection process if they had been controlled RFS, even though they may easily have had thick enough walls and enough preserved material either to not endanger wall thickness or to clean up in a machining process. The LMC concept allows the acceptance of holes and shafts that are more out of position than the original geometric tolerance would have allowed, and it does this on the basis that the shafts are bigger and the holes are smaller than the least material condition. They have more material and therefore have to be less well located. If the need is simply that all features have no walls that are too thin or have enough material to clean up during a machining operation, the LMC principle may be employed without adversely affecting functionality.

The least material condition symbol is usually used to preserve a sweep boundary--a constant boundary of virtual condition--bigger than the biggest hole and smaller than the smallest shaft.

For example:

Ø.250 - .260 hole

⌖	Ø.014 Ⓛ	A	B	C

FORMULA: The actual produced size plus the allowed geometric tolerance at that size equals the sweep boundary/virtual condition.

Ø.260 + .014 =	Ø.274	Ø.255 + .019 =	Ø.274
Ø.259 + .015 =	Ø.274	Ø.254 + .020 =	Ø.274
Ø.258 + .016 =	Ø.274	Ø.253 + .021 =	Ø.274
Ø.257 + .017 =	Ø.274	Ø.252 + .022 =	Ø.274
Ø.256 + .018 =	Ø.274	Ø.251 + .023 =	Ø.274
		Ø.250 + .024 =	Ø.274

As the hole moves or leans out of perfect position, it has the effect of widening the cylindrical boundary the hole sweeps over at true position by the amount that it moves or leans. It may not measure outside of its size limits of Ø.250 - .260, but the effect of having less than perfect position or orientation widens the area the hole covers centered about that true position location and, therefore, thins wall thickness. It is widened by the amount the hole moves or angles away from perfect. The result is that the hole must not violate that boundary of Ø.274 if material integrity and part strength are to be maintained.

If a mating boundary must also be calculated, as well as a sweep boundary when the least material condition symbol is used, it is easily done. The geometric tolerance allowed to the smallest hole size is subtracted from that MMC size to determine the worst mating boundary. For example: Ø.250 - .024 = Ø.226 worst mating boundary for the geometric control given the hole in the example used in this section.

A sweep boundary could be used to calculate a minimum wall thickness. If, for example, a minimum wall thickness needed to be calculated to a plane datum such as datum B or C, half the sweep boundary of Ø.274 is a radius of .137. This .137 can be subtracted from the basic dimension that emanates from the datums and locates the center of the hole's true position (to begin the process).

REGARDLESS OF FEATURE SIZE AND REGARDLESS OF MATERIAL BOUNDARY

The regardless of feature size concept (RFS) is used and its symbol implied when the geometric tolerance is to remain the same no matter what size hole or shaft is produced within its limits of size. The geometric tolerance never changes with the size of the feature it controls. The size of the feature must be met and the geometric tolerance must be met as separately verifiable requirements. If no MMB or LMB symbol is used after a datum feature of size, the Regardless of Material Boundary (RMB) symbol is implied.

The RMB concept is often used after datum features of size where balance is important--for example, on parts that rotate and also for parts that need a uniform seal.

Spinning parts could be functionally endangered if the tolerance of location, such as centering, were allowed to vary with the size of the feature. As with the MMB symbol (to say that, for example, as a hole grows or a shaft shrinks that it can be further off center on a spinning part), this could adversely affect a part's balance. Likewise, the regardless of feature size (RFS) implied symbol should be considered after the geometric tolerance for **very fast spinning parts**, but **especially the RMB symbol** should be implied after datum features of size for such parts and also for parts that **must seal uniformly**, instead of specifying the MMB or LMB symbols.

The RFS implied symbol says individual features gain no bonus tolerance as they vary in size, and the RMB implied symbol allows no datum feature shift. Datum feature shift may force some in-tolerance parts to assemble only off center. The RMB implied symbol after datum features of size is the only datum feature modifier that allows all parts accepted, as meeting their geometric tolerances, to be assembled on center and therefore to spin and balance better and to uniformly seal.

The MMB and LMB symbol after datum features of size will not control balance as well as RMB on rotating parts. The MMB symbol will not control sealing as well as the RMB implied symbol and the LMB symbol will control sealing, but not control uniform sealing. In other words, with the LMB symbol, the parts will seal better on one side than on the other.

If function must be preserved in such a way as to have mating features assemble in a specific way, such as a pin more in the center of a mating hole than the MMC symbol would maintain (since the MMC symbol allows the hole to move out of its perfect position more as the hole grows and the shaft to move out of its perfect position more as the shaft shrinks), then the implied RFS (regardless of feature size) concept should be used.

Where the MMB symbol (if used after datum features of size) allows a shift of feature patterns making it necessary to shift a part's weight off center of the datum axis to assemble the mating parts, the RFS concept (if used after datum features of size) allows no pattern shift as the datum feature departs from its MMC or LMC. This means that the only parts the inspector can accept are parts that can, at least, be assembled on center. Although it does not assure that the parts will be assembled on center in a balanced condition, it does assure that the parts accepted for assembly can be assembled on center or pushed back to center. This makes any balancing procedures that are subsequently attempted easier.

It is also worth noting that, in most cases, the use of the RFS concept protects the same worst mating boundary as does the MMC concept and the same worst sweep boundary for wall thicknesses and material preservation as the LMC concept.

The RFS concept would protect the minimum wall thicknesses as well as the LMC concept if two other rules did not interfere. One rule is that with the RFS concept there is no envelope of perfect form required at LMC, but there is one required at MMC. The second rule is that positional controls applied at RFS are inspected by verifying the location of the axes or center planes of their mating size. For example, a gage pin of the maximum inscribed cylinder generates an axis that must reside within the tolerance zone for a hole that is positioned. As is shown in FIG., the result is a thinner wall if the hole is produced large but curved.

The following calculation is based on the dimensions shown incomplete figure below. This figure uses a geometric control for a hole as follows:

Ø.250-.260

⌖	Ø.014 Ⓜ	A	B	C

However, for the following calculation we will assume the feature uses the RFS method of tolerancing as follows:

Figure 30

Ø.250-.260

⌖	Ø.014	A	B	C

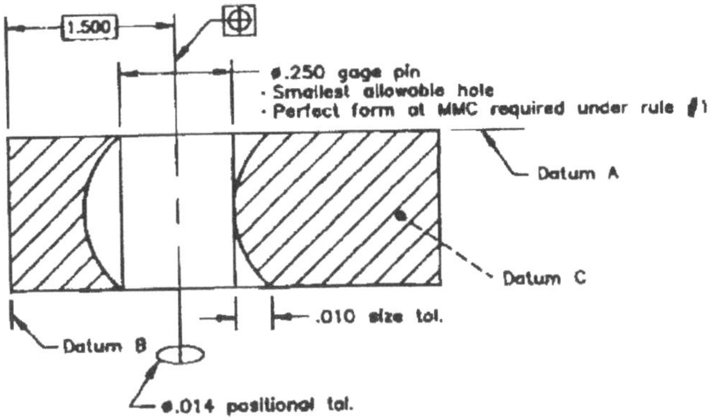

1.500 = basic dim.
-.142 = half boundary
1.358 = min. wall thickness to datum plane B

Ø.250 + .010 + .010 + .014 = Ø.284
worst outer boundry

or

.125 = half the MMC envelope
+.010 = size tol. all on one size on bottom as depicted
.135
+.007 = half the positional tolerance
.142 = half worst outer boundary

For example: **Ø.250-.260 hole**

⌖	Ø.014	A	B	C

Ø.250 - .014 = Ø.236 worst mating boundary (same as if the MMC symbol was used)

Ø.260 + .024 = Ø.284 worst sweep boundary (same as if the MMC symbol was used)

So, while it is important to note that the RFS concept preserves balance better than either the MMC or LMC symbol concepts, protects mating as well as the MMC symbol concept, and protects wall thicknesses and material preservation as well as the MMC symbol concept (but not as well as the LMC concept), it does create tighter, more restrictive tolerances on the average produced part. This runs the risk of increasing the cost of the overall product and should be used only when it is determined the use of the MMC or LMC concepts may endanger part function.

Material Condition Symbols and Concepts Explained

The following series of illustrations show the calculation of inner and outer boundaries using the Regardless of Feature Size Concept, the Least Material Condition Concept and the Maximum Material Condition Concept. It follows each with an explanation of what each concept may be used to achieve.

These explanations serve only as a brief introduction to the calculations of 'boundaries'. Modifiers (material condition symbols) are defined and a few uses for each are given. For a more in depth explanation of these concepts, see the section on Inner and Outer Boundary calculations.

If the hole is produced within its limits of size, but also moves out of its perfect location, inner and outer boundaries are created. These boundaries determine what the hole will mate with and what the minimum wall thicknesses are to the datum features and other surfaces.

The hole's smallest inner boundary is considered its worst mating condition. It determines what size shaft on a plate will mate with the hole at that location while seated on datum A. The hole's outer boundary is often used to calculate minimum wall thicknesses to the edges of the part.

Boundaries:

The concepts of boundaries generated by the collective effects of size and geometric tolerances are often referred to as simply inner boundaries and outer boundaries, but they are also called virtual conditions and resultant conditions. Any of these boundaries that are constant are defined as virtual conditions and the "worst case" of non-constant boundaries are resultant conditions.

Feature control frames that use maximum material condition symbols after the geometric tolerances generate constant (virtual condition) boundaries on the inner side for holes and on the outer side for shafts. Any feature control frames that use least material condition symbols after geometric tolerances generate constant (virtual condition) boundaries on the outer side for holes and on the inner side for shafts.

In all of these conditions, the resultant, or "worst case" for the non-constant boundaries are generated on the opposite side of the virtual conditions. The regardless of feature size concepts, when used on holes or shafts in the feature control frames, generate only "worst case boundaries" (resultant conditions) for the non-constant boundaries.

FIGURE 31 [Holes-MMC Concept]

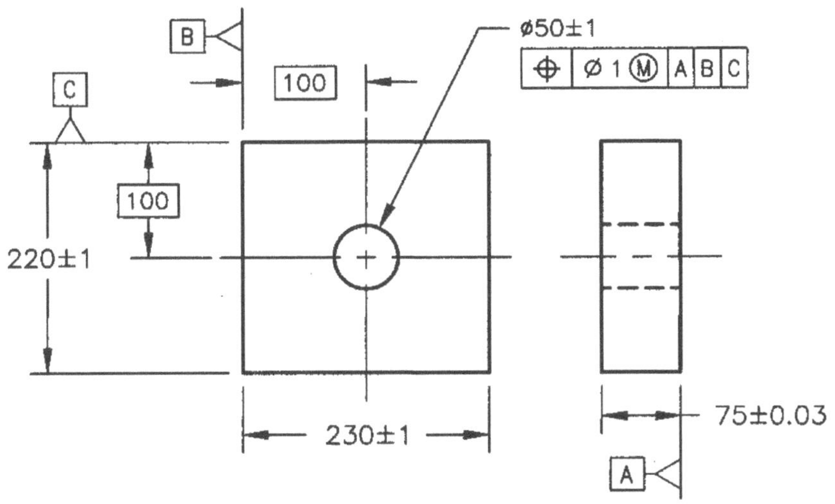

Ⓜ

- MMC = Maximum Material Condition
- Smallest hole
- Largest shaft
- Most material
- Heaviest feature
- Used on mating features
- Allows non-uniform airspace between mating features
- Allows fixed size gages to be used
- Generates constant boundaries (virtual conditions) that can be represented by fixed size gaging elements (pins, holes, slot widths, tab widths)

HOLE

Size		Geo. Tol.		Virtual Condition (constant boundary)
Ø49	-	1	=	Ø48) → worst case inner boundary
Ø50	-	2	=	Ø48)
Ø51	-	3	=	Ø48)

HOLE

Size		Geo. Tol.		Resultant Condition (worst case boundary)
Ø49	+	1	=	Ø50
Ø50	+	2	=	Ø52
Ø51	+	3	=	Ø54 → worst case outer boundary

The next illustration shows possible hole locations. These locations create the inner and outer boundary worst case sizes as calculated for the hole referenced at MMC. These sizes were calculated in the last illustration for the position toleranced hole using the MMC modifier. The following pages will show the boundaries and how they are calculated for holes and shafts called referenced at:

Ⓢ = implied (Regardless of Feature Size)
Ⓛ = (Least Material Condition)
Ⓜ = (Maximum Material Condition)

Figure 32
BOUNDARIES

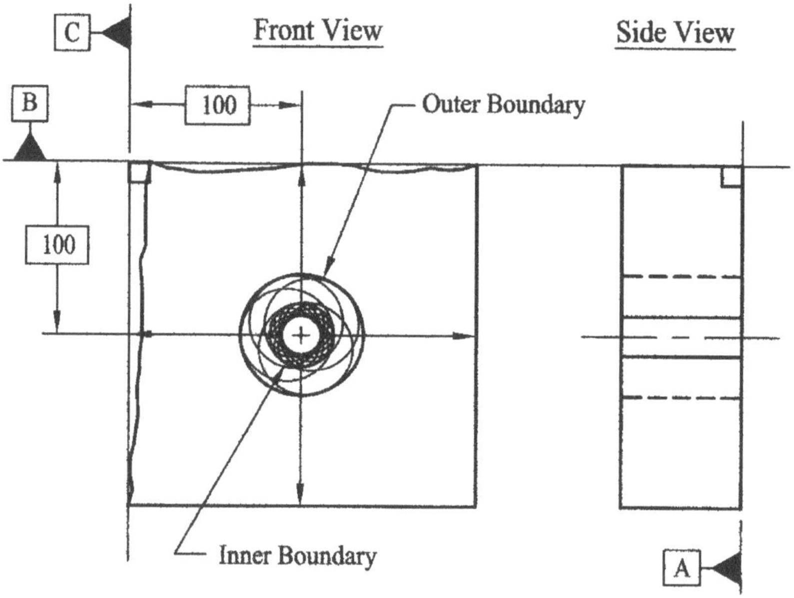

Constant vs. Non-Constant Boundaries

FIGURE 33 [Hole LMC Concept]

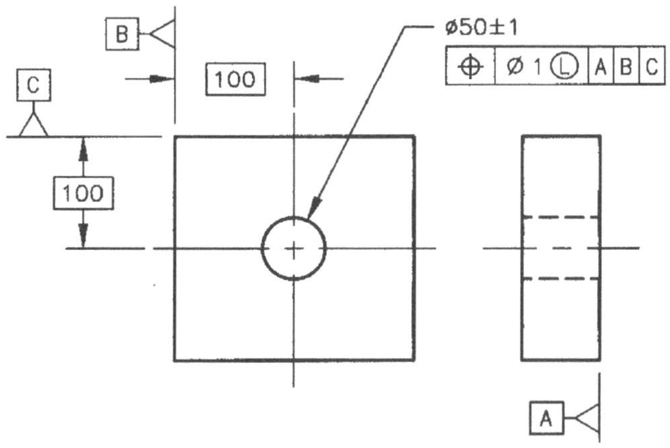

Constant vs. Non-Constant Boundaries

HOLE

Size	Geo. Tol.		Virtual Condition (constant boundary)
Ø51	+ 1	=	Ø52 → worst case outer boundary
Ø50	+ 2	=	Ø52
Ø49	+ 3	=	Ø52

HOLE

Size	Geo. Tol.		Resultant Condition (worst case boundary)
Ø51	- 1	=	Ø50
Ø50	- 2	=	Ø48
Ø49	- 3	=	Ø46 → worst case inner boundary

Ⓛ

- LMC = Least Material Condition
- Largest hole
- Smallest shaft
- Least material
- Lightest feature
- Used to preserve material for thicker walls
- Adds strength
- Allows feature sealing, but allows them to seal better on one side of the seal than the other. The seal is not uniform.
- Used on many casting drawings to allow surfaces enough material to be cleaned up by machine cuts.

FIGURE 34 [Hole RFS Concept]

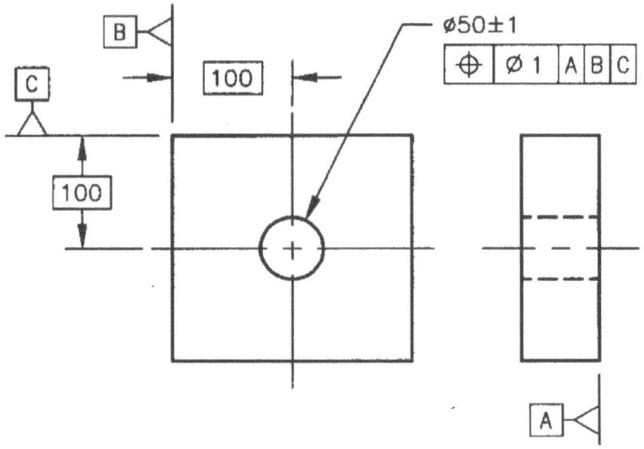

All Boundaries are Non-Constant in an RFS Call Out

HOLE

Size	Geo. Tol.		Inner Boundary
Ø49	- 1	=	Ø48 → worst case
Ø50	- 1	=	Ø49
Ø51	- 1	=	Ø50

HOLE

Size	Geo. Tol.		Outer Boundary
Ø49	+ 1	=	Ø50
Ø50	+ 1	=	Ø51
Ø51	+ 1	=	Ø52 → may be worst case

FIGURE 35 Worst Case When the Derived Median Line is Out-of-Straightness

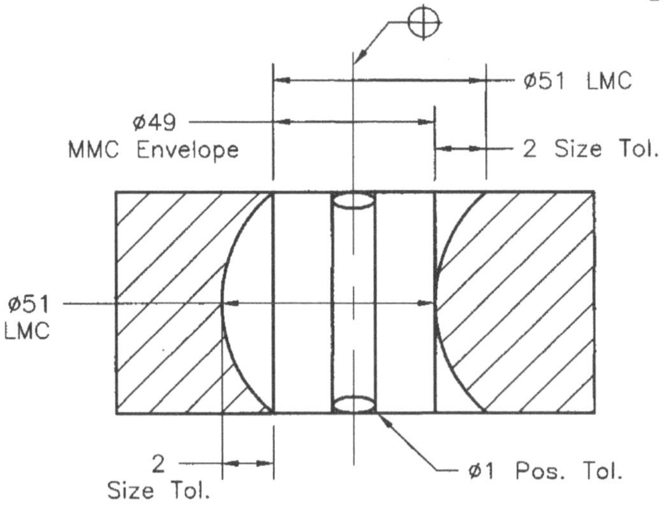

Ø49 + 2 + 2 + 1 = Ø54 worst case outer boundary

Only if the hole has a significant depth might this derived median line curvature (out-of-straightness) be a major factor. For very thin parts, such as sheet metal, it is probably not of concern in these analyses. In fact, many would contend that a banana-shaped hole is not likely to occur on most products. Therefore, for the analysis in this book, axially out-of-straight holes will not be mentioned.

Ⓢ implied

- RFS = Regardless of Feature Size
- No bonus tolerance = geometric tolerance remains the same regardless of the feature (produced) size
- Used to preserve uniformity
  - Airspace
  - Balance
- Used on spinning parts
- Used to protect uniform fit of seals

FIGURE 36 [Shafts-MMC Concept]

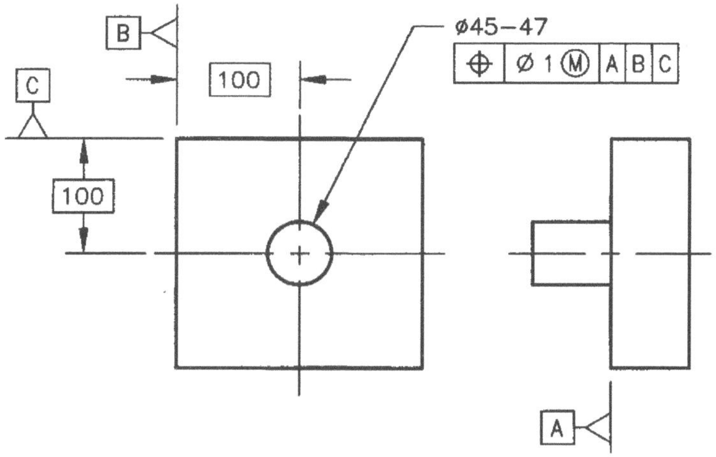

Constant vs. Non-Constant Boundaries

SHAFT

Size	Geo. Tol.		Virtual Condition	
Ø47	+ 1	=	Ø48	) → worst case outer boundary
Ø46	+ 2	=	Ø48	)
Ø45	+ 3	=	Ø48	)

SHAFT

Size	Geo. Tol.		Resultant Condition	
Ø47	- 1	=	Ø46	
Ø46	- 2	=	Ø44	
Ø45	- 3	=	Ø42	→ worst case inner boundary

FIGURE 37 [Shaft LMC Concept]

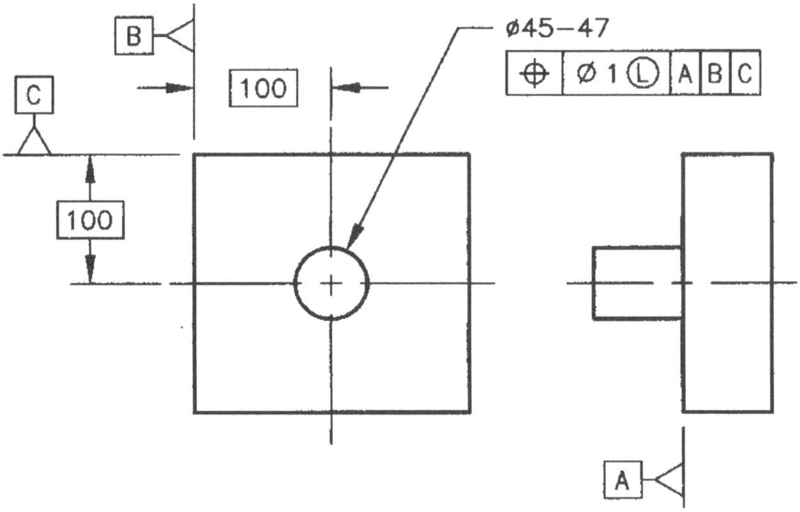

Constant vs. Non-Constant "Worst Case" Boundaries

SHAFT

Size	Geo. Tol.		Resultant Condition
Ø45	+ 1	=	46
Ø46	+ 2	=	48
Ø47	+ 3	=	50 → worst case outer boundary

SHAFT

Size	Geo. Tol.		Virtual Condition
Ø45	- 1	=	44) → worst case inner boundary
Ø46	- 2	=	44)
Ø47	- 3	=	44)

Creating an Equal Bilateral Toleranced Dimension From Virtual and Resultant Conditions for Figure 6

	Resultant Condition Shaft	=	Ø50
+	Virtual Condition Shaft	=	Ø44
	Sum		Ø94

	Resultant Condition Shaft	=	Ø50
-	Virtual Condition Shaft	=	Ø44
	Difference	=	Ø6

So,

$$\frac{Ø94}{2} = Ø47 \quad \& \quad \frac{6}{2} = 3$$

Therefore, expressed as an equal bilateral toleranced dimension = Ø47±3.

FIGURE 38 [Shaft RFS Concept]

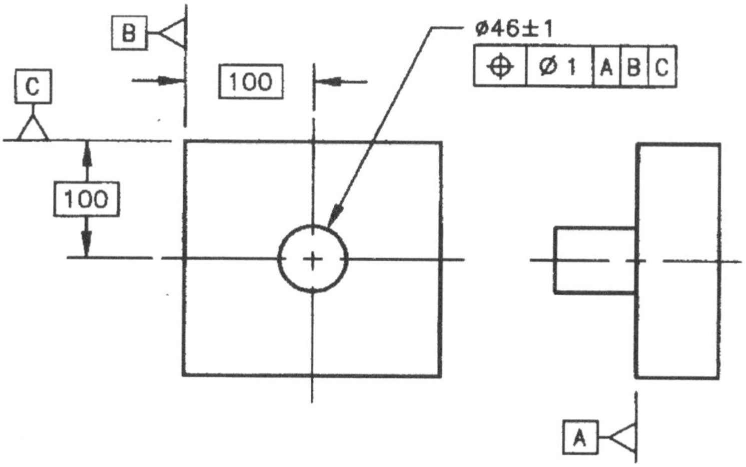

SHAFT

Size	Geo. Tol.		Outer Boundary
Ø45	+ 1	=	Ø46
Ø46	+ 1	=	Ø47
Ø47	+ 1	=	Ø48 → worst case

SHAFT

Size	Geo. Tol.		Inner Boundary
Ø45	- 1	=	Ø44 → worst case (maybe)
Ø46	- 1	=	Ø45
Ø47	- 1	=	Ø46

Creating an Equal Bilateral Toleranced Dimension from Inner and Outer Boundaries

	Outer Boundary Shaft	=	Ø48
+	<u>Inner Boundary Shaft</u>	=	<u>Ø44</u>
	Sum		Ø92
	Outer Boundary Shaft	=	Ø48
-	<u>Inner Boundary Shaft</u>	=	<u>Ø44</u>
	Difference	=	Ø4

So,

$$\frac{\text{Ø}92}{2} = \text{Ø}46 \quad \& \quad \frac{4}{2} = 2$$

Therefore, expressed as an equal bilateral toleranced dimension = Ø46±2.

The Ø44mm is identified only as ***maybe*** the worst case inner boundary. Regardless of feature size conditions have an oddity that must be considered in order to determine the absolute worst case inner boundary.

Figure 39

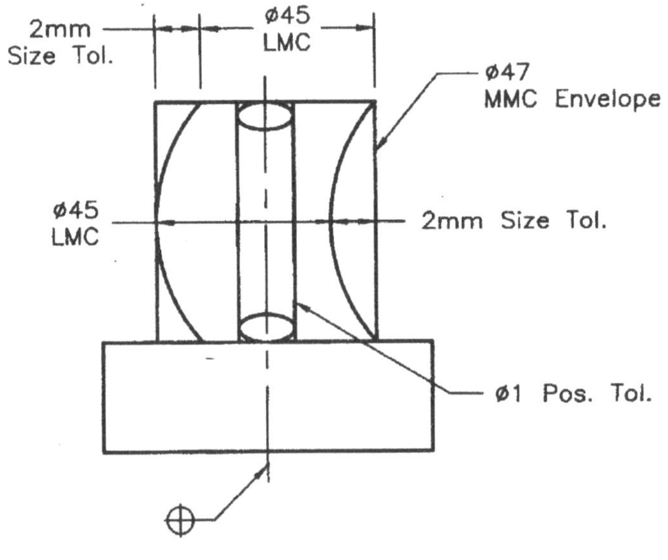

Ø47-2-2-1 = Ø42 worst case inner boundary
[with the derived median line out-of-straightness considered]

Only if the shaft has a significant length might this derived median line out-of-straightness be a major factor. For very short shafts, it is probably not of concern in the analysis. The analyses in this book will not consider this problem, but if your products run the risk of banana-shaped shafts, you may wish to use the above illustration to calculate the worst case inner boundary of your shafts.

Chapter 5

Form

•Flatness
•Straightness
•Cylindricity
•Circularity

Chapter Objectives

Readers will learn:

1. Flatness of a surface, flatness on a unit basis and flatness of a derived median plane; what they mean, the tolerance zone configurations and how they are measured.

2. Straightness of a surface, straightness of a derived median line; what they mean, the tolerance zone configurations and how they are measured.

3. Cylindricity of a surface; what it means, the tolerance zone configuration, what it controls and to within how much, how it is measured and how it relates to flatness.

4. Circularity, what it means, the tolerance zone configuration, what it controls and to within how much and how it is measured.

Form

There are **four geometric characteristic symbols** in the category called **form**. They are:
Straightness, Flatness, Circularity, Cylindricity

These symbols are used in feature control frames to tolerance defined 2-D and 3-D elements of part feature configurations. They are not related to datums, but are often used to tolerance the shape of primary datum features.

Two of the most common geometric characteristics used on primary datum features are flatness (for planar seating surfaces) and cylindricity (for cylindrical mating, alignment features). These are 3-dimensional form controls. Straightness, if used on a surface, and Circularity, which can only be used as a surface control, are considered 2-Dimensional control.

Flatness

Flatness used as a 3-D surface control, constructs a tolerance zone that consists of two parallel planes separated by the geometric tolerance. The entire surface being toleranced must have all its elements between these two planes. For example:

FIGURE 40 [Flatness of a Surface]

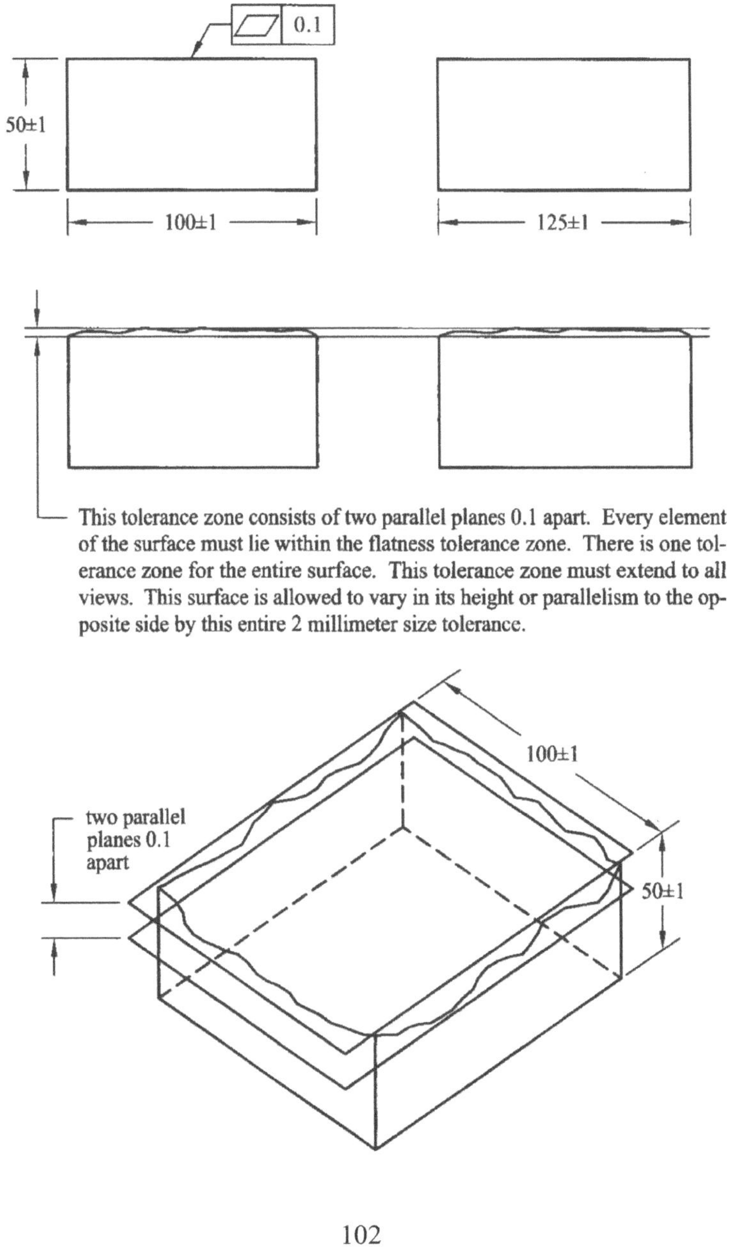

In this type of surface control, no datums are referenced. It is a 3-D form control that doesn't try to relate the orientation or location of the zone to datums. Since the surface is planar (not a diameter), no diameter symbol is appropriate in the feature control frame. It is a surface control, not an axis control. The use of a diameter symbol in a feature control frame denotes the control of an axis. Since this flatness control is just trying to tolerance the shape of a single surface, it is not a feature of size. It has no maximum or least material condition. Therefore, no material condition symbol is appropriate to use in the feature control frame. So, the only items to be found in the feature control frame are the flatness geometric characteristic and a geometric tolerance.

Measurement can be done by scanning the surface to determine if all points on the surface are within the tolerance zone. This can be done by manually orienting the surface to obtain optimal results as shown in FIGURE 41, or by allowing a computer program to merge the tolerance zone with the variables data collected by a machine like the coordinate measurement machine (CMM).

FIGURE 41 [Measuring flatness on a surface plate with a dial indicator and stand]

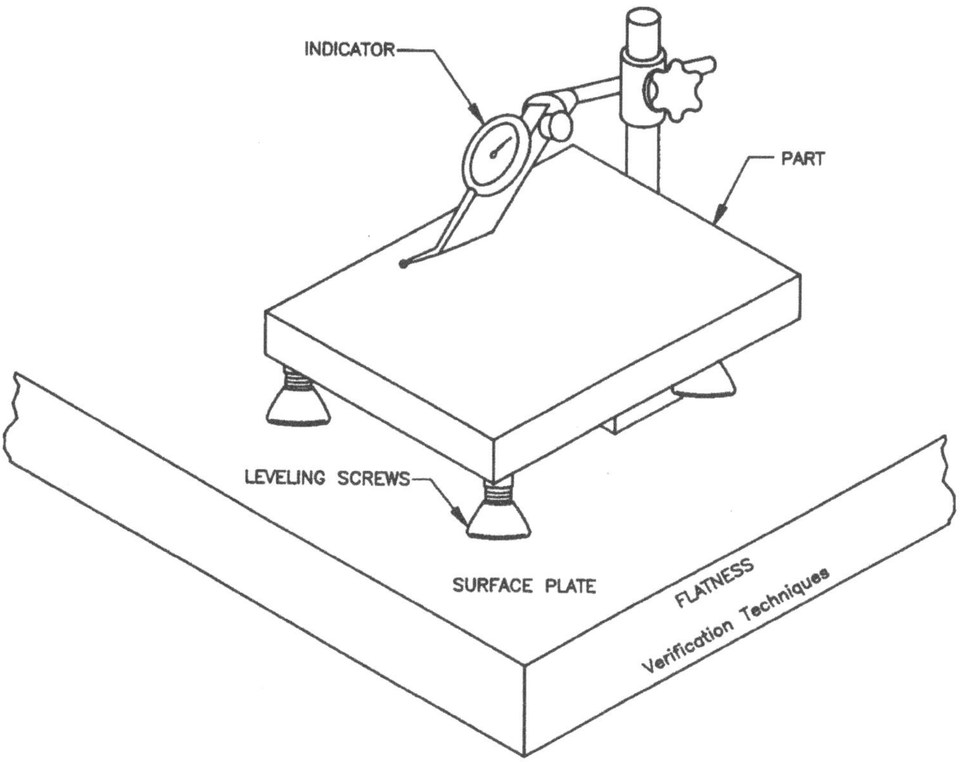

Sometimes, flatness is applied on a per unit basis. This would allow a tolerance that is so much per the unit specified. This type of constraint is generally used to refine an overall control of the entire surface. For example:

⏥	0.5
	0.1 / 20 X 20

This means the entire surface must have all its elements capable of fitting between 2 parallel planes that are 0.5mm apart. The refinement requirement tells us that every 20x20mm square on the surface must be flat to within 0.1mm. This prevents any abrupt surface variations within these 20x20mm squares. The entire surface could have a uniform curvature of 0.5mm, but no pits or bumps greater than 0.1mm.

FIGURE 42 [Flatness on a Unit Basis or Rate Basis]

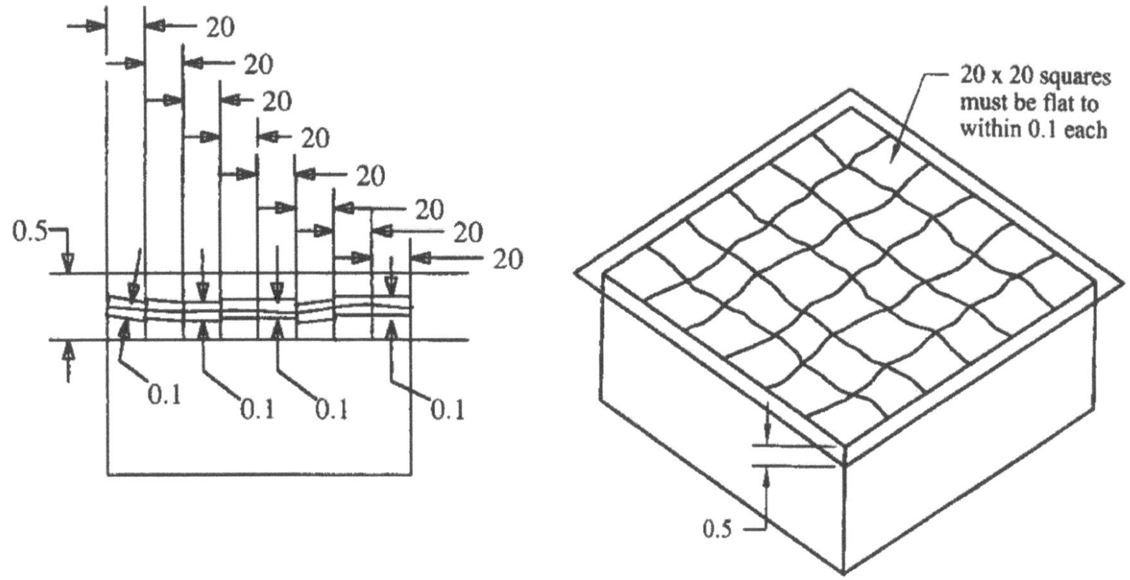

In the ASME Y14.5-2009 standard on Dimensioning and Tolerancing, flatness of the derived median plane replaced straightness of the derived median plane. It appears as shown in FIGURE 43.

FIGURE 43 [Flatness of the Derived Median Plane]

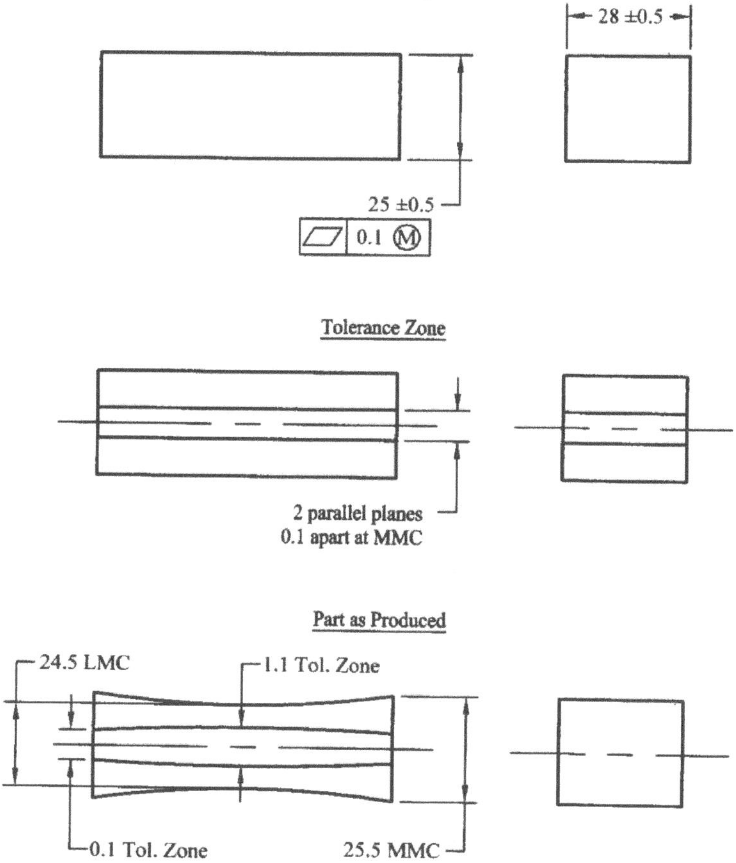

FIGURE 44 [Measurement of Flatness of the Derived Median Plane]

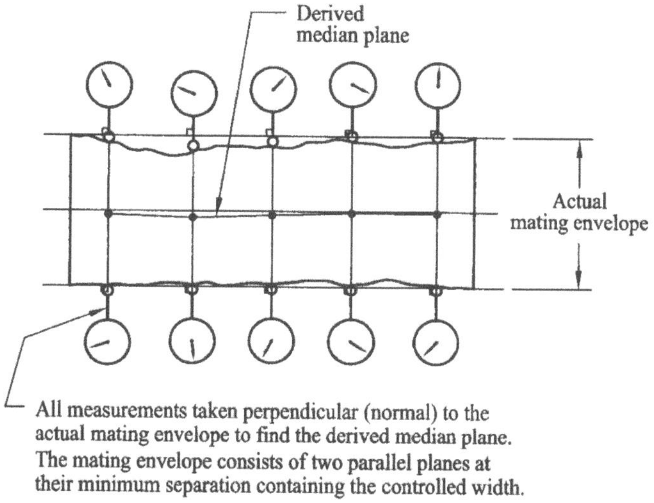

The tolerance zone for the flatness of the derived median plane control shown in FIGURE 43 and being verified in FIGURE 44 consists of two parallel planes that are 0.1 apart. It must contain the derived median plane. The maximum material condition symbol allows the flatness tolerance zone to grow locally as the part shrinks (is produced smaller than the MMC) locally.

As FIGURE 43 shows, the flatness tolerance zone may grow locally to a maximum of 1.1 in width if the controlled feature is produced at the least material condition of 24.5 at that location. This tolerance compliance is measured by probing points normal to the actual mating envelope. The actual mating envelope would consist of two parallel planes (much like vise jaws) that have closed on the width of the part (that is being controlled by this flatness requirement). The minimum separation of these planes would constitute the actual mating envelope and measurements are taken perpendicular to these planes on opposite sides of the part (as shown in FIGURE 44). These readings are then averaged to find the derived median plane. The derived median plane must reside within the flatness tolerance zone for the part to be accepted.

The methodology used to average the readings in FIGURE 44 would be to take each reading on the bottom of the part and subtract it from the reading directly above it on the top. Divide the results by 2. Then graph the results. Since there is no datum to orient or center the tolerance zone to, the collected and graphed median points may be manipulated to their optimal location to fit within the tolerance zone.

Since this flatness control has referenced its tolerance at MMC, a functional gage may be used to verify compliance. The gage would consist of two parallel rails that are 25.6 apart (the virtual condition size). Since Rule #1 is overridden when flatness is applied to the derived median plane, the produced part would first be measured for size violations at cross-sections. If no size limits are violated and the part is received by the functional gage measuring the flatness control, the part is adjudged to be in compliance with these requirements.

Another was to eliminate Rule #1 is to use the Independency Symbol (See Example on page 107). This would also say that perfect form at MMC is not required.

The following are more exceptions to Rule #1. Features do not fall under Rule #1 that:

1) are considered still in the as purchased condition stock size. These would be controlled by industry standards governing the production of the stock.
2) features that are subject to free state variation in the unrestrained condition.
3) features that used the **(ASME Y14.5-2009 and ASME Y14.5-2018) Independency symbol** next to the size dimension. See the FIGURE on the next page.
4) are dimensioned with AVERAGE dimensions (dimensions with the abbreviation AVG. after them).
5) use an applicable geometric tolerance such as Straightness of the Derived Median Line or Flatness of the Derived Median Plane (really a flatness type of control of the middle of a part width). These controls are capable of eliminating the envelope of perfect form at maximum material condition.

The measurement tool most approximating the MMC envelope is called a GO gage. Any of the five conditions described above and the Independency example shown on page 107 would eliminate the need to use a GO gage to measure their maximum material condition. Their MMC measurement would be reduced to a cross-sectional measurement, such as is conducted by a micrometer for a shaft, or using a coordinate measuring machine to probe points directly opposite one another in a hole.

Example: [Principle of Independency]

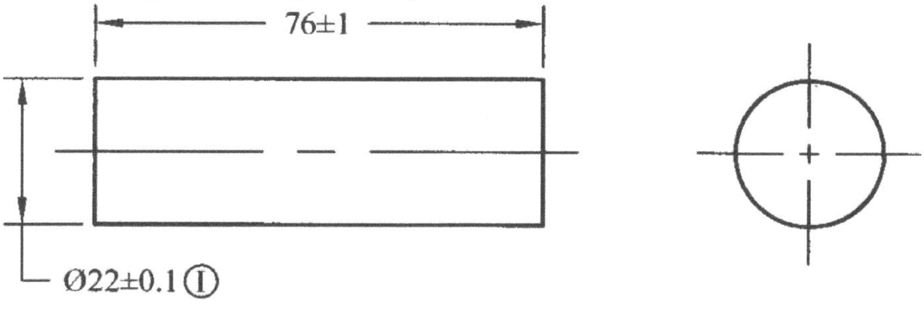

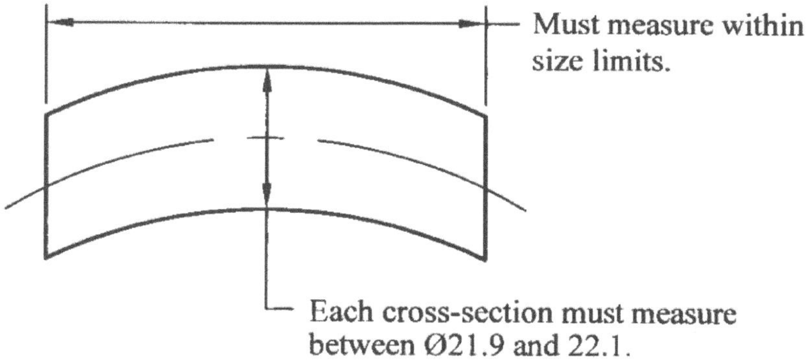

FIGURE 45 [Straightness of a Surface]

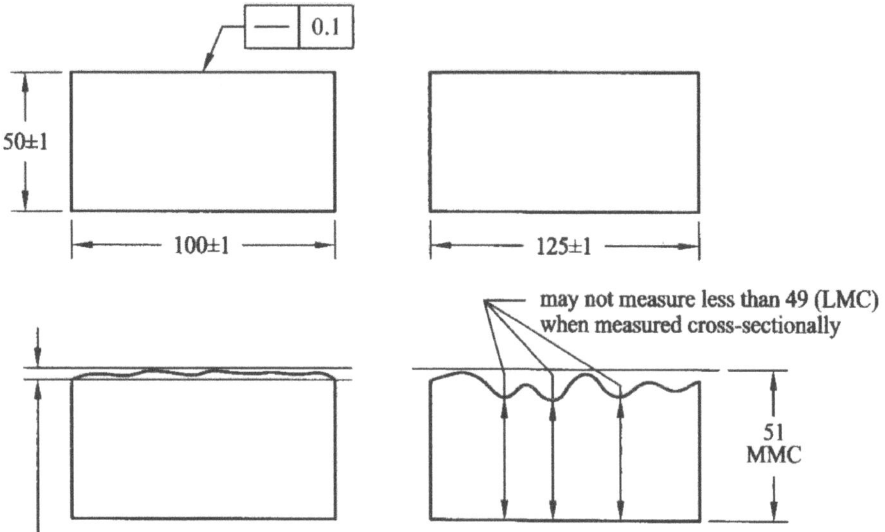

This tolerance zone consists of two parallel lines 0.1 apart. Every element of the individual line being inspected must lie within the straightness tolerance zone. These tolerance zones must exist in the plane of the view the straightness of the surface control is depicted in. If the surface spans the size tolerance of 49 to 51 millimeters in height, the line elements are allowed to differ in height and angle to one another by this entire 2 millimeter size tolerance.

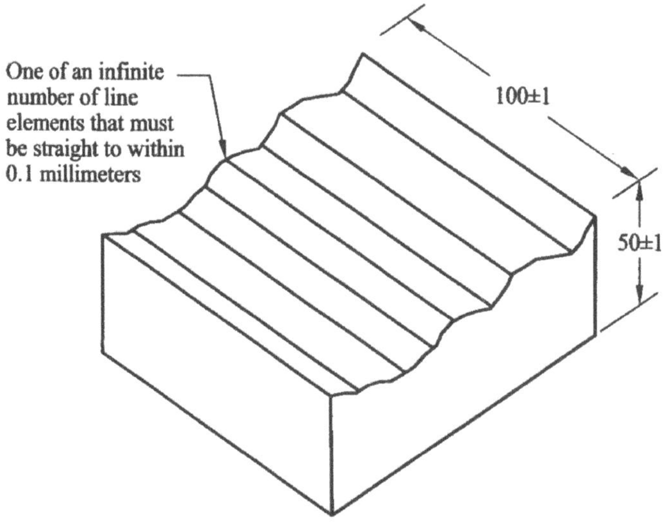

Straightness is one of the oddest geometric characteristic symbols/concepts in the Y14.5 standard. It can be used to control:

- line elements on a surface
- the axis of a diameter where it is referred to as 'Straightness of the Derived Median Line' (on the premise that all axes are, by definition, perfectly straight)
- Straightness of the Derived Median Plane' was changed in Y14.5-2009 to 'Flatness of the Derived Median Plane' (because a center plane is perfectly flat).

If used as a surface control, it states that each line element measured in the plane of the view the control is depicted, must be independently verified as residing within a tolerance zone that consists of two parallel lines separated by the amount of straightness tolerance. An example is a cone such as shown below.

FIGURE 46 [Straightness of Surface Line Elements on a Cone]

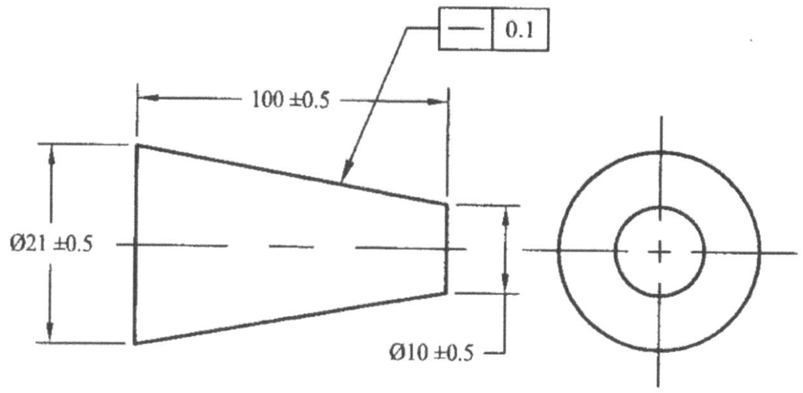

The straightness tolerance zone would consist of two parallel lines 0.1mm apart.

FIGURE 47 [Tolerance Zone is Two Parallel Lines 0.1 Apart for Each Line Element]

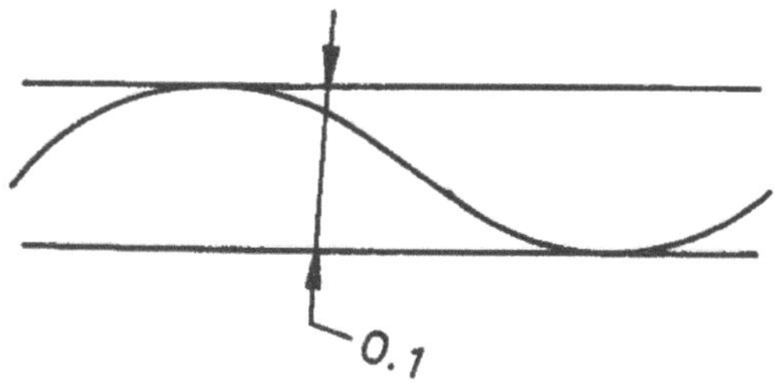

Each longitudinal line element would be independently verified, so the conical shape could be badly formed. For example, an end view of a compliant produced part could look like FIGURE 47.

FIGURE 48 [Two-dimensional line element control]

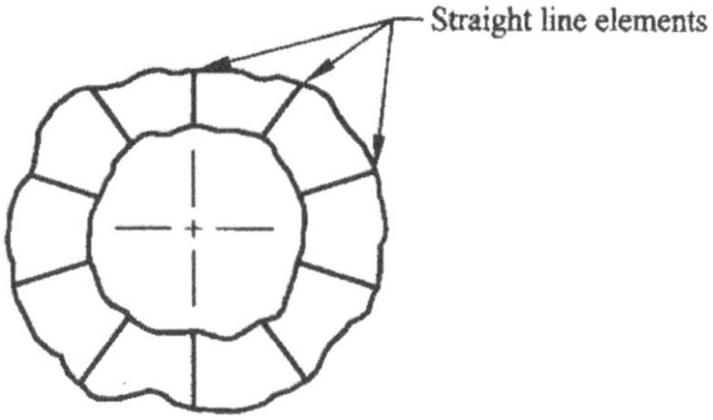

FIGURE 49 [Measuring straightness of Line Elements on a Cone]

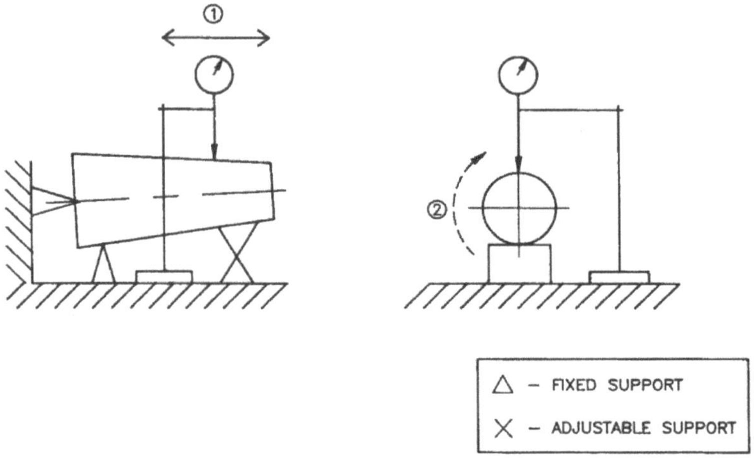

As long as no size dimensional tolerances have been violated and each line element fits within its own straightness tolerance zone (consisting of two parallel lines 0.1 apart), the part passes all requirements. No surface control, such as flatness, circularity, cylindricity, runout, total runout, profile of a line, profile of a surface or other controls that, when used to control shape of a surface (rather than an axis or centerplane), is allowed to use a maximum material or least material condition symbol associated with the geometric tolerance. Straightness, when used as a surface control, may not use one of these modifiers either. Also, no datum feature may be referenced in a straightness feature control frame. Yet, the equivalent of an implied datum is invoked in every surface straightness control. That implied orientation of measurement is the plane of the view in which the straightness control is depicted.

In line elements, only the plane of that view must meet the straightness requirement. Other views and line elements taken in their plane are uncontrolled by the straightness of the surface control. So, like a corrugated piece of sheet metal, the line elements may be quite straight in one plane but do not need to be straight in others. The applicable plane is solely dictated by the view the straightness control is shown in.

FIGURE 50

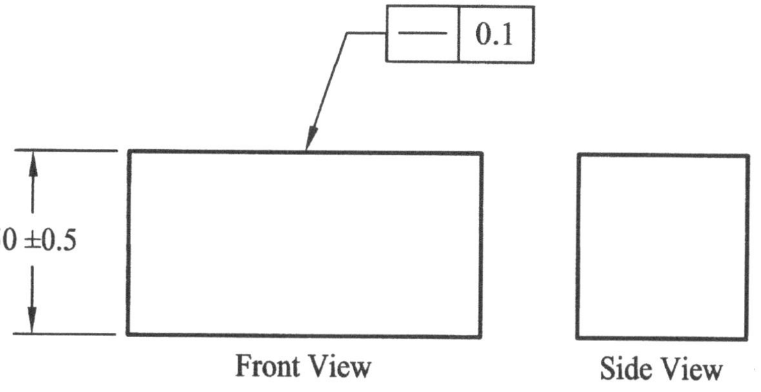

As Drawn

In FIGURE 51, this part complies with all requirements shown in the drawing specification.

FIGURE 51 [As Produced]

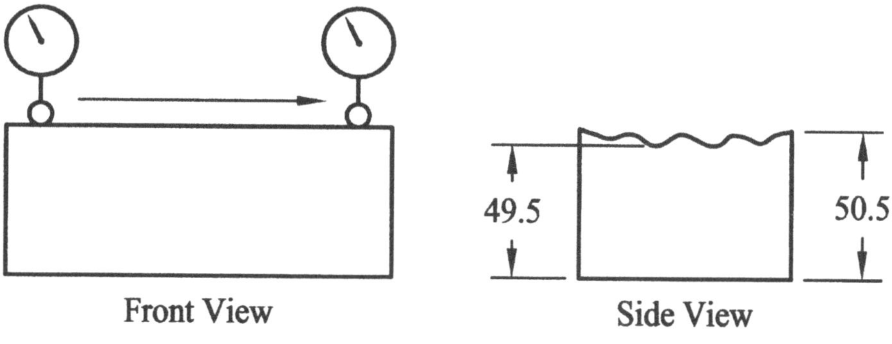

Straightness of a derived median line is a control of the middle of a diameter (its axis). For example, see the following figures.

FIGURE 52 [Straightness of a Derived Median Line]

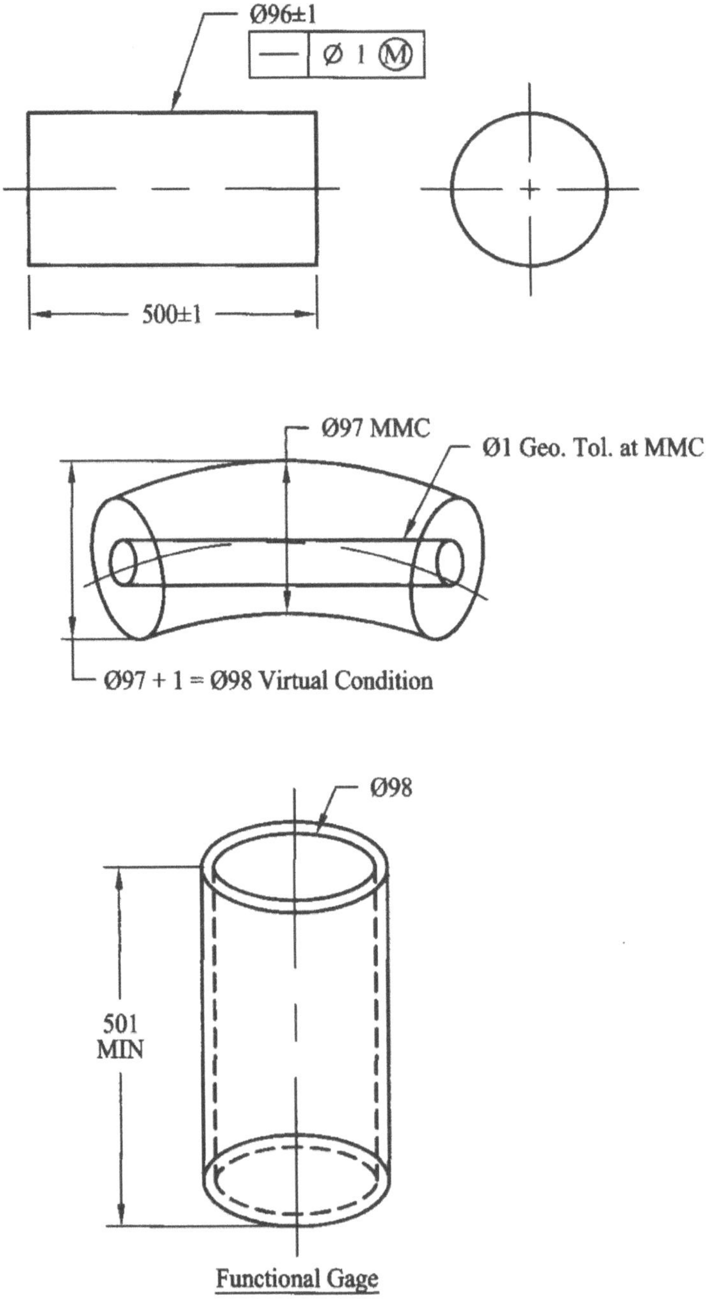

FIGURE 53 [Waisted-shaped part below]

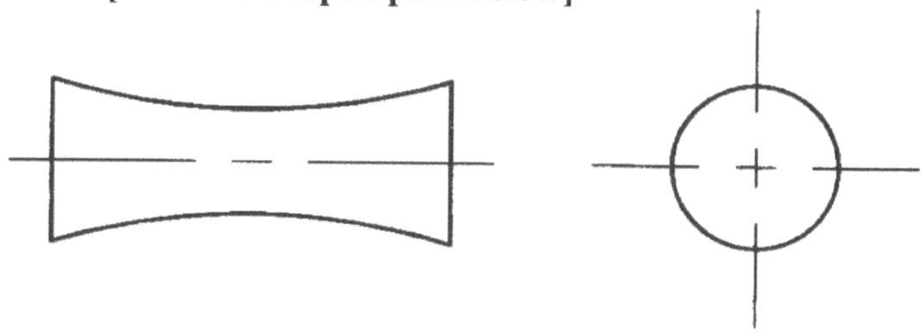

This part's surface is not straight, but its axis is straight.

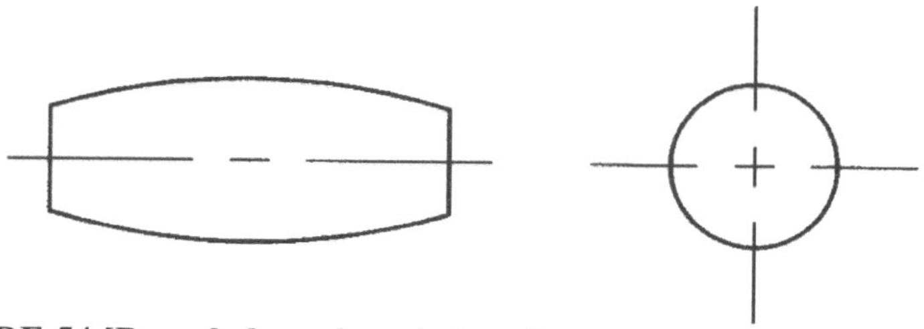

Likewise, this part's surface is not straight, but its axis is straight.

FIGURE 54 [Barrel-shaped part above]

But the following part has neither a straight surface nor a straight axis (derived median line).

FIGURE 55 [Banana-shaped part below]

Straightness of a derived median line, when used on a cylindrical feature, begins with a cylindrical tolerance zone in the middle of the part. For example:

FIGURE 56

As specified:

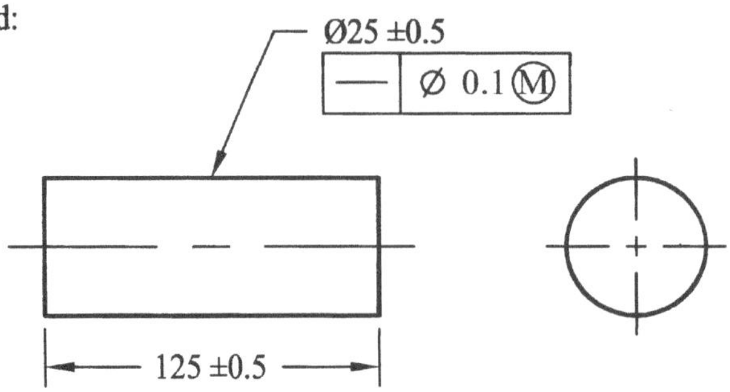

Tolerance zone

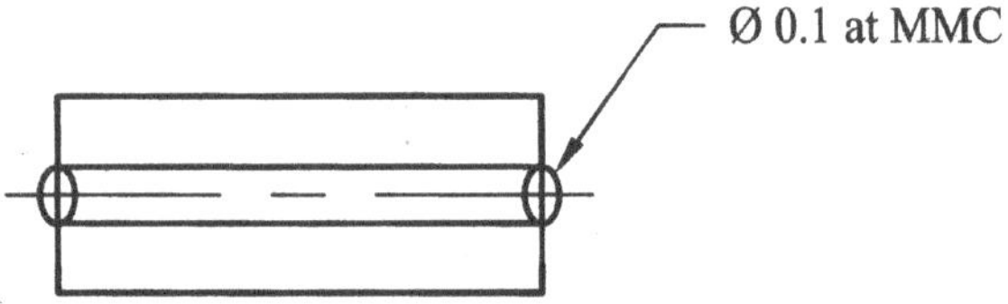

The cylindrical straightness tolerance zone is allowed a 'bonus tolerance' since the maximum material condition symbol is specified after the geometric tolerance. Oddly enough, unlike bonus tolerances calculated for position, perpendicularity, parallelism and angularity (on holes and shafts), this bonus tolerance is calculated on a local size. If the shaft shown in the previous illustration is produced smaller than its MMC anywhere on its diameter, the straightness tolerance grows by the same amount only at that location.

So, it's possible that what started out as a perfectly cylindrical straightness tolerance zone may end up bulging with bonus tolerance in localized areas. For example, see FIGURE 57.

Measurements for straightness of the derived median line are taken normal to the actual mating envelope. **For a shaft**, the median line is determined by taking measurements along the diameter normal to the minimum circumscribed cylinder. **For a hole**, these measurements would be taken normal to the maximum inscribed cylinder.

FIGURE 57

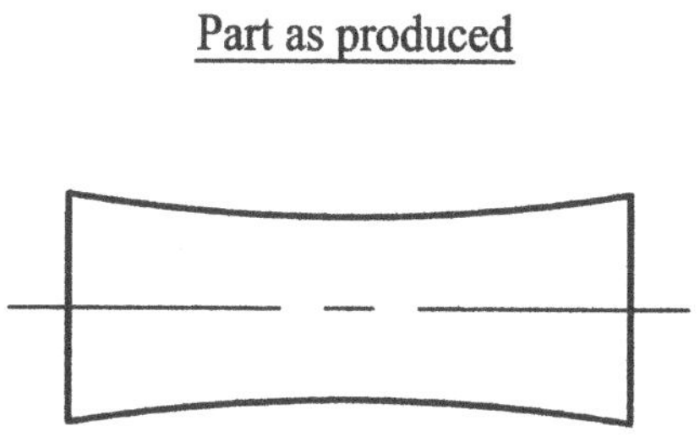

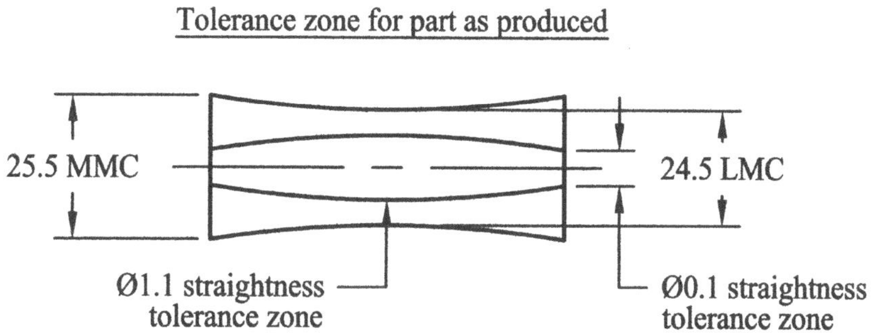

FIGURE 58 [Differential Measurements]

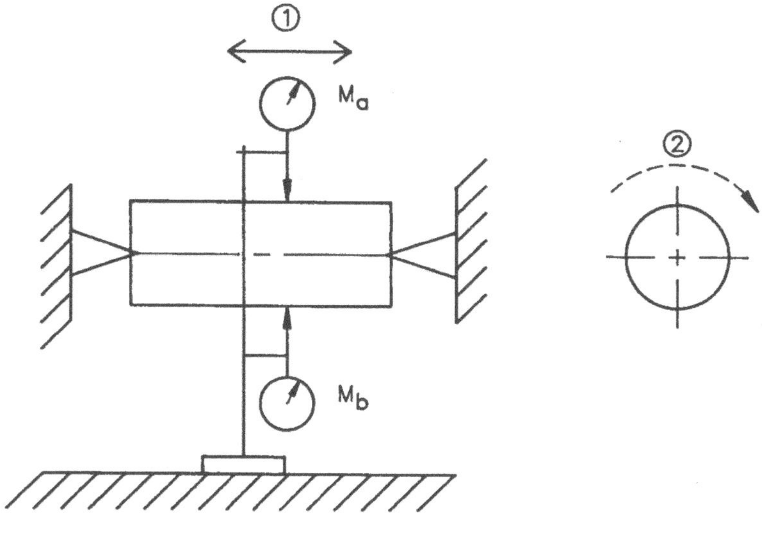

Radial deviation from a perfectly straight axis is = Ma minus Mb divided by 2. The results should be graphed for a visual display of compliance capability.

Another measurement possibility would be to use a functional gage. One such correctly designed, dimensioned and toleranced gage would be:

FIGURE 59 [Functional Gage]

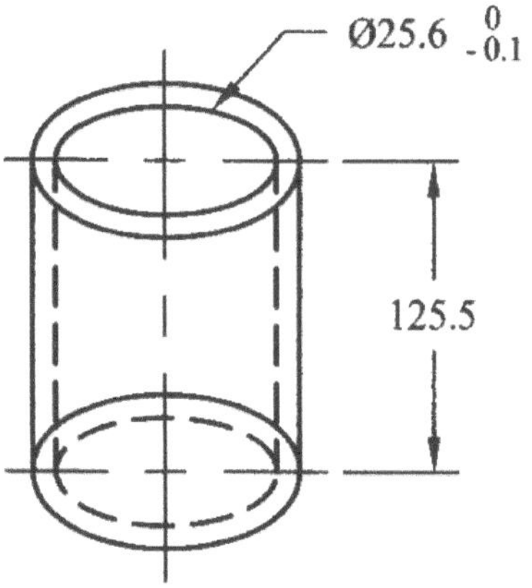

If the part fits into the functional gage and meets all of its separately verified size limits, the part is accepted as in compliance.

Another oddity of the straightness of the derived median line control is that it changes how the size of the part is verified. Instead of the part being measured for size compliance for fitting into an MMC envelope and LMC measured at cross sections, both MMC and LMC are verified with cross sectional measurements. Perfect form at MMC requirements imposed by Rule #1 are released. The part must simply measure cross-sectionally no larger than MMC and no smaller than LMC.

Straightness of a derived median line per unit length is a tool allowed by the Y14.5 standard. It is used to prevent an abrupt out of straightness variation of the feature over a short defined length. In this application, it is important to state a total allowed out-of-straightness control as well as the unit length control. The overall control prevents a possible large, trigonometrically induced, out-of-straightness variation for the overall length of the feature. In this control, RFS may be implied or MMC or LMC specified. If specified at MMC, functional gaging is possible with one gage each for the total straightness control and the unit length straightness control.

Figure 60 [Straightness of the Derived Median Line Per Unit Length]

Cylindricity

Like flatness, cylindricity is a 3-dimensional form control. And like flatness, cylindricity is a good control to use on a primary datum feature. In fact, the tolerance zones have a lot in common. The tolerance zone for flatness is two parallel planes separated by the geometric tolerance and the tolerance zone for cylindricity is two concentric cylinders separated by the geometric tolerance. One can almost see the flatness tolerance zone curling up and becoming the cylindricity tolerance zone.

FIGURE 61

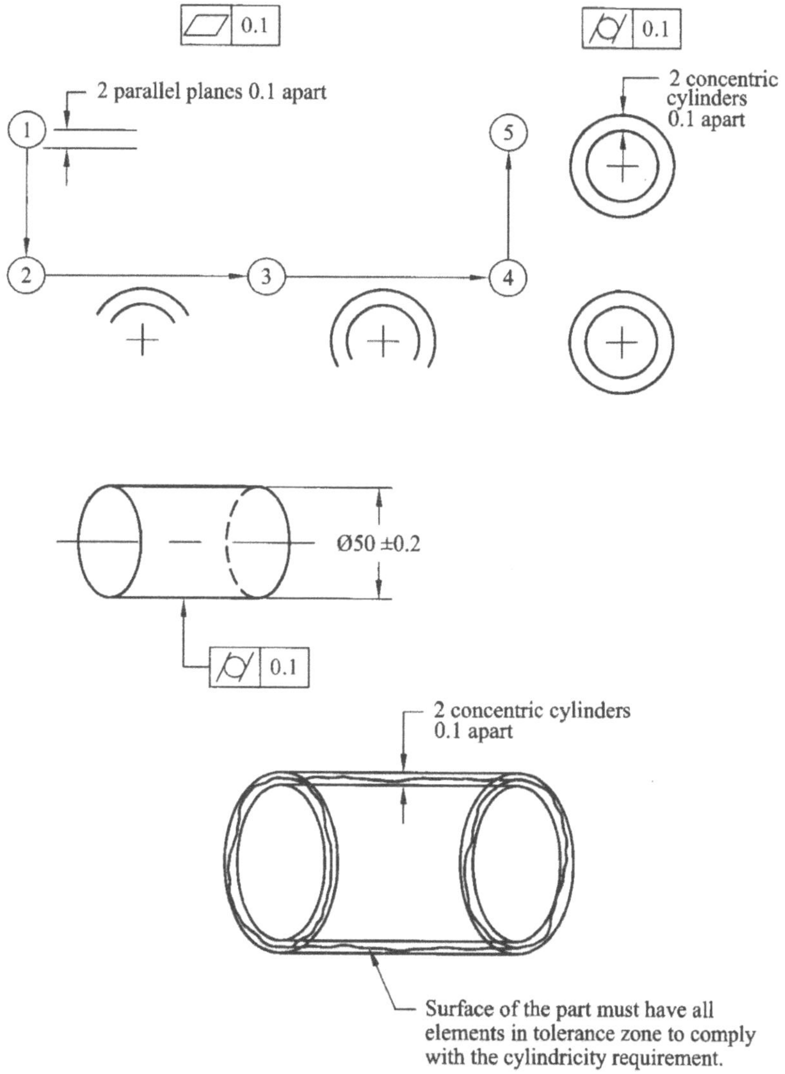

Cylindricity controls:

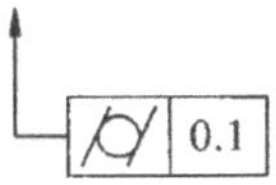

Surface flaws may not be greater than:

- Pits = 0.1
- Bumps = 0.1
- Flats = 0.1
- Ovals = Ø0.2
- Surface straightness = 0.1
- Axis straightness = Ø0.1
- Taper = 0.1 per side (Ø0.2)

Cylindricity can be described as a 3-dimensional form control of circularity, straightness and taper. As described in the chart above, the tolerance applies as two concentric cylinders radially separated by the amount of cylindricity tolerance that is shown in the feature control frame. It is meant to refine control of the surface form that size tolerances on the diameter would otherwise control (under Rule #1).

The size limits on the diameter are separately verifiable requirements. So, the diameter must not violate an envelope of perfect form at MMC (generally thought of as simulated by a GO gage). The diameter may not measure outside the LMC at cross-sections on the part (which can be measured with a NOGO gage or a simulation of the NOGO gage with any tool capable of measuring at cross-sections).

Separate from those diameter measurements, the part is ideally centered on the cylinder's own axis, rotated continuously about that axis and probed for surface flaws such as pits, bumps, flats, ovals, surface out-of-straightness, axis out-of-straightness and taper. If measured with a dial indicator, the full indicator movement (FIM) may not exceed the cylindricity tolerance in the feature control frame. The FIM in 360° revolutions of the part while the indicator is moved longitudinally down the surface (from end to end) may not be greater than the cylindricity tolerance when directly compared. Rotating the cylinder about its own axis is a convenience to determine cylindricity compliance. In reality, the axis is not important to this measurement. The inspector is just trying to verify that all points on the cylinder's surface lie between two concentric cylinders that are radially separated by the cylindricity geometric tolerance.

Cylindricity differs from circularity in that it controls surface taper (preventing conical shapes that exceed cylindricity tolerances) and surface straightness (limiting barrel shapes and waisted shapes to within the cylindricity tolerances). Circularity does not control surface taper or surface straightness.

Cylindricity differs from total runout in that cylindricity is measured from and centered around its own axis, whereas total runout is centered to a datum axis. This is why total runout, although it controls 3-dimensional form, is mainly considered a balance control, while cylindricity just controls 3-dimensional form.

Cylindricity is a surface control in that it shapes the feature surface. It will, therefore, not utilize a diameter sign inside the feature control frame. The diameter symbol, when used, implies control of an axis. Cylindricity is not a relationship control, so no datum features are referenced. Surface controls (such as cylindricity) are not allowed to utilize a maximum material condition symbol associated with the specified geometric tolerance. So, there will never be a growth of the tolerance zone ("bonus tolerance") for surface controls.

FIGURE 62 [Cylindricity Measurement Set-Ups-Method 1]

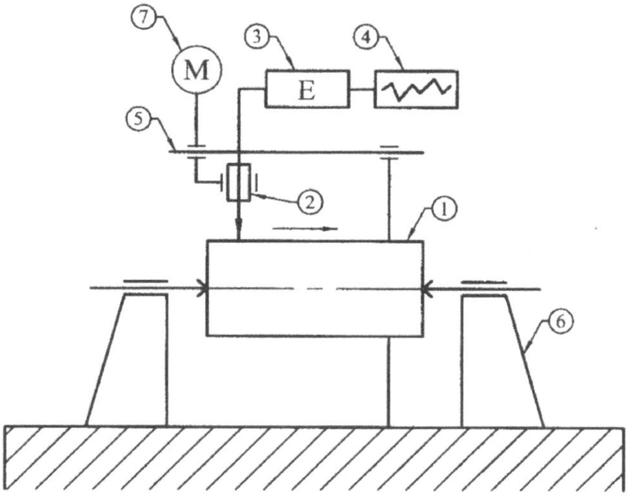

1) Specimen
2) Electric Sensor
3) Electronic Unit
4) Linear Printer
5) Slide Bar
6) Tip Support
7) Motor

Position the specimen between the tips. Make sure that the slide bar is parallel to the line connecting the two fastening tips. Scan the cylindrical surface to be measured continuously with a constant scanning rate according to the generatrix method. Draw up the measured profiles as a linear diagram.

Note: Webster's New World Dictionary defines Generatrix as a point, line or plane, whose motion generates a line, surface or solid.

FIGURE 63 [Cylindricity Measurement Set-Ups-Method 2]

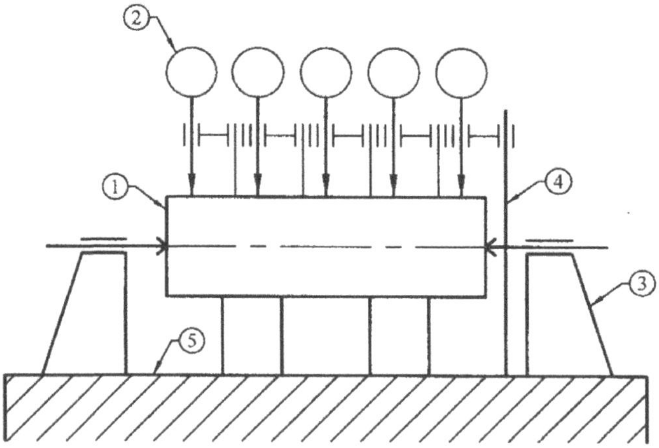

1) Specimen
2) Length Measuring Device
3) Tip Support
4) Measuring Supports
5) Test Plate

Position the specimen between the tips. A length measurement device is assigned to each test point on the generatrix to be measured so that all test points of a generatrix are determined simultaneously. Set all length measurement devices to the same indicated value with the help of the reference cylinder. Scan the cylindrical surface to be measured,

Circularity

FIGURE 64 [Circularity of a Surface—aka Roundness]

Design Requirements

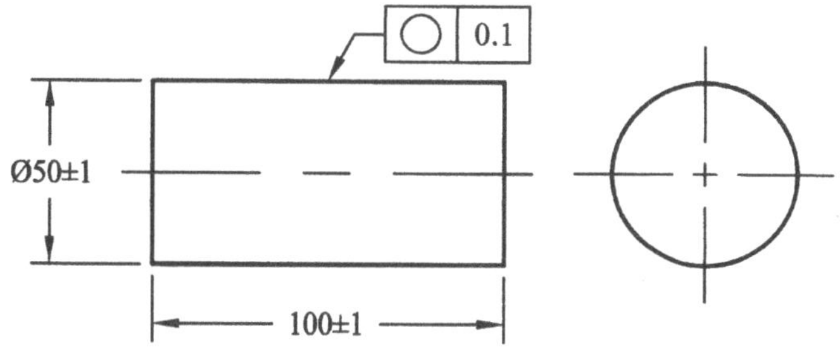

Possible Produced Configuration

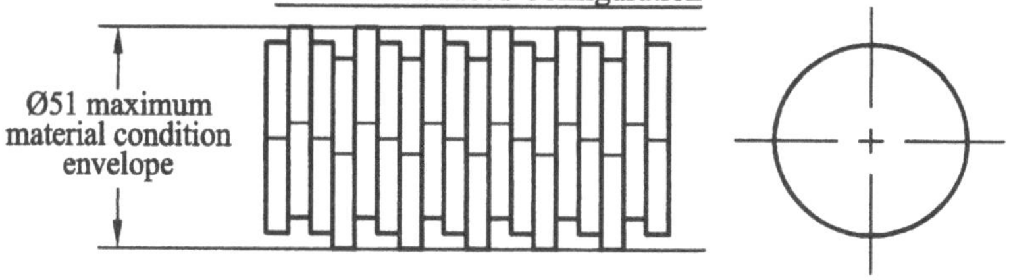

One might think of the Circularity tolerance as controlling a stack of coins that is allowed to be mis-stacked anywhere within the size limits of Ø49 to Ø51mm. Each circular cross section of the cylinder is one coin in the stack. Each cross-section must be circular (round) to within the 0.1mm circularity tolerance. The circular cross-sections may vary in size allowing taper and out-of-straightness as long as the size tolerance of Ø49-51mm is not violated. A circularity control of 0.1mm allows; pits to be no deeper than 0.1, bumps to be no taller than 0.1, flats to be no deeper than 0.1 and ovals to vary (on each coin in the stack) by no more than a diameter of 0.2.

Possible Configurations that would meet the Circularity requirement of 0.1

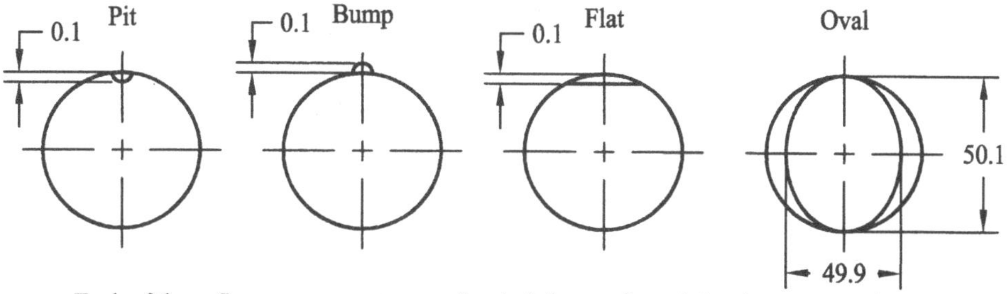

Each of these figures represents one of an infinite number of circular cross-sections for each cylinder and one of an infinite number of surface flaws possible.

The surface is comprised of an infinite number of circular cross-sections (like coins in a stack). The inspector is responsible for every cross-section of the surface being constrained within the circularity tolerance. The inspector will zero out the dial indicator (or probe) for each new circular cross-section measured. Inspectors will measure the cross-sections as described in number and distribution in the Measurement Plan for the workpiece. If no measurement plan exists, the inspector will measure enough cross-sections to convince himself that, had all cross-sections been measured, the workpiece would have proven compliant with the circularity requirement (or until a cross-section is found to be out-of-tolerance).

Some of the measurement uncertainties to be aware of:

- probe size
- establishing circle centers
- filtering of data (effects)
- number of circles probed
- number of points probed on each circle and
- temperature deviations from 20° C (68° F).

According to the ASME B89.3.1 Standard on the Measurement for Out-of-Roundness, unless otherwise specified, the centers used to establish an axis for the measurement of circularity are determined by the minimum radial separation circles able to contain the circular cross-section being measured. This minimum radial separation center (MRS) is established each time a new cross-section of the surface is measured.

Other methods of establishing an axis (or center point) for each cross-section may be specified. A very common method often specified is a least squares circle center (LSC). This LSC center is the center of a circle derived from the sum of the squares of the radial ordinates of the measured polar profile. This is the LSC which has the minimum value.

Another method that may be specified is the maximum inscribed circle center (MIC). This is the center of that which is the largest circle that can be inscribed (expanded into) within the measured circular cross-section's polar profile of a hole. If a shaft is under test, this method would change to the minimum circumscribed circle center (MCC). This is the center of the smallest circle that can be circumscribed about the measured circular profile cross-section being measured.

Circularity is a 2-dimensional form control of a surface. It controls pits, bumps, flats and oval shapes on cylinders, cones and any circular feature.
For example:

○	0.2

A circularity tolerance of 0.2 creates a tolerance zone that consists of two concentric circles 0.2 apart radially. Examples of acceptable part surface configurations:

- allows pits no deeper than 0.2
- allows bumps no taller than 0.2
- allows flats no deeper than 0.2 and

- allows ovals measured in one direction to be no greater or less than a diameter of 0.4 different when measured in another 90º direction.

All tolerances of size limits for MMC and LMC are separately verifiable. All tolerances of circularity must, therefore, be contained within the limits of size. When inspecting a circularity geometric characteristic tolerance, the full indicator movement (FIM) (formerly known as total indicator readout, TIR) must not exceed the tolerance in the feature control frame (the circularity tolerance) when compared directly. All FIM readings should be taken while rotating 360º around a stabilized axis of the single circular cross-section being tested at that time.

FIGURE 65 [Examples of Acceptable Part Surface Configurations]

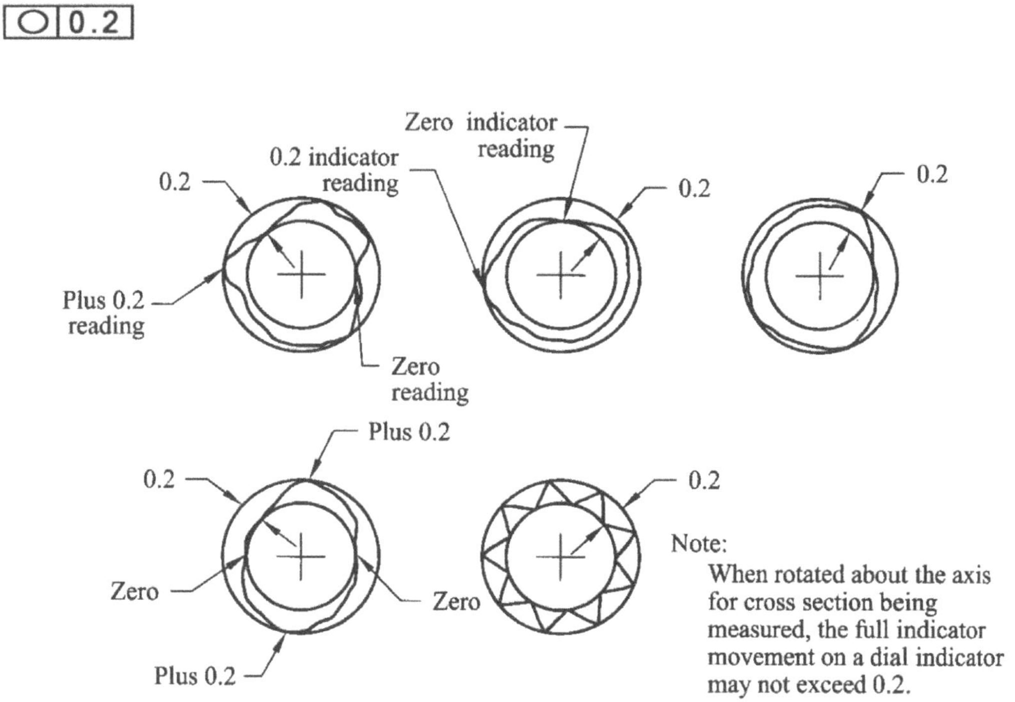

FIGURE 66 [Measurement for Out-of-Roundness using a Roundness Checking Machine]

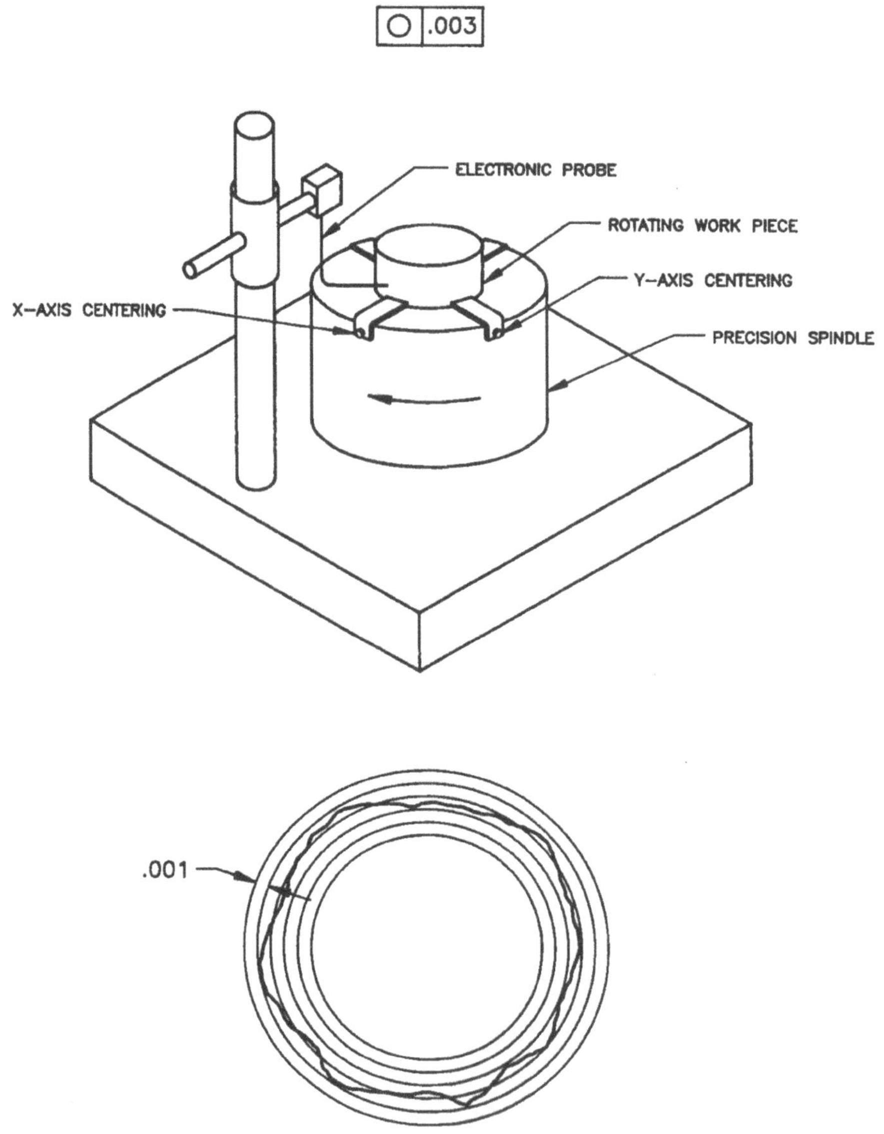

The part is placed on the machine table. Each circular cross-section is centered before it is measured. The probe is zeroed for each new measurement of a circular cross-section. The part is rotated under the electronic probe. The image of each circular line element being tested is transferred to a polar graph. The graduations of the polar graph may be anything appropriate to the measurement. The graph depicted above for FIGURE 66 is .001 per space between the circles. Since the circularity tolerance shown at the top of the figure is .003, each circular line element tested may span no more than 3 circles and comply with the tolerance. The circular line element shown above spans 2 circles. It is round to within .002. The inspector is responsible for all circular line elements on the cylinder.

FIGURE 67 [Measurement of Circularity using Bench Centers and Dial Indicators]

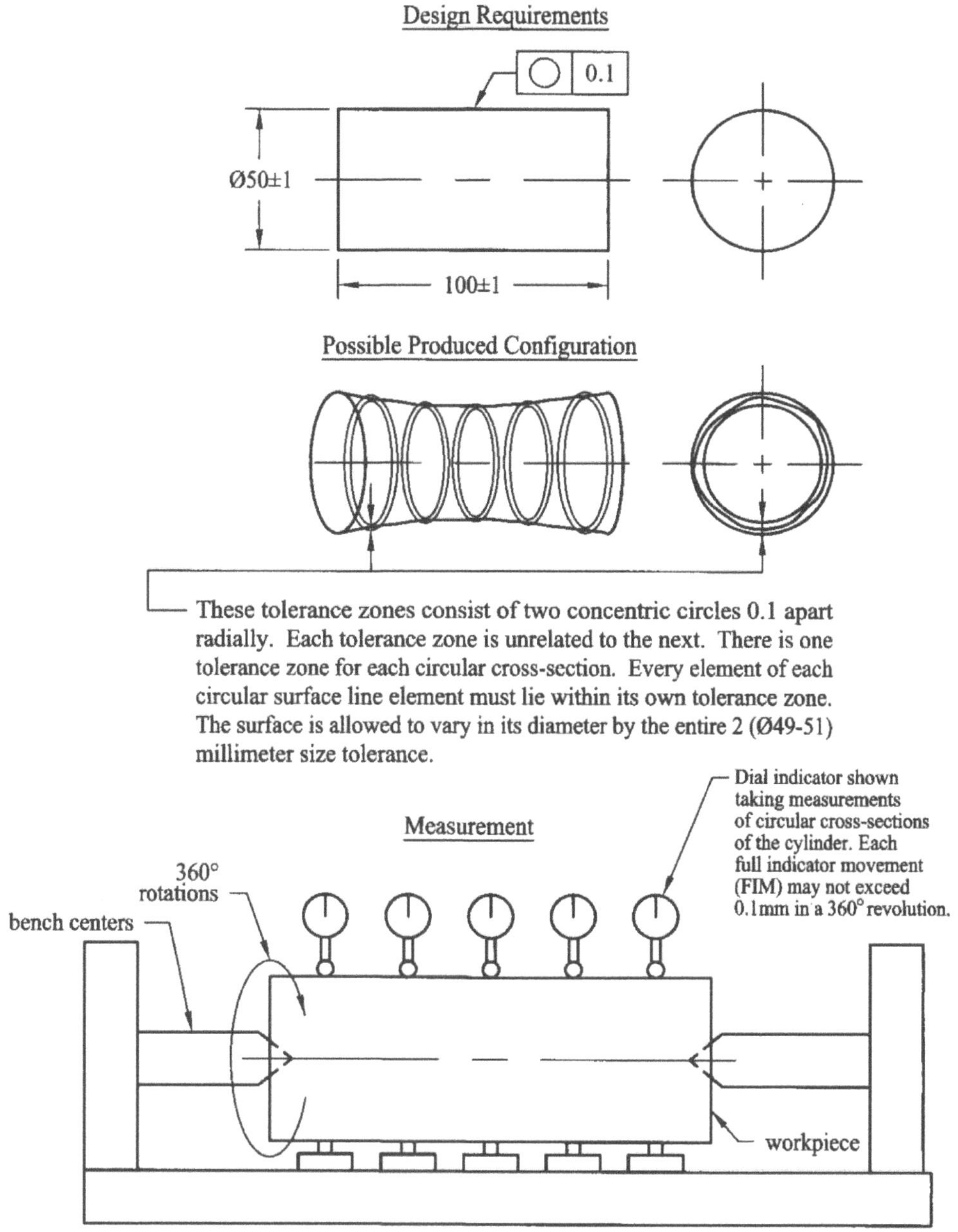

The measurement technique shown in FIGURE 67 would only be acceptable if the axis of the cylinder was straight. Technically, each circular cross-section is supposed to be centered to its optimal center point and then measured for circularity. The dial indicator is zeroed for each circular cross-section's measurement.

Circularity to Control Spherical Diameters

If a size dimension is given as a basic dimension on a spherical diameter, profile of a surface may be used to create a 3-dimensional tolerance zone to control its form. However, if the size dimension is shown as a plus and minus toleranced dimension, 3-dimensional form may be controlled on a sphere in a couple of ways:

1) The size limits can be used to control the form of a sphere per the "Perfect Form at MMC is Required" rule (Rule #1). According to this rule, the sphere (being a regular feature of size) may only depart from perfect form as it departs from its maximum material condition. So, the size tolerance is the form tolerance.
2) Another method to control the 3-dimensional form of a sphere is to control the form of every circle in all views that forms the sphere. This can be done by applying circularity to the sphere as follows:

FIGURE 68 [Spherical Diameters Controlled with Circularity]

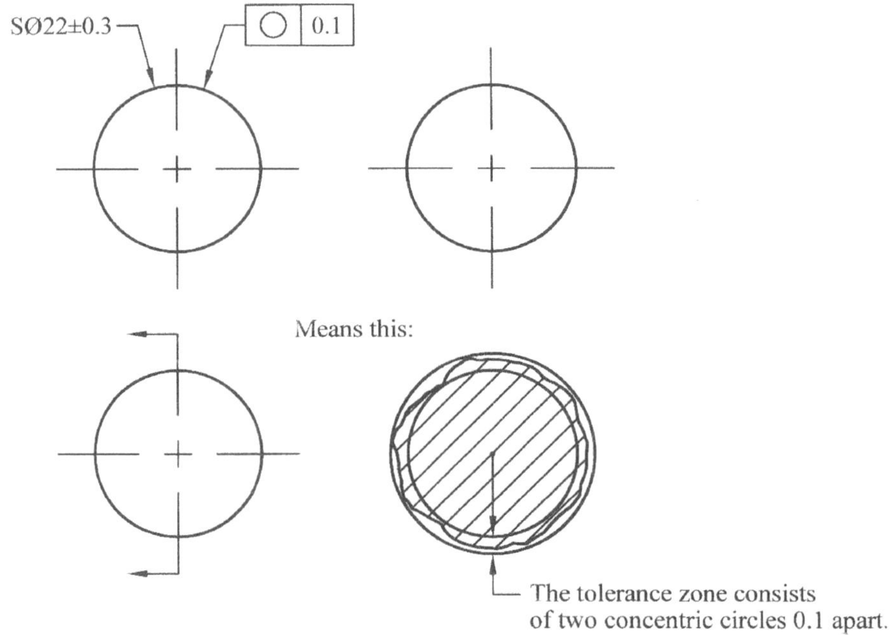

Each circular line element of the sphere must have all of its surface elements residing within its tolerance zone. Each circular line element is independently verified. Each circular element of the surface in a plane passing through a common center must not violate the circularity tolerance zone. Size limits of the sphere are verified separately. The circularity tolerance is not additive to the size limits. No diameter signs, material condition symbols or datum features references may be used inside of the circularity feature control frame.

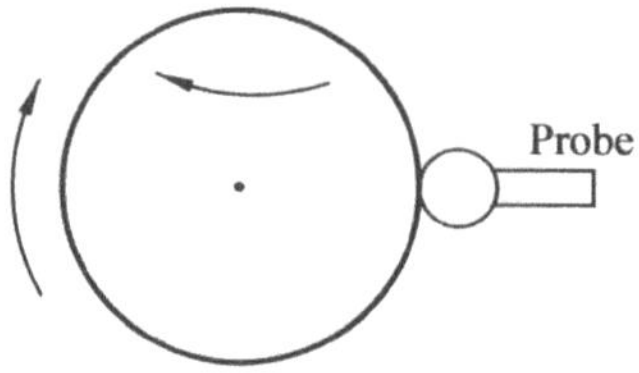

Part is rotated in all directions in full 360 degree revolutions. The FIM may not exceed 0.1.

Average Dimensions

When AVG is added after a dimension, it means that feature is to be verified as having dimensions that must only average within the given limits. There are tape measurements that may be taken about the periphery of the feature that will give an average of the surface diameter. If opposing points checks are chosen instead, then a minimum of four are required at each cross section of the surface. This AVG method is chosen by designers when they are certain that if a part averages within a dimensional range, then it will either function as is, of may be distorted (restrained) during assembly to conform to needed tolerances.

FIGURE 69 [Average Dimensions]

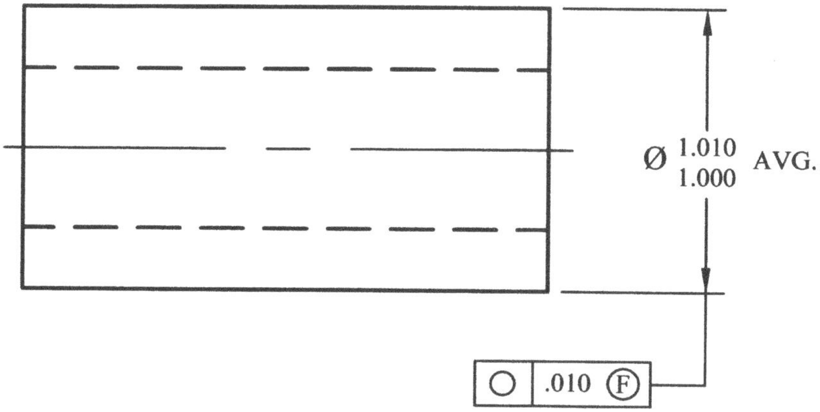

The roundness tolerance is used to (among other things) set the maximum upper limit size in the free state for any part made and also the minimum lower limit size in the free state (unrestrained) for any part made. Besides lending other control, such as form, the geometric (circularity, in this case) tolerance may also be used to recheck the average dimension. For example, if you take two size checks (at least four at each cross section is the usual minimum to establish an average for that cross section) and get, say, 1.003 on the first and 1.009 on the other, and then average them:

1.003
+1.009
2.012 divided by 2 = 1.006 AVG

Because 1.006 is the average, one must not exceed a maximum feature dimension of 1.006 plus the .010 roundness tolerance or 1.016" or a minimum feature dimension of 1.006" minus the .010 roundness tolerance or .996".

Note: No dimension on any of the parts checked may exceed:
1.010+.010 or 1.020 high limit, or 1.000 -.010 or .990 low limit.

○	.010 Ⓕ

This states that all measurements are taken while the part is unrestrained in the free state. This FREE STATE symbol is generally used when it has been specified in a general note that UNLESS OTHERWISE SPECIFIED ALL TOLERANCES ARE MEASURED IN THE RESTRAINED STATE.

Chapter 6

Orientation

•Orientation - an Overview
•Parallelism
•Perpendicularity
•Angularity
•Shifting vs. Growing Tolerance Zones

Chapter Objectives

Readers will learn:

1. Orientation controls; types, mathematical similarities; how they are used alone or as refinements of more powerful controls, the meaning of combined controls and how to read the controls as sentences.

2. Parallelism of surfaces, axes and centerplanes; what they mean, the tolerance zone configurations and how to measure them.

3. Coordinate measuring machines; least squares center vs. maximum inscribed cylinder center.

4. Tangent plane parallelism; tolerance zone configuration, what has to reside within it and how to measure it.

5. Perpendicularity of surfaces, axes and centerplanes; what they mean, the tolerance zone configurations and how to measure them.

6. Perpendicularity as a refinement of position.

7. How to read the geometric tolerances as sentences.

8. Angularity of surfaces, axes and centerplanes, what they mean, the tolerance zone configurations and how to measure them.

9. Tangent plane angularity; tolerance zone configuration and what has to reside within it.

10. Angularity of a line element.

11. Orientation of centerplane controls. The concept of unoriented, unlocated mating envelopes.

Orientation – an Overview

Mathematically speaking, perpendicularity, angularity and parallelism are all the same control. In fact, in the *Y14.5.1 Standard on Mathematical Definition of Dimensioning and Tolerancing Principles*, one algorithm (mathematical formula) defines all three. Whether used as a surface control, an axis of the actual mating envelope control, a centerplane of the actual mating envelope control, a line element or a tangent plane control, perpendicularity, angularity and parallelism tolerance an angle.

They just tolerance different angles and sometimes even that wanders into a gray area. When multiple angles are being toleranced by the same control, angularity may be used as a generic orientation symbol. It is allowed to tolerance any angle, including those usually reserved for perpendicularity and parallelism. Normally, however, perpendicularity is a control of 90° angles. Parallelism is a control of 0° angles. Angularity is control of all other angles. The angles they tolerance are always basic angles toleranced by these geometric characteristics in feature control frames. The tolerance is always in millimeters or inches, never in degrees.

If used as a line element control, besides tolerancing the angle, all three geometric characteristics control straightness of the line element. If used as a surface control on a planar surface, perpendicularity, angularity and parallelism control the flatness of that surface as well as the angle. The following controls only control the angle, not the shape: 1) tangent plane of the surface, 2) axis of the actual mating envelope and 3) center plane of the actual mating envelope.

The benefit of using three symbols instead of one to control orientation is the specificity of immediately conveying which angle it is that is being toleranced. The only point of confusion it causes is when one feature control frame has to tolerance multiple angles to more than one datum. As mentioned above, the ASME Y14.5 2009 and the ASME Y14.5 2018 standards specifically allow angularity to be used to tolerance any angle, including those situations where the datums referenced may require multiple angles to be controlled, even if the relationships to be held include perpendicularity and/or parallelism. To begin with, though, let's go over some simple and common uses of orientation controls.

Parallelism

If used to control planar surfaces, the orientation controls tolerance an angle and also flatness. With parallelism, a size tolerance may come into play. For example, see FIGURE 1.

In this type of usage, the size tolerance of plus or minus 1 millimeter on the 50 millimeter dimension already controls the parallelism between the two surfaces to within 2 millimeters. Therefore, any parallelism feature control frame tolerance would have to be smaller than 2 millimeters in order to steal the job of tolerancing parallelism from that size tolerance. This parallelism tolerance, shown as 0.1 millimeters, is much smaller than the 2 millimeters. It acts as a refinement that creates a parallelism tolerance zone that consists of two parallel planes 0.1 apart. These planes are perfectly parallel to datum plane A. Datum plane A is formed by (a minimum of) the three highest points on datum feature (surface) A.

See FIGURE 70 [Parallelism of a Surface]

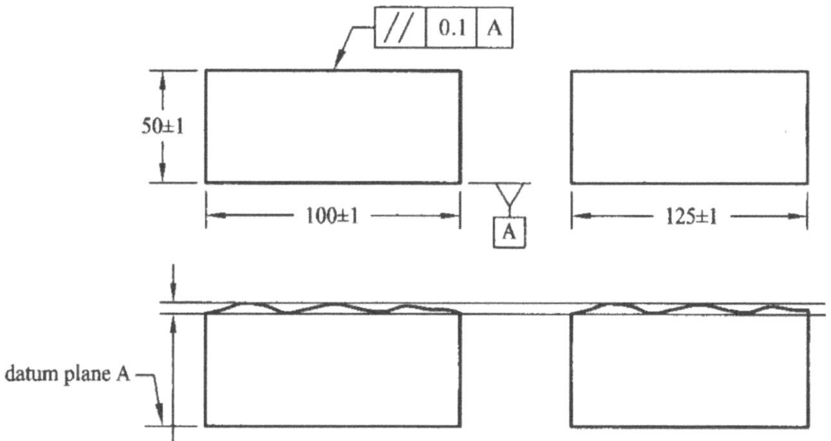

This tolerance zone consists of two parallel planes 0.1 apart that are also parallel to datum plane A. Every element of the surface being constrained must lie within the parallelism tolerance zone. There is one tolerance zone for the entire surface. This tolerance zone must extend to all views. The surface is allowed to vary in its height to the opposite side by the entire 2 (49-51) millimeter size tolerance.

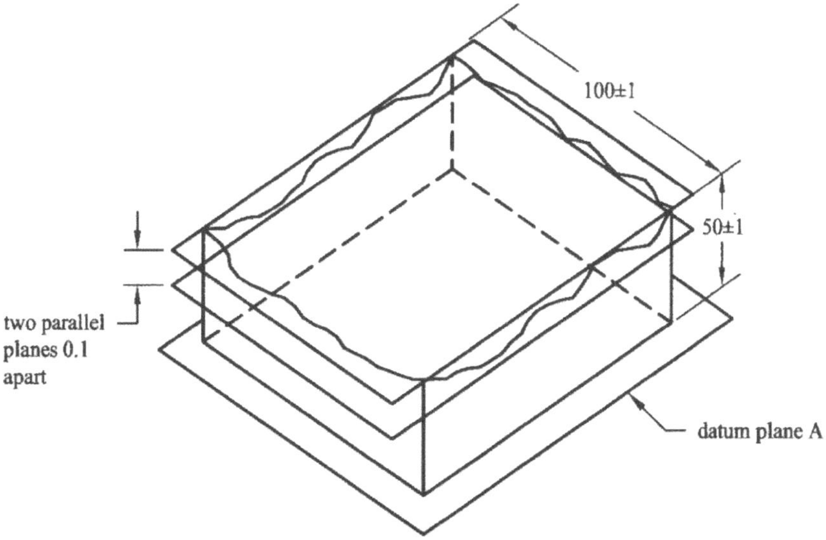

Measuring this type of parallelism on a planar surface is easy to do and makes the control understandable.

FIGURE 71 [Measuring Parallelism]

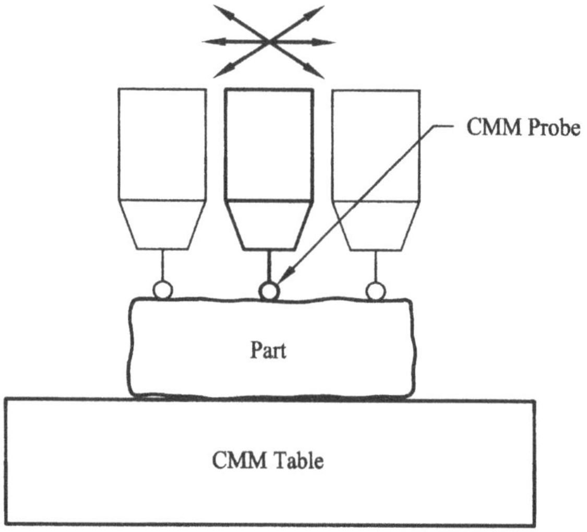

The part may be placed on a coordinate measurement machine table on the primary datum feature. The table becomes the datum feature simulator. The table may be probed to establish the datum plane in the computer memory. The surface being controlled for parallelism is then probed. A constant contact scanning probe is preferred to allow as many points as possible to be represented. Remember, this is not just a control of the angle, it is also a control of flatness, a three-dimensional form control. Three-dimensional form control measurements require many points to be probed in order to achieve a high degree of confidence in the data.

According to a study done by Electronic Measuring Devices, Inc., in 1988, it requires an average of 191 points per square inch to achieve a 90% certainty in the collected data results when inspecting a simple flatness control. An average of 102 points per square inch were needed to achieve a 75% certainty of the actual value of flatness. The lesson here is that methods that collect relatively few points are not recommended for inspecting controls that specifically tolerance or include an aspect that tolerances three-dimensional form.

By comparison, relatively few points are generally needed to achieve a high degree of certainty when inspecting controls of only angle or location. Part of the reason for this is the concept of the "mating envelope" (or "mating size"). The concept of "mating size" applies to inspecting position of axes or centerplanes, perpendicularity of axes or centerplanes, parallelism of axes or centerplanes and angularity of axes or centerplanes. The concept of "mating size" or "mating envelope" tells us that when we are inspecting these controls, we should probe a substitute for the feature that generates the axis or centerplane rather than the feature itself. See FIGURE 72 as an example.

FIGURE 72 [Mating Envelope]

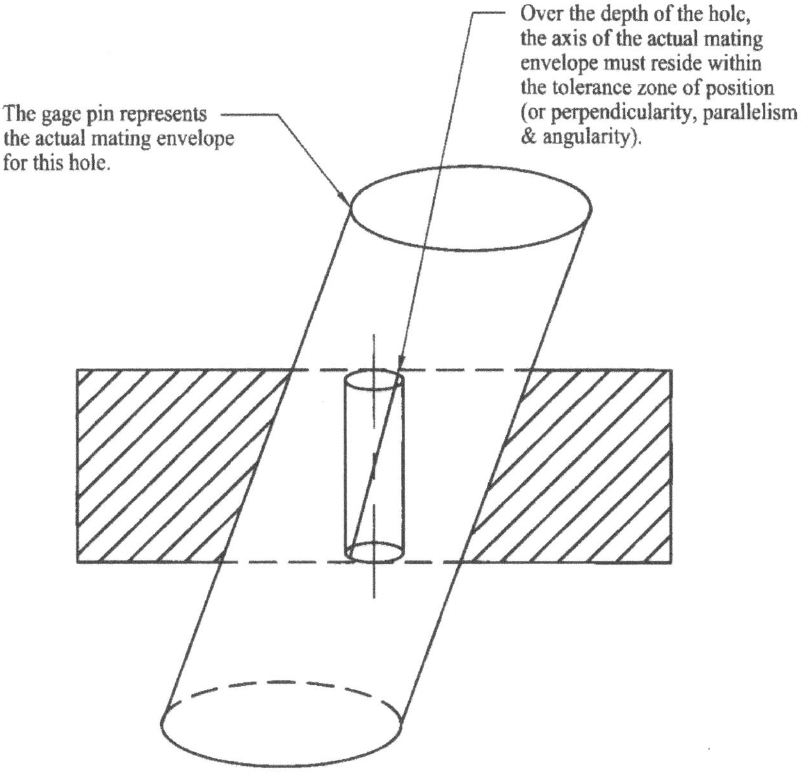

The axis of the actual mating envelope that exists at whatever angle or location the hole is produced, must (throughout the depth of the hole only) reside within the tolerance zone to comply with a tolerance of position, perpendicularity, parallelism, or angularity. If a substitute gage pin for the hole is probed, since that gage pin is the maximum inscribed cylinder, it touches the high points within the hole, represents the worst size for fitting something into the hole and is then assessed for how much it is out of position, perpendicularity, parallelism, or angularity. This is what these controls, when applied to the axis of a hole, require. That is why probing discrete points within the hole will not reap the theoretically correct results. For example, see FIGURE 73.

FIGURE 73 illustrates a very realistic, often practiced, method of inspecting position, perpendicularity, parallelism, or angularity of the axis of a hole. It is, however, not going to commonly give the same results as probing the gage pin substitute for the hole. Most software programs would take all points probed within the hole and give equal weight to the high points, low points and all other points probed, then through the use of an algorithm give the CMM operator an average axis for the probed points.

FIGURE 73 [Inspecting Position of a Hole]

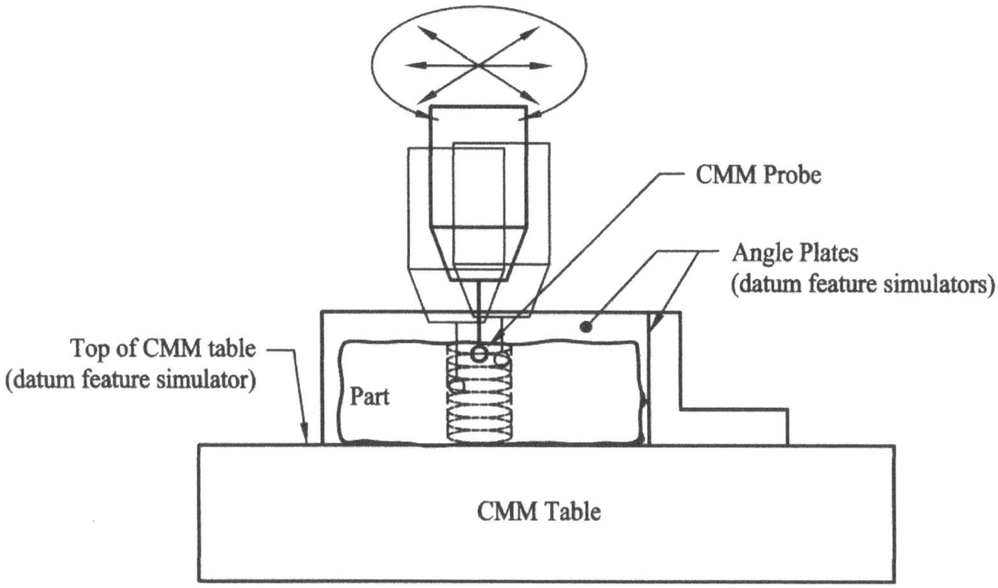

A common program is least squares center, which is like an average of all points probed. Unfortunately, although expedient, this does not comply with the theoretical concepts of correct compliance with the inspection of these geometric characteristics. Whereas, the probing of a gage pin representing the maximum inscribed cylinder or actual mating envelope at whatever angle or location the feature is produced, complies with the theory quite well. This concept of measuring the actual mating envelope is shown for parallelism in FIGURE 74.

FIGURE 74 [Inspecting Parallelism of a Hole]

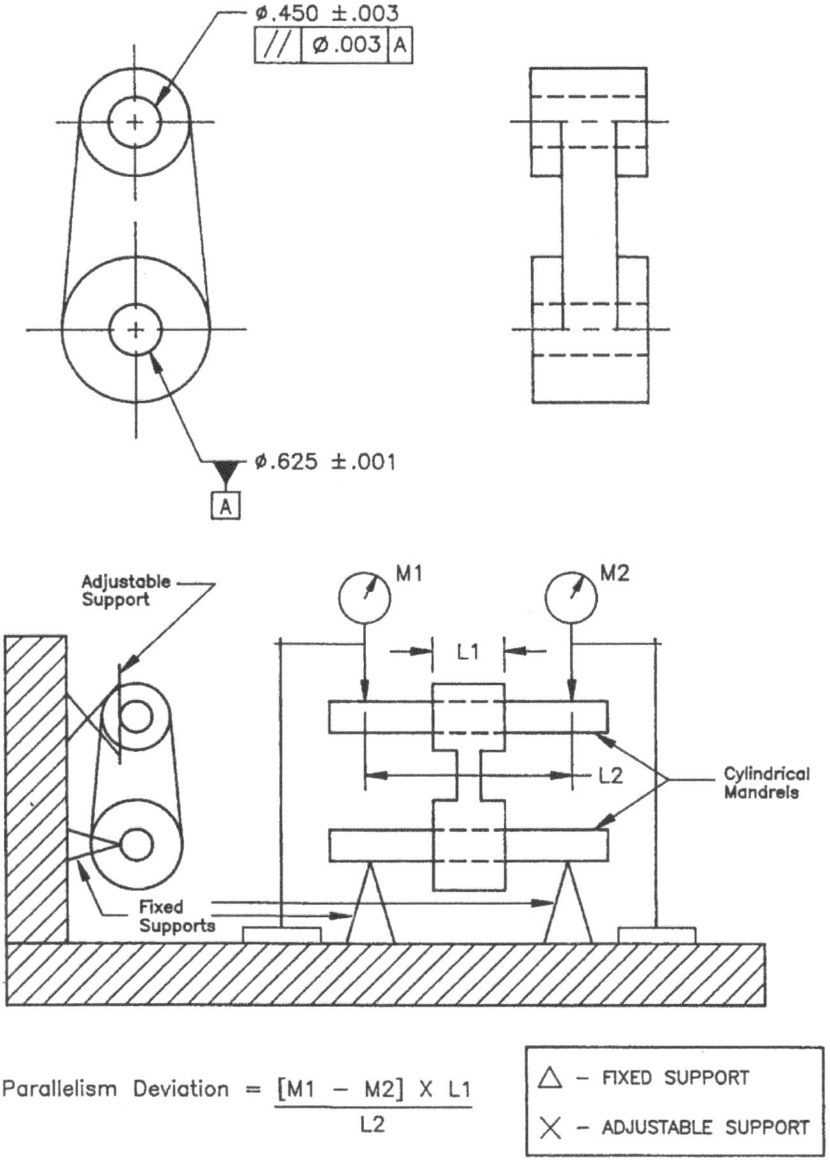

For the final determination of compliance with this parallelism control, the part is measured as shown above, the variables data inserted into the formula and the results compared to the parallelism tolerance in the feature control frame. Since no maximum material condition symbol has been used after the geometric tolerance, no additional tolerance based on the size of the produced hole (bonus tolerance) is considered. Likewise, since no maximum material boundary symbol has been used after the datum feature, no pattern/datum feature shift is considered. If desired, the collected data may be graphed to give a visual display of the out-of-parallelism.

FIGURE 75 [Tangent Plane Parallelism]

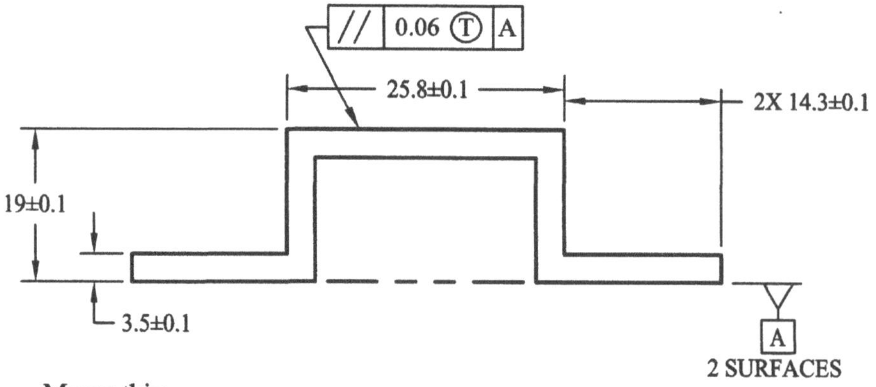

Means this:

The circled T inside the feature control frame means the Tangent Plane is controlled.

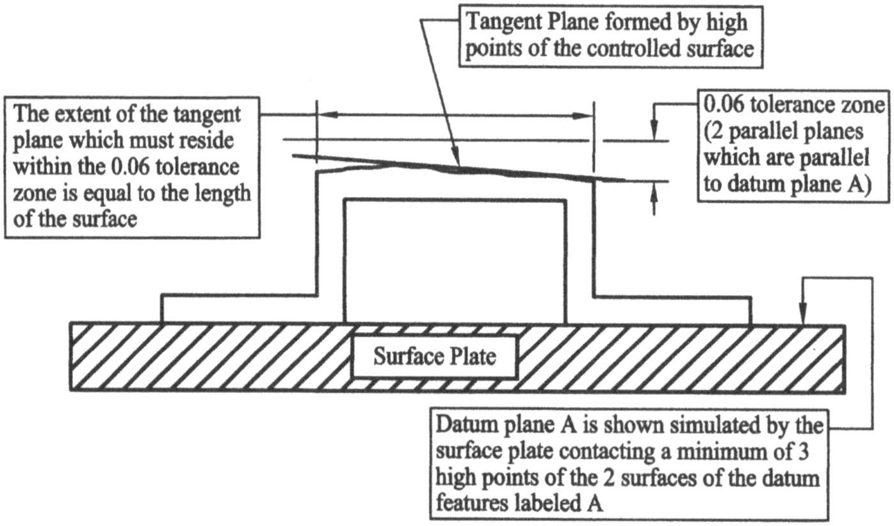

Possible measurement procedure and uncertainties:

The 0.06 tolerance zone will contain the Tangent Plane (which touches the surface high points) in order to comply with this parallelism control. The tolerance zone is 2 parallel planes 0.06 apart and parallel to datum plane A. As shown above, the Tangent Plane can be simulated by placing a flat plate on the surface high points. The plate is then probed to determine if the Tangent Plane's parallelism exceeds 0.06. If a plate is used and the top of the plate is probed (instead of the bottom which touches the surface high points) the plate must have very parallel top and bottom surfaces. Any out of parallelism that exists between the top and bottom of the plate must be taken into consideration and compensated for. Tangent Plane controls, if used for parallelism, perpendicularity or angularity, only control the orientation of the plane formed by the surface high points. Tangent Plane controls are unlike the use of regular surface orientation controls. In a regular orientation of a planar surface control, all elements of the surface must actually be within the tolerance zone for the part to comply with the specification. In these instances, since the tolerance zone consists of two parallel planes that are the geometric tolerance apart, flatness is controlled. A Tangent Plane control does not control surface flatness because all elements of the surface do not have to reside within the tolerance zone for the part to comply with the tolerance. Only the Tangent Plane of the surface must be within the tolerance zone to comply with the specification.

Perpendicularity

Perpendicularity is the same concept as parallelism, just at a different angle--90° as opposed to 0°. Feature control frames for perpendicularity of a surface use a geometric characteristic symbol, a geometric tolerance an at least one datum reference. As an example, see FIGURE 76 below.

FIGURE 76 [Perpendicularity of a Surface]

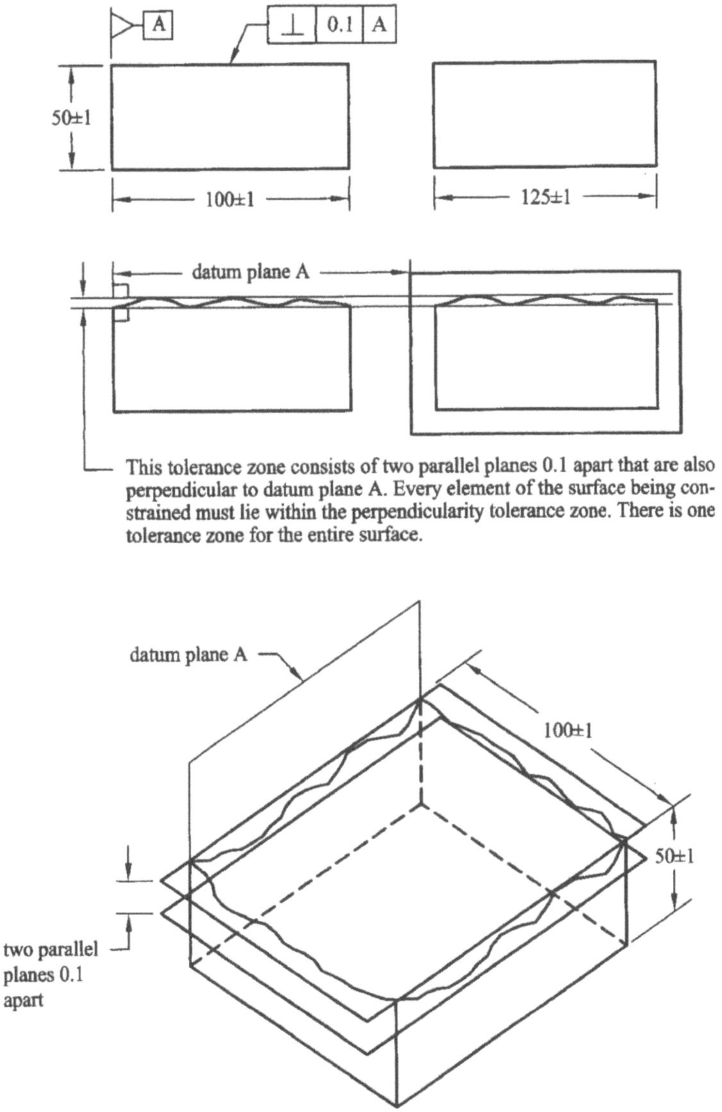

It creates a tolerance zone that consists of two parallel planes separated by the geometric tolerance. This tolerance zone is perpendicular to the primary datum that is referenced. It is a zone that may be 90° to a datum plane, a datum axis or a datum centerplane. If necessary, more than one datum is referenced (and often is). In this case, the surface being controlled is planar and, therefore, flat and not a feature of size. So, no material condition symbol may be used that is associated with the geometric tolerance. The datum feature being referenced may be planar as shown in this example, in which case no material condition symbol may be associated with the datum feature referenced. However, if the datum feature is a feature of size, such as a hole, slot, shaft or tab, the datum feature may be referenced at MMB or at LMB.

To measure perpendicularity of a planar surface to the datum plane or datum planes referenced, the part must be manually oriented to the datums (as shown the in the following illustration).

FIGURE 77 [Inspecting Perpendicularity]

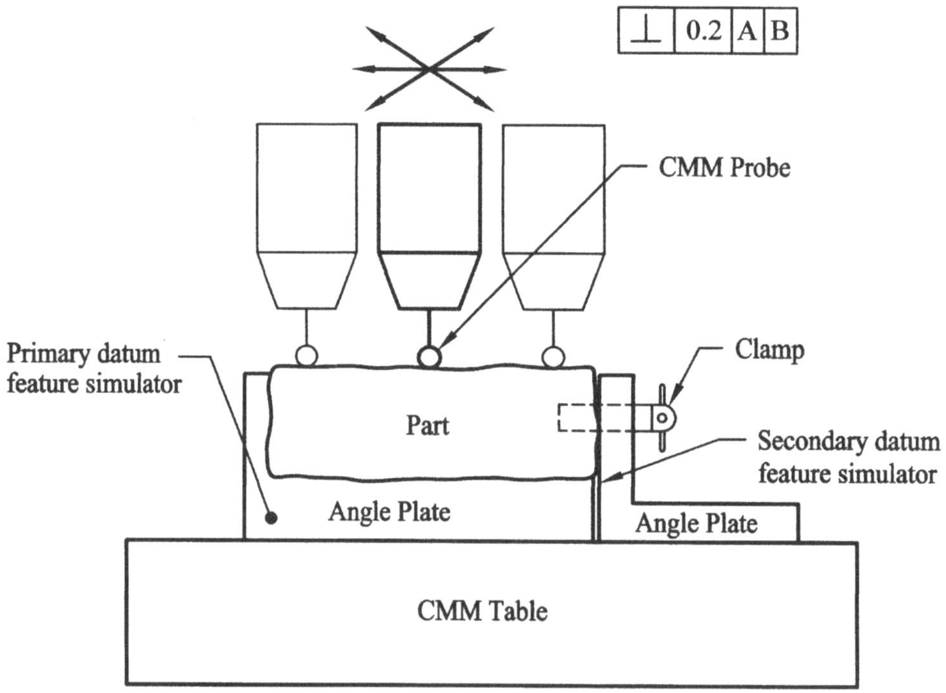

FIGURE 78 shows a part being inspected in a more 3-dimensional view.

FIGURE 78 [Inspecting Perpendicularity]

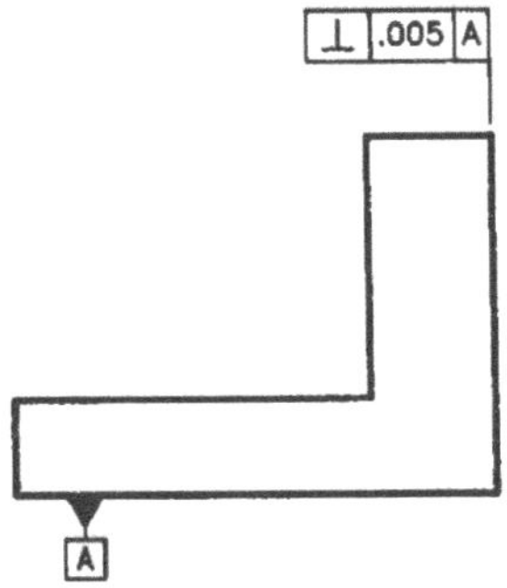

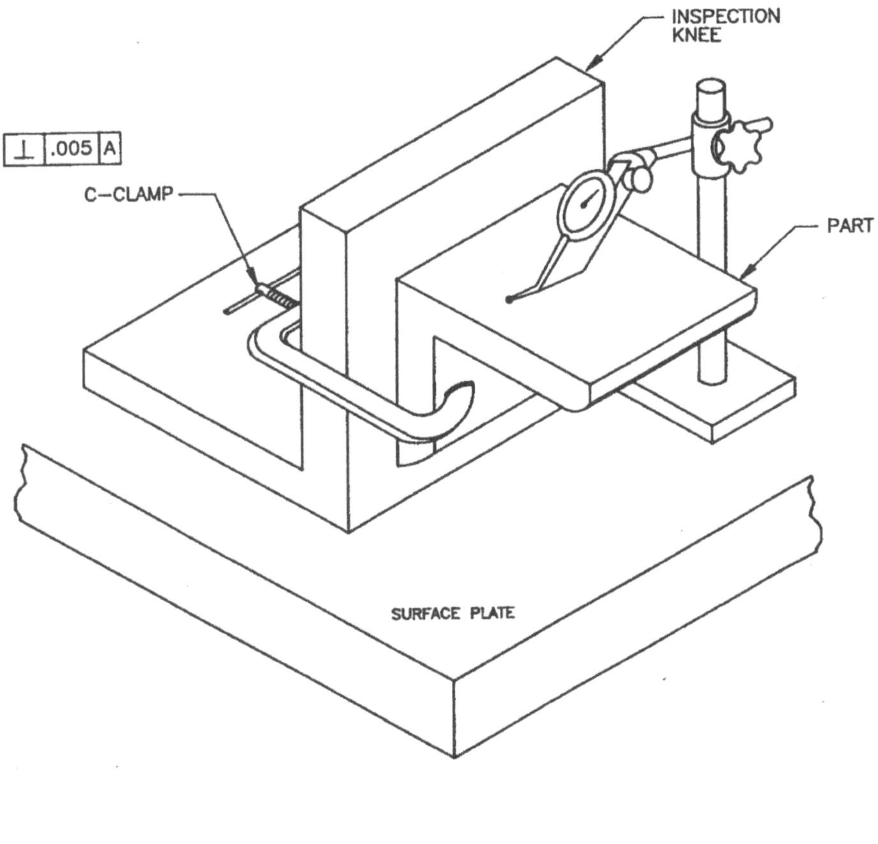

The part may, instead, be oriented to the datums by computer program.

Another way that perpendicularity may be used is to orient the axis of a hole or shaft to the axis of another hole or shaft. See FIGURE 79. In this illustration, the axis of the Ø6.3 hole is first positioned with a relatively loose tolerance (0.4 at MMC) to all of the datums. This makes certain the hole may not move away from true position as defined by its intersection of datum axis A and the 40 millimeter basic distance from datum plane B. Datum plane C is included in the position control simply to allow one set-up to be used when inspecting this position control and the subsequent refinement of perpendicularity. The part may be set-up one time, measured, and the collected data used to determine compliance with both controls.

FIGURE 79 [Perpendicularity of an Axis to a Datum Axis]

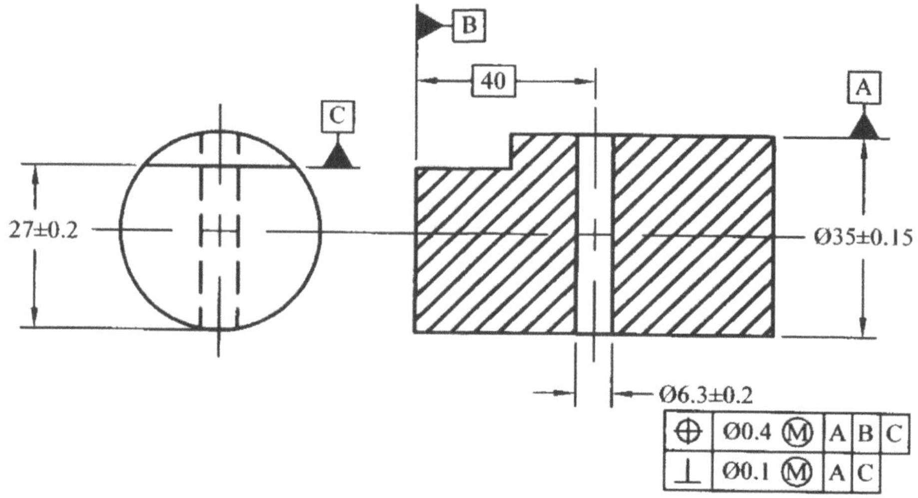

This can be done even though datum feature B is not included in the perpendicularity. Datum feature B, when used in the position control, is a secondary datum feature, but a tertiary datum plane (since datum axis A is comprised of the first two planes in the datum reference frame). As a tertiary distance datum plane, B receives only one point of contact and as an orientation datum is rendered useless. Therefore, only datums A and C control orientation. Since the axis of the hole is perpendicular to both A and C, it is the smaller orientation tolerance of Ø0.1 (at MMC) perpendicularity that ultimately controls the angles of the hole.

The following illustration shows the two tolerance zones, their relationship to the datums and to each other. The smaller zone may partially vacate the larger zone. But since the actual hole axis must be in both zones to comply with both controls, the Ø0.1 smaller zone must at least have a tangent line in common with the larger Ø0.4 zone in order to comply with the requirements.

FIGURE 80 [Perpendicularity of Axis to Axis Tolerance Zones]

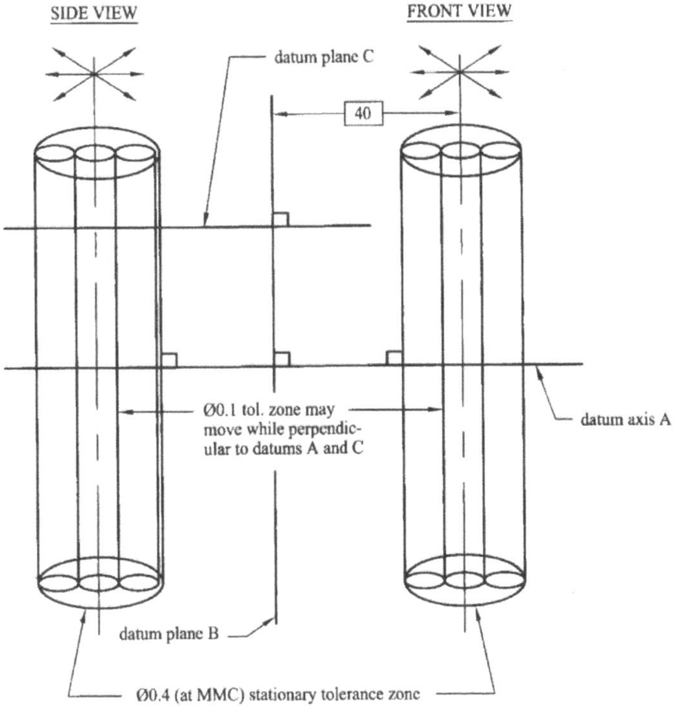

The position tolerance zone is stationary, locked in at a basic location intersecting datum axis A, 40 millimeters from datum plane B while 90° to datum axis A and datum plane C. The smaller Ø0.1 (at MMC) perpendicularity tolerance zone may move in any direction but must maintain perfect perpendicularity to datum axis A and datum plane C. The actual hole axis must be produced residing in both zones for the part to pass inspection. When two controls are used on the same feature, they often influence one another, but are read as separate statements. For example:

⌖ = Position
Ø = of the axis
0.4 = must be held to within 0.4 mm
Ⓜ = if the hole is produced at its maximum material condition of Ø6.1
A = to datum axis A for intersection and perpendicularity
B = and to B for a 40mm distance
C = and to C for angular orientation (perpendicularity in this case)

Then, and in addition to that;
⊥ = Perpendicularity
Ø = of the axis of the hole
0.1 = must be held to within 0.1mm
Ⓜ = if the hole is produced at its maximum material condition of Ø6.1
A = to datum axis A (for perpendicularity)
C = and to datum plane C (for perpendicularity).

The smaller 0.1 tolerance zone must maintain perfect perpendicularity to datums A and C. But it may move around, even may move partially outside of the larger 0.4 tolerance zone. It must maintain at least a small portion inside (a tangent line element would do, in theory) of the 0.4 (at MMC) tolerance zone, since the axis of the maximum inscribed cylinder of the hole must be in both tolerance zones to be in compliance with these controls.

The larger 0.4 (at MMC) tolerance zone is locked into true position by its relationship to all of the datums and the strengths of the position geometric characteristic symbol. It is a stationary tolerance zone with all six degrees of spatial freedom eliminated (3 rotational and 3 linear translations).

FIGURE 81 [Perpendicularity of Planar Surface to a Datum Axis]

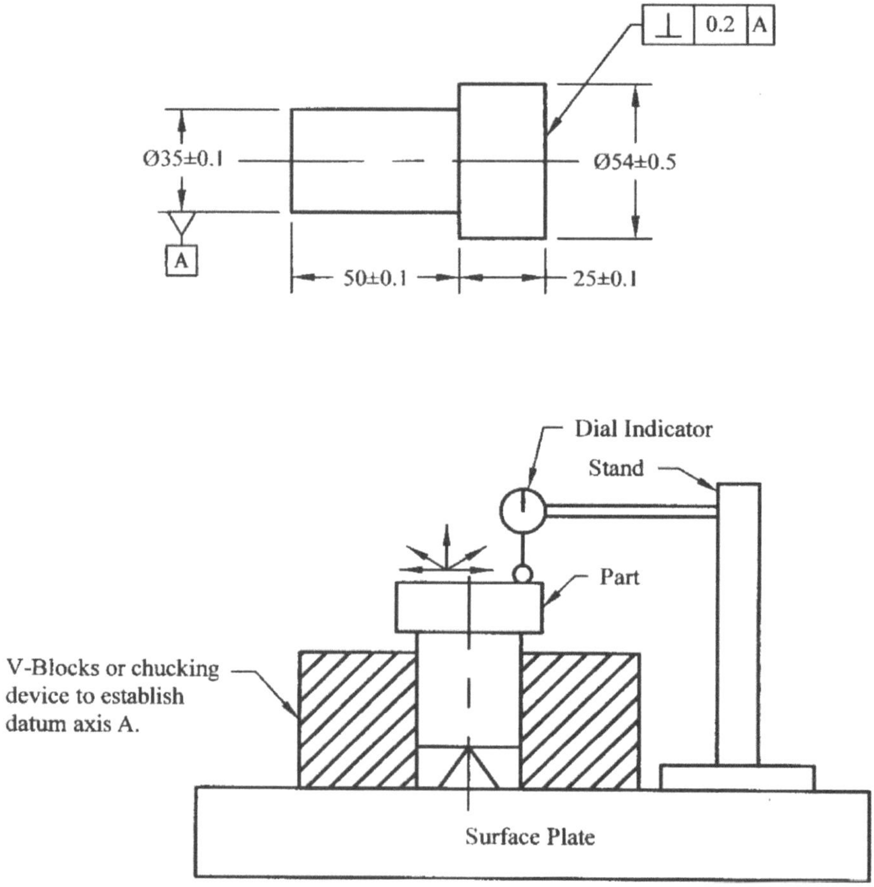

The dial indicator traverses the surface in all directions. The full indicator movement may not exceed 0.2 millimeters for the surface to comply with this perpendicularity control. This not only controls the 90° angle, but also the flatness of the surface.

FIGURE 82 [Perpendicularity of a Centerplane to a Datum Plane]

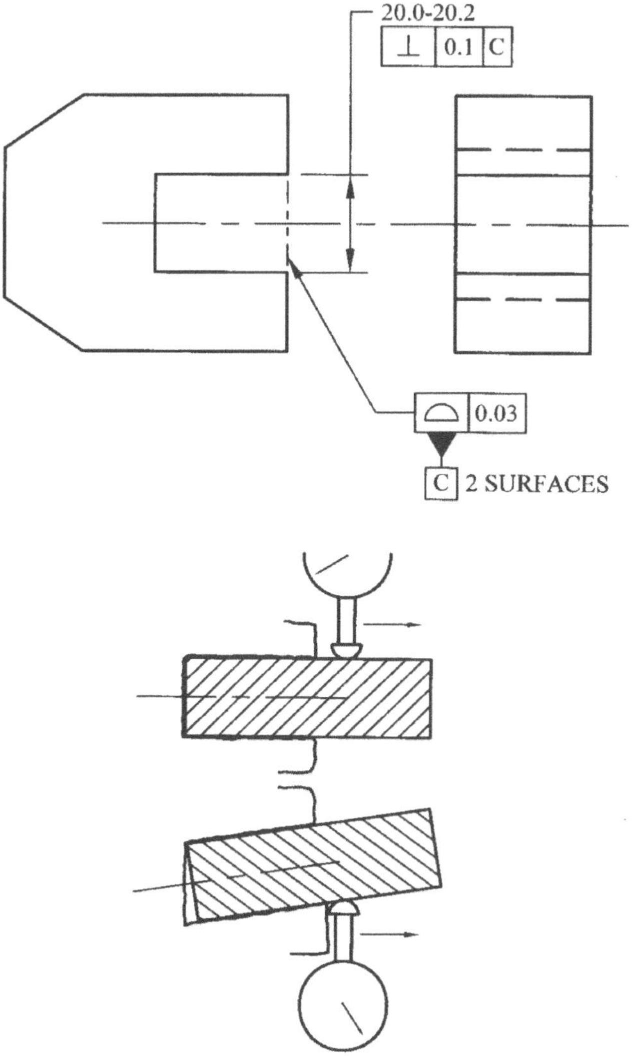

This perpendicularity control is used to orient the slot centerplane to the datum plane. The centerplane of the slot is constructed by the largest gage block that can be inserted into the slot. This is to simulate the actual mating envelope for the slot. The actual mating envelope is unoriented when simulated to determine its deviation from a perfect 90 degree angle from datum plane C. Once the gage block is inserted into the slot, the probe travels longitudinally along the block. The probe must maintain a 90 degree angle to the datum plane. It should travel a distance that is equal to the depth of the slot. If the full indicator movement exceeds the perpendicularity tolerance, the feature fails the requirement.

Unless an exception is invoked, features of size, such as slot widths, tabs, holes, shafts and spheres are technically all actual mating envelope measurements, if they are controlled by perpendicularity, angularity, parallelism and position.

Angularity

Like the other orientation controls, angularity can be used to orient a surface, an axis or a centerplane. In some instances, it can (like parallelism and perpendicularity) be used to control line elements and tangent planes. The angles that it focuses on are those not controlled by perpendicularity and parallelism. However, as mentioned earlier in this unit, it is often used as a generic orientation symbol where it tolerances more than one angular relationship that may also include perpendicularity or parallelism. One of the most common uses of angularity is to orient and make flat a single surface. For example:

FIGURE 83 [Angularity of a Surface]

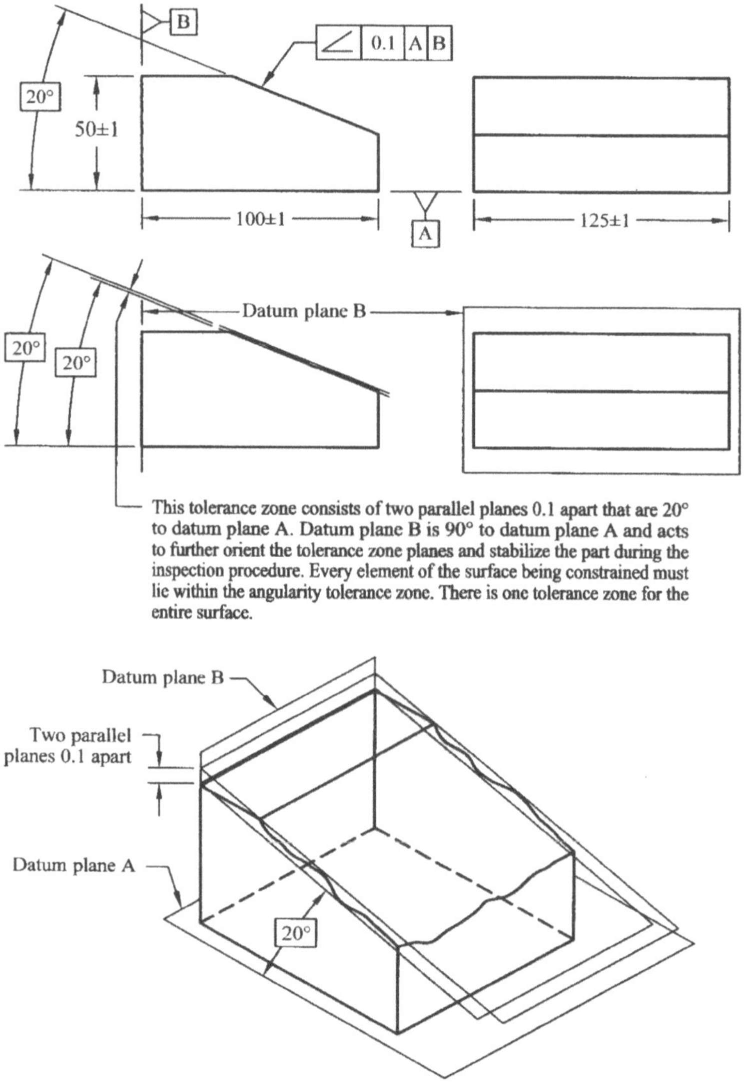

In these instances, one or more datum references must be used in the angularity feature control frame. This type of control can be inspected in a wide variety of ways including computer assisted variables data collectors such as a coordinate measuring machine. One of the oldest and most reliable inspection techniques involves a surface plate (or machine table), a sine plate or sine bar and a dial indicator. For example:

FIGURE 84 [Inspecting Angularity]

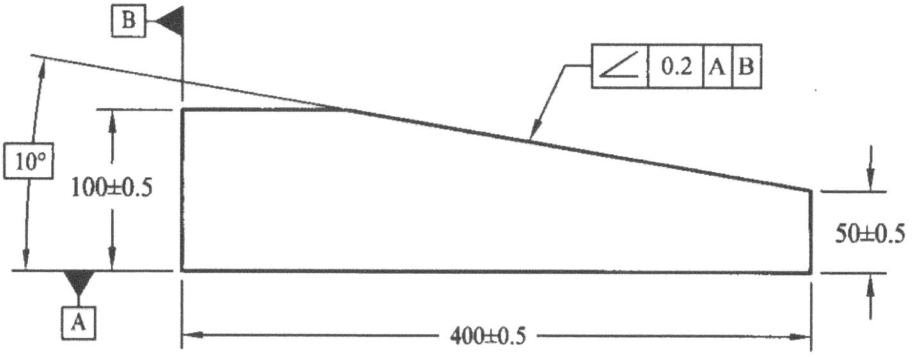

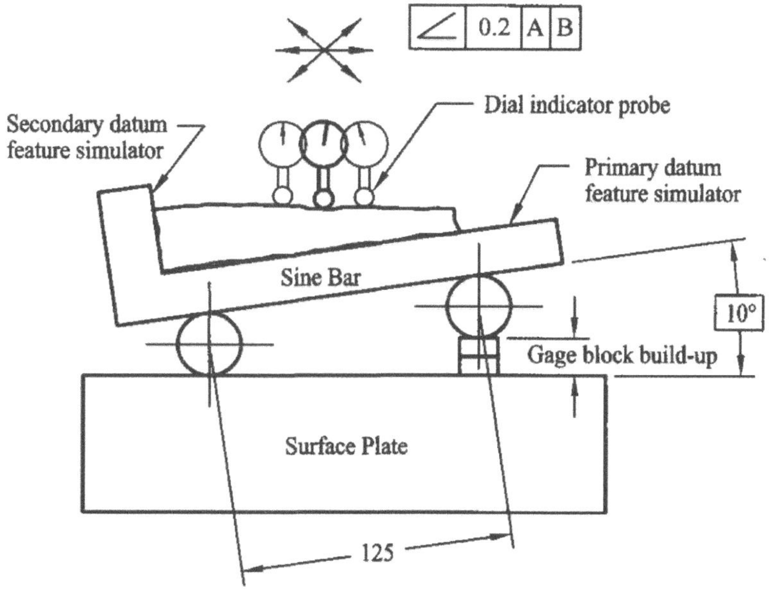

FIGURE 85 [Angularity of a Tangent Plane]

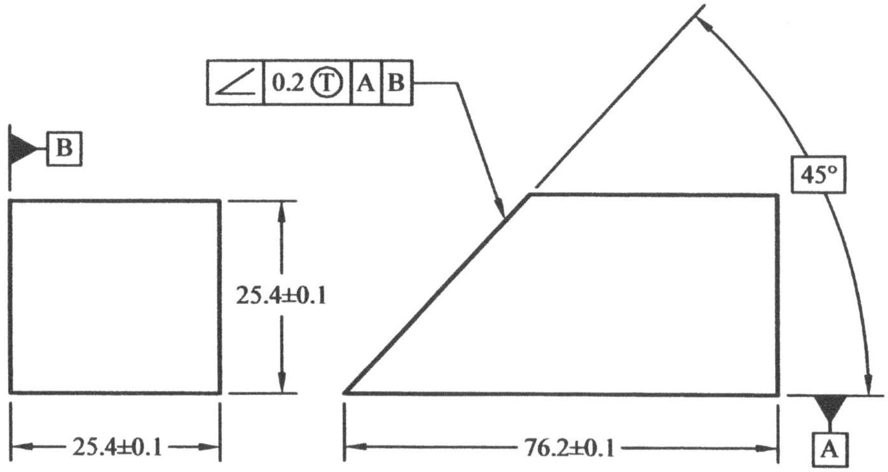

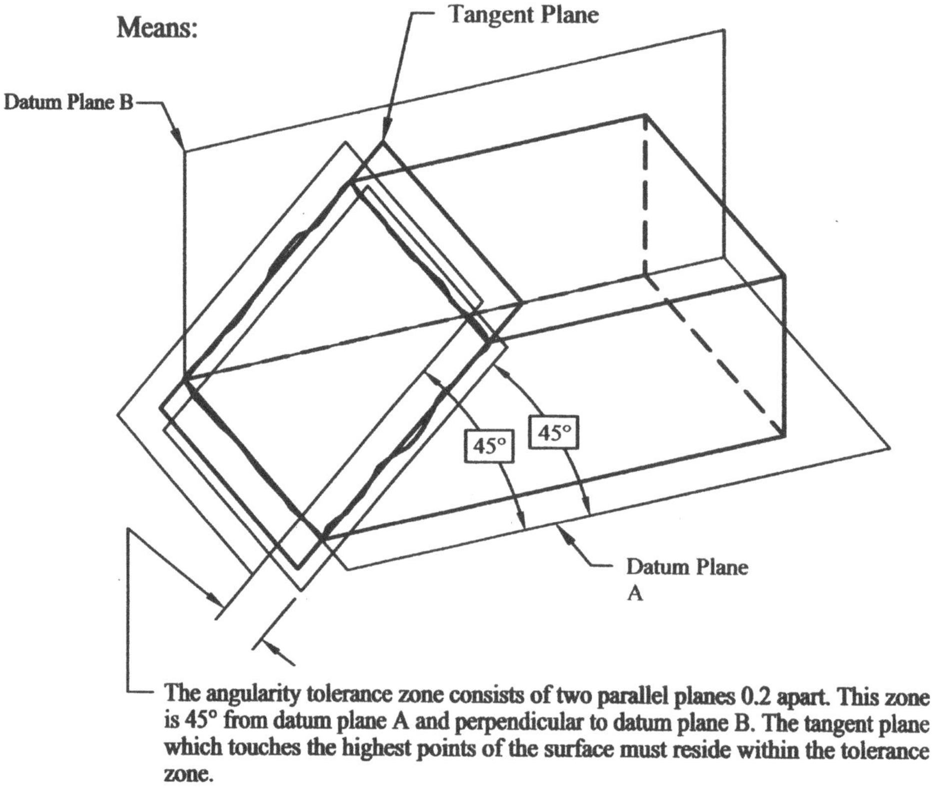

FIGURE 86 [Angularity of a Line Element]

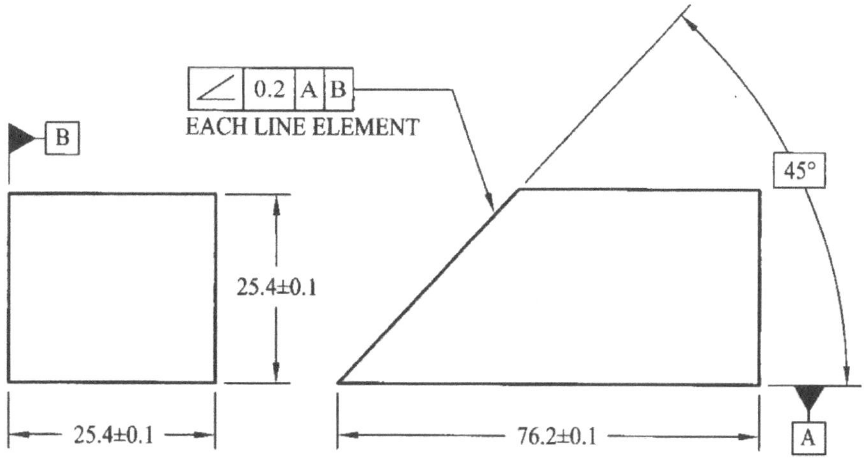

Means:

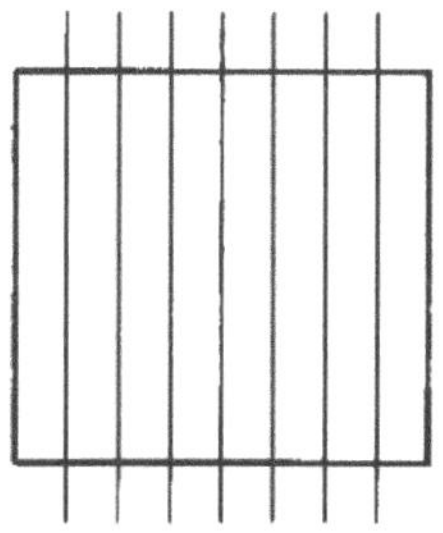

Each line element has its own tolerance zone. Each tolerance zone is independently verified as containing its line element.

All tolerance zones consist of two parallel lines that are 0.2 millimeters apart, oriented at the basic 45° angle from datum plane A and parallel to datum plane B.

Angularity as a Refinement of Position

Angularity is often used on holes and shafts as a refinement of the orientation aspect of a position tolerance. It's like writing a technical paper where one starts with a broad general statement and then refines it in subsequent paragraphs.

The broad general statement given in FIGURE 87 says that the hole must be on the part. It must hold all basic location dimensions as well as the specified and implied basic angles to within a diameter of 0.5 if the hole is produced at maximum material condition.

Then, in the refining angularity feature control frame, it states that the specified and implied basic angles to A and C must be held to within a diameter of 0.25 if the hole is produced at maximum material condition.

FIGURE 87 [Angularity as a Refinement of Position]

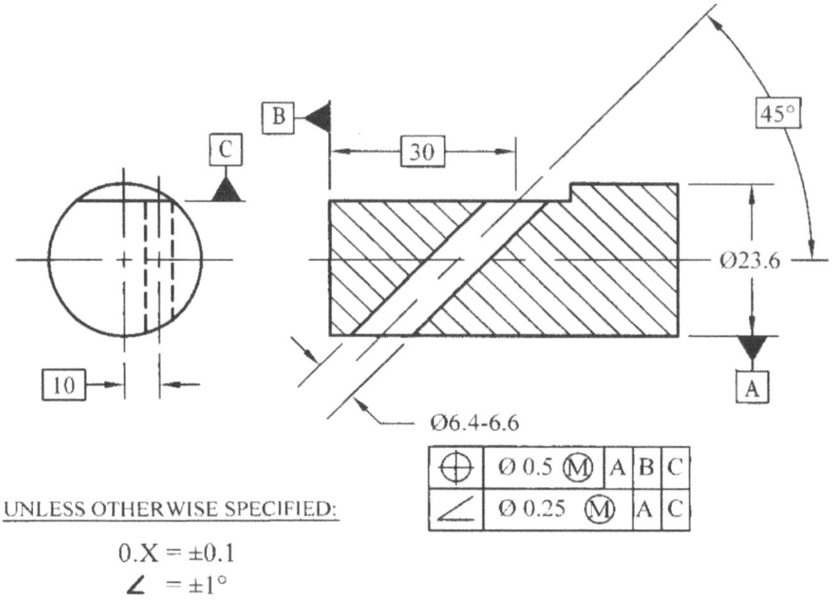

As you can see, the hole is loosely toleranced for its location on the part with a position tolerance. This feature control frame can be read:

⊕ = Position
⌀ = of the axis
0.5 = must be held to within 0.5
Ⓜ = if the hole is produced at its maximum material condition of 6.4
A = 45° and 10 millimeters from datum axis A
B = 30 millimeters from datum plan B
C = and oriented 90° in the left side view to datum plane C.

In this datum reference frame, datum axis A forms the primary and secondary datum planes. That means that secondary datum surface B forms the tertiary datum plane. Datum surface C forms a fourth datum plane to orient the two planes that cross (at a 90 degree angle) to create the primary datum axis A. The 0.5 (or larger with bonus tolerance) tolerance zone is perfectly oriented and located from this axis and planes. The axis of the maximum inscribed cylinder of the hole must reside in the tolerance zone to be in compliance with this control. Only then is the angularity control looked at. It says:

∠ = the angularity
⌀ = of the axis of the hole must be held in a cylindrical tolerance zone that is a diameter of 0.25
Ⓜ = if the hole is produced at MMC (of ⌀6.4)
A = that is 45° to datum axis A
C = and 90° to datum plane C as depicted in the left side view.

The fact that the angularity geometric characteristic symbol is used to control two different angles, 45 degrees to the primary datum and 90 degrees to the secondary datum makes a lot of people uncomfortable. It is widely adopted that when more than one angle must be toleranced with one control, and these angles fall under more than one of the orientation symbols, that the relationship to the primary datum suggests which geometric characteristic symbol is used. The reader of the drawing is expected to look at the angles in the part model or on the part drawing and deduce which angles are being toleranced.

Lately, a movement has started that says when multiple angle categories are to be toleranced with one geometric characteristic symbol, the angularity symbol is used. The simple fact is that as long as the reader of the drawing is made aware that the datums referenced hold multiple angles and these angles (whatever they are) are being held by one tolerance zone oriented to both of these datums, the goal of clarity of intent has been met. In this instance, the tolerance zones in the figure under discussion would be as follows:

FIGURE 88 [Tolerance Zones]

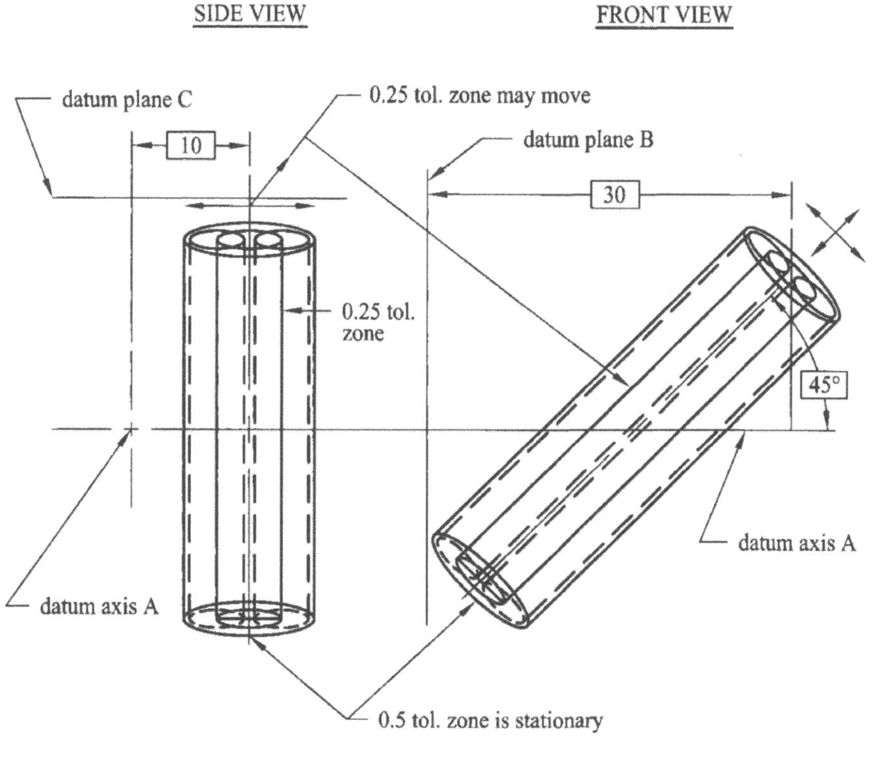

The smaller 0.25 orientation tolerance zone may move but must hold its basic angles to datums A and C. The larger position tolerance zone may not move or rotate. It is stationary and must maintain its basic angles and distance from datums A, B and C. The maximum inscribed cylinder axis of the hole must reside within both the 0.5 (at MMC) and the 0.25 (at MMC) tolerance zones to comply with these requirements.

Chapter 7

Profile

•Profile of a Surface
•Profile of a Line
•Composite Profile
•Composite Profile vs. Two Single Segment Profile
•Coplanarity
•Continuous Feature of Size
•Conicity
•Dynamic Profile

Chapter Objectives

Readers will learn:

1. Profile of a surface, profile of a line, composite profile, two single segment profile, coplanarity, continuous feature of size and conicity; their tolerance zone configurations and how to measure them.

2. All around controls, all over controls, unilateral tolerances, the unilateral/unequal tolerance symbol, equal bilateral tolerances, unequal bilateral tolerances, non-uniform zones, showing extent of control, and ASME vs. ISO tolerance zones.

3. How to tolerance complex curved surfaces to mate using unilateral profile tolerances.

4. How to sequentially apply and read feature control frames for perpendicularity, position and profile of a surface.

5. Composite vs. two single segment profile of a surface tolerances; their similarities, differences, strengths, weaknesses and uses.

6. Relating patterns of features using 3 levels of profile controls.

7. Locating planar surfaces from one another using profile of a surface.

8. Dynamic Profile

Profile of a Surface
...Basic Dimensions and the Tolerances that Apply to Them

Basic dimensions are boxed dimensions such as: 100 .

These dimensions may be used to describe size, shape (form), angles or location. They are targets to shoot for; in some cases, the ideal size, shape, angle or location at which a feature may be produced. Unlike a plus or minus toleranced dimension, they are toleranced indirectly. Instead of tolerancing the imaginary dimension, the real feature's deviation from the basic dimension is toleranced.

FIGURE 89 [Lawn Sprinkler Part]

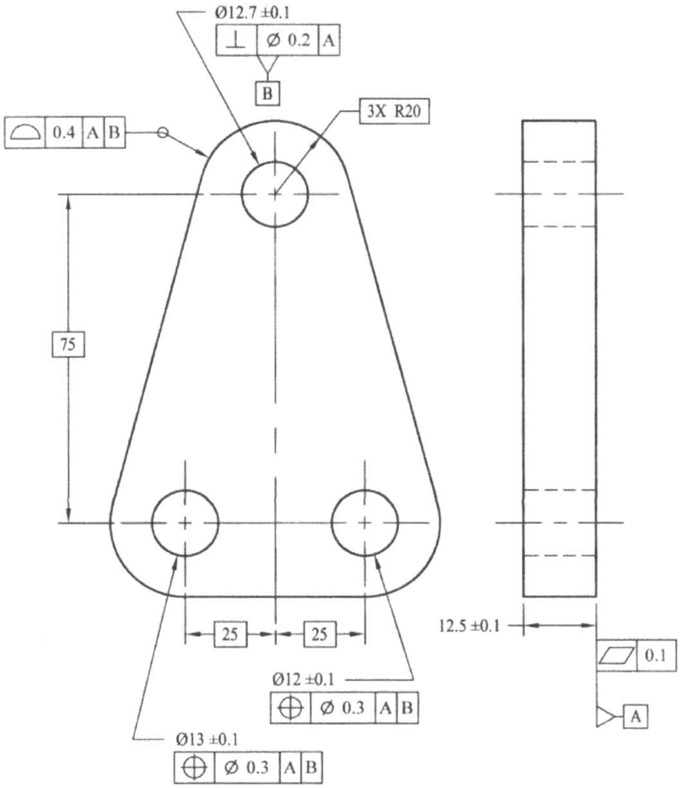

The illustration of the lawn sprinkler part shows basic dimensions to describe the perfect size, three-dimensional form (shape), angle and location of the part. The datum features A and B are used to stabilize the part and generate a primary datum plane A (for perpendicularity) and datum axis B for location of the other two holes and the surface of the part's periphery (profile of a surface).

The illustration of the lawn sprinkler part's tolerance zones shows each of the holes at the bottom is ideally 75mm down from datum axis B and 25mm measured left or right of the B hole's axis. The basic dimensions show us the target to shoot for, but the positional tolerance zones of a diameter of 0.3mm shows us how much the axis of each of these holes may stray from that target. As long as the axis of the maximum inscribed cylinder of each hole is produced inside of its 0.3 positional tolerance zone, it meets the location requirement.

FIGURE 90 [Lawn Sprinkler Part Tolerance Zones]

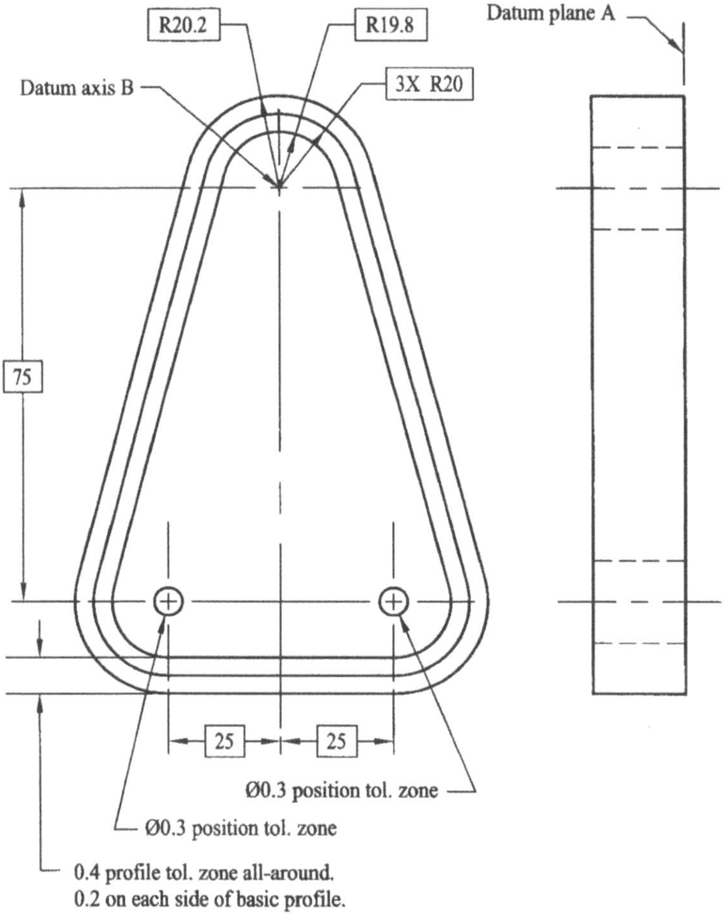

The perfect profile configuration is defined by the basic dimensions. The size, shape, angles and location of the surface is defined by basic dimensions of three radii of 20mm which are located by the basic distance dimensions from datum axis B of 75, 25 and 25. The tolerance zone of profile of a surface shows the surface must remain within a 0.4 total wide zone that is equally distributed on both sides of the configuration defined by the basic dimensions. The profile tolerance zone is perfectly perpendicular to datum plane A. As long as the produced surface of the profile does not violate the 0.4 tolerance zone, the geometric tolerance requirement has been met.

There are only four things that geometry can control. These are size, 3D form (shape), angles and location. This profile tolerance, as depicted, controls all four of these requirements within the 0.4 tolerance zone.

The statement that is often made is that basic dimensions have no tolerances. A healthier way to think of that is that the tolerance "that applies" to basic dimensions is found in feature control frames. This part contains those tolerances in the position and profile controls shown on the field of the drawing. Sometimes these geometric tolerances are embedded into the CAD model.

FIGURE 91 [Profile of a Surface All Around – Equal Bilateral]

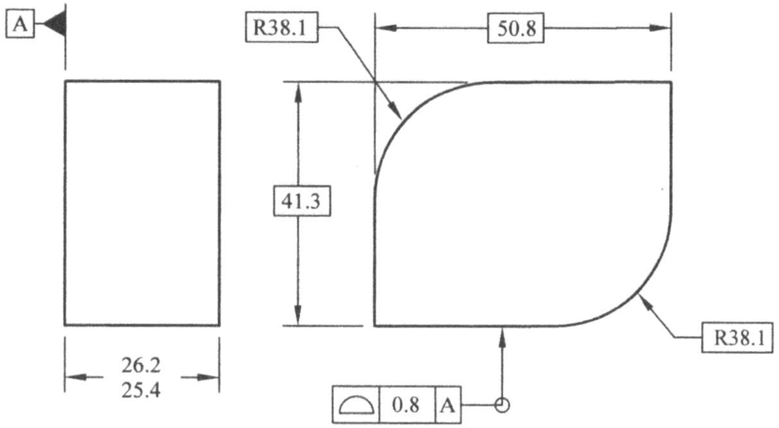

Meaning

COMPARATOR CHART

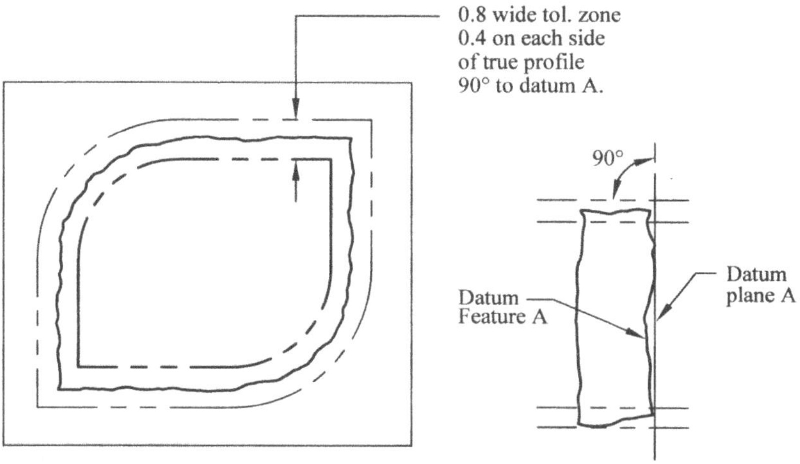

The designer of this part had, at some point, decided what the perfect part would look like (its configuration) and how big it should be (its size). Then, the tendency is to step away and think the job is complete. Someone from manufacturing usually inserts a bit of reality into the situation and says, "We can't make it exactly like that. How far from perfect may it deviate before it fails to function?" In other words, tolerance the thing!

Many designers, feeling they have more important projects to tackle, or lacking the knowledge to appropriately tolerance the part, tend to either slap some ill-conceived tolerances on the part or relegate the tolerancing to someone else not as familiar with the part's function. It sometimes happens that a few parts are made, then inspected, and the largest deviations from perfect are recorded as the tolerances allowed on the drawing (sort of tolerancing, after the fact).

Plus or minus tolerancing may be used, but with complex features that have many angles, radii, sizes or even locations, the accumulation of error this creates from angle to angle and radius to radius (and so on) can allow manufacturing to create a part that conforms to the drawing requirements but doesn't even remotely resemble the configuration shown on the drawing. It is similar to the situation often created when a line of holes is located by dimensioning and tolerancing each hole from the previous hole. The accumulation of error may not even require the last hole to reside on the part (if it wasn't for the hole violating its size requirements), to say nothing of the fact that these holes would stand very little chance of aligning with related mating holes or shafts on another part in the assembly.

To avoid this type of problem on this part, the dimensions of 50.8, 41.3, the two radii of 38.1 and the two 90° angles shown all around in one view have been made to be basic dimensions. This means they will draw their tolerances from an appropriate feature control frame (a profile control, in this case) and never from the general tolerances most often shown in the drawing's title block. The fact that the tolerance is all around in this view, as designated by the circle on the elbow of the leader line from the feature control frame to the part, creates a non-accumulating tolerance of one continuous zone. This zone follows the configuration drawn; and, since no phantom line is used, the zone is considered equal bilateral, 0.4 on each side of the basic profile given. Only at the 90° corners does a hypotenuse-type effect create a distance which is greater, from sharp point to sharp point, than 0.4 outside and 0.4 inside the basic profile.

With profile, use of a datum or datums is optional; but, in this case, a seating surface has been used as a primary datum feature for part stability and repeatability during inspection and also to reflect the way the part seats in the assembly. The actual surface of the part, if it resides within the tolerance zone, will be controlled for size, form (including surface flatness, where appropriate), perpendicularity of one surface to another as shown and, because of datum A, line element perpendicularity of the profile to datum plane A.

This tolerance zone can easily be simulated in an optical comparator overlay created specifically for this part (see "Means This" portion of FIGURE 91) and, subsequently, inspected on an optical comparator, perhaps similar to the one shown in the next illustration, when properly seated on datum feature A. A strong light source strikes the part and its enlarged image is shown on the screen to which the appropriately enlarged transparency overlay of the tolerance zone has been clamped. If the image can be manipulated (while maintaining the required datum relationship) to fall within the tolerance zone, the controlled feature is on its way to acceptance. The thicker the part, the more the possibility increases that pits in the surfaces or the out-of-perpendicularity of the line elements to datum A may not have been reflected in the screen image and must be inspected. These items, as well as the part's thickness, should be investigated before finally accepting the part.

FIGURE 92 [Optical Comparator]

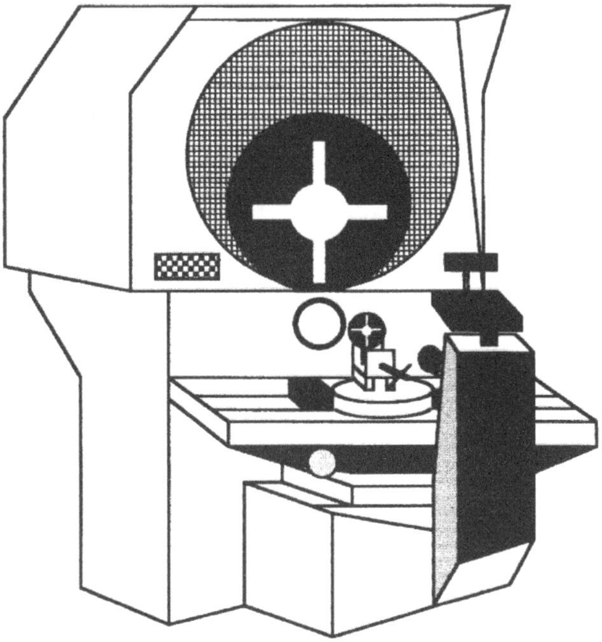

The example in FIGURE 93 shows the preferred method of displaying an unequal bilateral profile tolerance zone per *ASME Y14.41 Digital Product Definition Data Practices standard, the ASME Y14.5-2009 and the ASME Y14.5-2018 standards*. The circled 'U' symbol means either unilateral or unequal. We know which by the tolerance value that is shown after the circled 'U'. In these cases, the tolerance value shown after the circled 'U' is the portion of the total profile tolerance that is outside of the basic profile.

This example shows a profile tolerance that is unequal. It has a 0.2 profile tolerance outside of the basic profile. That leaves 0.6 of the total 0.8 profile tolerance inside of the basic profile. The number shown after the circled U represents the amount of tolerance that is allowed "growth of material". The remainder is the allowed shrinkage of material. "Growth of material" means that a shaft (male feature) gets bigger, but to grow material into a hole (female feature), it would have to get smaller.

If the entire 0.8 appeared after the circled 'U' symbol, the profile tolerance would have been plus 0.8 and minus zero--a unilateral profile tolerance. Likewise, had the tolerance depicted after the circled 'U' been zero, it would have been a profile tolerance of plus zero and minus 0.8. This use of the circled 'U' is the preferred method, rather than using a phantom line to show the direction of the profile tolerance zone.

FIGURE 93 [Preferred Method for Unequal or Unilateral Profile Tolerancing per ASME Y14.5-2018]

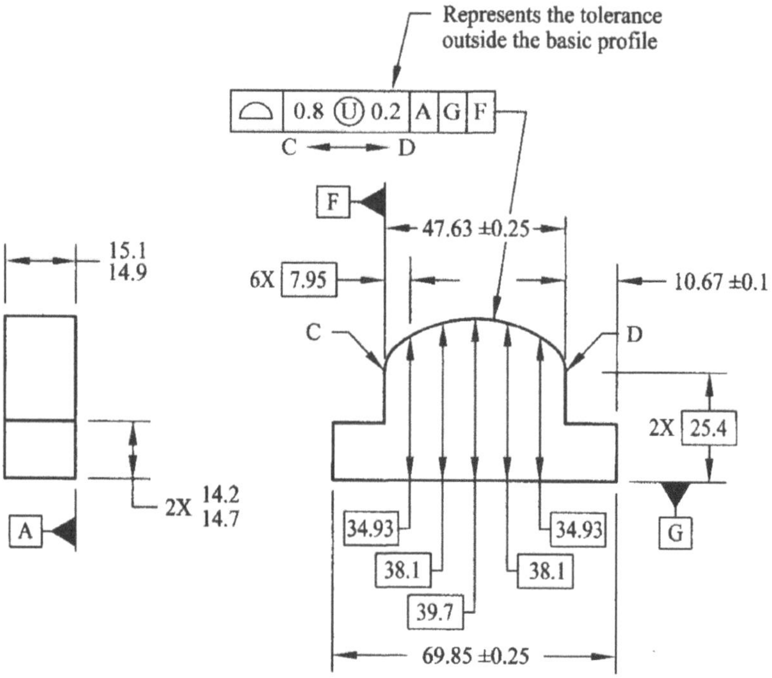

FIGURE 94 [Unequal Bilateral Profile Tolerancing-Alternate Method]

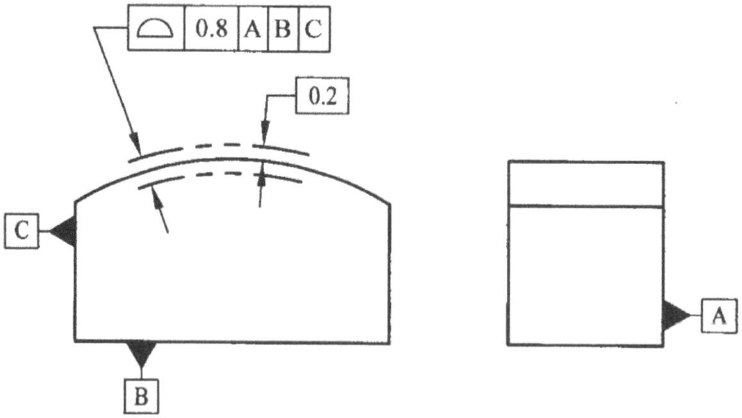

This is an ASME Y14.5 supported alternate method example of an unequal bilateral profile tolerance. Since 0.2 of the profile tolerance is shown outside of the basic profile, the remainder of the 0.8 profile tolerance zone resides inside of the basic profile in FIGURE 94.

FIGURE 95 [Profile of a Surface]
Figure 1 – Equal Bilateral
Figure 2 – Unilateral (Inside)-Alternate Method to Figure 8-5
Figure 3 – Unilateral (Outside) – Fully Dimensioned

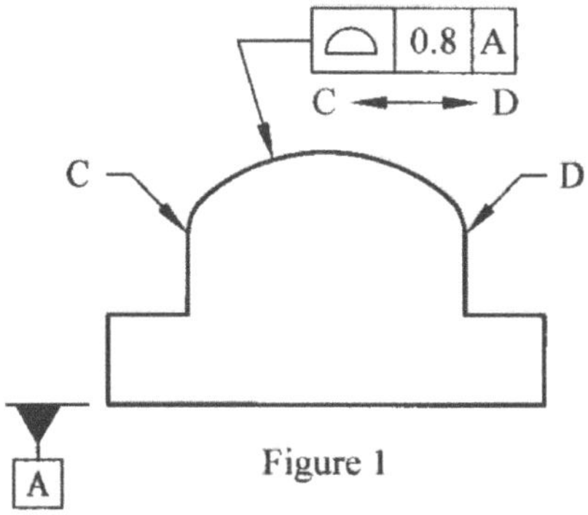

Figure 1

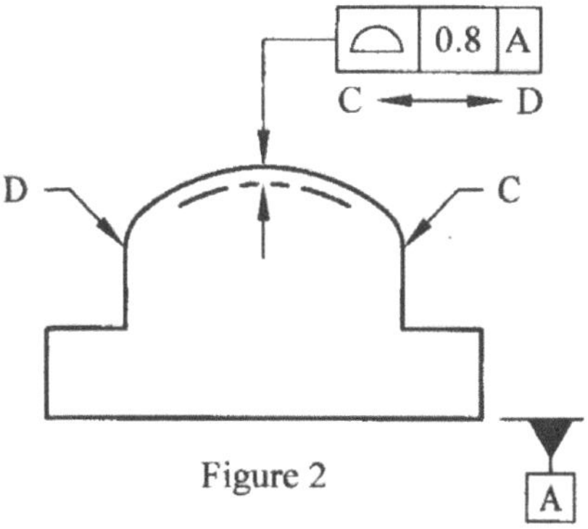

Figure 2

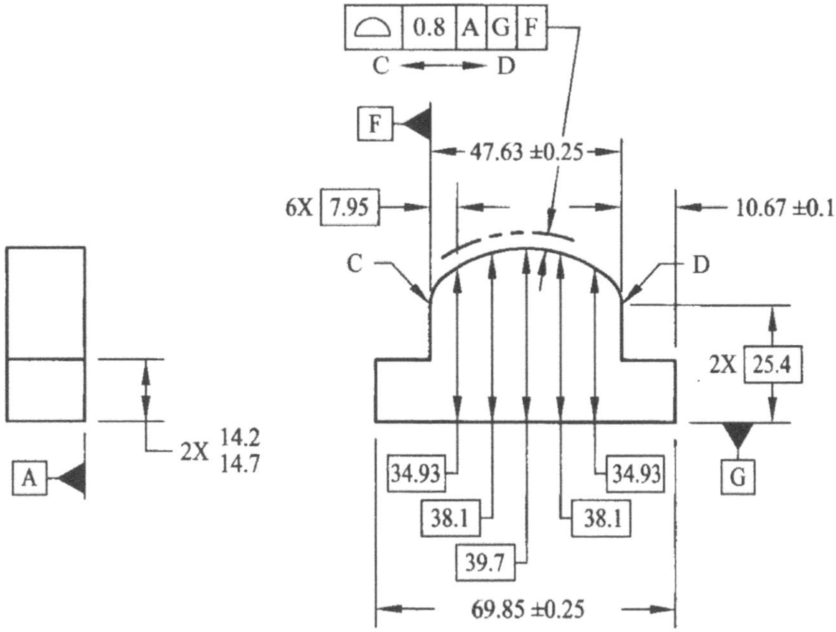

Figure 3

In FIGURE 95, the three examples show how to designate the direction and extent of a profile tolerance. Figures 1 and 2 are not dimensioned to show a less cluttered explanation, but would, of course, be fully dimensioned on the finished drawing. Figure 3, although incomplete, gives an idea of one method in which the basic profile may be dimensioned.

In Figure 1 of FIGURE 95, an equal bilateral profile tolerance is indicated by a leader line and accompanying arrowhead extending from the feature control frame to the basic profile. The profile tolerance uses half of the total profile tolerance (shown in the feature control frame) on each side of the basic profile. The "between symbol", a line with arrowheads on each end extending between two letters (in this case, C and D), shows where the profile tolerance zone begins and ends. Where the letters are not shown, both on the drawing of the part and below the feature control frame, the profile tolerance extends until a sharp corner is shown, signifying that the controlled surface ends (unless the “all around” or “all over” symbol is shown). See the end of this section for illustrations that use the new symbol for “all over”.

To comply with the profile tolerance, the entire surface under control must reside within the tolerance zone. In Figure 2 of FIGURE 95, the profile tolerance zone is bounded on the outside by the basic profile and is bounded on the inside by a profile uniformly 0.8 smaller. The surface may not extend outside (grow beyond) the basic profile (which is fully defined on the finished drawing) but may extend inside the basic profile by 0.8. This is designated through the use of a phantom line shown inside of the basic profile. A line extends from the feature control frame and points at the basic profile with an arrowhead. An opposing line, and its accompanying arrowhead, point at the phantom line. The phantom line may be shorter than the controlled profile zone, since it is only to show the direction of the tolerance. It is the job of the letters (C and D, in this case) to show the beginning and end of the tolerance zone. Below this profile feature control frame, it is shown in symbology that the profile tolerance zone extends between line elements C and D.

In Figure 3 of FIGURE 95, this more fully dimensioned example shows the profile tolerance zone is bounded on the inside by the basic profile and is bounded on the outside by a profile uniformly 0.8 larger. The surface may not extend inside (shrink beyond) the basic profile but may extend outside the basic profile by 0.8. This has the effect of controlling size, form, orientation and surface location deviations allowed from the basic profile shown. This control is designated in this example through the use of a phantom line shown outside of the basic profile. A line extends from the feature control frame and points at the phantom line with an arrowhead. An opposing line, and its accompanying arrowhead, point at the basic profile.

As in Figure 2 of FIGURE 95, the phantom line is used to show whether the tolerance zone is unilateral inside or unilateral outside (outside in Figure 3 of FIGURE 95, inside in Figure 2 in FIGURE 95), and the letters C and D, both on the drawing and below the feature control frame, show the extent of the tolerance zone. Of course, letters other than C and D may be used, and, in fact, an entire profile of a part may be segmented out into separate regions, each of which is given a different profile tolerance as appropriate.

Notice in these examples that where segments of a profile have different tolerances (are not designated all around ), the extent of each profile tolerance is designated by the use of reference letters. These letters show the limits of each profile tolerance's influence on the part.

The profile tolerance zone can be described as two uniform boundaries within which all elements of the line (for profile of a line) or surface (for profile of a surface) must lie. The dimensions that define the basic profile of the feature (from which the uniform boundaries emanate) may be basic radii, basic coordinate dimensions or basic angular dimensions. Formulae are sometimes also used.

If plus and minus tolerances or limit dimensions are given to a part for sizes of features while separately identifying a profile as basic, the part must stay within those size limits. However, if the part has no such tolerance of size on opposing features and only the basic dimensions identify the basic profile and the size, then the profile tolerance in the feature control frame tolerances both profile and size. So, form/profile and size of the part's features are first defined by the basic part profile and then toleranced by the feature control frame.

Both profile of a line and profile of a surface may be used on the same surface. If this is done, the profile of a line tolerance must be the smaller of the two (in order to have any meaning). Profile must be shown in an appropriate view to show the desired basic profile. In gaging a profile tolerance, the feature should, whenever possible, be gaged normal (perpendicular) to the basic profile. Several instruments used to check profile are optical comparators with appropriate comparator charts/overlays, templates with dial indicators, and computerized Coordinate Measuring Machines. [Note: When profile tolerance involves a sharp corner of a part, the tolerance zone must extend to the intersection of the part's boundary lines.]

The profile of the surface in FIGURE 96 must first be defined by basic dimensions. The profile tolerance zone will follow the same basic configuration as these basic dimensions define. One tolerance zone exists for the entire surface under control. The zone consists of two uniform boundaries 0.04 apart and 0.02 on each side of the basic profile (since this is shown as an equal bilateral tolerance zone). As explained in ISO (international) standards, a tolerance sphere of a diameter of 0.04 may be thought of as passing through the basic profile with its radius 0.02 on each side to form the tolerance zone. The surface being controlled must reside within the profile tolerance zone to be acceptable. Generally, the basic dimensions of such a control will emanate (or be traceable) from the datum reference frame (more complete than shown here). Such a control would (under those circumstances) control the size, form, orientation and location of the surface being profiled to the datums. A profile template of the perfect profile configuration may be used, as shown, for comparison of the produced surface being controlled.

Once the part/object being inspected is set up on the datum(s), the surface is indicated/probed for a record of deviations from the perfect profile as simulated by the profile template. The copy tip, which contacts the template, becomes the zero point or master against which readings on the indicator are compared. If deviations registered are greater than 0.04 or 0.02 on each side of the perfect profile from the master form template, the surface is considered in violation of its profile of a surface tolerance. Although not shown in this example, the indicator should consistently be normal to the desired geometry of the controlled surface.

It is important to note that material condition symbols are inappropriate next to the controlled feature's geometric profile tolerance and should not be used. However, it is common and often functionally appropriate for datum features of size used in the feature's control to be referenced with the MMB (or LMB) material condition symbol. This MMB symbol does not allow a growth in the profile tolerance zone but does allow a shifting (movement) of the profile zone as the datum feature of size is produced at sizes other than MMB (and still within allowed size limits).

FIGURE 96 [Inspecting Profile of a Surface]

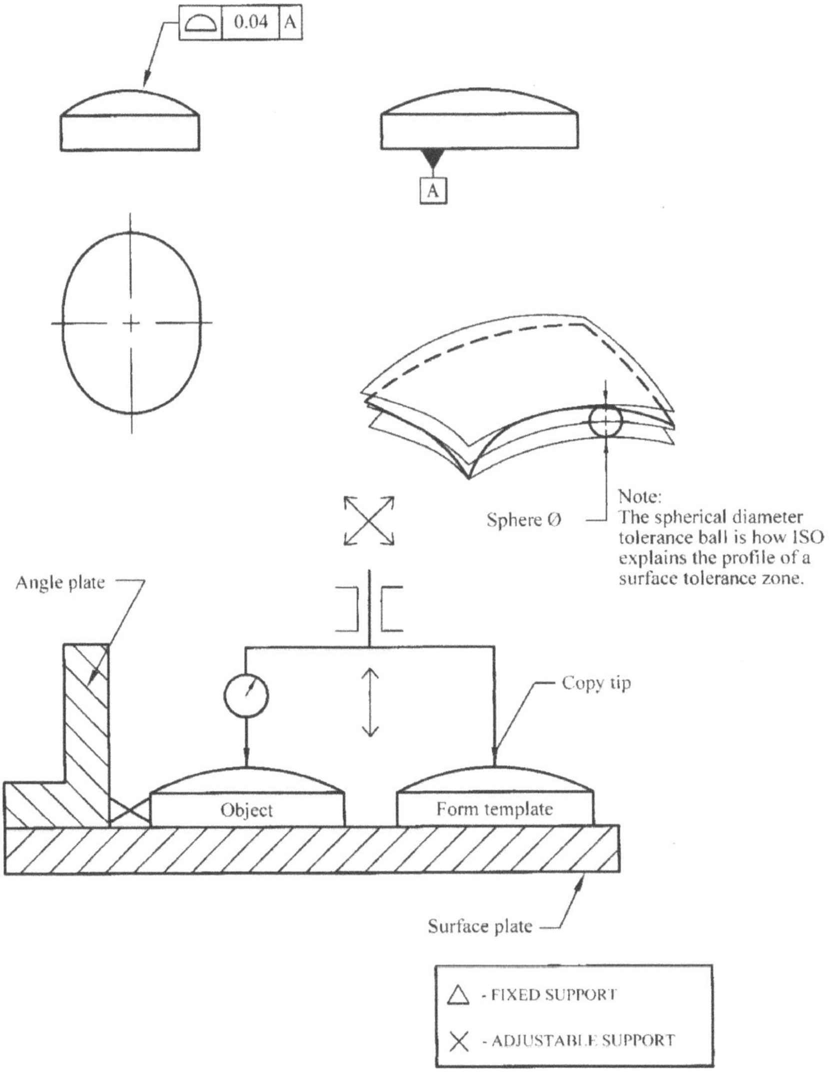

Ideally, profile measurements are taken normal to the basic geometry with the probe always perpendicular to the surface of the part. In FIGURE 96, consideration must be given to the error that could be introduced if the probe takes on any other angle to the surface being measured.

FIGURE 97 [Profile of a Surface All Over Tolerance Zones]

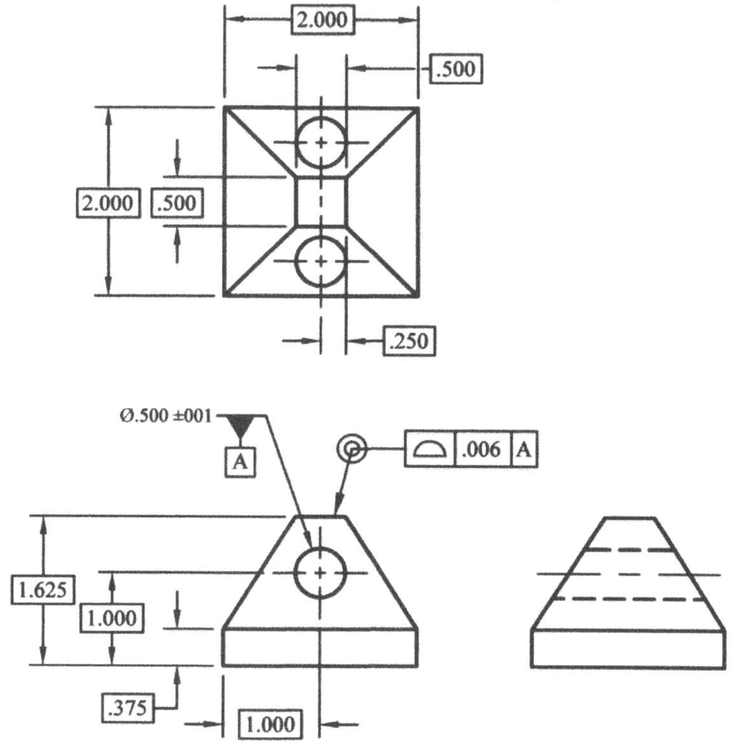

FIGURE 98 [Means This-Applies in All Views Simultaneously]

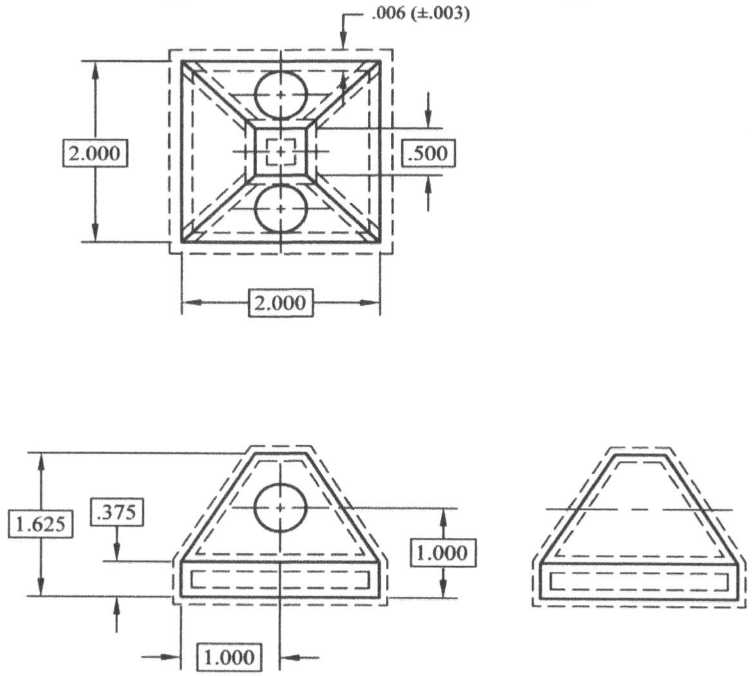

FIGURE 99 [Profile of a Surface All Over Shown on a GD&T Sheet]

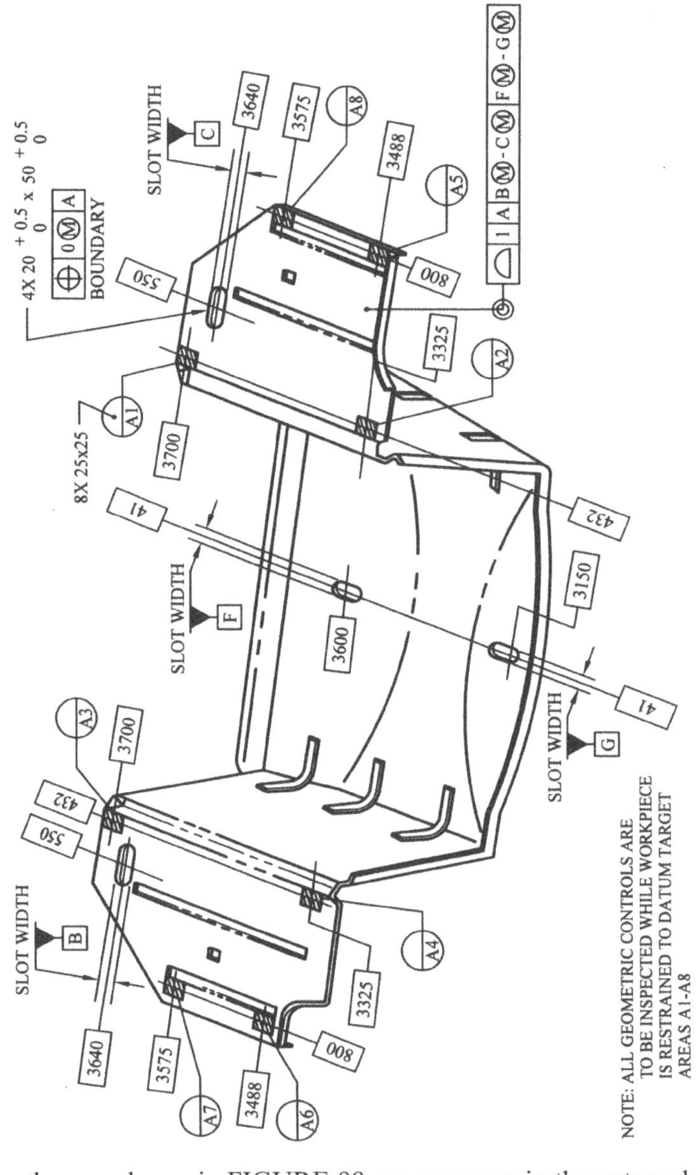

GD&T sheets, like the one shown in FIGURE 99 are common in the automobile industry. They show an isometric view of the part and all of the GD&T. Any dimensions that are depicted are optional. A full definition of the part may be found in the CAD model. Some of the dimensions shown in this drawing are "body line' dimensions that relate the panel to the rest of the vehicle. Often, the CAD model is downloaded into a Coordinate Measuring Machine's software. The software builds a tolerance zone that is plus or minus 0.5 millimeters around the CAD model. The part can be fixtured on the CMM table, then the profiled surfaces scanned and all collected points compared to the CAD model to determine if they fall within the Profile of a Surface All Over continuous tolerance zone. This tolerance zone applies in all views.

FIGURE 100 [NONUNIFORM Profile of a Surface Tolerance Zones]

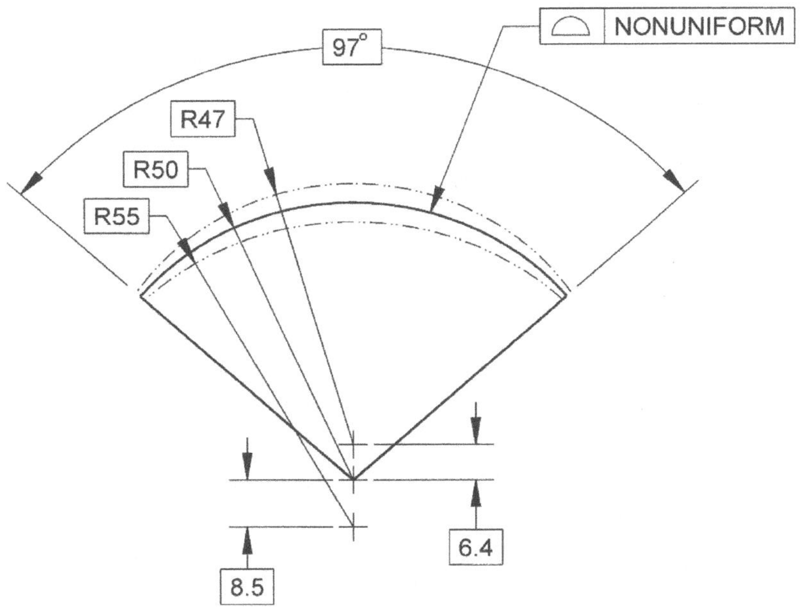

A nonuniform profile tolerance zone can be used to smooth the transition between profile segments. Instead of abrupt surface variations being allowed at the end of one profile tolerance zone and the beginning of another, the zones can now be shown to narrow at each end. The differences in the radii, taken in conjunction with the offset of the radii (as shown in the figure above) are considered when calculating the profile tolerance. The extent of each profile tolerance zone must be made clear using the usual methods, such as between symbols.

If needed, a NONUNIFORM profile tolerance may also be used for Profile of a Line.

Y14.5-2018 introduced the non-uniform profile tolerance zone's use of numerical values inside of the feature control frame, to define the proportional variation from one value to another, between two specified locations on the feature being tolerance. The tolerance values are related to the specified locations on the toleranced feature by the letters shown on each side of the "between" symbol below the feature control frame. So, the term NONUNIFORM (as shown in Figure 100) can replace these tolerance values inside of the feature control frame when the profile boundaries are given in the part model data.

From-To Symbol. Another symbol introduced in the Y14.5-2018 standard is one that indicates that a specification transitions from one location to a second location. The leader line from the feature control frame is directed to the segment of the feature to which that tolerance applies. The illustration that follows this text shows the "from-to" symbol indicating tolerance transitions that shows where one profile tolerance begins and ends, then shows where another profile tolerance begins and ends. This can be used when specifying a profile tolerance that varies in size from one place to another. See figures 101 and 102.

FIGURE 101 [NONUNIFORM Profile Tolerance]

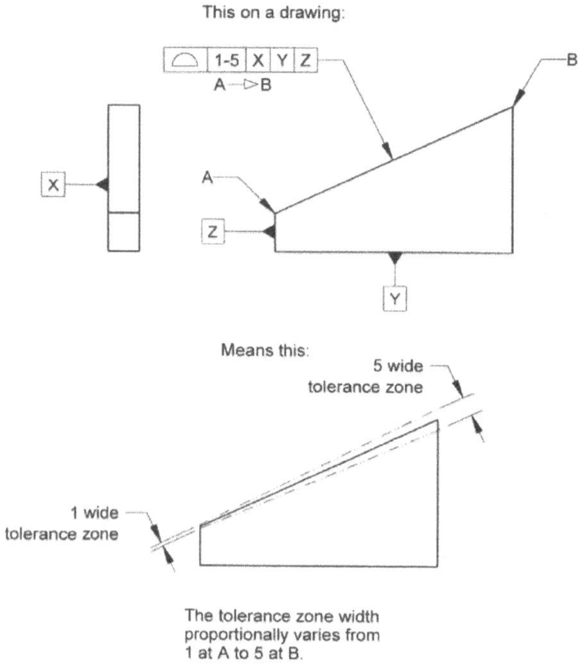

FIGURE 102 [NONUNIFORM (Variable) Profile Tolerance]

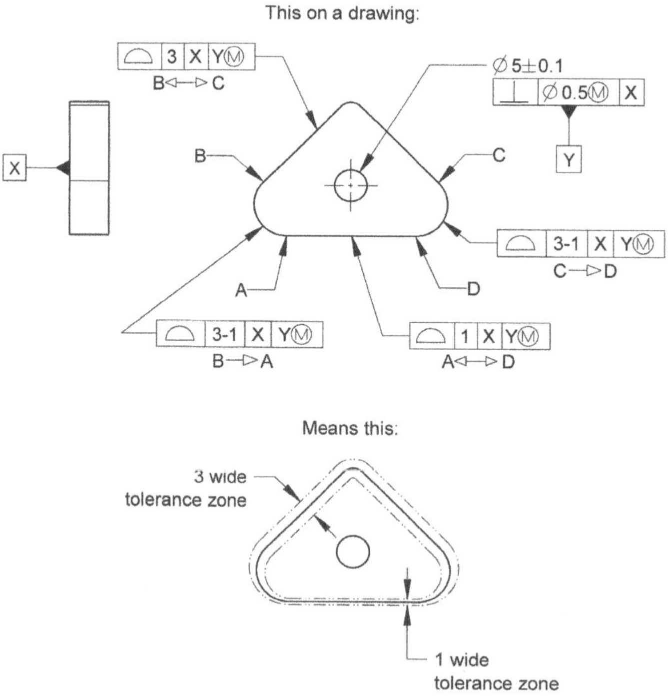

Dynamic Profile Tolerance Zone Modifier

The humble name of a new concept for Profile is "Dynamic Profile". This principle is invoked using a symbol that looks much like a Delta symbol Δ within a profile control (See figure). This concept was specifically developed to divorce the control of form and size. So, when a profile control that applies ALL AROUND or ALL OVER would normally control size and form to within the same tolerance, this control allows the size to change within limits given in a larger (regular) profile tolerance, while the form is controlled within the smaller Dynamic Profile tolerance. It may be used with Composite Profile or Multiple Single Segment Profile controls.

FIGURE 103 [Dynamic Profile in a Composite Profile Control]

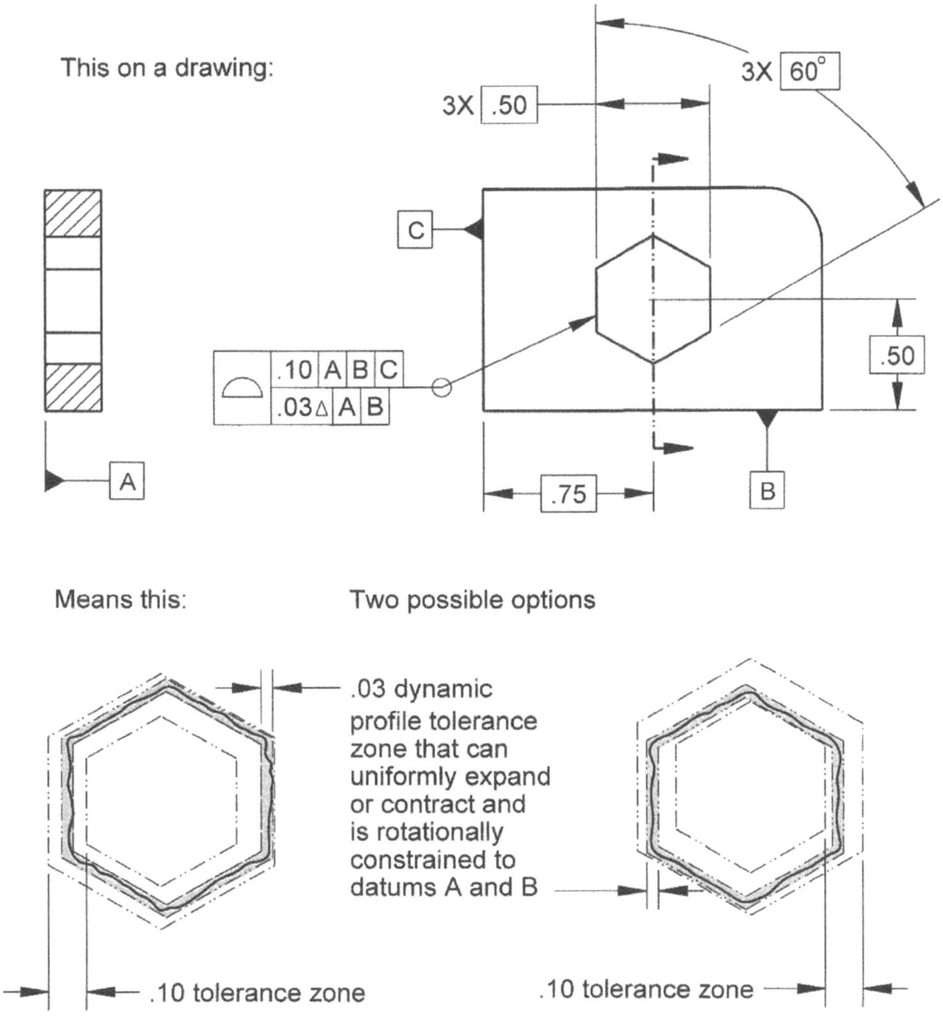

FIGURE 104 [Dynamic Profile in a 2-Single Segment Profile Control]

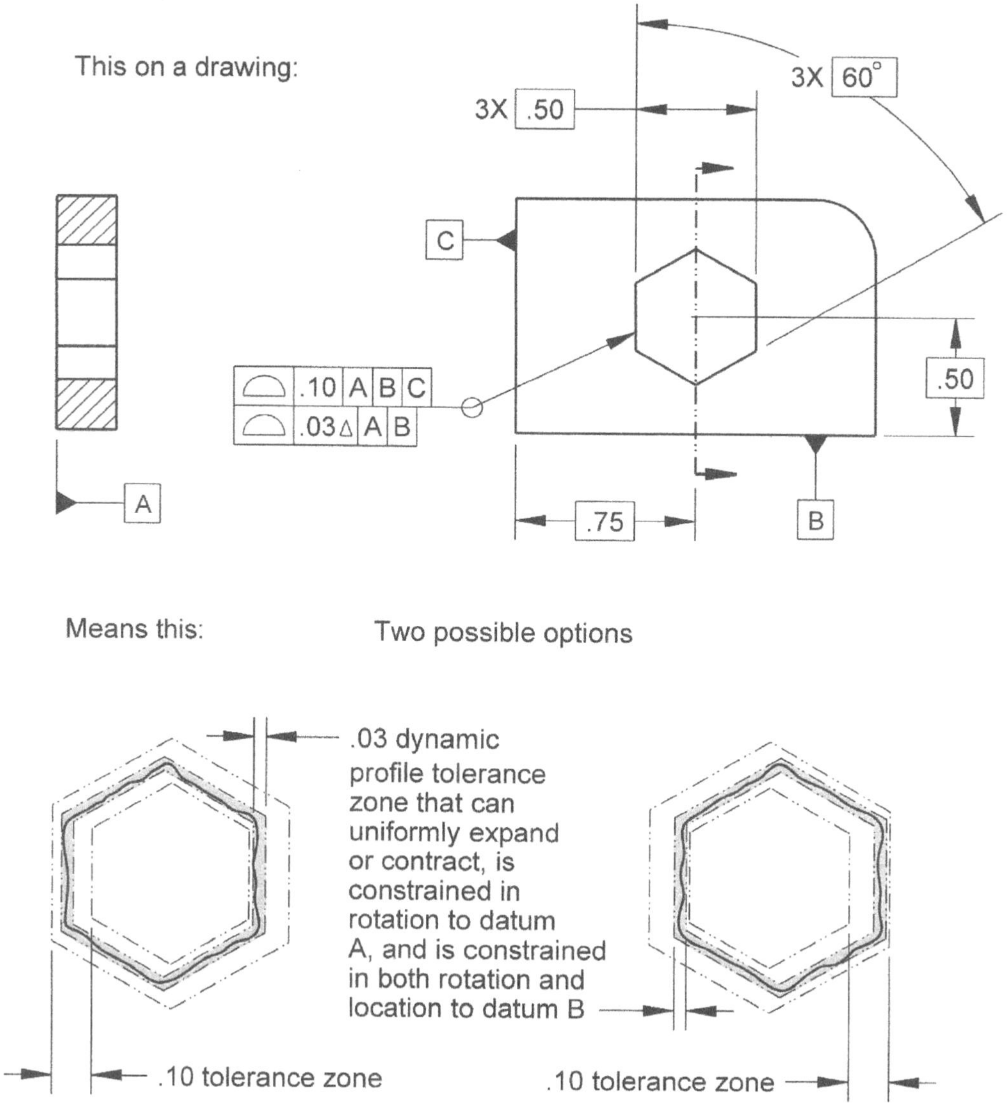

FIGURE 105 [Profile of a Line – Unilateral Inside]

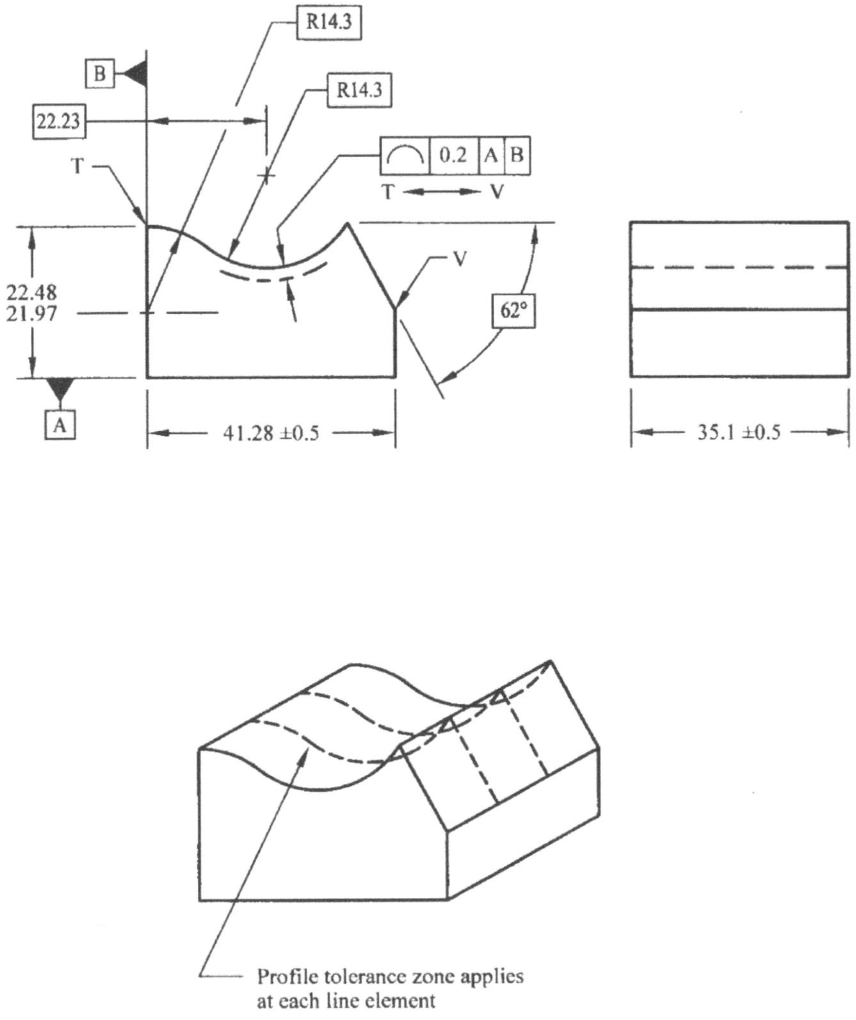

This illustration deals with the geometric characteristic Profile of a Line. For any profile of a line characteristic that references datums, each cross section of the surface under control in the plane that is appropriately oriented to the datums generates a line element. Every line element of the produced surface that is within the extent of the Profile of the Line control (in this case between T and V) must lie within a tolerance zone that consists of two uniform line boundaries that are the geometric tolerance apart (in this case 0.2). These tolerance zone line elements follow the basic dimensions that define the line profile.

Profile of a Line is a two-dimensional control of surface line elements as opposed to Profile of a Surface, which is a three-dimensional control of the entire surface (or portion of the surface) being toleranced.

As stated, this Profile of a Line control applies to line elements between T and V. The datums used do not control the height or location of the line elements from datums A and B, but do control the orientation of these line elements. This control has the effect of slicing up the surface into an infinite number of line elements. Each of the line elements must also reside, at their left most edge of the front view, somewhere between the 21.97 and 22.48 millimeters dimensions. The Profile of the Line tolerance of 0.2 is not additive to these dimensions.

Both the requirements of size and geometric tolerance must be respected and complied with. However, the shape of each line element must simply fall within its own 0.2 profile tolerance. This could allow abrupt surface variations from line element to line element. Each profile tolerance zone may begin at a different height within the 21.97 and the 22.48 range. The datums referenced will stop the tolerance zones for all of the line elements from angling, since all of these geometric tolerance zones will be perfectly oriented to the datums and, therefore, to one another.

As the 0.2 tolerance zones for each line element maintain their orientation to the datums A and B, the usable portion of the tolerance zones will diminish in size as they approach the size limits of 21.97 and 22.48, until at those sizes, none of the profile tolerance zones are usable. This is because the size limits must not be violated. This profile of a line control is used to control the form and orientation of each line element.

In FIGURE 106, once the profile under control is defined by basic dimensions, the tolerance zone follows the same basic configuration as the profile. One tolerance zone exists for each line element of the surface. Each zone consists of two uniform line elements 0.6 apart that forms a boundary for each line element of the surface which may not be violated. A profile template of the perfect profile configuration may be used, as shown, for comparison of the produced surface being controlled.

Once the object/part being inspected is set at an optimum orientation for each line element, that line element is indicated for a record of deviations from the perfect profile as simulated by the profile template. The copy tip which contacts the template becomes the zero point or master against which readings on the indicator are compared. If deviations greater than 0.6 from the master are registered, the line element is considered in violation of its profile of a line tolerance.

FIGURE 106 [Inspecting Profile of a Line]

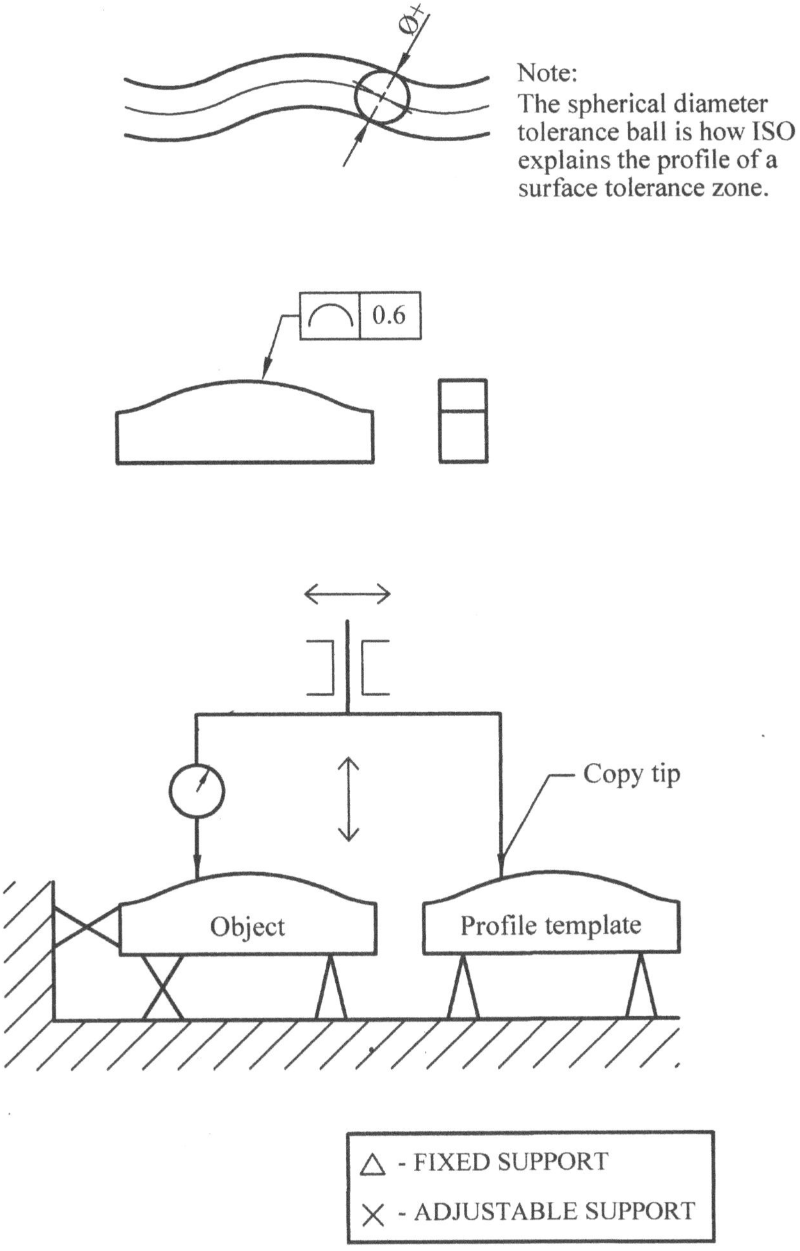

Although not shown in this example, the indicator should consistently be 90° to the desired geometry of the line element under test. As with all profile measurements, if the probe is other than normal to the surface of the part, cosine error may be introduced into the measurement.

FIGURE 107 [Mating Parts Using Unilateral Profile of a Surface Tolerancing-Part 1]

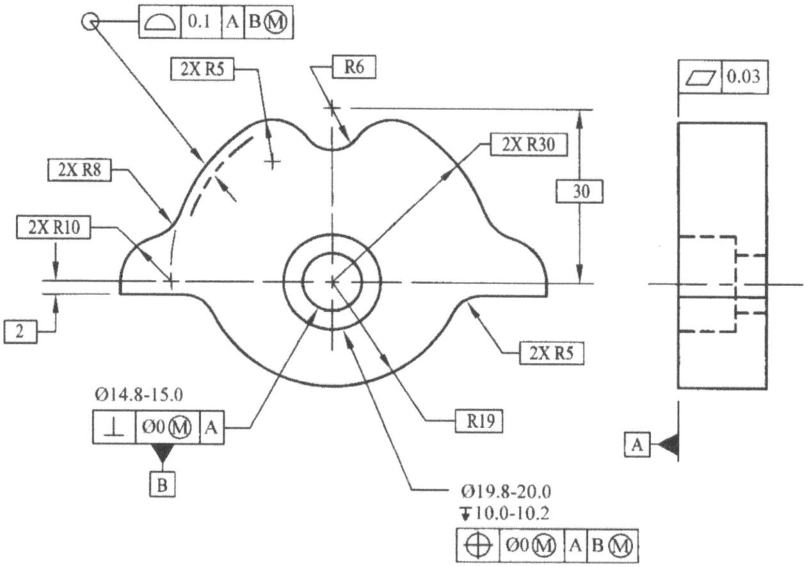

FIGURE 108 [Mating Parts Using Unilateral Profile of a Surface Tolerancing-Part 2]

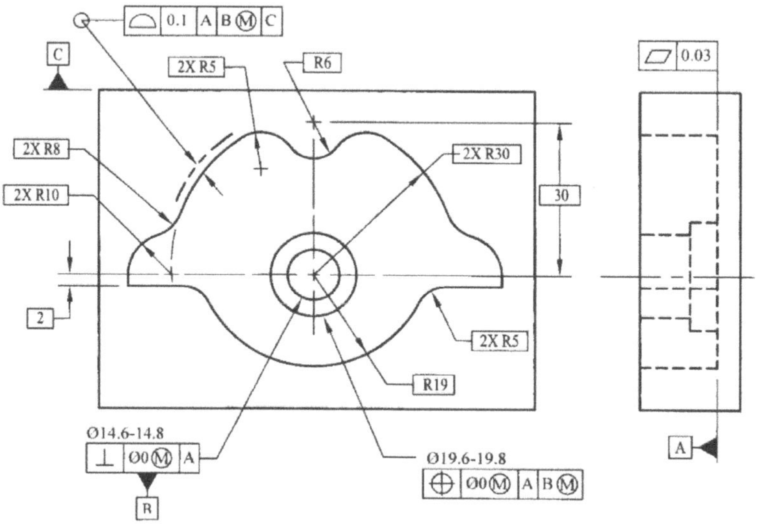

These illustrations show the power and versatility of profile of a surface. **There are only four things that can be controlled geometrically (with geometry). They are:**

- **3-D form**
- **size**
- **angle**
- **location.**

These all-around profile of a surface controls tolerance all four. These illustrations also show how phantom lines can dictate the side of the basic profile to which a profile tolerance zone applies.

If, as shown in FIGURE 107, the phantom line is depicted to the inside of the basic configuration (as defined by basic dimensions), then the profile tolerance is all inside. The surface may shrink, pit, wave or display any aberration, as long as it does not violate the unilateral inside tolerance zone which consists of two uniform boundaries that are 0.1 apart. The outer boundary is the basic profile shown on the drawing. When the profile tolerance zone traverses around a sharp corner, the zone extends to the point of intersection. This has the effect of creating a hypotenuse at the sharp corners of the part's profile tolerance zone.

As can be seen on FIGURE 108, the phantom line is depicted on the outside of the mating cavity. It shows that the profile tolerance is unilateral outside and the cavity is therefore only allowed to grow. These are mating parts. This type of tolerancing allows us to use all the same basic dimensions to define the configuration of both parts. We then know that since the oddly configured shaft can only shrink and the oddly configured hole can only grow, the worst fit they would have, if the parts are within tolerance and assembled optimally, is a line-on-line fit.

The most uniform clearance (with the datums aligned) we would see is the sum of the profile tolerances, which is 0.2. If a line-on-line fit was undesirable, then the only thing that needs changing on the drawings would be to either shrink slightly the basic size that defines the oddly configured shaft or to increase the basic size of the oddly configured hole. In other words, either make the shaft smaller or the hole bigger in their geometric definition to which the profile of the surface tolerance is then applied.

STEP 1:

Since both parts use the same type of linear progression of geometric tolerances, we will go over FIGURE 108 only. This part begins by assigning a primary datum feature that is the bottom of the cavity. This is the surface that the mating part seats on. Since one of the major criteria for selecting datum features is to select as a primary datum the surface to which the part bolts, on which the part seats or that dictates the angle at which the two parts will assemble, this makes a very good primary datum feature. Ideally, it should be assigned a flatness control. This would be Step 1. This surface must be flat to within 0.03mm.

STEP 2:

Select a secondary datum feature that is a mating and/or alignment feature. The 14.6-14.8 shaft mates with a hole on the other part and makes a convenient origin of measurement from which to locate the larger 19.6-19.8 shaft and the profile of the cavity.

So, this shaft is made datum feature B. It is given a perpendicularity control to relate it back to datum plane A. This control can be read as:

⊥	=	Perpendicularity (the 90° angle)
∅	=	of the axis of this diameter (a shaft)
0	=	must be held perfectly
Ⓜ	=	if it is produced at MMC (∅14.8), but could be allowed to be out of perpendicularity a maximum of ∅0.2 if the shaft is produced at the LMC diameter of 14.6
A	=	to A.

This shaft then becomes datum feature B.

It could also be read as: **"Perpendicularity of a diameter that mates while the mating part seats on A."**

STEP 3:

The position control used on the ∅19.6 to 19.8 shaft references only datums A and B, so it is Step 3. Since the larger diameter shaft is depicted as perpendicular to A and coaxial to (shares the axis of) B, these will be the relationships that the positional control will tolerance. It can be read as:

⌖	=	Position
∅	=	of the axis of the ∅19.6-19.8 shaft
0	=	must be held to a minimum tolerance of zero
Ⓜ	=	if it is produced at ∅19.8 (its MMC) but could receive as much as ∅0.2 bonus tolerance if produced at ∅19.6 (its LMC)
A	=	to hold its perpendicularity to datum plane A
B	=	and its coaxiality to datum axis B
Ⓜ	=	but could receive an allowable additional shift away from axis B as feature B departs in its produced size from 14.8 to 14.6 (to a maximum shift of ∅0.2 if B does not use it up by being out of perpendicularity to A).

Then datum plane C is assigned to act as an angular orientation datum to stop the rotation of the profile of the cavity around the axis of B. Datum plane C will act to orient the cavity and our measurement of the cavity from the secondary datum axis B. This sets us up to begin Step 4.

STEP 4:

This is the profile of the surface all-around control. The profile's size, shape and location have all been defined with basic dimensions (as well as some implied basic 90° angles). The profile feature control frame will tolerance all of these basic dimensions to within a unilateral (all growth as shown by the phantom line) tolerance of 0.1mm. It can be read as:

⌓	=	Profile of the surface all-around in this view
0.1	=	must be held to within 0.1 of the basic profile defined by the basic dimensions shown in the unilateral outside growth tolerance zone
A	=	while the zone is held perpendicular to A

B	=	located from datum axis B
Ⓜ	=	but could receive an allowable additional shift away from axis B as feature B departs in its produced size from 14.8 to 14.6 (to a maximum shift of ∅0.2 if B does not use it up by being out of perpendicularity to A).
C	=	while measured along the intersecting (at 90°) planes that create the axis of B, which are oriented by datum plane C (as defined by a minimum of the two highest points of contact on C).

The planes that construct the axis of B must be either perpendicular or parallel to datum plane C. Since both the position and profile controls reference the A datum and B at MMB, when the shift allowed by the BⓂ occurs, both features must shift together, if at all.

FIGURE 109 [Another Example of Mating Parts Using Unilateral Profile Tolerancing]

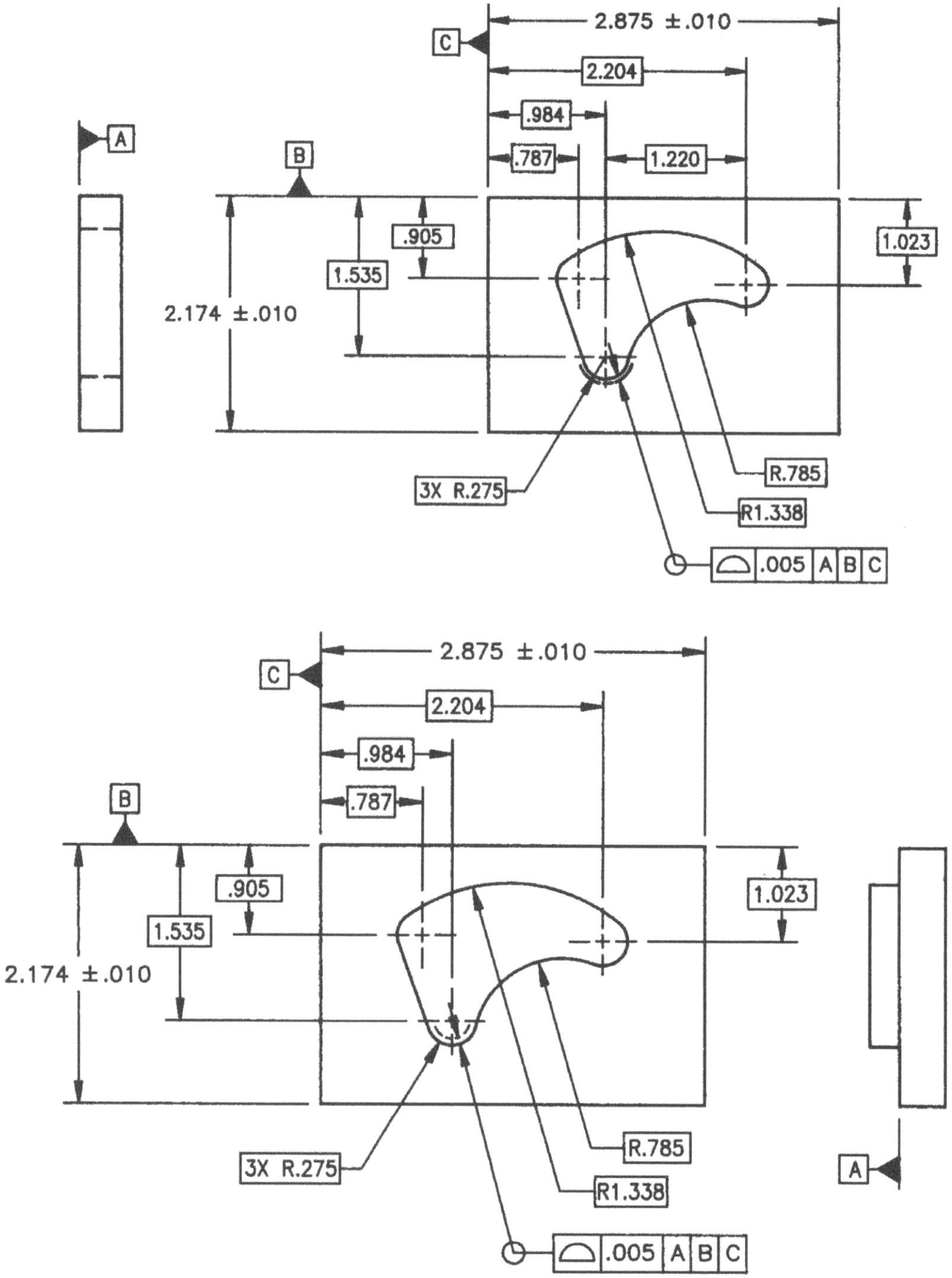

FIGURE 110 [Composite Profile Tolerancing]

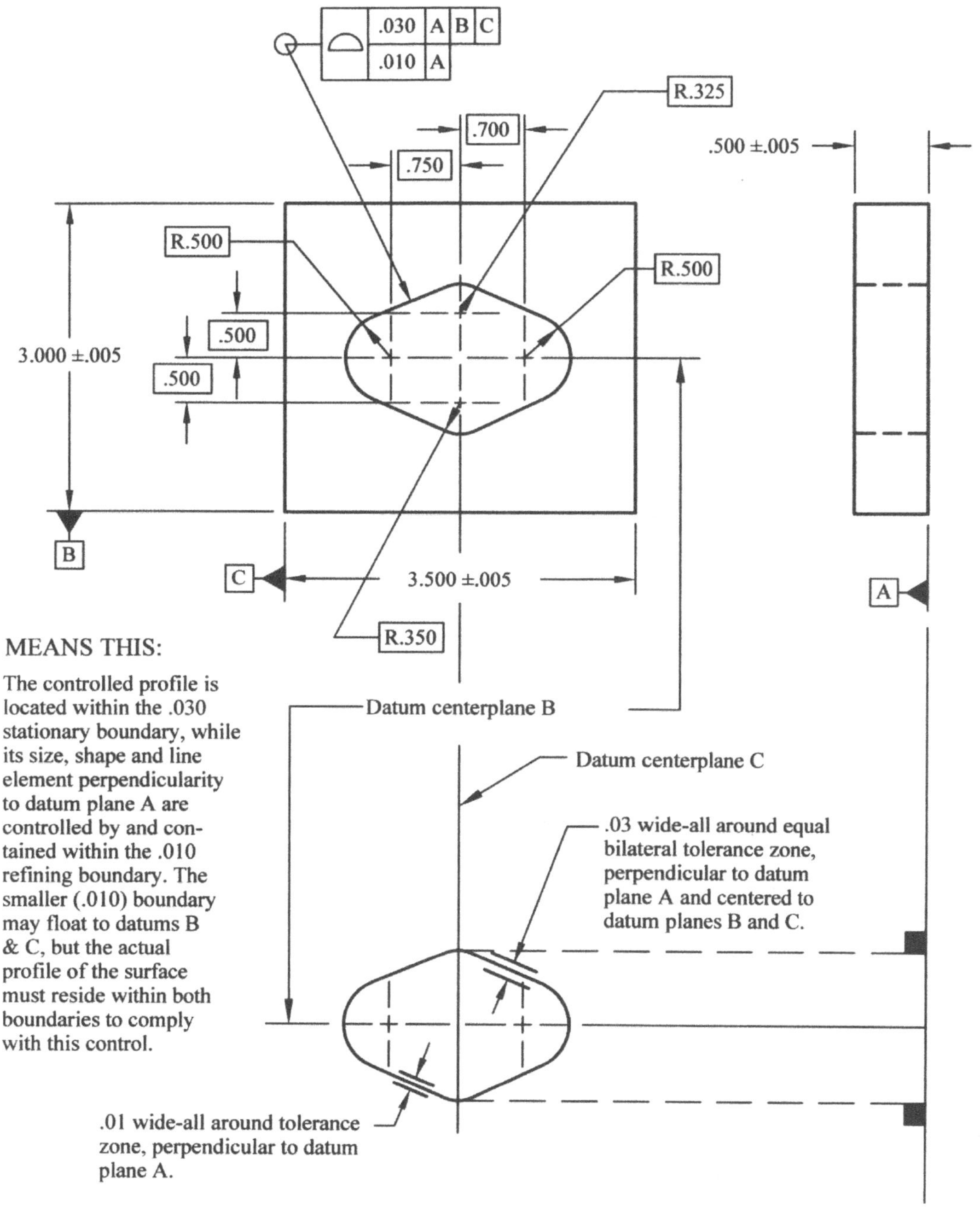

This example of composite profile tolerancing is used to show a need for a tighter tolerance of .010 on the size, shape and perpendicularity (to datum A) for the oddly-configured hole than its location tolerance of .030 to datum centerplanes B and C.

FIGURE 111 [Composite Profile Tolerancing]

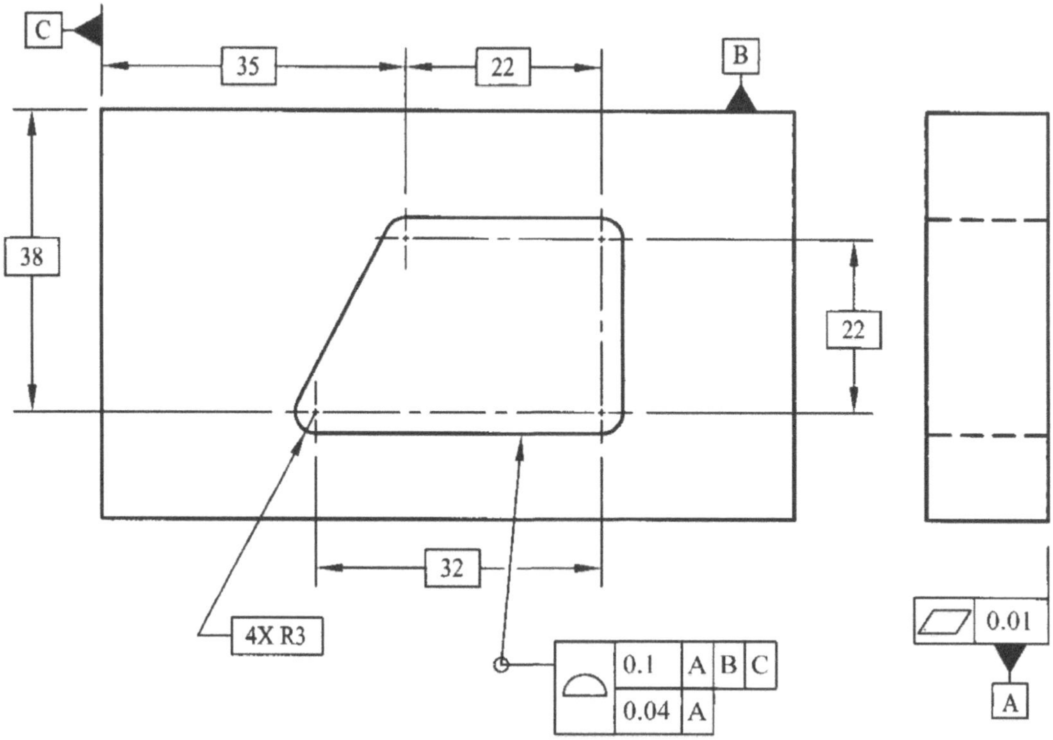

This composite profile control has two separately verifiable levels of geometric tolerance. The upper level of profile tolerance of 0.1mm locates the oddly shaped hole, while the lower level of control tolerances to within 0.04 per surface the size and shape (3-D form) of the oddly configured hole. Since the primary datum A is repeated in the lower level control, the perpendicularity of the surface of the hole is held within the 0.04 tolerance zone which is oriented to A.

The 0.04 level of tolerance controls the 3-dimensional form of the hole automatically since that is its main function. The fact that this profile of a surface tolerance extends all-around in this view means that it will also control the size of the oddly shaped hole. The size tolerance is 0.04 per surface. So, if we wanted to know how that affects the basic dimensions on the part, we could just add them up and apply the profile tolerance of 0.04 per surface. For example, see Figures 112 and 113.

FIGURE 112

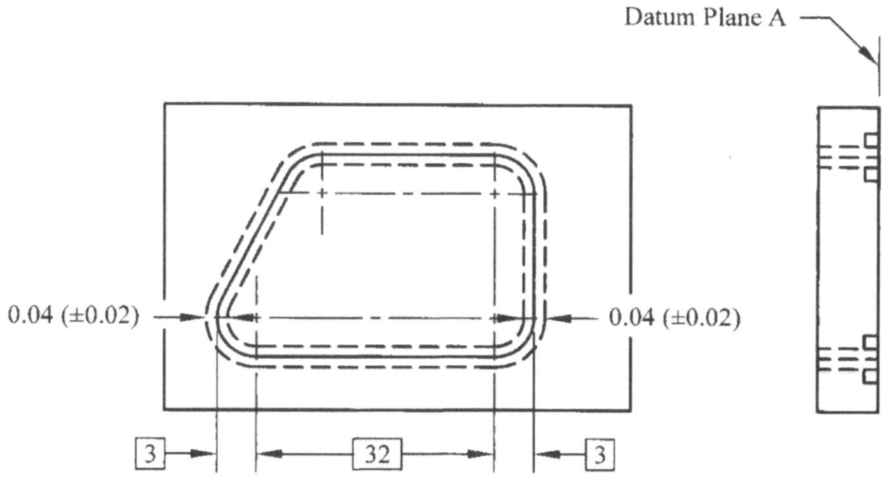

0.02 + 3 + 32 + 3 + 0.02 = 38.04 MAX
38 - 0.02 - 0.02 = 37.96 MIN

In the other direction:

FIGURE 113

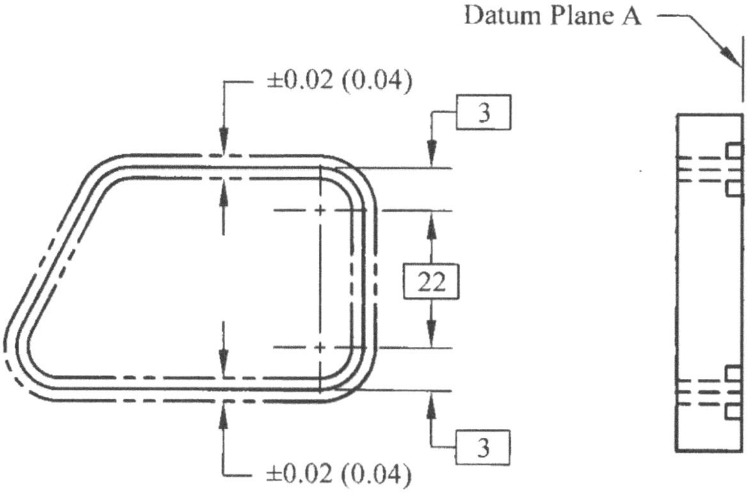

0.02 + 3 + 22 + 3 + 0.02 = 28.04 MAX
28 - 0.02 - 0.02 = 27.96 MIN

Since the 0.04 all around tolerance zone is perfectly oriented 90° to datum plane A, the 0.04 tolerance also controls the perpendicularity of the oddly-shaped hole.

Since all that geometry can control is size, 3D form, angles and location, we have to deduce what is left for the 0.1mm (upper level) profile control to do. The only remaining degrees of spacial freedom are rotation of the hole to datum plane B and location to datum planes B and C. We might, in a very simplistic way, view the 0.1mm profile tolerance as allowing rotational and locational deviations from the 38mm dimension from B and the 35mm dimension from C. See the illustration below.

FIGURE 114

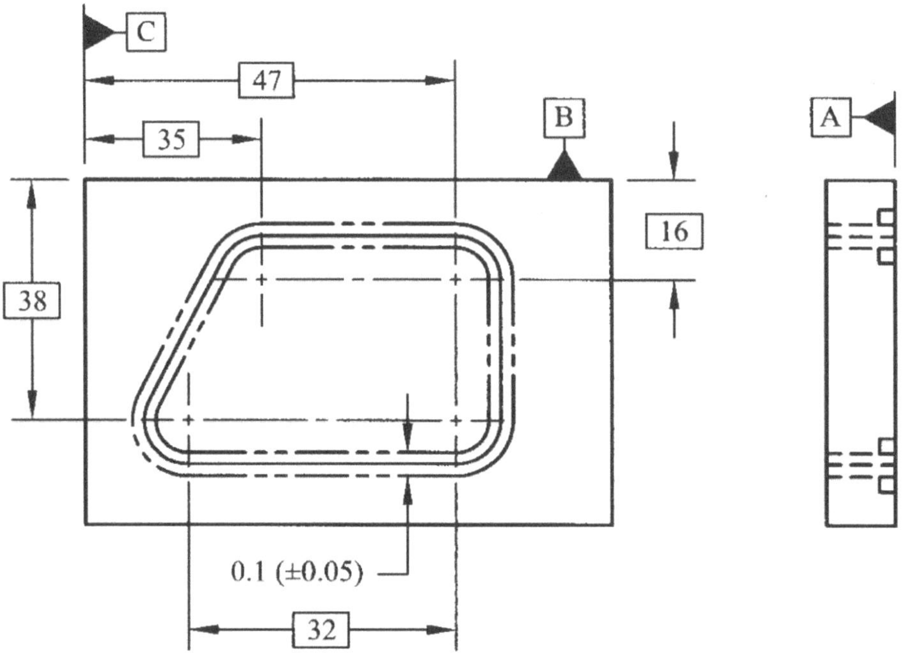

The 0.04mm tolerance does everything else. Since this is a composite profile of a surface tolerance (one profile symbol straddling multiple levels of control), it falls under the rule that says any datums used in other than the first level (with the 0.1) are not capable of tolerancing location to datums. So, even if datum B was brought down and referenced after the 0.04 and datum A, datum B would only be capable of stopping the rotation of the 0.04 tolerance zone to datum plane B. Location of the hole in the direction of B would still be toleranced by the 0.1mm listed in the first level of profile tolerance.

If a tighter location tolerance was desired in the direction of B than in the direction of C, a 2-single segment profile control would be needed. It would look like this:

⌓	0.1	A	B	C
⌓	0.04	A	B	

This would tighten the amount that the oddly configured hole could deviate from the 38mm dimension from B as well as how much the hole could rotate to datum B. Whereas, a similar composite profile control like this:

⌓	0.1	A	B	C
	0.04	A	B	

could only tighten the rotational control to datum plane B. The movement in the direction of B for the hole would still be toleranced by the 0.1mm.

This composite profile control above can be read as two separate sentences. The upper level control can be read as:

⌓ = Profile of the surface

0.1 = must be held to the basic dimensions shown to within ±0.05mm (a total wide zone of 0.1) for size and shape

A = and to A for perpendicularity

B = and B for location and rotation

C = and C for location.

It does everything to within 0.1mm. But then the lower level control steals some of these jobs by tightening the tolerance on these items to within 0.04mm. It can be read as:

⌓ = Profile of the surface is refined to

0.04 = within ±0.02 (a total wide zone of 0.04) of the basic dimensions for size and shape (overriding the upper level's control for size and shape/3D form control)

A = and to A for perpendicularity (or while measured oriented perpendicular to A) (overriding the upper level's control for perpendicularity tolerance to A)

B = and to B for rotation only (overriding the upper level's control for rotation to B).

Composite Profile vs. Two Single Segment Profile Controls

To explain in more depth the similarities and the differences between composite profile and two (separate) single segment profile controls, let's extend the discussion about the part shown in FIGURE 111. For example, a composite profile control, such as is shown in that figure, but with no datum listed in the lower level control:

⌓	0.1	A	B	C
	0.04			

In the upper level control, datum A is used to control perpendicularity and datums B and C are for location. The meaning is exactly the same as two single segment profile controls such as:

⌓	0.1	A	B	C
⌓	0.04			

In both cases, the upper level control acts to confine the location to datums B & C and the orientation (perpendicularity) to datum A of the oddly configured feature, while the lower level control acts to define the limits of the feature's size and shape/form/profile. Even if the two types of controls allow the primary datum to appear in the lower level controls, such as:

⌓	0.1	A	B	C
	0.04	A		

Composite Profile Tolerance

and

⌓	0.1	A	B	C
⌓	0.04	A		

Two Single Segment Profile Tolerances

the controls remain identical in meaning. The addition of datum A in each of the lower level controls simply means that the job of refining the perpendicularity tolerance of the line elements of the oddly configured profile now falls to the lower level control.

Since the tolerance in the lower level of each control happens to be smaller, it is the one that more closely controls the angle to A and, of course, still controls the size limitations and the shape/form/profile tolerance. Since datums B and C have not been brought down into the lower level control (with the tighter tolerance), the relationship to B and C is controlled by the upper level control that includes B and C (with the looser tolerance).

In this example, the difference in meaning between composite profile tolerancing and two single segment profile tolerancing arises only when one or more of the datums used for location (B and C) in the upper level control are brought into play in the lower level control, such as shown next.

⌓	0.1	A	B	C
	0.04	A	B	

versus

⌓	0.1	A	B	C
⌓	0.04	A	B	

Composite Profile Tolerance **Two Single Segment Profile Tolerances**

These two controls are **<u>not</u>** identical in meaning. The composite profile control uses datums in the lower control only to refine the angular relationship of the oddly-configured feature to those datums already used in the upper level control. It may only use datums in the lower control in the same order as they have been used in the upper level control. Two single segment profile tolerances may use datums in the lower level control that have been used in the upper level control in the same order or may use them in a different order. For example:

⌓	0.1	A	B	C
⌓	0.04	C	A	

In addition to the versatility of being able to use the same datums in a different order, the lower level control in two single segment profile tolerances may use entirely different datum features than the upper level control (something composite profile tolerancing is not allowed to do). For example:

⌓	0.1	A	B	C
⌓	0.04	D	E	

At any rate, the datums used in the lower-level control for two single segment profile retain their full implication and are <u>not</u> limited to an angular tolerance refinement of the controlled feature. A datum which has a basic dimension originating from it to locate the feature retains that locational relationship in a refining level control. Datums used in a lower-level control with a tighter tolerance of profile refine the relationship between that datum and the controlled profile for angle or location, not just for angle--as is the case with composite profile tolerancing.

In the example shown below for the composite profile control,

⌓	0.1	A	B	C
	0.04	A	B	

the upper-level control creates a stationary total-wide, all-around tolerance zone which confines the movement of the controlled feature surface. The surface of the controlled feature must reside within the 0.1 tolerance zone which is located as shown by the basic dimensions from datums B and C and perpendicular to primary datum plane A. The smaller lower-level, total-wide, all-around tolerance zone of 0.04 may float in its location to datums B and C but must maintain its perpendicularity to datum A and its angular orientation (rotational relationship) to datum B. The actual surface of the controlled feature must reside within both the 0.1 and the refining 0.04 tolerance zone to be in compliance with this composite profile control.

If, instead, the example was of two single segment profile controls, such as:

⌓	0.1	A	B	C
⌓	0.04	A	B	

the upper-level control would create a stationary, total-wide, all around tolerance zone which (like composite profile) confines the movement of the controlled feature surface. The surface of the controlled feature would have to reside within the 0.1 tolerance zone which is located as shown by the basic dimensions from datums B and C and perpendicular to primary datum plane A.

The smaller lower-level, total-wide, all-around tolerance zone of 0.04 may float in its location only to datum C, but must maintain its locational relationship as shown by the basic location dimensions to datum B (since B is listed in the lower-level refining control which also refines size/shape/profile) and its perpendicularity to datum A. This also has the effect of (by default) refining the rotational relationship to tertiary datum plane C. [This is a more restrictive control than the composite profile control described.] The actual surface of the controlled feature must reside within both the 0.1 and the refining 0.04 tolerance zone to be in compliance with these two single segment profile controls.

FIGURE 115 [Extension of Composite Profile of a Surface Concept]

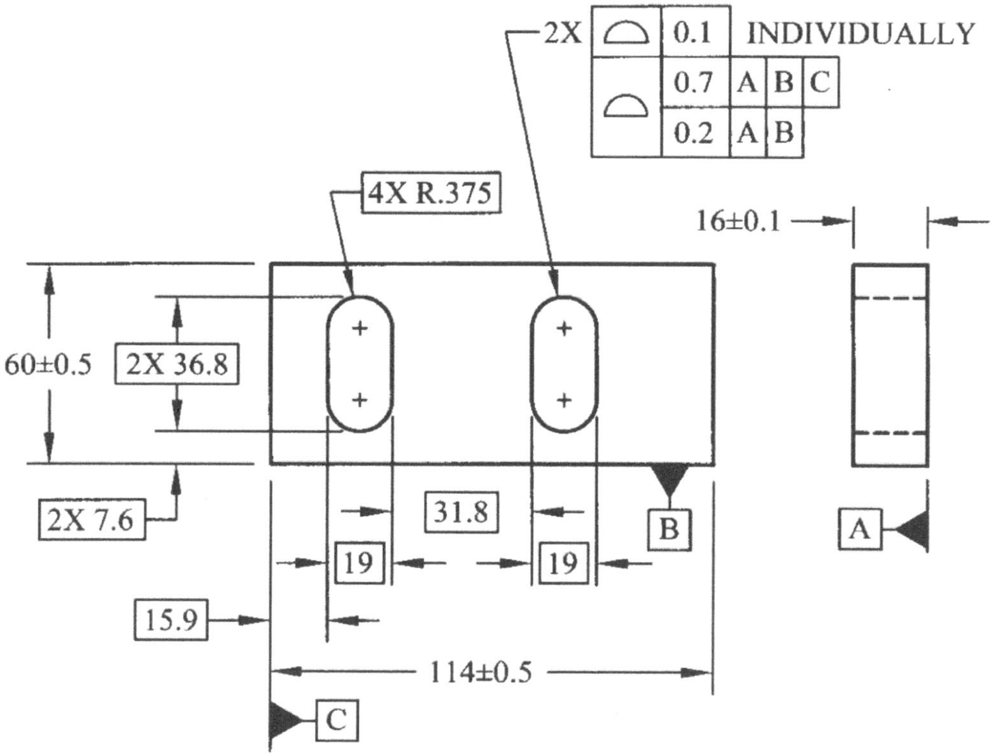

This illustration shows that profile of a surface can be used to locate patterns of features. The example above depicts a situation wherein the INDIVIDUAL form of the elongated holes is more important than their location to each other or their relationships to the datums referenced. These geometric controls do the following:

1) The size and form of the profile of each of the elongated holes is held to within a tolerance zone that is 0.1 wide and follows the basic configurations for size and shape and holds these dimensions to within plus or minus 0.05 per surface.
2) The location to datums B and C (while the part is seated on datum feature A) is held to within 0.7. These tolerance zones are perpendicular to datum plane A and located the basic distances depicted on the drawing from datum planes B and C. This location tolerance zone is plus and minus 0.35 per surface (an equal bilateral tolerance zone).
3) The location of the elongated holes to each other is held to within what the 0.2 tolerance zones dictate. These 2 tolerance zones are basically located from one another, perpendicular to datum plane A and oriented as a pattern to datum plane B.

Coplanarity

FIGURE 116 [Continuous Feature of Size]

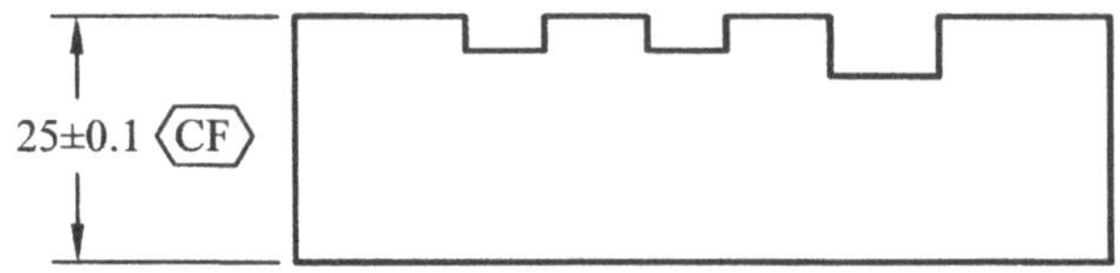

MEANS THIS:

GO GAGE to inspect the MMC of the Continuous Feature

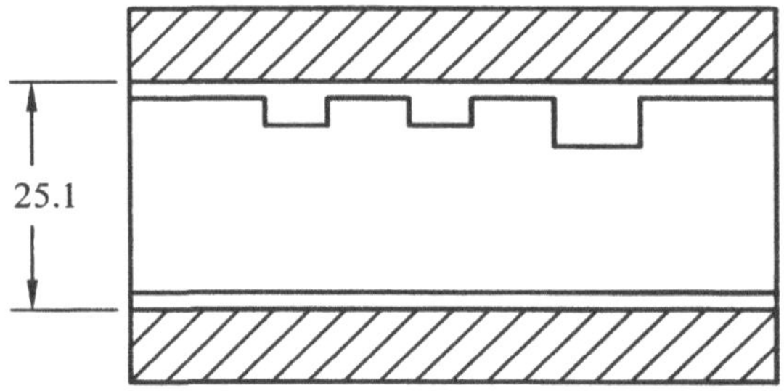

If the entire Continuous Feature fits into the GO GAGE it proves that the MMC has not been violated. This means that the size tolerance not only controls the form of the part features, but it also controls their coplanarity to each other. This treats these features as though they are one continuous feature without interruptions. LMC is still inspected at cross-sections.

When a surface is entirely interrupted by, for example, grooves (as shown in FIGURE 116), it becomes more than one surface. Once it is classified as more than one surface/feature, simple form or orientation controls often become ineffective. These controls, such as flatness and parallelism, are unable to create one tolerance zone within which multiple surfaces must reside. One way to apply these controls indirectly is to build one continuous MMC envelope for multiple interrupted features. To do this we can use the Continuous Feature of Size symbol. If used and associated with size dimensions the Continuous Feature symbol allows one MMC envelope of perfect form to contain all features of that size. This would apply to multiple features that share the same widths, but are interrupted. It would control these (width) features of size within one MMC envelope for; flatness, straightness and parallelism. The size tolerance would become the form tolerance. In the case of multiple, interrupted cylindrical features of the same diameter and sharing the same axis; roundness, straightness, taper, coaxiality and straightness of the axis would be controlled as though the multiple features were one continuous diameter. This concept first introduced in the ASME Y14.5-2009 standard on Dimensioning and Tolerancing may be used in place of geometric characteristic symbols that would normally terminate control when a surface is interrupted.

In FIGURE 116 this symbol is used to apply Rule #1, which allows size tolerance to control surface form (in this case flatness, straightness and parallelism) to interrupted features of size. As shown, the symbol for continuous feature is used next to the size dimension.

In the case of coplanarity, profile of a surface, when used in this type of situation, has always been capable of creating one tolerance zone within which more than one surface must reside. See FIGURE 117 for an example of profile of a surface to control coplanarity.

However, per ASME Y14.5-2018 standard, the coplanarity control can be invoked using the Continuous Feature (CF) symbol, much as ISO uses the Common Zone (CZ) symbol. It states: The symbolic means for identifying **two or more interrupted features or** interrupted regular features of size as a single feature or feature of size is shown. The continuous feature symbol shall either be applied to a size dimension of an interrupted regular feature of size, directed to a geometric tolerance for an interrupted surface, or to a datum feature symbol applied to interrupted features. When using the continuous feature symbol, the extension lines between the features may be shown or omitted; however, extension lines by themselves do not indicate a continuous feature. The number of surfaces that are included in the continuous feature such as 2X or 2 SURFACES may be added beside the CF symbol.

FIGURE 117 [Continuous Feature used for Coplanarity]

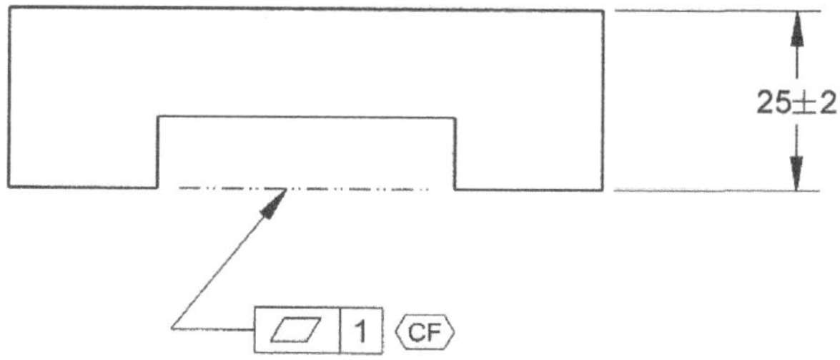

Means This:

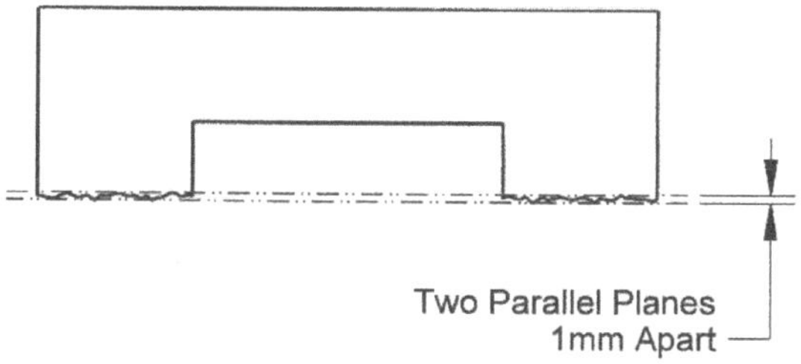

Figure 117 shows the Continuous Feature symbol being used to control coplanarity combined with a flatness geometric characteristic. This is new per ASME Y14.5-2018.

FIGURE 118 [Coplanarity Using Profile of a Surface]

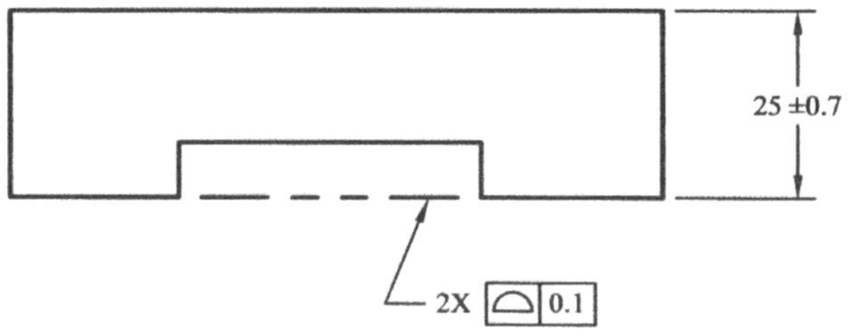

MEANS THIS:

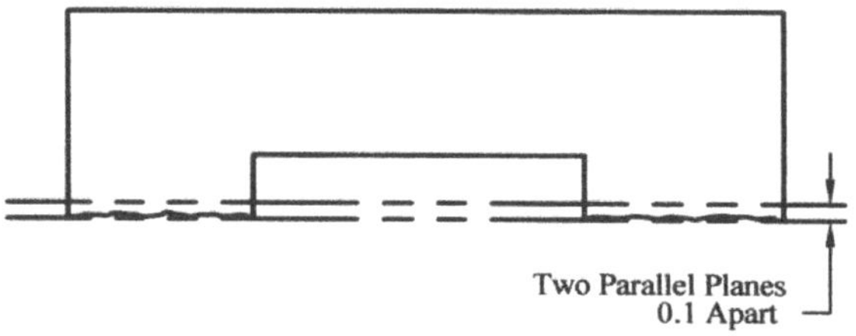

As shown FIGURE 118, the tolerance zone of profile consists of two parallel planes 0.1 apart within which both surfaces must reside to conform. If in conformance, both surfaces will not only be straight, flat and parallel to one another to within 0.1, but they will also be coplanar to within 0.1. If the tolerance on the 25 millimeter dimension was as tight as the profile tolerance and the Continuous Feature of Size symbol was used, the profile tolerance could be eliminated. The size tolerance would, if that tight (a total of 0.1), control all of the same factors that the profile tolerance currently controls.

That is not the case in FIGURE 118. The size tolerance is a total of 1.4 (plus or minus 0.7). If not for the profile control, this would allow one of the surfaces to be offset from the other by 1.4. The profile of the surface control allows the part to uniformly grow or shrink as long as it does not exceed its size limits and keeps the two surfaces under control (flat and straight, while also parallel and coplanar to each other) to within 0.1.

More control is possible. For example, if a datum feature was added as follows:

FIGURE 119 [Coplanarity and a Parallelism Datum Reference]

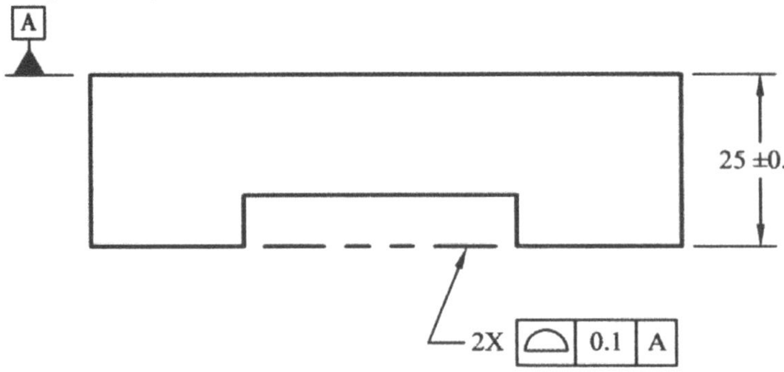

The controlled surfaces are now flat, straight, coplanar and parallel, not only to one another, but also parallel to datum plane A.

Even more control is possible. For example, in FIGURE 120, a basic dimension was added to replace the toleranced height.

FIGURE 120 [Profile of a Surface Controls Coplanarity, Parallelism and Location]

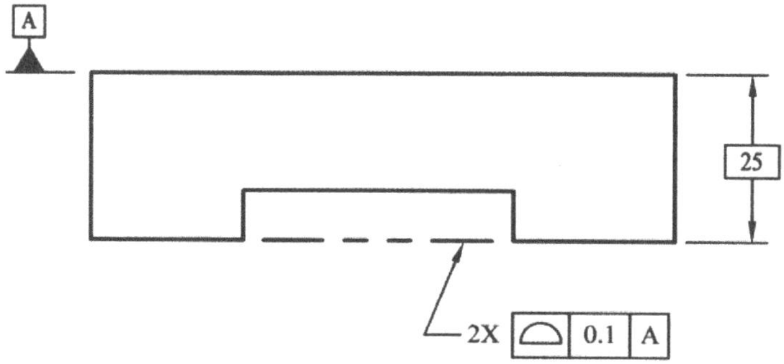

The controlled surfaces are now flat, straight, coplanar and parallel to one another and also parallel to datum plane A and located to datum plane A. The tolerance zone is located equally distributed about the basic dimension of 25mm located from datum plane A. Both surfaces may be no farther than 25.05 from datum plane A and no closer than 24.95.

An equivalent control could use a dimension origin symbol instead of a profile control and a datum feature. For example:

FIGURE 121 [Dimension Origin Symbol]

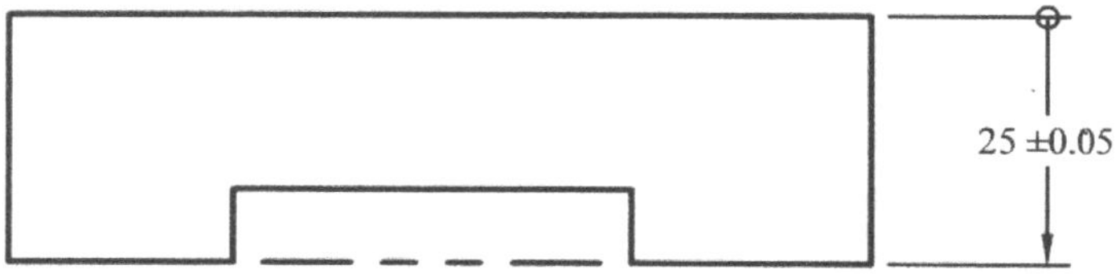

This is a simple looking control, and yet, it does what the basic dimension of 25mm with a profile control and a datum feature would do. The tolerance zone for both is as follows:

FIGURE 122 [Figures 120 and 121 Mean This]

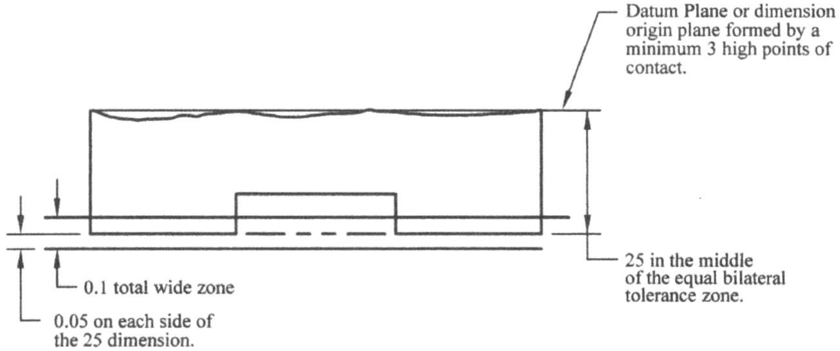

FIGURE 123 [Coplanarity]

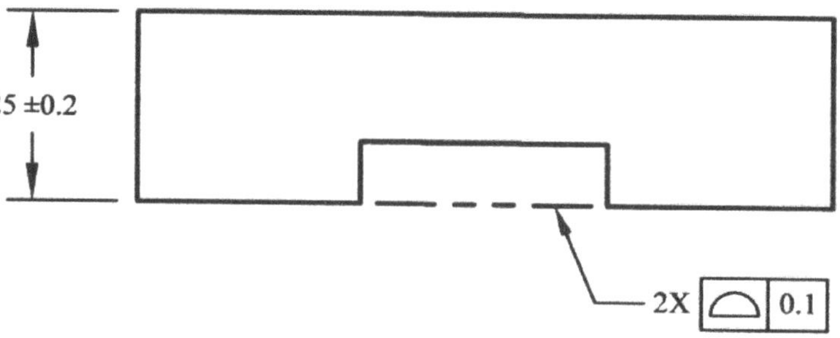

FIGURE 123 above shows an example wherein the size tolerance is larger than the required coplanarity of the two surfaces on the bottom. Therefore, an augmenting profile of the surface control has been added to give the required coplanarity control of 0.1. It forms a tolerance zone consisting of two parallel planes 0.1 apart within which all elements of both surfaces must reside. The control on Figure 35 allows the part to taper from top to bottom by the entire size tolerance (25.2 to 24.8), while keeping the two lower surfaces coplanar by 0.1. For example, the part produced could look like this:

FIGURE 124 [Figure 123 Means This]

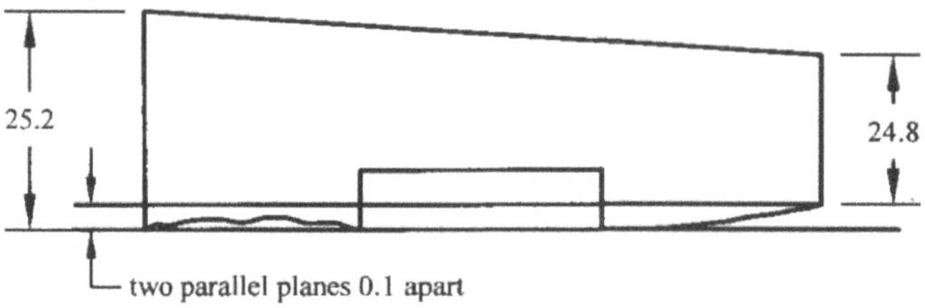

FIGURE 125 [Coplanarity and Parallelism]

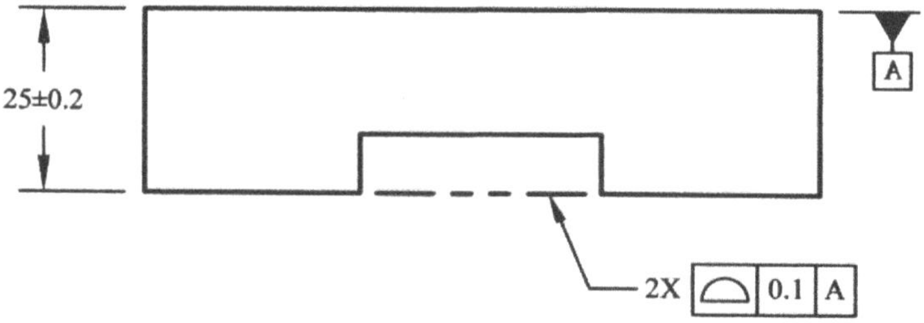

FIGURE 125 above adds an additional requirement of parallelism of the tolerance zone to datum plane A. The inclusion of datum feature A into the profile of the surface control, while retaining the liberal size tolerance of 25 plus and minus 0.2, makes certain both controlled coplanar surfaces on the bottom are not only flat, straight, parallel (to one another) and coplanar (to one another) to within 0.1 as they were in FIGURE 123, but also parallel to datum plane A to within 0.1.

The control on FIGURE 125 allows the part to grow or shrink within the size limits of 25.2 to 24.8. Although one part could be produced at 25.2 and another part at 24.8, these sizes may not be spanned on a single part. The datum plane referenced in the coplanarity control requires the tolerance zone of two parallel planes 0.1 apart to be parallel to datum plane A. For example, FIGURE 126 shows what the produced part could look like.

FIGURE 126 this: .....or this:

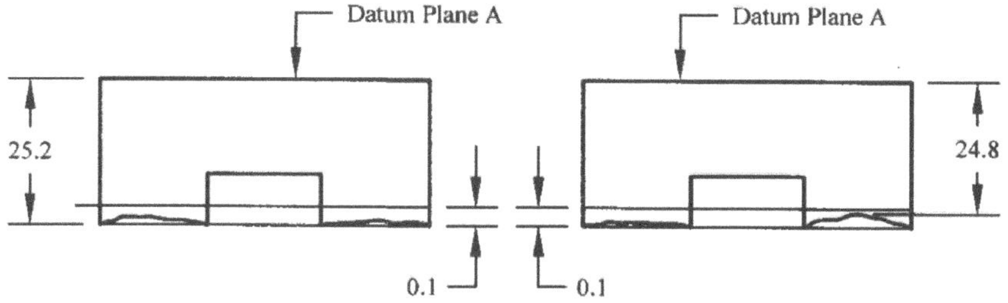

But, the 25.2 and the 24.8 dimensions may not co-exist on the same part and be in compliance with the specifications.

FIGURE 127

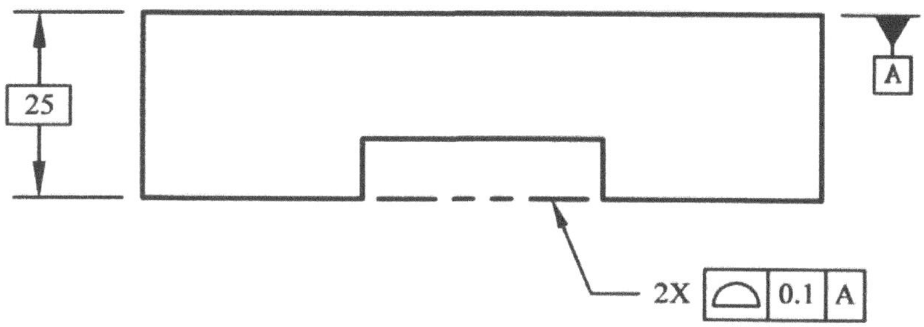

FIGURE 128 [Means This in its Effect on Size]

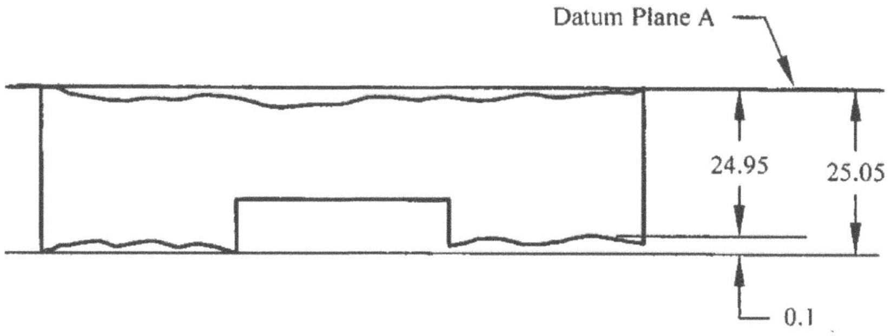

The two surfaces on the bottom must be coplanar to within 0.1.

FIGURE 127 above changes the size dimension to a basic dimension 25. This, in combination with the inclusion of datum feature A in the profile of the surface control, accomplishes all that FIGURE 125 does and, additionally, ensures that the location of the two controlled surfaces is 25

plus and minus 0.05 from datum plane A. Note that datum feature A desperately needs a flatness control. It is currently completely uncontrolled.
The following FIGURES 129, 130 and 131 show a variety of ways to control the location and coplanarity of planar surfaces to each other.

FIGURE 129

Assembly

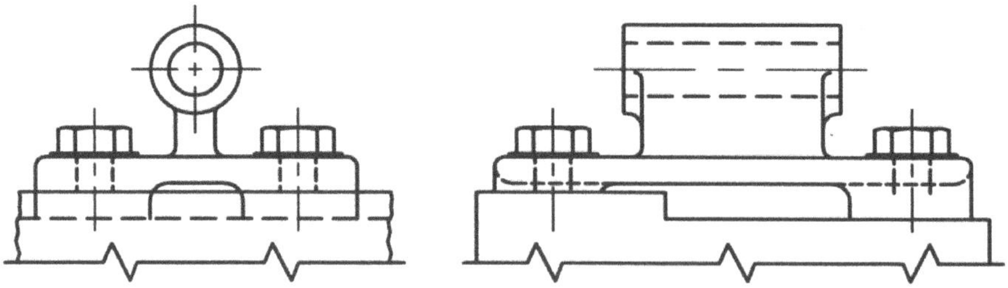

Detail drawing callout

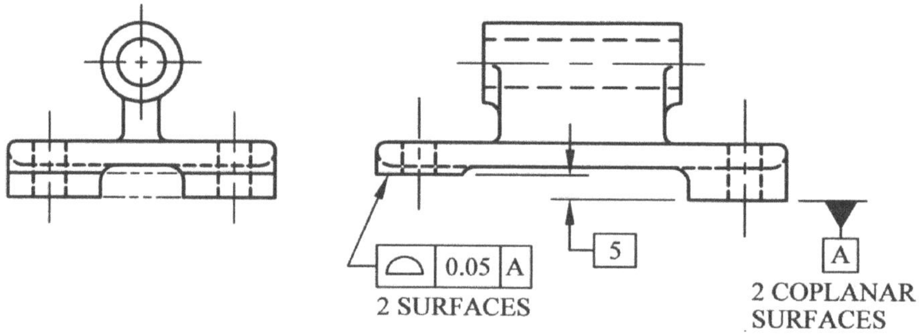

Means this:

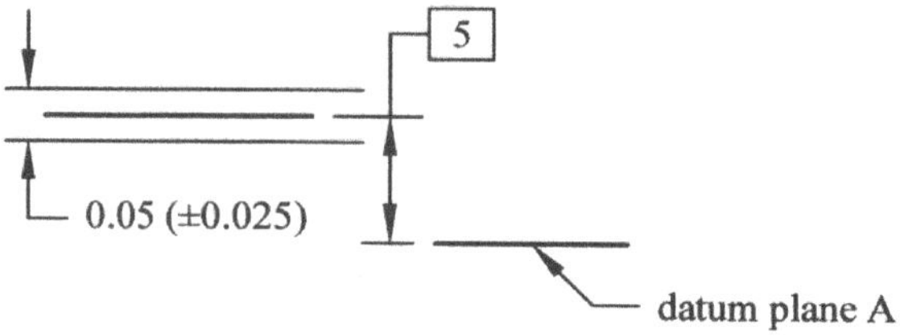

FIGURE 130 [Detail Drawing Callout]

Detail drawing callout

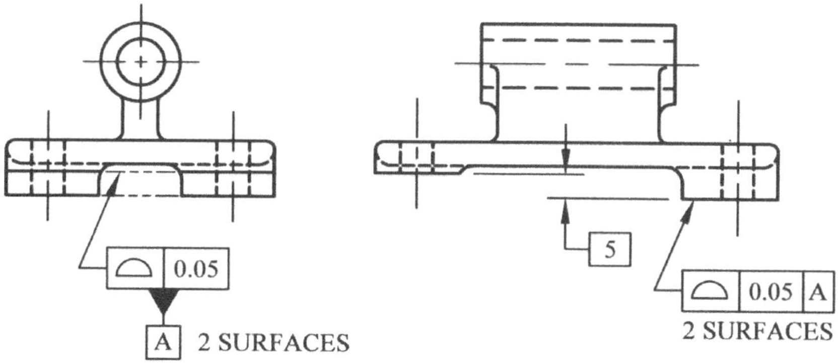

Means this:

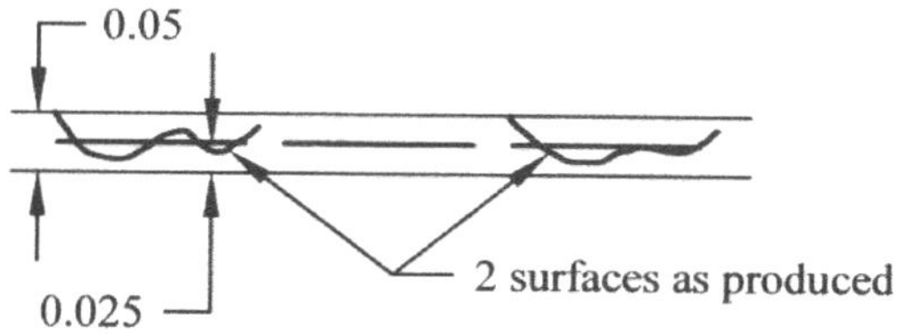

Then this:

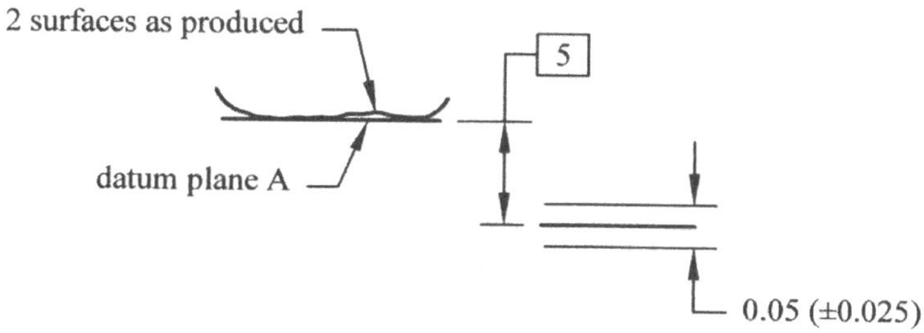

FIGURE 131 [Detail Drawing Callout]

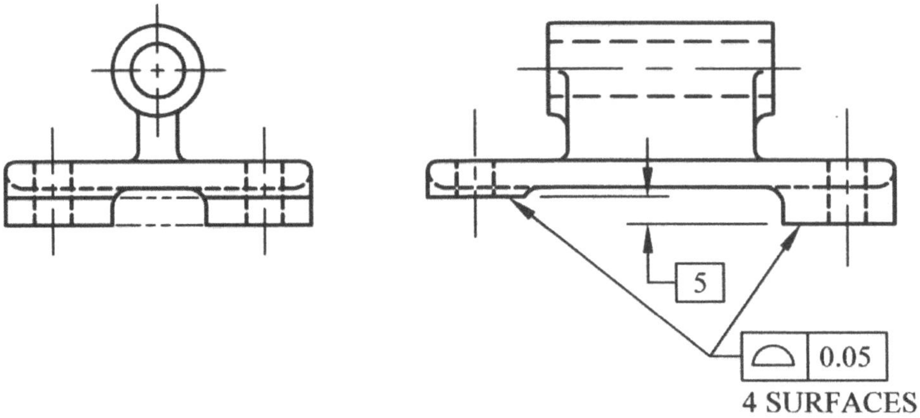

Means this:

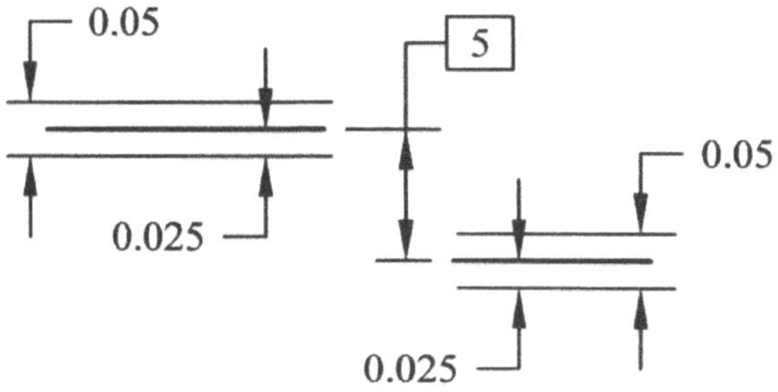

Conicity

FIGURE 132 [Conicity - Profile Tolerance for a Conical Feature • Individual Feature]
OPTION 1

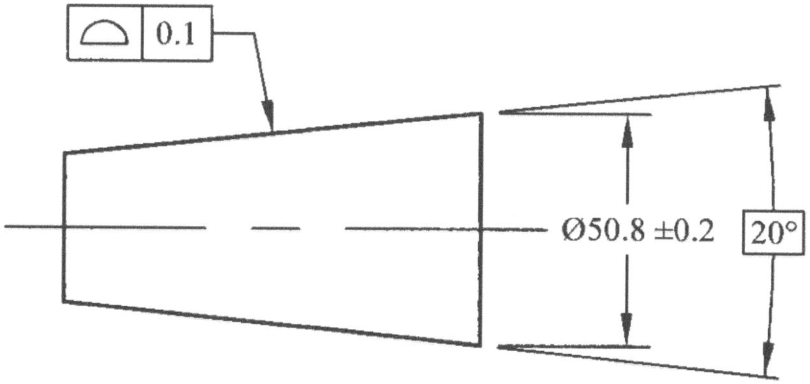

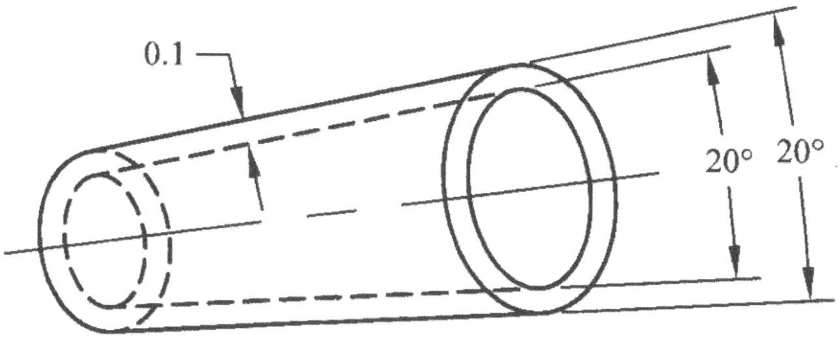

In Figure 132 Option 1, conicity is a 3-dimensional surface control for cones. The surface of the cone must lie within a tolerance zone consisting of two concentric cones that are, in this case, at a 20 degree basic angle and 0.1 apart radially. Size tolerance is separately controlled by the 50.8 plus and minus 0.2.

FIGURE 133, Option 2, allows the profile of a surface conicity tolerance to control not only the shape of the conical surface, but also its size. By making the 50.8 dimension basic, its tolerance is taken from the profile tolerance of 0.1 radially. On the diameter at that end, the size limits are approximately 50.7 to 50.9. Conicity is held to within the 0.1 tolerance zone.

FIGURE 133 [Conicity - Profile Tolerance for a Conical Feature • Individual Feature]

Option 2

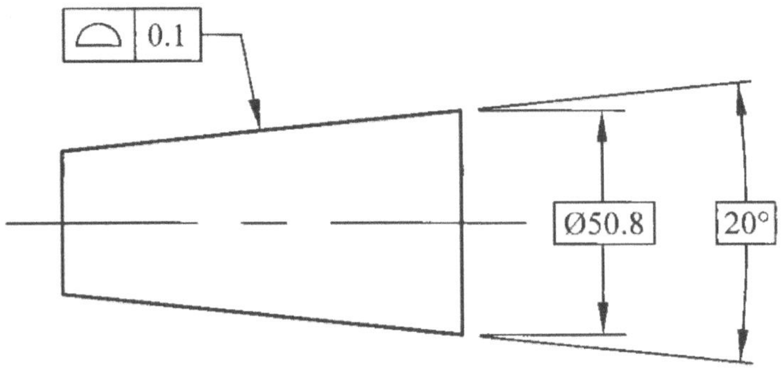

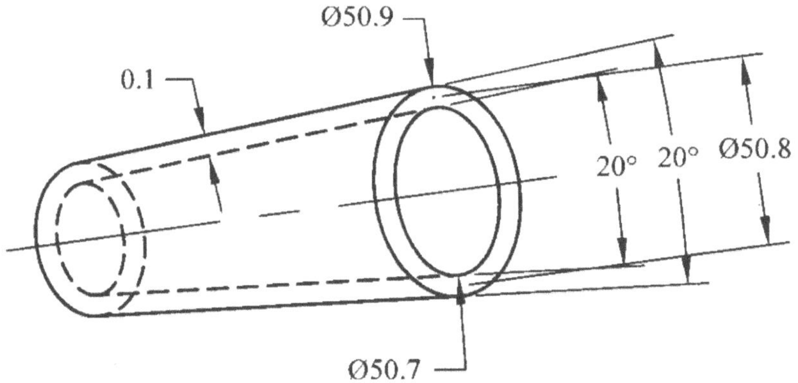

Note: The purpose of this illustration is to show that profile of a surface can be used to control not only the conical shape of this part but also its size. This is done by making the size dimension (50.8) a basic dimension. **But since measurements for profile are taken normal to the desired geometry of the part, the resulting size dimensions shown of 50.7 to 50.9 are only approximate. This slight discrepancy is known as cosine error.**

FIGURE 134 [Conicity - Profile Tolerance for a Conical Feature • Related Feature]
Option 1

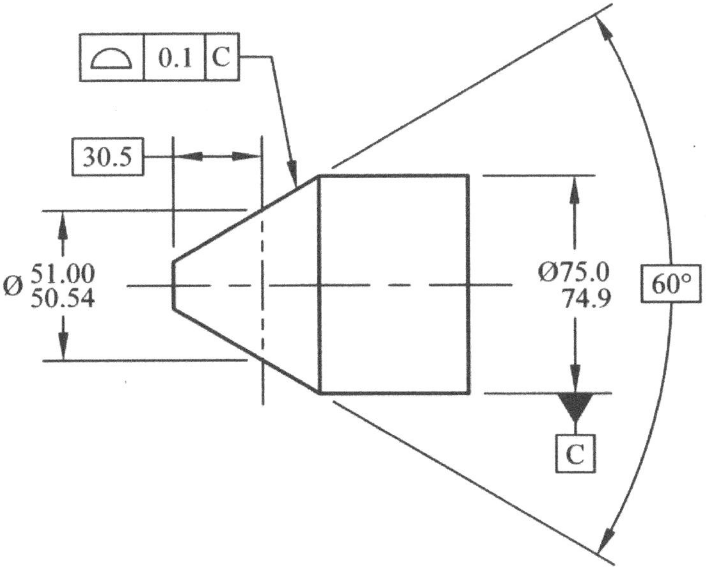

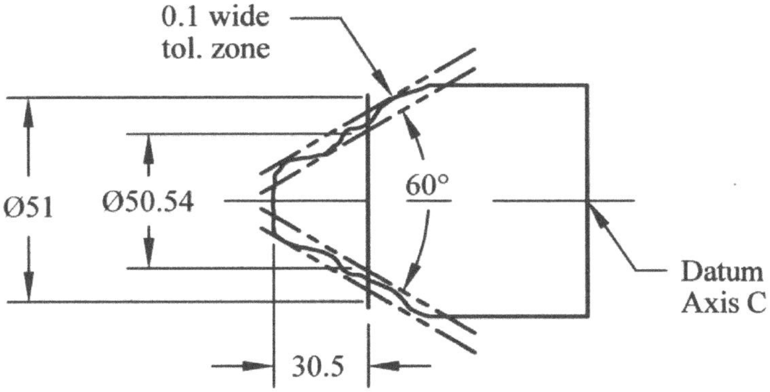

Datum feature C is referenced in this conicity control to make the cone concentric to datum axis C. The conical surface may be off center of datum axis C no more than a radius of 0.05 and still be in compliance.

FIGURE 135 [Conicity - Profile Tolerance for a Conical Feature • Related Feature]
Option 2

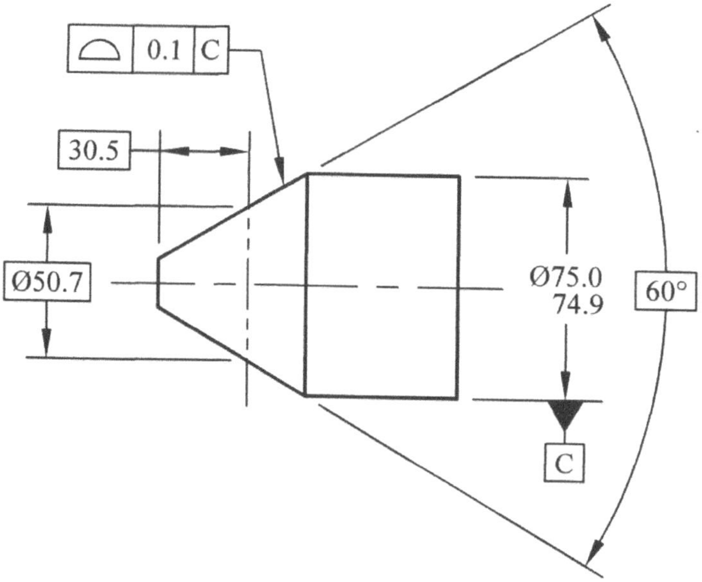

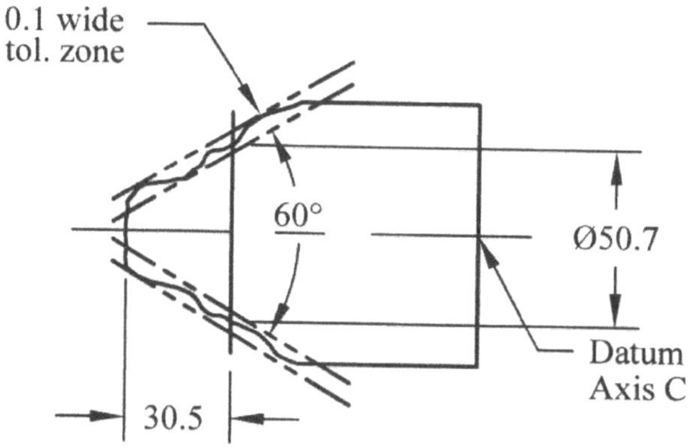

By making the 50.7 dimension a basic dimension, the size of the cone is toleranced by the profile control.

FIGURE 136 [Conicity - Profile Tolerance for a Conical Feature • Related Datums • Releasing Degrees of Freedom]

The following illustration shows a conical datum feature. In the position feature control frame, the spatial degrees of freedom that each datum reference constrains are listed in brackets after the datum reference. Datum feature A generates an axis (through the center of the cone) and a plane (at the apex of the cone) that would normally control the special degrees of freedom for the x, y and z linear movements, as well as the u and v rotations. But, in this feature control frame, since z does not appear within the brackets after datum reference A, the z location is no longer controlled from the apex of the cone. Therefore, datum B has been added to the feature control frame to control the distance in the z direction. Since w (the rotational degree of freedom around the Z axis) is unnecessary to stem, it is not included in any brackets.

Notice that when the degrees of freedom that each datum constrains are to be spelled out in the feature control frame, the datum axes are shown on the part representation. When this is done, datum axes X, Y and Z are shown using capital letters and the degrees of freedom in the feature control frame are shown using lower case letters. Degrees of freedom letters are only allowed to appear in the feature control frame and within the brackets, while the datum axes appear outside the feature control frame on the illustration of the part.

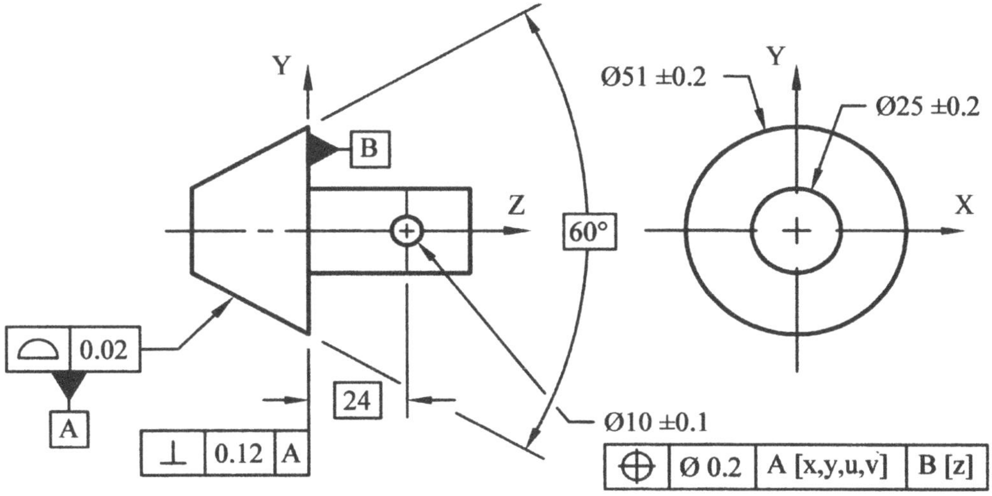

SYMMETRICAL PARTS

Even though the Symmetry geometric characteristic symbol was discarded in the Y14.5-2018 standard, there is an old symbol that has been used for a long time in the drawing standards to indicate that only half of a part was displayed and that the half not shown is exactly the same. This is not a tolerance. It is only a way to show the desired part is meant to be identical the half that was pictured. Here is a part that uses the symbol to convey that the part is intended to be symmetrical to the half that is shown:

FIGURE 137 [Symmetrical Parts]

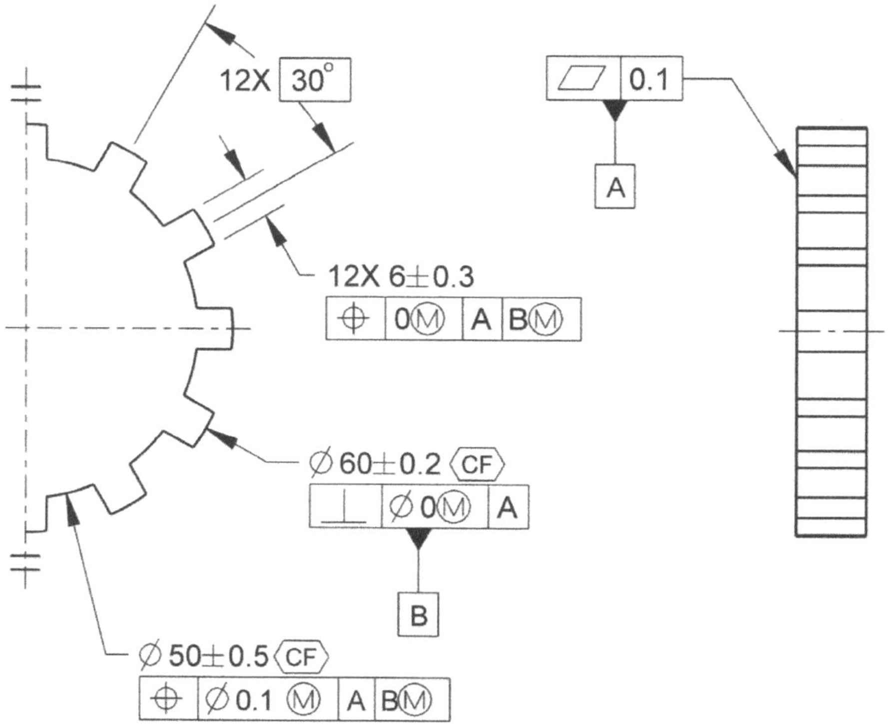

Even though the tabs on this part have been positioned, this is an illustration of a part that very much resembles a gear. Complex, newly designed, critical gears may need to specify all facets of the gear reside within a surface controlling tolerance zone. This chapter in the book is about Profile. To achieve the critical configuration of such a part, Profile of the Surface may be the best control to use. It would control the size, shape, angles and locations in the view shown containing the "Symmetrical Part" symbol, and replace all of the other geometric characteristics in that view. However, this approach, **if used as an all-around control**, would require one profile tolerance for all features.

IMPLYING A MANUFACTURING SEQUENCE ON COMPLEX PART CONFIGURATIONS

Using Compound Pattern Datums and Profile

The part in FIG. 138 mates with a similarly amoeba-shaped cavity that contains 7 pins/shafts. These shafts insert into the holes on this part while the cavity entirely encompasses the outer periphery of the external amoeba.

The manufacturing facility has requested we create a series of feature control frames that suggest a manufacturing sequence. The part will be punched out of a blank rectangular piece of sheet metal in three separate presses. Each press will either punch out one of the hole patterns or the outside periphery's profile. The manufacturing sequence will be suggested by the datums, or lack thereof, in each feature control frame. If a control uses datums, then that implies the datums have been created/punched out in a press prior to the feature or features controlled to that datum or datums.

The 4-hole pattern is labeled datum D. The holes are positioned and use datum A only. It is read, "The axes of these Ø.250 - .260 holes are positioned to one another and must maintain their basic hole-to-hole location dimensions and their perpendicularity to datum plane A perfectly if produced at maximum material condition (a Ø.250)." It also implies that one should not produce the holes at their Ø.250 size. If produced at that size, the required perfection of hole-to-hole location and perpendicularity to datum plane A would be impossible with which to comply.

If, instead, the holes are produced at a larger size (within their limits of size), an additional tolerance of position is gained. A maximum of Ø.010 of positional tolerance is available to each hole if the hole is produced at Ø.260 (least material condition). Since both the 3-hole pattern's positional control and the profile control use datum D, the 4 holes (datum pattern D) must precede them. The 4-hole pattern is punched out (created) in the first press.

Four real holes are manufactured. These holes are imperfect but, hopefully, good enough to meet their size requirements and not violate the perfect imaginary boundaries of virtual condition. These virtual condition boundaries consist of four cylinders (a diameter of .250 of an inch) with their axes located exactly as dictated by the basic hole-to-hole dimensions displayed on the drawing and perfectly perpendicular to datum plane A. These virtual condition boundaries have a central axis which is datum D; but, for all practical purposes, it is the axes of the boundaries that serve as the origin of measurement for features controlled to datum pattern D. The concept of a central axis for datum D is more difficult to deal with in reality than the task of establishing datum pattern D for purposes of fixturing and measurement.

• DATUM REFERENCE SYSTEMS BUILT ON ONE ANOTHER
• LOCATION AND PROFILE USING PERMANENT DATUM REFERENCES

FIGURE 138

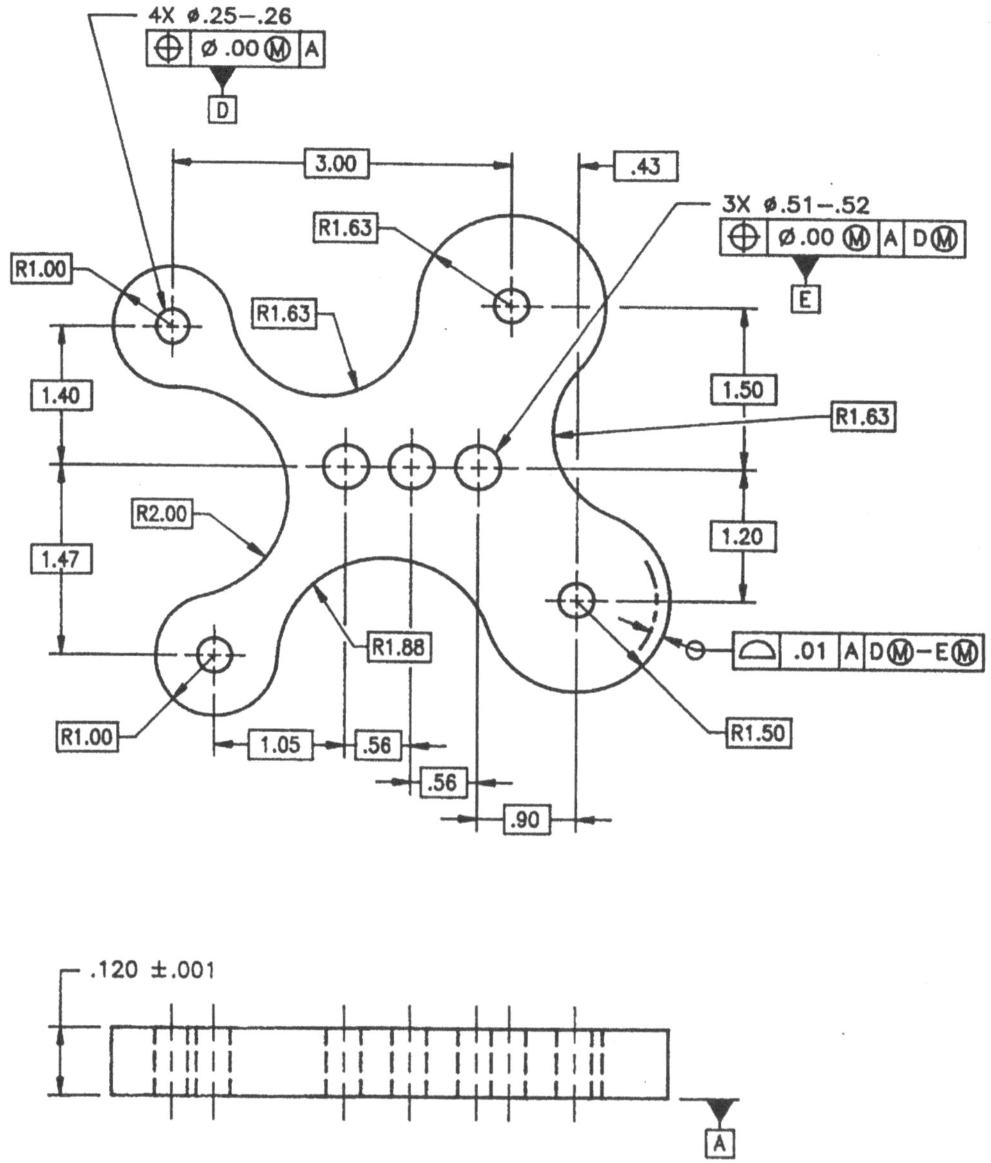

For example, if we merely visualize the second pressing operation, in the bed of the press exist four pins .250 in diameter, perpendicular to the bed and separated by the basic dimensions shown on the product drawing. If I place the still-rectangular part into the press, slip the 4 holes punched in press one down over the 4 pins in press two and bottom out in the bottom of the press' bed, I have accomplished two things: 1) I have fixtured the part by establishing datum plane A and datum pattern D, and 2) I have inspected/ gaged the position of the 4 holes to one another and datum plane A.

The second pressing operation is implied to be based on datums that exist at that time--datums A and D. Since the 3-hole pattern is positioned to datum plane A and datum pattern D, the second press punches out the 3 holes Ø.510 - .520. They form datum pattern E. Again, datum pattern E generates a central axis dictated by the axes of the individual holes' virtual condition boundaries located at their true position from datums A and D. Just as before, the datum axis is difficult to pinpoint and measure from, but that task is not necessary. We must establish the pattern datum, if it is subsequently listed in a feature control frame, and gage the features controlled from the pattern datum. To do this, we don't have to find the centroid axis E. As a matter of practicality, once the pattern datum is established, we can measure/gage the features controlled to it by using the physical embodiment of datum pattern E--the fixture/gage pins. These pins are represented at their true position to datums A and D and made at Ø.510 (their virtual condition size).

There remains one press and one feature control frame. The outside of the part is profiled to datum plane A for line element perpendicularity and located from compound pattern datum D-E invoked at maximum material condition. Any deviations allowed in the basic size, form and location of the profile are stated in the final feature control frame. This part may only shrink from these basic dimensions and relationship as the profile tolerance is unilateral inside. The reason for this is that the mating cavity uses all the same basic dimensions and relationships to shafts (instead of holes) with the same virtual condition boundaries. It, however, is profiled unilateral outside (the cavity may only grow). This will give a mating minimum gap (worst case) of zero and a maximum gap (not including pattern shift) of the .010 on this part plus the allowed profile tolerance on the mating cavity. If one were to try to accomplish a similar mating design condition for irregular shapes such as these, without basic dimensions from specified datums and feature control frames of position and profile, it would be infinitely more difficult, if possible at all. Plus and minus tolerancing for all this geometric definition is anemic by comparison.

We still must complete the final pressing operation. To do this, in the bed of the last press we must establish the compound pattern datum D-E at maximum material/virtual condition. To do this, we will have seven pins--four at Ø.250 and three at Ø.510. These pins are perpendicular to datum plane A represented by the press bed and separated by the basic dimensions taken from the part drawing representing the true position of the holes being fixtured and simultaneously gaged. If the holes pass over the virtual condition pins and seat on the datum feature simulator for datum A, the part is fixtured for the final pressing operation to profile the part; and, the holes have been gaged in their relationship of perpendicularity to datum A and their position to one another.

Once the part is fixtured, the final press is activated and creates the profile of the periphery. The sequence of events is complete and the part finished. Three geometric definitions are toleranced in three feature control frames, sequentially built on one another and accomplished in three punch presses. The manufacturing operations are implied by this type of sequential feature definition. Of course, the manufacturing facility does not have to follow the implied sequence and can choose an entirely different procedure; but, certainly, one has been implied. The final part, no matter how it has been manufactured, must comply with the tolerances shown on the drawing.

In reality, another entirely different approach could have been used and given us a part that would conform to the same geometric definition. For example, see FIG. 139.

FIG. 139 implies a part produced in one press--not three. There are no controls that imply they are complied with first. No leaders, so to speak. All holes are positioned and the part outside profiled as a simultaneous requirement. The interrelationships are created by this requirement for implied simultaneous creation and required simultaneous inspection of all features defined by the three controls. They are all part of one pattern, interrelated to one another and to datum plane A.

In both of the drawings shown, the profile tolerance will not be allowed a growth. It is a surface control. No surface controls are allowed a bonus tolerance (a growth in the geometric tolerance zone) as the controlled feature changes size (grows or shrinks). The profile of the surface tolerance is defined as .010 and remains at that regardless of the feature profile produced size. However, the circled M that appears after the datum features D and E in Fig. 1 shows the inspectors that a pattern shift is allowed for the profile as datum features D and E depart from their maximum material condition. Also, the simple fixturing of both parts in the two illustrations with all holes controlled at MMC shows that each hole may drift from its true position as it grows.

FIGURE 139

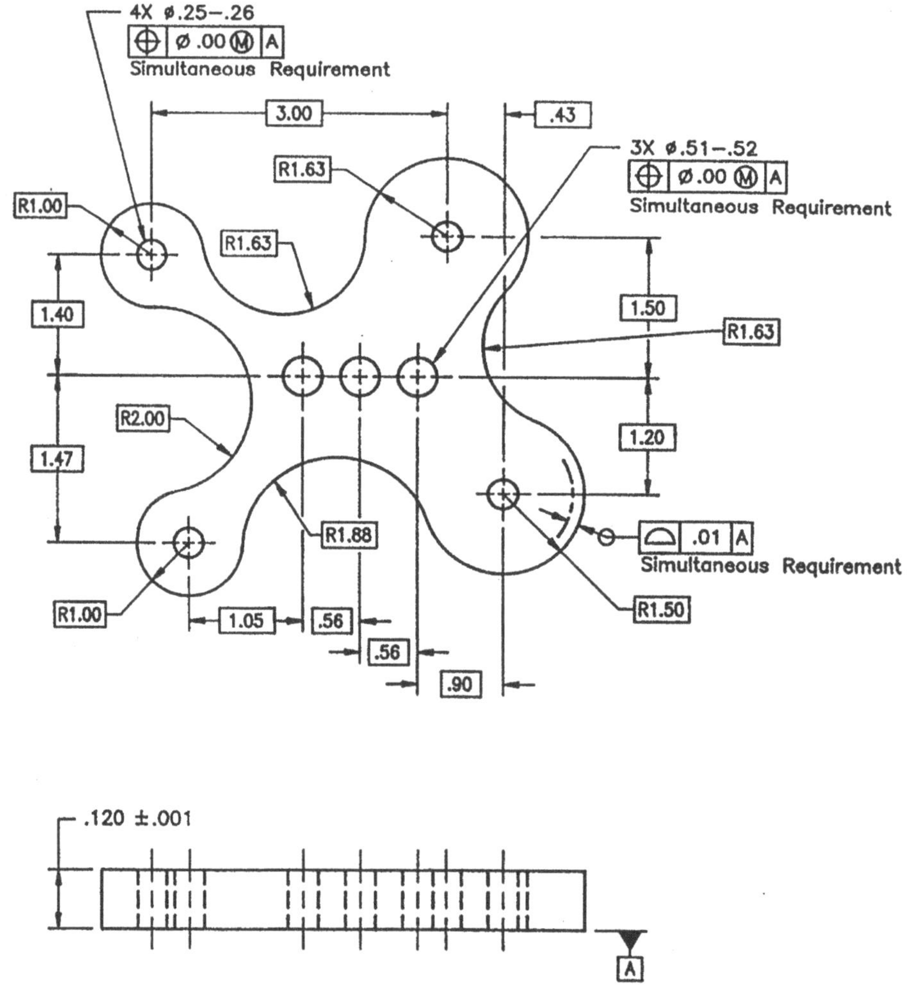

For the second illustration (FIG. 139), the allowed shift of profile is less blatantly called out but still logically deduced. The holes are positioned to one another at MMC. The axes of the virtual condition boundaries are the origin of measurement, but they are imaginary. They are located at true position--but to one another. If the holes grow, they may move away from true position by the amount of their growth.

Still, who is to say where the virtual condition axes reside. If the holes all grow and all move in the same direction, the virtual condition boundaries may be viewed as at that new location. If this is true, the result is that the profile, which is simultaneously measured from the axes of the virtual condition boundaries, is viewed as having shifted with the movement of the virtual condition boundaries. Therefore, the profile experiences the same advantage as the pattern shift allowed the profile in the first illustration (FIG. 138) when the datum features were controlled at MMC in that profile control.

The controls shown in both illustrations create the same tolerance zones with the same compliance requirements, while each implies a different manufacturing procedure. Inspection of either part may be dealt with to within various levels of satisfaction using an optical comparator with a comparator chart (overlay/template) made specifically for this part, or templates with fixtures and dial indicators, or fixtures and CMM probing with appropriate software. And to a certain degree, the mating functional requirements are capable of being inspected adequately with a receiver-type functional gage that consists of seven virtual condition sized pins located at their true position in a cavity that is the size and configuration shown by the basic dimensions on the external profile. If our part drops into the cavity, clears the pins (sized at 4X Ø.250 and 3X Ø.510) and datum feature A contacts the bottom of the cavity (with 3 high points of contact on datum feature simulator A), all controls have been inspected except for the internal boundary of the profile tolerance zone.

Chapter 8

Runout

•Circular Runout
•Total Runout
•Continuous Feature of Size

Chapter Objectives

Readers will learn:

1. Apply circular runout to diameters to achieve balance and two-dimensional form.
2. Measure circular runout on diameters with full indicator movement at circular cross-sections.
3. Apply total runout to diameters to achieve balance and three-dimensional form.
4. Measure total runout on diameters with full indicator movement traversing the entire surface.
5. Apply geometric tolerances on round parts in a sequential fashion with escalating power.
6. Apply, define and set-up compound datum features used for total runout.
7. Apply and measure circularity, cylindricity, circular runout and total runout on the same part.
8. Define a ball bearing assembly.
9. Apply and measure perpendicularity and total runout of a planar surface to a datum axis.
10. To utilize the continuous feature of size symbol for control of total runout.
11. To apply runout tolerances to assemblies.

Runout

There are two types of Runout. One is circular runout. The other is total runout.

Circular Runout

Circular runout is a two-dimensional surface control. It can be used on a planar surface to control the shape of the circular line elements of that surface and their relationship to a datum axis. That is not the way it is most often used, though. Most often it is used on diameters, to control the circularity (roundness) of circular line elements (pits, bumps, flats and ovals) and their coaxiality to a datum axis. Through its control of circular elements on the surface it controls the axis of the diameter. The axis is controlled for straightness and how far off center of the datum axis it may be. Separate from any runout controls that are inspected, the size limits of the part must be inspected and verified as being in compliance with the design specifications.

The control shown in FIGURE 140 depicts a geometric tolerance of 0.2. This means that if the circular line elements are inspected while rotating around the datum axis established by datum feature A, the full indicator movement (FIM) may not be greater than 0.2. The datum axis is established by the minimum circumscribed cylinder about datum feature A.

Once the axis of datum feature A is established, perhaps as shown, with a collet-type chucking device, a dial indicator can be placed on one of the circular cross-sections of the diameter being controlled by runout. The part is rotated about the datum axis 360 degrees and the FIM recorded. If the full indicator movement is not greater than 0.2, that circular cross-section has proven to be within the runout tolerance and another circular cross-section is chosen. The dial indicator is reset at zero while in contact with the circle to be tested. The part is rotated 360 degrees, the FIM recorded, and again compared to the runout tolerance of 0.2. If the FIM does not exceed what directly appears inside of the circular runout feature control frame, that circle is adjudged to be in compliance and another circular cross-section chosen to inspect. This is done over and over again, until the inspector is convinced that, had all circular cross-sections been inspected, all would have been within the tolerance. Of course, if any of the cross-sections has a FIM that is greater than the runout tolerance, the part is rejected as being out of tolerance.

FIGURE 140 [Circular Runout of a Cylindrical Surface to a Datum Axis]

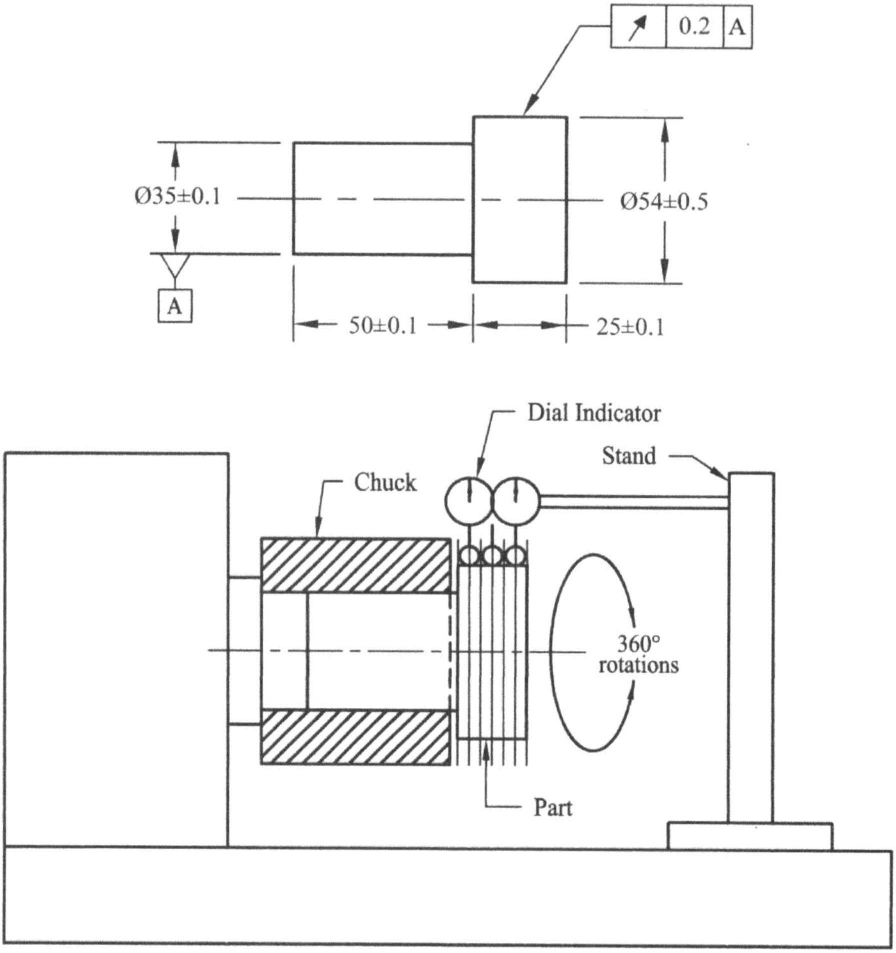

The dial indicator remains stationary for every 360° revolution. The surface has an infinite number of circles, so the inspector inspects enough of them to convince the inspector that the part is either within tolerance or is not within tolerance. Each circle receives a new zero setting on the dial indicator. The surface is inspected as though it was a stack of coins each sharing the same center. The full indicator movement of each may not exceed 0.2 millimeters for each circle under test to comply with this circular runout geometric tolerance. This controls the circularity (2D form) and concentricity (centering for balance).

If, as shown, the circular runout tolerance is 0.2, it prevents any pits in the cylindrical surface from being deeper than 0.2, any bumps from being taller than 0.2, any flat from being deeper than 0.2, any oval from measuring bigger in one direction than another by a diameter of 0.4, the center points of the circles from being off center of the datum axis by more than a radius of 0.1 and the entire axis of the controlled diameter from bowing by more than 0.2. Circular runout does not control taper at all. This job is relegated to the diameter's size tolerance. When used on diameters, think of circular runout as having the same shaped tolerance zone as circularity (roundness), two concentric circles that are the geometric tolerance apart. Each circle on the diameter has its own tolerance zone, which is why the dial indicator is re-zeroed each time a measurement is taken.

If used on a cylindrical surface, one must be cautious to notice the diameter's size tolerance. If the size tolerance is tighter on its control of the entire diameter than the circular runout tolerance is, then the size tolerance controls the surface's form and all the runout tolerance controls is concentricity (how far off center of the datum axis the controlled feature's axis my stray). This is true for circular and total runout controls.

The difference between circularity and circular runout is that circularity has no datum axis it can reference. So, if you think of circularity as a stack of coins, each coin with its own tolerance zone, the coins could be stacked with each coin off center of the next, because each tolerance zone could have a new center. This is not the case with circular runout. Since circular runout must reference a datum axis, each tolerance zone for every coin in the stack must be centered on the same datum axis. That does not prevent the diameter from looking like an hour glass, a barrel or a nose cone as long as the size limits are not violated.

If circular runout of 0.2 had been used on a planar surface, instead of a diameter, it would have controlled the circular line elements of that surface from having a full indicator movement of greater than 0.2. Each circular line element of the planar surface would be measured individually with each receiving a new zero on the dial indicator. This prevents pits and bumps of each line element from being greater than the circular runout tolerance and also prevents more material from being on one side of the datum axis than the other side by more than 0.2. This is a kind of balance control that limits the wobble of the surface, which is why the part is ideally inspected while rotating. However, it must be remembered that flatness of the planar surface is not controlled by circular runout. The surface could look like a dish or sphere and still pass the inspection with each circular line element not exceeding its FIM. This is because each circular line is inspected with a new zero.

Total Runout

Total runout is a three-dimensional surface control.

FIGURE 141 [Total Runout of a Cylindrical Surface to a Datum Axis]

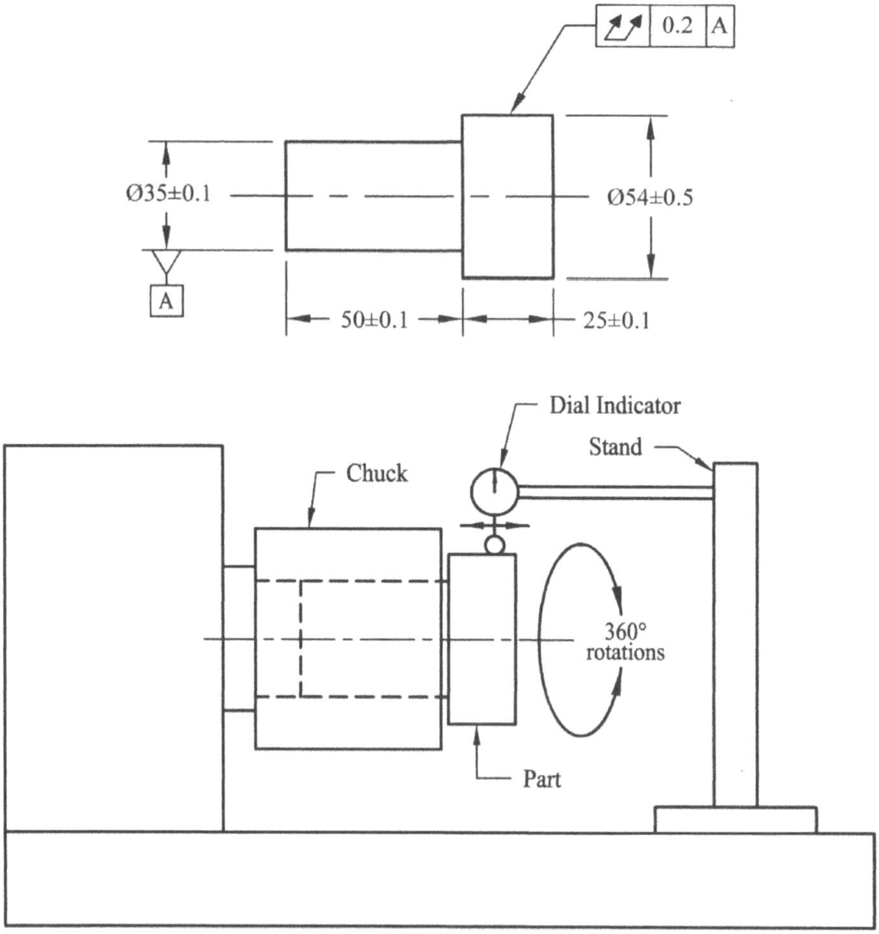

The dial indicator traverses the surface in the longitudinal direction while the chuck rotates in continuous 360° revolutions. The full indicator movement may not exceed 0.2 millimeters for the surface to comply with this total runout geometric tolerance. This controls the cylindricity (3D form) and concentricity (centering for balance).

When inspecting a diameter, such as the one shown in the example above, it must separately be ascertained that the size of the diameter has not exceeded its tolerances. It must also be determined whether the size limits are tighter than the total runout tolerance (not the case in the illustration above). If that was the case, then the total runout tolerance only controls the concentricity of the diameter to the datum axis. In these situations, the size limits would control the three-dimensional form of the diameter. So the roundness, straightness and taper usually controlled by total runout would be controlled by the diameter's size tolerance, if the size tolerance was tighter than the total runout tolerance. If, however, the size tolerance on the diameter being controlled by total runout was looser than the total runout tolerance then the roundness, straightness and taper of the diameter would be controlled by the total runout tolerance.

Looking at FIGURE 141, one can see that the size tolerance on the 54-millimeter diameter is plus or minus 0.5 (a total of 1 millimeter). Since the total runout tolerance is less than 1 millimeter, it controls the three-dimensional form of the surface. So, with this control of total runout of 0.2, pits can be no deeper than 0.2, bumps can be no taller than 0.2, flats can be no deeper than 0.2, ovals can be no larger in one direction than another by a diameter of 0.4, the part can taper by no more than 0.2 per side (a diameter of 0.4), the axis can not bow by more than 0.2 total, the axis can't be off center by more than a radius of 0.1, and the parallelism of the axis may not be out more than a total of 0.2.

Measurement of the total runout control shown in FIGURE 141 illustrates that once the datum axis has been established as the minimum circumscribed cylinder axis of the datum feature A, the part is rotated about that datum axis. The dial indicator is brought into contact with the larger diameter's surface and zeroed out. The part rotates quickly, while the dial indicator is moved down the surface of the 54-millimeter diameter (longitudinally) slowly. This is to bring as many points on the surface into the measurement procedure as possible. In all those revolutions, while moving the dial indicator from end to end, the full indicator movement may not exceed the 0.2 total runout tolerance.

Both circular and total runout controls are surface controls and therefore not allowed to use a diameter sign in the feature control frame. A diameter sign is used only to denote a direct axis control, while runout controls the axis indirectly by controlling the surface. No material condition symbols are allowed inside of a runout feature control frame. A datum axis must be referenced in all runout controls. It need not be the only datum referenced, in that planar datums may accompany the datum axis referenced. If this is done, the datum axis is most commonly the secondary datum referenced. If it is primary, the datum axis would most likely overcome all geometric characteristics of any planar datum listed as secondary (for example to control the angle of measurement).

All runout controls are thought of primarily as balance type-controls and therefore implied to be measured while the part is rotating around the datum axis.

FIGURE 142 [Linear Product Definition and Establishing a Compound Datum Axis]

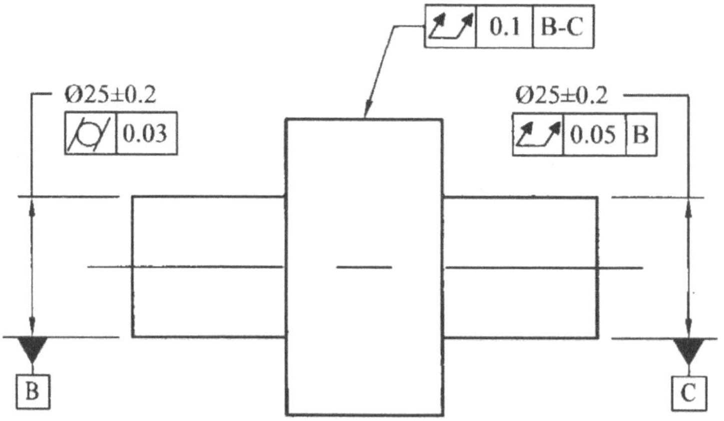

B-C means: Diameters B&C should be chucked simultaneously to establish the datum axis common to both. This technique is difficult to accomplish but better complies with the theory of compound datum establishment than putting the part between centers.

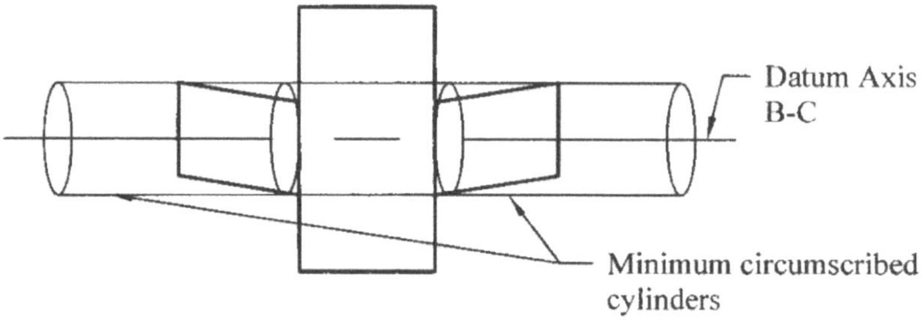

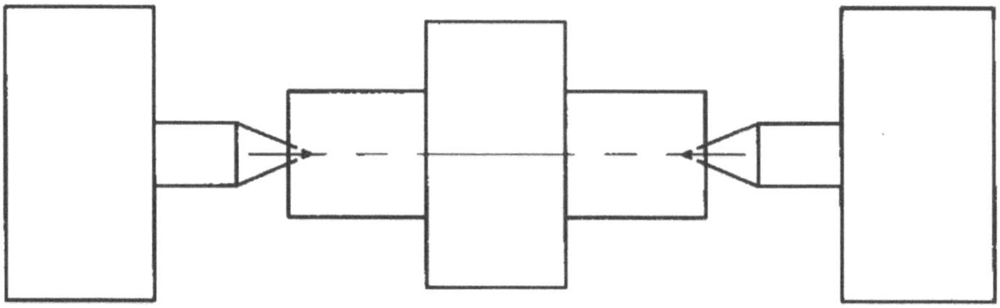

Centerdrilling the datum features and putting the part between centers would only comply with the specification for the establishment of compound datum B-C if the axis created is the same as the axis established by simultaneous chucking. Still it is often easier to accomplish, and would provide a rough simulation of the desired axis. Another option to simultaneously chucking up on datum feature B and C would be to put them into V-Blocks.

FIGURE 143 [Defining and Measuring Round Parts]

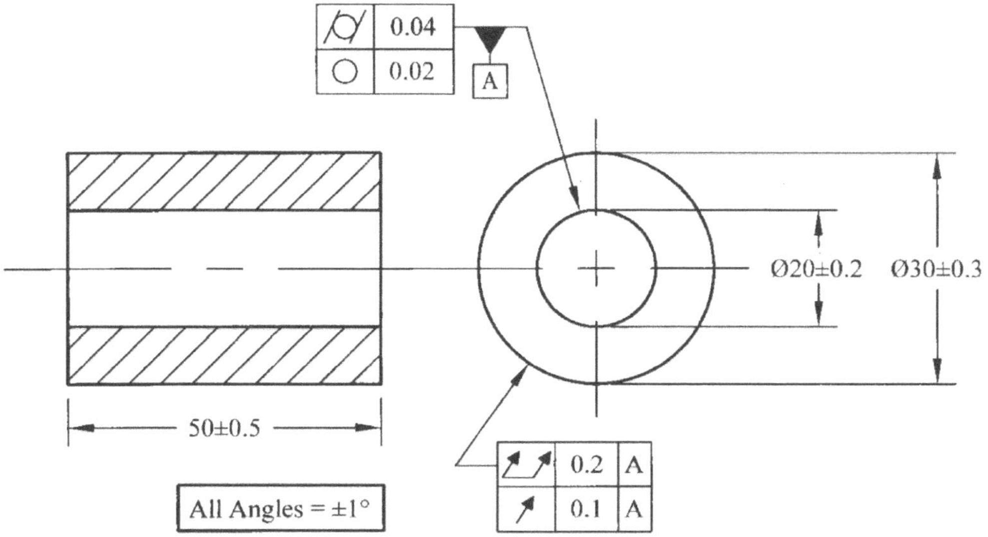

The setup shown below is to inspect the circularity and cylindricity controls.

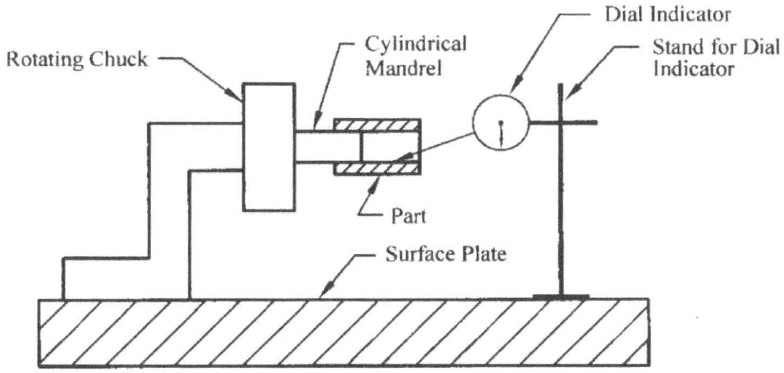

The set-up below is to inspect the runout controls.

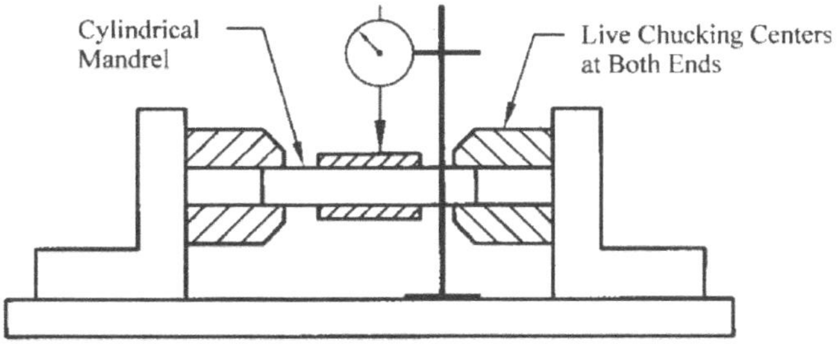

FIGURE 144 [Defining and Measuring Runout with Compound Datum Features]

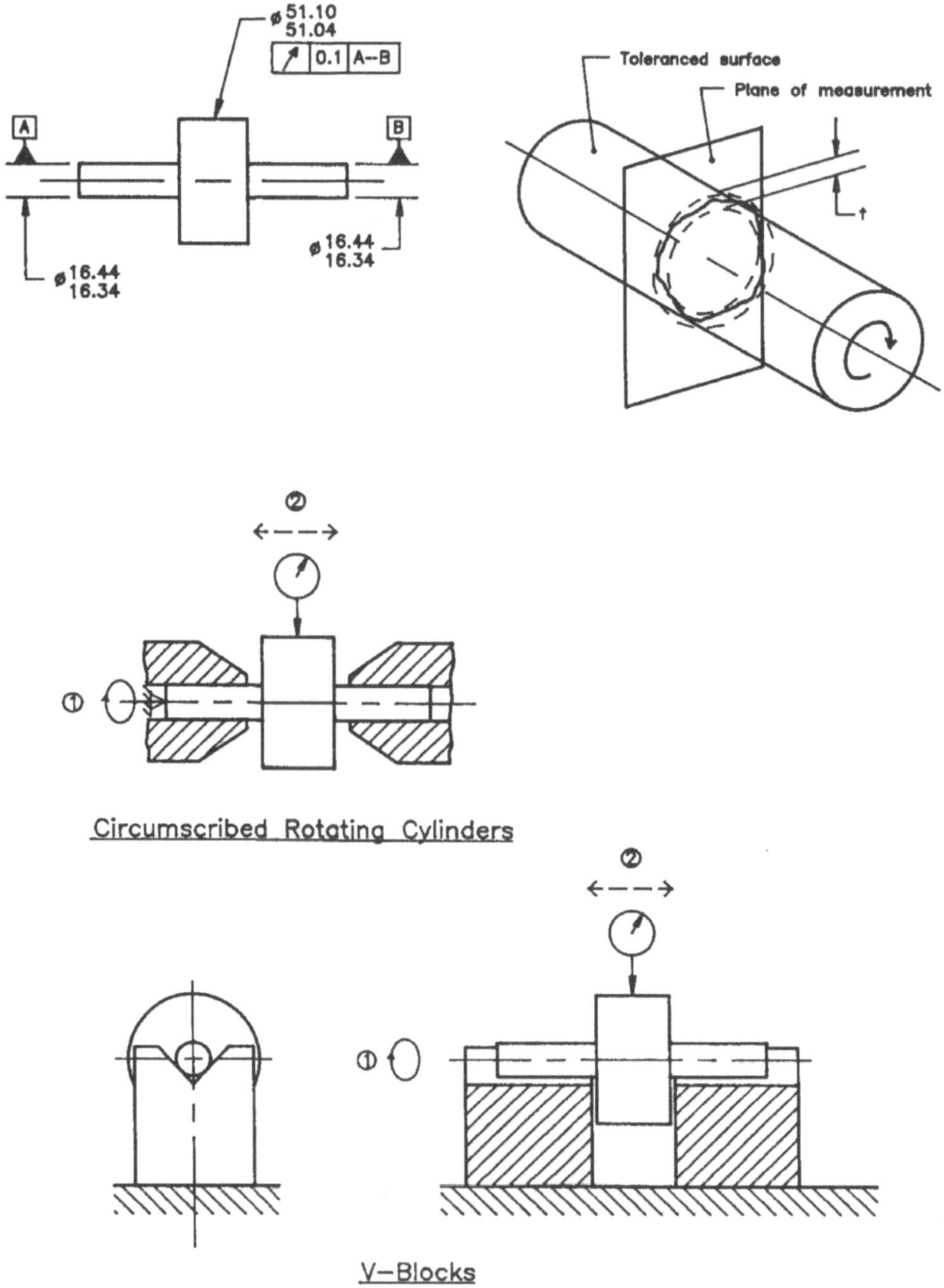

FIGURE 144 shows circular runout controlling the roundness and coaxiality of a surface to a datum axis. This datum axis is created by two coaxial diameters. An infinite number of circles comprising the constrained feature exist. Runout is measured at a sufficient number of these to convince the inspector that, had he inspected them all, the part is either in compliance with the runout requirement, or that it is not. For each circle, the tolerance zone consists of two concentric circles that are 0.1 apart radially and centered on the datum axis. The tolerance zone implies that no compliant circle may have a pit that is deeper than 0.1, no bump that is taller than 0.1 and no flat that is deeper than 0.1. However, the size tolerance on the diameter is 51.04 to 51.10, a difference of 0.06. Since this is a tighter control of form than the circular runout control, the form of the surface is controlled by the size tolerance. No surface variations from a perfect cylinder may be greater than a diameter of 0.06. The runout tolerance is relegated to controlling only how far off center the center of each circle may be. The axis of each circle may not be off center of the datum axis by more than 0.05 (half of the 0.1 runout tolerance).

Two ways are shown to establish the datum axis from the two coaxial datum features. Coaxial circumscribing rotating cylinders are preferred, but may be difficult to simulate. Vee-blocks may be used for a rough check if it is known that the datum features are sufficiently well formed and coaxial as to make the error they introduce negligible.

Another method used to measure runout controls to axes constructed by compound datum features is simply to centerdrill each diameter and put the part between bench centers. These centers will only give an accurate reading of a feature's runout if the axis they construct is essentially the same axis as would have been derived from the twin circumscribed cylinders method. If not, false readings may occur. Also, if the centers are not very coaxial to one another, bench or machine centers may wear the surfaces unevenly once inserted, in an effort to wear them into better coaxiality. This could cause a variety of problems, including a sloppy or loose fitting fixturing of the part. Problems such as these could lend to greater measurement error. See FIGURE 145.

FIGURE 145 [Problems Substituting Center Drilled Holes for Datum Feature Diameters]

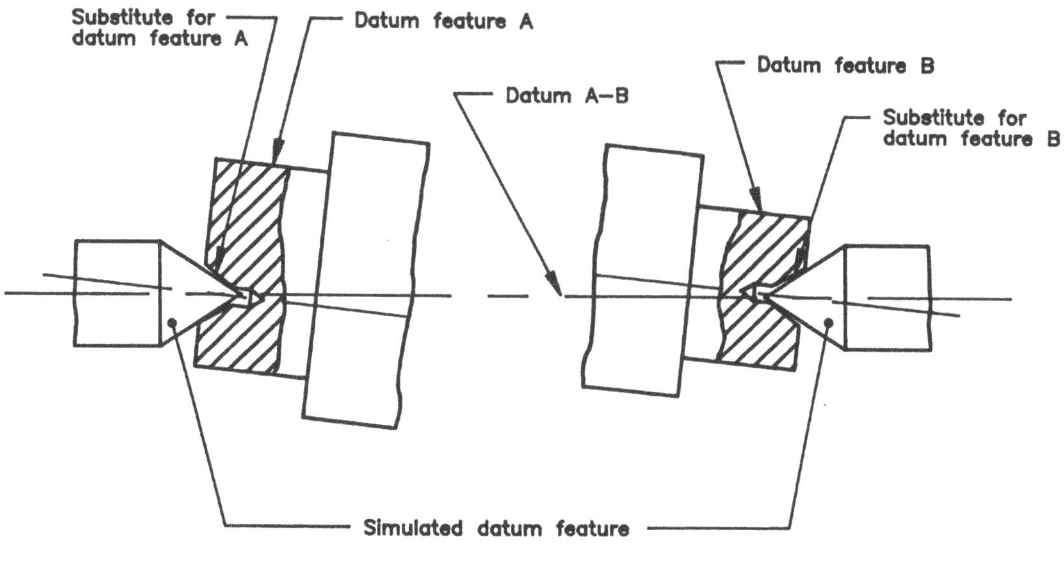

FIGURE 146 [Ball Bearing Ring Assembly]

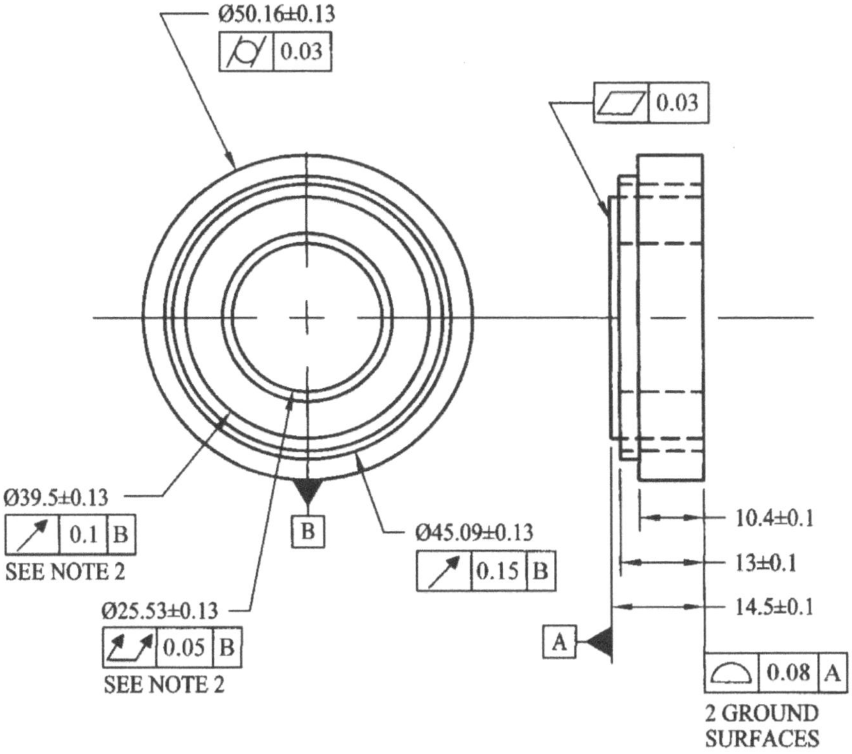

NOTES:

1. TO BE INTERPRETED PER ASME Y14.5M.
2. TO BE INSPECTED WHILE DATUM FEATURE B IS STATIONARY AND THIS CONTROLLED FEATURE ROTATES ON ITS BEARINGS.
3. UNLESS OTHERWISE SPECIFIED ALL ANGLES ARE ±1 DEGREE.
4. CAUTION: THIS PART IS FRAGILE AND MAY BE DAMAGED IF NOT HANDLED WITH CARE.
5. THIS DRAWING IS FOR THE PURPOSE OF INSPECTING THE ASSEMBLED BALL BEARING RING ONLY. DETAIL DRAWINGS OF EACH PART IN THE ASSEMBLY ARE AVAILABLE FOR PIECE PART INSPECTION.

FIGURE 147 [Runout and Total Runout Used on a Variety of Features]

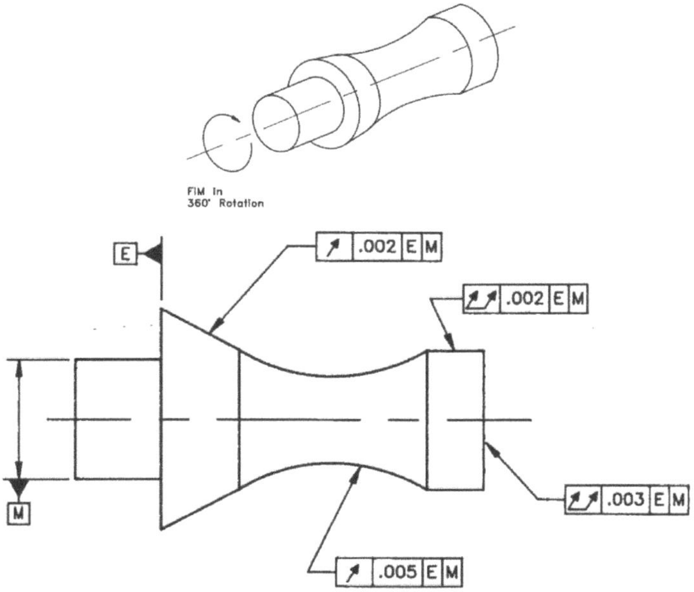

FIGURE 148 [Continuous Feature of Size Symbol Used Instead of Total Runout]

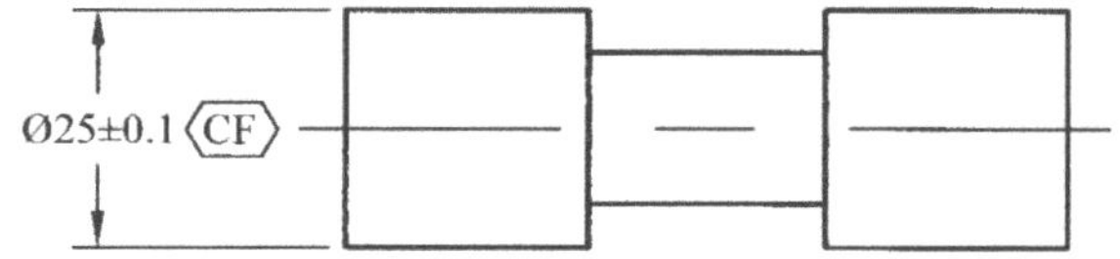

MEANS THIS:

GO GAGE to inspect the MMC of the Continuous Feature

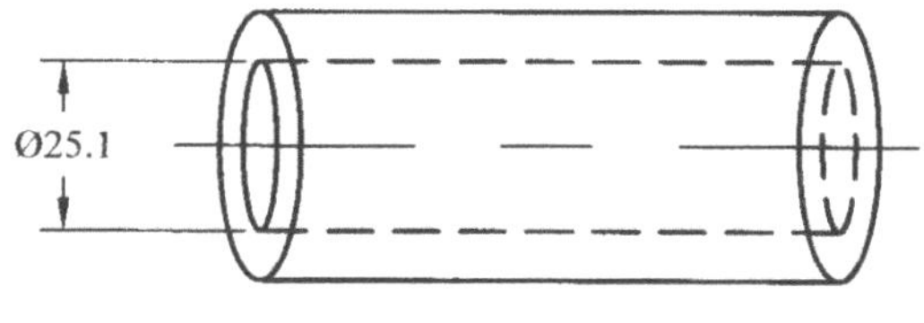

If both of the 24.9-25.1 diameters of the Continuous Feature fit into the GO GAGE it proves that the MMC has not been violated. This means that the size tolerance not only controls the form of each diameter, but it also controls the coaxiality of the diameters to each other. This treats the two diameters as though they were one continuous feature without interruptions. LMC is still inspected at cross-sections.

FIGURE 149 [Perpendicularity of a Planar Surface to a Datum Axis]

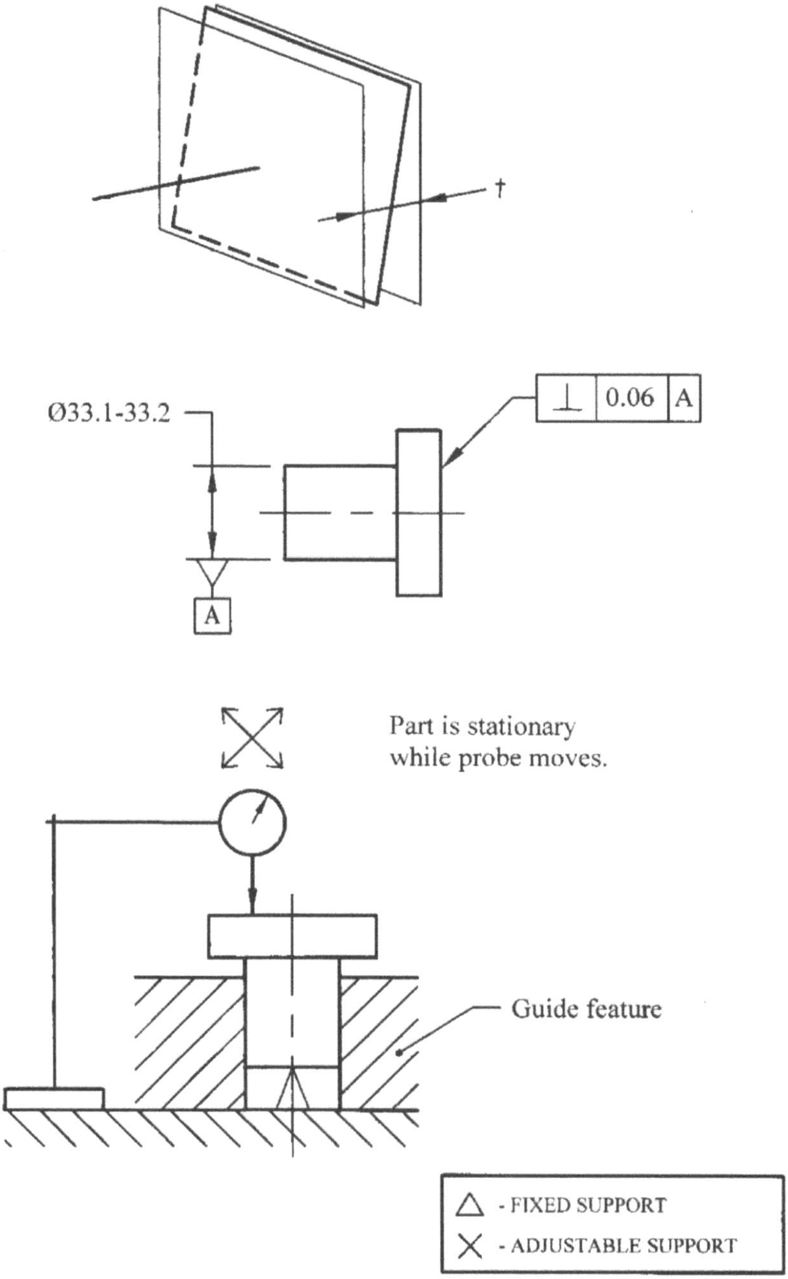

Total runout can be used on planar surfaces to control flatness and perpendicularity of the surface while being measured in a rotational state. In fact, the only difference between total runout and perpendicularity if both are used on a planar surface and related to a datum axis is that total runout is implied to be measured while rotating about the datum axis and perpendicularity is implied to be measured while the part is stationary but perpendicular to the datum axis.

In the case of total runout, the part is chucked up onto the datum feature diameter to generate a datum axis with the minimum circumscribed cylinder. Then the dial indicator is mounted 90 degrees to the datum axis, the part is rotated continuously about the datum axis and while the indicator is in contact with the surface being controlled, the indicator is moved from the outside edge of the surface to the middle of the surface. In all those revolutions, as the part is moved across the planar surface, the FIM may not exceed the total runout tolerance. To achieve the greatest degree of certainty about the validity of the measurement, the part should be rotated rapidly and the dial indicator should be moved slowly from edge to middle. In this way, many points on the surface being inspected will be tested.

In FIGURE 149, entitled Perpendicularity of a Planar Surface to a Datum Axis, datum feature A is held in a chucking device or between two Vee-blocks (marked Guide Feature) while supported at an appropriate height. This simulates the minimum circumscribed cylinder for the datum feature diameter. An indicator and its holding stand are shown as a possible method of determining compliance with the perpendicularity tolerance. All are mounted, as shown, on a surface plate. The tolerance zone consists of two parallel planes which are perpendicular to the datum axis (as would be a similar control using the geometric characteristic to total runout of a surface to a datum axis).

To verify compliance for perpendicularity, the indicator travels over the entire controlled surface registering any full indicator movement. If the full indicator movement does not exceed the tolerance of perpendicularity given in the feature control frame, it is assumed that all elements of the surface are within the geometric tolerance zone. This controls flatness, straightness and perpendicularity of the surface to the datum axis.

As depicted in FIGURE 150, Inspection of a Total Runout of a Planar Surface, this tolerance zone is the same as perpendicularity of a planer surface to a datum axis would generate. The difference is that perpendicularity is measured while the part is stationary. With total runout, the part rotates.

FIGURE 150 [Inspection of a Total Runout of a Planar Surface]

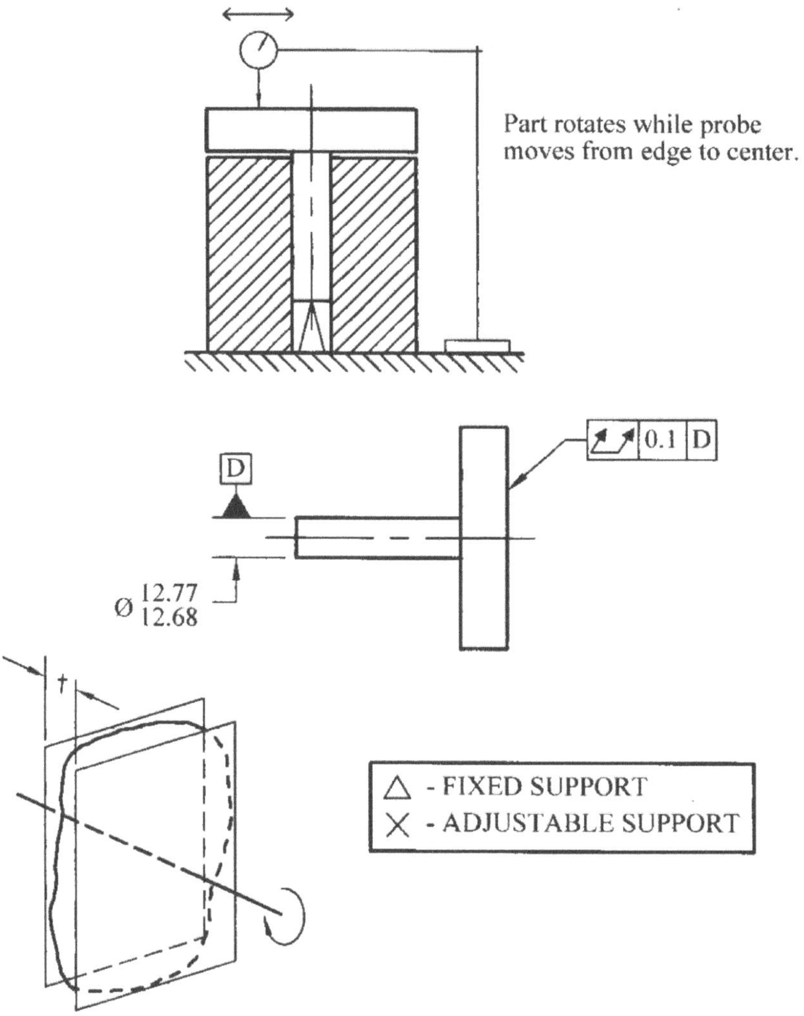

FIGURE 151 [Total Runout of a Planar Surface to a Datum Axis]

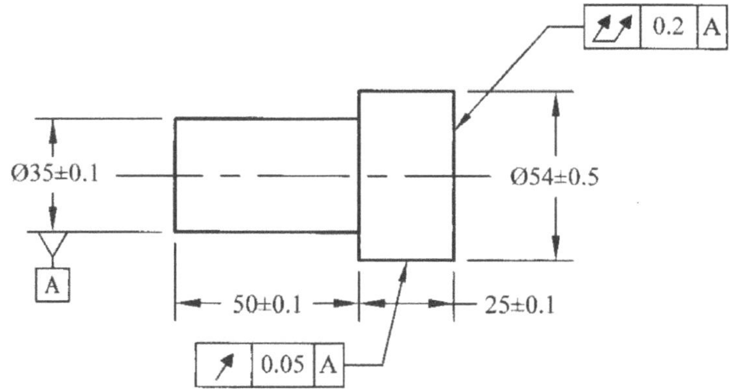

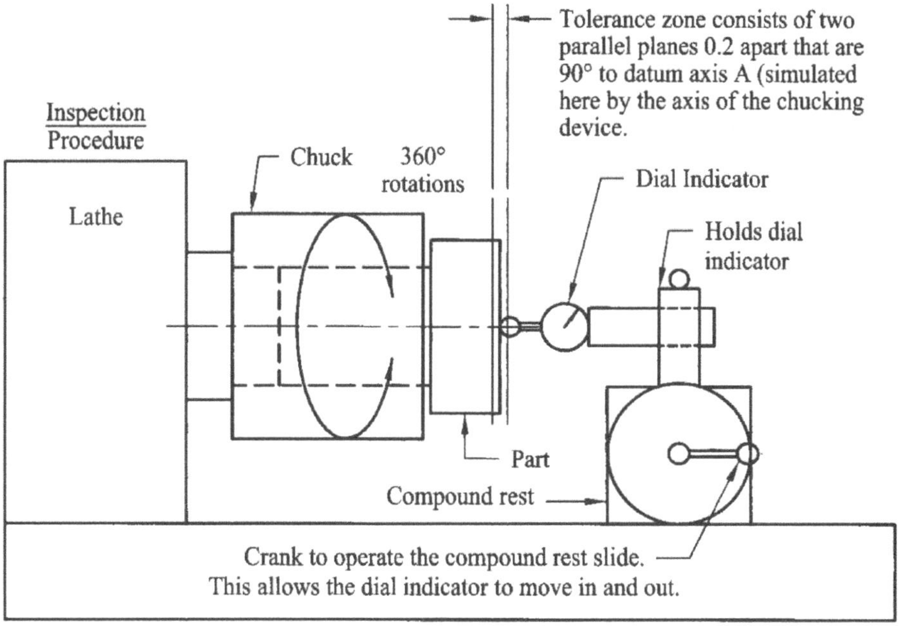

The dial indicator traverses the surface in an in and out direction while the chuck rotates in continuous 360° revolutions. The full indicator movement may not exceed 0.2 millimeters for the surface to comply with this total runout geometric tolerance. This controls the flatness (3D form) and perpendicularity (for balance). It makes certain a more equal amount of material exists on either side of the datum axis. It is similar to a perpendicularity control, but is inspected in a rotational state.

As shown in Figure 152, ASME Y14.5-2018 allows runout controls to be related to datum features in an assembly that would not generally be considered to generate a datum axis to which the constrained feature is coaxial. These runout tolerances are related, in these situations, to an axis of rotation while fixed at a basic angle to the datum reference frame and the assembly is fixed in translation and rotation to the datum reference frame. In Figure 152, the datum features do not establish the location of the axis of rotation. Instead, they fix the orientation of the axis to be certain that the feature being constrained with a runout tolerance is compliant at the assembly level. If necessary, the axis of rotation my be labeled for clarity, preferably with a note, such as, "AXIS OF ROTATION".

FIGURE 152 [Runout Tolerance Shown at the Assembly Level]

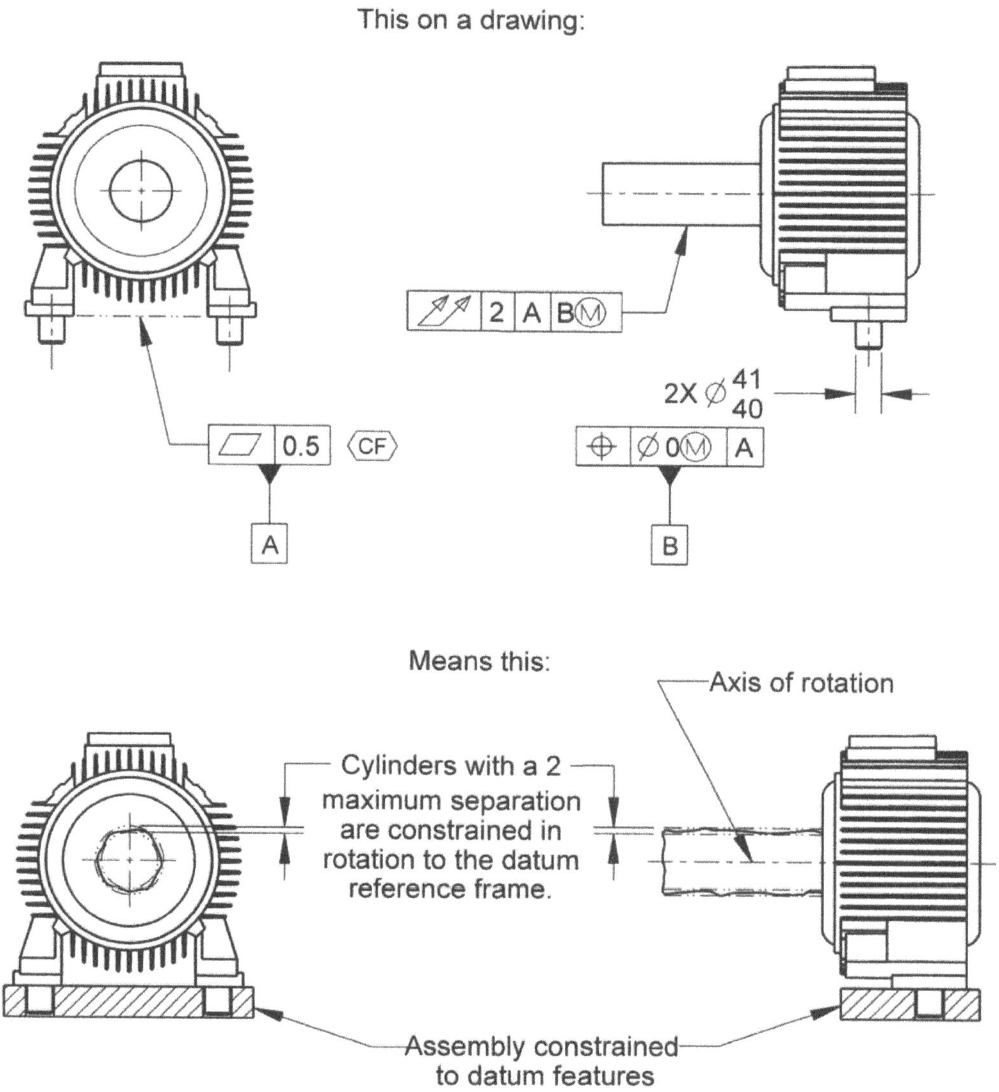

DIFFICULTIES IN PROPER DATUM FEATURE SIMULATION
(Perfection is God's Domain)

A somewhat irate caller phoned to say he was upset about the explanation given in the Y14.5 Standard for the establishment of a compound datum axis (or, as it's sometimes referred to, a simultaneous datum). An example is shown below.

FIG. 153

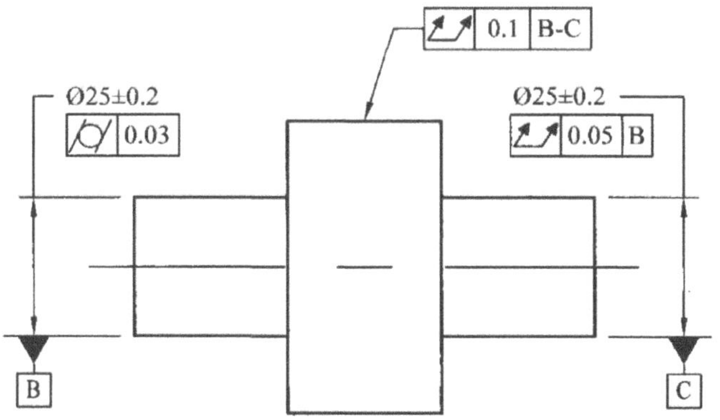

B-C means: Diameters B&C should be chucked simultaneously to establish the datum axis common to both. This technique is difficult to accomplish but better complies with the theory of compound datum establishment than putting the part between centers.

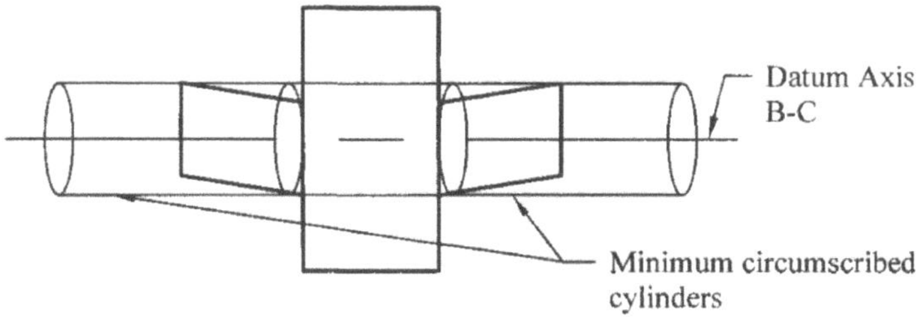

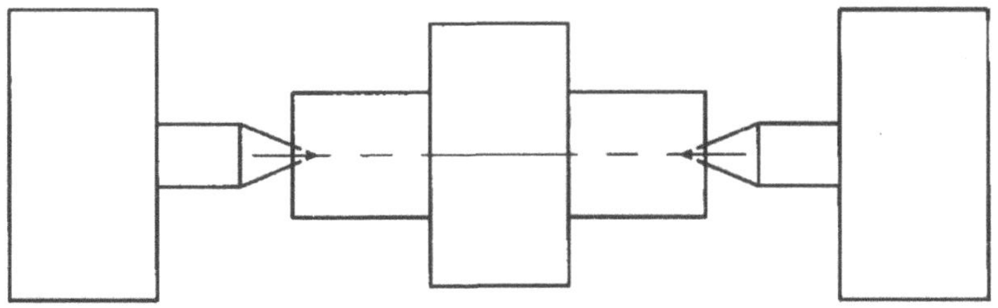

Centerdrilling the datum features and putting the part between centers would only comply with the specification for the establishment of compound datum B-C if the axis created is the same as the axis established by simultaneous chucking. Still it is often easier to accomplish, and would provide a rough simulation of the desired axis. Another option to simultaneously chucking up on datum feature B and C would be to put them into V-Blocks.

The inspector complained, "This type of control calls on me to establish an axis in a way that I cannot comply with."

The theory is explained as a common axis formed by both features simultaneously. Minimum circumscribed cylinders contacting features B and C would give this axis. Chucking both simultaneously is one way to think about it. The axis achieved is then rotated about and deviations from perfect (in this case) total runout (cylindricity and concentricity) are recorded with full indicator movement. Chucking both datum features simultaneously isn't easy.

Fig. 154

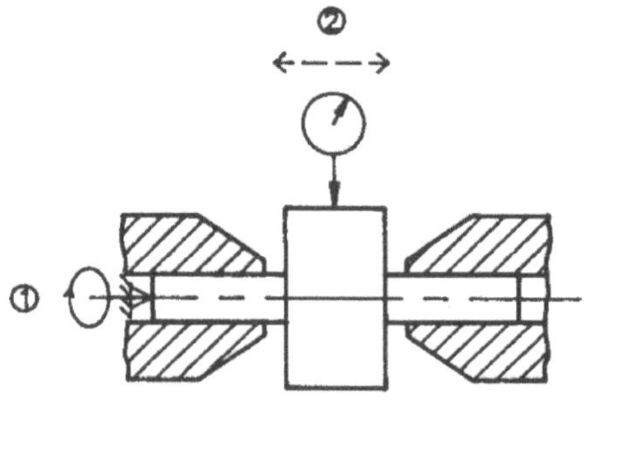

Circumscribed Rotating Cylinders

"What about Vee-Blocks?", I said?

Fig. 155

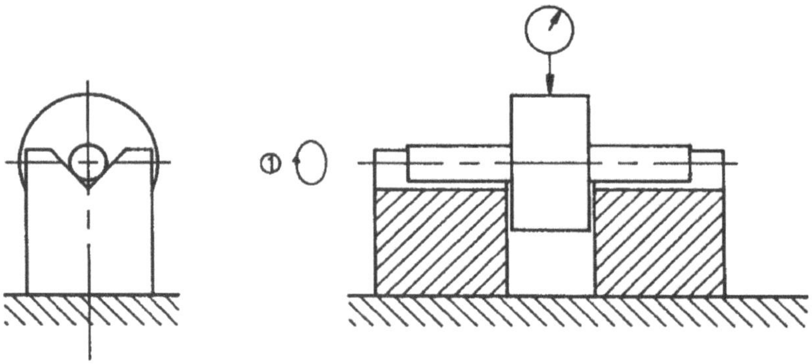

V-Blocks

"Well, you know," he said, "if the datum features aren't perfectly round, the axis you're measuring from will move up and down as the part is rotated."

This business of simultaneous datums has a lot of people angry.

"With coordinate measuring machines, some people indicate them in while the part is fixtured center drilled hole to center drilled hole. The CMM will do a swirlie or traverse a helix longitudinally, while the part rotates, on both datum features to construct a compound datum axis.", I said. The inspector said he could do that, but it wouldn't be perfect. "And it wouldn't hit the highest points on the surfaces. And even if it did, the software has a filtering algorithm that would eliminate the highest points and the lowest points and give us something like a least squares center."

Similarly, center drilling each separately, then putting them between centers and using a dial indicator to construct the dual simultaneous datum feature axis offended his sensibilities. It is understandable. It's less correct than indicating them in with trial and error, or the CMM doing a swirlie on the datum feature diameters. Both are done. Still, as the man said, "It isn't perfect!" No kidding! We live in an imperfect world.

He was asked, "What if only one datum feature was used in the control?"

"One diameter?"

"Yes, only B, for example. What would you do to establish that axis?" he was asked.

"Grab it," he said.

"With what?"

"A chuck," he said.

"What kind?"

"Three jaw," he said

"Not good enough," was replied. "It would not give the axis of the minimum circumscribed cylinder."

"Ten jaws", he said.

"Not good enough. Even ten jaws wouldn't encompass all elements of the surface. Even a collet would miss a few that peeked out between its seams.

At that point, the message was becoming clear. Just as when ANSI B4.4, The Inspection of Workpieces standard, relates the theory of GO gages, then subsequently says even though the theory calls for it, "a full form gage is not always necessary or used," "if it is known that, with the . . . processes used, the error . . . is so small that it does not affect the character of the fit of the assembled workpieces."

Let's face it, when measuring force is explained, it says "all measuring operations given in or in any way associated with this standard are understood as being referred to a zero-measurement force." All readers understand some force is applied, but the error is ignored or compensated for. Compensation is a tricky, nasty business, and most people pretend the error isn't there or ignore it because they believe that it really isn't harming or negating the data being collected. And, in many cases, they are absolutely correct.

When an inspector checks any surface geometric control with a dial indicator or probe, they are expected to touch every element of the surface. And, although theoretically possible, no one has ever done so. They check as many elements as they need to, and assure themselves that, had they checked them all, all would have been within the tolerance zone. It's a confidence builder. They want to be able to tell themselves, "No one out there is going to find an element of this surface that falls in the forbidden zone." They know they are responsible for them all, but, as the inspector, they cannot afford to spend the rest of their life, and possibly the lives of their descendants, checking one part. So. . . they settle. There are recently developed guidelines for certain part configurations that help them decide when to settle. But, let's face it, they still settle.

Once they realize what they're doing is not perfect, not exactly complying with the theory of the control for set-up or inspection--ever, the whole picture comes into a better perspective. Just as a designer realizes he/she must tolerance a part because one can never make perfect parts, the inspectors must realize the inspection set-up is full of errors. No piece of equipment they use is perfect--just some are better than others. There are no perfect set-ups. But they try. The inspectors examine their guidelines, standards, environment, equipment and budget, then hopefully infuse some knowledge of part function and come up with the best they can do that day under the given circumstances. It's judgement and professionalism and fear.

They know if they settle for too much imperfection, eventually it will come home to roost, laying goose eggs in the dollar sign column of their company's account book. Business goes away in one form or another if they don't try hard enough to comply with known requirements and theoretically perfect set-ups or if they don't have a firm grasp of them to begin with. Things can afford to be perfect only if they are imaginary. In the harsh light of reality, these theories of perfection sometimes merge poorly with the capabilities of our environments. Still, inspectors continue to keep in mind those unattainable goals and strive to achieve them as best they can.

The inspection procedure will never be perfect; but in order to even aspire to perfection, one must first know what perfect is. It's something to keep in mind--the bull's eye to shoot for. It reassures them that they have tried to do the right thing--their very best.

DIMENSIONING AND TOLERANCING APPROACHES

COMMON TOLERANCING METHODS

Objectives: Readers will learn the most common tolerancing methods for the majority of product configurations in the world.

Although features of size (holes, shafts, etc.) lend themselves to many different ways to geometrically define their desired position (orientation and location) and the allowable deviation from that described perfection, dimensioning and tolerancing for location has a few frequently used and most often appropriate formats. This chapter will explore and explain a few of these commonly used tolerancing schemes. One of the most favored is shown below:

FIG. 156

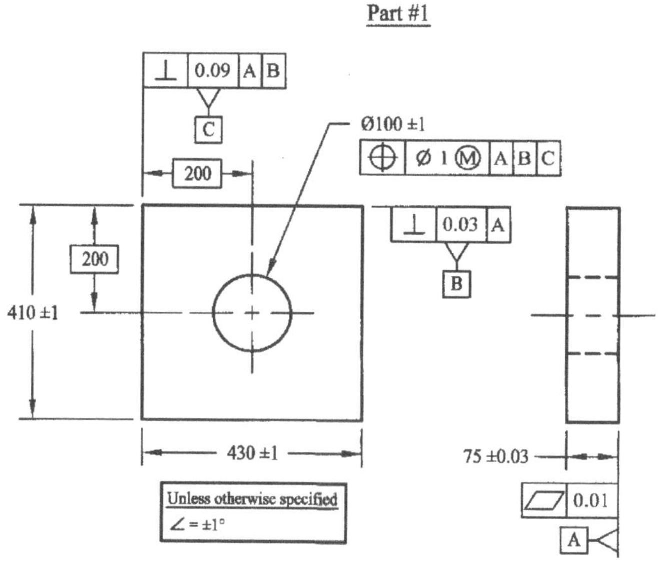

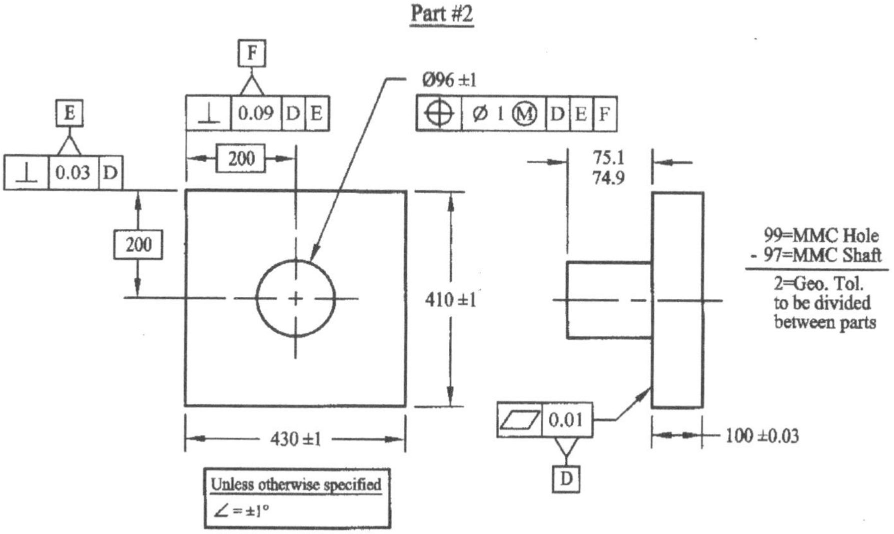

The first figures' tolerancing method shows hole on one part and a shaft on the mating part dimensioned from plane datums, 3 mutually perpendicular planes formed from the high points of 3 datum features. The high point contact required to form these planes is a minimum of three from the primary, two from the secondary, and one from the tertiary. It is a popular tolerancing format because besides fulfilling design requirements, the features are often easily accessible to manufacturing and inspection personnel. **The tolerancing scheme is:**

1) Choose a planar primary datum feature based on seating surfaces at assembly. This is for perpendicularity of feature axes. Consider giving it a flatness control. If not, size limits will have to control flatness. This size and form tolerance control is explained extensively in ASME Y14.5 (The American National Standard on Dimensioning and Tolerancing). The reference to size tolerance refers to the size limits assigned between this planar surface and its opposing side on the part.
2) Choose two location datums from which basic dimensions will be assigned. Give the secondary datum feature a perpendicularity control to the primary datum plane. Give the tertiary datum feature a perpendicularity tolerance to the primary and secondary datum planes, in that order. This can be done either on the field of the drawing (as these illustrations show) or in the general tolerance block, or in the drawing notes.
3) Position all features of size in the pattern(s) back to the 3 datums chosen.

FIG. 157: Part 1 is a Cavity with 4 Pins. Part 2 is a Disk with 4 Holes.

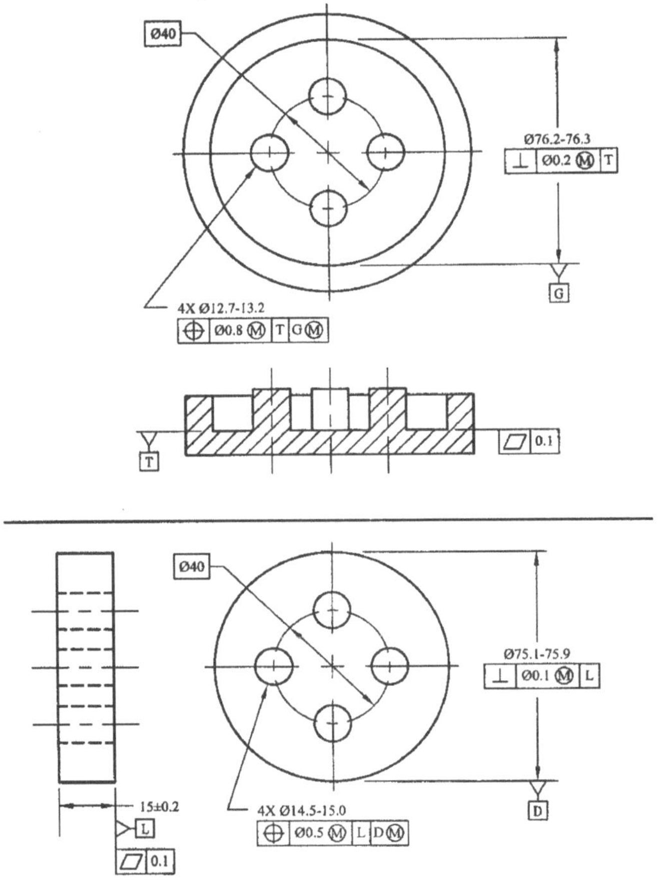

The tolerancing method shown in the second set of mating part figures is very versatile and sound for most cylindrical parts with feature patterns located from a functional diameter. **The tolerancing scheme is:**

1) Choose a primary datum feature that is a planar surface. Consider a flatness control for this surface to allow it to seat on the mating feature without rocking. (If no flatness control is specified, the size limits of that surface in combination with the opposing side of the part will control flatness to within specified size tolerances under Rule #1 in ASME Y14.5.)
2) Choose a secondary datum feature of size (a functional diameter, in this case). Consider a perpendicularity control back to the primary datum. As is used in this figure, a tolerance of zero at MMC is often appropriate.
3) Position all features in the pattern back to the chosen primary and secondary datums.
4) Although not appropriate on the part shown, if rotation of the pattern must be kept to within the specified tolerance to another feature on the part---such as a flat or slot, or other significant feature, it may be added as a third datum feature in this control. For example, see the third figure.

Figure 157a: [Part 2 Shifting Boundaries Using Perpendicularity and Position]

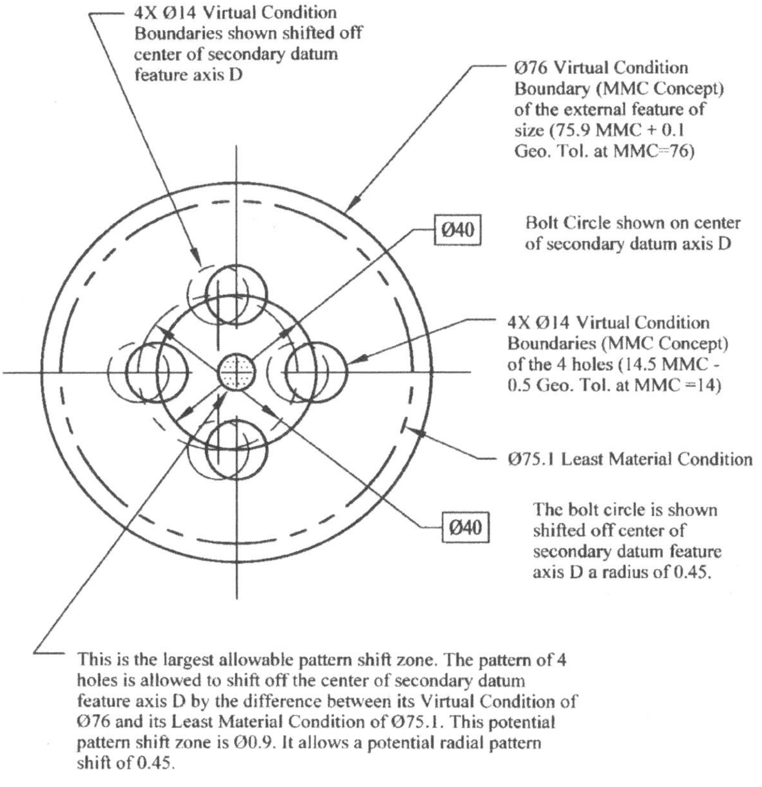

Since it is the datum feature that acts as the origin of measurement for the 4 hole pattern, the holes are seen as if they, as a group, have shifted in their relationship to the axis of secondary datum feature D (the actual diameter as produced) as shown above. In theory, it is viewed as the datum feature axis being allowed a departure from the datum axis, and in assembly this entire part may have to be shifted off the center of the imaginary datum axis to assemble.

Virtual condition boundaries (MMC concept) are generated by the collective effect of the maximum material condition of a feature of size and the geometric tolerance applicable at that size. The surface of the feature, such as the 4 holes shown on the part illustrated in FIGURE 157-Part 2, must not violate the virtual condition boundary (Ø14) in order to comply with the individual positional tolerance given to it. The pattern of the 4 virtual condition boundaries theoretically begins perfectly centered to the secondary datum axis D (20mm from center), but because of the Maximum Material Boundary symbol used after datum feature D in the 4 holes' positional control, the pattern may shift as datum feature D departs from its virtual condition boundary of Ø76.

As FIGURE 157a shows, this pattern shift/datum feature shift is only allowed while maintaining the relationship between the four virtual condition boundaries of 90 degrees to one another (and the primary datum) and their Ø40mm distance from hole to hole. Since datum feature D may be produced as small as Ø75.1 and could, in theory, be perfectly perpendicular to datum plane L, the maximum datum feature shift that may be experienced by the 4 virtual condition boundaries is a Ø0.9mm. If the full amount of datum feature shift is allowed, the pattern of 4 holes may be shifted off the center of secondary datum feature axis D radially an amount equal to one-half of the datum feature shift zone (which is a Ø0.9) or a radius of 0.45. This has the effect of allowing the 4 virtual condition boundaries to deviate from their 20mm basic dimension distance from the secondary datum feature axis of D by the amount of radial shift.

This part mates with a part consisting of 4 pins in a cylindrical cavity. The datum feature shift/pattern shift allowed by the Maximum Material Boundary symbol used after datum feature D in the position control allows some parts to pass inspection that can only be assembled off center of the mating cavity. Possible negative effects of this are imbalance for spinning parts and non-uniformity of fit (unequal airspace between datum feature D and its counterpart cylindrical cavity). This shift of the 4 holes as a pattern is in addition to the individual tolerance of position around the Ø14mm virtual condition boundaries afforded to each of the holes.

Size limits of each hole's MMC and LMC are verified separately from the conditions discussed here.

Regarding Figure 158: This third datum feature (shown in the third figure) actually creates a fourth datum plane for angular orientation of the pattern of 4-holes. One of the planes crossing at the axis of the secondary datum feature of size passes through the center of the keyway/slot. It says, in essence, "while measuring the location of these features from the secondary datum axis, measure along the two planes that form that axis in a fashion that is oriented perfectly to this fourth plane." In this example, the angular orientation plane is generated by the centerplane of the keyway. The desired orientation to that plane will be dimensioned basic or drawn so that the desired orientation is clear (for example, drawn at a 90° angle or right on the centerline of the secondary datum). In this example, of the two planes crossing at and creating datum axis D, one will pass through the center plane of the keyway (and become the same plane as the centerplane of the keyway). This will say that measurements of the position of the holes in the gear will be taken perpendicular to datum plane A, located from the axis of datum B while oriented to the centerplane of C.

This third datum feature which generates the fourth datum plane (which merges with and becomes the same as one of the planes crossing at the axis of datum B) is often thought of as a machinists' stop to "clock" the measurements taken from datum axis B or to butt up against and stop part rotation during production, and is used by inspectors to set the orientation of measurement--not to be measured from, but to be oriented to, while actually measuring from the location datum axis.

This angular orientation datum, may be generated by a variety of features such as flat, holes, slots, tabs, shafts, etc.

FIG. 158 Keyway Used to Give Angular Orientation

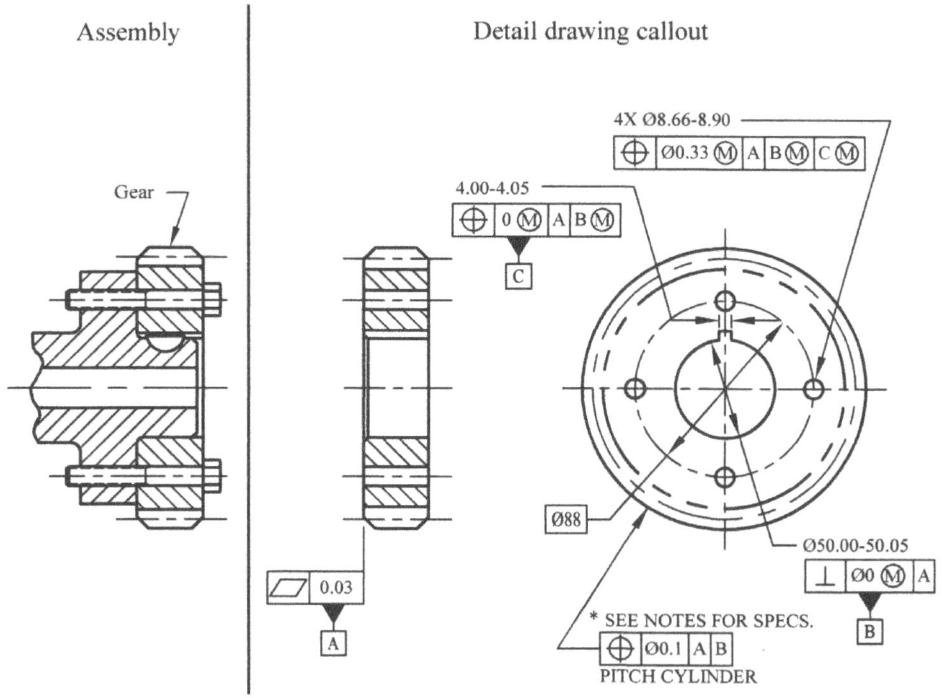

Regarding FIG. 159, FIG. 160 and FIG. 161:
These tolerancing schemes are all the same. Only the part configurations are different. **The tolerancing scheme is:**

1) Choose a significant feature of size on the part from which to measure other features of size. Make it a datum feature (called datum feature D in the Figure 4 example given and datum feature D in the next three examples).

2) Position this feature from the outside of the part, usually to three planes formed by three datum features (called datum features A, B, and C in the examples given). The primary datum feature in Figures 159-161 is for perpendicularity. The primary will often be for perpendicularity and the secondary and tertiary for location.

3) The general tolerance note of "UNLESS OTHERWISE SPECIFIED all angles equal plus or minus 1 degree" will tolerance the relationship between datum features A, B and C.

FIG. 159

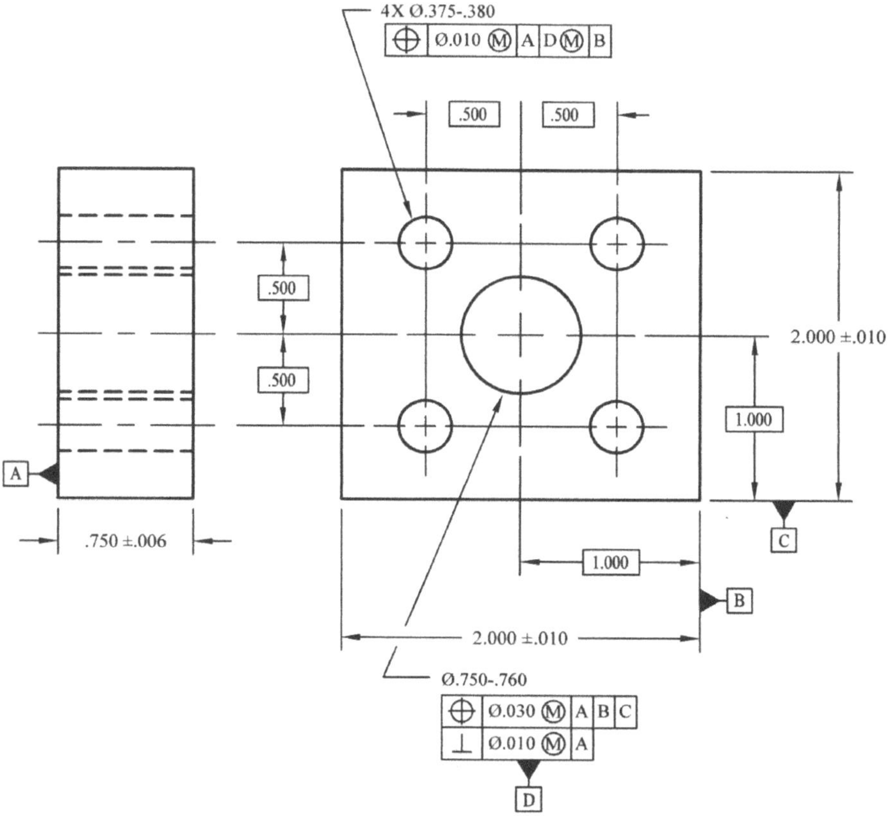

.X = ±.06
.XXX = ±.005
∠ = ±1°

FIG. 160

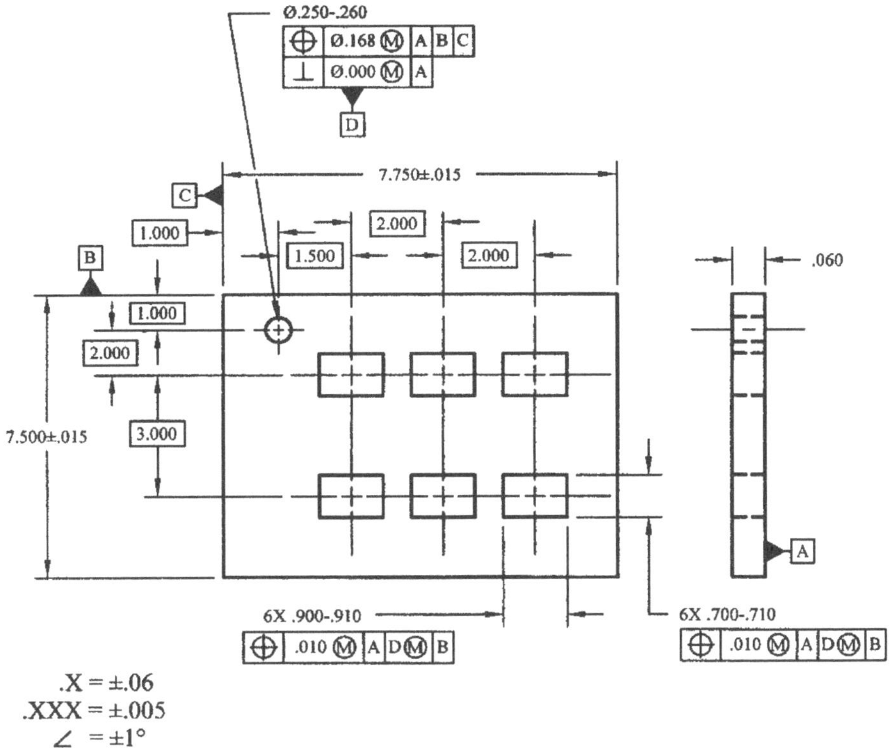

FIG. 161

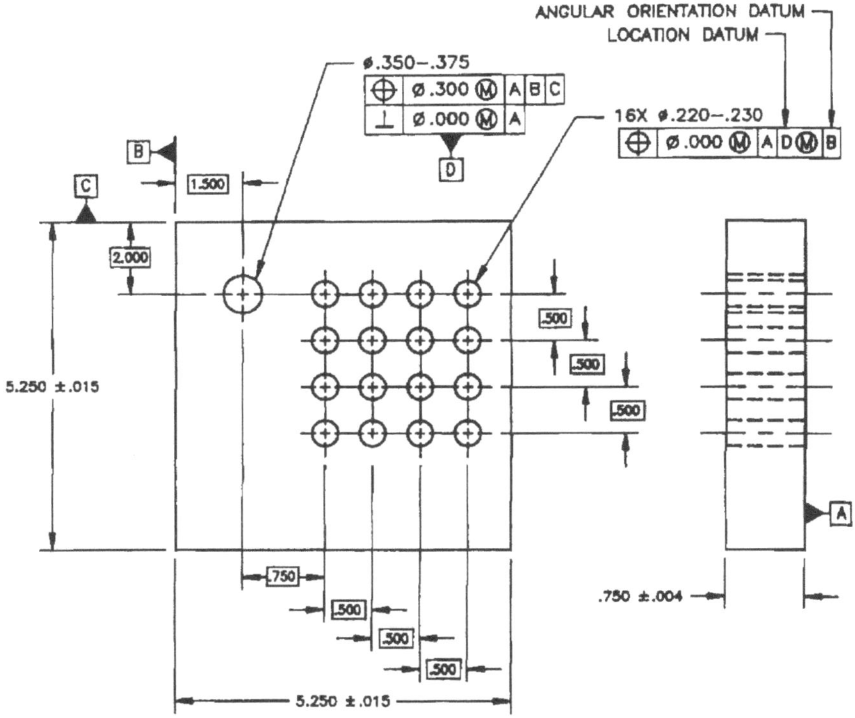

4) Consider a flatness control for the primary datum feature. The size tolerance of the part width is used, in this case to control flatness under Rule #1.

5) Consider perpendicularity controls for the three planar datum features to one another. Often, the angular tolerance in the general tolerance block will suffice here.

6) Consider a refinement of perpendicularity for the datum feature of size (called datum feature D in all three of these drawings). Place this perpendicularity refinement beneath the positional control.

* The positional control is usually a large, cost-saving tolerance.

** The MMC concept virtual condition of the perpendicularity tolerance tells what size mating feature virtual condition will assemble at that orientation. It is the functional mating size. The positional virtual condition of D is not. It is merely as stated---a cost-saving factor.

7) Position all features of size in the pattern(s) back to the same primary datum feature (A in these cases) for perpendicularity, then to the secondary datum feature of size (D in these cases) for location, then add an angular orientation datum (B in these cases) for protection of pattern rotation.

FIG. 162

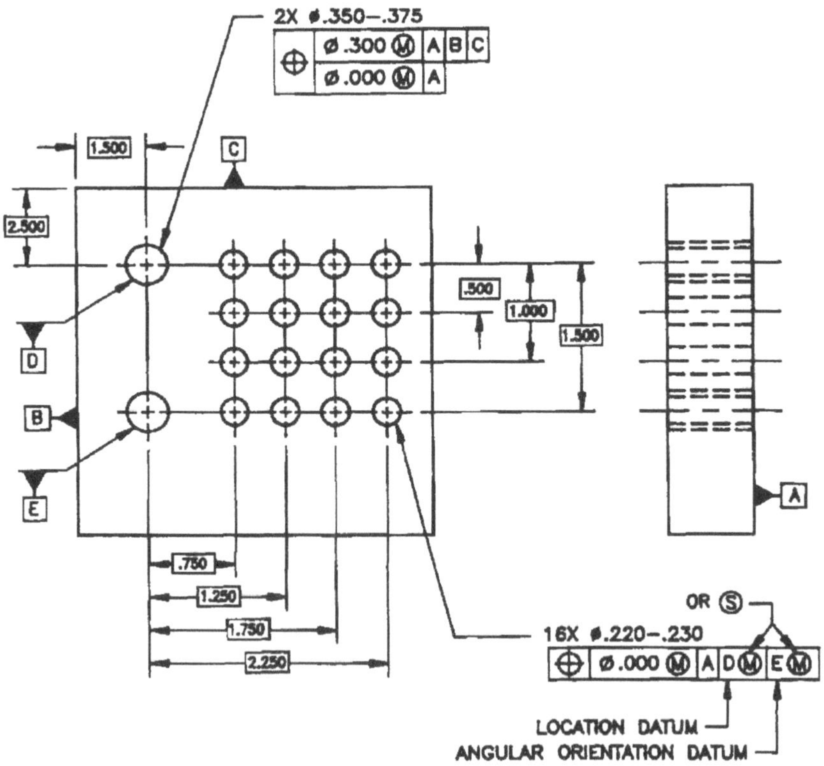

The tolerancing scheme in FIG. 162 is the same as the one used for FIG. 161 and explained in the previous section. The only difference is the pattern rotation is protected to a datum feature of size with its own tolerance of location on the part. This means the pattern of 16 holes is oriented to the outside edges of the part (actually to the datum planes B and C) only through the common plane formed between datums D and E.

Datum feature E is allowed a movement of a diameter of .300 (.150 radially) or more if datum feature E grows toward .375. Datum feature D is allowed a similar tolerance of movement which could occur in an opposite direction. The plane that might be struck between the two for pattern orientation could be between .300 (at the maximum material condition of .350) and .325 (at the least material condition of .375) away from perfect orientation to the datum plane B used for the same pattern in the drawing for FIG. 161.

FIG. 163

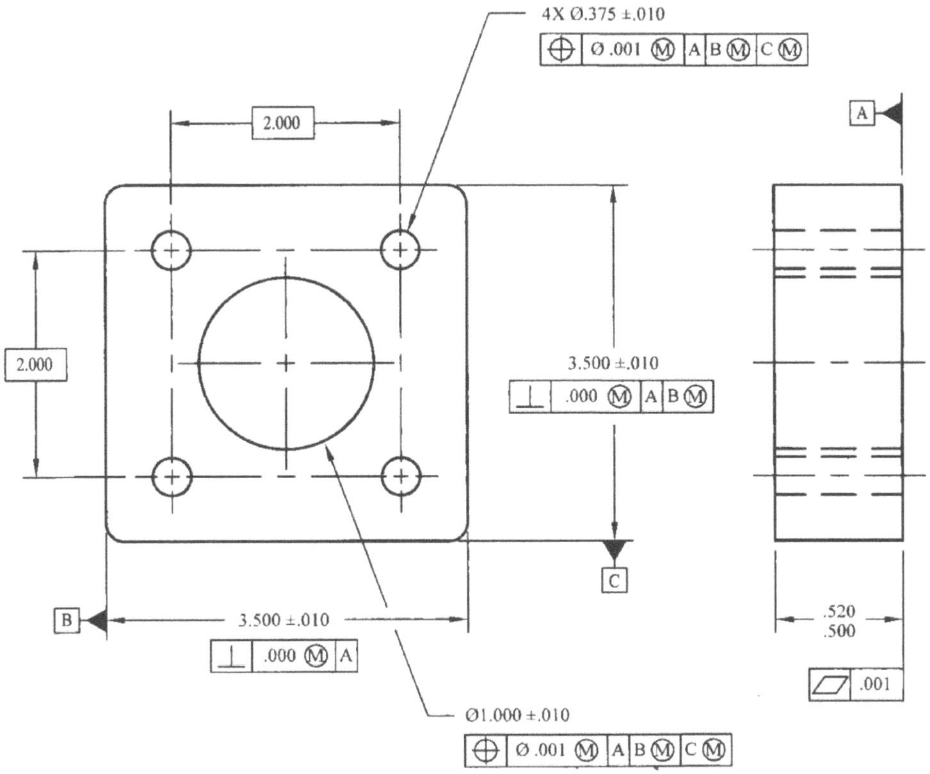

.X = ±.06
.XXX = ±.005
∠ = ±1°

This tolerancing format uses a primary datum feature for perpendicularity and a secondary and tertiary datum feature of size for location. The only unique quality of this tolerancing scheme is that the secondary and tertiary datum features of size are both formed by two parallel planar surfaces. Each set of parallel planar features generates a centerplane from which the features in the pattern are dimensioned. In this case, the center larger hole is supposed to be a basic dimension of zero from these centerplanes.

The 4 holes in the smaller hole pattern are dimensioned 2.000 inches from one another on the drawing; but since their feature control frame clearly states they are to be positioned from datums B and C for location, and the holes are shown centered about those planes, it is exactly the same as if the holes had been shown offset from the centerplanes of B and C by 1.000 in each direction.

All three planes (A, B, and C) are formed by the feature high points in order of precedence as shown in the feature control frames, and all three planes per the definition of a datum reference frame are mutually perpendicular even though the real features they are formed by won't have perfectly perpendicular sides.

Or, if each was considered as primary, instead of one primary, one secondary and one tertiary, the centerplanes of primary plane B and primary plane C wouldn't be exactly perpendicular. When these planes are considered in their interrelationship and given an order of precedence as A primary, B secondary, and C tertiary, each plane generated takes its orientation not only from its real datum feature, but also from the previously established plane. This assures it will be perpendicular to the other datum plane or planes in the reference frame.

Geometric tolerances are then established from the "true positions" set up by the datum planes and basic dimensions shown from them. **This tolerancing scheme is:**

1) Choose a primary datum feature. Give it a flatness control given previously described considerations.

2) Choose a secondary and tertiary datum feature of size.
   a) Control perpendicularity between primary, secondary and tertiary features.
   b) There are many ways to control the perpendicularity of one datum feature to another in this scheme. One is to give a zero at MMC perpendicularity tolerance of datum feature B to A, then give a zero at MMC perpendicularity tolerance of datum feature C to A primary and B at MMC secondary. This creates an envelope of perfect form and orientation at MMC by MMC. In other words, it ties the MMC envelopes to one another and to A.

   c) Use tolerance block tolerances. (Ex: ANGLES ±1°)

   d) Use notes. (Ex: Unless otherwise specified, all dimensions relate to datum A primary, datum B secondary, and datum C tertiary.)

   e) Use perpendicularity tolerances besides zero at MMC.

3) Position all features in the pattern(s) to the datum reference frame chosen.

FIG. 164

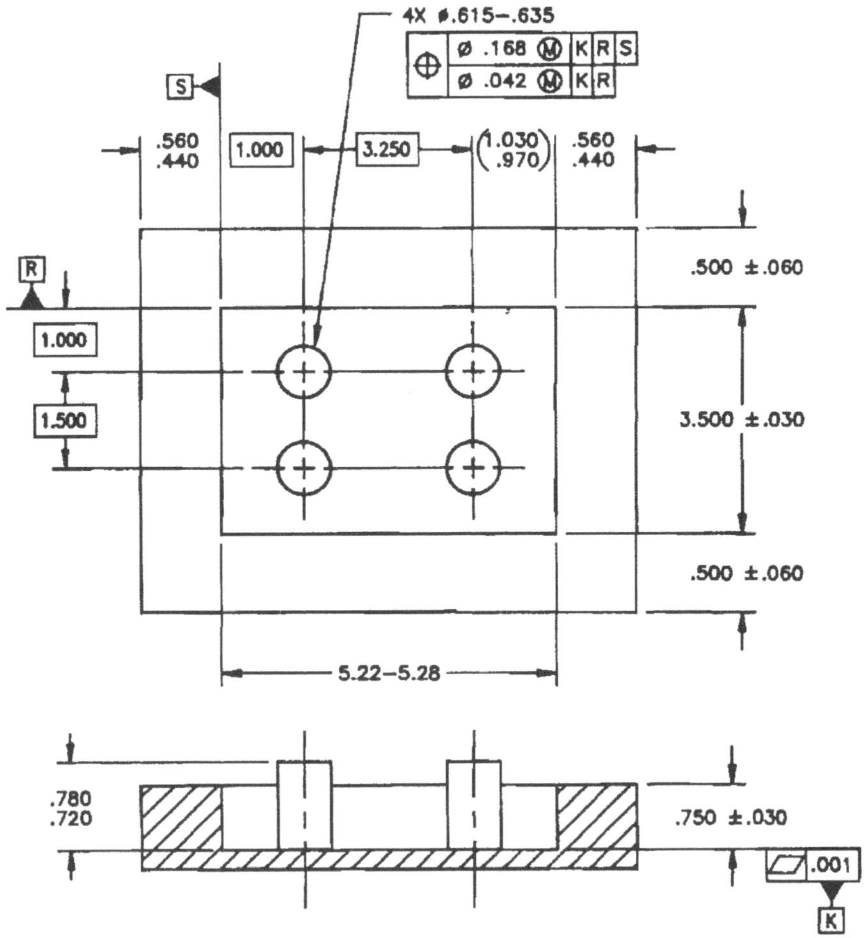

[Note: Unless otherwise specified, all dimensions relate to datums K primary, R secondary and S tertiary.]

Composite positional tolerancing as shown in FIG. 164 has the capability of controlling the overall location of the pattern of features to within a loose tolerance to datums and, at the same time, to maintain a tighter feature-to-feature tolerance.

The upper tier of tolerance creates stationary tolerance zones centered around true position as defined by the basic dimensions as located from the datums listed. The main purpose of this tier is to lower part costs. The upper tier is not what makes the part function. In reality, in this situation, if the tolerancing engineer were willing to replace the tolerance in the upper tier with the smaller tolerance in the lower tier, the lower tier could be eliminated entirely without the loss of part functionality. Therefore, we learn it is the tolerance in the lower tier that makes the part work. The tolerance in the upper datum-to-pattern tier in composite positional tolerancing merely makes the part cost less than having a one tier conventional positional tolerance using the tolerance given in the lower, feature-to-feature tier.

The datums listed in the lower tier, although not the main function of the tier, limit the allowed deviation from perfect orientation of the features within the pattern to those datums. In this case, the primary datum listed (K) limits the allowed out-of-perpendicularity of the features and attempts to insure that when this part is assembled with its mating part that the primary datum feature will seat well on the primary datum feature used for the same purpose on the mating part.

The secondary datum feature listed (in this case R) stops the pattern rotation of the smaller feature-to-feature tolerance zones which float within the larger zones. It limits the orientation of the float of the smaller zones, which are connected as though by a wire framework of the feature-to-feature basic dimensions. The secondary datum feature in the lower tier acts as an angular orientation datum which says that while the smaller zones may float within the larger stationary zones, they may float only in a fashion that maintains the pattern of the tolerance zones' orientation (angle) to R.

In this case, perfect parallelism and perpendicularity of the tolerance zone framework is maintained. This allows each individual feature (shaft) to move in its smaller tolerance zone. Each of these zones is connected to the other smaller zones by the basic framework (basic feature-to-feature dimensions). The framework itself may move within the larger stationary zones (maintaining the pin axis location within both zones) but datum K in the lower tier allows the tolerance zone feature-to-feature framework to move only while perfectly perpendicular to datum K. Likewise, repeating the secondary datum feature, R in this case, allows the feature-to-feature tolerance zone framework to move while perfectly perpendicular and parallel to datum R. This means any out-of-perpendicularity to K or pattern rotation to R must occur within the smaller zones.

Although somewhat complicated at first glance, this tolerancing scheme is very useful in lowering part cost by allowing larger pattern movement to datums while maintaining tighter feature-to-feature relationships and still maintaining the smaller tolerances of orientation to the datum reference frame. **The tolerancing scheme is:**

1) Choose a primary datum feature. Consider a flatness control for it.

2) Choose secondary and tertiary planar datum features. Control the perpendicularity of the 3 datum features to one another through the use of title block angular tolerances, perpendicularity or profile controls applied directly to the field of the drawing, or through the use of drawing notes such as: "Unless otherwise specified, all dimensions relate to datums K primary, R secondary and S tertiary."

3) Specify a large cost-saving positional tolerance in the upper tier of the control to control the pattern movement to the three specified datum planes.

4) Specify a smaller function preserving tolerance in the lower tier of the control to limit the feature-to-feature movement.

5) Consider specifying an orientation datum feature in the lower tier for refinement of perpendicularity.

6) Consider specifying an angular orientation datum in the lower tier to refine the pattern rotation to within the smaller feature-to-feature tolerance.

[Steps 5 and 6 limit the pattern float of the feature-to-feature tolerance zone's framework to be properly oriented (for example, perpendicular and/or parallel to the specified datums).]

FIG. 165

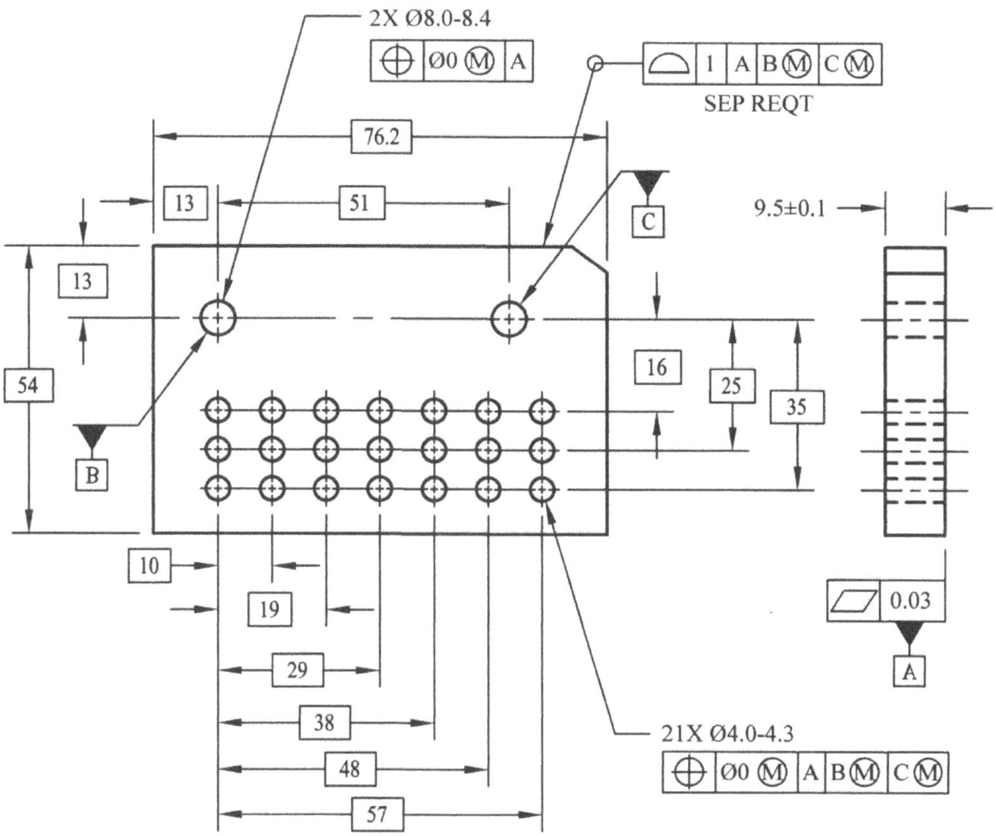

FIG. 165 shows a planar surface acting as the primary datum feature and holes acting as the secondary and tertiary datum features. **The tolerancing scheme is:**

1) Choose a primary feature that is a planar surface. Consider a flatness control for this surface to allow it to seat on the mating feature without rocking. If no flatness control is specified and the part is rigid, the size limits of that surface in combination with the opposing side of the part will control flatness to within the specified size tolerances under Rule #1 of ASME Y14.5.

2) Choose a significant set of features of size (holes, in this case) from which to construct the location and orientation of the secondary and tertiary datum planes. Call out one of the holes as a secondary datum feature from whose axis measurements of controlled features will originate. Call out the other hole as the tertiary datum feature through whose axis one of the two planes that construct the secondary datum feature's axis will pass.

This gives an orientation to both planes that construct the secondary datum axis. These planes will act as the origin of measurement for subsequently-controlled features.

3) Position these datum features to one another. If they are holes or shafts of the same size limits (a pattern of two), the positional control need only include a position symbol, a diameter sign, a geometric tolerance, a material condition (specified or implied) and (if functional for seating) the primary datum feature.

4) All other features of size, such as holes and shafts, are positioned to the established primary, secondary and tertiary datum features (which generate the datum reference frame).

5) The part profile is dimensioned with basic dimensions back to the datums, then a profile of the surface tolerance is applied to this part profile using the established datums.

Chapter 10

Position with Fixed Fastener Assemblies And Projected Tolerance Zones

•Tolerancing Mating Parts in a Fixed Fastener Assembly for Position Tolerancing; GO Gages; Reading Feature Control Frames; Projected Tolerance Zones for Position Tolerances and How to Measure Them; Referencing Datum Features at MMC/Datum Feature Shift and What it Means

Chapter Objectives

Readers will learn:

1. To tolerance a fixed fastener assembly with appropriate geometric characteristic symbols.
2. To distribute geometric tolerances between mating datum features, clearance holes and threaded holes that use projected tolerance zones.
3. The questions to ask to select the optimal datum features for an assembly.
4. How to visualize Rule #1 using the GO gage principle.
5. How to read feature control frame symbols as a sentence.
6. Three options to prevent binding when threaded holes and screws are involved in an assembly.
7. How to measure a projected tolerance zone.
8. What the maximum material condition symbol after a datum feature of size means.

Position with Fixed Fastener Assemblies and Projected Tolerance Zones

FIGURE 166 [Tolerancing Mating Parts in a Fixed Fastener Assembly]
Consider these parts cylindrical.

Assembly

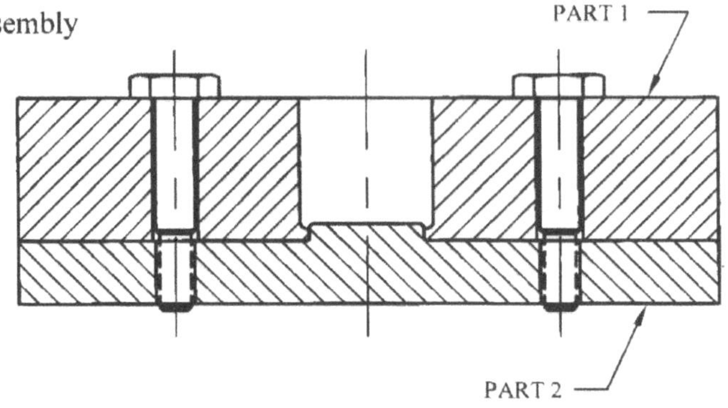

PART 1

Detail drawing

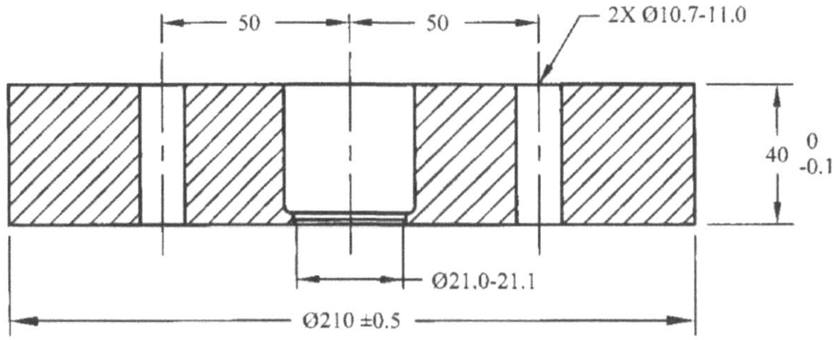

PART 2

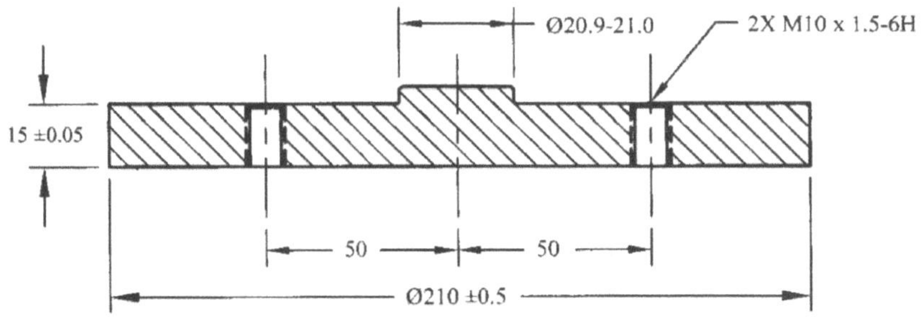

The assembly drawing shows two parts bolted together by screws that pass through clearance holes on Part #1 and are screwed into the threaded holes on Part #2. This situation is a classic fixed fastener assembly condition. These parts follow a tolerancing scheme that is common to many assemblies.

STEP 1:

Find the primary datum feature by asking; "What surfaces seat?" or "What surfaces need the most physical contact when assembled?" or "What surfaces dictate the angle at which these two parts will assemble?" or simply, "What surfaces are we trying to bolt to?" The answer to these questions should all be the same; "The bottom of Part #1 and the top of Part #2." These surfaces then become the primary datum features for each part. One has been named datum feature A (on Part #1) and the other, datum feature Z (on Part #2).
These surfaces are planar in shape and the next question should be; "What characteristic of geometry must each possess to seat well without rocking (too much)?" The answer has to be a geometric characteristic which is not related to a datum. Because, in this linear progression of geometric definition, these datum features are first. They can't be related to each other because these detail parts will most likely end up being produced separately. The producers of one part may never even see the other part's detail drawing. And, these primary datum features can't be related to anything else because each is first (primary) on its own part. There is nothing else to relate them to at this stage of the definition.

So, the answer to the question is flatness, a geometric characteristic often assigned to planar primary datum features. We may want to look at the thickness tolerance on each part because under Rule #1, the size tolerance controls the form of a rigid regular feature of size. Part #1 has a thickness tolerance, so let's examine that. The thickness is 40 plus zero and minus 0.1.

Rule #1 says that if a regular feature of size is produced at its maximum material condition (40 in this case), it must have perfect form (straightness, flatness and parallelism in this case). This assumes every two opposing points on the part thickness would measure (with a probe or micrometer type measurement) 40 mm.

And if it was true, the form of the 40mm thickness would have to be perfect to comply with the size requirement under Rule #1. This rule is based on the principle of the GO Gage invented by Taylor in 1905. So, it assumes the part produced at MMC (40) would only GO into the GO Gage (a couple of rails set at a distance apart of 40mm) if the part had perfect form. For example:

FIGURE 167 [GO Gage]

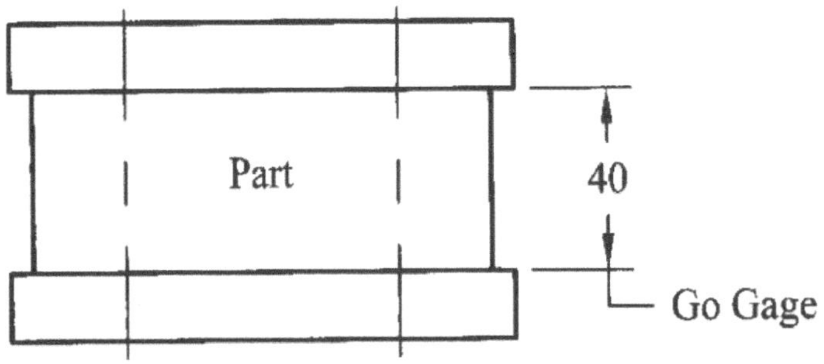

— , // & ⏥ = O at MMC

By that same token, if produced smaller than MMC (40), the part's surfaces may have less than perfect form to a maximum deviation of the entire size tolerance of 0.1. For example:

FIGURE 168 [GO Gage]

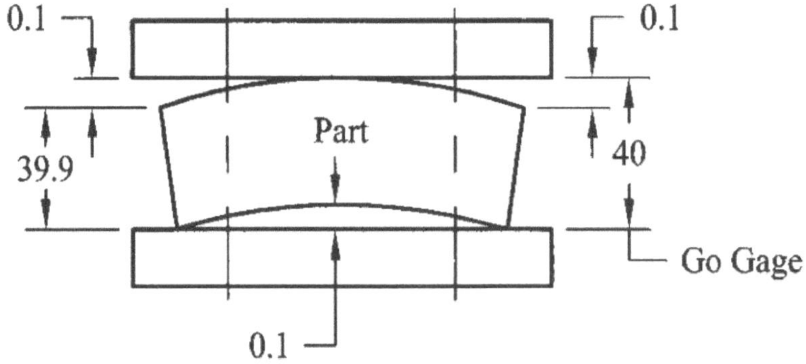

If every cross-sectional measurement of the thickness of the part measures 39.9 (the LMC), then the part may be out of flatness, straightness and parallelism by the entire 0.1. Although this may appear solely on one side as a pit 0.1 deep (if the opposite side is perfectly flat), the figure shows the entire 0.1 existing on both the top and bottom simultaneously if the part has a bowed centerplane.

As one can see, the tolerancing specialist thought this was too much for the primary datum feature to be allowed to be out-of-flatness, so this configuration was disallowed by the subsequently imposed flatness requirement of 0.03 of both datum features A and Z. Now, the maximum each primary datum feature may be out of flatness (rock or wobble) is 0.03. Together, during assembly, the maximum they may rock on one another is the sum of their flatness tolerances of 0.03 plus 0.03 or 0.06.

FIGURE 169 [Step 1] Detail Drawing Callouts
Consider these parts cylindrical.

PART 1

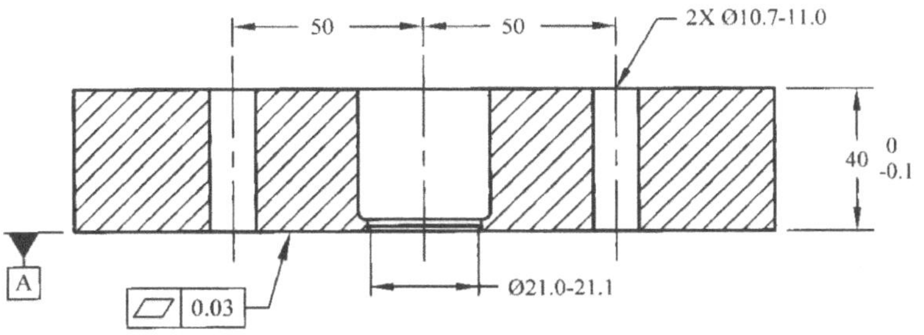

PART 2

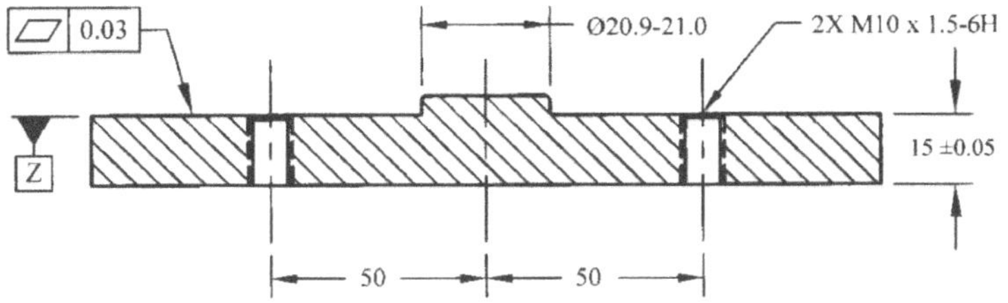

STEP 2:

A secondary datum feature must be selected. Since the primary datum features have a lot of surface area and will stabilize the parts when holding/measuring other geometric relationships for angles and location, the secondary datum feature need not have as much stabilizing ability. It will be to locate other part features from, while the primary controls the perpendicularity of those features.

The secondary datum features will be chosen on the basis of; "What features mate and/or align the assembly?" We can see that the center bore in Part #1 mates with the shaft in the center of Part #2. But, likewise, the two Ø10.7-11.0 holes mate with the 2 threaded holes on Part #2. It is very attractive to use the single mating features as datum features, since it is easier to find the axis of a single feature than a pattern of features. But, let's determine what aligns the assembly before we make the final decision on our secondary datum feature.

To determine which features control the alignment, let's look at the fit of the features from part to part and determine which is the tightest. The tightest fit from part to part is most likely to align the parts when they assemble. The fit on the center diameters has a minimum size tolerance difference of Ø21-21 or 0. The maximum clearance is Ø21.1-20.9 or 0.2. The fit of the screws into the clearance holes is 10.7-10.0 = 0.7 minimum or 11.0-10.0 = 1.0 maximum. Since the threaded holes will act as vehicles to receive the screws and move them around, the threaded holes and screws will merge. For the purposes of this exercise (and in general tolerance analysis), we will refer to the combination of the threaded holes and screws as "mounted screws."

Clearly, the fit between the center hole on Part #1 and the center shaft on Part #2 is tighter than between the mounted screws and their clearance holes. So, the alignment is more likely done by the single mating features in the middle of each part. They are easier to measure from, have a tighter fit, and, therefore, do the alignment. They will become the secondary datum feature on each part. We must tolerance their relationship to their primary datum feature and give each a name.

The hole on Part #1 is called datum feature E and the shaft on Part #2 is called datum feature D. Since E is depicted perpendicular to A, and D is depicted perpendicular to Z, we know they are an assumed 90° dimension. But, these relationships must be toleranced. We could use a general note that says something like, "Unless otherwise specified, all angles are plus and minus 1 degree." But, that would mean if E or D was subsequently referenced in a feature control frame at MMC, a virtual condition (the collective effect of size and geometric tolerance from a feature control frame), would be impossible to calculate, which means fixtures or gages that needed to simulate E or D at their virtual condition (MMC concept) could not be built. So, in this case, it is better to assign each a perpendicularity tolerance. Since the MMC of both E and D is Ø21, their difference is zero. This zero at MMC tolerance is assigned to both perpendicularity tolerances. The perpendicularity control on Part #1 can be read as:

⊥ = Perpendicularity
Ø = of the axis of the diameter
0 = must be held to a zero tolerance
Ⓜ = if the feature is produced at the MMC size of Ø21
A = to A (or while seated on A).

FIGURE 170 [Step 2] Detail Drawing Callouts
Consider these parts cylindrical.

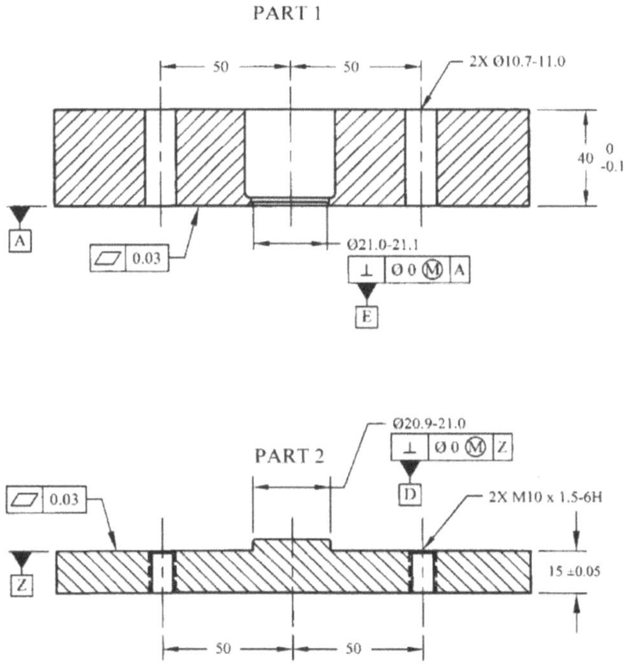

STEP 3:

The next step on each part is to tolerance the 2-hole pattern. The important relationships that we must tolerance are:

a) Hole to hole distance (100mm)
b) Perpendicularity to the primary datum
c) Distance out from the axis of the secondary datum (50mm).

Since the maximum material condition of the clearance holes is 10.7 and the maximum material condition of the mounted screws is 10, the difference of 0.7 is divided between the two parts. Threaded holes are generally perceived to be more difficult to manufacture than clearance holes. Also, threaded holes do not gain an easily quantifiable amount of bonus tolerance because of the effects caused by the angles of the threads and the tendency of the pitch cylinders of the screw and the threaded hole to center one another. The bonus tolerance on a threaded hole is negligible. Still, the use of the MMC modifier allows the inspector to reap a small tolerance benefit if gages with fixed size threaded gage pins are used.

Potentially, the clearance holes could be produced at sizes larger than the MMC of Ø10.7 and gain additional position bonus tolerance (a maximum of 0.3 if the holes are produced at Ø11.0). The threaded holes can't count on a meaningful bonus tolerance. Since ease of manufacture and the potential gain of bonus tolerance work in favor of the clearance holes, a larger portion of the available Ø0.7 position tolerance should be assigned to the threaded holes. This will give them a better chance of passing inspection than if the tolerance was equally divided between the threaded holes and clearance holes. The proportions of this division begins with the statement; "Give more tolerance to the features that are difficult to manufacture." Then it proceeds to: "How much more difficult to manufacture are they?" There is no set rule on dividing up tolerances in this manner, just judgment and knowledge of the product and manufacturing. In this case, the threaded holes have been given 0.4 of the 0.7, with the remaining 0.3 allotted to the clearance holes.

We can estimate that if a machinist shoots for the mean clearance hole size of 10.85, then the average bonus tolerance will be about 0.15. If we add this to the 0.3 we assigned them, the clearance holes will average a position tolerance of 0.45. The threaded holes will average the 0.4 we assigned them and from bonus tolerance a negligible, non-quantifiable wiggle. Therefore, it is likely, under these assumptions, the average position tolerance for the clearance holes and threaded holes for the finished products will be roughly equal. Again, if that seems to run against the premise of assigning more tolerance to the set of holes that are hardest to manufacture, we can reassess our tolerance distribution and divide up the 0.7 differently.

Each set of holes is related back to the primary and secondary datum references on each part; to A and E on Part #1 and to Z and D on Part #2. The 50mm dimensions are converted to basic dimensions. Since these parts mate and don't rotate, balance is not an issue. So, the datum features of size (mating features D and E) may be referenced at MMC. This will provide additional tolerances drawn from D for the shifting of the pattern of threaded holes as a group (to a maximum of Ø0.1; the difference between 21.0 and 21.1 provides a maximum shift of that pattern from datum feature axis E of Ø0.1). This just means that, although the position tolerance zones for the holes on each part will maintain their 100mm distance from one another, they might not be 50 millimeters from their datum feature axes. As a pattern, the zones could shift off center left or right (in the view shown) by a maximum of radius of 0.05 (a Ø0.1).

The clearance holes' position feature control frame can be read:

The clearance holes' position feature control frame can be read:

⌖ = Position
Ø = of the axes of these diameters
0.3 = must be held to within a diameter of 0.3 (per hole) to each other (100mm)
Ⓜ = if produced at MMC
A = and to A for perpendicularity (by orienting the Ø0.3 tolerance zones 90° to A)
E = and to E for distance (50mm) by locating the Ø0.3 zones 50mm from E
Ⓜ = if E is produced at its virtual condition to A. But the pattern of tolerance zones may shift (as a group while maintaining their 100mm from each other) off the axis of datum feature E by its growth. Any datum feature shift may be negated by the use of E's growth as perpendicularity tolerance.

The threaded holes' position control is read in a similar way but it must be remembered the MMC modifier used after the 0.4 is not going to provide one-for-one bonus tolerance since it is based on growth of the hole's pitch cylinder. It does, however, allow the inspector to use fixed size threaded gage screws (M10) for inspection. These zones use a projected tolerance. So, the sentence (position feature control frame) would be read with the extra phrase: **"...while the zones are projected (above the primary datum) out of the hole 40mm (the maximum thickness of the mating part in this instance)."** This 40mm is representative of the maximum height that the body of the screw projects from the hole. Remember also, it is the axis of the pitch cylinder of the threaded hole that is being projected and that must reside in the projected positional tolerance zone.

FIGURE 171 [Step 3] Detail Drawing Callout

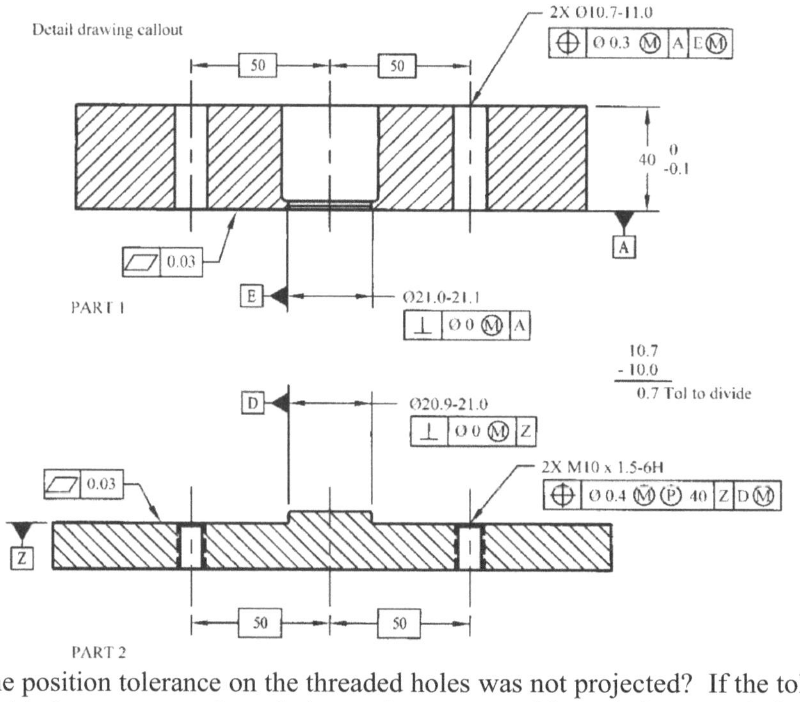

But, what if the position tolerance on the threaded holes was not projected? If the tolerance zone on the threaded hole was not projected, the mating part would need clearance holes that have a virtual condition that is larger than the current version (10.7 MMC minus 0.3 geometric tolerance at MMC = 10.4 virtual condition).The diameter of 10.4 mating clearance hole virtual condition is so small as to cause binding when the screws are inserted and attempts are made to screw them into the threaded holes.

The illustration that demonstrates this condition (FIGURE 172) assumes the two parts are 40 millimeters thick to keep this calculation simple. Since Part #1 is actually thicker than Part #2, the interference problem would be more severe. If it was of a thickness that was less than the depth of the threaded hole, the interference problem would lessen but still need to be calculated. With the numbers shown in FIGURE 172, the mating clearance hole would have to have a virtual condition size of at least 10 + (3 x 0.4) = 11.2 diameter. Since the mating holes aren't even allowed to be made at a diameter of that size, the parts do not fit together even if they don't use all of the position tolerance.

There are three solutions to the problem:

1) Change the clearance hole diameter to be, for example, a diameter of 11.5 – 11.8 and keep the position tolerance of 0.3. This would give the clearance holes a virtual condition (worst mating condition) of 11.5 minus 0.3 = 11.2. But the head of the screw may not have enough land and may even drop right into the hole. It might be necessary to use washers to make this viable.
2) Decrease the position tolerance on the threaded hole to be extremely smaller. This would lessen the interference, but it would dramatically increase the cost of Part #2.
3) As the original requirement shows, just use the projected tolerance zones on the threaded holes. This means that there is no position tolerance zone inside of the threaded hole at all.

   The entire projected position tolerance is outside of the threaded hole. It reflects the maximum height that the screw would have to project from the hole in order (in this case) to pass through the mating part (40 millimeters). This approach actually makes it easier to inspect the threaded hole's position tolerance by taking a threaded gage pin and screwing it into the hole.

   Then the position tolerance is inspected by probing the pin at its base nearest the threaded hole and also over the height of the projection (40 millimeters) to determine if both circles generate a center point that resides within the 0.4 tolerance zone. A gage pin with a thread that is split up the middle is an option that may be desirable in order to allow the pin's pitch cylinder to contact the hole's pitch cylinder. This will act to negate any wiggle that may have been experienced while the pin is probed. See FIGURE 173.

FIGURE 172 [Position Tolerance on the Threaded Holes that was Not Projected]

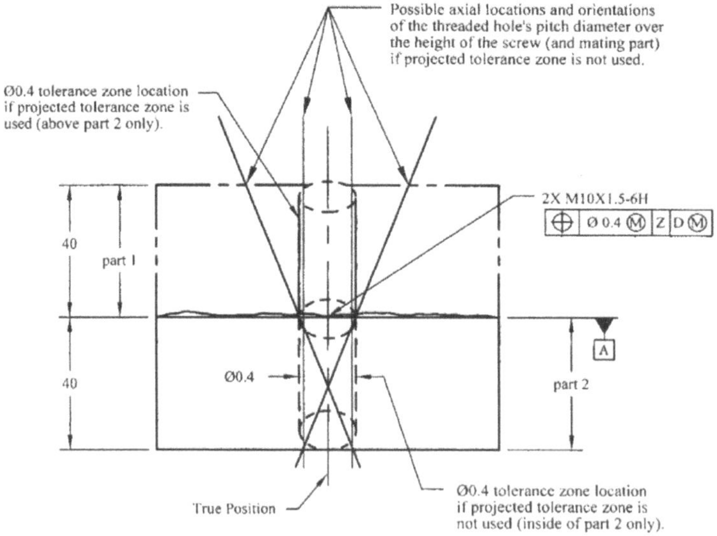

FIGURE 173 [Inspecting Threaded Holes with a Projected Position Tolerance]

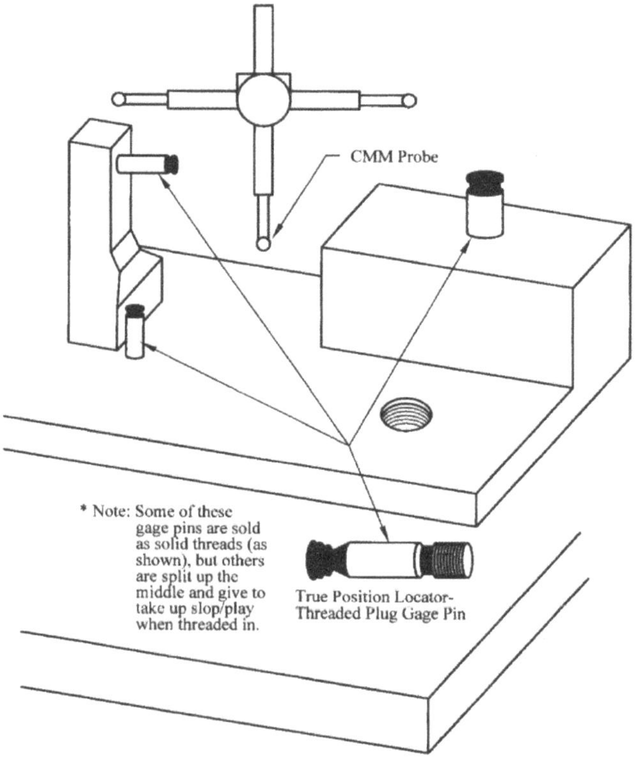

See FIGURE 173 for the inspection of a completely different part that also has threaded holes. It is shown ready to be inspected with a CMM (coordinate measuring machine) and threaded gage pins. Threaded plug gages are often used to inspect the position of threaded holes that have projected tolerance zones. The pin is screwed into the hole, then the CMM probe determines the hole location over the height of the tolerance projection. For this inspection procedure, it is helpful to have threaded gage pins that either expand or contract to meet the size of the hole's pitch diameter. That way, the gage pin doesn't wobble when being probed.

FIGURE 174

What does the MMC symbol after the feature E allow?

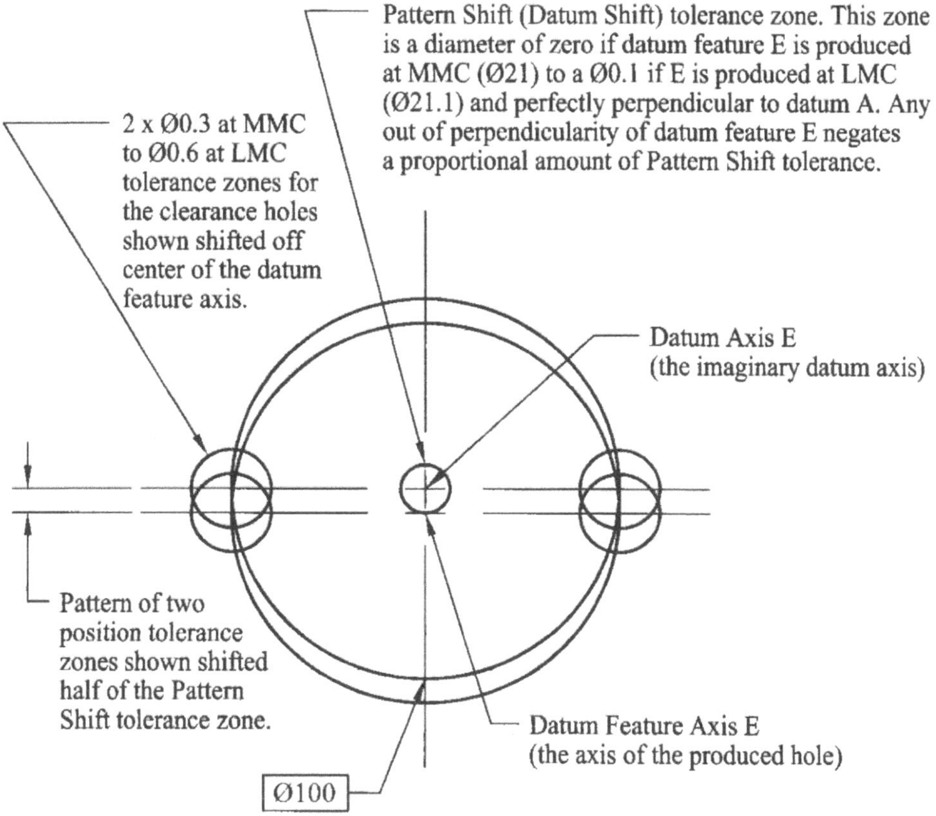

In FIGURE 174, since it is the datum feature that acts as the origin of measurement for the 2-hole pattern, the holes are seen as if they, as a group, have shifted in their relationship off the axis of secondary datum E (the actual diameter as produced).

In theory, it is viewed as the datum feature axis being allowed a movement away from the imaginary datum axis. In assembly, a part accepted with this datum feature shift may have to be shifted off the center of the imaginary datum axis (represented by the center of the assembly) in order to mate. See FIGURE 175 for clarification.

FIGURE 175

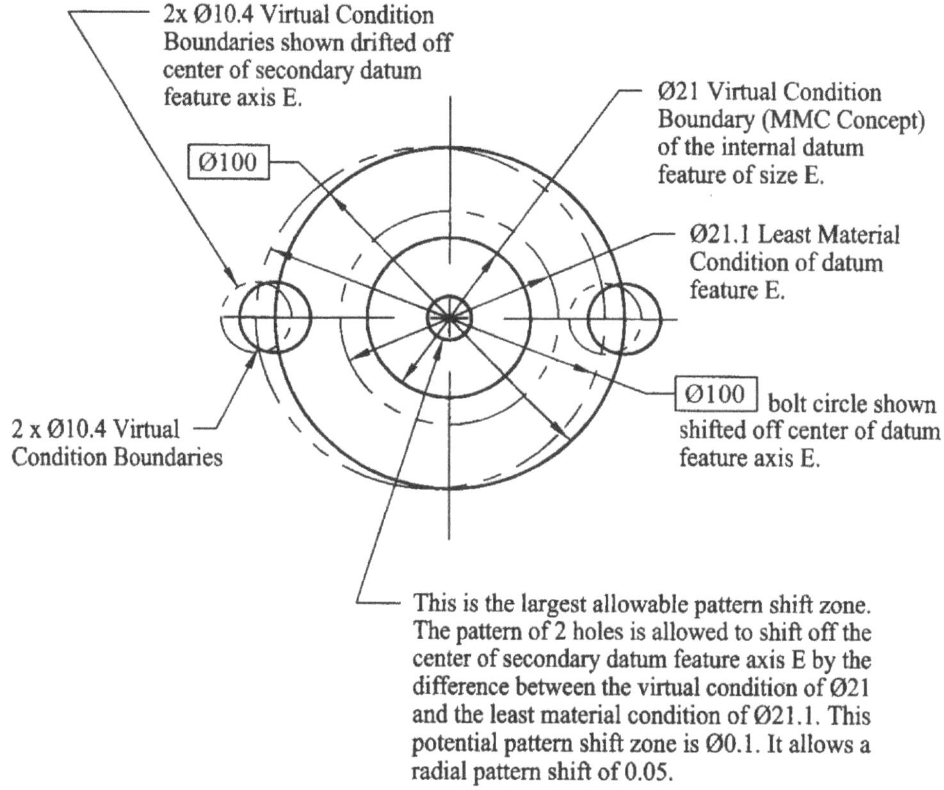

STEP 4:

Step 4 would require a coaxiality-type control be applied to the outside diameter of each part. These controls would reference the same datums as the position control on each part. To center the outside diameter of each part to the datums shown, a control such as runout, total runout, concentricity or position could be used. This last control would act to complete the Geometric Tolerancing Scheme.

Its tolerance would be calculated to be large enough to reduce production costs but small enough to preserve a robust wall thickness between the hole pattern and the outside diameter of each part.

FIGURE 176 [Step 4]

Detail drawing callouts

PART 1

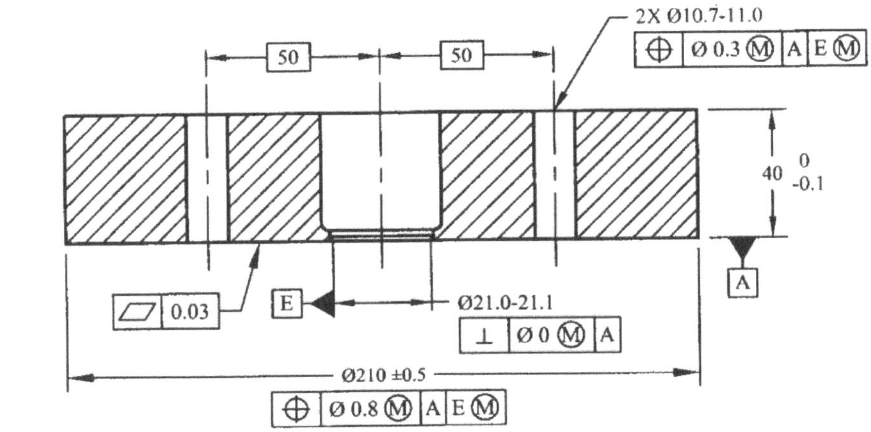

PART 2

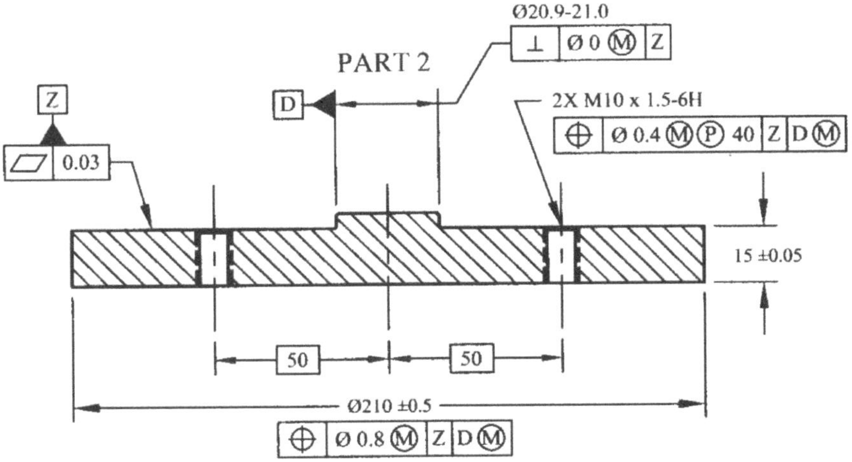

FIGURE 177 [The Final Assembly]

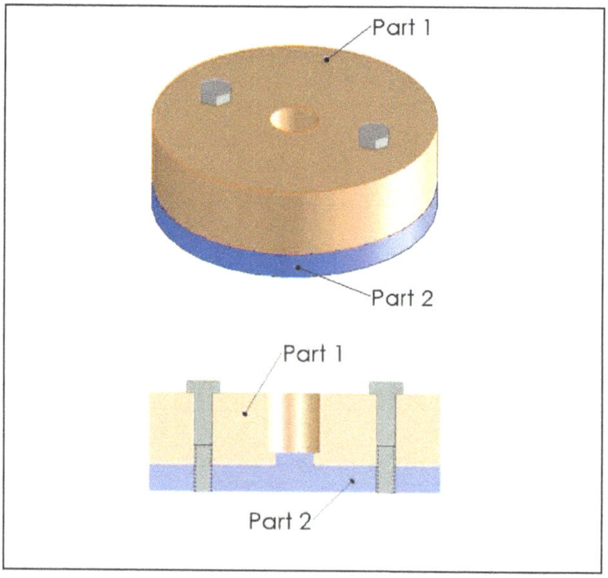

SolidWorks Corporation

Clearance Hole Sizes When Projected Tolerance Zones Are Not Used

When the projected tolerance zone concept and symbology are rejected for use on threaded holes, a substitute option must be selected to prevent binding or interference. One option to reduce interference is to reduce position tolerances. This will usually increase manufacturing costs. Another option is to increase the clearance hole size.

In order to calculate the size of the clearance holes necessary to allow assembly in a fixed fastener situation when threaded holes or tight-fitting shafts such as dowel pins are involved, the following formula may be used:

H = maximum material condition of the clearance hole
F = maximum material condition of the fastener/screw/dowel pin
T_1 = position tolerance of clearance hole
T_2 = position tolerance of threaded hole
D = minimum depth of engagement of fastener/screw/dowel pin
P = maximum projection of fastener/screw/dowel pin out of the tight fitting hole

The following calculation would substitute for the projected tolerance zone used in **FIGURE 176** and determine the maximum material condition of the clearance holes.

$$H = F + T_1 + T_2 (1 + 2P/D)$$

$$H = 10 + 0.3 + 0.4 (1 + 2 \times 40/14.95)$$

$$H = 10 + 0.3 + 0.4 (1 + 5.351)$$

H = 10 + 0.3 + 0.4 (6.351)

H = 10 + 0.3 + 2.5404

H = 12.8404

So, if a projected tolerance zone was not used in **FIGURE 176**, the clearance hole would have to be dimensioned to have a maximum material condition of a minimum of a diameter of 12.8404 to avoid interference. If this option is chosen instead of a projected tolerance zone, care should be given to the amount of contact the head of the screw will have with the mating part. If insufficient to bear the load, a washer may be required.

If 12.8404 was rounded off to 12.84, the following might be used on the clearance holes.

2X ⌀12.84-13.14

⌖	⌀0.3 Ⓜ	A	E Ⓜ

FIGURE 178

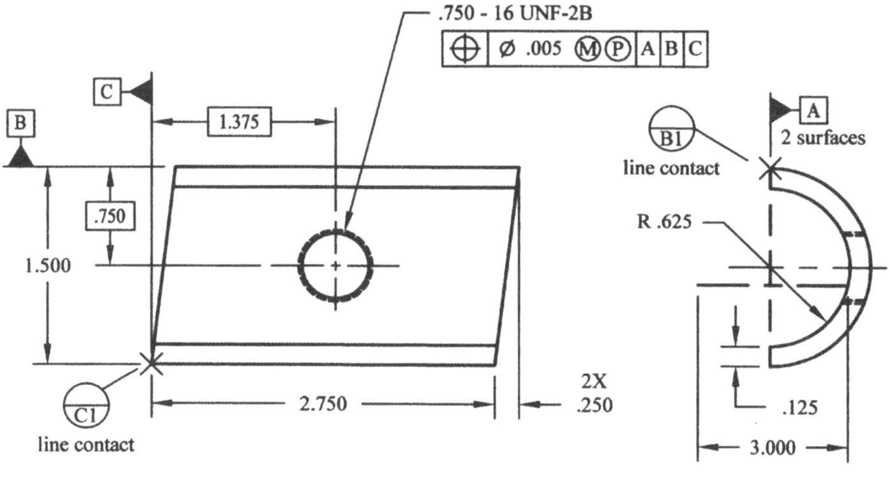

.XXX = ±.005
∠ = ±1°

Sometimes, it is difficult to determine the origin and direction of a projected tolerance zone. In these instances, an option is available which allows the height of the projected tolerance zone to be removed from the feature control frame. It is, instead, placed next to a chain line on the field of the drawing. The chain line is placed next to the center line of the hole outside of the part. An example of this is shown in FIGURE 178. The projected tolerance zone is 3.000.

Projected Tolerance Zone

In ASME Y14.5M-1994, the wording for what a projected tolerance zone means wasn't significantly different than it was in ASME Y14.5-2009 or in Y14.5-2018, although some clarification is given as it pertains to a composite positional tolerance (that the projected tolerance zone only applies in the level of control the projected tolerance zone symbol is shown in). The ASME Y14.5-2018 standard has not changed the projected tolerance zone concept at all.

The significant passage in all three Y14.5 standards is identical. The passage, in part, reads, "The specified value for the projected tolerance zone is a minimum and represents the maximum permissible mating part thickness, or the maximum installed length or height of the components, such as screws, studs, or dowel pins." "The direction and height of the projected tolerance zone are indicated as illustrated." It shows two illustrations. They are similar to the following illustration.

FIGURE 179

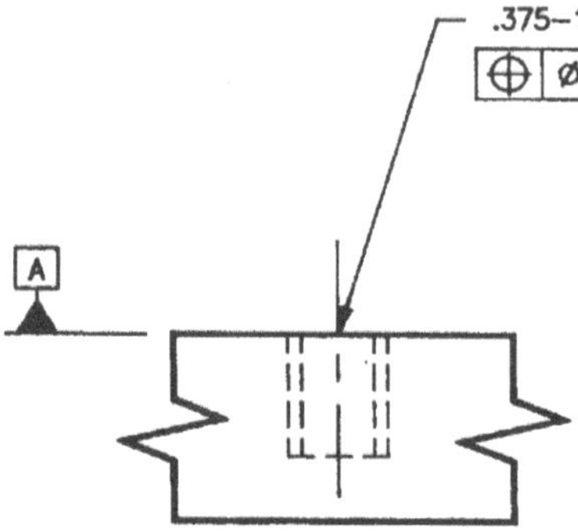

The section in both versions of Y14.5 goes on to say: "The minimum extent and direction of the projected tolerance zone are shown in a drawing view as a dimensioned value with a heavy chain line drawn closely adjacent to an extension of the center line of the hole." This passage first was used to introduce the optional way to show a projected tolerance zone that makes more specific where the zone begins and ends.

FIGURE 180

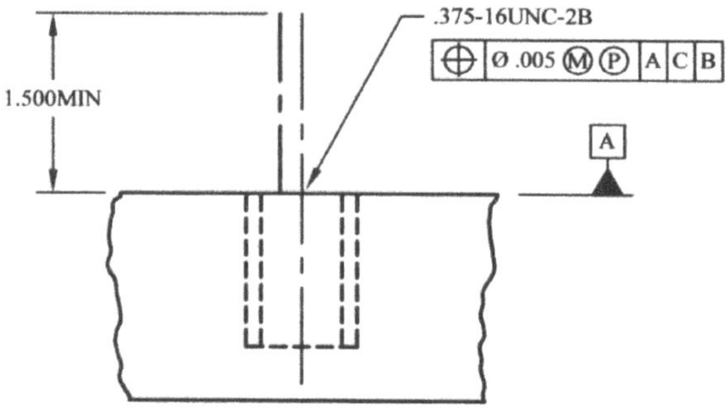

Notice that the abbreviation for the word "Minimum" (MIN) is placed next to the 1.500 dimension to show the minimum height that the position tolerance zone is to be projected out of the threaded hole above datum plane A.

In Y14.5-2009 and the Y14.5-2018 standards, the same text passage is used, but shows the following type of illustration:

FIGURE 181

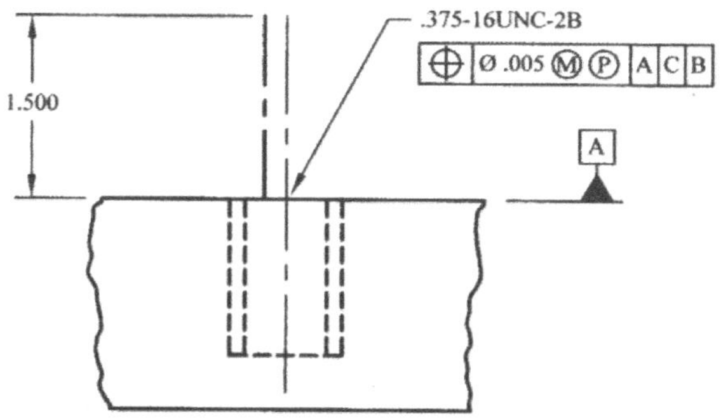

The only difference between Y14.5-1994 and the more recent 2009 and 2018 revisions of the Y14.5 standard is in the illustrations. The words given in the text of the standards would seem to support either illustration.

In the Y14.5M-1994 standard, it was never stated that the abbreviation MIN was to be used next to the height of the projected tolerance zone, but the illustration showed it.

In the Y14.5-2009 and the Y14.5-2018 standards, it does not state that the abbreviation MIN is not to be used next to the dimension that shows the height of the projected tolerance zone, yet the abbreviation is not shown. It is also curious that in the list of changes made from Y14.5M-1994 to Y14.5-2009 given in the 2009 standard, no mention of this is made.

In either case, the meaning is the same. The meaning is shown in the following illustration.
FIGURE 182

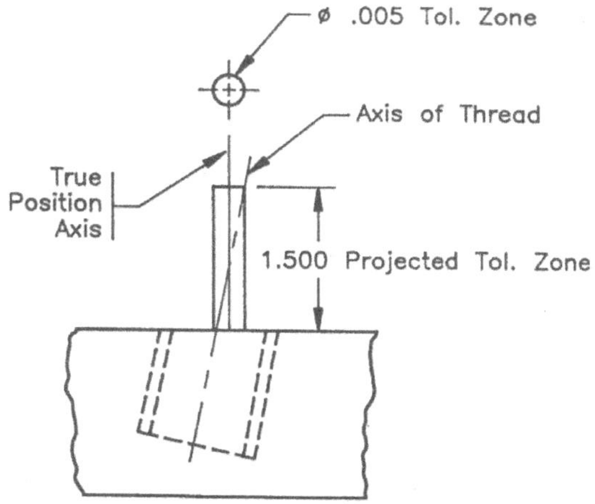

Threaded Holes Treated as Screws to Calculate Virtual Condition

The virtual condition of a threaded hole controlled with perpendicularity or position with a projected tolerance zone at MMC should be calculated exactly as though it were a screw. Use the MMC size of the screw that will fit into the hole, then add that MMC to the geometric tolerance (perpendicularity or position) to get the virtual condition (MMC concept). (Note: This also works for RFS conditions to calculate the worst mating boundary size.)

This virtual condition may be used to build a functional gage to check the tapped hole. The functional gage can consist of a plate to represent the A datum. It will have a hole in it the calculated virtual condition size. If the control is true position instead of perpendicularity, the axis of the gage hole will be at true position as simulated from the datum planes that make up the complete datum reference frame. With the actual part in the datum reference frame, if the screw made for the tapped hole can pass through the hole in the gage and screw into the actual part, the part is good.

The question is often asked as to whether the RFS or MMC concept should be employed in threaded holes. When a screw is inserted into a threaded hole, there is some "play" (slop, if you will) during assembly. This slop between the screw and threaded hole is the result of the pitch diameter of the screw being smaller than the pitch diameter of the threaded hole, and this slop or "play" helps us to more easily assemble the parts. Hence, there is more slop if the class of fit is a 1B instead of a 2B, and even less for a class 3B threaded hole. The larger the pitch diameter of the hole, the more "play" (slop) we derive.

If the "play" is utilized to help us assemble the products, it should be allowed as additional tolerance during the inspection procedure. This is bonus tolerance, and using the MMC concept in the control recognizes that. If you can't measure the bonus tolerance, the dimensioning and tolerancing engineer can suggest gaging procedures that will automatically accommodate the allowed minimum bonus tolerance available for that hole (for example, functional receiver-type gaging).

If you do not wish to utilize the bonus tolerance because you can't quantify its exact number of thousandths of an inch or portions of a millimeter, that is your decision. Use of the RFS symbol in the control will negate it in the inspection procedure. Still, for a tolerancing engineer (as merely a describer of physical phenomena) to say it doesn't exist by putting in an RFS modifier instead of an MMC is not recommended. The RFS symbol implies that the hole contracts to meet the screw size--and that is rarely true.

In visualizing the act of a screw and threaded hole assembling, don't view the additional "play" (as the hole's pitch diameter grows) as a tolerance zone growth, rather see it as allowing you to use the available "play" to push the screw over into the projected tolerance zone of the hole. In inspection, see the gage screw as entitled to the same advantage. Because, however it is viewed, additional tolerance exists even if it is small and we can't ascribe a number to it. As tolerancing engineers, we are tellers of physical truths. **A feature control frame is nothing but a sentence in the form of symbols strung together like words and phrases.** If we lie in a geometric control, the result is usually either: 1) parts that work are rejected, or 2) parts that don't work are accepted.

Importance of Formulae for Floating Fastener Situations or Fixed Fastener Situations

The formulae for floating and fixed fastener situations are important in so much as they allow mating parts to have size and positional tolerance in combination so as to create a situation wherein the worst mating conditions of the mating parts will be compatible.

For example, in the floating fastener cases, the main purpose of inputting the MMC of the holes and the MMC of the fasteners into the formula to calculate the holes' positional tolerance is to create a MMC concept virtual condition (a boundary or tube) equal to or greater than the MMC of the fastener that will fill it. This virtual condition boundary will run through the thickness of both parts and no special or extra data must be inserted into the formula. It works for any number of holes and any number of hole patterns. The important factor in whether or not the parts will mate at assembly is the positional MMC concept virtual condition boundary being compatible on each part for every respective hole each fastener fits through.

With fixed fasteners, the principle is the same. If the worst mating conditions are compatible (with help from projected tolerance zones, where applicable), the parts will mate at assembly.

Projected Tolerance Zones for Tight Fitting Holes

The reason we use projected tolerance zones on tight fitting holes is to assure that once the shaft is pushed into the hole, it won't lean too far to increase its effective mating size and not allow it to mate with (fit inside of) the hole(s) on the mating part(s). The fit of the shaft, although tight fitting into one part, is not necessarily a tight fit into the other (mating) part.

FIG. 183 PROJECTED TOLERANCE ZONES
(Verification Without Functional Gaging)

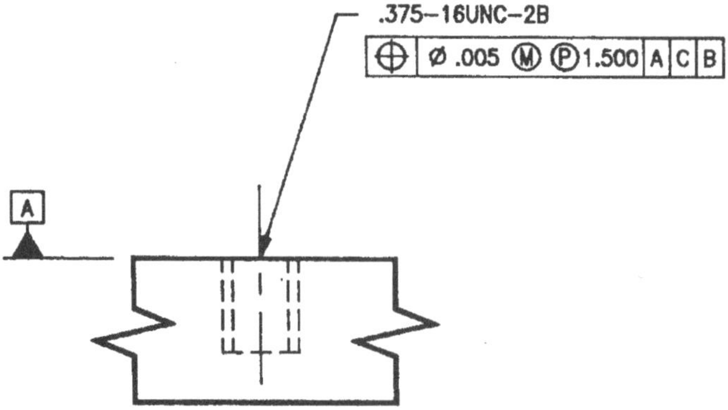

Besides the use of functional gaging to check the location of this threaded hole, verification can also be accomplished through the use of a pin threaded on one end (to the depth of the hole) and with a smoothly ground (1.5" long) periphery on the other. The threaded end could be screwed into the full depth of the controlled hole, the part oriented to and positioned from datums A, C, and B and the axis of the 1.5" length checked with conventional methods for location to determine if the axis of the hole, as dictated by the pitch cylinder, when projected over the height specified [1.5 inches], exceeds the confines of the .005 diameter tolerance zone.

Tight Fits for Features with a Liberal Positional Tolerance

The question sometimes arises as to the correct procedure to designate tight fitting, and perhaps loosely located features such as dowel holes.

ASME Y14.5 recommends the use of projected tolerance zones. However, when doweling two or more parts together (where tight fits and the coaxiality of the holes on one part to respective holes on the other parts are very important, but location of the holes on the parts can be loosely controlled other than that), other methods might be beneficial.

When not using projected tolerance, probably one of the easiest and most clear procedures is to show the holes in an assembled view. With the two or more parts shown together, the mating features can be given very tight size limits with one feature control frame with perhaps a very loose positional control. In these situations, MMC would probably be the most logical material condition because an RFS condition would limit any additional positional tolerance as the features departed from MMC. The difference between MMC and LMC (possible bonus tolerance) is very small anyway; and, when you consider our criterion is loose location (other than coaxiality) of the dowel holes, then MMC makes even more sense. One important criterion would be to clarify that the features are to be treated as one. A note such as "MATCH DRILL and REAM" would act as clarification.

FIG. 184

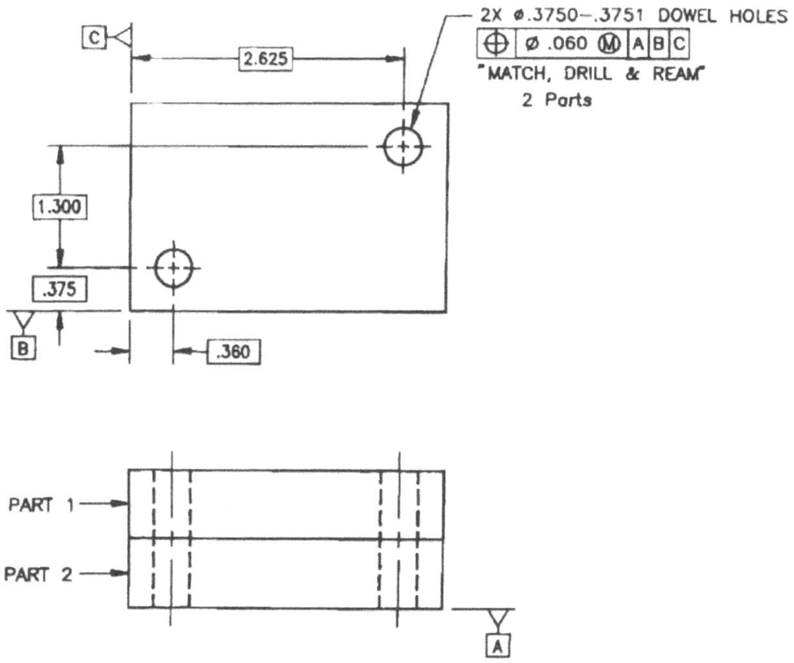

Another way to define such a situation is shown below.

FIG. 185

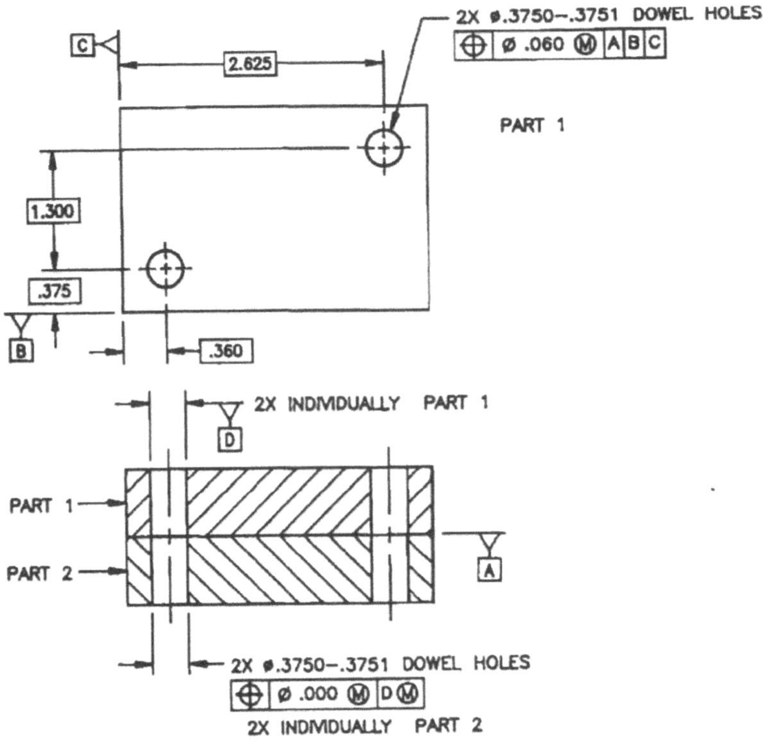

Unless notes were added, this control would leave it up to the logic of the manufacturer that holding such a tight coaxiality (from the holes on Part 1 to the holes on Part 2) would be much easier if the holes were "MATCH DRILLED and REAMED".

The use of projected tolerance zones for the type of situations previously shown adds very little, if any, benefit to features on parts that are match drilled and reamed. However, if a tight fit is needed of the dowel pin into holes on only one of the parts, while a loose fit of that dowel into the holes on the second part is desired, projected tolerance zones would be very beneficial to control holes on the first part. These features need not be match drilled, but virtual condition boundaries must be created that assure mating.

A LOGICAL APPROACH TO PART TOLERANCING

LINEAR SEGMENTED THINKING

FIG. 186

Example: A Crankshaft Sub-Assembly

Linear segmented thinking is the conscious act of defining a part (or parts that interact with one another) one element at a time. This is done by asking questions about the parts that are basically how the part features physically act or interact. One must focus on the feature one is defining, thinking in depth about it only and relating it only to features that have been defined prior to it. If nothing else has been defined because it is first, then that feature can only be considered for a form control.

In this way, one works one's way through a part definition as though through a story, leaving no doubt as to the beginning, middle and end. It is possible that one part may interact with several components and would, therefore, have several stories to tell, but each would follow the same rules of definition. If the stories relate to one another because they have common components, such as common datums, or one defines controlled features within one story that become the datums for the next, then the stories must be interrelated with a specific sequence of events.

If this sequence of events also describes a viable manufacturing sequence, then one has created a well-defined, fully-functional and easily-producible product--the goal of the tolerancing engineer. For example, see FIGURE 186 and 187.

This is a crankshaft sub-assembly. It will be used to demonstrate this process. Notice how the parts within the assembly interact. For this exercise, pretend nothing else exists other than what you see here. Once the parts are designed for geometric configuration and features sized, as shown here, one must ask, "What do I do first?"

FIG. 187

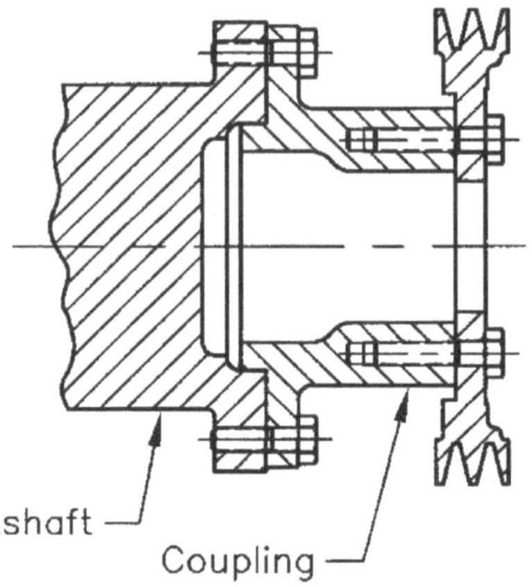

FIG. 188

Detail drawing callouts

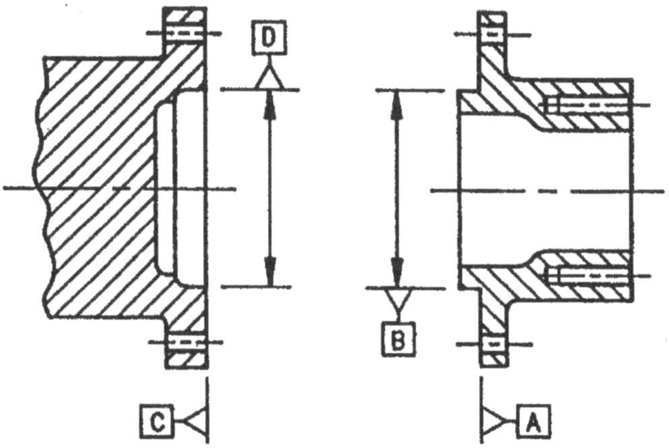

FIG. 189

Detail drawing callouts

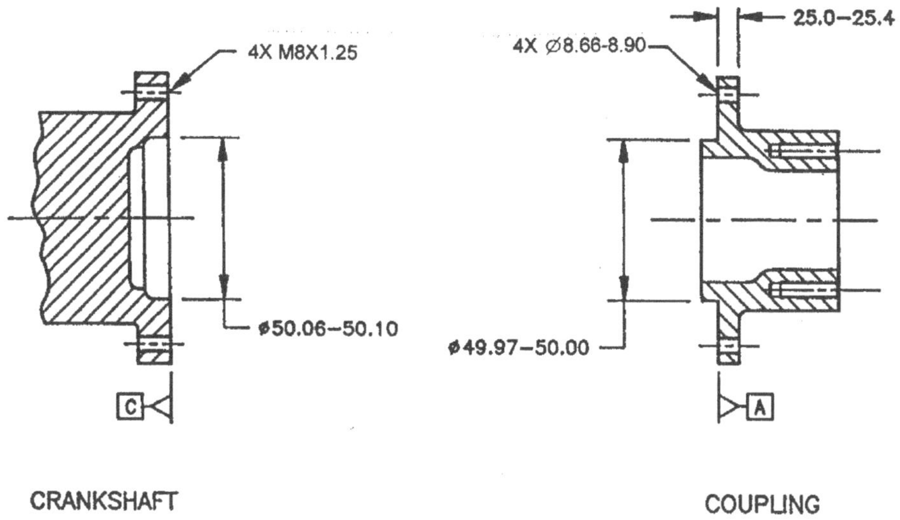

Another word for "first" is "primary". We must first choose the primary datum feature on both parts. In order to do this, we must ask two questions:

1) When the parts assemble, what surfaces need the most physical contact?
2) What surfaces/features actually will create the angle at which the subsequently related part features will function?

These two questions are important because if a planar surface, for example, needs a minimum of 3 high points of contact, and it is not chosen as a primary datum feature, its chances of receiving such contact and bearing the load when bolts are tightened diminish greatly. The chances of some other unsuitable feature, such as a shaft or the bolts, being side loaded, increase. Also, the primary datum feature creates the angle at which features controlled to it are created. Whether this angle is one of perpendicularity, parallelism or angularity is determined by which datum feature is chosen as primary. Sometimes a primary datum feature does double duty and controls not only angle but also distance/location. But make no mistake; it always, at the least, creates the angle of the features controlled to it.

So, we must look to the assembly drawing to see what surfaces must seat and what features must clear/mate with one another. It is clear that the center diameters mate but need not touch. They need not contact; therefore, they do not create the angle. The seating surfaces will be given names through the use of the datum feature symbol. We will call one datum feature A and its seating surface datum feature C. These are the beginnings of the geometric definition of each piece, the beginning of the story of each--the primary datum feature from which other features will take their angle and the angle at which each part will rotate. If they were not chosen as primary, in a worst-case situation, the part surfaces would not seat at assembly. If one were to look at surfaces A and C, a point of contact would be seen to touch, but not three.

The surfaces would be at an angle to each other, the threads of the bolts visible through the seam. If more force was then applied to the bolts in an attempt to close the gap, side loading of the bolts and/or the central external diameter would occur. This could cause them to fatigue, break or strip. This is a mistake we will avoid, by making A and C primary datums.

We must focus our attention on these surfaces and ask ourselves what is expected of them. They must seat on one another without rocking too much. Has either been controlled with a surface form tolerance? Though not flamboyantly evident, one has. Datum feature A has, in combination with its opposing side, been given a size tolerance. Since no flatness tolerance appears directly in a feature control frame on datum feature A, under Rule #1 in ASME Y14.5, the size tolerance controls the surface form. Datum feature A must be flat to within the difference between 25.0 and 25.4. A tolerance of flatness of 0.4 has been imposed on datum feature A.

We must look to the size limits we will give to the features that will be subsequently controlled to datum feature A. Since this is a tightly toleranced part with tightly controlled features, it is logical that the primary datum feature have a tight enough tolerance to make it sufficiently reliable from which to orient and measure these subsequent features. Since 0.4 is not sufficient to limit the rock when compared to the tolerances given these features of size, we should tighten it. We could simply tighten the size tolerance between datum feature A and the opposing size of the part. In this situation, however, it would be more cost effective to leave that alone and refine the flatness of datum feature A with a feature control frame tolerance. This tolerance must take into consideration the balance between a perfectly flat, impossible-to-produce surface and one that is so out-of-flat as to endanger functionality but cost very little to produce. One must discuss with manufacturing personnel the capability and cost to produce a certain amount of flatness on the surface and weigh that against the overall product cost and functional requirements. It does no good to produce a part no one can afford to buy. In this case, let us assume the balance of all these requirements led us to a flatness tolerance of 0.01 and apply it to both datum features A and C.

If they are primary, the first element in the creation of the geometric definition of each part, we must then choose a second, or secondary, datum feature for each. The secondary datum feature should create a plane or an axis from which to locate subsequent features. The secondary datum feature should be functional and representative of a mating feature or alignment feature. The central diameters that interface/mate fulfill these criteria. We will name them datum features B and D. Datum feature B goes into/mates with datum feature D.

A geometric control must be considered for datum features B and D. Let us contemplate datum feature B. It must be considered for form control. But currently, the form is being controlled by Rule #1 in ASME Y14.5. If produced at 50 (the maximum material condition), the diameter must be perfect in roundness, straightness and taper (cylindricity). The same is required of its counterpart mating datum feature D at a size of 50.06. Since 50 with perfect form will fit into 50.06 with perfect form, the form control is sufficient and there is no reason to refine it.

So, we must look for another suitable control. Some might suggest position or concentricity or runout, but this is not possible at this stage of creation/part definition. Since only datum feature A exists at this time on the part as the first defined feature and we are now defining datum feature B, the only interrelationship B may share is with A. This relationship is one of angle, not location. The angle is one of simple perpendicularity of an axis (that of datum feature B) to a plane (that formed by datum feature A). So, datum feature B is assigned a perpendicularity control back to A. Simultaneously, datum feature D is assigned a perpendicularity control back to C.

If you are looking at the control assigned to datum feature B, the control can be read, "The axis may be out of perpendicularity 0.03 millimeters if produced at maximum material condition to datum plane A" or "Perpendicularity within 0.03 is to be held for a diameter that mates while the part is seated on datum A (or C)". The tolerance has been calculated using the fixed fastener formula of the maximum material condition of the hole (Ø50.06) minus the maximum material condition of the shaft (Ø50.00) equaling the geometric tolerance (0.06) to be divided between the mating features. If one of the features is more difficult to make perpendicular than the other, then the tolerance is divided unevenly, giving a greater portion of the tolerance to the one most difficult. In this case, we have determined they are approximately equal in difficulty, so the 0.06 has been divided equally and datum feature B assigned half (0.03). What you do to one feature on one part, you attempt to do to its counterpart feature on the other. So, datum feature D is assigned a similar control using the counterpart of datum feature A (which is datum feature C) from which to hold its perpendicularity.

FIG. 190

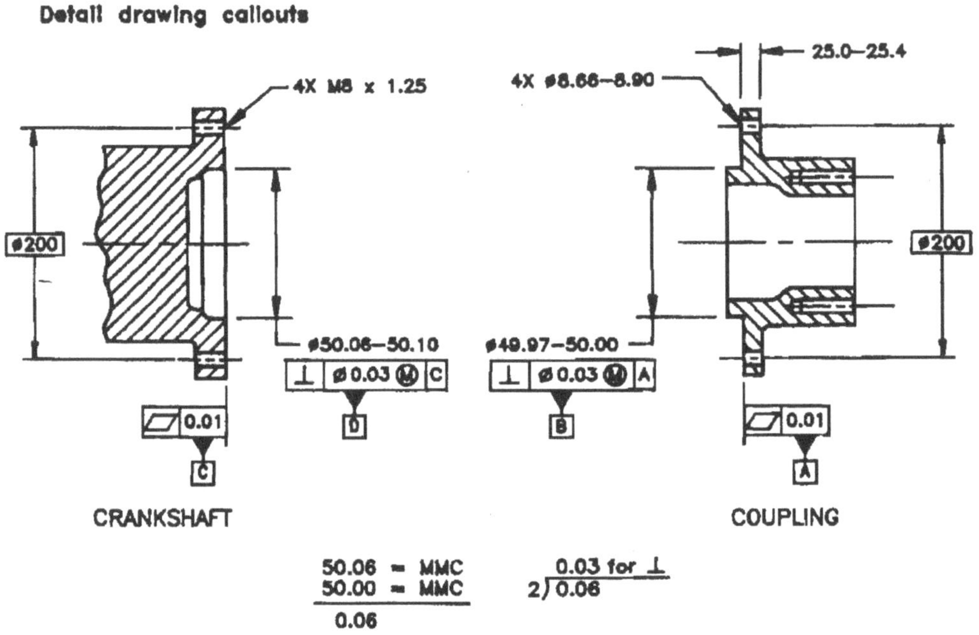

Now that the datum features primary and secondary have been chosen, assigned and geometrically controlled and interrelated on each part, we can determine if any other datums are necessary. In this example, we are only going to take the dimensioning and tolerancing up to the point where the crankshaft and coupling will bolt together; therefore, the bolt holes are the only remaining features to be defined. If the holes are defined to existing datums, that will be sufficient to stop all degrees of freedom for these features except pattern rotation around the axes of either datum B or D. Since no distinguishing features, such as keyways or flats-on-the-periphery exist on these parts, it will not be necessary to create a datum feature to stop this pattern rotation. Therefore, sufficient datum features exist to complete the exercise.

At this point, since the bolt holes have not been defined, we have accomplished the following with existing controls:

1) The flatness controls on datum features A and C will allow these surfaces to seat on one another without rocking too much.

2) The perpendicularity controls on datum features B and D will allow B to insert into D. These, in tandem with the flatness controls, assure us this will happen while datum features A and C seat without rocking too much.

Now, in order to bolt the parts together, we must define the bolt circle as a basic diameter from hole to hole; or, as an equivalent alternative, we may give a basic dimension from the center of datums B and D to each hole. The holes are then ready to be toleranced in a feature control frame from the existing datums. To start, we will look at the clearance holes. The four 8.66 - 8.90 holes need to be given a tolerance for (1) their angle from the primary datum A and (2) their distance from the secondary datum B.

Since a geometric characteristic symbol like concentricity or runout would assume the controlled holes individually share the same axis as datum B, these controls are not suitable. The axis of datum B is shared by the imaginary bolt circle of the hole pattern, but that is not what is being toleranced. The language described in Y14.5 tolerances real features--not imaginary things. The holes are real and a distance from the axis of B. The basic diameter of the bolt circle is an imaginary target from which each of the holes may stray. The feature control frame we are about to create will define the allowed tolerance of the deviation from the perfect angle and location the holes may experience.

Position is a geometric characteristic strong enough to control angle and distance from the existing datums and more appropriate than any of the others. Therefore, position will be used. A statement such as, "Position of diameters that mate while the part is seated on datum feature A and the holes are centered about the axis of datum feature B while B spins" is constructed through the use of geometric symbology. Or, an equivalent statement, such as "The axes of these holes may be out-of-perfect position a certain amount if they mate at maximum material condition to datum A for perpendicularity and B for location regardless of the size of datum feature B" is constructed with the same symbology. For example:

4X ⌀8.66–8.90

⌖	⌀0.22 Ⓜ	A	B

This control can be read, "Position of the axes must be held to within a diameter of 0.22 if produced at maximum material condition (8.66) to A for perpendicularity and B for location regardless of material boundary (RMB)."

What is done to one set of features is attempted for the counterpart set of features with which they interface. Therefore, the M8 threaded holes get a similar control, such as:

This control can be read, "Position of the axes of the threaded holes' pitch diameters must be held to within a diameter of 0.44 at MMC while the positional tolerance zones are projected outside of the part 25.4mm, perpendicular to datum plane C and located from axis D regardless of material boundary (RMB).

The threaded holes don't receive a one-for-one bonus tolerance for positional tolerance as do the clearance holes because of the centering effect as the pitch diameters of the threaded holes and screws engage. Although some additional play or bonus tolerance exists as the pitch diameters of the screws shrink and the pitch diameters of the holes grow, it is a small amount and difficult to quantify. Therefore, the threaded holes are at a disadvantage to the clearance holes. For this reason, we have assigned more of the available geometric tolerance to the threaded features. For example:

MMC Holes	=	8.66
MMC Screws	=	8.00
Geo Tol. to be divided	=	0.66
Threaded Holes	=	0.44
Clearance Holes	=	0.22

The tolerance assigned to the threaded holes must now be put into a cylindrical zone that resides outside of the part. As is discussed in the section of this book on fixed fastener situations for threaded holes, the tolerance zone is projected to the position in space which represents the portion of the body of the screw which is not threaded into the tapped hole. Since in this situation the heads of the screws will rest on the top of the mating part, the positional tolerance zone is projected a minimum of the maximum thickness of that part, which is 25.4. The portion of the tolerancing discussed in this step is displayed in the last figure of this section.

Notice that a bolt circle diameter of 200mm basic has been added to the final illustrations. Also, a size and a circular runout control has been added to each outside diameter. The allowed deviation from the 200mm basic bolt circle is toleranced by the position control on each part. The runout control is to make certain the outside diameters are spinning true, at the proper angle and coaxiality within the specified runout tolerances. The sizes of the outside diameters and their runout tolerances were determined by setting an acceptable wall thickness on each part and testing the limits of how far off center the diameters may be before an unacceptable amount of vibration is experienced.

A good drawing is easy to follow. It is like a map with a point of origin, a destination and specific interconnected features/pathways to get you there. You never feel lost reading it because one control feeds directly into the next until completion of the product definition. These steps are best created by focusing as much attention as possible on the features you are defining at that moment, making certain they are complete and correct. Only when you are certain that step is finished and your direction is correct do you move on to define the next feature or pattern.

FIG. 191

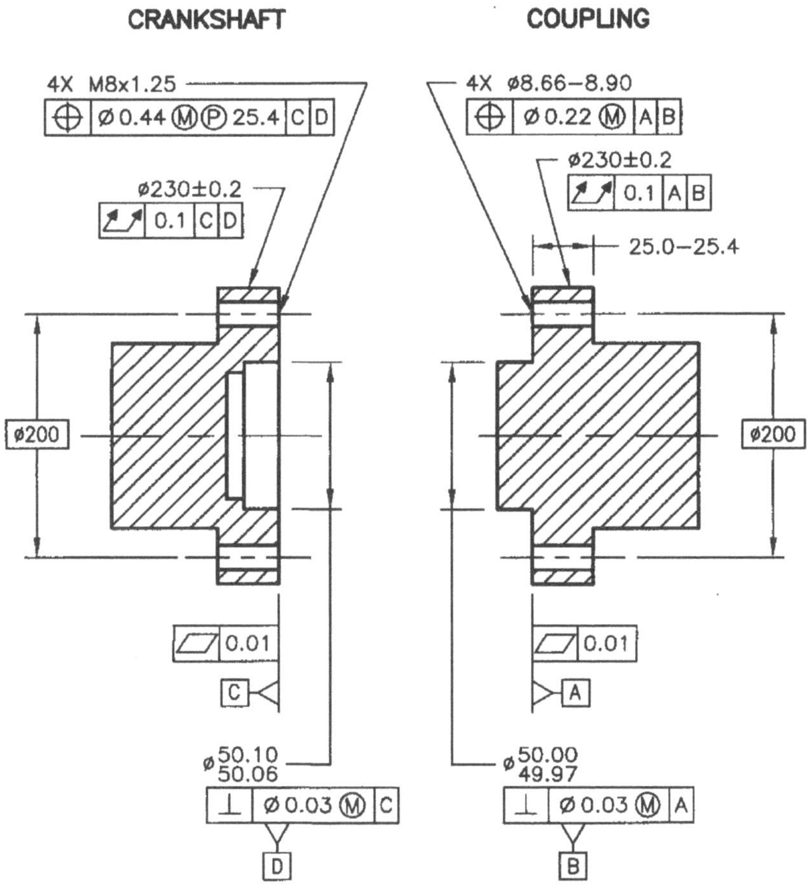

DOING SOMETHING VS. DOING NOTHING

During a recent workshop I was faced with an individual I felt I had encountered at many other companies in my travels. He wears a different face each time. Sometimes he is a she, at times older and at others younger, but the mentality remains the same.

I had just spent a significant span of time explaining the workings, interrelationships and other pertinent facts concerning a product, with the goal in mind to involve the group in a discussion about what dimensioning and tolerancing was needed. So, with all necessary groundwork laid, I said, with a great deal of anticipation, "Okay, what should we do first?"

The immediate response I received came from the front row. "That depends on the function of the part," he screeched, as though his very existence was at stake. "We can't touch that part. The designer is the only one that can decide these things!"

I smiled. "I am the designer," I said. "I have just spent a great deal of time telling you exactly what the function of the part is and how it relates to the assembly. You have an assembly drawing in front of you, showing exactly how the part interrelates. You have a detail drawing of this part. I will give you any additional information you think you need to begin suggesting tolerancing controls to use. Let's just discuss it. If you'll suggest something, the group will go over the advantages and disadvantages of the control and from that decide if it is the best in this situation and for that feature. If it is, we will use it and go on to control other features until the part is complete." I smiled again reassuringly and waited for him to make a suggestion.

He did. His suggestion was, "This is not possible! If we change anything, we could be inflicting unknown difficulties into the design. It's best to just leave it alone and let the people in manufacturing make it. They will know what we need."

"Through psychic guidance?" I asked. "We are the ones who know what it needs. You know the function. The part needs to be dimensioned and toleranced. For goodness sake, **JUST DO IT!**" Normally, I am not the excitable type, but this "Anything but action", "Excuses, excuses and abuses", "We can't do anything about that because" stuff is about to grind me down.

I know this guy has made a career explaining to his peers and bosses why things are so bad and why nothing can be done about it, while his competitors have seen his non-action as an opportunity to design and build better - and at a lower cost.

I often put up slides of a company's products (after a few days of training them on generic products) and generate discussions about what is good, what is not so good, and what is blatantly incoherent and in conflict with the functional requirements. Some people amaze me with their knowledge and flexibility. It doesn't matter to them what is on the current drawing or model, whether they have created it, or it was drawn by someone sitting to their right or by someone who retired 20 years ago. I can look into their eyes - so filled with new knowledge - and tell that all the lights are on and that everyone is home. They are writing furiously on the drawing copy they have in front of them and verbally making suggestions to lead the conversation. They are blending the recently studied examples, folding them into their products right before me, applying what they have learned at breakneck speed to rectify that which they can now easily see was not in the best interest of their products and their company and themselves.

And I also see others. The looks on their faces horrified! Years of side-stepping and loud talk, of pretense and ignorance being washed away in the cleansing light of new-found knowledge. These are the ones that would rather retire than learn, destroy colleagues rather than have them shine, and build the walls in the scrap area higher rather than change the process. They are insecure little people with more power and bluster than they are entitled to, standing on the road of progress, mean and afraid. **They are in the way.**

I don't want to hear why it can't be done, only ideas and suggestions as to how it can. I want healthy, cooperative discussions - not arguments. Give me creativity and inspiration. I don't want to hear that the problem is too big and the people and resources too small to solve the difficulties. I need. . . **we all need** . . . tenacity and dedication, open-mindedness and deductive reasoning. **We need to work together**, not be threatened by one another.

I have seen the panic in their eyes when the solutions to age-old (and what in their minds have seemed to be unsolvable) problems just naturally float to the top in discussions in my classroom, inspiration striking the person next to them - the ones with the lights in their eyes - and yet eluding them completely. They want to protect what was theirs, the bliss of their ignorance, their position as top intimidator, the loudest yeller, the one with the boss's ear. And yet, they are helpless as their power melts before them, their protests drowned out by the ideas to improve, the zeal to correct, the willingness to adapt and the insight to apply displayed by their peers - whom they once considered less than their equals and are quickly, and for all to see, becoming the future of the company. Their voices, once so strong and confident in ignorance and negativity, becoming tinny and whinny sounding until they are nothing but a whisper, a Neanderthal grunt ignored or finally recognized for what it is, what it has always been.

These are wonderful and sad situations, like watching a vile, loathsome creature sinking in quicksand, throwing them a rope only to see them shrink from it as though you had thrown in a venomous snake. They sink lower and lower, until they are hard to see at all. But they are not gone. And yet, if we are lucky, we will never hear from them again.

Chapter 11

Tolerancing Mating Parts in a Floating Fastener Assembly (Two Single Segment Positional Controls)

•Floating Fastener Assembly Condition
•Two Single Segment Position Controls

Chapter Objectives

Readers will learn:

1. How to convert a floating fastener assembly with radial hole patterns from plus and minus tolerances to geometric tolerances.

2. How to assign the optimal datum features.

3. How geometric tolerances and datum features look when assigned to a CAD model.

4. How to tolerance two-single segment position tolerances to mating parts in an assembly.

5. How to hold tight tolerances for important functional relationships and loose tolerances for less important relationships.

6. How datum shift due to maximum material condition symbols used after multiple datum features on multiple parts accumulate and affect an assembly.

Floating Fastener Assembly Condition

The two part assembly shown in FIGURE 192 must mate with the following requirements:

1) The inner shaft must go into the outer shaft.
2) The planar end of the 37.95-38.00 inner shaft must bottom out in the 38.05 – 38.12 outer shaft's hole.
3) The 6.25 clearance holes in both parts must align well enough to insert the 6.0 bolt. This bolt and its accompanying nut are used to bind the parts together. Since the bolt is not fixed at any angle by either part's 6.25 holes, this is considered a floating fastener assembly condition.
4) The title block tolerance for angles unless otherwise specified is plus or minus 1 degree.

FIGURE 192 [Illustration of Assembly and Parts to be Toleranced]

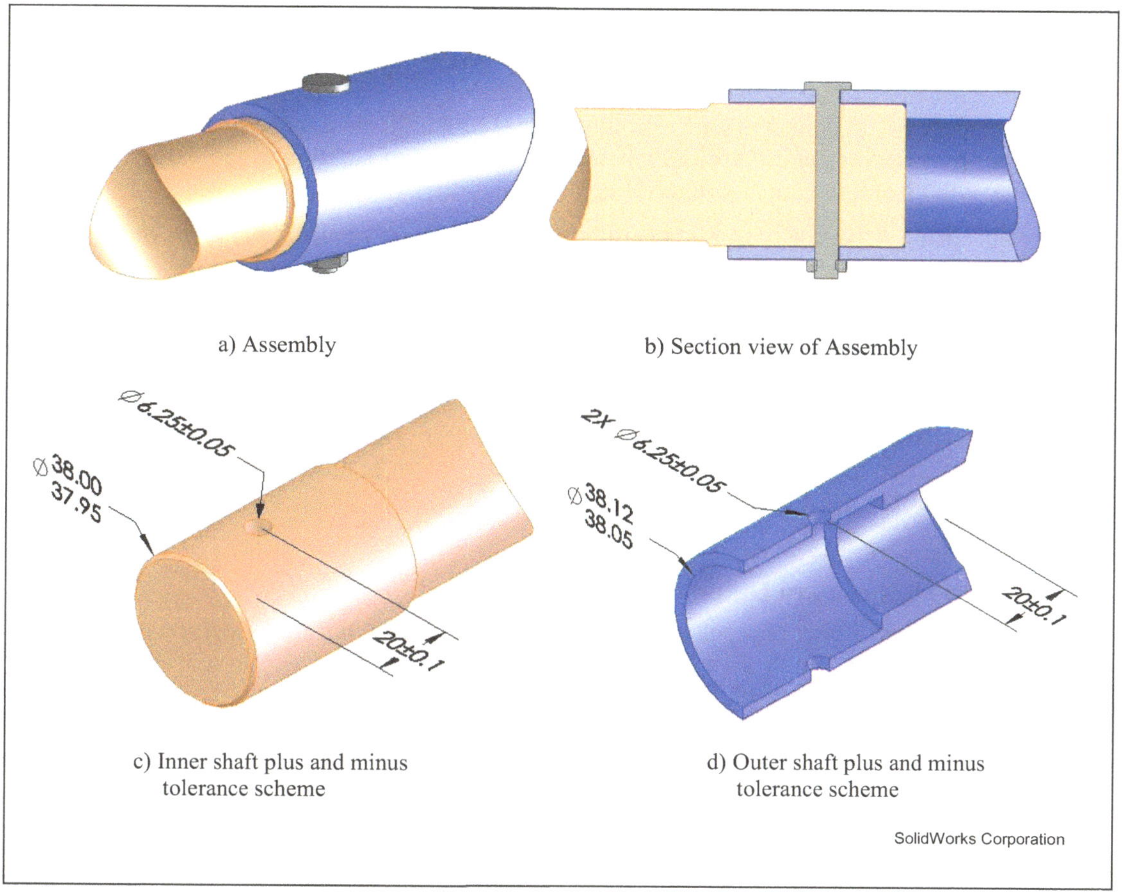

SolidWorks Corporation

Selecting the Primary Datum Features

It is clear that the designer wanted the 6.25 holes on these parts to intersect either the axis of the 37.95-38.00 or the axis of the 38.05-38.12. But, since there is no tolerance stating how well they must intersect, the drawing is considered incomplete. Unless such a tolerance is displayed in accordance with the principles of the ASME Y14.5 standard on Dimensioning and Tolerancing, the parts could easily be manufactured and the features not align well enough to mate.

To begin, we will choose a primary datum feature on the basis of what features a) mate, b) align the parts to one another, c) have enough surface area to stabilize the parts in the way that they function, and d) have a precise enough form to allow measurement data to be repeatable from measurement to measurement, no matter how many times the part is set-up.

a. The 37.95-38.00 diameter shaft mates with the 38.05-38.12 hole, so it qualifies as a primary datum feature under this criterion. But, the 6.25 holes also mate with the bolt, so they also qualify as a possible datum feature.

b. The mating features with the tightest fit from part to part act to align the two parts in the assembly. Under this criterion of alignment, only the 37.95-38.00 diameter shaft and the 38.05-38.12 hole that it mates with qualify as the alignment features. Their fit is tighter than the 6.25 holes and the bolt they mate with.

c. The 37.95-38.00 diameter shaft and the 38.05-38.12 hole have more surface area than the other part features and therefore have a better chance of stabilizing the parts in the way they will assemble than any other selection possibility.

d. Under Rule #1 in the Y14.5 standard, the size tolerance controls the form of a shaft or hole. Therefore, the shaft and hole with the tightest size tolerance must have the best form. The 37.95-38.00 diameter shaft and the 38.05-38.12 hole have a tighter size tolerance than the 6.25 holes. They are therefore required to have a more precise cylindricity (cylindricity controls roundness, straightness and taper).

So, the primary datum feature symbol will be assigned to the 37.95-38.00 diameter shaft and the 38.05-38.12 hole. In this case, we will call the primary datum feature on the inner shaft A and the primary datum feature on the outer shaft B. See FIGURE 193.

Assigning the Secondary Datum Feature

The axes of the 6.25 holes on each part will be held perpendicular and intersect the primary datum axis. The type of control capable of tying the hole axis to the datum axis for angle and location is a position geometric characteristic.

Next, we will move on to control the movement of the 6.25 holes longitudinally up and down the datum axis. If the holes get too close to the front of the part, the wall may be too thin. Another concern is that we may not get enough engagement of the parts in the assembly to stabilize them or to allow the end of the inner shaft to bottom out in the bottom of datum feature B.

FIGURE 193 [Assignment of the Primary Datum Features]

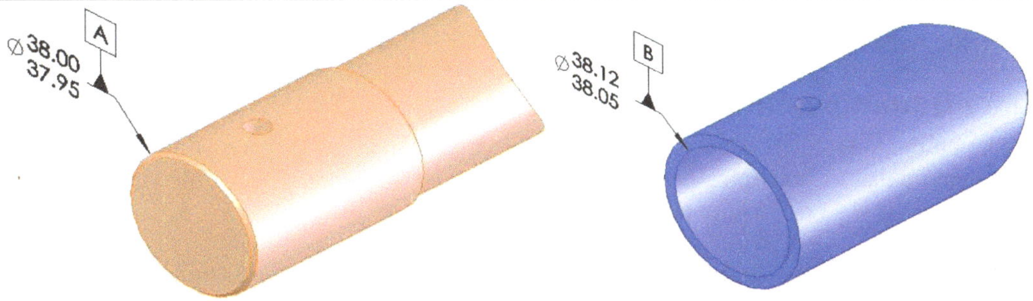

a) Inner Shaft

b) Outer Shaft

SolidWorks Corporation

Currently, the 20mm dimension on each part in FIGURE 192 tries to control the distance of the 6.25 holes to the end of the inner shaft and to the bottom of the datum feature B hole on the outer shaft. This implies a datum/origin of measurement, which is against the Y14.5 rules. All datums must be specified. But, more importantly, the tolerance on that distance is one we could control with the position tolerance on the 6.25 holes. That would allow us to inspect the entire position of the 6.25 holes on each part in one set up.

To this end, we will make the secondary datum feature the end of the inner shaft and the bottom of the B hole on the outer shaft. The 20mm dimensions must be converted to basic dimensions for their tolerance to be given in the position control that will be assigned to the Ø6.25 holes. The secondary datum feature is assigned as datum feature D on the inner shaft and datum feature C on the outer shaft. The 20mm toleranced dimensions are changed to basic dimensions of 20. See FIGURE 194 below.

FIGURE 194 [Assignment of the Secondary Datum Features]

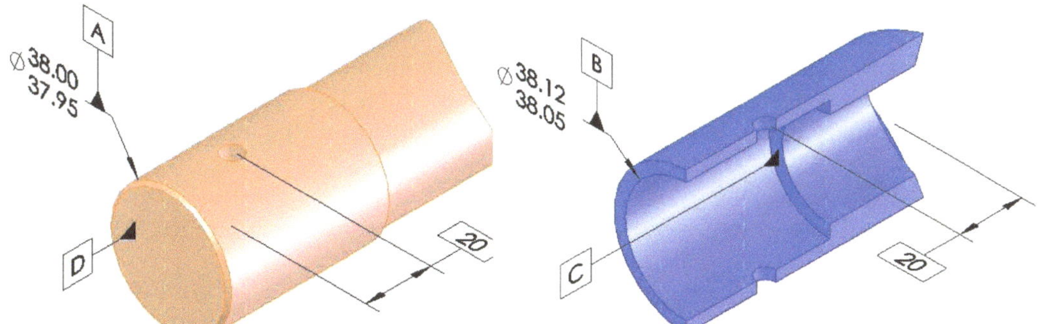

a) Inner shaft

b) Outer shaft

SolidWorks Corporation

It is important to control the angle of the secondary datum feature on each part to the primary datum axis, but since the angular tolerance of plus or minus 1 degree in the title block (not shown in these illustrations) will do this well enough, no perpendicularity geometric characteristic will be utilized.

Tertiary Datum Feature

If the hole on the inner shaft is positioned to datum axis A, and datum plane D, it will be required to be perpendicular to and intersect the datum axis, and be located from the datum plane to within whatever is assigned as the position tolerance.

That would still allow the hole's axis to rotate around the datum axis. But since there is no keyway or flat or other significant feature on the outside diameter of the part, it wouldn't be possible to tell that it had rotated. If such a significant feature did exist, we would assign it as the tertiary datum feature and reference it in the position control to limit that rotational degree of spatial freedom. Since that rotational degree of freedom is not a concern on either part, a tertiary datum feature is not required.

Assigning a Position Tolerance

The 6.25 holes on each part will now be positioned to the datums in the order that they were assigned.

This tolerance will be calculated using the floating fastener formula. It says:

6.2	=	MMC of the holes (smallest allowable hole size)
- 6.0	=	MMC of the Bolt (largest allowable shaft size)
0.2	=	Position tolerance for every hole on both parts

Since it is important that the tolerance on the 20-millimeter dimensions also be controlled to within the calculated 0.2, we will be able to control all facets of the position of the 6.25 holes with one position feature control frame on each part.

If the 20-millimeter dimensions did not control one part from bottoming out in the other (D against C) while the 6.25 holes aligned and received a bolt, then two feature control frames could have been used. One position control would be stacked on top of the other with a tighter tolerance to the primary datum in one level of control and a looser tolerance in the other level to the secondary datum.

In this case, one level of position tolerance suffices for each part as shown in FIGURE 195.

FIGURE 195 [Assignment of the Geometric Position Tolerance on the 6.25 Holes]

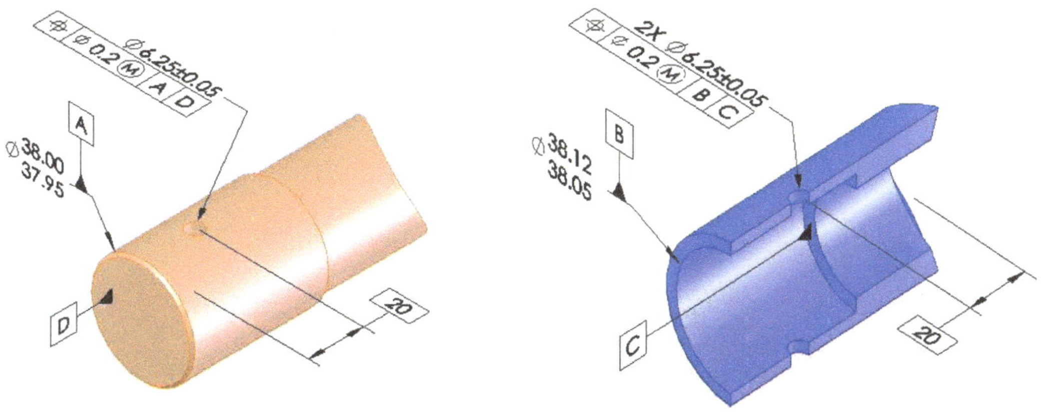

a) Inner shaft b) Outer shaft

SolidWorks Corporation

Floating fastener assemblies can consist of two or more parts with any number of corresponding non-threaded, clearance holes that go through the parts. These parts are to be fastened together with separate screws and nuts. The screws are considered floating fasteners in that they are not fixed at any orientation by the holes. They "float" through the parts and then the parts are bound together by applying nuts to the threaded portion of the screws. The danger that fixed fastener assemblies may encounter of needing a projected tolerance zone to provide protection for the angle of the threaded hole being exaggerated over the height of the screw as it projects from the threaded hole is not a factor in floating fastener assemblies.

The floating fastener formulas are as follows:
T = position tolerance to be fully applied to each clearance hole on both parts
H = MMC hole
F = MMC fastener/screw

$T = H - F$ $H = T + F$ $F = H - T$

These formulas can be used to calculate tolerance to be assigned to holes in the assembly, or, if that information is given, the hole MMC or even the closest applicable fastener MMC size can be calculated for use. If the tolerance size is sought and both parts have holes with the same size limits, then the formula $T = H - F$ only has to be used once to calculate the position tolerance for all clearance holes under consideration on both parts. However, if the hole size limits are different on each mating part, the formula must be used separately for each hole of a different size.

Tolerancing Mating Parts in a Floating Fastener Assembly (Two Single Segment Position Controls)

FIGURE 196

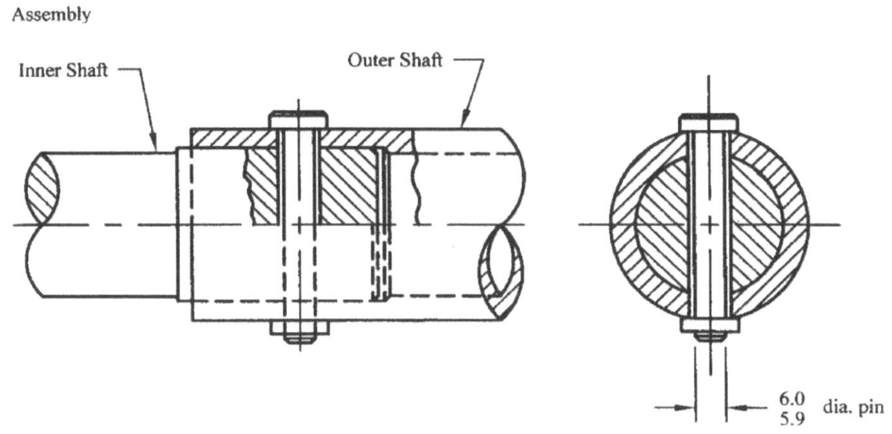

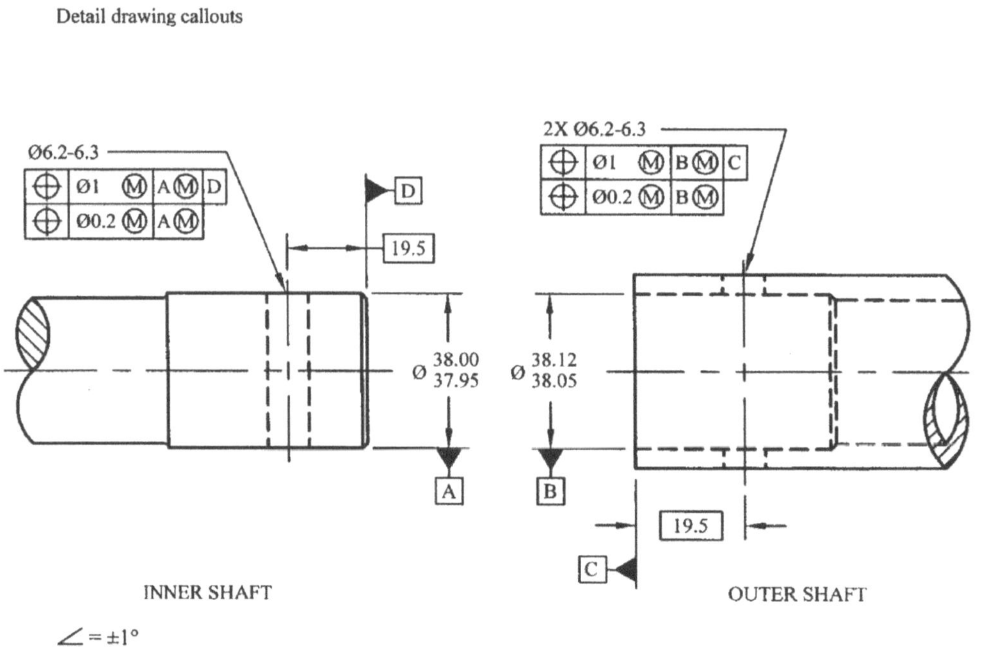

The assembly shown in FIGURE 196 is slightly different in its requirements than was the simpler single position control depicted in the previous section of this text. These illustrations show an assembly drawing of an inner shaft mating with an outer shaft. When they assemble, a 5.9-6.0 bolt is used to bind the parts together. The bolt floats through the 6.2-6.3 clearance holes on both the inner and outer shaft, then a nut is used to complete the assembly. The end of the inner shaft (datum feature A) does not seat (bottom out) in the bottom of the hole (datum feature B). So, the bottom of the hole will not become a datum feature. Datum feature C is more accessible.

Ideally, datum features should be representative of mating features, alignment features or seating features (or a combination of these attributes). We can see here that the primary datum feature on each part mates (with the primary datum feature on the other part) and aligns the assembly. It would be nice if the secondary datum feature also fit the criteria of a mating, seating or alignment feature, but that is not the case in this situation.

The clearance holes on both parts must receive a bolt with a diameter of 5.9 to 6.0. The parts are bound together by a nut. Since the bolt floats through all of the clearance holes on both parts and is not fixed at a particular angle by any threaded holes or tight-fitting holes, this is a classic floating fastener assembly.

We have decided to position the 6.2 to 6.3 holes loosely from the end of each part, but tightly to A and B. This can be done, as shown, in a two-single segment positional control. The upper level segment positional control on each part is to orient the part to the datum axis A or B while measuring distance from the end of the part (D or C). This 1mm position tolerance at MMC applies only to the relatively unimportant relationships (since the tighter 0.2 position tolerances steal the job of tolerancing the important relationships). The 1 at MMC tolerances the 19.5 basic distance from the end of the part to the 6.2-6.3 holes. All other requirements are held to within the tighter 0.2 at MMC positional tolerance that appears in the lower level segment of the control. This tolerance was calculated using the floating fastener formula of:

6.2	=	MMC hole
- 6.0	=	MMC shaft/screw
0.2	=	Geometric tolerance for all holes

The tolerance is used to locate and orient the holes' axes to either datum axis A or datum axis B. The upper segment tolerance was chosen based on manufacturing capability for an affordable cost and to be certain the hole did not move so far from the end of the part that D bottomed out in the outer shaft's hole (B) before the 6.2-6.3 holes had a chance to align. Since the 6.2 - 6.3 hole is centered to datum axis A to within 0.2, we can calculate the minimum wall thickness between the hole and the O.D. fairly easily. The LMC of A is 37.95. The outer boundary of the 6.2 - 6.3 hole is:

6.30	=	LMC
+ 0.30	=	Geometric Tolerance at LMC
6.60	=	Outer Boundary
+0.05	=	Datum feature shift due to A µ Reference
6.65	=	Hole Outer Boundary with Shift

So,

37.95	=	LMC Shaft A
- 6.65	=	Hole O.B. with Shift
31.30	=	Material
2		
15.65	=	Minimum wall thickness

Likewise, minimum wall thickness to the end of the part can be calculated using the upper segment tolerance of 1mm.

6.3	=	LMC Hole
+1.1	=	Geometric Tolerance at LMC
7.4	=	Outer Boundary Hole

Note: A at MMB has no effect on distance to D.

So,

19.5	=	Basic Dimension
-3.7	=	½ of 7.4
15.8	=	Minimum Wall to Plane D

Then we must determine the effect of the general tolerance of ±1° on the 90° angle between the 38mm O.D. and the end of the part. If that was about 0.64, our final minimum wall thickness would be:

15.80		
-0.64		
15.16	=	Minimum Wall

Since the wall thicknesses are all robust enough to maintain the strength and durability of this part, we have to assume the 1 mm at MMC positional tolerance to D and the 0.2 at MMC positional tolerance to A is acceptable (in that respect). The worst mating condition for the holes on both parts is calculated with a virtual condition formula using the 0.2 tolerance. So,

6.2	=	MMC of all clearance holes
-0.2	=	Geometric Tolerance at MMC
6.0	=	Virtual Condition (to A or B)

Since the MMC of the screw/bolt is 6.0, we can see that the worst fit is a line fit, which is acceptable in this case. Now, let's take a look at the assembly conditions and tolerance zones we have generated and what they mean. FIGURE 197 explains the non-uniformity of fit allowed in the assembly by using the maximum material boundary symbol after datum features A and B.

FIGURE 197

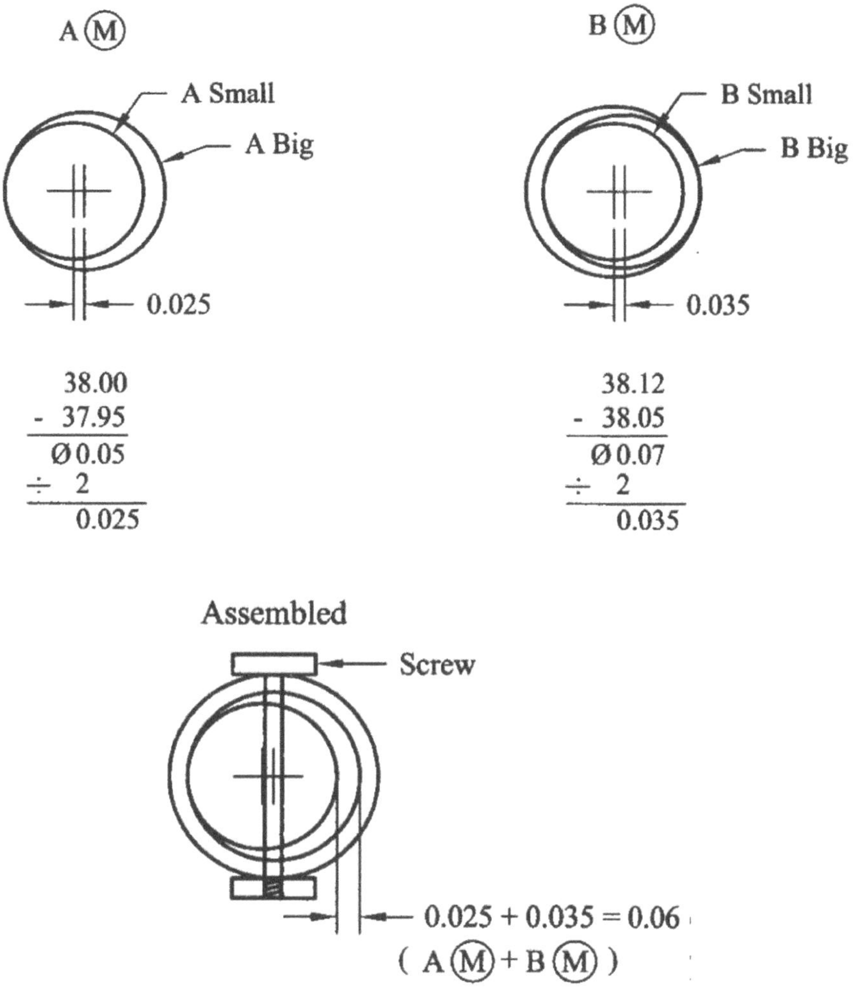

As shown in Figure 197, use of the Maximum Material Boundary (MMB) symbology after datum features A and B allows an additional unequal air space between A and B at assembly. In fact, this non-uniformity of fit may be required to assemble the parts. If the MMB symbology had not been used after datum features A and B, the assembler would have been able to assemble all parts with equal airspace between A and B.

In FIGURE 196, the parts would have functioned just as well with one level of position tolerance, but would have been more expensive to manufacture since the tighter Ø0.2 at MMC tolerance would have been held to all datums. For example:

Ø6.2-6.3

⊕	Ø0.2Ⓜ	AⓂ	D

Inner Shaft

2X Ø6.2-6.3

⊕	Ø0.2Ⓜ	BⓂ	C

Outer Shaft

The false belief seems to be that multi-level/segment controls are more expensive than single level/segment controls. But the truth is the reason behind the majority of multi-segment controls, especially those that don't list a different set of datums in each segment, is to loosen the tolerance to the less important datums only listed in the upper segment, while holding the more functionally important relationships to within the tighter tolerance listed in the lower segment control(s). Granted, these multi-level controls may require a higher degree of GD&T knowledge to interpret but should be worth it to allow inspectors to buy a larger portion of the parts that are produced. These additional parts that are acceptable under the multi-segment geometric tolerances are just as functional and should, therefore, be passed on to assembly by the inspectors.

FIGURE 198

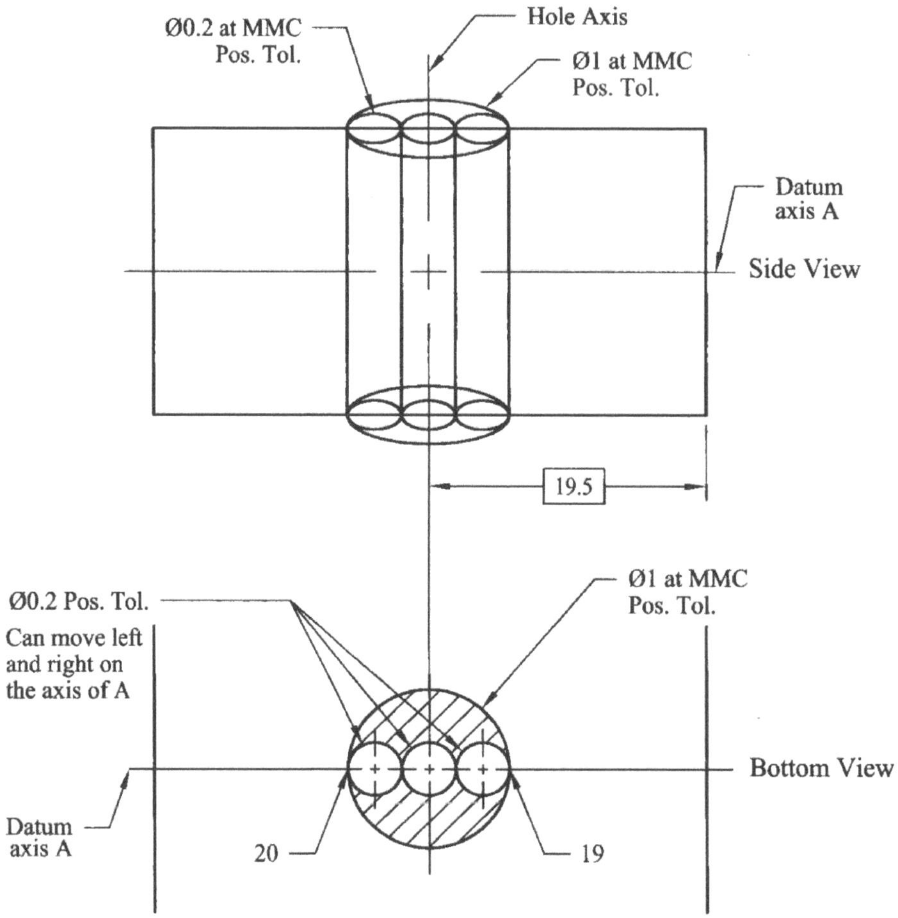

This illustration shows the meaning of the two single segment position tolerances used on the **inner shaft** in the FIGURE 196 detail drawing.

Chapter 12

Why use GD&T?

•Composite Position Tolerancing on a Floating Fastener Assembly Part

•Minimum Wall Thickness Calculation using the Pattern Locating Tolerance Zones in a Composite Position Tolerancing

•Composite Tolerancing with Three Levels of Control

•Differentiation of Similar Features

Chapter Objectives

Readers will learn:

1. How to convert a plus and minus toleranced part to a composite positional tolerance.
2. To interpret plus and minus tolerancing and its ambiguities.
3. To convert rectangular tolerance zones into cylindrical tolerance zones.
4. The advantages of specified datum features implied datum features.
5. How to calculate and test geometric tolerances for composite position controls.
6. How to calculate a minimum wall thickness.
7. How one position tolerance can function as well as two, but cost more.
8. How to apply and interpret composite positional tolerancing with three levels of control.
9. How to differentiate between features of similar size and configuration.

Why use GD&T?
Composite Positional Tolerancing on a Floating Fastener Assembly Part

I WONDER AS I WANDER

Years ago, I gave a workshop at a New England company that shall go nameless. The engineering manager did not meet me at the entrance on the first day as my itinerary indicated he would. After repeated calls and paging from the guard shack, the guard finally went to his office and brought him to me. He did not act happy, but I attributed it to personal problems. I had to run to keep up with him as he guided me to the room where the workshop was to be held. I peered into the cramped, closet-like room and saw 20 chairs and no tables.

"We will have about 30 people running in and out of here. Ten will have to stand," he said
.

I nodded. "For four days?"

He nodded. "The old engineering manager arranged this workshop before he was promoted to the company's headquarters. I'm the engineering manager now, and I've decided we will hold the workshop because of our contract to do so, but we aren't going to use this stuff." He was smiling now, having gotten to air his point of view.

"But it's good stuff," I said. He sneered at me as though he smelled something bad. "Does that mean we can't have more chairs?", I asked.

His face got red. "I have real work to do." He spun on his heels. "I can't be bothered with this," he said as he disappeared down the hall.

I looked around. There was no audio-visual equipment, no board or chart pad to write on and people were beginning to arrive. I pictured violent acts taking place, the weaker attendees being forced to relinquish their seats by burly guys with one eyebrow. I asked the early arrivals for help and things began to work out. We got the needed equipment and chairs - even the luxury of tables for them to sit at.

However, I noticed the engineering manager lurking in the halls, passing by the open door, peeking around corners and other curious behavior while we held the sessions, incredulous, I'm sure, that we were overcoming the barriers he had so carefully constructed.

He heard the course participants discussing the topics at break times, and feeling their enthusiasm and knowledge levels growing, he decided on another tactic. He would ambush the group during each subsequent break, berating their enthusiasm and belittling their knowledge. I heard him ranting, "We're not going to use that junk around here!"

Still, the people were very bright and took to the topic and unbridled encouragement without a hint of reluctance or shame.

The creature in the hall was losing momentum to the crowd. His hair fell into disarray and his shirttail wasn't properly tucked in. I noticed a frenzied look overtake his face and his movements about the hallway took on a scurrying mode, like a scared creature looking for a gathering wherein he might be welcomed. But he was having more and more trouble getting people's attention. Even waving his arms didn't seem to help much. He went home sick before lunch on the third day.

When I called for him two weeks later, I was told he was no longer an employee of the company. The new engineering manager, one of the course graduates, assured me implementation of the course principles were proceeding smoothly - and that the old engineering manager was assumed lost during a storm.

"A storm?" I asked.

"Yes," he said with a hint of amusement. "It took place in the Vice President's office a few days after you left."

Why use Geometric Dimensioning and Tolerancing?

In 1973, an example similar to the following illustration was put into the ANSI Y14.5 Standard.

FIGURE 199

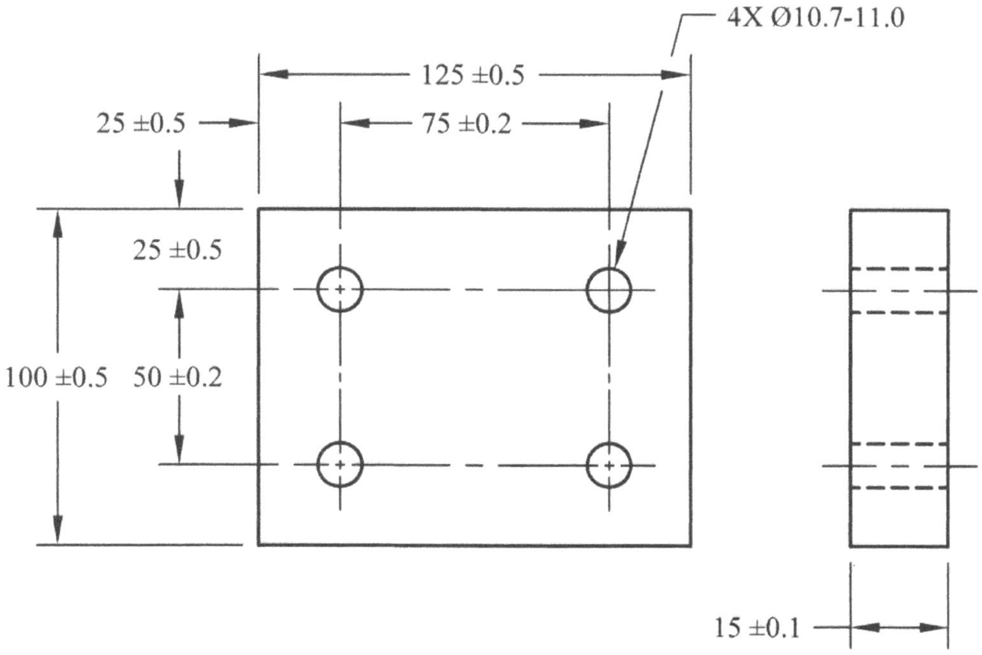

Three interpretations of the tolerance scheme were given that were popular at the time. One interpretation of the hole location tolerance zones is that four zones are generated that measure 1 x 1 millimeter for each hole location from the edges of the part. In addition to those tolerance zones, four more are generated that measure 0.2 x 0.2 for the hole to hole locations. That interpretation is as follows:

FIGURE 200 [Interpretation #1]

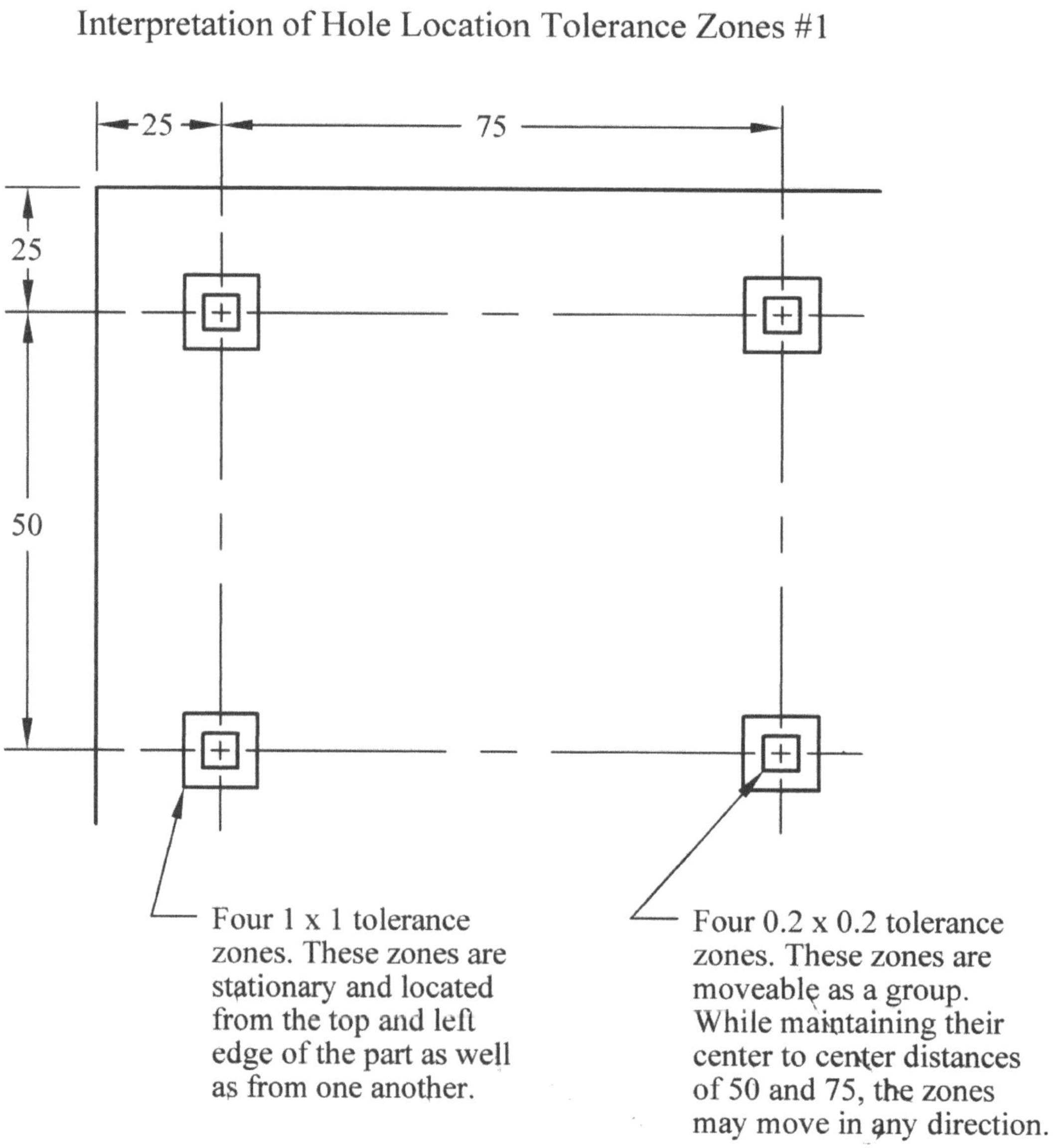

For the part to meet this interpretation of the drawing requirement, all hole's axes must be produced to reside within both the 1 x 1 and the 0.2 x 0.2 tolerance zones.

Another interpretation is that the upper left-hand hole in the front view has a tolerance zone that is 1 x 1 millimeter, and that the other three holes have tolerance zones that are 0.4 x 0.4. That interpretation is shown below.

FIGURE 201 [Interpretation #2]

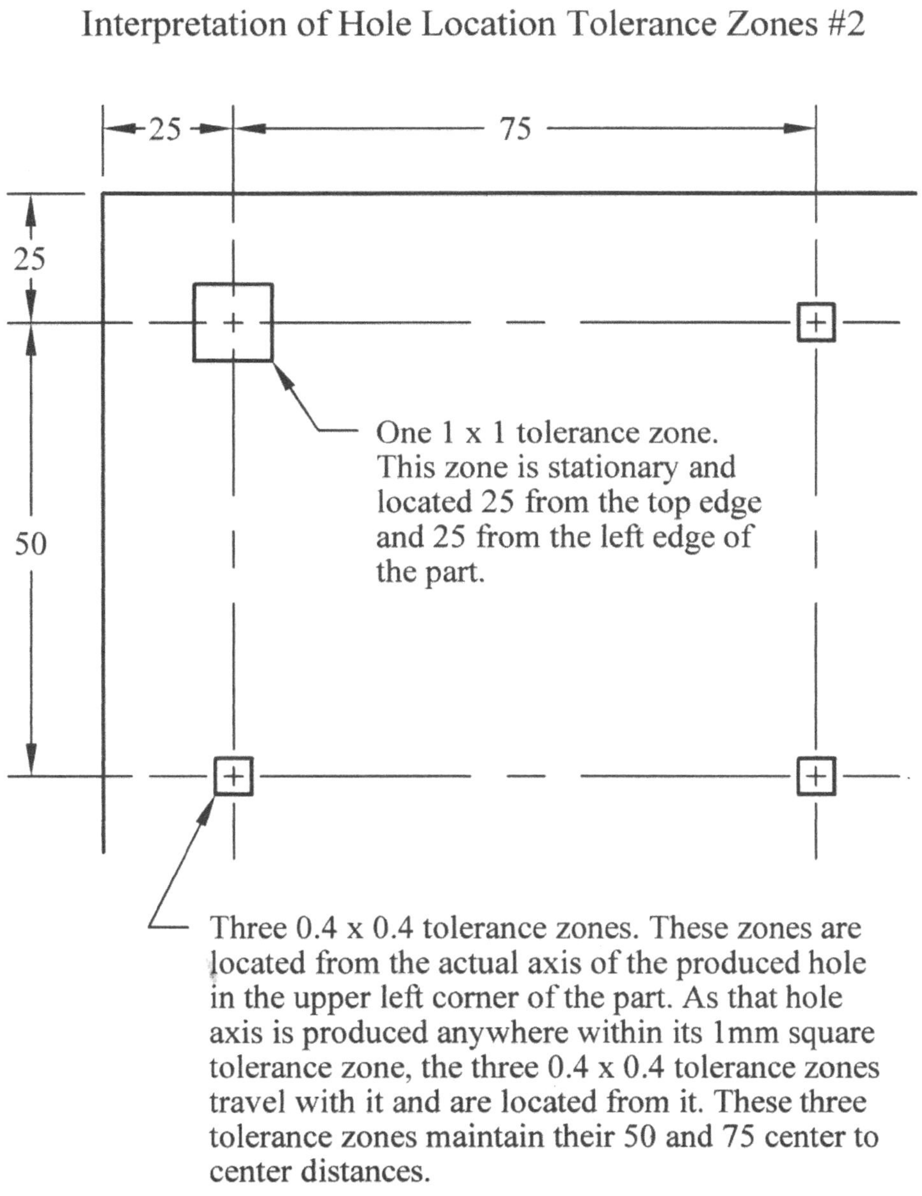

For the part to be in compliance with this interpretation, each hole must be produced with its axis residing within its respective tolerance zone.

Another, more controversial, interpretation is that the upper right-hand hole is not measured from the upper left and top edges of the part at all. Instead, this interpretation is that those edges are measured from the hole. This interpretation has its basis in industries that do not like to use the edges of a part as datum features. P.C. boards and automobile panels, for example, rarely use the part edges as datum features. This may be either because those edges are inherently unstable (as is the case with flexible parts) or that this approach does not reflect how the part is manufactured.

In this interpretation, since the hole generates an axis, which is the intersection of two planes crossing at a 90-degree angle to one another, the interpretation is often as shown in Figure 4. Measuring along planes crossing at a 90-degree angle, they surmise, creates a tolerance zone that consists of two parallel planes 1 millimeter apart for each edge and these zones are 90Δfrom each other.

FIGURE 202 [Interpretation #3]

Interpretation of Hole Location Tolerance Zones #3

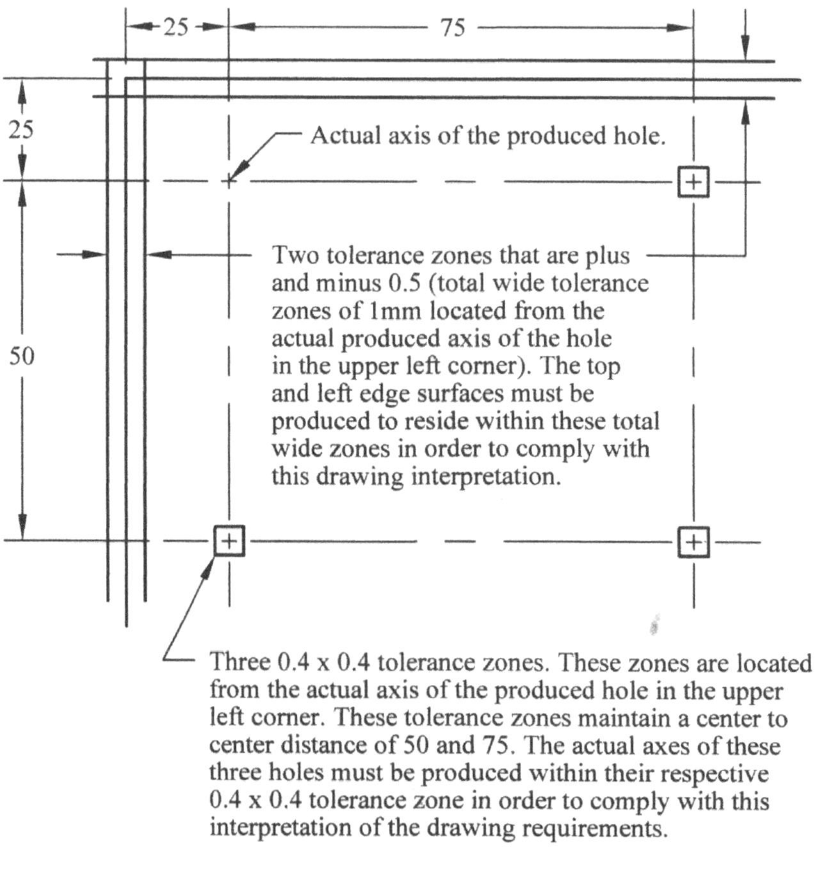

The point the Y14.5 committee was trying to make was that drawings like this are ambiguous. They have no one correct meaning. When people saw the multiple interpretations for this type of drawing in the Y14.5 standard, instead of getting the point, many began to argue about which interpretation was correct.

So, when the Y14.5 standard was revised and reissued in 1982, these illustrations were replaced with a rule. The rule states that all datum features must be specified. They may not be implied. So, drawings like this are not in compliance with that rule.

In the following steps we will convert this drawing into one that is compliant with Y14.5. We will start by adding the datum features to the part as shown in FIGURE 5.

Step 1:
The primary datum feature is the surface that bolts to and seats on the mating part. We will call this surface datum feature A.

Step 2:
The secondary datum feature will be the longest surface from which a dimension originates to locate the first hole in the pattern. This will be B.

Step 3:
The tertiary datum feature will be the remaining surface from which a dimension originates to locate the first hole in the pattern. This will be C.

FIGURE 5

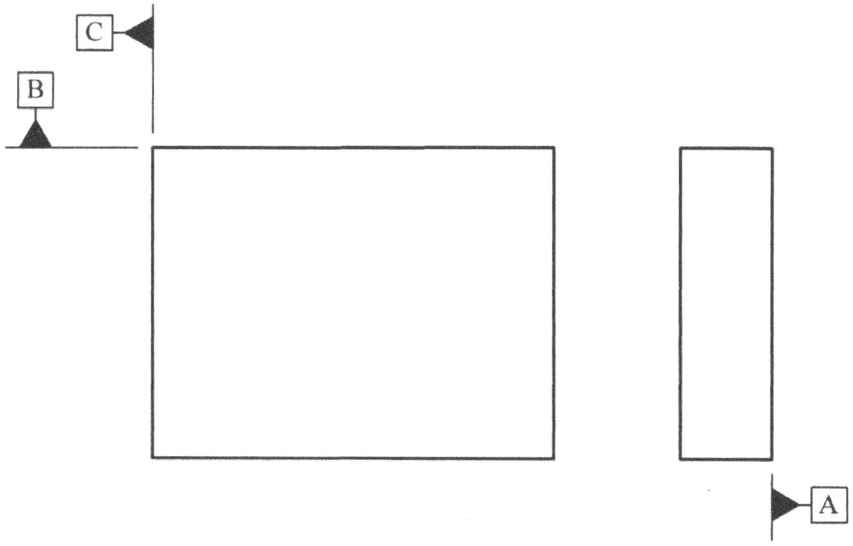

Step 4:
The next step in converting this part to a position control is to make all the dimensions that locate the holes basic dimensions. This means that these dimensions will be targets to shoot for. The tolerances will be transferred into a position control in a later step and become the basis for the allowed deviation of the axes of each hole from true position as defined by the basic dimensions. See FIGURE 203 for the basic dimensions that show true position (the perfect hole locations).

FIGURE 203

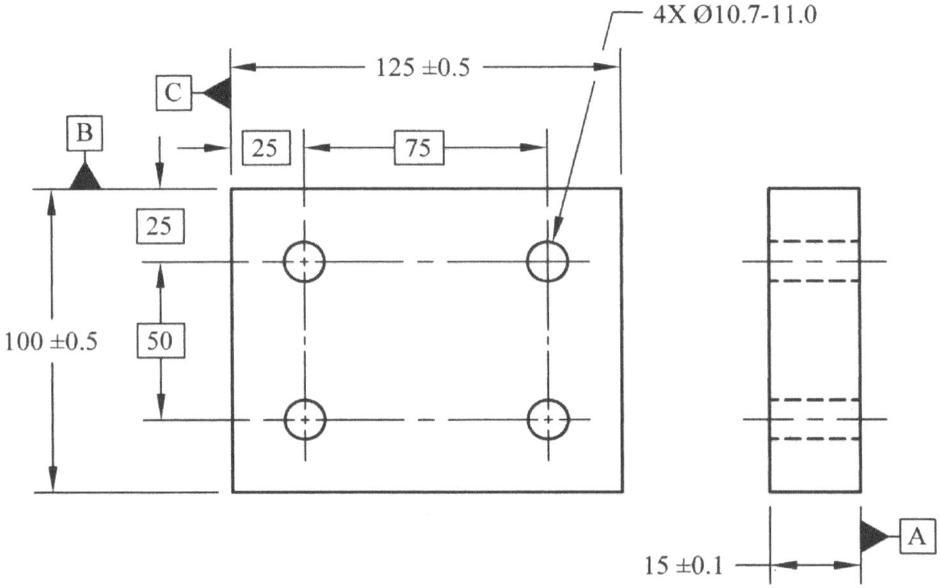

Step 5:
The original plus and minus toleranced drawing had a tighter tolerance between the holes than to the edges of the part. In converting to a position tolerance, this affords us the opportunity to use a composite position tolerance.

Composite position tolerancing will hold a tighter tolerance between the holes than to the edges of the part. The premise of this type of control is to calculate the tolerance from hole to hole that will allow these clearance holes to align with the clearance holes in the next part of the assembly well enough to pass 10-millimeter screws through and use nuts to bind the parts together. This is known as a floating fastener assembly.

To begin with, let's calculate this hole to hole tolerance with the floating fastener formula.

	Ø10.7 =	MMC Clearance Holes (smallest hole size)
-	Ø10.0 =	MMC Screws (largest screw size)
	Ø 0.7 =	Geometric Tolerance for All Holes

This is the tolerance that will be found in the functional lower level position control. This tolerance will apply to each hole in its allowed deviation from the 50 and 75 millimeter dimensions between the holes.

The tolerance for the datum-to pattern relationship is a number that can be selected on a purely cost-saving basis. This will be a larger tolerance and will apply to the 25 millimeter dimensions from datums B and C. Since this tolerance is going to be converted from the original 1 millimeter square tolerance zones, it will have to be tested to determine the minimum wall thicknesses from the holes to the edges of the part. The conversion follows.

The minimum circumscribed cylinder about a 1mm square is 1.4. So, our new position tolerance will be a diameter of 1.4 to the datums A, B and C. The hole-to-hole tolerance (and a refining tolerance to A for perpendicularity) will be Ø0.7 as calculated above with the floating fastener formula.

FIGURE 204

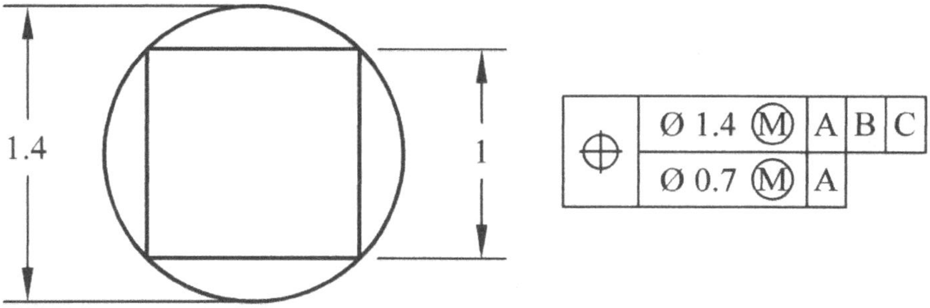

<u>Step 6</u>:
The calculated tolerances for the hole-to-hole and datum-to-pattern tolerances are transferred into the position control. The upper level control tolerances the 25 millimeter dimensions from datums B and C to within a Ø1.4 at MMC. The lower level control tolerances the 50 and 75 millimeter dimensions between the holes to within a Ø0.7 at MMC for each hole.

The four Ø1.4 tolerance zones are stationary, locked in to the basic dimensions from datum planes B and C (with datum A acting to orient the zones). The four Ø0.7 tolerance zones are locked into position from one another and perpendicular to datum plane A, but may float as a group in their location. Since the actually produced axis of each hole must reside within both of its tolerance zones to be in compliance, the smaller Ø0.7 zones may not float entirely outside of their respective Ø1.4 zones.

See the following finished conversion in FIGURE 205.

FIGURE 205 [Completed Conversion to Composite Position Tolerancing]

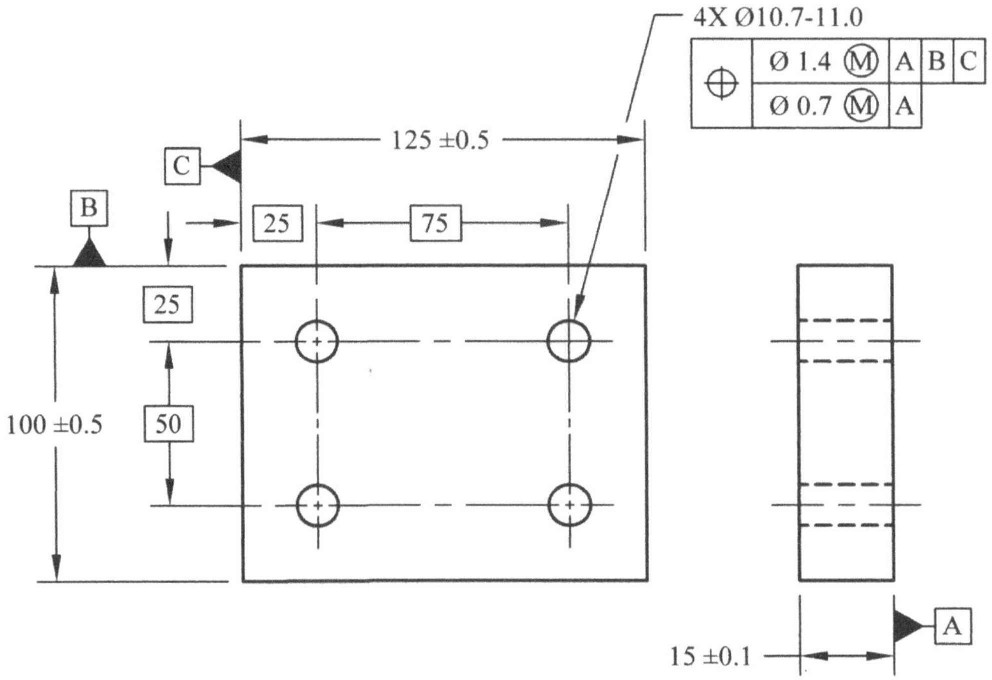

UNLESS OTHERWISE SPECIFIED
ALL ANGLES ARE ±1°

The position tolerance zones for the finished composite tolerancing requirement are shown in FIGURE 206. For the part to meet this drawing requirement, all hole axes must be produced to reside within both the Ø1.4 at MMC and the Ø0.7 at MMC tolerance zones.

FIGURE 206 [Interpretation of Hole Location Tolerance Zones]

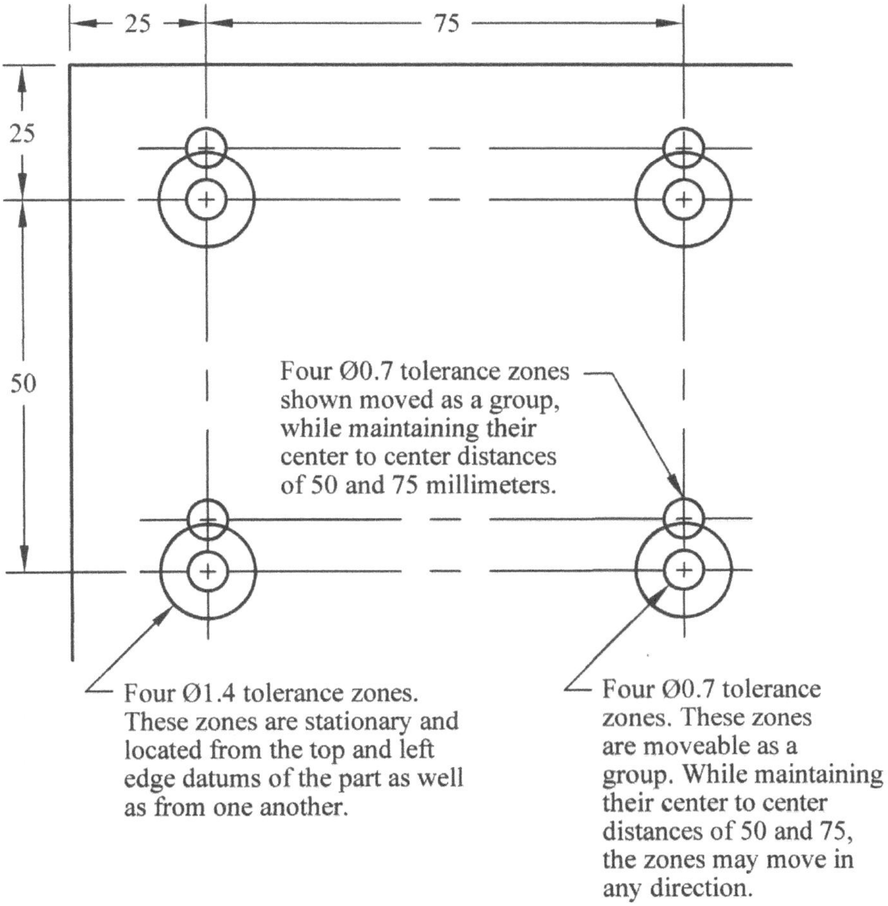

Minimum Wall Thickness Calculation using the Pattern Locating Tolerance Zones in a Composite Positional Tolerance Figure 292

As mentioned, the position tolerance of a Ø1.4 at MMC must be tested to determine if it allows the hole pattern to be out of position so much as to allow it to endanger the wall thicknesses between the holes and the outside edges of the part.

The factors that will affect the wall thickness between the holes and the edges of the part are:

1) The least material condition size of the holes (the largest holes).
2) The position tolerance to the datums assigned at MMC.
3) The "Bonus Tolerance" allowed to holes produced at LMC.
4) The dimensions to the edges from the holes.
5) The flatness of the edges of the part.
6) The tolerance on the angles of the part surfaces.

So, a good formula that includes these factors would be:

	11.0 =	LMC Holes
+	1.4 =	Geometric Position Tolerance that applies at MMC
+	0.3 =	Bonus tolerance allowed an 11 millimeter hole because it is 0.3 larger than MMC
	12.7 =	Outer Boundary of the holes

12.7 divided by 2 = 6.35

6.35 = ½ the Outer Boundary of the hole

	25.00 =	Basic Dimension from datum plane B
-	6.35 =	1/2 Outer Boundary of the holes
	18.65 =	Minimum Wall Thickness between the edge of the hole and datum plane B

Since the surface called datum feature B is real, it may be out of flatness. Its flatness is controlled by the size tolerance on the overall dimension of 100 plus and minus 0.5 in the direction of datum feature B.

If the part was produced at 100.5 at its highest points and the surface opposite datum feature B was perfectly flat, datum feature B could be out of flat by the full-size tolerance of 1 millimeter (plus and minus 0.5). This would thin the wall between the holes along datum B and the surface B by the entire 1 millimeter.

So, 18.65 minus 1 = 17.65. However, if datum feature B was controlled by a geometric tolerance such as perpendicularity, it would override the 1 millimeter of tolerance we subtracted from 18.65 and control the flatness of datum feature B. It would also control the last factor in this calculation which is the amount the surface called datum feature B could lean away from datum plane B. This perpendicularity tolerance would be subtracted from the 18.65, instead of the 1 millimeter of tolerance to determine the minimum wall thickness to datum surface B.

However, we instead have a general tolerance note on the drawing for all angles shown at the bottom of the finished drawing that says; UNLESS OTHERWISE SPECIFIED, ALL ANGLES ARE TOLERANCED TO WITHIN PLUS OR MINUS 1 DEGREE. This makes the situation different. Given these circumstances, the 1 millimeter of size tolerance would be used to subtract from the 18.65 for the flatness control on datum feature B and the 1 degree would be used to calculate how much the surface B could lean away from datum plane B in toward the holes. For example, in this case, since the part is 15.1 thick, 1 degree over that thickness would thin the wall by about 0.26 millimeters. So, the minimum wall thickness with that as a factor would be 17.65 minus 0.26 = 17.39 to datum feature B.

If using the plus or minus 1 degree in calculating the minimum wall thickness to datum feature C, we would have to remember that it is 75 millimeters from datum B to the second hole. This would impact the wall thickness because datum surface C could lean away from datum plane C. It only must touch the datum plane with one point of high point contact. Since the math constant is 0.425 (rounded off) per 25 millimeters per 1 degree, it would impact the wall thickness by about 0.425 times 3 = 1.275 millimeters. So, the minimum wall thickness from the edge of that hole to datum feature C (not datum plane C) would be approximately 17.65 minus 1.275 = 16.375 (the thinnest wall yet).

That exhausts the minimum wall thickness calculations to the datum features. Let's examine the non-datum features. We will leave out the impact of the 1 degree angular tolerance we have been speculating about for this discussion. It would appear as though the wall thickness to the non-datum features on the opposite sides of datum features B and C will be the same. We can see the minimum wall from the hole closest to the opposite side of the part from datum features B would be calculated by the following factors; 100 minus 0.5 = 99.5 (LMC) and 99.5 minus the 75 millimeters of basic dimensions = 24.5. 24.5 minus 6.35 (1/2 of the outer boundary of the holes) = 18.15 minimum wall thickness. Even subtracting out the 0.26 for the 1Δangle would still reap a positive and robust wall thickness.

We can see from these calculations that the use of the Ø1.4 at MMC position tolerance was safe and saved us from having to hold the tighter functional tolerance of Ø0.7 at MMC to datums B and C. The Ø0.7 at MMC tolerance (which makes it possible for this part to mate with a similar part with the same clearance holes using M10 screws) had to be held from hole to hole and to datum A (on which the part seats in assembly). But it did not have to be held to datums B and C which do not have to align with the edges of the mating part. If they did have to align with the edges of the mating part, the tighter Ø0.7 at MMC tolerance would have had to be held from hole to hole and to all datums.

That control would have looked like: 4X Ø10.7-11.0

⌖	Ø0.7 Ⓜ	A	B	C

This control would have worked as well as the composite position tolerance, but would have been more restrictive and therefore, probably more expensive.

Minimum Wall Thickness Calculations

In the right side view of the detail drawing below, what are the minimum and maximum wall thicknesses between the edge of the hole and the left edge of the part?

FIGURE 207

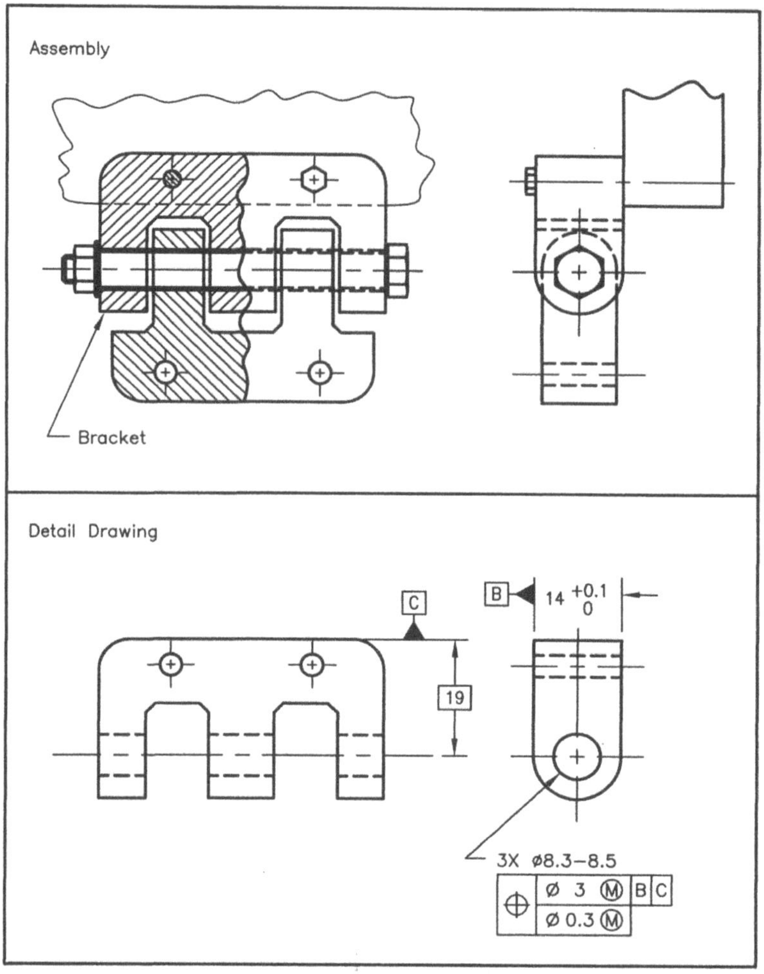

One formula that lends itself to calculating the minimum wall thickness from the surface of the holes to the left (or right) side of the part in the right side view of the detail drawing is as follows:

Step 1:

8.5	=	LMC of the holes
+ 3.0	=	Geometric Tolerance at MMC
+ 0.2	=	Bonus tolerance potential from hole size limits (8.5-8.3=0.2)
11.7	=	Outer Boundary of the holes

Step 2:

14.1	=	MMC of Datum Feature B
-11.7	=	Outer Boundary of the holes
2.4	=	Material on both sides of the holes

Step 3:

2.4 = Material on both sides of the holes divided by 2 = 1.2 Min. Wall to the surface high point

Step 4:

1.2 = Min. Wall to the surface high point minus the size tolerance (that controls the flatness) on the 14.0 to 14.1 dimension of 0.1 (pit in surface) = 1.1 Min. Wall thickness to the surface low point

A Tolerance Stack-Up Analysis Approach to the Same Problem

FIGURE 208

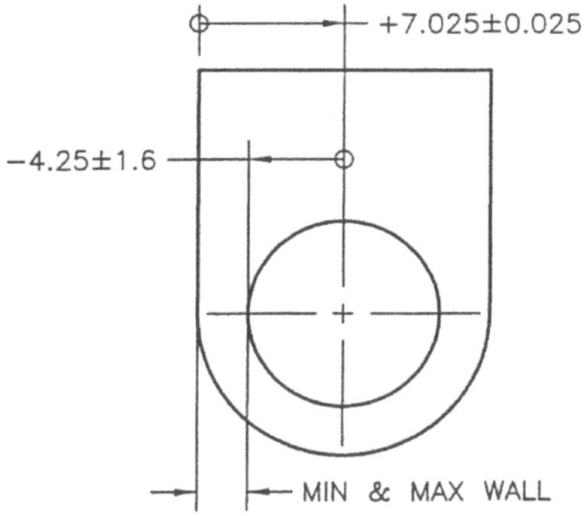

8.5 = LMC Hole	8.3 = MMC
+ 3.2 = Geo. Tol. at LMC	- 3.0 = Geo. Tol. at MMC
11.7 = Outer Boundary of Hole	5.3 = Inner Boundary of Hole

11.7 = Outer Boundary of Hole	11.7 = Outer Boundary of Hole
+ 5.3 = Inner Boundary of Hole	- 5.3 = Inner Boundary of Hole
17.0 = Sum	6.4 = Difference

$\frac{17}{2}$ = 8.5 Mean Dimension $\qquad$ $\frac{6.4}{2}$ = 3.2±Tol.

Ø8.5±3.2 divided by 2 to convert to a radius equals R4.25±1.6

Datum feature B;

14.1 = Outer Boundary of B	14.1 = Outer Boundary of B
+ 14.0 = Inner Boundary of B	- 14.0 = Inner Boundary of B
28.1 = Sum	0.1 = Difference

$\frac{28.1}{2}$ = 14.05 and $\frac{0.1}{2}$ = 0.05 equals 14.05±0.05

14.05±0.05 divided by 2 to convert to a radius equals R7.025±0.025

-	+	± Tol	
	7.025	0.025	
-4.250		1.600	
-4.250	+7.025	1.625	Totals

7.025 - 4.250 2.775	2.775 - 1.625 1.150 = MIN Wall	2.775 + 1.625 4.400 = MAX Wall

This would seem to be the end of the analysis. But, in this case, another look at the part is necessary. With a regular feature of size as the primary datum feature, an odd situation is possible.

In our analysis above, we divided every factor into radii and calculated the MIN and MAX WALL from the radial factors. This is sufficient for the MAX WALL. However, all of the size tolerance can gravitate to one edge of datum feature B.

The illustration below shows the results.

FIGURE 209

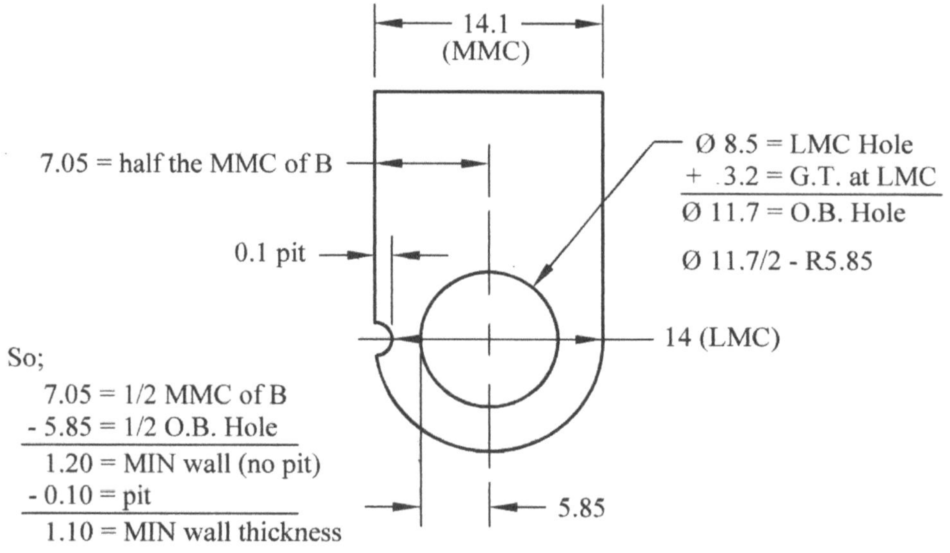

The second analysis just shows the importance of judgment and imagination in any analysis and why tolerance stack-up analysis is so hard to get right when pat routines are followed without every possibility being considered.

FIGURE 210 [Composite Tolerancing with Three Levels of Control]

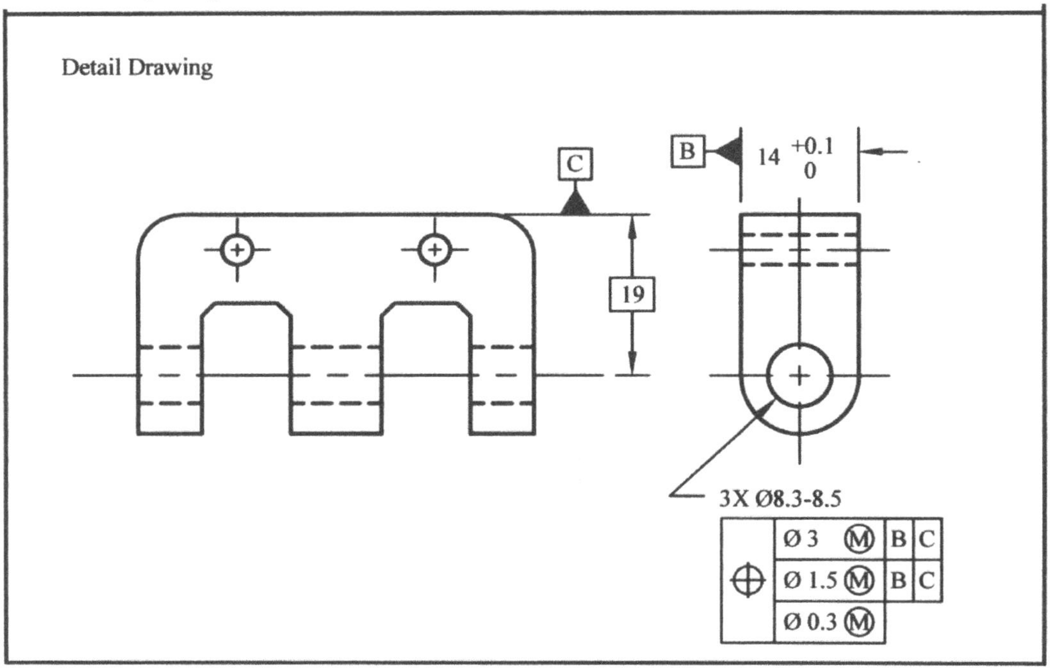

The control shown in FIGURE 210 illustrates a concept first presented in the ASME Y14.5-2009 standard. This three level Composite Positional Tolerancing control means the following:

1. The top level control locates the three holes to datums B and C. Regardless of the material boundary of datum feature B, the three holes must be centered to datum centerplane B and hold the 19 millimeter basic dimension from datum plane C to within a diameter of 3 millimeters if the holes are produced at their maximum material condition.

2. The middle level control is ultimately only holding the orientation of the three holes (as a pattern) to datum centerplane B and datum plane C to within a diameter of 1.5 millimeters if the holes are produced at their maximum material condition.

3. The third level control is aligning the three holes' axes to one another for coaxiality to within a diameter of 0.3 millimeters if the holes are produced at their maximum material condition. It is worth noting that no datum feature references are used here, in that, none are necessary to position the three holes to each other. This tolerance is the one that allows the 8 millimeter screw to pass through the three holes. It generates a virtual condition (inner boundary) for the three holes that is calculated by subtracting the 0.3 millimeter diameter position tolerance from the 8.3 millimeter maximum material condition of the holes. 8.3 minus 0.3 equals an 8 millimeter diameter virtual condition. This is the worst mating condition and is compatible with the 8 millimeter screw that must pass through. Since 8 millimeter screws are actually a little smaller than 8 millimeters, there is even a little clearance built into this assembly condition.

FIGURE 211

Differentiation of Coaxial Holes
Composite Positional Tolerance

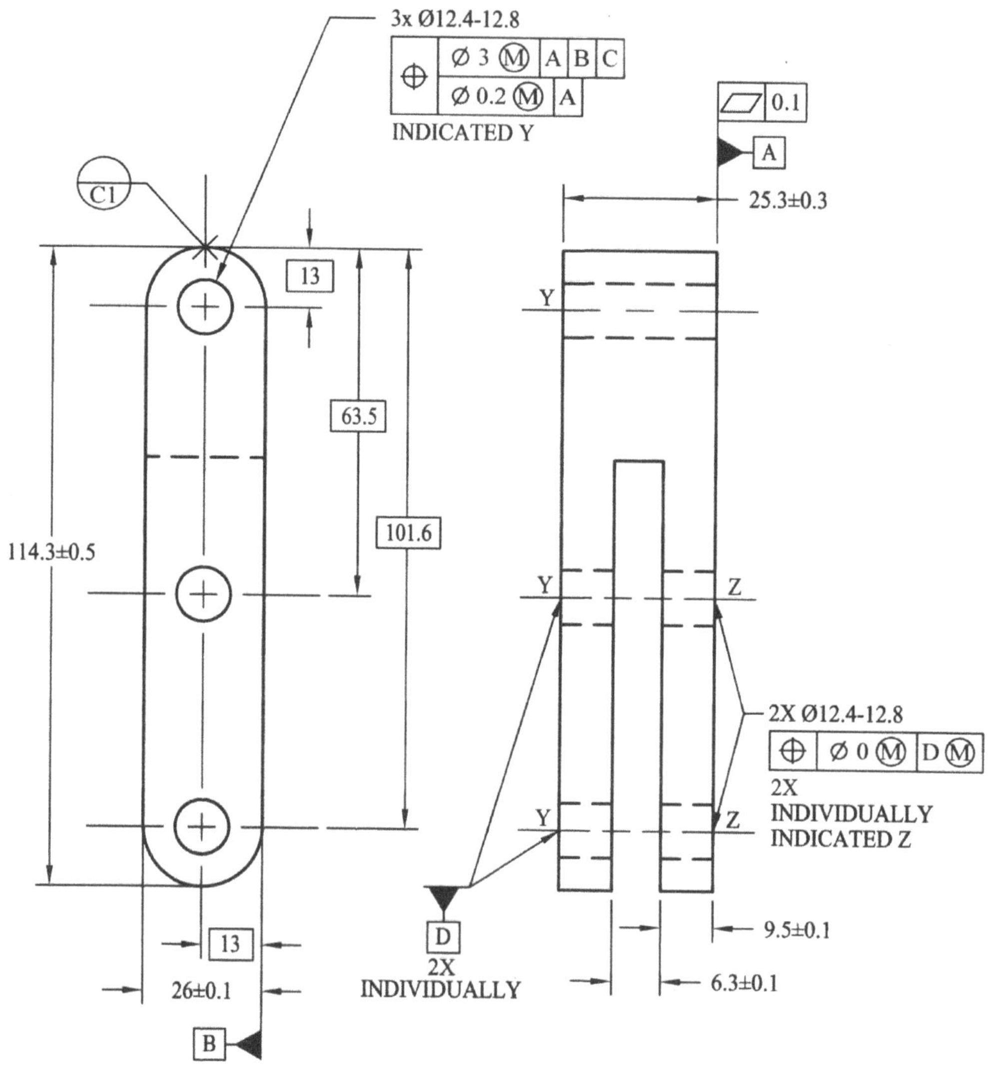

Chapter 13

Composite vs. Two Single Segment Positional Tolerancing

•Composite Positional Tolerancing vs. Two Single Segment Positional Tolerancing Controls

•One Level Positional Tolerancing vs. Composite Positional Tolerancing and Simultaneous Requirements

•Two Single Segment Position Controls

Chapter Objectives

Readers will learn:

1. The differences between composite positional tolerances and two-single segment positional tolerances.
2. What the tolerance zone limitations are for composite and two-single segment controls.
3. How Simultaneous Requirements do and don't apply to composite positional tolerancing.
4. How to combine single features with patterns of features in composite controls.
5. How to specify and interpret composite and two-single segment controls on parts with radial hole patterns.
6. How to refine geometric controls to be more cost effective.

Composite Positional Tolerancing vs. Two Single Segment Positional Tolerancing Controls

FIGURE 212 [Composite vs. Two Single Segment Position Tolerancing]

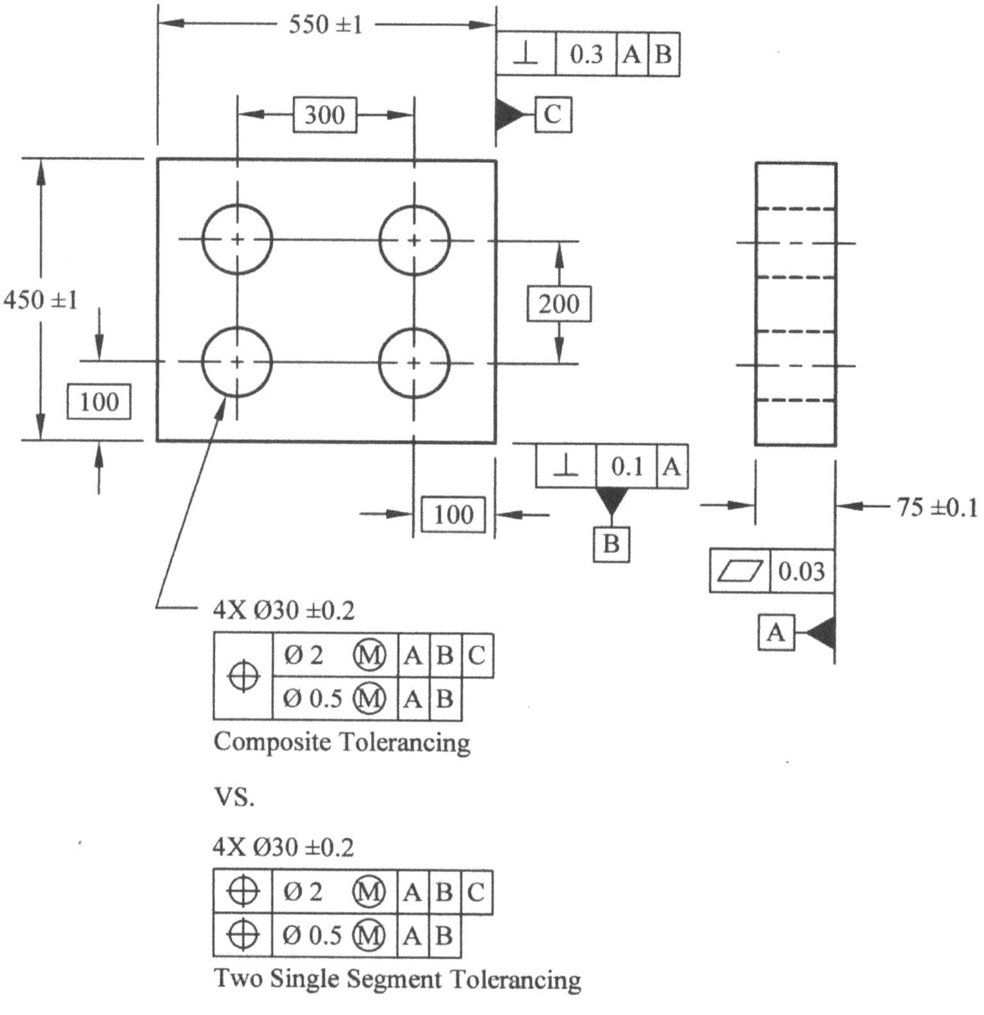

When the 1994 version of the ASME Y14.5 Standard on Dimensioning and Tolerancing was issued, a major distinction was made between composite positional tolerancing and two single segment positional tolerancing controls.

In FIGURE 212, datum feature B is a planar surface from which a basic dimension of 100 millimeters originates to locate a pattern of holes. Depending on which of the two controls are shown (composite position tolerancing or two single segment position tolerancing), the interpretation would be quite different.

Composite Positional Tolerance

⌖	Ø 2 Ⓜ	A	B	C
	Ø 0.5 Ⓜ	A	B	

2 Single Segment Positional Tolerance

⌖	Ø 2 Ⓜ	A	B	C
⌖	Ø 0.5 Ⓜ	A	B	

The essential element of holding the positional relationship between features within a pattern is kept by both. The smaller tolerance between the two levels of control is the one that (at least) refines the positional tolerance of the features to each other (the hole to hole tolerance, for example). Also, if no datums are included in the lower level control, or if the only datum included happens to be the primary datum used in the upper level control and that is a datum used to control perpendicularity, then the two types of control [(1) composite and (2) two single segment position], are identical in meaning. However, if a datum is used in the lower level control that is either not used in the upper level control or in a different order than it is used in the upper level control, or if the datum used in the lower level control is one used for location in the upper level control, the differences become evident.

First, whether we are talking about profile or position, there are some rules that govern a composite tolerance that do not apply to a two single segment control. In a composite position tolerance, only datums used in the upper level (Pattern Locating Tolerance Zone Framework--PLTZF) control may be used in the lower level (Feature Relating Tolerance Zone Framework--FRTZF) control. And, if used in the FRTZF, they must be used in the same order as they were used in the PLTZF control above. For example, a primary datum in the upper level (Pattern Locating Tolerance Zones) control may not become secondary in the lower level (Feature Relating Tolerance Zones) control.

With two single segment positional tolerance controls, the datum features used in the lower level control do not have to be in the same order or even be the same datum features used in the upper level control. In fact, since the two controls are entirely separate positional tolerance controls for the same pattern of features, the use of datums, or the lack thereof, is completely up to the discretion of the tolerance assigner (the tolerancing engineer).

In a composite positional tolerance control, any datums used in the lower level (Feature-Relating Tolerance Zones) control are used to refine ORIENTATION ONLY to those datums. Angle tolerance refinement is the only use for datums in the lower level of a composite positional tolerance control. Datums used for orientation/angle in the upper level (PLTZF) control refine orientation/angle in the lower level (FRTZF) control. Datums used for location in the upper level (PLTZF) control refine only orientation/angle to that datum if used again in the lower level (FRTZF) control. This, of course, is in addition to what might be considered the main purpose of the lower level (Feature-Relating Tolerance Zone Framework) control and that is to refine the relationship between all features within the controlled pattern (for example, hole-to-hole, or pin-to-pin tolerance).

With two single segment positional tolerance feature control frames, a datum used in the upper level positional tolerance control may have that relationship refined in the lower level control, or an entirely different set of datums may be used. If datums used in the upper level control are also used in the lower level control (and in the same order), they refine the tolerance on whatever the datum was used for in the upper level control. Orientation/angular datums refine tolerance on the angle and location datums refine tolerance on the location to those datums.

The following examples show some similarities and some differences between these types of controls.

FIGURE 213

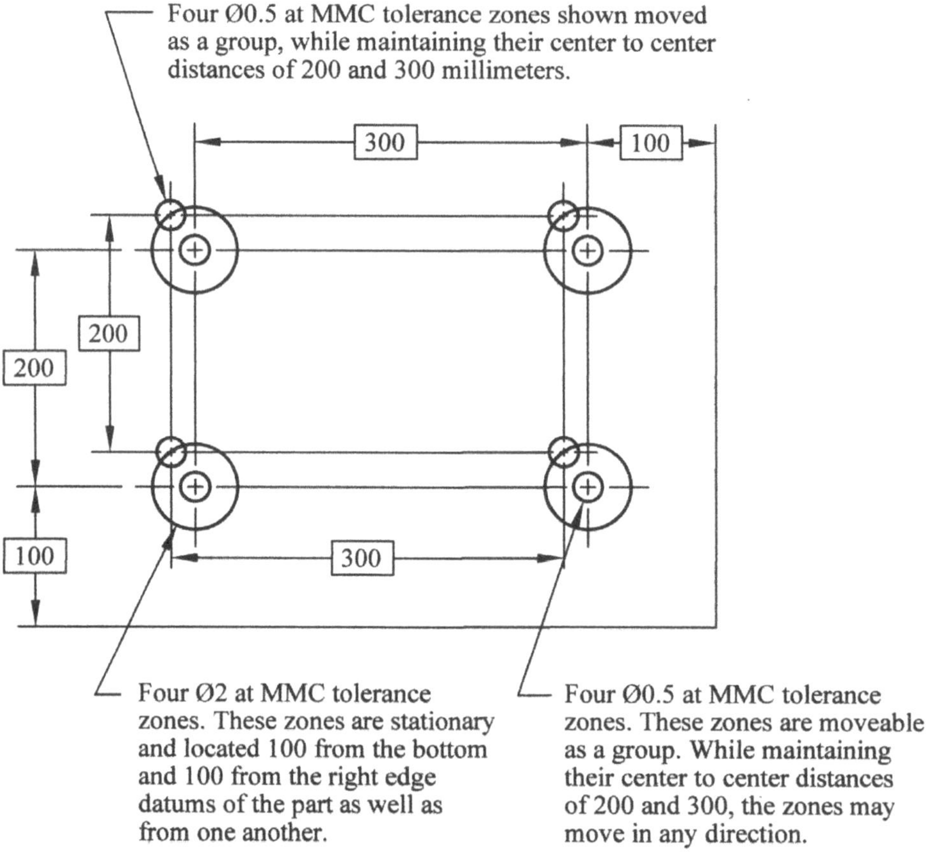

For the part to meet either drawing requirement shown below, all hole axes must be produced to reside within both the Ø2 at MMC and the Ø0.5 at MMC tolerance zones. These two controls, as shown, are identical in meaning.

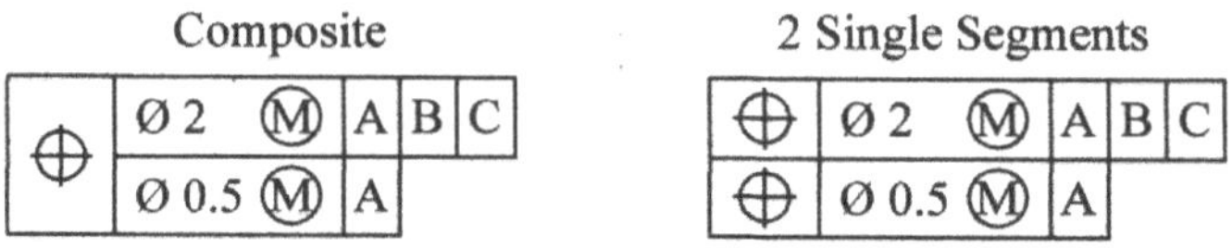

FIGURE 214

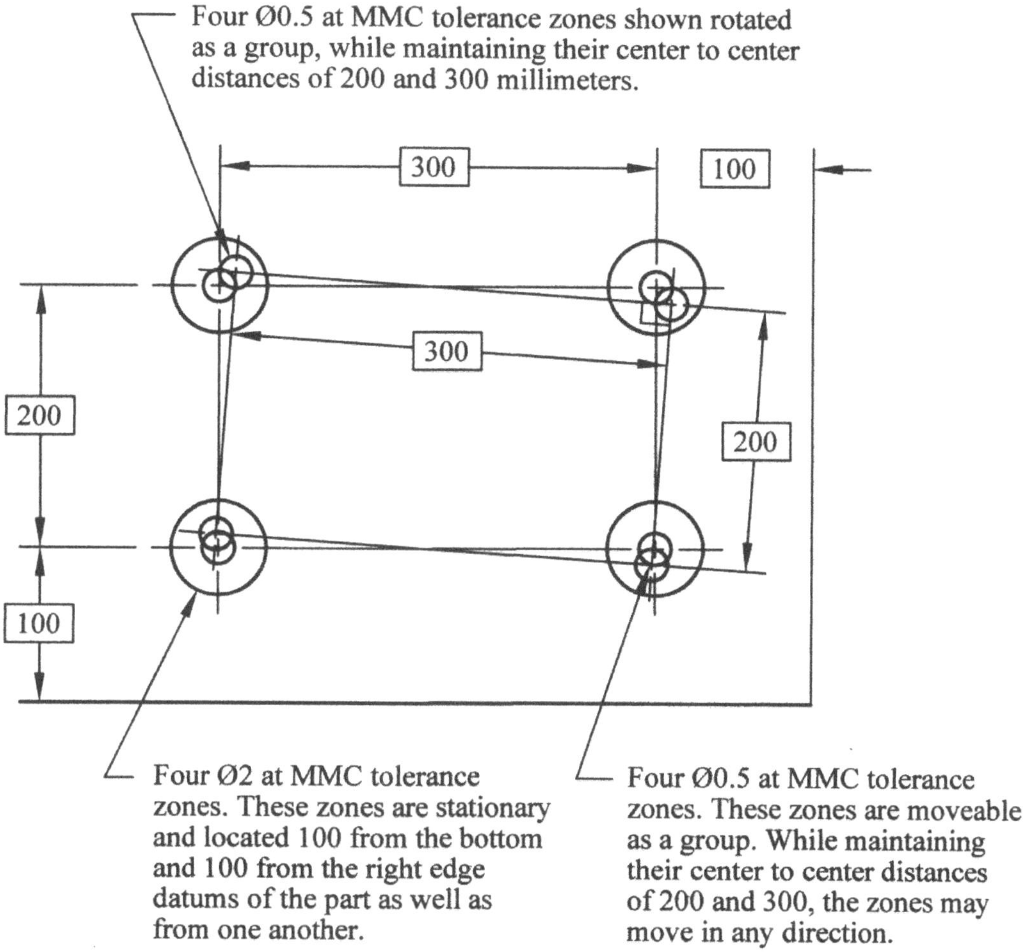

For the part to meet either drawing requirement shown below, all hole axes must be produced to reside within both the Ø2 at MMC and the Ø0.5 at MMC tolerance zones. These two controls, as shown, are identical in meaning.

Composite

⌖	Ø 2 Ⓜ	A	B	C
	Ø 0.5 Ⓜ	A		

2 Single Segments

⌖	Ø 2 Ⓜ	A	B	C
⌖	Ø 0.5 Ⓜ	A		

FIGURE 215

For the part to meet the drawing requirement shown below, all hole axes must be produced to reside within both the Ø2 at MMC and the Ø0.5 at MMC tolerance zones.

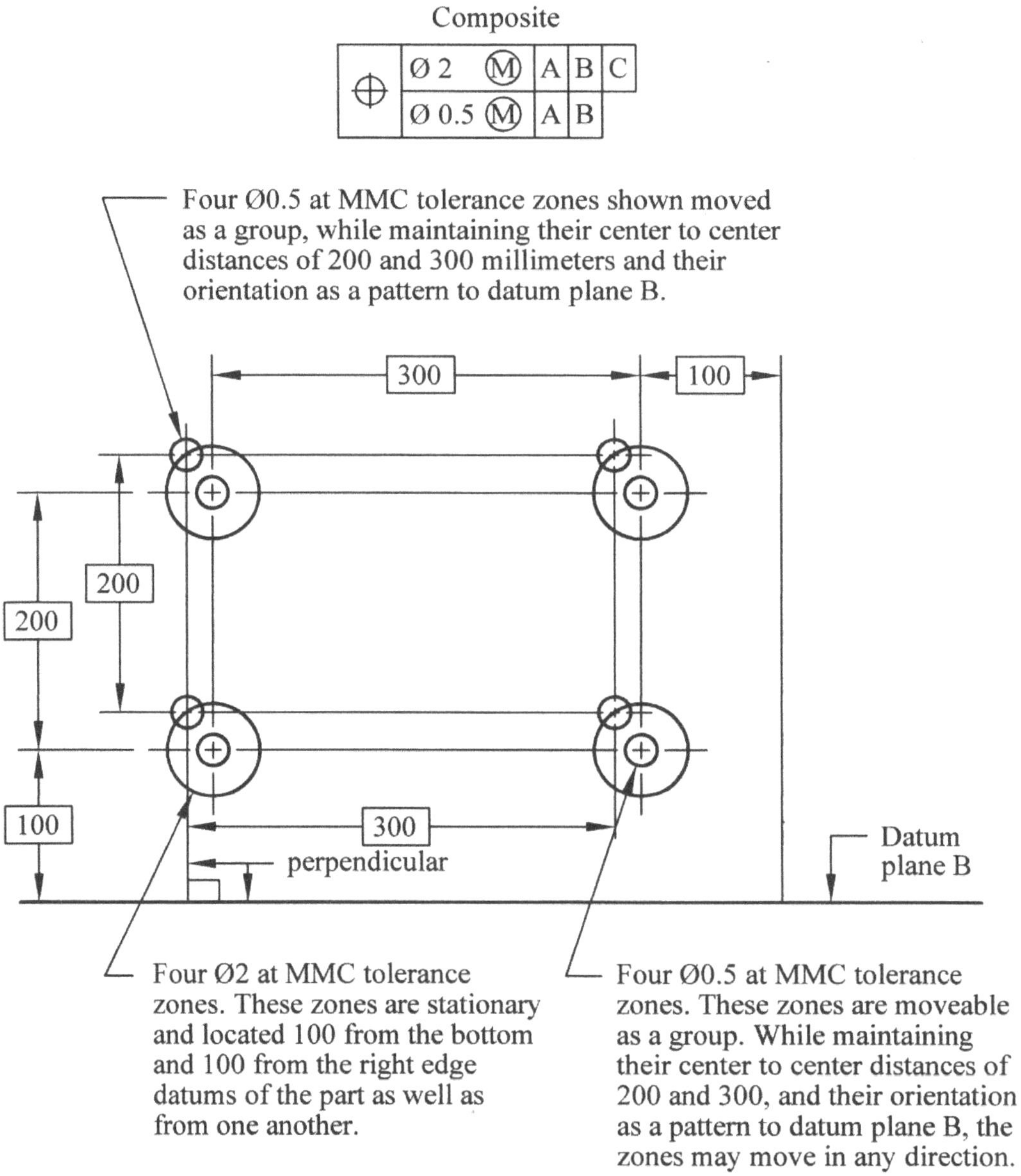

FIGURE 216

For the part to meet the drawing requirement shown below, all hole axes must be produced to reside within both the Ø2 at MMC and the Ø0.5 at MMC tolerance zones.

2 Single Segments

⌖	Ø 2 Ⓜ	A	B	C
⌖	Ø 0.5 Ⓜ	A	B	

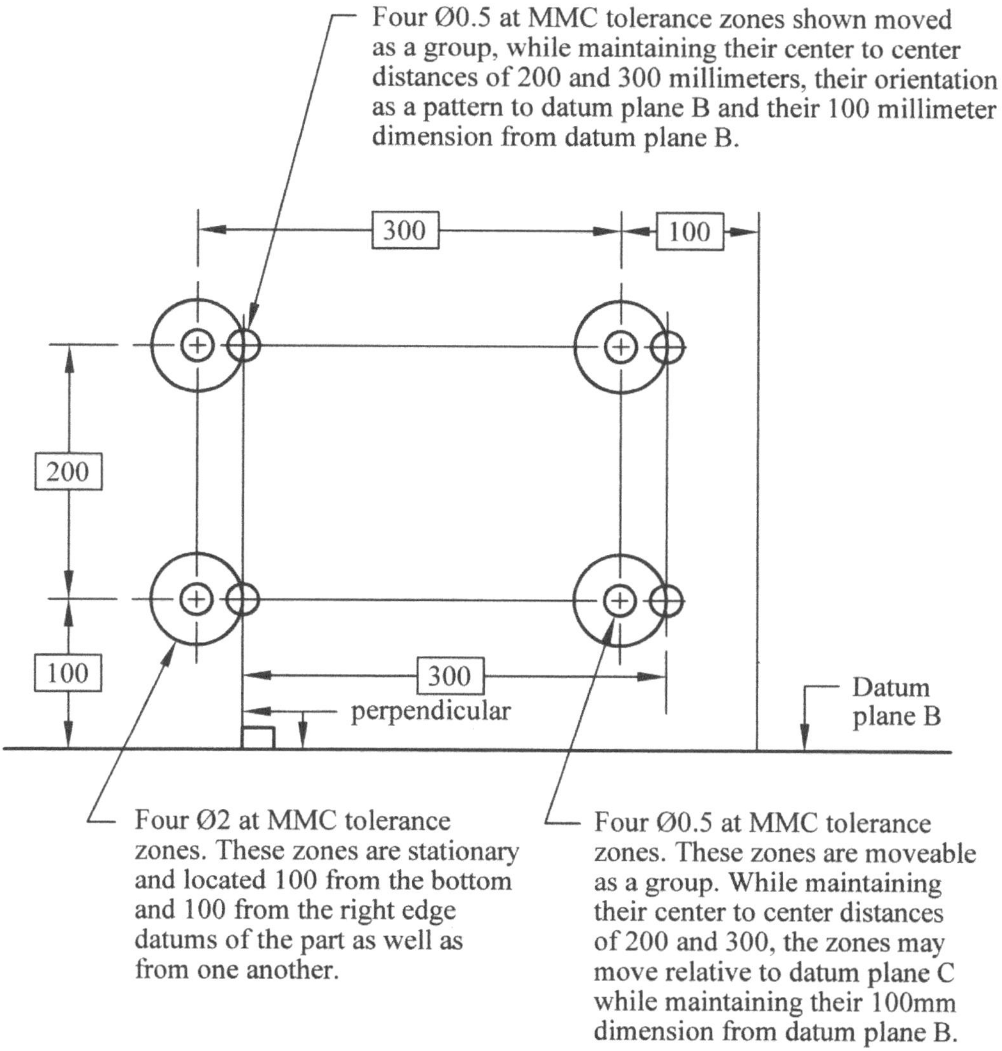

One Level Positional Tolerancing vs. Composite Positional Tolerancing and Simultaneous Requirements

One Level Position Controls

When more than one feature or pattern of features is located by basic dimensions relative to the same datums referenced in the same order of precedence, they are considered one pattern of features. This statement is true if the datum features are:

1) not subject to size considerations, or
2) if they are datum features of size and they use the same material condition symbols after the datum features of size.

This concept of multiple patterns of features being considered one pattern is known as the simultaneous requirement rule, or in gaging, it is called the simultaneous gaging requirement rule. This means that one set-up or a single functional gage is required to determine compliance with the positional tolerance of all features within the combined patterns. This rule also applies to profile controls or a combination of profile and position controls.

If, for example, the primary datum feature rocks on the primary datum plane, only one rocked orientation may be used to verify tolerance compliance for all features that are a simultaneous requirement. It also means that any pattern shift (or datum feature shift as it is sometimes called) allowed by datum features of size referenced at MMC or LMC would have to be experienced by all features within the combined patterns (by the same amount and in the same direction), or not experienced at all. If treated as a separate requirement or separate gaging requirement, each pattern would be allowed to rock in different ways or shift in different directions and still be accepted as within tolerance. Conceivably, more parts would pass inspection if this were allowed, but some accepted parts may not assemble with unified mating threaded hole patterns or unified mating pin patterns. If a separate requirement is desired for features normally considered a simultaneous requirement, a local note may be used beneath the feature control frame that states SEP REQT.

Composite Positional Tolerancing

For a composite position or profile control, the previous statements in this section apply only to the upper segment pattern locating tolerance zones. Multiple features or patterns of features using the same location datum reference frame are only considered one pattern of features (a simultaneous requirement) when inspecting the larger upper level tolerance (the datum-to-pattern Locating Tolerance Zone Framework). **The simultaneous requirement rule does not apply to the lower segment control (hole-to-hole Feature Relating Tolerance Zone Framework) for multiple patterns of features.** When inspecting feature-to-feature requirements (feature relating tolerance zone frameworks), each pattern must be considered separately.

This rule may be changed through the use of a local note that states SIM REQT (simultaneous requirement) listed after (to the right of) the lower segment of each applicable composite control.

These lower segment controls may also be changed to a simultaneous requirement through the use of one composite control with multiple leader lines, each leading to separate features or patterns you want to combine. As shown in the following illustration, it is sometimes useful to clarify the total number of features within the final simultaneous pattern through the use of a local note. In this method, the hole sizes are listed on the drawing, separated from the composite positional control. This method is especially useful when single holes are to be combined with other single holes or with patterns of holes, since the use of composite tolerancing is not appropriate for use on a single feature.

FIGURE 217 [Composite Position Tolerancing and Simultaneous Requirements]

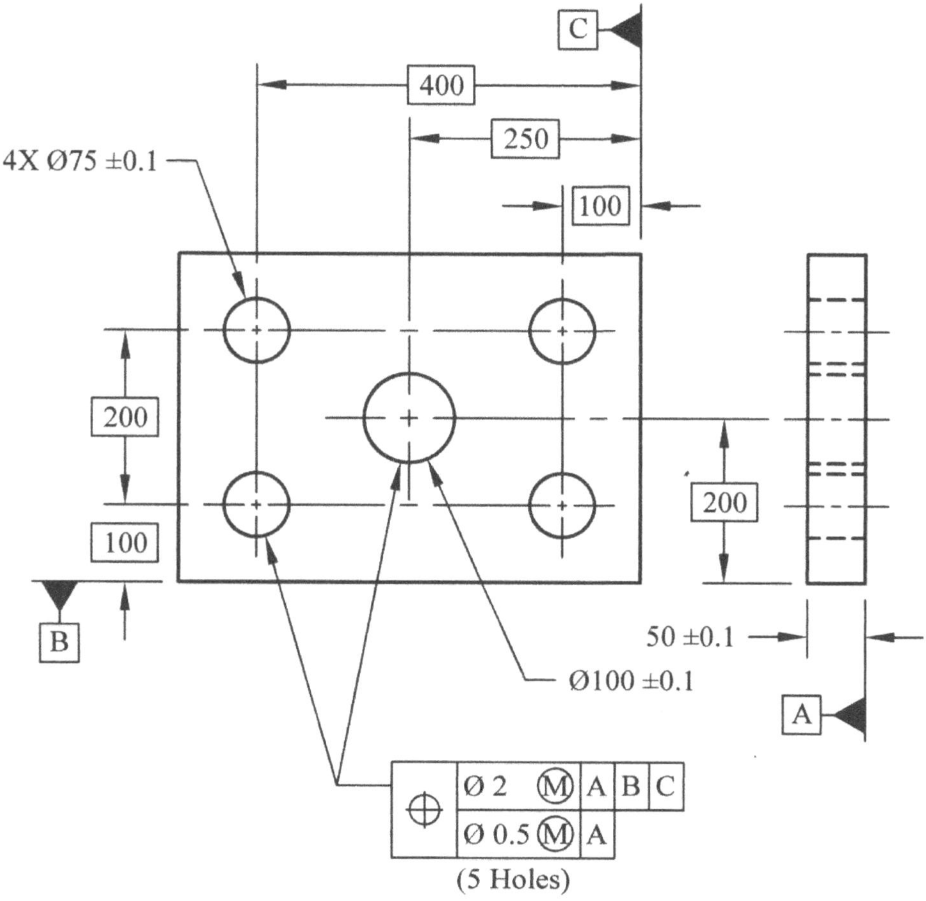

FIGURE 218 [Composite Positional Tolerancing Feature Control Frames]

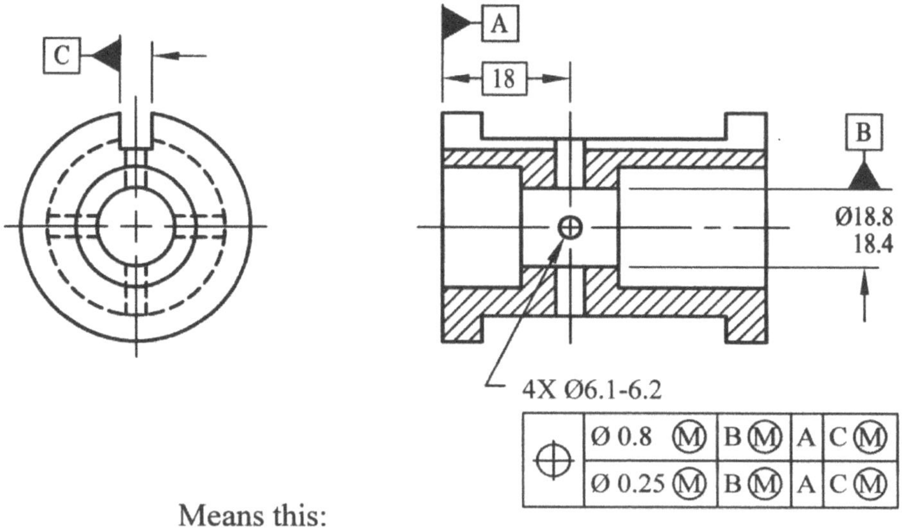

Means this:

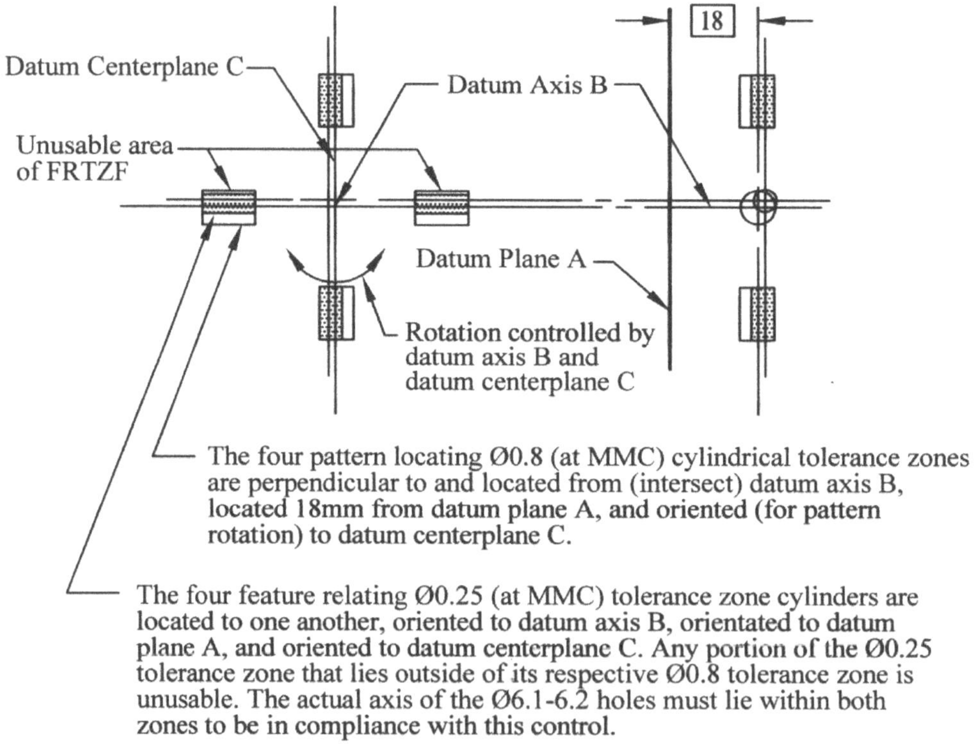

Only one possible displacement of the feature-relating tolerance zone framework (FRTZF) relative to the pattern-locating tolerance zone framework (PLTZF) is shown above.

If the geometric tolerance said only

4X ⌀6.1-6.2

⌖	⌀0.8Ⓜ

this would hold the relationship between the four holes to ⌀0.8Ⓜ per hole. But the tighter tolerance of ⌀0.25Ⓜ overrides that.

4X ⌀6.1-6.2

⌖	⌀0.25Ⓜ

This tighter position tolerance controls the hole-to-hole relationship within the four hole pattern because it is the tighter of the two tolerances.

If it said

4X ⌀6.1-6.2

⌖	⌀0.8 Ⓜ	BⓂ
	⌀0.25 Ⓜ	

this would mean:

1) the four holes were held to one another to within ⌀0.25Ⓜ.
2) the location and perpendicularity of the holes to datum axis B would be held to within ⌀0.8Ⓜ.

If it said

4X ⌀6.1-6.2

⌖	⌀0.8 Ⓜ	BⓂ
	⌀0.25 Ⓜ	BⓂ

1) the hole to hole tolerance would still be ⌀0.25Ⓜ per hole.
2) the intersection of each hole's axis to datum axis B would be held to within ⌀0.8Ⓜ.
3) the BⓂ in the level of control with the tighter ⌀0.25Ⓜ would apply to each hole's axis perpendicularity to datum axis B.

   By definition, any location datums brought from the upper level (⌀0.8Ⓜ control) down to the lower level (⌀0.25Ⓜ) control are only able to control orientation (angles only), not location.

   So the control of ⌀0.8Ⓜ would still apply to the location of the holes, tolerancing their intersection to datum axis B.

If it said

4X Ø6.1-6.2

⊕	Ø0.8 Ⓜ	BⓂ	A
	Ø0.25 Ⓜ	BⓂ	

or

4X Ø6.1-6.2

⊕	Ø0.8 Ⓜ	BⓂ	A
	Ø0.25 Ⓜ	BⓂ	A

1) the hole to hole tolerance (within the four hole pattern) would be held to within ∅0.25 per hole.
2) the location/intersection of each hole to datum axis B would be held to within ∅0.8Ⓜ.
3) the perpendicularity of each hole's axis to datum axis B would be held to within ∅0.25Ⓜ.
4) the tolerance on the 18mm dimension from datum plane A would be ∅0.8Ⓜ.

Even though datum plane A appears in one of the two lower level controls with a tolerance of ∅0.25Ⓜ, but not in the other, these composite position controls are identical in meaning. In the feature relating tolerance zone framework (lower level) control, location datums, such as A, are not capable of tolerancing location dimensions, such as the 18mm. That means that whether datum A is referenced with the ∅0.25Ⓜ or not, it does not control location (just as B does not control location in this level of tolerance).

Some might surmise that datum A listed in the lower control would refine the parallelism, of the holes to datum plane A, but orientation to B in that level already has done that. By virtue of the fact that datum plane A is (by definition) perpendicular to datum axis B, any hole axis that is perpendicular to B is automatically parallel to secondary datum plane A to within ∅0.25Ⓜ. That is true whether or not datum A is referenced in the level of control containing the ∅0.25Ⓜ tolerance.

So, if datum A is referenced in the level of control containing the ∅0.25Ⓜ tolerance, datum A means nothing—just as if it wasn't there. The logical question is then, "Why is A there?" The answer is so that we can reference C. For example:

4X Ø6.1-6.2

⊕	Ø0.8 Ⓜ	BⓂ	A	CⓂ
	Ø0.25 Ⓜ	BⓂ	A	CⓂ

There is a rule.

The rule is that in composite position tolerancing only datum features referenced in the <u>pattern locating tolerance zone framework (upper level)</u> control may be referenced in the <u>feature relating tolerance zone framework (lower level)</u> control. And, any datum referenced in the feature relating tolerance zone framework control must be used in the same order as they were referenced in the upper level (in the pattern locating control). The feature relating control (lower level) doesn't have to reference any datums. But, if it does, they may not be different than were used in the pattern locating control (upper level), and they may not be used in a different order.

In our composite position tolerance, we want to control the rotation of the four hole pattern as they intersect the axis of B. We want them oriented to the slot labeled datum feature C. We would like the hole pattern's ⌀0.25Ⓜ position tolerance zones to be either perpendicular or parallel (as depicted) to the centerplane of slot C while the ⌀0.25Ⓜ zones are centered on datum axis B.

Since we may not reference C in the feature relating level of control (lower) unless it already appears in the same order in the pattern locating control (upper), it is listed up top as a place saver so that it can be brought down to the lower control. This will comply with all rules—keep one set-up for inspection to inspect both levels of position tolerance and keep the four hole pattern from rotating away from their depicted relationship to datum centerplane C by more than the ⌀0.25Ⓜ tolerance allows.

Two Single Segment Position Controls

FIGURE 219

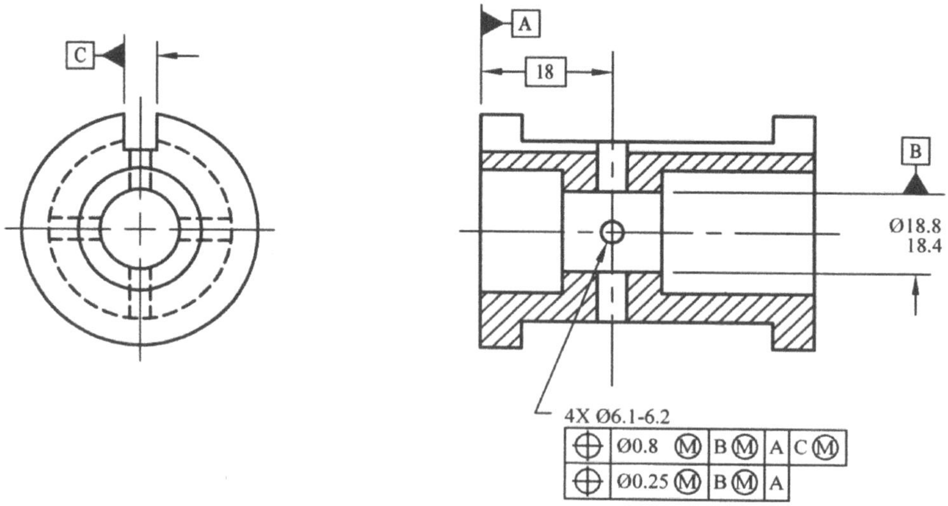

Means this:

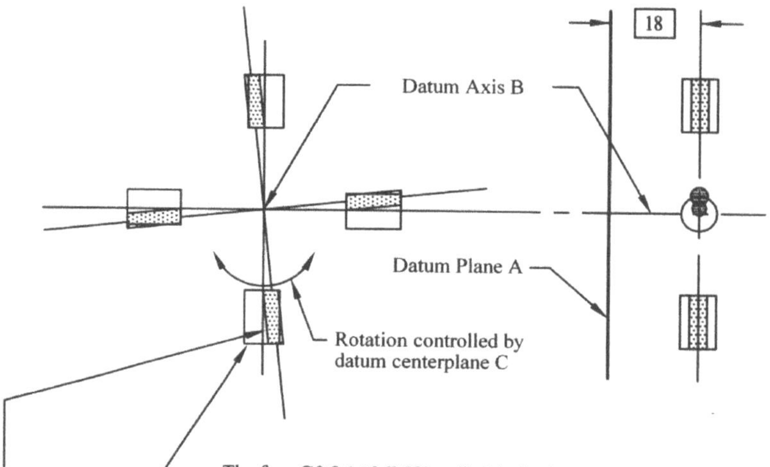

The four Ø0.8 (at MMC) cylindrical tolerance zones are perpendicular to and located from (intersect) datum axis B, located 18mm from datum plane A, and oriented (for pattern rotation) to datum centerplane C.

The four Ø0.25 (at MMC) tolerance zone cylinders are perpendicular to and located from (intersect) datum axis B, and located 18mm from datum plane A. Any portion of the Ø0.25 tolerance zone that lies outside of its respective Ø0.8 tolerance zone is unusable. The actual axis of the Ø6.1-6.2 holes must lie within both zones to be in compliance with the control.

The most important thing to remember about a two single segment position control is that whatever jobs the control with the tightest position tolerance bids on, it gets. For example, if the control was

4X Ø6.1-6.2

⌖	Ø0.8 Ⓜ
⌖	Ø0.25 Ⓜ

without any datums referenced, the only job being bid on by both levels of control is the hole to hole tolerance relationship. But since the lower level control has a tighter tolerance (∅0.25Ⓜ versus ∅0.8Ⓜ), it wins the job of tolerancing the relationship of the holes to each other within the pattern. So, the tolerance on the perpendicularity and intersection of the holes' axes to one another is controlled to within the smaller position tolerance of ∅0.25Ⓜ. The ∅0.8Ⓜ tolerance has no job at this stage.

If the control was:

4X Ø6.1-6.2

⌖	Ø0.8 Ⓜ	B Ⓜ
⌖	Ø0.25 Ⓜ	B Ⓜ

the hole to hole tolerance within the four hole pattern is still being held by the ∅0.25Ⓜ tolerance. But now, the perpendicularity of each hole to datum axis B and the intersection/ location of each hole to datum axis B is being held to within the ∅0.25Ⓜ tolerance. If this was a composite position control, the location of the four holes to datum axis B would be held to within the ∅0.8Ⓜ tolerance since in composite position tolerancing datums used in the lower level controls lose their ability to control location. That is not true for two single segment position controls. So, location (perpendicularity and intersection) of all holes to datum axis B is held to within the tighter ∅0.25Ⓜ tolerance. Again, the position tolerance of ∅0.8Ⓜ has no job.

If it said:

4X Ø6.1-6.2

⌖	Ø0.8 Ⓜ	B Ⓜ	A
⌖	Ø0.25 Ⓜ	B Ⓜ	A

every relationship explained to this point would still hold true. The hole to hole tolerance and the perpendicularity and intersection of the holes' axes to datum axis B is being held to within the ∅0.25Ⓜ. In addition, that, with datum A now listed in both levels of position, the distance from datum plane A of 18mm would now be toleranced by the tighter ∅0.25Ⓜ position tolerance. In a two single segment position tolerance, the level of control with the tightest position tolerance applies to any relationship it references. It is capable of holding tighter hole to hole tolerances and orientation and location tolerances to the datums referenced in that level of control. In this case, it has referenced every relationship that the looser ∅0.8Ⓜ tolerance has referenced. So, again, there is no relationship that is toleranced by the larger ∅0.8Ⓜ position tolerance. It has no job yet. All jobs are done better and held tighter by the ⌀0.25Ⓜ position tolerance.

But, if it was:

4X Ø6.1-6.2

⌖	Ø0.8 Ⓜ	B Ⓜ	A	C Ⓜ
⌖	Ø0.25 Ⓜ	B Ⓜ	A	

since the level of position with the tightest tolerance does not reference datum C, it does not take the job from the upper level control. The rotation of the four-hole pattern, which clocks (orients) the holes to centerplane C as depicted in the drawing, is held by the ⌀0.8Ⓜ tolerance.

Datum centerplane C is called an angular orientation datum. Had C been referenced in the lower control with the tighter ⌀0.25Ⓜ tolerance, the upper level tolerance of ⌀0.8Ⓜ would have nothing that it applied to tighter than the ⌀0.25Ⓜ tolerance and, therefore, would be taken off the drawing. In a two single segment position tolerance, if the tighter tolerance level wants to do all of the jobs done by the looser level tolerance, the looser level control is wiped out. For example,

this:

4X Ø6.1-6.2

⌖	Ø0.8 Ⓜ	B Ⓜ	A	C Ⓜ
⌖	Ø0.25 Ⓜ	B Ⓜ	A	C Ⓜ

means this:

4X Ø6.1-6.2

⌖	Ø0.25 Ⓜ	B Ⓜ	A	C Ⓜ

FIGURE 220 [Assembly Drawing]

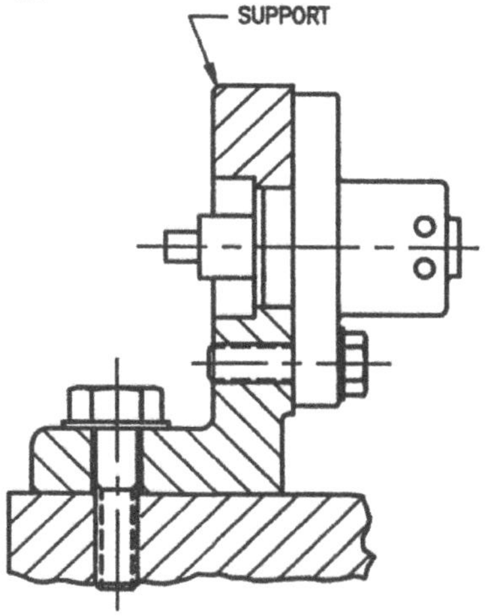

FIGURE 221 [Detail Drawing]

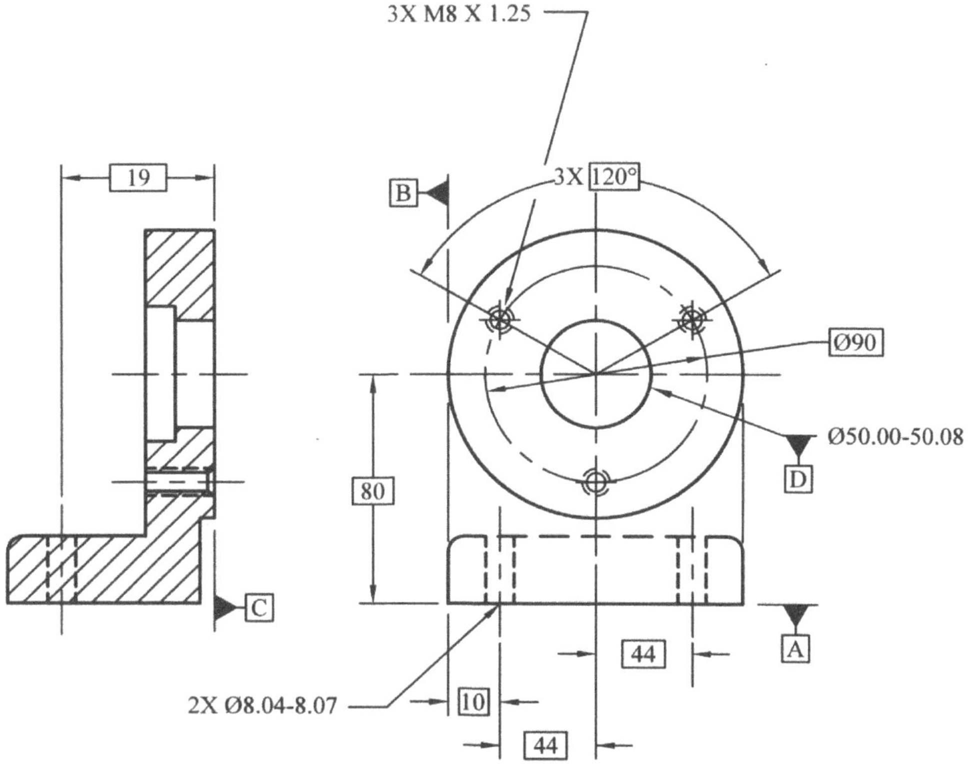

The problem of defining a part in a functional manner is compounded by the ever-present fear that when such a definition is founded, it will be difficult to produce and, therefore, prohibitively expensive. This problem can often be solved by simply taking it on in a two-step approach:

Step 1: Create a functional definition that ties down everything within an equal tolerance to all datums in the control.

Step 2: Disassociate the less important relationships in each control from the tight tolerance that makes the controlled features function. Some-times two controls are less costly than one.

The assembly drawing shows two separate components interacting with the part to be toleranced. Datum features have already been chosen. Some are good, while others are mediocre. We will work with these to show how to eventually disassociate the tight tolerances that make the features mate/function from the datums simply meant to keep the features being controlled on the part. Looser tolerances will eventually be assigned to the datums of lesser functional importance. But this will not occur until we have found an initial part definition and are ready to refine and optimize it.

Let us say the part is received by the machinist as a rough casting. The surfaces that serve as seating surfaces will be machined first and subsequently used on which to stabilize the part while machining other features. Looking at how the component parts seat on our part, we should ask what surfaces need the most contact. These surfaces will serve as primary datums and, if possible, machined first. For example, one interfacing part seats on datum feature A and one on datum feature C. If we wish them to be machined, we must give them an appropriate geometric control to impart that knowledge to the machinist.

Likewise, we must decide if we wish to imply a machining sequence. If not, both datum features A and C will be given a flatness control and the tolerance on the perpendicularity relationship they share relegated to the general angular tolerance note on the drawing that usually resides with the title block tolerances. If, instead, an implied machining sequence is desired, one of the datum features (either A or C) will be given a flatness control and the other a perpendicularity control back to the one that has been made flat. In this case, we will give datum feature A a flatness control and make datum feature C perpendicular to A. Although the machinist has the latitude to improvise, this symbology, once on the drawing, implies a sequence of events.

Datum feature A will be machined first, then datum feature C will be machined perpendicular to A.

We must then decide what will be implied as machined next. Since datum feature B touches no surface on either of the interfacing parts, we will not machine it, nor give it a feature control frame tolerance. It will simply comply with the title block's "Unless otherwise specified, all angles are equal to plus or minus one degree" tolerance. (The plus or minus one degree is given as an example only. Each part is equipped with suitable default tolerances to cover one's anatomy.) In the preliminary geometric controls given, not machining a rough cast surface like datum feature B and subsequently invoking it as an origin of measurement will cause some consternation. However, refining controls given in step 2 (the optimization) will divorce tight tolerances from this datum feature.

The central hole called datum feature D plays an important role in allowing one of the component parts to mate with the part we are defining. Since the 3-hole pattern of threaded holes works directly with and will be subsequently defined and measured from datum D, we will define D next. There are usually a couple of options in each product definition. We can measure from the outside of the part and work our way into the internal features, or we can begin with the internal features and work our way out, defining the external features from the internal. In this instance, we will do a little of both. We will measure datum feature D from the established outside datum surfaces and then measure the 3 threaded holes from datum D. As you can see, we are focusing our attention on only the one component part that interfaces with datum feature D and the 3 threaded holes first. When that is finished, we will switch our attention to the 2 clearance holes that will tie our part to the larger part that seats on datum feature A.

Since this first control for datum feature D is only a gross geometric definition, we will choose an all-encompassing control. Positioning the hole will allow us to use datums that give it a tolerance on its angle and location. It is a diameter that mates while seated on datum feature C. And to tie it loosely to the desired location on the part, we will use datum features A and B. Since datum feature A is machined and B is not, we will push B as far back in the control as possible. This will make B a tertiary datum feature, requiring only 1 point of high point contact. The control reads:

Ø50.00 - 50.08

⌖	Ø0Ⓜ	C	A	B

The zero positional tolerance was necessary because it is known that the shaft that inserts into datum feature D has a virtual condition of Ø50.00. Remember, this is only a preliminary control and will be refined. The refining control must also keep in mind that datum feature D can use only the features that have been defined prior to it and none that will be subsequently defined from it. For the refinement, we must recognize a much larger tolerance can be used to some of the datums than from others.

In this example problem, we will use a diameter of 1 millimeter of tolerance to the datum features that are relatively unimportant to the function of the feature being controlled. Since the purpose of this larger tolerance is to lower part cost, make the features more producible and protect wall thickness, calculations must be performed to determine that this feature, and those controlled to it, do not break out/endanger part integrity and wall thickness. This technique is discussed in the section on wall thickness calculations and will not be discussed here. The formulas given in that section can also be worked in reverse, a desired wall thickness decided on and inserted as the final minimum wall thickness answer. Then, the formula is worked from that end forward to calculate the positional tolerance that gives that result.

The chosen preliminary geometric control says, "The axis may be out of position a diameter of zero if the hole is produced at maximum material condition to datum C for perpendicularity and datums A and B for location". Since the maximum material condition symbol is most often used on features that mate, saying as the holes grow and shafts shrink they are easier to fit together and, therefore, may be less geometrically perfect, the circled M can often be read, if used in a positional control, as "Position of a feature that mates".

In this case, a diameter is being positioned, so the control can be read, "Position of a diameter that mates" and, when the datums are added, "Position of a diameter that mates while the part is seated against C and the diameter located from A and B". The refining control will first loosen all relationships to a diameter of one millimeter, then refine the important functional relationship(s) back to the original tighter geometric tolerance (in this case, zero at MMC). For example:

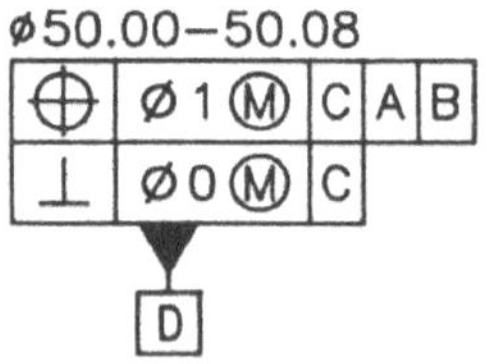

The only surface that is contacted (at this stage of the part definition) by the mating component part that interfaces with datum feature D is datum feature C. Therefore, it is the only relationship that must be held to within the tighter functional/mating tolerance of zero at MMC. Since the relationship between D and C is one of perpendicularity only, that is the appropriate symbology for the refining control. It says, "Perpendicularity to within zero at MMC of a diameter that mates while the part is seated on datum feature C", or "The axis may be out of perpendicularity zero if the hole is produced at maximum material condition (Ø50.00) to datum plane C". This feature (datum feature D) has been defined, then optimized in its definition. We are now ready to move on and define the 3-hole pattern in a similar manner.

The 3-hole pattern, when properly defined, will complete our journey in defining the features of our part necessary to allow it to mate with one of the other component parts. The 3 holes must keep the proper 90mm bolt circle distance and 120° angles from each other, as well as a relationship to datums selected; therefore, the position geometric characteristic is chosen. These are diameters which will need cylindrical tolerance zones to encompass their axes; therefore, the diameter symbol will be included in the feature control frame.

These controlled axes are generated by pitch diameters that mate and, although the amount of additional (bonus) tolerance gained is small (not one-for-one like a clearance hole derives) and not easily quantifiable, it exists (as the pitch diameter of the hole increases) and makes part assembly easier, so the maximum material condition symbol will be included. The mating component part seats on datum feature C; therefore, this surface needs more contact than any other with this component part.

To achieve that, surface C will be the primary datum feature. The 3-hole pattern works directly with datum feature D and must maintain a direct interrelationship with D in order to assemble. Therefore, the datum axis D will act as the origin of measurement in both the X and Y planes for the 3-hole pattern. In other words, datum feature D will act as the secondary datum feature but will generate two planes at 90° to one another from which the 3 holes will be measured. With datum features C and D in the control, three mutually perpendicular planes exist. But, because two of these planes are generated by an axis at the center of a hole, the two planes have orientation only to one another and datum plane C. Unless we are willing to relegate the rotation tolerance of the 3 holes, as a pattern, around the axis of D to the angular tolerance in the drawing's general tolerance note (in the drawing's title block), we must include another datum feature.

This datum feature will generate a plane that will give the two planes of datum axis D an orientation and, consequently, orient the 3-hole pattern to be measured from that axis and along its crossing planes. This will control the rotation of the pattern to the datum plane generated by this third datum feature (fourth datum plane). With the given datum features, either A or B may be chosen to serve this purpose; but, since A is a machined surface and B is not, A seems the better choice. It is flatter and, therefore, more repeatable. This plane receives 2 points of high point contact when the part is set up for measurement of the 3-hole pattern. The holes are not measured from datum plane A but rather from datum axis D while along planes that are perpendicular to datum plane C and at the appropriate angle to datum plane A. Datum feature D, being a datum regular feature of size that mates at the same time (with the same component part) as the 3 holes, will receive a maximum material boundary modifier. This will allow a pattern shift of the 3 holes, as a group, as datum feature D grows from its maximum material boundary, as long as D maintains its perpendicularity to datum plane C.

Any pattern shift/datum feature shift gained by the growth of D is negated by an equal departure of D from perfect perpendicularity to datum plane C. So, strictly speaking, the pattern shift of the 3 hole pattern away from the center of D is gained as datum feature D grows outside of its maximum material boundary (MMB), which is a 50 millimeter diameter cylinder that is perfectly perpendicular to datum plane C. The preliminary control will appear as follows:

3X M8 etc.

⌖	Ø0.02Ⓜ	C	DⓂ	A

The tolerance of Ø0.02 was simply derived through the use of the fixed fastener formula. Since the clearance holes in the mating component have a maximum material condition of 8.04 diameter, and the maximum material condition of the bolt is 8.00 millimeters, the tolerance to be divided between related threaded and clearance holes on each part is the difference of 0.04. In this case, we have chosen to divide the tolerance equally and give each threaded hole half of the available geometric tolerance, which is 0.02. However, in many cases where the clearance holes have a large size tolerance that is potentially available to them as positional bonus tolerance, a larger portion of the total geometric tolerance than just half should be considered for the threaded features. Even though the threaded features have a maximum material condition symbol next to their geometric tolerance, the centering effect of a threaded feature tends to negate most of its bonus tolerance. The threaded features may need more than half of the total original geometric tolerance to be divided between them and the clearance holes to even things out and make the threaded holes more producible. They start out at a distinct disadvantage to the clearance holes in the area of potential bonus tolerance.

The tolerance zones must not be inside of the threaded holes but rather projected outside of the part entirely at a height above the part that reflects how far the bodies of the mating screws protrude above the surface of datum plane C. In this case, since the head of the bolt will rest on the top of the mating part, that height of the projected tolerance zone happens to also be the maximum thickness of the mating part (25.4). To accomplish this, the feature control frame is changed to appear as below:

3X M8

⊕	Ø0.02Ⓜ Ⓟ25.4	C	D Ⓜ	A

This may be read a number of ways. For example: "The axes of the holes' pitch diameters may be out of position to one another a diameter of 0.02 if produced at maximum material condition while the zones are projected above and perpendicular to datum plane C 25.4 millimeters and located from datum axis D if datum feature D is produced at maximum material boundary while the hole pattern is properly oriented to datum plane A." Another possible reading is: "Position of diameters to one another (within 0.02 per hole) that mate, with all tolerance zones projected above datum C 25.4 millimeters, while the part is seated on C, the holes located from D while D mates and the hole pattern properly angled from A."

This preliminarily completes the definition of the 3 threaded holes sufficiently to allow them to mate with their interfacing component part. However, this definition can be improved for ease of manufacture by loosening the rotational relationships to datum plane A. The component part that seats on datum C, whose pin inserts into datum feature D and whose clearance holes align with our threaded holes, doesn't have much to do with datum A. If more rotation of the 3-hole pattern were allowed, it would not affect the part function at all. Yet, if we simply eliminate datum A from the control, the threaded hole pattern's angular relationship to it would be controlled by the title block's general tolerance note for all otherwise untoleranced angles. It is possible that tolerance may be inappropriate--too tight or loose. If one wishes to specify a certain tolerance that is different, it may be done by defining the relationship of the hole pattern in two levels of control. This is called a 2 single segment position control. Let us speculate we have calculated an allowed pattern rotation to datum A that would be most appropriate at a diameter of 1 millimeter. Yet, all other interrelated tolerances must be held as tightly as before. We may state this as follows:

3X M8 x 1.25

⊕	Ø 1 Ⓜ Ⓟ25.4	C	D Ⓜ	A
⊕	Ø0.02 ⓂⓅ25.4	C	D Ⓜ	

The order of the datums in the upper level control is important. A rotational datum such as A can be misconstrued as a locational datum unless it follows sufficient datums that have already given the controlled features location in both the X and Y directions. The 3-hole pattern may now rotate around the axis of datum D as a unit within stationary tolerance zones of diameters of 1mm. But they may not depart individually from their basic defined relationships to one another, datum C, or datum D more than a diameter of 0.02 per hole. The control is now optimized--cost effective and functional. It also may shed some perspective on the myth some hold to be true--that two geometric controls are more expensive than one. As has been shown here, that simply is not factual.

We can now switch our attention to the other mating component and the 2 clearance holes on our part that will help bind the parts together. The 8.04 - 8.07 holes must be at least held 88mm apart and perpendicular to the seating surface A. Since the holes must reside within the confines of our part, the pattern must be related to datums that represent the part. There are many options, but to begin we will create a general preliminary statement using half of the available geometric tolerance available to holes on both parts. Let us assume an 8mm shaft or screw. Half the available geometric tolerance would be 0.02. (8.04 - 8.00 = 0.04 and 0.04 divided by 2 equals 0.02). Again, the same considerations mentioned when tolerancing the 3 threaded holes should be considered--saving a greater tolerance for the threaded holes, when appropriate, because of their disadvantage to clearance holes in gaining quantities of bonus tolerance. Once the tolerance is calculated, the preliminary control could appear as:

2X Ø8.04 - 8.07

⌖	⌀0.02Ⓜ	A	C	B

or

2X Ø8.04 - 8.07

⌖	⌀0.02Ⓜ	A	C	DⓂ

or

2X Ø8.04 - 8.07

⌖	⌀0.02Ⓜ	A	C	D

The choice between datums B and D is most difficult. At this stage of the definition, datum B looks unattractive because of the tight geometric tolerance and the roughness and resultant non-repeatability of a high point plane formed by the cast surface B. Although datum feature D doesn't have a roughness problem, if used, it would require centering the 2 hole pattern to datum hole D. Centering can be difficult and slow without fixtures; and, since the 2 holes have little to do with D, it may not be worth the trouble. The 2 holes have little to do with datum feature B either. But, if B is used, it would require only a rail to push the surface against to establish a high point plane. Still, the surface is rough and what is the high point one time may crumble and not be the high point at the next manufacturing or measurement procedure. If D is used, the maximum material boundary symbol may be used after it or the regardless of material boundary implied.

Although the correct syntax would be D regardless of material boundary, since functionally the size of hole D has nothing to do with the location of the 2 holes, before using the RMB concept one should ask, "Will the MMB concept hurt the part's functionality?" In this case, the use of the MMB concept after datum D in the control would allow an additional shift of the 2 hole pattern to one side or the other a maximum of half of the actual growth of D. Since the size tolerance of D is so small (50.00 - 50.08), the additional side shift of the hole pattern of half of 0.08 seems safe in terms of potential damage to wall thickness integrity. Cost savings are also small, but fixturing and gaging may be simpler if MMB is used over RMB.

Having considered all the possibilities, we choose. Let us say we choose B over D because the next step of optimizing the control will divorce the tight tolerance from B and replace it with a liberal tolerance. So, we will use a control that states, "Position of diameters (to one another to within 0.02 per hole) that mate while the part is seated on A and the pattern is measured from C and B." or "The hole axes may be out-of-position a diameter of 0.02 if the holes are produced at maximum material condition holding perpendicularity to datum A and location/distance from datums C and B."

Datum A is the obvious choice as primary because it is the surface that needs the most contact, the seating surface in the interrelationship between our part and the component part that uses the 2-hole pattern. Datum C is chosen over B because it is a flatter, machined surface and B is not. Still, the control can be refined by allowing a larger tolerance for the less important relationships than for the important ones. The component mating part interacts directly with each of the 2 holes and datum feature A, but does not touch or align with either C or B. Therefore, the relationship (tolerance) can be loosened to C and B. Let us assume a wall thickness calculation and other considerations will allow this looser tolerance to be a diameter of 1mm for holes produced at a diameter of 8.04. The control may now appear as:

2X Ø8.04–8.07

⌖	Ø 1 Ⓜ	A	C	B
⌖	Ø 0.02 Ⓜ	A		

It is important that the lower level refining control use position as the geometric characteristic symbol. Perpendicularity would not be a strong enough symbol to maintain the 88mm dimension between the holes. So our lower level control can be read, "Position of diameters (to within Ø 0.02 per hole) to one another, if produced at maximum material condition, while the part is seated on A" or "The axes of these holes may be out-of-position to one another and out-of-perpendicularity to datum plane A, a diameter of 0.02 per hole, if produced at maximum material condition". Even at this level of refinement, some of the produced parts would function under the worst possible conditions but still be in violation of the features' size limits. Therefore, a further refinement is possible. One may ask, "What is the smallest clearance hole an 8mm bolt that has been threaded into a hole allowed a positional tolerance of a diameter of 0.02mm (the amount we reserved for the threaded holes on the mating component part) could hope to fit within?" If the answer to that question is 8.02mm, then the control can be stated as:

2X Ø8.02 - 8.07

⌖	Ø0.98 Ⓜ	A	C	B
⌖	Ø0 Ⓜ	A		

This refinement has the advantage of maintaining the same wall thick-nesses and worst mating boundaries as the previous controls, while allowing the inspectors to accept more of the produced functional parts. This should reduce the overall cost of the accepted parts. Should the holes be produced at 8.04 (the old MMC), the bonus positional tolerance gained would again raise the upper level positional tolerance to a diameter of 1mm and the lower level positional tolerance to a diameter of 0.02mm. So, things have changed for the better in terms of functional produced parts accepted as within limits by inspection, but have remained the same in the functional worst-case boundaries created for wall thickness calculations and mating conditions.

In summary, a part can be defined in logical steps, sequential in their description of features, based on interrelationships, one built on another, presupposing the existence of one feature while defining the next logical interdependent feature or pattern.

The following example shows a somewhat preliminarily complete version of the tolerancing procedures discussed in this section.

FIGURE 222 [Finished Drawing with Multi-level Geometric Tolerances]

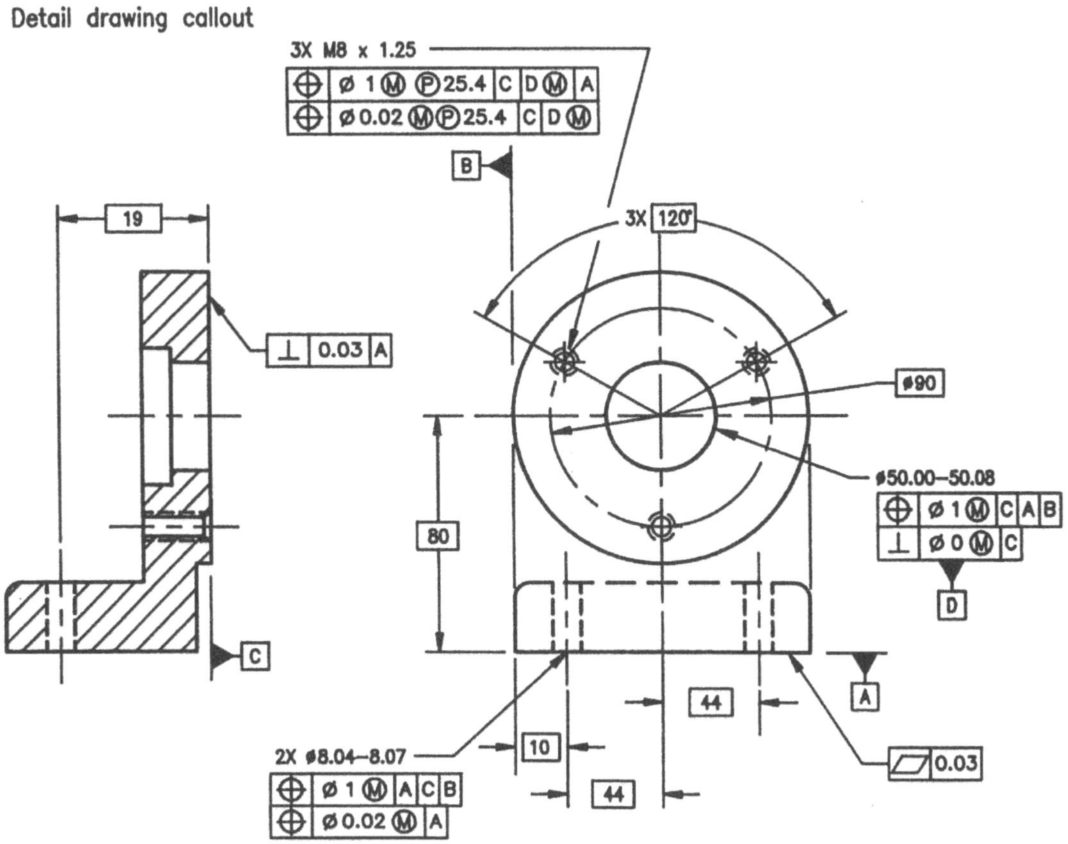

Chapter 14

Position Boundary Concept

•Position Boundary Concept
•Tolerance Zone vs. Boundary Concept Explanation

Chapter Objectives

Readers will learn:

1. The positional boundary concept and how to apply it oddly configured "closed loop" features.

2. How to position elongated holes.

3. How to gage elongated holes.

4. How to position hoses, pipes and tubes using positional boundary.

5. Notes common to hoses, pipes and tubes.

6. To apply the positional boundary concept to an oddly configured hole, make it a datum feature and represent it in a functional gage at the correct virtual condition size.

7. What to do when gage measurements conflict with Coordinate Measurement Machine measurements.

8. How to create odd tolerance zones for cylindrical holes.
   -rectangular tolerance zones
   -bi-directional positional tolerancing, polar coordinate method.

9. Calculating the Applicable Boundary Size

10. Tolerancing a Gear

Position Boundary Concept

The Y14.5 Dimensioning and Tolerancing Standard has for the last several revisions depicted examples of elongated holes being located by basic dimensions from specified datums and toleranced for this location with position feature control frames. But it was always different from other positional controls in a couple of ways. First, until the 1994 revisions, the position section in Y14.5 described position as being used only on features of size. Features of size were defined as anything cylindrical, spherical or any two parallel planar surfaces.

An elongated hole was clearly a more complicated shape than any of these and seemed to violate the statement that only features of size were to be positioned. Also, a definition of the position tolerance zone for an elongated hole was not given and the words in that section of the standard instead said that a "boundary" was formed that the surface of the hole may not violate.

This boundary was the shape of a perfect elongated hole, but at a size generated in each direction by subtracting the applicable position tolerance from the produced size of the hole. If referenced at MMC, the boundary was the virtual condition of the hole calculated by the formula; MMC minus Geometric Tolerance at MMC equals virtual condition. If referenced at regardless of feature size (RFS), the size became harder to calculate but still fell under the concept of mating size. The mating size of the elongated hole was to be used to determine the hole size and the geometric tolerance (without bonus tolerance) was then subtracted to determine the appropriate "boundary" size that the surface of the hole could not violate. The mating size of an elongated hole would, in theory, be the size of a perfectly shaped elongated gage pin that expands into the hole and engages all high points within the hole in both directions simultaneously.

Once this boundary size is determined, the boundary is represented either with a gage pin located at true position of the hole or a simulation of the boundary by other methods such as a computer simulation of the gage pin. If a physical gage pin is used, the part is put onto the gage and the datum features on the part seated on, aligned with or mated with the datum feature simulators on the gage (as appropriate). Then the gage pin is attempted to be inserted into the hole being gaged and if it is received by the hole, it is assumed to have met its position tolerance. Size tolerances are, as always, must also be verified.

If a computer simulation of this procedure is used, for example, using a coordinate measuring machine, points are probed within the hole usually after the datums have been established and the surface points probed determined to either violate or not violate the "positional boundary" generated by the computer software. No tolerance zone was explained by Y14.5 and, therefore, only the boundary method could be used to properly inspect the hole location.

A discussion about positioning elongated holes ensued prior to the Y14.5M–1994 Standard wherein some said positioning of elongated holes should be made illegal, and examples in Y14.5 changed to reflect all elongated holes must be toleranced using a profile control. Most hated that idea, wanting instead to be allowed to use position on not only elongated holes, but also on other oddly configured features of any closed-loop variety, like D-shaped holes or even amoeba-shaped holes (or shafts).

This is the concept that was passed, but it was done with the provision that an oddly configured feature that was not a common cylinder, sphere or width would be positioned under the "boundary concept." No tolerance zone would be described for these features in the Y14.5 Standard (but the tolerance zones for oddly shaped features would be the same shape as the perfectly shaped feature itself).

The "boundary" method can, in these instances, be clarified as applicable by writing the local note "BOUNDARY" beneath the position feature control frame. **Although in ASME Y14.5-1994 the local note "BOUNDARY" was required, in ASME Y14.5-2009 and ASME Y14.5-2018, it is an option.**

The positional boundary concept would require the oddly configured closed-loop feature (hole or shaft of any shape) to not violate the boundary which is the collective effect of the actual mating size and the applicable position tolerance. They would be subtracted to get the boundary for a hole. They would be added to get the boundary for a shaft. Again, the terms shaft and holes as used here are generic and represent any female (hole) closed-loop feature and any male (shaft) closed-loop feature no matter what their configuration.

Now, some of these features are designed to be oddly configured enough that they must first be given basic dimensions for any size and shape. These basic size dimensions must then be toleranced by a profile feature control frame. This will set the MMC and LMC of the feature by the extremes allowed by the profile tolerance zone. Then the positional boundary tolerance is used to control the location of the feature. Two different geometric tolerances are required in this situation. Profile controls size and shape. Position controls angle and location.

Some features though, like elongated holes, can be controlled for size and shape by plus and minus tolerances. And, positional boundary is used to control angles and location.

FIGURE 223 [Elongated Holes with Bidirectional Positional Tolerances]

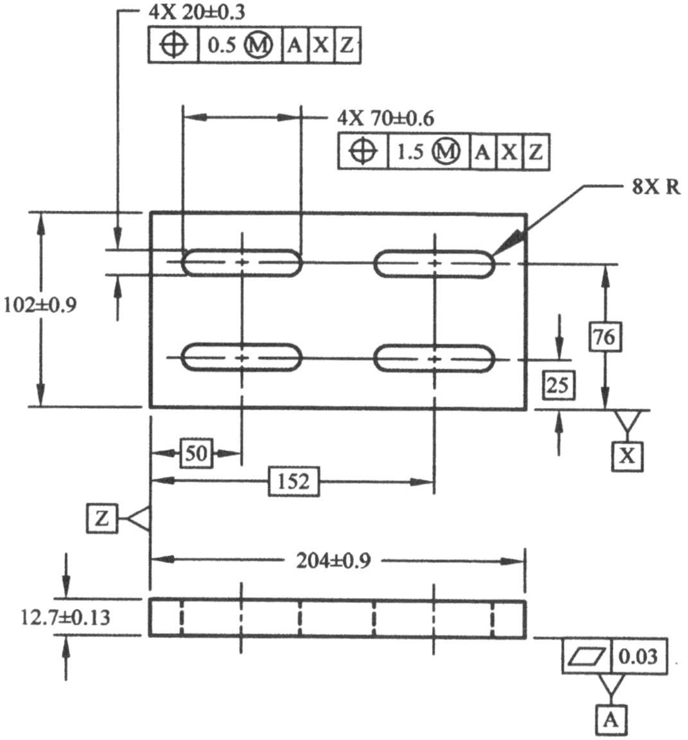

FIGURE 224 Means this: Four elongated virtual condition boundaries are generated at true position. The surface of each hole may not violate its virtual condition boundary. Each perfectly imaginary virtual condition boundary is 19.2 by 67.9. These boundaries may be simulated by computer software or functional gage pins.

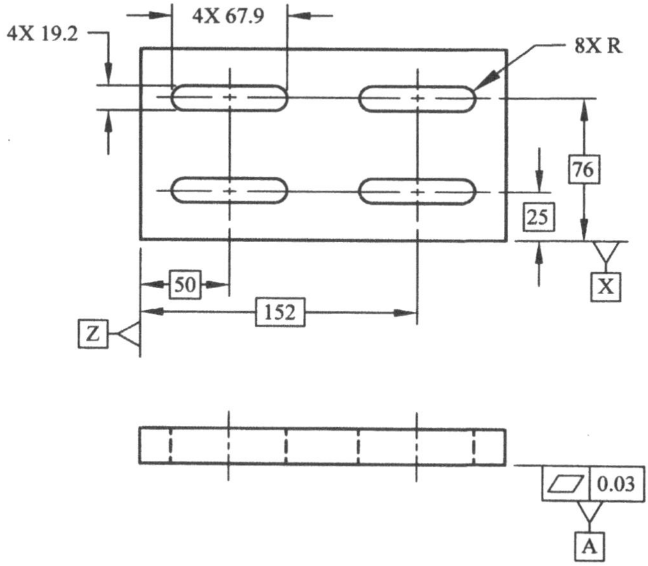

FIGURE 225 [Gage for Elongated Holes with Bidirectional Positional Tolerances]

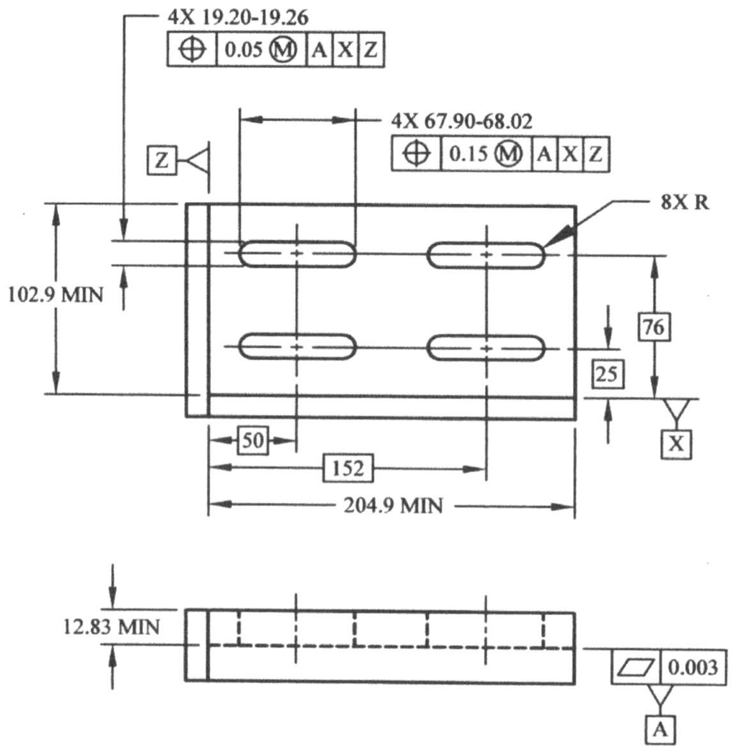

Another such feature that plus and minus tolerancing can be used on to control size and positional boundary can be used to control angle and location is a hose. Like a hose, a tube or pipe can have the same qualities, which is to be nominally round at cross-sections. This makes it a perfect candidate to use the concept of positional boundary. Each cross-section can be measured to determine whether or not the plus and minus size tolerance has been met, then the entire feature being controlled can be measured for violation of the positional boundary generated by the combined effect of size and geometric (positional boundary) tolerance.

FIGURE 226 [Positional Boundary Concept Used on Hoses, Pipes and Tubes]

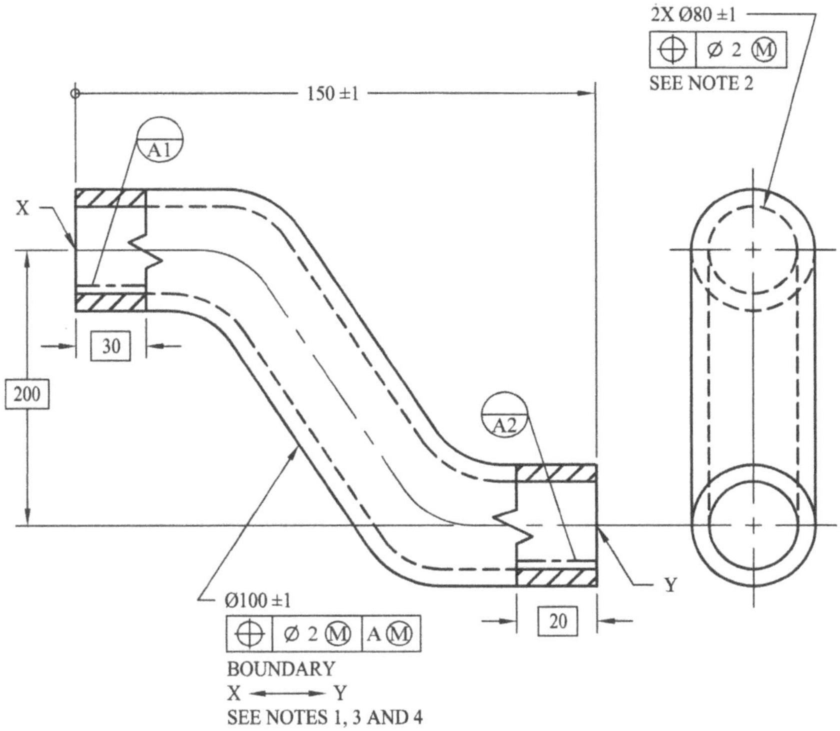

NOTE 1: FOR COMPLETE DEFINITION OF THIS PART REFER TO THE CAD MODEL.

NOTE 2: THIS POSITION CONTROL APPLIES TO DATUM TARGET DIAMETERS A1 AND A2 ONLY. IT POSITIONS THEM TO EACH OTHER.

NOTE 3: UNLESS OTHERWISE SPECIFIED, CONSIDER ALL CAD DIMENSIONS AS BASIC.

NOTE 4: DATUM AXIS A IS FORMED HALFWAY BETWEEN THE VIRTUAL CONDITION BOUNDARIES OF DATUM TARGET DIAMETERS A1 AND A2.

In these situations, a CAD model usually locates all points on the axis through a variety of methods to show what the perfect part would be. The diameter being controlled is then depicted with its plus and minus tolerance. The positional boundary on this part would follow the basic configuration but at a diameter of 101 (MMC of O.D.) + 2 (Geo.Tol. at MMC) = Ø103. This Ø103 boundary would be located from datum axis A. A functional gage could be created to measure this positional boundary. Size violations would be separately verified.

The functional gages for hoses, pipes and tubes are sometimes referred to as "Trough Gages" or "Worm Gages" in that they can end up looking like a trough that the hose, pipe or tube is dropped into to determine compliance. In this case, the trough would follow the basic configuration shown in the CAD model, be a diameter of 103mm and have a couple of attached plug-in shafts to represent datum feature A. These plugs would be 200mm from center to center and both would be a diameter representing the virtual condition size of datum target diameters A1 and A2.

Since the diameters designated as datum target diameters A1 and A2 are both Ø80 plus and minus 1 and positioned to each other to with a positional tolerance of 2 at MMC, their virtual condition as holes is Ø79 (MMC) – 2 (Geo.Tol. at MMC) = Ø77. So the plugs that are inserted into datum target diameters A1 and A2 simultaneously to fixture the part are represented at a diameter of 77mm. The use of datum features in the positional boundary control for the outside diameter is optional. This control would be equally sensical without them. If they were omitted from the feature control frame, the trough gage would not alter in size or configuration expect for the exclusion of the plugs that would have represented the datum features.

FIGURE 227 [Positional Boundary Concept used on Hoses, Pipes and Tubes]

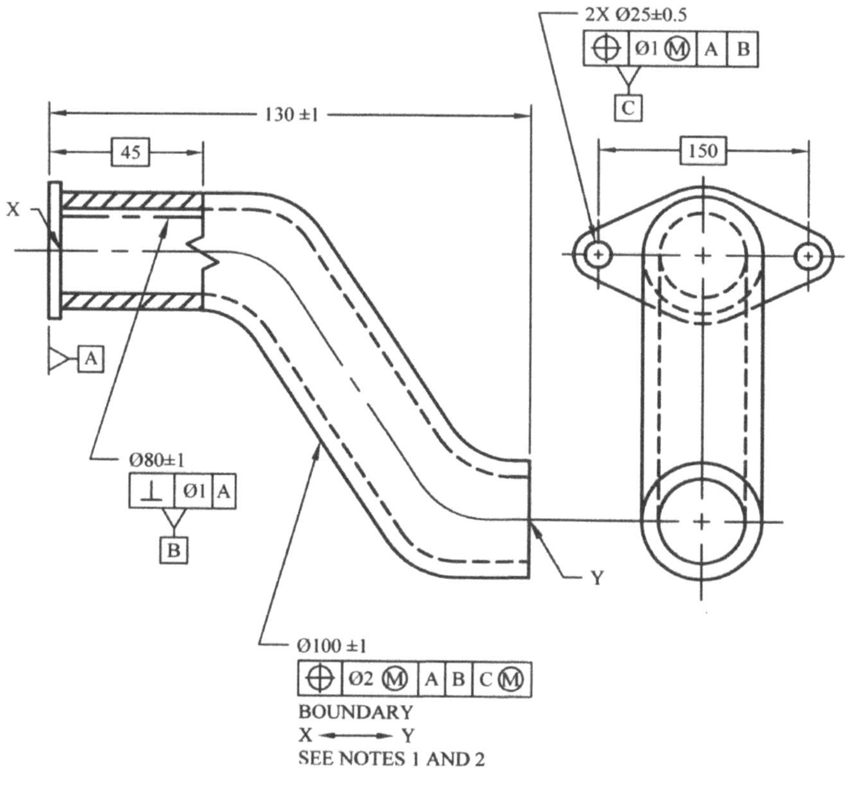

NOTE 1: FOR COMPLETE DEFINITION OF THIS PART REFER TO THE CAD MODEL.

NOTE 2: UNLESS OTHERWISE SPECIFIED, CONSIDER ALL CAD DIMENSIONS AS BASIC.

FIGURE 227 is another example, not dramatically different, but with a different datum structure and a configuration that is still nominally round. Since this configuration is not sufficiently complex at each cross-section to warrant a combination of profile and positional boundary to be used to define it, once the datum features are established and controlled, positional boundary is used on the controlled tube and simply references those datums.

FIGURES 228 and 229 show a feature that is sufficiently complex to need a profile of a surface control to define its size and shape. It is stabilized for measurement by use of a primary datum feature which also acts to control perpendicularity at this level of control. Note that even though Figures 228 and 229 use the positional boundary control, the local note "BOUNDARY" has been omitted from beneath the feature control frame. This is to point out that ASME Y14.5-2009 and ASME Y14.5-2018 has made this note optional.

FIGURE 228

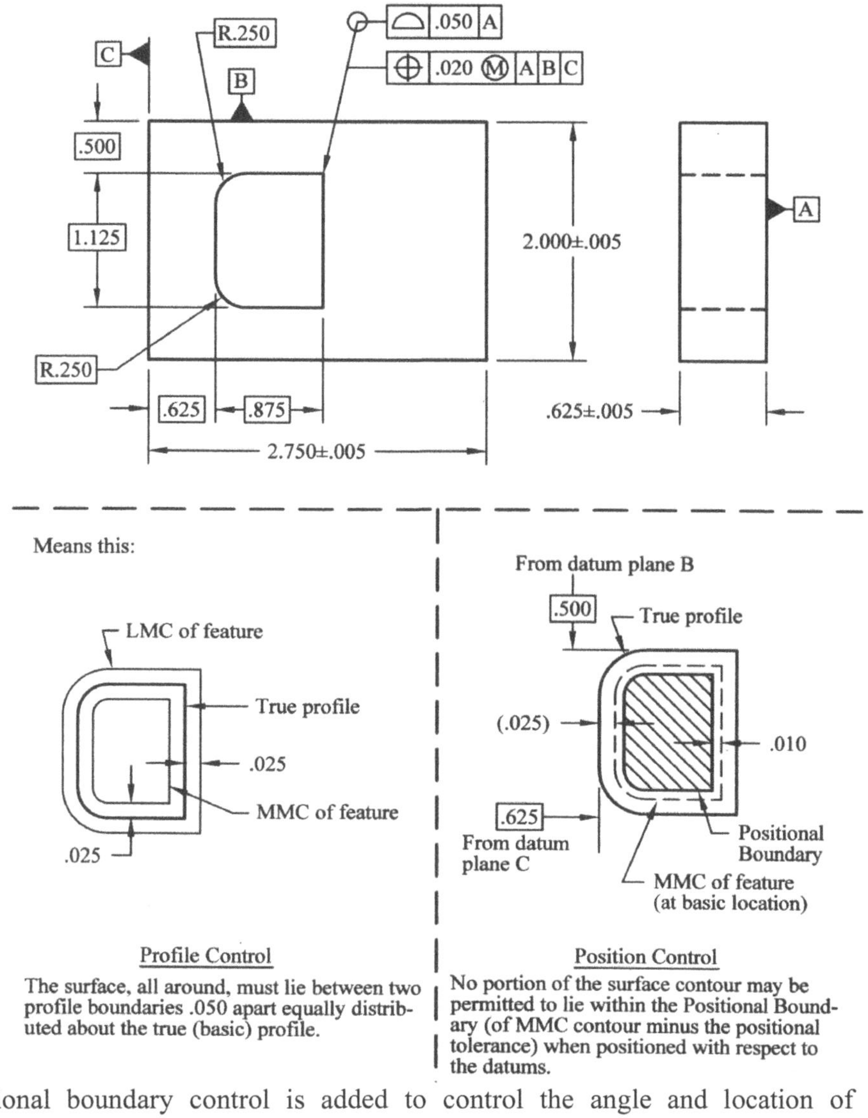

The positional boundary control is added to control the angle and location of the oddly configured hole to the datums. Whichever control (profile or position) has the tightest tolerance will control the perpendicularity to datum plane A, since both reference the A datum. Still, both controls need datum A to better stabilize their measurement.

FIGURE 229 [Positional Boundary – Workpiece and Gage]

Workpiece

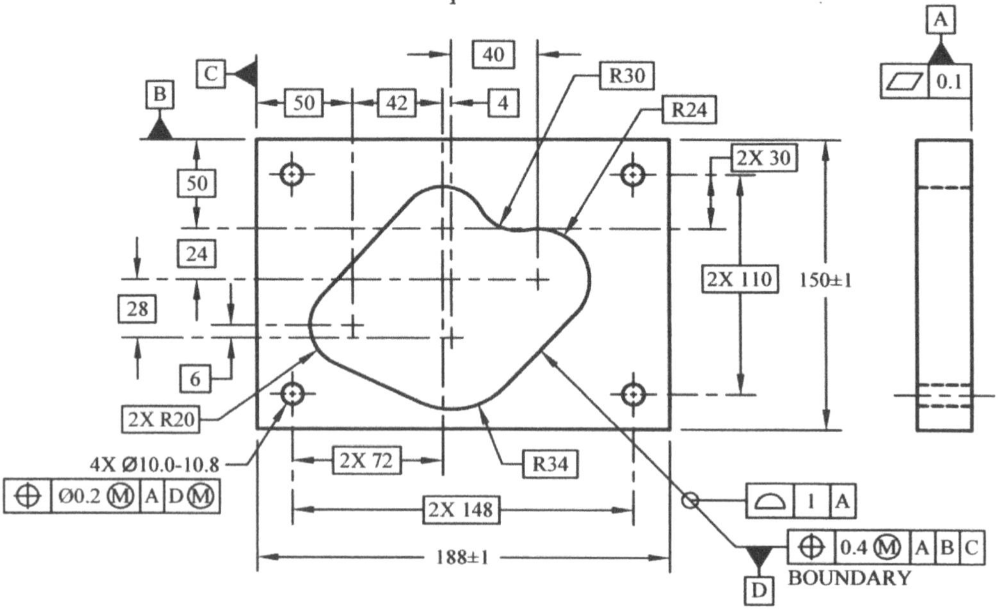

Gage

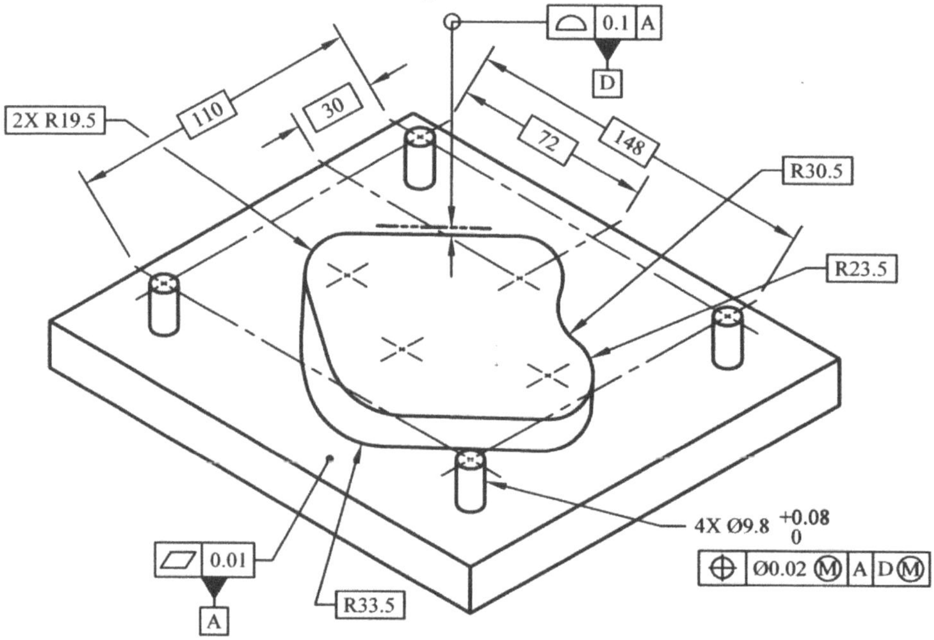

In Figure 229, **Datum Feature Simulator D is produced at virtual condition to Datum A.** Since datum feature D on the part has two virtual conditions to datum plane A, the pertinent Maximum Material Boundary (MMB) must be utilized for the gage design. We must consider that the profile of a surface control sets the maximum material condition of the irregular closed feature while oriented to datum plane A.

This means that the irregular closed (hole) feature must have perfect perpendicularity to datum plane A when produced at MMC to pass this requirement.

Although the position control does allow movement of 0.4 at MMC to datums B and C, the use of datum A in that control is for part stabilization. It does not allow an additional perpendicularity tolerance of surface elements for the hole. If the hole was out of perpendicularity to datum plane A the sum of the two controls of 1.4, the irregular hole would violate the profile of a surface control (measured while the part seats on datum A). Also, the 4-hole position control is verified with a gage that represents datum feature D at its Maximum Material Boundary (MMB) using D's profile tolerance and its relationship to datum A only (since A is the only datum feature preceding D at MMB in the position control for the 4-hole pattern). D is simulated at its applicable MMB as dictated by the profile control which references A. This is because when gauging the 4-hole pattern, the position of datum feature D has already been verified. In Figure 229, the applicable MMB to simulate datum feature D is also its MMC size as dictated only by the profile of a surface tolerance. D's other MMB, to verify the position of D itself, combines the profile and position tolerance to be calculated). The gage depicted in Figure 229 is not trying verify D's position tolerance. That has already been done.

For all these reasons, the size of the depicted gage's datum feature simulator to represent D to inspect the 4-hole pattern (to datums A and D at MMB) is the irregular feature's 'basic size' minus the profile tolerance of 0.5mm per side. Since the profile tolerance is plus and minus 0.5 at the surface, each surface of the hole is allowed to grow material by 0.5. Overall, since the surface loops around and closes up, the effect would be a grow material of the basic size by the full 1 millimeter (0.5 on each side of the irregular hole). When a hole experiences growth of material, it gets smaller. This dictates the size of the gage pin simulating datum feature D. Some of the oddly configured hole's radii get bigger and some get smaller to grow material.

Tolerance Zone vs. Boundary Concept Explanation (Tolerance Zone vs. Surface Methods of Measurement)

For many years, it was thought that if a part was measured for verification of a position tolerance, it did not matter whether the axis or centerplane of a regular feature of size was determined to reside within the positional tolerance zone or the surface of the feature was determined not to violate the virtual condition boundary. Both methods of verification were believed to give equivalent results. If a hole's axis was in the position tolerance zone, it was assumed the hole would fit over its gage pin and vice versa. It was not considered that a hole, if produced with extreme surface aberrations (for example, was very out-of-round) that one measurement method might accept the part and the other reject it.

FIGURE 230 shows that it is possible for a hole to fit over its functional gage pin produced at the virtual condition **boundary** size for the hole (in this case a diameter of 9.3) and not have its axis within its position **tolerance zone**. This illustration has exaggerated the shape of the hole to show the problem. It shows that a hole that is inspected in the proper way on, for example, a coordinate measurement machine (CMM). The CMM would try to prove that the maximum inscribed cylinder (MIC) for the hole (the largest gage pin that can be inserted into the hole) has its axis in the position tolerance zone.

In this illustration, the MIC for the hole has an axis that is far outside of the position tolerance zone. This position tolerance zone is located at the perfect position of the hole, which is perfectly perpendicular to datum plane A, 30.16 millimeters from datum plane B and 19.05 millimeters from datum plane C. The size of the position tolerance zone is dependant on the size of the MIC. If the hole's MIC is larger than a diameter of 9.4 (the maximum material condition of the hole), the position tolerance zone will be larger than a diameter of 0.1, by the difference. The MIC may be located anywhere and at any angle in the hole. Still, it is easy to see that the axis of the MIC can be outside of the position tolerance zone (as is the one depicted in this illustration). While this would result in the rejection of the hole, the functional gage pin accepts the hole as being within its position tolerance. As shown, the functional gage pin (located at the perfect hole position) is a diameter of 9.3 and fits into the hole.

The rules of inspecting position tolerance are not always followed. But, even if the hole was inspected by probing the hole (wrong methodology), instead of probing the maximum inscribed cylindrical gage pin that fits into the hole (right methodology), the results from this and what the functional gage pin at a diameter of 9.3 would accept could be quite different.

Rule Change: In the **ASME Y14.5-2009** standard for features of size that reference the MMC, LMC or RFS modifier, if the results of verification of the position tolerance zone and the verification of the virtual condition boundary differ, in that one method rejects the part and the other accepts the part, the verification of the virtual condition boundary acts as the final word on whether the part is good or bad. In the **ASME Y14.5-2018** standard, the virtual condition boundary method (surface method) of measurement takes precedence over the tolerance zone method only on features controlled using the MMC and LMC modifiers. If the RFS modifier is used on the features being inspected, the tolerance zone method of measurement takes precedence and acts as the final arbiter on geometric tolerance compliance (good vs. bad).

FIGURE 230 [Tolerance Zone vs. Boundary Concept]

Figure 1:

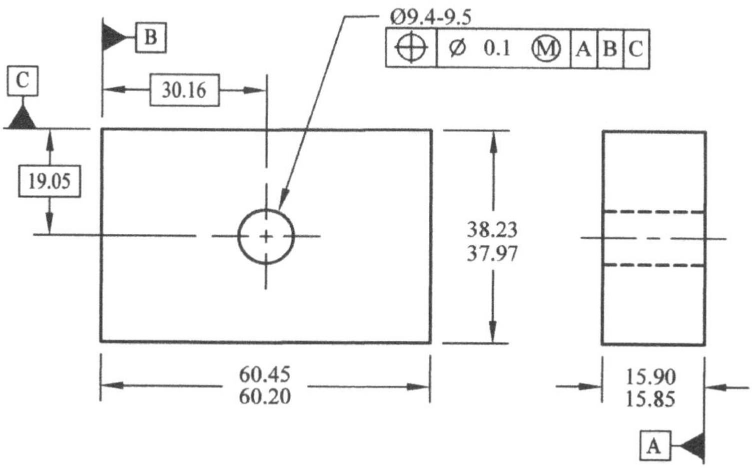

Figure 2:

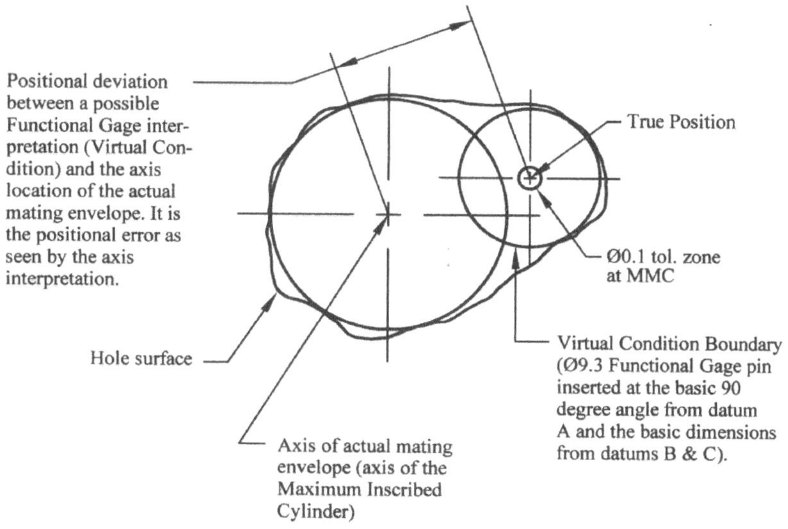

Gages check boundaries. CMM's most commonly check tolerance zones.

The "boundary method" takes precedence over the "tolerance zone method" of position tolerance zone verification when features being measured are referenced at MMC.

Gages win under the MMC concept.

In practical measurement terminology, the surface (virtual condition boundary) method shall take precedence where the tolerances are applied at MMC or LMC. If the part is inspected entirely correctly by probes meant to determine if the position tolerance zone has been violated and the part checks bad, but the part is accepted by a properly designed functional gage, then the part is adjudged to be a good part.

However, per ASME Y14.5-2018, where tolerances are applied at RFS, the surface method is applicable. The axis or center plane must reside within the positional tolerance zone to be adjudged to be in compliance.

These statements are true for the inspection of features of size (holes, shafts, tabs and slots) for not only position, but also for perpendicularity, parallelism and angularity.

Odd Tolerance Zones for Cylindrical Holes

Bi-Directional Positional Tolerancing (Rectangular Coordinate and Polar Coordinate Methods)

Rectangular Coordinate Method

Sometimes, it is functional to assign a feature of size more tolerance of position in one direction than another. In these instances, more than one feature control frame is required. For example, a feature using the rectangular coordinate method:

FIGURE 231 [Rectangular Coordinate Method]

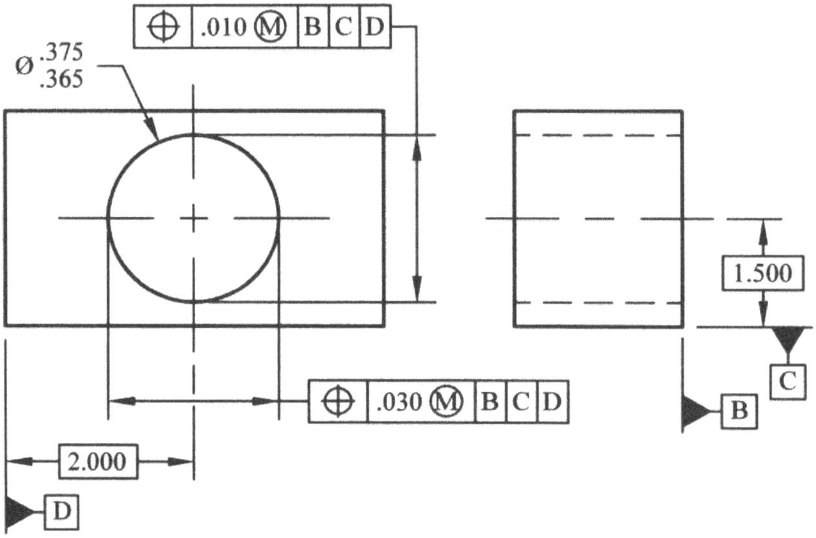

The feature is located by basic dimensions from the specified datums. The datums and their order of precedence do not change from one feature control frame to the other. Only the tolerance is different. No diameter symbol is given or proper in the description of the positional tolerance zone. The tolerance zone is not cylindrical, but rather two sets of parallel planes. One set is perpendicular to the other. In two dimensional terms, the tolerance zone can be described as rectangular. True position lies at its center. In these instances, it is sometimes desirable to refine the perpendicularity with an additional feature control frame. For example, see FIGURE 232.

FIGURE 232 [Rectangular Coordinate Method with a Perpendicularity Refinement]

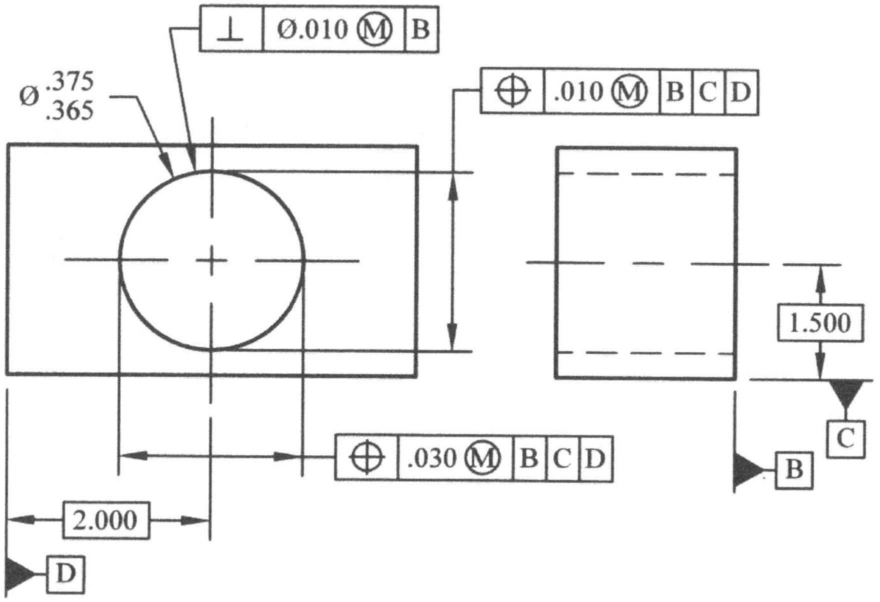

As you can see in this example, the perpendicularity refinement matches the smaller positional tolerance. This is to keep the feature from being allowed to lean farther in one direction than another. To assure this, a cylindrical tolerance zone for perpendicularity is used.

Polar Coordinate Method

Bi-directional positional tolerancing using the polar coordinate method differs somewhat from the rectangular coordinate method. FIGURES 233 and 234 show an arc being struck from datum axis C that has a radius of 2.500. As is reflected in the feature control frame, a tolerance zone of .015 width is required. Therefore, additional arcs are struck .0075 on either side of the true position arc of 2.500. This constructs the .015 tolerance zone. The plane, 72° from datum plane A, crosses the true position arc and creates a true position axis for the hole. Two parallel planes are simulated parallel to the true position centerplane that is 72° from datum A. These parallel planes are .030 apart and .015 on either side of the feature true position axis. This completes the true position tolerance zone in which the axis of the actual feature must lie.

The perpendicularity control further defines the feature. It refines the perpendicularity of the feature to a cylindrical tolerance, matching the smaller positional tolerance. This keeps the feature from being allowed to tilt farther in one direction than another. This type of positional control has been proven practical in some controls of gear mounting centers.

FIGURE 233 [Bi-directional Position Tolerancing, Polar Coordinate Method]

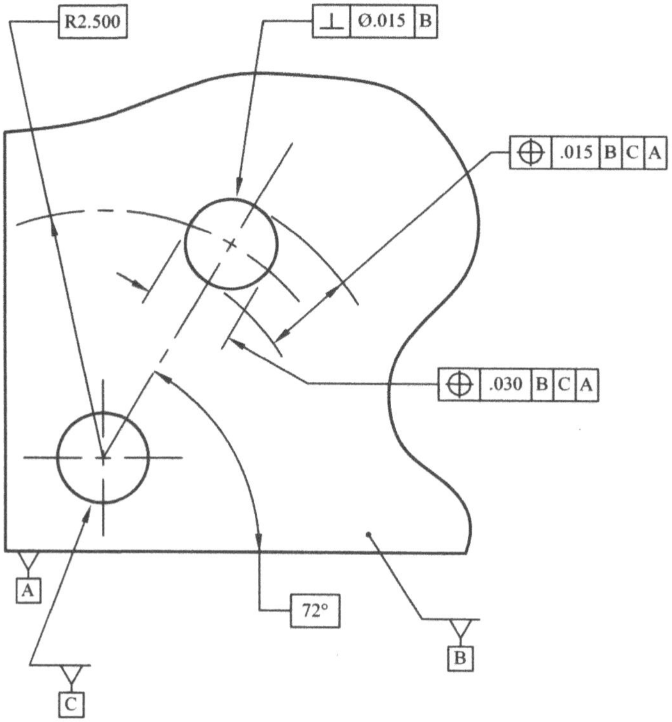

FIGURE 234 [As drawn above, means this]

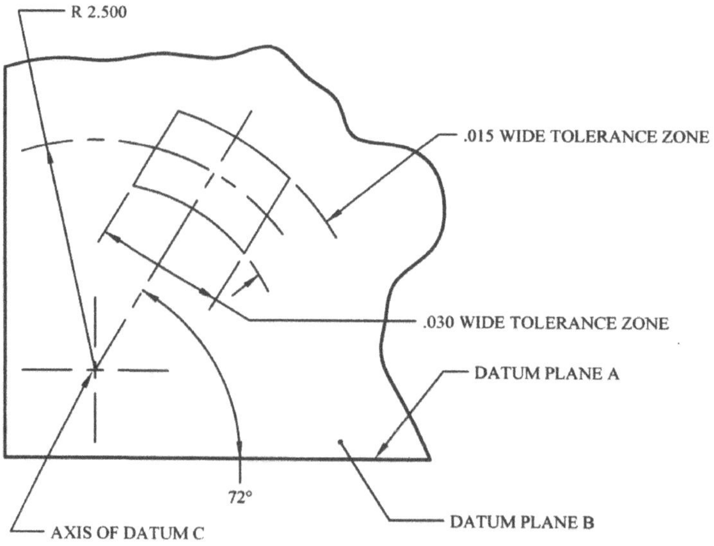

FIGURE 235 [Correct Maximum Material Boundary Calculations]

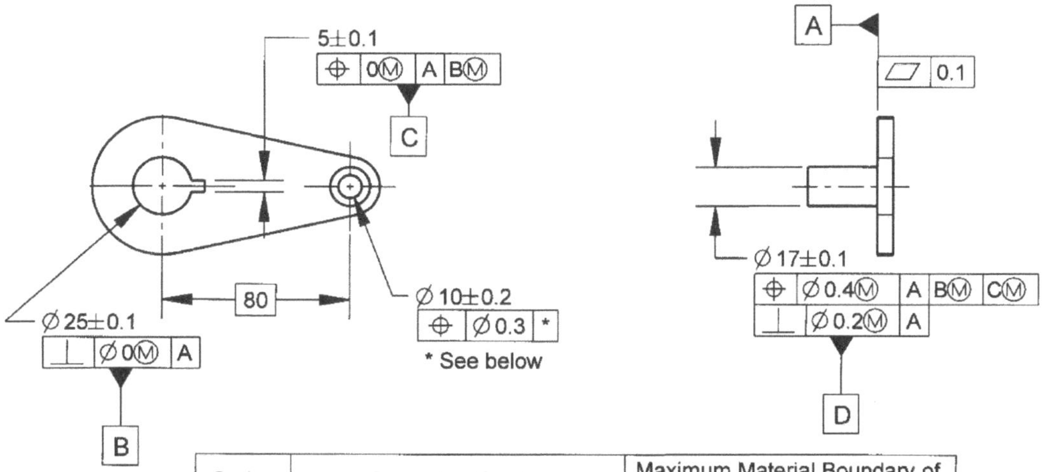

Option	Geometric Control	Maximum Material Boundary of datum feature D
(a)	⌖ Ø0.3 DⓂ	17.1
(b)	⌖ Ø0.3 A DⓂ	17.1+0.2 = 17.3
(c)	⌖ Ø0.3 A BⓂ DⓂ	17.1+0.4 = 17.5

FIGURE 236 [Correct Least Material Boundary Calculations]

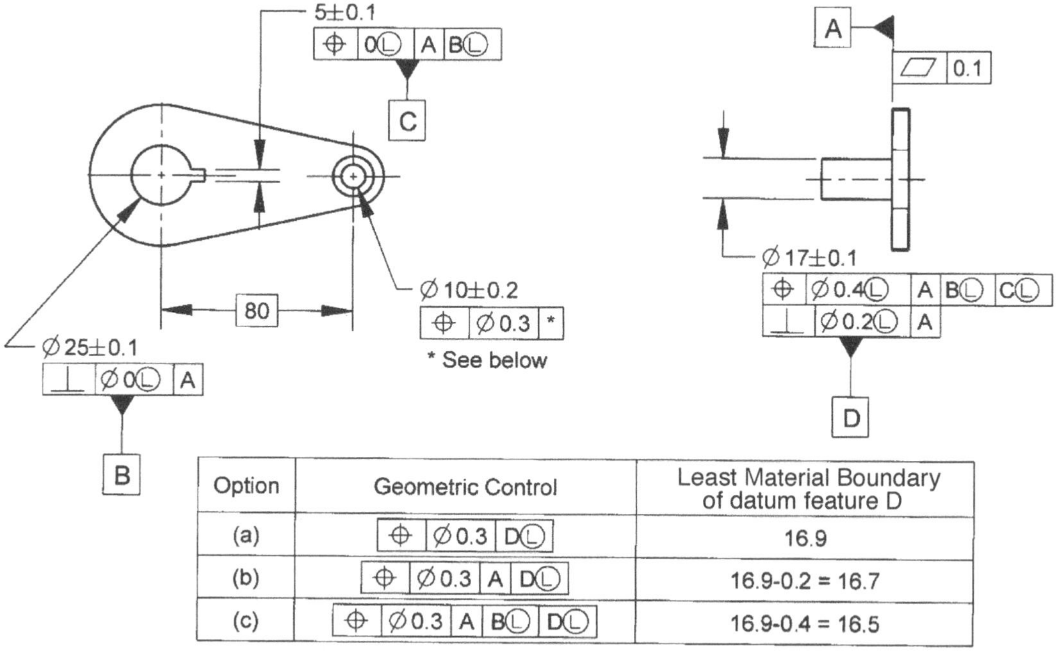

Option	Geometric Control	Least Material Boundary of datum feature D
(a)	⌖ Ø0.3 DⓁ	16.9
(b)	⌖ Ø0.3 A DⓁ	16.9-0.2 = 16.7
(c)	⌖ Ø0.3 A BⓁ DⓁ	16.9-0.4 = 16.5

Geometric Tolerancing a Gear

Figure 237

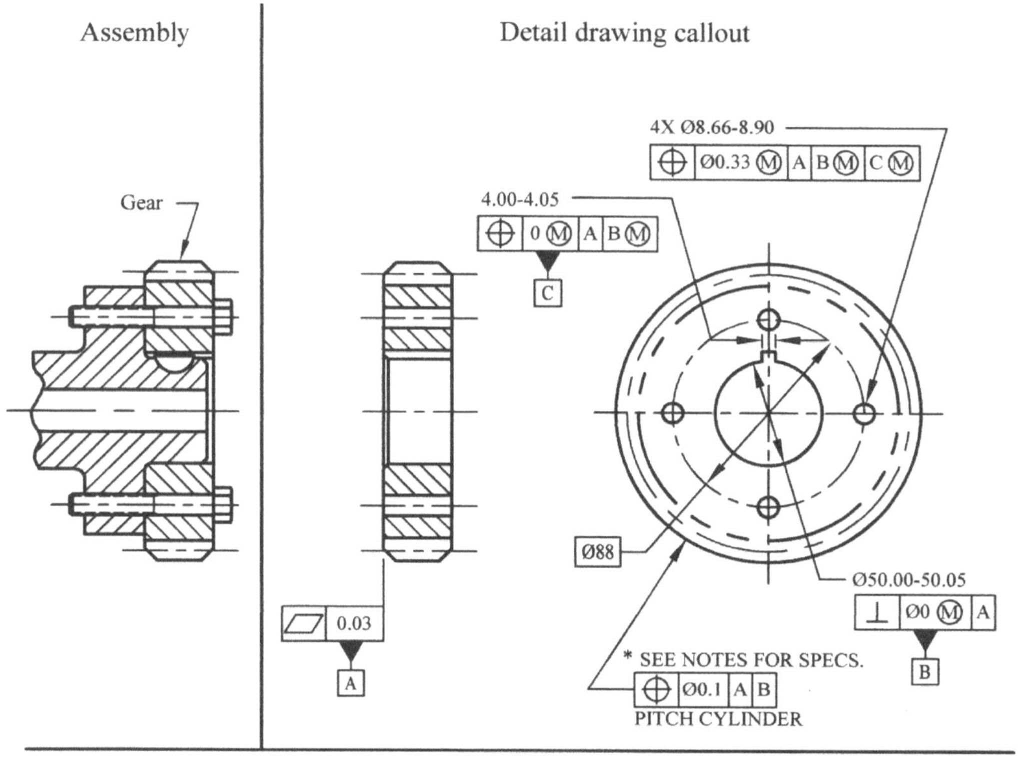

8.66 = MMC
\- 8.00 = MMC
0.66 = Tol. to be divided

$$2\,\overline{)0.66} = 0.33$$

50 = MMC Hole
\- 50 = MMC Shaft
0 = Tol. (Allowance)

This gear follows the same tolerancing technique as the automobile panel and the P.C. board. All of the parts have merely taken a piece of the technique used by the first part shown in this section; a rectangular plate with five holes. The technique is altered in each case to fit the situation but the alteration is small.

STEP 1:

This gear begins by choosing a seating surface as a primary datum feature. This surface seats the gear against the next part in the assembly so that the clearance holes can be used to bolt the gear into place. This seating surface needs the most contact and dictates the angle at which the two parts will assemble. It must seat without rocking so much that it side loads the shaft that fits into the 50.00-50.05 hole or the screws that fit into the four clearance holes.

Since it is a planar feature, this primary datum feature is given a flatness control to limit the amount it may rock. This flatness tolerance is chosen based on two major factors; a) cost of manufacture at an affordable price for the customer, and b) reducing the rock to an amount that runs little risk of side loading the mating shaft and screws to within a significant amount.

STEP 2:

A unique feature that both mates with and aligns the next part in the assembly with the gear is chosen as a secondary datum feature. This feature is the 50.00-50.05 center hole. The relationship between this hole and the primary datum plane is then toleranced. This relationship is one of perpendicularity. This control says, **"Perpendicularity of the axis of this diameter must be held perfectly (if it is produced at its maximum material condition) to datum plane A."**

Or,

"Perpendicularity of a diameter that mates while the part seats on A (is to be held to a zero tolerance at MMC)." This tolerance is a minimum of zero for a 50mm hole, but it would grow to a maximum of 0.05 if the hole was produced at the least material condition of 50.05. This mating and alignment hole is then made the secondary datum feature. It has been named datum feature B. The perpendicularity tolerance, in this case, has been based on the formula:

MMC Hole - MMC Shaft	or	MMC Hole - Virtual Condition Shaft
Geometric tolerance to be divided between the two		Geometric Tolerance for the hole

Either way, in this situation it ends up being 50 minus 50 equals zero. Zero tolerancing at maximum material condition is a good method of tolerancing that will be further explained in Step 4. In this Step 2, however, the part producer must simply be aware that manufacturing this hole at the MMC of 50 is not a good idea in that it would be entitled to no perpendicularity tolerance. The larger the hole is produced, the greater the perpendicularity tolerance to a maximum of 0.05 at 50.05.

STEP 3:

Aligning the four-hole pattern with the keyway can be done in a variety of ways. One of the clearest is to simply make the keyway the tertiary datum feature and then reference it in the position control used on the four-hole pattern.

First, we recognize the function of 4.00-4.05 keyway is to clock or act as a timing feature when trying to get the four 8mm screws to pass through the clearance holes on this gear and thread into the threaded holes on the next part in the assembly. This makes it the perfect feature to use on the four clearance holes to establish their angular orientation and prevent the pattern from rotating out of alignment with the threaded holes.

To do this, a keyway is designed into both the gear and the next part in the assembly. The key (rectangular block/shaft) that slides between them will be used to align the clearance holes and threaded holes. The keyway is first related to the primary and secondary datums. The keyway's relationship to the primary datum plane is one of perpendicularity. Its relationship to the secondary datum axis is one of location, since the centerplane of the keyway must pass through the axis of datum B. Both of these relationships (perpendicularity and location/intersection) will be handled by a position control.

This control can be read: **"Position of the centerplane must be held to within a zero tolerance if the keyway is produced at its maximum material condition (of 4.00) to A for perpendicularity and B for location/intersection."** The entire width of the keyway is labeled as datum feature C. This means the centerplane will generate datum centerplane C. (See illustration of Step 3 for an explanation of the MMB symbol used after datum reference B.)

FIGURE 238 [Illustration of Step 3]

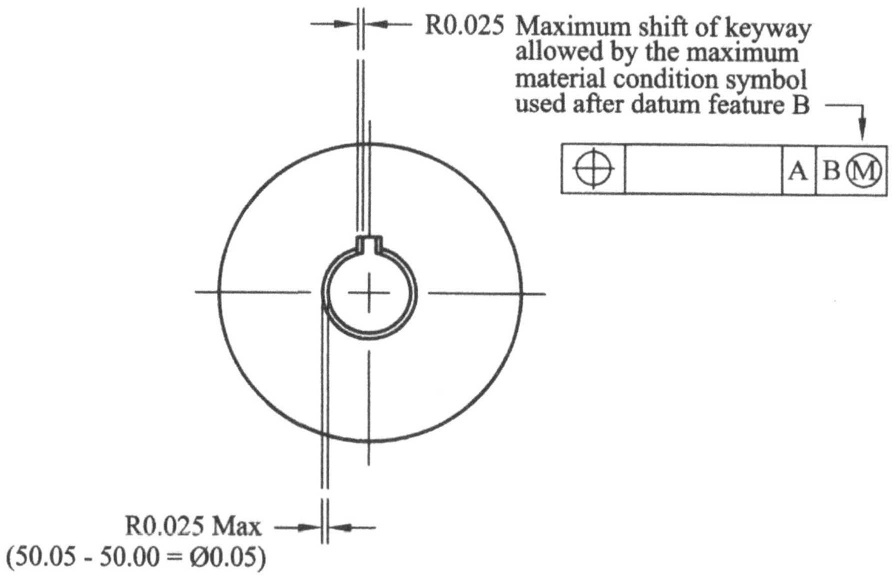

In Step 3, specifying the MMB symbol after datum feature B allows the keyway to shift off the center of B by a maximum of a radius of 0.025 if feature B is produced at a size of 50.05. This would mean that to get this part to align and mate with its mating part, the entire weight of this part may have to be shifted back by a maximum of this 0.025. This is generally perceived as lending to the imbalance of rotating assemblies. If this gear was rotating at a high rate of revolutions per minute, this could adversely affect its ability to function. To stop this additional shift and better balance the part, the maximum material boundary symbol may be omitted after datum feature B in this position control. If it is, however, it should also be omitted in the position control on the four-hole pattern.

STEP 4:

The four-hole pattern is now related to all three datum references. The four holes are positioned to each other, then to A for perpendicularity, then to axis B for distance/location (44mm from the axis of B), then to centerplane C for angular orientation (clocking).

We wouldn't want the four holes being allowed to shift away from the axis of B without the keyway shifting in the same direction, by the same amount. If this were allowed to happen (one shifting and the other not), we would find it impossible (in worst case situations) to assemble the parts. We would be able to get the four screws to bind the parts together, but could not get the key into the keyways. Or, we could get the key into the keyways, but not the screws into the holes. So, if an MMB symbol is used on one, it should be used on the other. And, if it is not used on one, it should not be used on the other.

FIGURE 239 [Illustration for Step 4 of the Datum Feature Shift Allowed B Referenced at MMB]

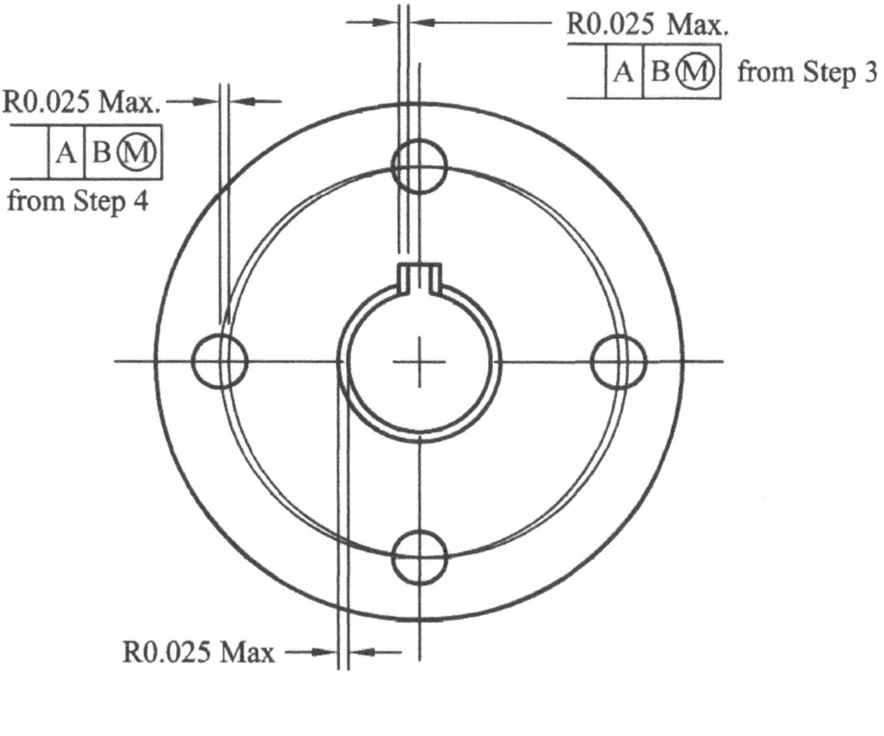

FIGURE 240 [Illustration for Step 4 to Show Additional Pattern Rotation Allowed by Datum Feature C Referenced at MMB]

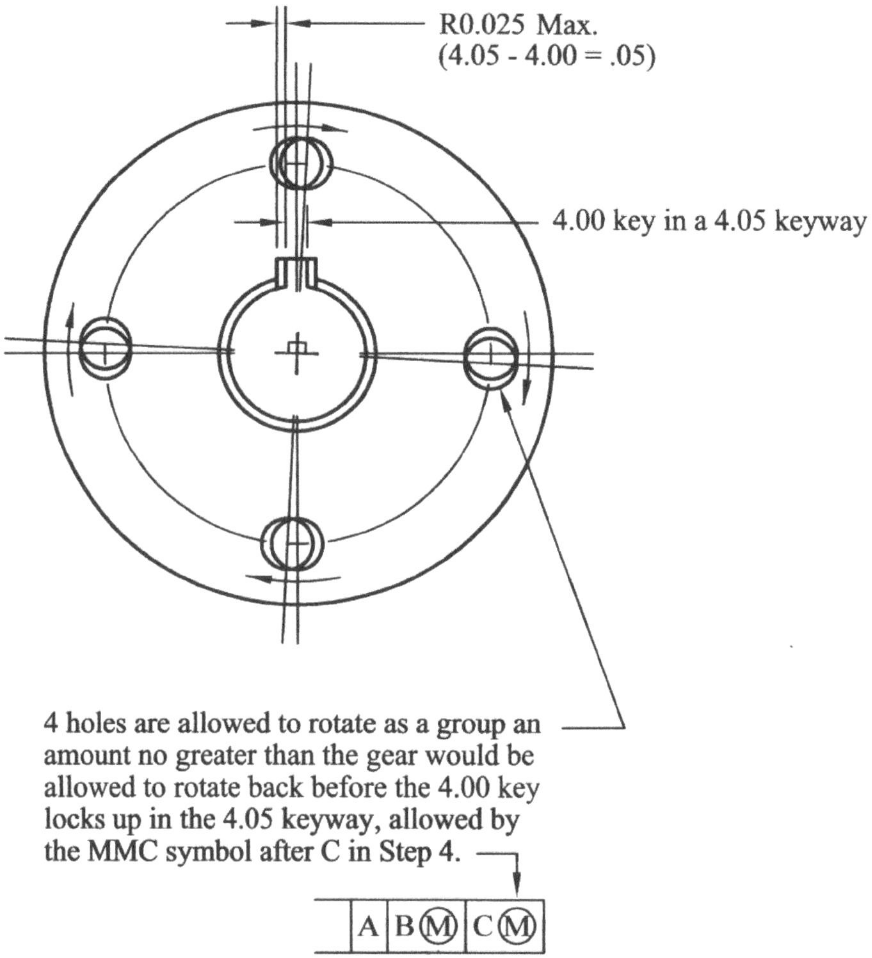

It is also worth considering an alternate Step 4 that zeros out the geometric (position) tolerance. Currently, the control reads:

4 X Ø8.66-8.90

⌖	Ø0.33 Ⓜ	A	B Ⓜ	C Ⓜ

This tolerance was calculated using the same formulas used in Step 3. This would allow the acceptance of a variety of produced parts, some of which follow.

Size		Geometric Tolerance Used		Virtual Condition Boundary
8.90	-	0.57	=	8.33
8.88	-	0.55	=	8.33
8.86	-	0.53	=	8.33
8.80	-	0.47	=	8.33
8.78	-	0.45	=	8.33
8.76	-	0.43	=	8.33
8.74	-	0.41	=	8.33
8.72	-	0.39	=	8.33
8.70	-	0.37	=	8.33
8.69	-	0.36	=	8.33
8.68	-	0.35	=	8.33
8.67	-	0.34	=	8.33
8.66	-	0.33	=	8.33

But, any diameter produced below Ø8.66 would be rejected by the inspector. Many of these would be just as functional as the others that were accepted. Some examples of rejectable features that would function just as well are:

Size		Geometric Tolerance Used		Virtual Condition Boundary
8.65	-	0.32	=	8.33
8.64	-	0.31	=	8.33
8.63	-	0.30	=	8.33
8.60	-	0.27	=	8.33
8.55	-	0.22	=	8.33
8.50	-	0.17	=	8.33
8.45	-	0.12	=	8.33
8.40	-	0.07	=	8.33
8.38	-	0.05	=	8.33
8.35	-	0.02	=	8.33
8.34	-	0.01	=	8.33

The infinite extension of this logic is 8.33 using zero positional tolerance would still generate a worst mating condition of 8.33. So, if additional weight is not a major factor (these smaller holes weigh a little more), and all we care about is the parts fitting together, then Step 4 should be changed to read:

4 X Ø8.33-8.90

⌖	Ø0 Ⓜ	A	B Ⓜ	C Ⓜ

STEP 5:

Most often gears are controlled for their configuration by a gear chart. These are often standardized for a product with tolerances. If this is not enough specificity, all dimensions on the gear may be made into basic dimensions and toleranced with a profile of a surface control. Datums may also be referenced for angle and location relationships (including angular orientation/clocking). But, most often, this is not done. The configuration is controlled and toleranced by the gear charts and the angle and location relationships are toleranced with position (as shown on this drawing). When this is done, the portion of the gear being positioned must be specified. This is most often the functional pitch cylinder.

For an illustration of Step 5, see the positional control on the pitch cylinder in Figure 1.

Threads, Gears and Splines

Screw threads, gears and splines generate axes that may be difficult to establish during inspection and, therefore, are sometimes avoided for use as datum features. Screw threads, when given an orientation or location control, it is understood that the application applies to the axis of its pitch cylinder. Likewise, when screw threads are used at datum features, the datum axis is derived from the pitch cylinder. Where an exception to this practice is desirable, a note such as MAJOR DIA. or MINOR DIA. must be specified beneath the feature control frame or beneath or adjacent to the datum feature symbol.

However, gears and splines have no implied portion of the feature that generates the axis to be oriented or located. Likewise, when used as a datum feature, no specific portion of the gear or spline is implied to generate the datum. Therefore, a note such as MAJOR DIA., MINOR DIA. or PITCH DIA. must be specified beneath the feature control frame or beneath or adjacent to the datum feature symbol.

Sequential Tolerancing Technique Using the Simultaneous Requirement Rule

Before we exit this chapter, let's see one more example of a tolerancing technique that uses a linear progression of geometric tolerances. It completes the part in a way that allows all who view the drawing to know where geometric definition begins and where it ends.

FIGURE 241

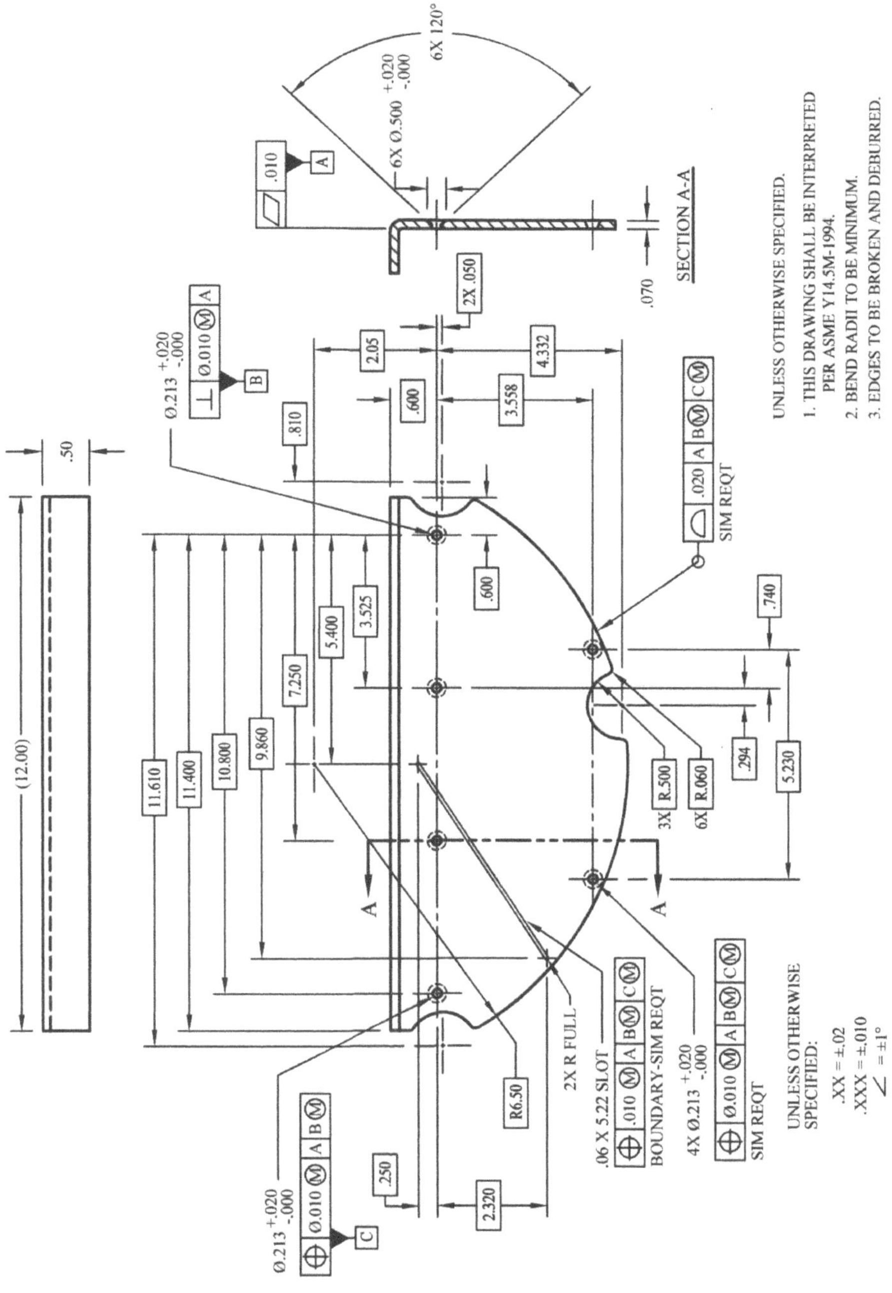

STEP 1:

Choose the primary datum feature as the surface that seats on the next part in the assembly. In this case, it is evident by the side that the holes are chamfered on, that the part bolts to what has been chosen as datum feature A. Datum feature A is then given a flatness control. Datum plane A will act as the perpendicularity datum for all subsequent features related to it.

STEP 2:

One of the holes within the hole pattern is selected as the secondary datum feature. It would be nice if this hole was unique, but since no unique feature is available that would make a great secondary datum feature, we will just select one hole within the pattern. This hole must be related to the primary datum plane. Their relationship is one of perpendicularity. So, datum feature axis B is held perpendicular to datum plane A. Datum axis B will act as the location datum for all subsequent controls on this part.

STEP 3:

Another hole within the pattern of holes is selected as the tertiary datum feature. Any hole will do, but the farther this hole is from datum feature B, the more stable the datum reference frame. Also, for ease of interpretation, it is optimal if this datum feature (hole) is in alignment with datum feature B in the plane of the front view of the part. Since datum axis C will act to orient subsequent measurements from datum axis B, the common plane between datum axes B and C (for angular orientation, anti-rotation) that gives us an angle of measurement (while measuring from datum axis B) is easiest to visualize if it is horizontal to the plane of the view. This hole is perpendicular to datum plane A and located 10.800 from datum axis B. To tolerance both these relationships, position tolerance is used on datum feature C relating it to A and B at MMB.

STEP 4:

Now that all of the datums needed to create a datum reference frame necessary to eliminate the six spacial degrees of freedom have been established, all subsequent round holes, the elongated slot and the part profile are related to them with either position tolerance or profile of a surface tolerance. This creates a SIMULTANEOUS REQUIREMENT that makes them part of the same pattern of features.

Remember, the rule is:

Simultaneous requirements apply to position and profile tolerances that are related to the same datum reference frame. The same datum reference frame is defined as using the same datum features referenced in the same order of precedence with same modifiers applied to the datum feature references. The default of simultaneous requirements requires that all applicable geometric tolerance zones are related to the same datum reference frame and are appropriately constrained as a single pattern to a single datum reference frame. This means that any datum feature shift allowed by the MMB symbols after the datum features must be experienced by all features within the pattern (as a group). It also means that the features must be measured in the same set-up. If a gage is used to measure the geometric tolerance, it must be one gage for all features within the pattern.

This is a very powerful technique and, as has been shown in this section, can be used on a wide variety of parts from many industries.

Chapter 15

Centerplane Datums

•Centerplane Datums – an Overview
•Centerplane Datums on Mating Parts

Chapter Objectives

Readers will learn:

1. To convert an assembly from plus and minus tolerancing to geometric tolerancing using center plane datums.

2. To apply a step by step procedure of assigning widths as datum features, relating all datum features to one another and positioning hole patterns to the established datums.

3. Calculating and assigning perpendicularity and position tolerances to a fixed fastener assembly.

Centerplane Datums – an Overview

At times, datum features that are most functional are formed by widths. As can be seen in the assembly drawing and the two detail drawings, Part #1 and Part #2 are toleranced in the ambiguous, obsolete fashion using plus and minus tolerances on location dimensions. In these seemingly simple drawings, we sometimes assume that these origins of measurement create a specific order of implied datum features. These implied datum features have no universal interpretation as to which features are the origin and have no order of reference. We don't know how to form our coordinate reference system, nor which specific features create the 3 mutually perpendicular planes which we would call a datum reference frame. Therefore, any measurements taken on the part tend to be crude and produce data that is grossly unrepeatable. Datum features must be specified on parts to rid us of (or at least greatly reduce) these ambiguities.
This section will undertake the task of converting this completely plus and minus toleranced part into a balanced mixture of plus and minus tolerancing, specified datum features, basic dimensions and geometric tolerances. This conversion should greatly increase measurement repeatability and reduce confusion about the functional requirements needed. It will allow parts produced within specified tolerances to assemble with such ease, as to be interchangeable with other parts produced run after run.

The following illustration shows the assembly drawing. For the detail drawings, see the next page.

FIGURE 242 [Assembly Drawing]

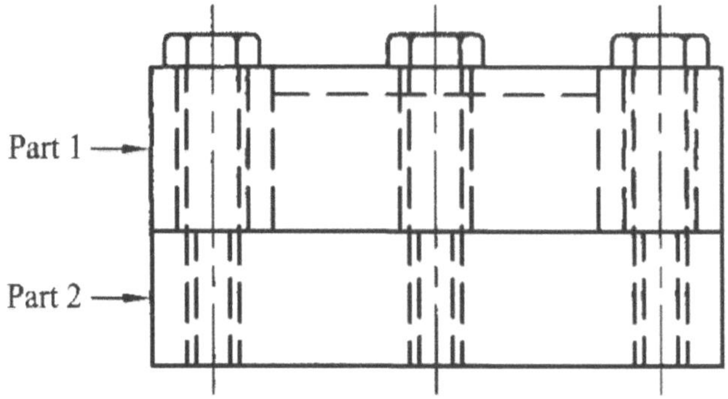

Part #1 has 6 clearance holes that receive 10 millimeter screws to bind Part #1 to Part #2. Part #1 also has a rectangular hole that mates with a rectangular shaft on Part #2.

FIGURE 243 [Part #1]

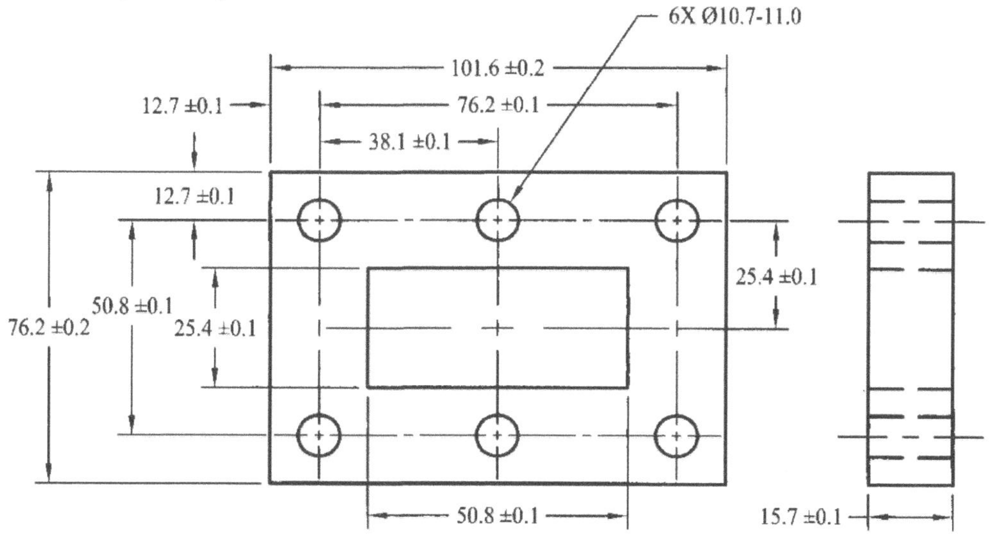

UNLESS OTHERWISE SPECIFIED ALL ANGLES ARE ±1°

This illustration shows Part #2. Part #2 has 6 threaded holes that are used to bind this part to Part #1 and a rectangular shaft that mates with the rectangular hole on Part #1.

FIGURE 244 [Part #2 Centerplane Datums on Mating Parts]

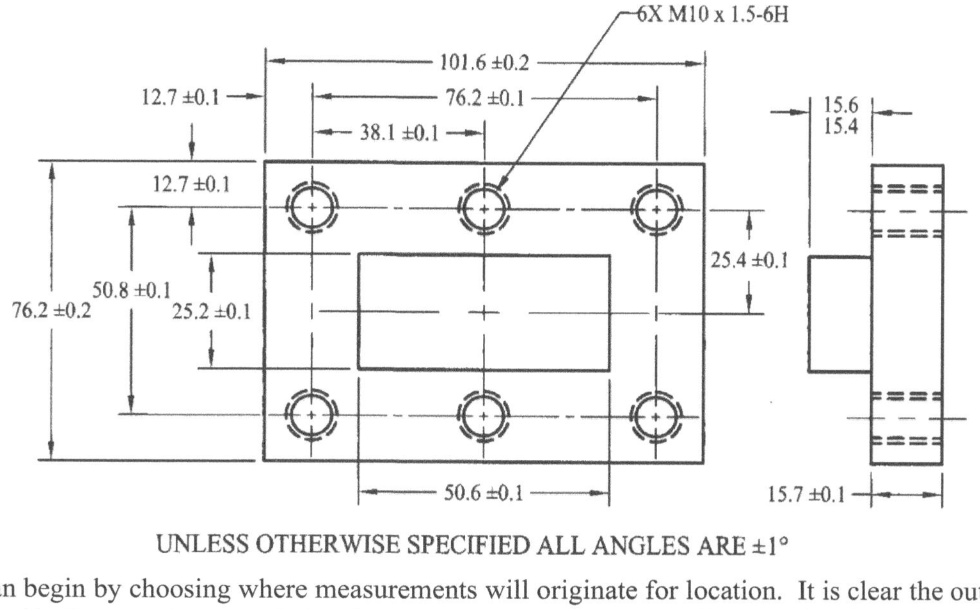

UNLESS OTHERWISE SPECIFIED ALL ANGLES ARE ±1°

We can begin by choosing where measurements will originate for location. It is clear the outside edges of both parts play no role in whether or not the two parts will assemble. So, although we could make such an approach work, the more functional and direct tolerancing scheme would begin in the middle of the functional mating widths.

Centerplane Datums on Mating Parts

For this case, we will go back to an assembly condition and describe the Geometric Tolerancing Scheme for both parts. This scheme consists of the following steps:

- To achieve the goal of measuring from centerplane datums we will eliminate the two 12.7 dimensions that lead us in from the outside edges of the part.
- All dimensions will be left intact with their accompanying plus and minus tolerances, except the dimensions that locate the holes from one another and the centerplanes. The 25.4 location dimension could either be converted to a basic dimension (of 25.4) like all other location dimensions, or it could be eliminated.

  Using centerplane datums, once the 50.8 dimension is converted to a basic dimension, the 25.4 becomes redundant. In this case, we will eliminate it to be consistent with Part #2. It's simply a personal preference decision with no technical difference.

  When datum features of size mate with similar features on the mating part, they will be referenced at MMC.

Step 1

Establish the primary datum features as the surfaces that seat on one another. Give them a flatness control.

Step 2

Establish the secondary datum feature of size as one of the widths that mate and has the most surface area. This datum feature will generate a secondary datum plane that is the centerplane of the width. This secondary datum feature will be controlled for perpendicularity to the primary datum plane only.

Step 3

Establish the tertiary datum feature of size as the other mating width. This datum feature will generate a tertiary datum plane that is the centerplane of the width. This tertiary datum feature will be controlled for perpendicularity to the primary and secondary datum planes.

Step 4

Position the holes on each part to the datums established in Steps 1 through 3.

Step 5

Position the outside widths on each part to the datums selected in Steps 1 through 3.

Step 6

Position the remaining outside widths on each part to the datums selected in Steps 1 through 3.

Step 1, Part #1

The primary datum feature is assigned as the seating surface in the assembly. It is given a flatness tolerance to limit its rock during inspection and assembly. All location dimensions become basic dimensions, and the redundant 25.4 location dimension is either eliminated or made a basic dimension.

Step 1, Part #2

The primary datum feature is assigned as the seating surface in the assembly. It is given a flatness tolerance to limit its rock during inspection and assembly. All location dimensions become basic dimensions.

Step 2, Part #1

The secondary datum feature is assigned as the 25.4 width. This will generate a secondary datum centerplane. The secondary datum feature will be controlled for perpendicularity to the primary datum plane.

The perpendicularity tolerance is established by subtracting the MMC of this hole width of 25.3 from the MMC of the mating shaft width of 25.3. The perpendicularity tolerance is zero at MMC. This could grow to as much as 0.2 at the hole width's LMC of 25.5. The perpendicularity tolerance zone confines the centerplane of the width between two parallel planes that are both perpendicular to datum plane A and are a distance apart that ranges from 0 at MMC to 0.2 at LMC.

Step 2, Part #2

The secondary datum feature is assigned as the 25.2 width. This will generate a secondary datum centerplane. The secondary datum feature will be controlled for perpendicularity to the primary datum plane.

The perpendicularity tolerance is established by subtracting the MMC of this shaft width of 25.3 from the MMC of the mating hole width of 25.3. The perpendicularity tolerance is zero at MMC. This could grow to as much as 0.2 at the shaft width's LMC of 25.1. The perpendicularity tolerance zone confines the centerplane of the width between two parallel planes that are both perpendicular to datum plane D and are a distance apart that ranges from 0 at MMC to 0.2 at LMC.

Step 3, Part #1

The tertiary datum feature is assigned as the 50.8 width. This will generate a tertiary datum centerplane that is perpendicular to both the primary plane and the secondary datum centerplane. All three planes are imaginary, therefore perfect. However, the real tertiary datum feature width will be controlled for perpendicularity to the primary and secondary datum planes.

The perpendicularity tolerance is calculated by subtracting the MMC of the mating shaft width of 50.7 from the MMC of this hole width of 50.7. The perpendicularity tolerance is zero at MMC to a maximum perpendicularity tolerance of 0.2 if produced at LMC. The perpendicularity tolerance zone confines the centerplane of the width between two parallel planes that are both perpendicular to datum plane A and datum centerplane B. They are a distance apart that ranges from 0 at MMC to 0.2 at LMC.

Since the secondary datum feature will be referenced at MMC an additional shift zone could form about the secondary datum centerplane that could be as large as another 0.2 (if it was produced at LMC). This would increase the possible out of perpendicularity tolerance between the secondary datum feature and the tertiary datum feature to a maximum of 0.4.

Step 3, Part #2

The tertiary datum feature is assigned as the 50.6 width. This will generate a tertiary datum centerplane that is perpendicular to both the primary plane and the secondary datum centerplane. All three planes are imaginary, therefore perfect. However, the real tertiary datum feature width will be controlled for perpendicularity to the primary and secondary datum planes.

The perpendicularity tolerance is calculated by subtracting the MMC of this shaft width of 50.7 from the MMC of the mating hole width of 50.7. The perpendicularity tolerance is zero at MMC to a maximum perpendicularity tolerance of 0.2 if produced at LMC. The perpendicularity tolerance zone confines the centerplane of the width between two parallel planes that are both perpendicular to datum plane D and datum centerplane E. They are a distance apart that ranges from 0 at MMC to 0.2 at LMC.

Since the secondary datum feature will be referenced at MMC an additional shift zone could form about the secondary datum centerplane that could be as large as another 0.2 (if it was produced at LMC). This would increase the possible out of perpendicularity tolerance between the secondary datum feature and the tertiary datum feature to a maximum of 0.4.

Step 4, Part #1

Position will be used to locate all clearance holes from one another and the datums. The holes will be held perpendicular to datum A, 25.4 millimeters (implied basic dimension) from datum centerplane B and 38.1 from datum centerplane C.

The position tolerances will be calculated by using the fixed fastener formula. The MMC of the clearance holes is 10.7 minus the MMC of the screws of 10 millimeters is a geometric tolerance to be divided between the two parts of 0.7. We will assign the threaded holes more than half of this tolerance on the basis of their difficulty to manufacture.

Step 4, Part #2

Position will be used to locate all threaded holes from one another and the datums. The holes will be held perpendicular to datum D, 25.4 millimeters (implied basic dimension) from datum centerplane E and 38.1 from datum centerplane F. The threaded holes will use a projected tolerance zone of 15.8 (the maximum amount of material the screws must pass through before reaching the threaded holes).

As was explained in a prior case, use of the projected tolerance zone prevents binding when the screws are threaded in.

Step 5, Part #1

The outside width of 76.2 will be positioned to the primary datum and then, in the same control, it will reference only the datum centerplane to which it is centered. The 76.2 dimension is centered to datum centerplane B. The maximum material condition modifier is not used because the outer edges of the part do not mate and the size of the datum features has no bearing on their location. This is also true for Step 5, Part #2, and both parts for Step 6. The tolerance will be chosen based on the fact that these outside dimensions are not functional and therefore, their tolerances can be large.

We must be cautious in not allowing a tolerance that is so big that it endangers the wall thicknesses.

Step 5, Part #2

The outside width of 76.2 will be positioned to the primary datum and then, in the same control, it will reference only the datum centerplane that it is centered to. The 76.2 dimension is centered to datum centerplane E. The tolerance will be chosen based on the fact that these outside dimensions are not functional and, therefore, their tolerances can be large.

We must be cautious in not allowing a tolerance that is so big that it endangers the wall thicknesses.

Step 6, Part #1

The outside width of 101.6 will be positioned to the primary datum and then, in the same control, it will reference only the datum centerplane that it is centered to. The 101.6 dimension is centered to datum centerplane C. The tolerance will be chosen based on the fact that these outside dimensions are not functional and therefore, their tolerances can be large.

We must be cautious in not allowing a tolerance that is so big that it endangers the wall thicknesses.

This step could have referenced datums A, B and C instead of A and C only. This would have squared the 101.6 width to A and B before centering it to C.

FIGURE 245 [Step 6, Part 1]

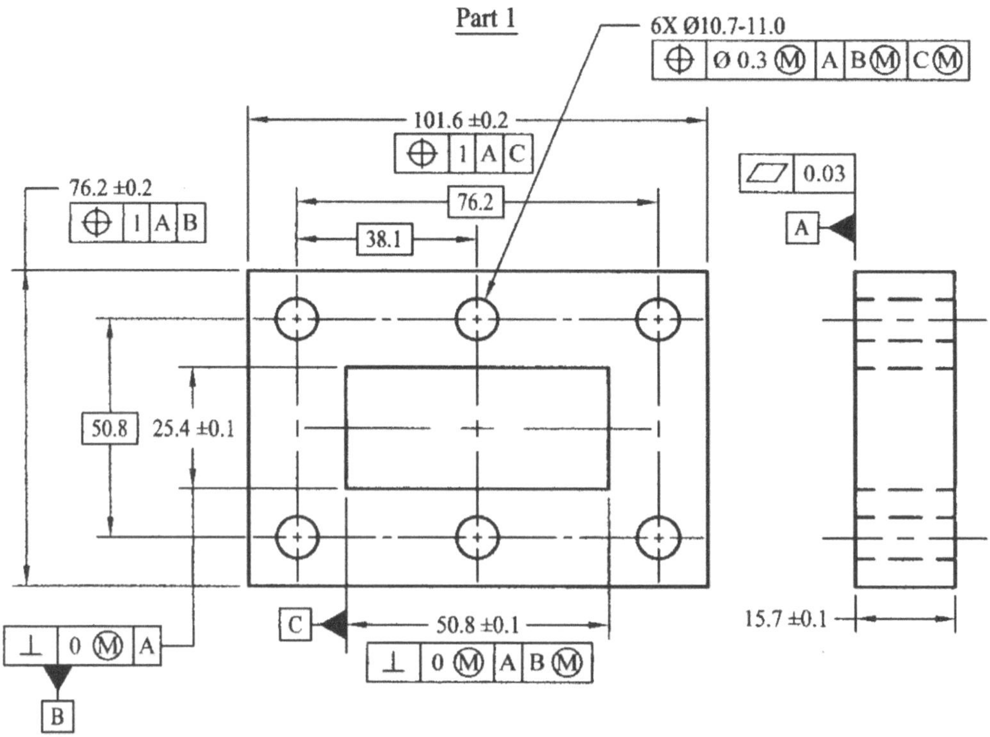

UNLESS OTHERWISE SPECIFIED ALL ANGLES ARE ±1°

Step 6, Part #2

The outside width of 101.6 will be positioned to the primary datum and then, in the same control, it will reference only the datum centerplane that it is centered to. The 101.6 dimension is centered to datum centerplane F. The tolerance will be chosen based on the fact that these outside dimensions are not functional and therefore, their tolerances can be large.

We must be cautious in not allowing a tolerance that is so big that it endangers the wall thicknesses.

This step could have referenced datums D, E and F instead of D and F only. This would have squared the 101.6 width to D and E before centering it to F.

FIGURE 246 [Step 6, Part 2]

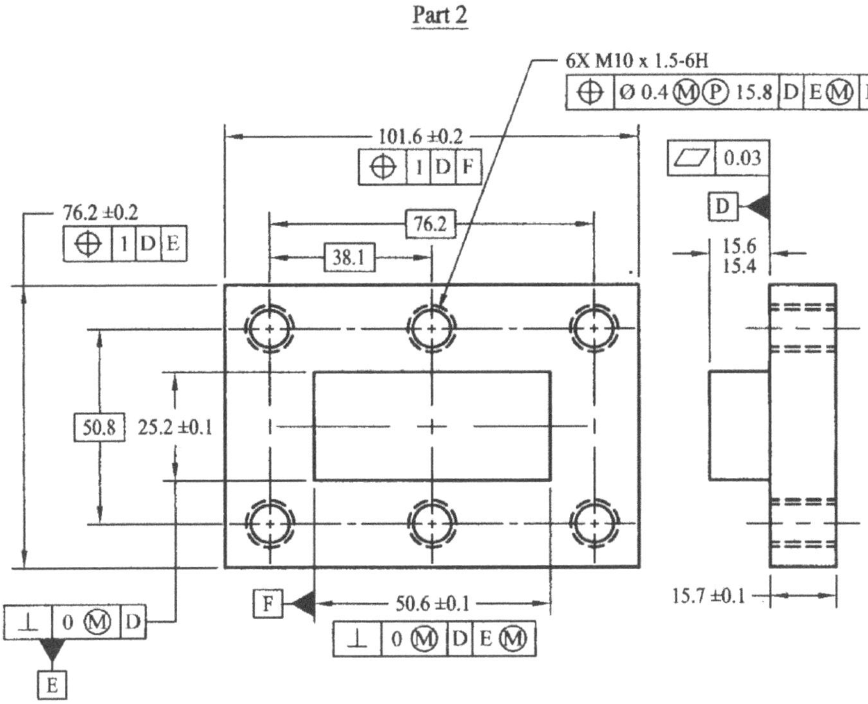

UNLESS OTHERWISE SPECIFIED ALL ANGLES ARE ±1°

Chapter 16

•Unique Effects on Controlled Features that Invoke LMC and LMB
•Wall Thickness Calculations
•Datum Feature Shift

Chapter Objectives

Readers will learn:

1. The unique effects of utilizing LMC and LMB concepts on design calculations and measurement procedures.

2. To calculate minimum wall thicknesses that include datum feature shift.

Unique Effects of LMC and LMB

Unique Effects on controlled Features that Invoke the LMC and LMB Concepts

Invoking the Λ symbol (and principle) by including it in the feature control frame for positional controls adds some advantages and also has some unique effects on the features of size being controlled with it. The most obvious effect is the increase in positional tolerance for features as their **actual minimum material envelope** departs in size from least material condition (LMC). For most non-threaded features, this growth in tolerance is proportional. In other words, as the feature's actual minimum material envelope departs in size from LMC, the positional tolerance zone grows an amount equal to that departure, allowing more movement for the axis or centerplane of the controlled feature. This symbol is often used to protect wall thickness. It is used where as much tolerance as possible is desired based on the produced size of the controlled feature, but where the use of the MMC symbol [or MMB symbol after datum features] would endanger the integrity (wall thickness/material strength) of the part. In certain cases, the RFS (regardless of feature size) and RMB (regardless of material boundary) concepts could be used to protect and control the part (and often is a more desirable control), but this would eliminate any additional geometric tolerance derived as the controlled feature was produced at various sizes (within size limits).

An interesting and often undesirable factor about using the LMC symbol is its effect of changing the way we measure size. **When the LMC modifier is used, the envelope of perfect form at maximum material condition is eliminated. MMC is then checked for violations at cross sections. LMC, however, which is normally checked for cross-sectional, 2 point opposing point size violations, is checked for violations of a newly created envelope at LMC (not something most inspection departments are used to dealing with).**

Another factor to give consideration is the calculation involved in figuring worst mating size boundaries so we can be certain mating parts will, indeed, mate at assembly. The worst mating size formulae listed below can be used; however, additional thought must be given to calculating the appropriate values to insert into each portion of the formula. Additional tolerance drawn from size will change the tolerance called for in the feature control frame at LMC to a larger amount applicable at MMC. The formulae are still:

INTERNAL FEATURES

MMC
\- Geo. Tol. applicable at MMC
Worst Mating Boundary Size

EXTERNAL FEATURES

MMC
\+ Geo. Tol. applicable at MMC
Worst Mating Boundary Size

These formulas apply no matter what material condition symbol/modifier is used in the feature control frame for controls such as positional tolerancing. The difficult factor in the formula is to insert the appropriate value of geometric tolerance applicable at the MMC size.

Identifying the MMC should be no difficult task; however, the positional tolerance that applies at MMC should not be confused with the positional tolerance that appears in the feature control frame. Remember that the geometric tolerance is applied at LMC.

The geometric tolerance at MMC will be that tolerance plus the difference between MMC and LMC in the size limits.

Wall Thickness Calculations

When we use geometric definitions, we employ many formulas to protect us from problems common to the task of creating designs for parts that will always mate at assembly (if produced per print), that have built-in repeatability factors, realistic tolerances based on part function and producibility, and countless other factors. One of the most important things we can do is to protect the parts we design for wall thickness/material strength. Wall thickness between the external features on a part and the internal features (holes, slots, etc.) are often critical to function. Minimum wall thickness between these internal and external features can be calculated (and usually quite easily) using, in addition to other factors, the calculations in this unit and the principles behind them. We often work with "worst case" situations.

We ask ourselves "What is the worst thing that could happen?", and we incorporate into our design those factors that will make that situation work (or we change the "worst case" to something we know will work). And, in doing so, any better situations that arise in the production of our designed parts, we know also will work.

One of the "worst case" situations we often use is to protect the ability of parts to mate at assembly. As you know, when an MMC modifier is used we call this "worst case" concept (the inner boundaries for holes and the outer boundaries for shafts) virtual condition. We design mating features so their worst-case mating boundaries are compatible; and, therefore, theoretically and statistically, we know that puts us in great shape for those features produced within these limits to virtually always mate at assembly.

As mentioned, another "worst case" formula centers around the concept of protecting each part independently to make certain no breakout or near breakout situations occur between external features and their internal features. No matter what the material condition symbol, this concept can be used. It can help us determine if features on our parts are in danger of breakout before the part is ever put into production. To be able to use a generic term, I will refer to these as boundaries, which will reflect a worst-case feature boundary for wall thickness.

Whether the MMC or the LMC concept is used in the feature control frame, the formulae are as follows:

INTERNAL FEATURES	EXTERNAL FEATURES
LMC	LMC
+ Geo. Tol. applicable at LMC	- Geo. Tol. applicable at LMC
Outer Boundary Hole	Inner Boundary Shaft

Sometimes, the minimum wall thicknesses are found by simply subtracting one boundary from another and dividing by two. At other times, more factors must be brought into play to determine this type of "worst case" situation. However, whatever the situation, a little practice with different part designs will soon make the important factors and ways of including them in this minimum wall thickness concept easy to identify and utilize.

The following example demonstrates this concept and other factors to determine minimum wall thicknesses. It is a good beginning for learning how to calculate any wall thickness problem. I call it the "Drinking Straw" problem. The premise is that most parts can be turned in drinking straws (or gun barrels) with a little thought.

FIGURE 247 (Inner and Outer Boundaries for LMC)

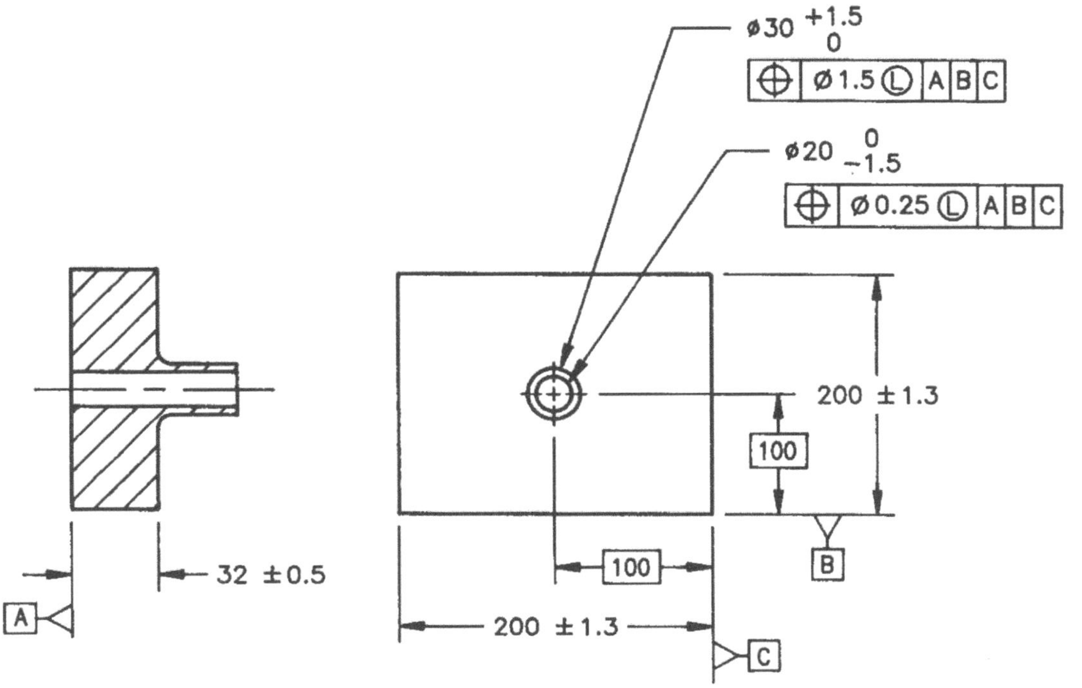

Shafts
30.0 = LMC Shaft
- 1.5 = Geo. Tol. at LMC
28.5 = Inner Boundary of Shaft

Holes
20.00 = LMC Hole
+ 0.25 = Geo. Tol. at LMC
20.25 = Outer Boundary of Hole

28.50 = Inner Boundary of Shaft
- 20.25 = Outer Boundary of Hole
8.25 = Material with Hole

8.25 divided by 2 = 4.125 Minimum Wall Thickness

Displacement/Datum Feature Shift

Does displacement of the feature pattern allowed because of the MMB symbol used after datum feature D endanger wall thickness to the O.D. of the part for this pattern controlled to the center datum feature of size? For an explanation and a calculation of the minimum wall thickness of the 4 holes to the O.D. of the part, see the paragraphs after FIGURE 248.

FIGURE 248 [Displacement and a Simultaneous Requirement]

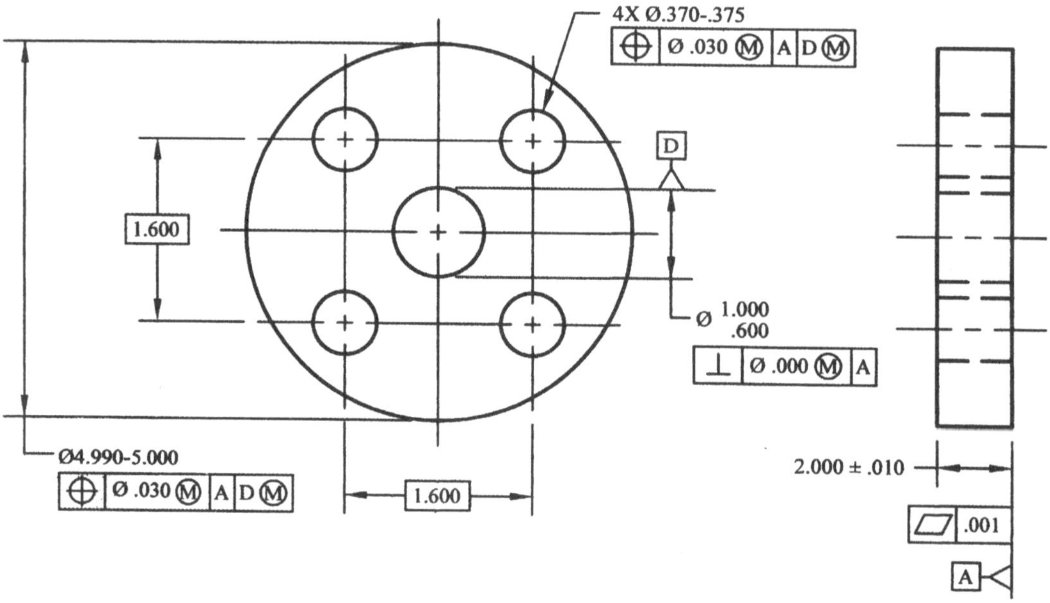

Minimum wall calculation:

LMC	4.990
-Geo. Tol. at LMC	- .040
Minimum Inner Shaft Boundary	4.950 = Inner Boundary of the O.D.

Because the 4 holes and the 4.990 - 5.000 O.D. are located to the same datums in the same order of precedence, they are considered one pattern, a Simultaneous Requirement; therefore, as D departs from MMC, they may shift--but all must shift in the same direction. So, if they all shift in the same direction, their relationship to one another does not change. Therefore, the thickness between the two is not further endangered.

4.950 = Inner Boundary of the Shaft

.375 = LMC
+.035 = Tol. at LMC
.410 = Outer Boundary of the (Hole)

1.600 = Bolt Square
x 1.4142 = Square Root of 2
2.26272 = Diagonal of Square (Bolt Circle)

4.950 = Shaft O.D. Inner Boundary
-2.263 = Diagonal of Square (Bolt Circle)

2.687 = Inner Boundary of the Shaft without the Bolt Circle
- .410 = Outer Boundary of the Hole
2.277 = Total Material

2.277 divided by 2 = 1.1385 Minimum Wall Thickness

FIGURE 249 [Datum Feature Shift with a Separate Requirement]

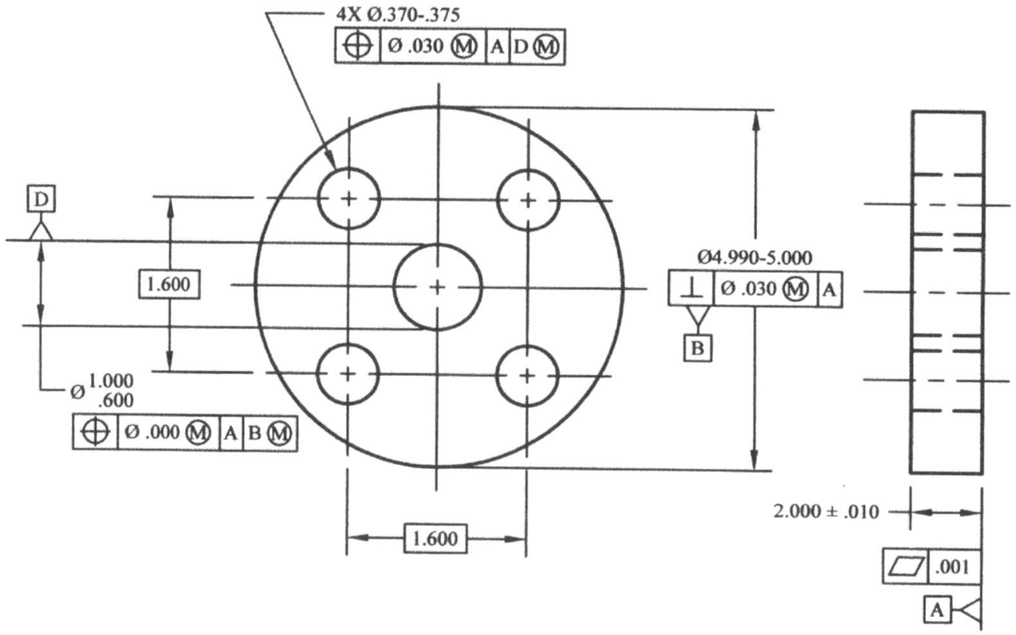

Is displacement a factor in minimum wall thickness of the 4-hole pattern to the O.D. of the part? I will explain and calculate minimum wall thickness of the 4 holes to the O.D. of the part below. I will use the term resultant condition to refer to the "worst case, non-constant boundary". In these cases, the resultant conditions will be the inner boundary of the O.D. and the outer boundary of the holes.

Answer: Yes

4.990 = LMC of O. D.	.375 = LMC of Holes
- .040 = Geo. Tol. at LMC	+.035 = Geo. Tol. at LMC
4.950 = Resultant Condition of O.D.	.410 = Resultant Condition of Holes

1.600 = Bolt Hole Square
x 1.4142 = Square Root of 2
2.263 = Diagonal of Square (Diameter of Bolt Hole Circle)

4.950 = Resultant Condition of O.D.
- 2.263 = Diameter of Bolt Hole Circle
2.687 = Resultant Condition of Shaft w/o Bolt Hole Circle

2.687 = Resultant Condition of Shaft w/o Bolt Hole Circle
- .410 = Resultant Condition of Holes
2.277
- .400 = Movement of D at LMC Allowed to O. D.
1.877

1.877
- .400 = Displacement (Datum feature shift Allowed)
1.477 = Total Material from D at Its As-Produced Location

1.477 divided by 2 = .7385 minimum wall thickness of 4-hole pattern to O. D. of the part (datum feature B)

Logic of adding displacement to 4 holes as a pattern

1. Datum feature D may be out of position a diameter of .400 to the outside diameter of the part if it is produced at its LMC of 1.000.

2. As datum regular feature of size D departs from its MMB of .600 to its LMC of 1.000, the pattern controlled to it may shift with the datum feature axis from the datum axis a like amount (one for one). This additional displacement is allowed due to the MMB symbol next to datum feature D in the positional tolerance for the 4-hole pattern.

FIGURE 250 [Using Datum Feature Shift in Calculating Minimum Wall Thickness]

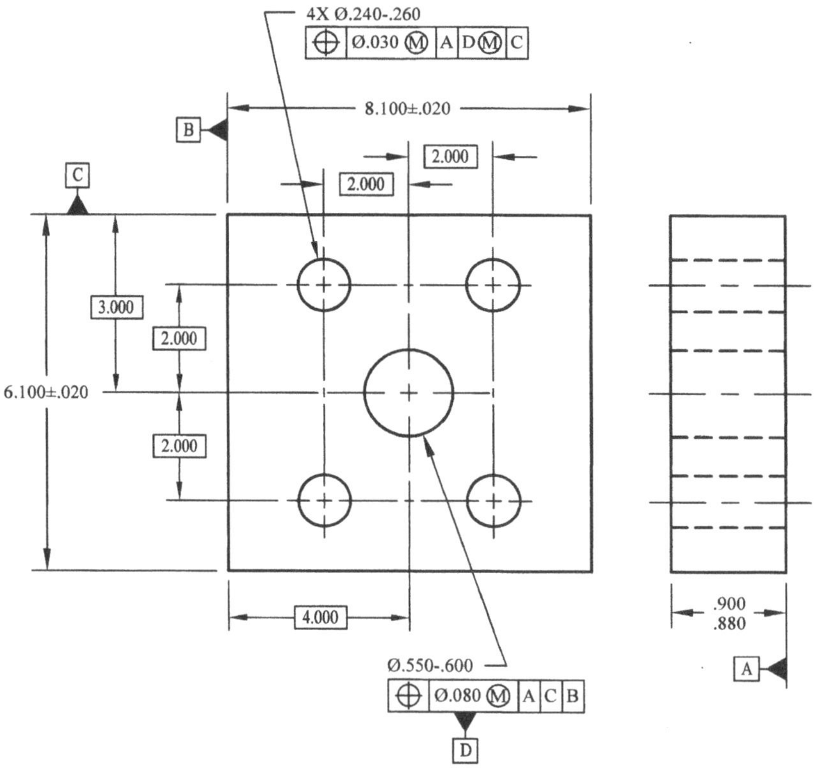

∠ = ±1/2°

Note: 1/2° over .900=.008 (rounded off)

What is the minimum wall thickness of the hole pattern to datum feature C? Answer:

.310 = Resultant Condition of Holes
+.130 = Allowed Movement of feature D at LMC to C

.440
+.130 = Displacement (Datum feature shift Allowed from D at its As-Produced Location
.570 = Resultant Condition with Movement of D from C and Pattern from D

$\text{Tan. } A = \frac{\text{opp.}}{\text{adj.}}$

$\text{Tan. } 1/2° = \frac{x}{.900}$

$.00873 = \frac{x}{.900}$

$.900\ (.00873) = x$

$.007857 = x$

.570 divided by 2 = .285 = Half Hole with All Movement One Way
1.000 = Basic Dimension
- .285 = Effective Half Hole with All Movement One Way
.715 = Basic Dimension - Effective Half Hole
- .008 = Perpendicularity of Feature C controlled by 1/2Δover .900
.707 = Minimum Wall Thickness without consideration of the flatness tolerance on datum feature C
- .040 = datum feature C's flatness tolerance as controlled by the 6.100 ± .020
.667 = Minimum Wall Thickness with all factors

Wall Thickness Calculations and Choosing the Pertinent Tolerances and Boundaries for Inclusion

Doing any tolerance analysis calculation depends a great deal on knowing which tolerances are to be included as pertinent and which tolerances are excluded as non-factors. Many times, just having a series of formulas that can be used time after time and altered to fit any given situation is the best way to begin. One such formula that can be used in its entirety, or in part, is illustrated for the part below.

FIGURE 251

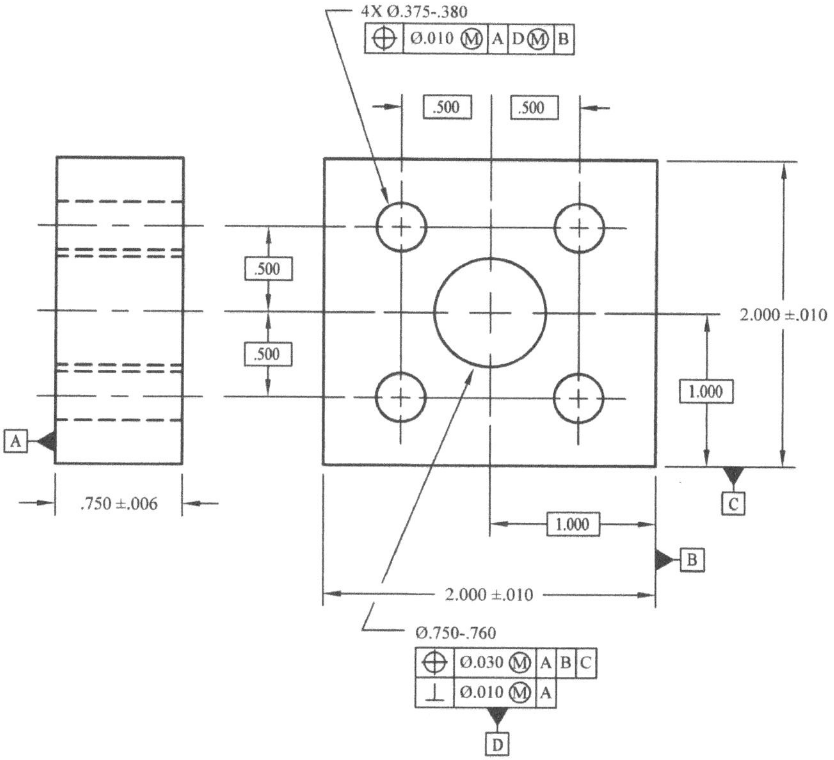

Calculating the minimum wall thickness of one of the small holes to datum feature B, many of the factors common to such problems can be assembled into a base formula. It is a challenging problem in that the four holes are not measured from datum feature B for distance. They are oriented perpendicular to datum plane A, located to datum axis D and the pattern oriented to datum plane B. This means the holes are located from the center hole, while measured along planes that are perpendicular to datum plane A and perpendicular or parallel to datum plane B.

Datum plane A, being a primary datum plane, would be established by a minimum of three high points of contact. Since datum axis D is generated by the secondary datum feature hole D, it creates the secondary and the tertiary datum planes.

An axis is the intersection of two planes crossing at a 90 degree angle to one another. That means that A and D alone create a coordinate system (datum reference frame) consisting of three mutually perpendicular planes. But it also means that one rotational degree of spatial freedom has not been eliminated until datum plane B is referenced to orient the two planes that cross at the axis of D. Datum plane B is known as an angular orientation datum. It is a planar surface that would be fitted to its datum plane with a minimum of two points of contact to construct a fourth datum plane. This plane could be thought of as a datum feature simulator in a gage or fixture that slides into contact with surface B. This datum feature simulator would be perpendicular to datum plane A, but not located from datum axis D.

To begin the calculation, we must know that the unknown being calculated is the minimum wall thickness between either of the small holes nearest to datum feature B. So, the small holes' outer boundaries must be calculated. This will be done by taking the largest hole (its Least Material Condition) and adding its positional tolerance (allowed movement) at LMC.

.380	=	LMC of the Four Holes
+ .015	=	Position Tolerance of the Hole at the Least Material Condition Size
.395	=	Outer Boundary of the Four Holes

To this number we must calculate, then add, the positional tolerance of datum feature D. Since the four holes are measured from datum axis D for location, if D is out of position, it takes the four holes with it. Datum feature D may be out of position to datum plane B by a .030 if it is produced at its Maximum Material Condition of .750. But it may move out of position a .010 more if it is produced at its LMC of .760. This is a total movement of a maximum of a .040 of datum feature D and consequently the four-hole pattern to datum B. So,

.395	=	Outer Boundary of the Four Holes
+ .040	=	Position Tolerance of datum axis D to datum plane B
.435	=	Outer Boundary of the Four Holes with D's Position Tolerance

To this, we must consider the additive effects of the MMC symbol after datum feature D in the four holes' position control. This means that the real hole's datum feature axis D may shift away from the imaginary datum axis D by the difference between the LMC of datum feature D and the pertinent inner boundary (virtual condition boundary) of datum feature D. Since D has two inner boundaries, we must choose the one that applies in this situation. **The key to this decision is to notice the four holes' position control and the datum references that precede the D referenced at MMC. Since the only datum that is referenced prior to the D at MMC reference is datum A, the inner boundary that applies is the one that is between D and A.** That means we do the following calculation.

.760	=	LMC of Datum Feature D
- .740	=	Inner Boundary of Datum Feat. D to Datum Plane A (.750 MMC minus .010)
.020	=	Datum feature shift/Pattern Shift Due to the MMC symbol after D

That means the .030 at MMC tolerance, which would be .040 at LMC is not used as a factor a second time in this calculation because of datum feature D's refinement of perpendicularity control to datum plane A. If no refinement of perpendicularity to datum plane A had been used on datum feature D, the .040 would have been used again in this calculation.

Therefore, we add the .020 to the outer boundary size of the holes within the four-hole pattern.

	.435	=	Outer Boundary of the Four Holes with D's Position Tolerance
+	.020	=	Datum feature shift/Pattern Shift of the Four Holes Due to MMB after D
	.455	=	Outer Boundary of the Four Holes with all Factors

Since this outer boundary is to be subtracted from the calculated basic dimension from datum B (1.000-.500=.500), and the basic dimension from datum D goes to the axis of the hole, the outer boundary of the holes must be divided by 2.

$\frac{.455}{2}$ equals .2275

.5000	=	Basic Dimension from Datum Plane B to Two of the Small Holes
-.2275	=	1/2 Outer Boundary of the Four Holes
.2725	=	Minimum Wall Thickness to Datum Plane B

Since datum Plane B is perfect, but datum feature surface B is not, we must now consider the aberrations that may befall that surface. First, it may be out of flat. Datum plane B hits the two highest points of datum feature B, but its low points are dictated by the overall size tolerance on the 2.000 dimension. The 2.000 dimension is $\pm$.010. If the side of the part opposite datum feature B was produced perfectly flat and the overall part size was 2.010, then datum feature B could experience the entire size tolerance as out-of-flatness tolerance.

So, .2725 minus .020=.2525 minimum wall thickness. The only other factor to consider on this part is the allowed out-of-perpendicularity tolerance of datum feature B to datum plane A. Since this tolerance is not given on the field of the drawing, we would have to go to the drawing's general tolerance notes to derive it. If that tolerance was, for example, $\pm$1 degree, we would calculate what that would be over the .756 maximum part thickness. This would be approximately .013. This would then be subtracted from .2705 to give us the minimum wall thickness from datum feature B to the holes nearest it.

.2525	=	Min. Wall Thickness to Datum Surface B without B's perpendicularity tol.
- .0130	=	1 Degree Over .756 Part Thickness
.2395	=	Minimum Wall Thickness for the Hole Nearest to Datum Feature B

Formulas like this one and the logic of including pertinent factors explored in this section can be invaluable for any tolerance analysis. They can also be used in other situations one may encounter with similar factors and tailored to fit the situation.

Single Part Analysis

Problem: Explain and calculate minimum wall thickness of the four holes to the O.D. of the part.

FIGURE 252

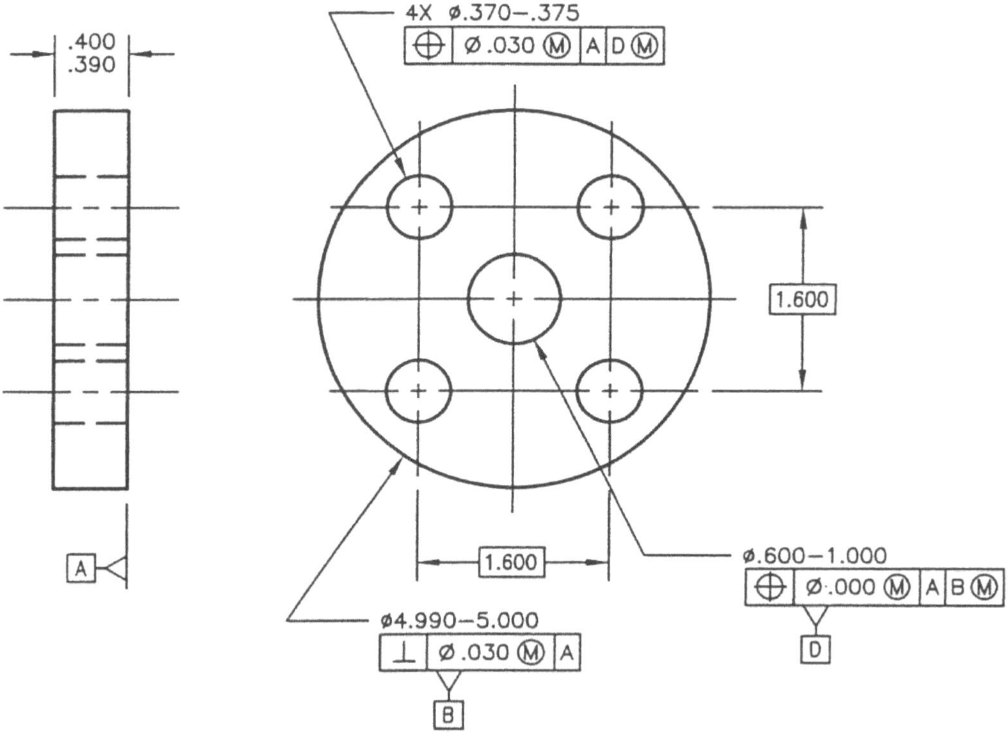

Answer to FIGURE 252.

Step 1:

.375	=	LMC Hole
+ .035	=	Geo. Tol. at LMC
Ø.410	=	Outer Boundary Hole
+ .400	=	Pattern Shift <Δμ>
Ø.810	=	Outer Boundary Hole with Pattern Shift
+ .400	=	Position Tol. of D
Ø1.210	=	Outer Boundary Holes with all Facets
R.605		

Step 2:

4.990	=	LMC Outside Diameter
- .040	=	Geo. Tol. at LMC
Ø4.950	=	Inner Boundary of Outside Diameter

$$\frac{\text{Ø }4.950}{2} = \text{R}2.475$$

Step 3:

1.600 = Basic Distance between Holes
x 1.4142) = The square root of 2
Ø 2.263 = Bolt Circle Diameter

R1.1315

FIGURE 253 [Circuit Graphed]

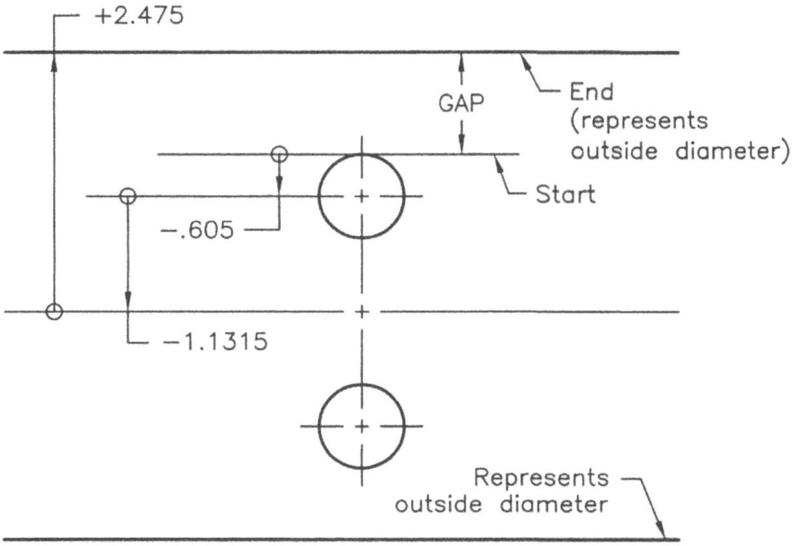

−	+	
.6050		
1.1315		
	2.475	
1.7365	2.475	Totals

2.4750
- 1.7365
.7385 = MIN WALL

Alternate Method Answer [Double check using diameter method.]

Step 1: Shaft

4.990 = Shaft
- .040 = Geo. Tol. at LMC
4.950 = Inner Boundary Shaft

Step 2: Bolt Square converted to Bolt Circle

1.600 = Bolt Square
x 1.4142 = The Square Root of 2
2.263 = Bolt Circumscribed Circle

Step 3:

4.950 = Inner Boundary Shaft
- 2.263 = Bolt Circle Diameter
2.687

Step 4: Hole

.375 = LMC Hole
+ .035 = Geo. Tol. at LMC
Ø.410 = Outer Boundary Hole

Step 5:

2.687
- .410
2.277

$\frac{2.277}{2} = 1.1385$

Step 6:

.400 = Accumulated Error from D (Position Tol.)
+ .400 = Pattern Shift of 4 Holes Due to D µ
Ø.800

$\frac{.800}{2}$ = R.400

Step 7:

1.1385
- .4000
.7385 = MIN WALL

. . . matches answer using Tolerancing Stack-Up approach.

Calculating Minimum and Maximum Wall Thicknesses with Profile Tolerance and Separate Requirements

Problem: Calculate the minimum and maximum wall thicknesses from the right side of the part in the front view and the 4.0 – 4.3 hole closest to it. Remember, since the profile control and the position of the 21 holes are Separate Requirements, they may shift in opposite directions when the effect of the B at MMB is considered.

FIGURE 254

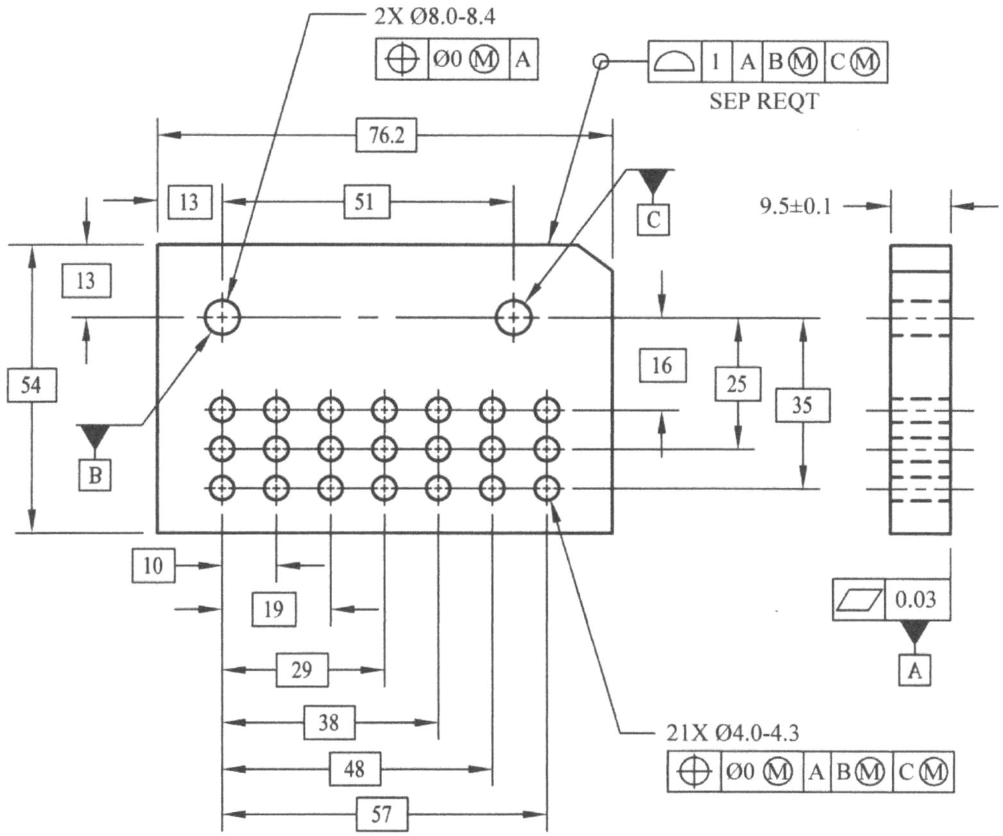

Answer to FIGURE 254.

Step 1:

4.3 = LMC Hole
+ 0.3 = Geo. Tol. at LMC

Ø4.6 = Outer Boundary
+ 0.4 = Pattern Shift (B µ)

Ø5.0 = Outer Boundary with Pattern Shift

4.0 = MMC Hole
- 0.0 = Geo Tol. at MMC

Ø4.0 = Inner Boundary
- 0.4 = Pattern Shift

Ø3.6 = Inner Boundary with Shift

5.0 = Outer Boundary
+ 3.6 = Inner Boundary
8.6 = Sum

5.0 = Outer Boundary
- 3.6 = Inner Boundary
1.4 = Difference

$\frac{8.6}{2}$ = Ø4.3 Mean

1.4 = 0.7 Plus and Minus Tol.

$\frac{Ø4.3}{2}$ = R2.15

$\frac{0.7}{2}$ = 0.35

So, R2.15 ± 0.35.

Step 2: Right Edge of Part from Datum B

76.2 = Basic Overall
- 13.0 = Basic Dimension Left Edge to B

63.2 = Basic Dimension right Edge to B

63.2⊥0.5 (Profile Tolerance)±0.2 (Pattern Shift-Separate Requirement)

63.2
+ 0.5
+ 0.2
63.9

63.2
- 0.5
- 0.2
62.5

63.9
+ 62.5
126.4

63.9
- 62.5
1.4

$\frac{126.4}{2}$ = 63.2

$\frac{1.4}{2}$ = 0.7

So, 63.2±0.7.

FIGURE 255 [Two-Single Part Analysis MIN and MAX GAP Right Edge]

Step 3:

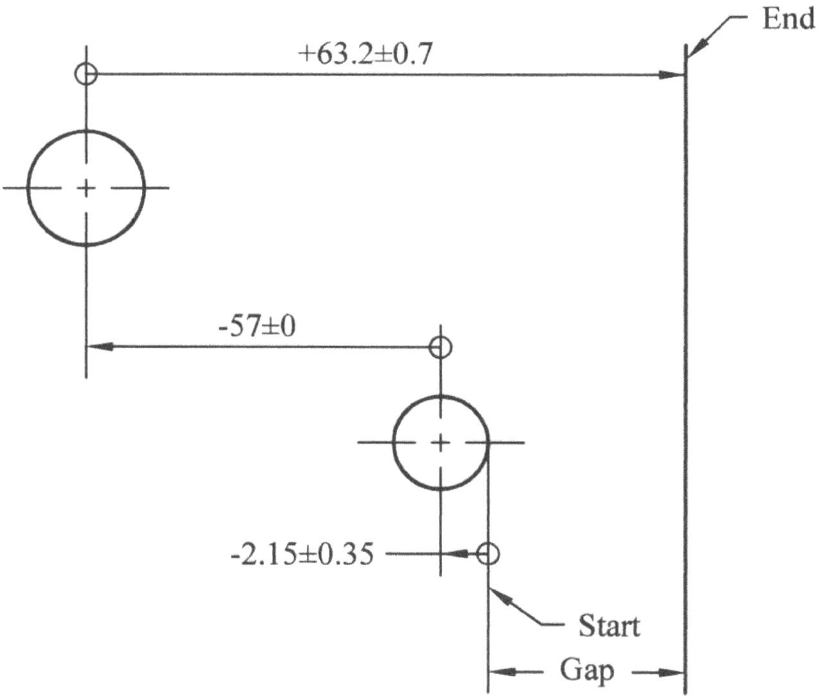

-	+	±Tol	
2.15		0.35	
57.00		0.00	
	63.20	0.70	
59.15	63.20	1.05	Totals

$$\begin{array}{r} 63.20 \\ -\ 59.15 \\ \hline 4.05 \end{array} \qquad \begin{array}{r} 4.05 \\ +\ 1.05 \\ \hline 5.10 \end{array} = \text{MAX GAP} \qquad \begin{array}{r} 4.05 \\ -\ 1.05 \\ \hline 3.00 \end{array} = \text{MIN GAP}$$

Chapter 17

Geometric Dimensioning and Tolerancing as a Language, Statistical Tolerancing, PC Boards

Chapter Objectives

Readers will learn:

1. How to read a feature control frame's symbology as a sentence.
2. How switching datums affects gage design, part fixturing and accumulated tolerance error.
3. How to visualize and calculate datum feature shift.
4. How to tolerance a PC board.
5. When a simultaneous requirement applies to single features or patterns of features and when they are a separate requirement.
6. Angular orientation datum features and sliding versus stationary datum feature simulators.
7. Profile tolerance zone and pattern shift.
8. How to apply and read sequential tolerancing schemes.
9. Statistical Tolerancing

GD&T as a Language

Geometric Dimensioning and Tolerancing is a language of linear logic. In every facet of the language, whether writing it, reading it or measuring the results, it must be done a piece at a time to do it well. It's like putting together a jigsaw puzzle.

You always keep in mind the picture on the front of the box, but you only have one piece of the puzzle in your hand at any time. So, you need to concentrate all of your skill on that piece to make it fit. You have the option of dropping it back into the pile, if you determine it wasn't the best choice at this time. But if you stick with it, your focus must be absolute. And most people don't focus well. They are diffuse in their thoughts, even chaotic. You can see it in their eyes, the way they skim over the design drawing, dwelling on nothing and, therefore, understanding and accomplishing very little.

To properly read a drawing, you must find the beginning. The beginning is usually the primary datum feature, often selected as such because it is the surface the next part in the assembly seats on, bolts to, derives the most contact from and, therefore, dictates the angle at which the two parts assemble. This primary datum feature is real and, therefore, flawed and is controlled for form.

If it is a planar surface, it is controlled for flatness, usually either by using a geometric characteristic symbol or a size tolerance that combines this surface with one that opposes it. Rule #1 can kick in and allow this size tolerance to control the form or it can be overridden by applying a geometric tolerance that controls the form better than the size tolerance. If this feature is cylindrical in shape, it must be controlled for cylindricity. And, again, we have the choice of using a geometric characteristic symbol or a tight enough size requirement to do the job.

Once the primary datum feature has been selected and controlled, the **secondary datum feature is selected. This selection is often based on choosing a mating and/or alignment feature, one that mates with and/or aligns with the comparable features on the next part in the assembly.** This feature is then related back to the primary datum. If that relationship is one of perpendicularity, and it often is, then a choice is made. Again, this is a choice between using a geometric characteristic symbol (perpendicularity) and a plus or minus tolerance on the angle. This can be deferred to a general tolerance note on the drawing. But it is usually easier to calculate assembly problems such as might be done in a tolerance stack-up analysis if geometric tolerances are used. **Once the secondary datum feature is chosen and related to the primary, it is decided whether or not a tertiary datum feature is required.** If so, it is related to the primary and secondary datums.

Then other features on the part are related to all three. And on and on. The point here is that this is **a linear progression. One step at a time is performed. Each step relates features to those controlled in previous steps.**

Tolerancing PC Boards

Either of the following approaches may be used for a variety of products and part configurations. For example, a P.C. board can use a shift variation of each tolerancing scheme. For example, see FIGURE 256.

FIGURE 256

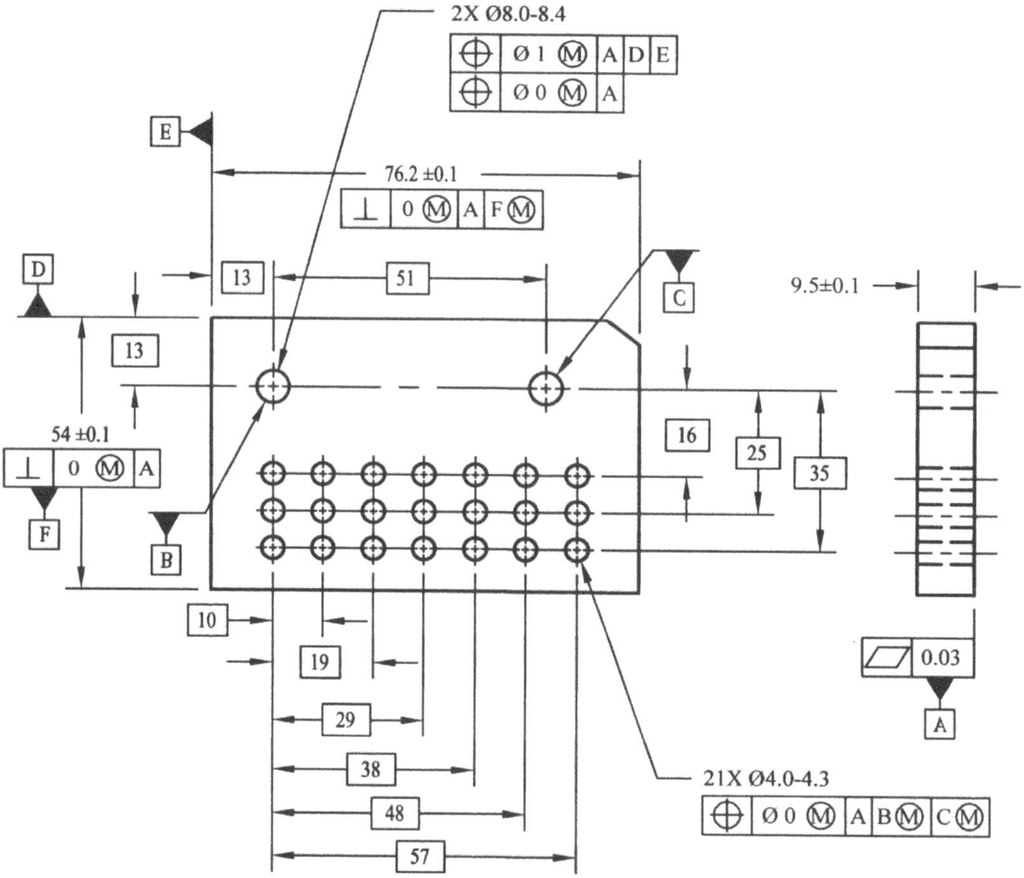

This is a common tolerancing approach used on the many parts throughout a wide variety of industries. Step 1 is to assign a primary datum to stabilize subsequent relationships and measurements. This surface should have sufficient surface area to achieve part stabilization and hopefully in a functional way. For example, choosing a surface that seats on the next part in the assembly would be ideal. This surface is given a 3-dimensional form control, in this case, a flatness control.

We know the important relationships to be held are between the 23 holes on the part. Since two of the holes are larger, with more surface area and stabilizing ability, they will eventually become datum features from which the other 21 holes will be measured. But first, we must loosely relate (position) the two holes to the edges of the P.C. board, then more tightly control their relationship (distance) to each other and to A (for perpendicularity).

So, a couple of edges of the board are selected to measure in from. Ideally, the longer of the two edges, with more surface area, will be selected as secondary (after A primary) and the other will be tertiary. The 2-hole pattern is first positioned loosely (to within 1mm at MMC) to these datums that have been named A, D and E.

Then in the second level of position of these 2 single segment position controls, the 2 holes are more tightly positioned (to within zero at MMC) to each other (tolerancing the 51mm basic dimension) and in their relationship to A (for perpendicularity), a refinement of portions of the upper level segment of this 2-segment control. The holes are then labeled as datum features B and C.

The 21-hole pattern is subsequently positioned to each other, then to A for perpendicularity, then to B for location (distance) in both the X and Y directions, then to C to act as the angular orientation datum. C makes certain one of the planes that constructs the axis of hole B passes through the axis of hole C. This gives us an angle of measurement that must be followed. We will measure the location of the 21 holes with our measurements oriented to A, while measuring from the axis of B along planes crossing at the axis of B, one of which passes through the axis of hole C.

This is the definition that the Y14.5 committee has always claimed. However, with the redefinition of the angular orientation datum feature in ASME Y14.5-2009 (and perpetuated in the ASME Y14.5-2018 standard) in cases like this, it is questionable. Datum feature C would be represented in theory by a stationary boundary located by the applicable basic dimension (51) from B. The effect on gages and fixtures would be that datum features B and C would be represented by gage pins at their applicable maximum material boundary (MMB). For the measurement of the 21-hole pattern, that means that B and C would be represented by gage pins that are a diameter of 8 millimeters and basically located from one another at 51 millimeters apart. The exact same representation would be true if both holes were called datum pattern B and the 21 holes located to A and B at MMB. This means that B and C are equally capable of locating and controlling the angular orientation. And, as with the datum pattern of two holes, the datum axis likely resides halfway between their maximum material boundaries (or gage pins). So, the location datum axis would not necessarily reside at the center of B with the angle of measurement dictated by the common plane between B and C. The common plane between B and C's maximum material boundaries would control location in one direction and angular orientation. But the other datum plane that creates the location datum axis would be at a 90-degree angle to the common plane between B and C's MMB's and would most likely lie halfway between those MMB's.

FIGURE 257

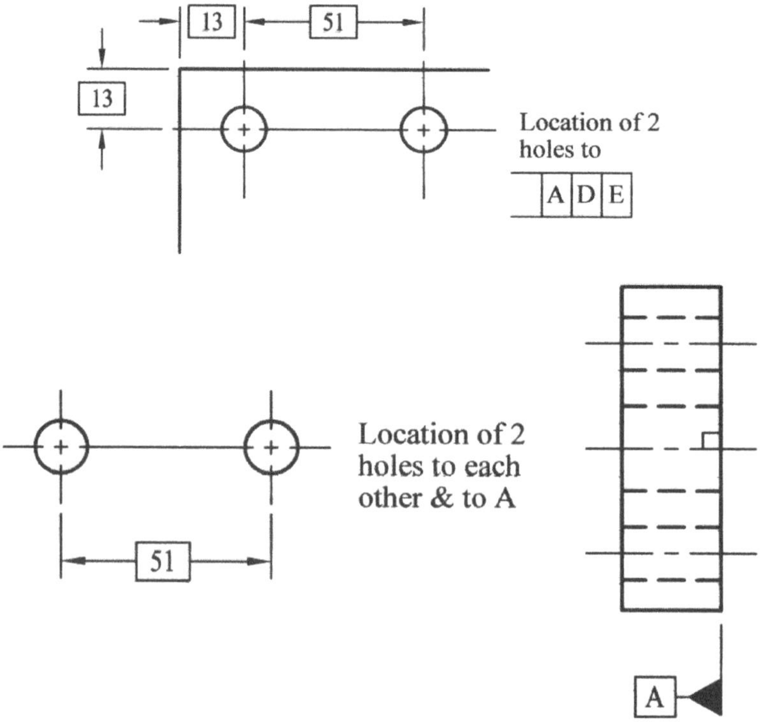

FIGURE 258

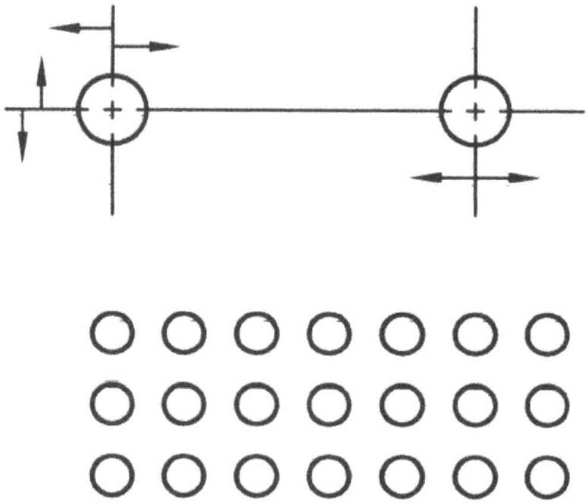

There are two perpendicularity controls on the outside widths of the part shown in this approach that just act to square up the part. They create an envelope of perfect orientation at maximum material condition. There are a couple of ways to understand these controls. The perpendicularity control of the centerplane of datum feature F gets represented in understanding the more complex perpendicularity control on the 76.2±0.1mm width. So, let's focus on the interpretation of this more complex control.

First, let's see what it is trying to control in terms of understanding the gage it would take to inspect it.

FIGURE 259 [Gage for Checking Perpendicularity of Part Widths]

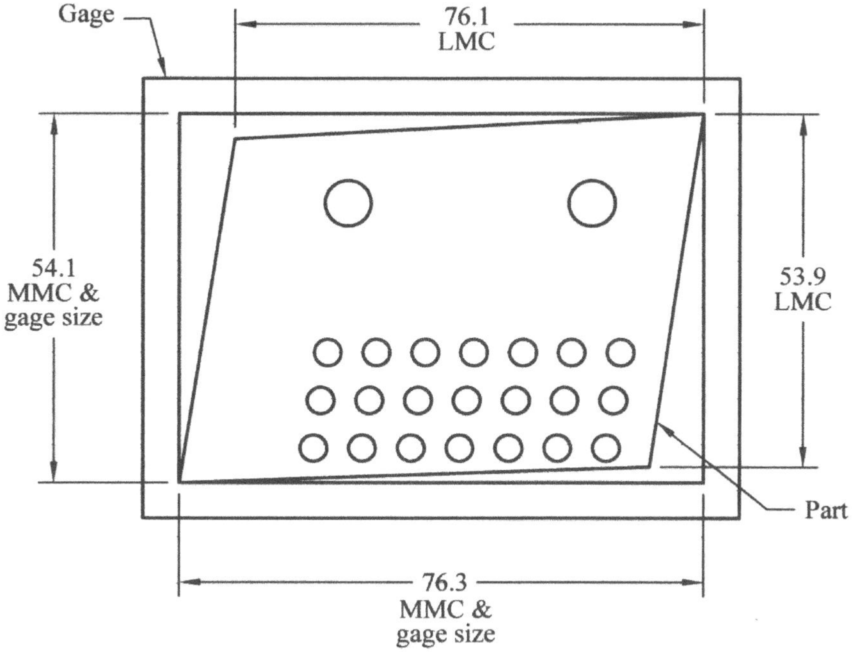

If the part is produced at MMC by MMC (54.1 x 76.3), it must have perfect perpendicularity to fit into the gage. But, if the part is produced at less than the 54.1 or the 76.3, it may be out of perpendicularity by the amount it is smaller in both directions. In terms of the tolerance zones this creates, the following illustration applies:

FIGURE 260 [Perpendicularity Tolerance Zones in Middle of Gage]

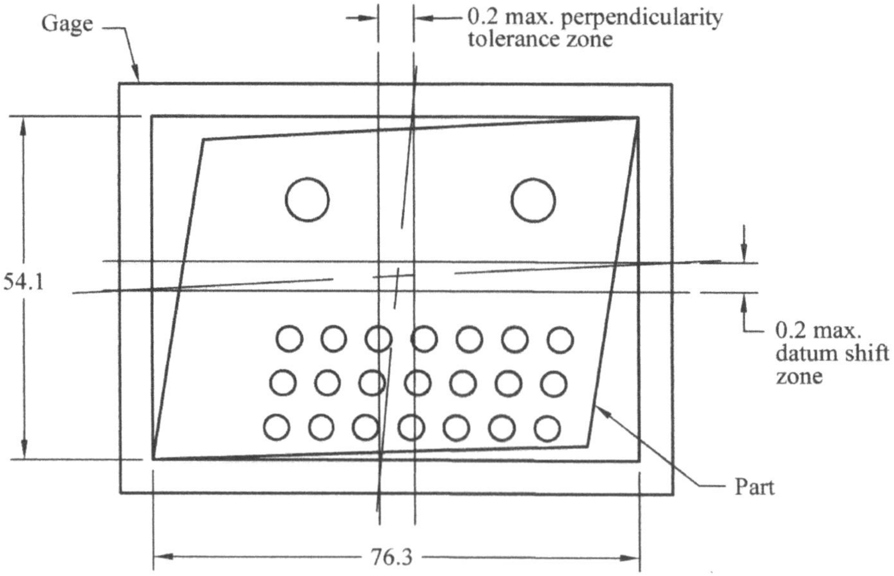

So, the maximum material condition symbol after the zero perpendicularity tolerance allows the zone to grow from zero at 76.3 to 0.2 if the part width is produced at 76.1 (LMC). Additionally, the maximum material boundary (MMB) symbol after the datum feature reference F allows a datum feature shift (or in this case, a datum lean) zone to grow from zero at the middle of the gage width in the 54.1 direction (at that size) to a maximum of 0.2 if the datum feature is produced at 53.9 (LMC). This is difficult to understand if we think of datum centerplane F to exist in the middle of the part. But the concept becomes clearer if we think of the datum centerplane F as existing in the center of the gage width's 54.1 size.

As the part's datum feature F shrinks, it may lean away from the center of the gage. So, in this particular instance, the perpendicularity between the 76.2±0.1 width's centerplane and the 54±0.1 width's centerplane could be out 0.2 plus 0.2 = 0.4. It is important that we not picture this as the tolerance zone simply growing to 0.4, but rather that it grew to 0.2 (76.3-76.1 = 0.2) and the datum feature centerplane leaned away from the imaginary datum centerplane (represented here by the gage width's centerplane) 0.2 (54.1-53.9 = 0.2). Each zone is a maximum of 0.2. There are 2 zones. One zone contains the centerplane of the 76.2±0.1 width, and the other zone contains the datum feature centerplane of the 54±0.1 width.

These perpendicularity controls would not have been necessary had we used the other tolerancing scheme that we explored on the last sample part that employed a profile all-around control.

Let's look at an example of this P.C. board that uses that approach:
FIGURE 261

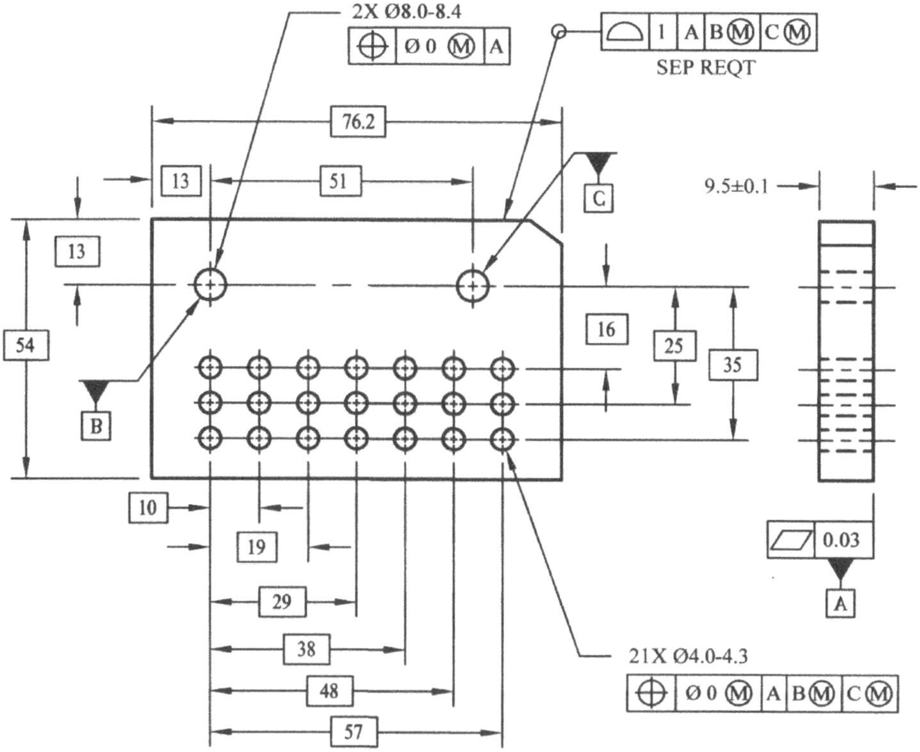

STEP 1:

Step 1 is assignment of a primary datum feature that will stabilize the board for subsequent relationships and their measurement. Ideally, this will orient the board in the way it seats in the assembly, but stabilization is the real key here. This surface is then controlled for flatness. The less flat it is, the less stable it is. We do not want our primary datum feature rocking while we try to hold and measure the relationship of the other features on the board to datum A and to each other while seated on A.

This control can be read, **"Flatness of this surface must be held between two parallel planes that are 0.03 apart."**

...or

"All points on this surface must reside between two parallel planes that are 0.03 apart to comply with the flatness requirement."

...or simply, **"Flatness of the surface must be held to within 0.03 mm."**

STEP 2:

Step 2 is a step that combines two controls into one step. It is mainly to tolerance the 51mm distance between the two holes that become datum features B and C. The size limits of the holes begin by saying 2x. In other words, there are two holes in this pattern. Therefore, the position control begins by tolerancing their relationship to each other. This hole-to-hole tolerance is Ø at

MMC (8) per hole, but could grow (as the holes grow) to a 0.4 (maximum diameter) at LMC (8.4). The only datum reference is A. The relationship between the holes and datum plane A is one of perpendicularity.

So, the position tolerance zones for each hole, which are a basic distance of 51mm from center to center, are also perpendicular to A. If the axes of the holes are in their position tolerance zones and the zones are 90° to datum plane A, then the holes will not only be their allowed distance from one other, but also perpendicular to A to within the allowed tolerance. For example:

FIGURE 262

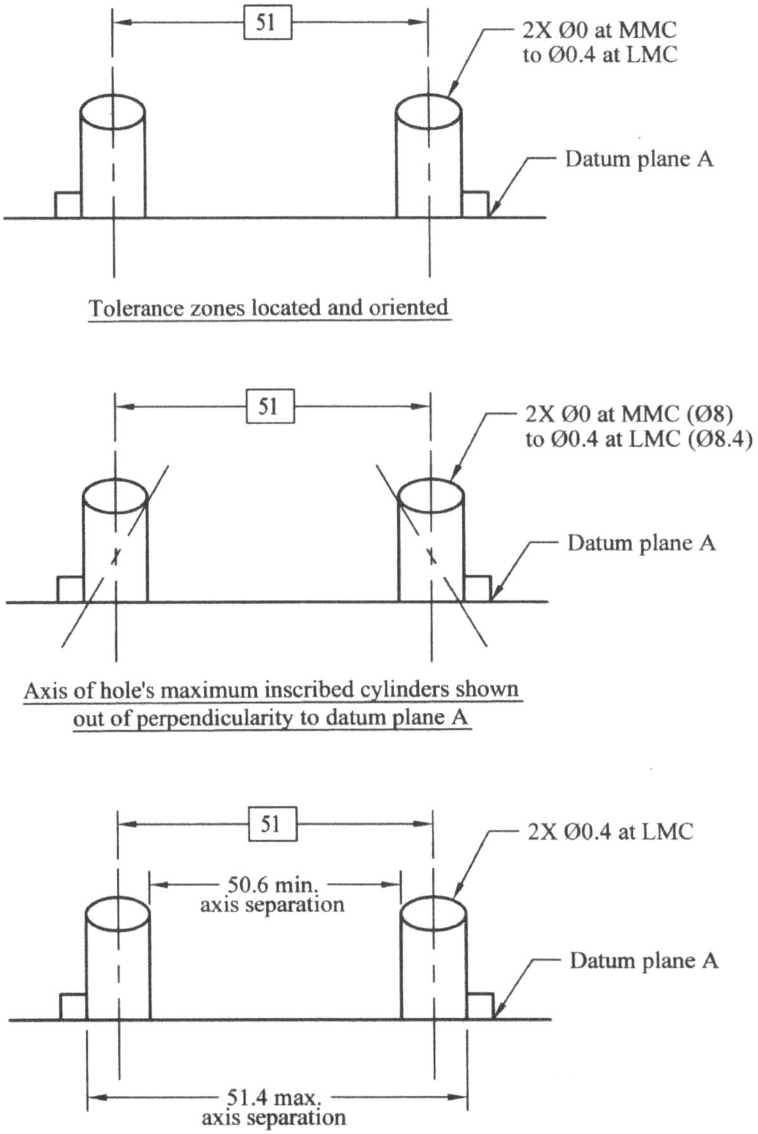

These relationships could have been controlled in a more sequential fashion and the same results accomplished. For example:

FIGURE 263

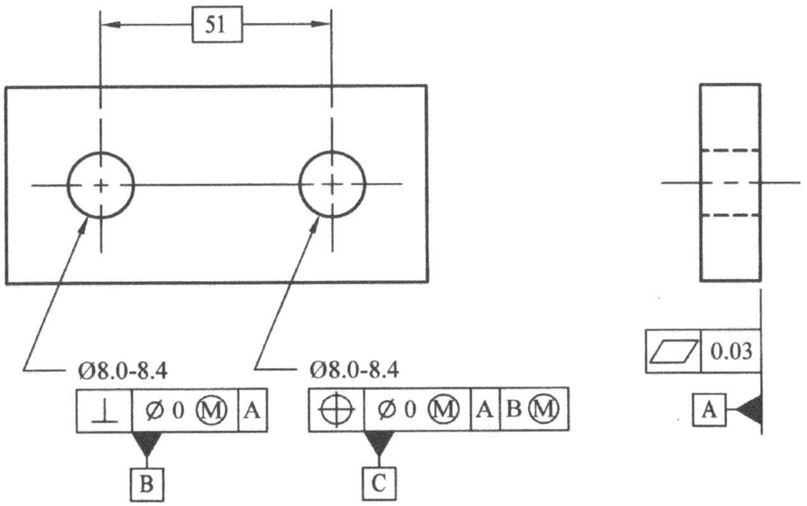

The tolerance zones on the previous page would also apply for the illustration above when assessing the control exerted by the positional tolerance controlling datum feature C. The only difference in the explanation would be:

FIGURE 264

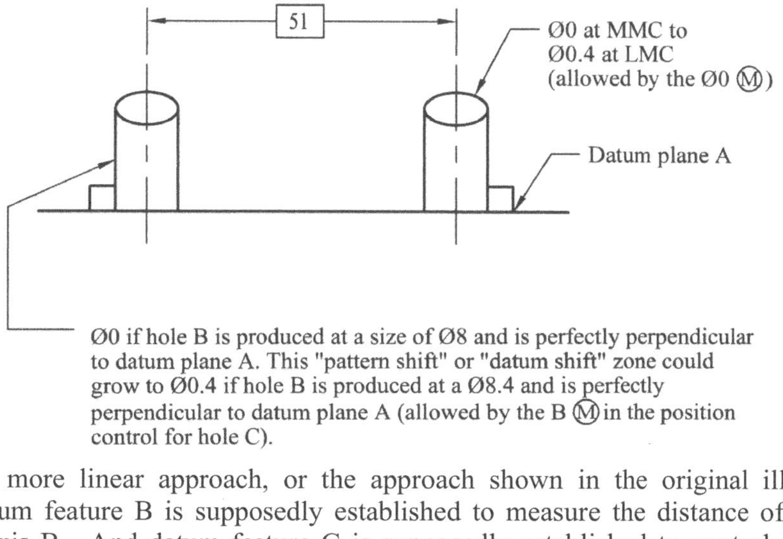

In either this more linear approach, or the approach shown in the original illustration, the secondary datum feature B is supposedly established to measure the distance of the 21 holes from datum axis B. And datum feature C is supposedly established to control your angle of measure (angular orientation). It orients the planes that cross at the axis of B (and create that axis). So, one of the planes that creates the axis of B passes through the center of C. The other plane that creates the axis of B is perpendicular to the plane that runs through the centers of B and C. In truth, B and C are equal in their ability to both locate and orient any features measured from them. However, if B and C were referenced at regardless of material boundary (RMB), wording in ASME Y14.5-2009 and ASME Y14.5-2018 would support B as the location datum and C as the angular orientation datum. Under the RMB concept Y14.5-2009 and Y14.5-2018 state that B would be expanded into with a maximum inscribed cylinder before C was utilized (expanded into) and therefore B would locate and C would just control angular orientation.

FIGURE 265

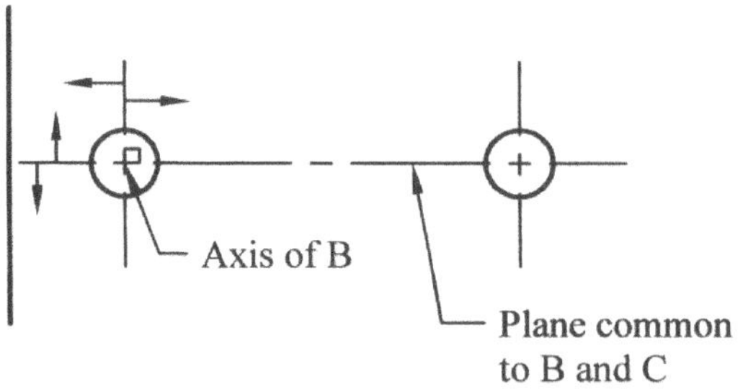

So, all measurements for the location of the 21 x Ø4.0-4.3 holes are taken from the axis of B along the planes oriented by C.

In the original control for these two holes,

2X Ø8.0-8.4

⊕	Ø 0 Ⓜ	A

it can be read: "Position of the axes of diameters (dimensioned 51mm basic from each other) must be held to zero if produced at MMC (Ø8.0) to Ø0.4 if produced at LMC (Ø8.4). These tolerance zones that are between Ø0 and Ø0.4 and 51mm apart are also perpendicular to datum A and, therefore, also control the perpendicularity to A within these tolerance zones".

Or

⊕ = Position
Ø = of diameters
Ⓜ = that mate
A = while the part seats on A.

Or,

⊕ = Position (defined by the 51mm basic dimension)
Ø = of the axes (of 2 diameters)
0 = must be perfectly held to each other
Ⓜ = if the holes are produced at their MMC (Ø 8)
A = and to A for perpendicularity (or while the part seats on A).

FIGURE 266 [Gage of Step 2 – Hole to Hole and Tolerance Zones]

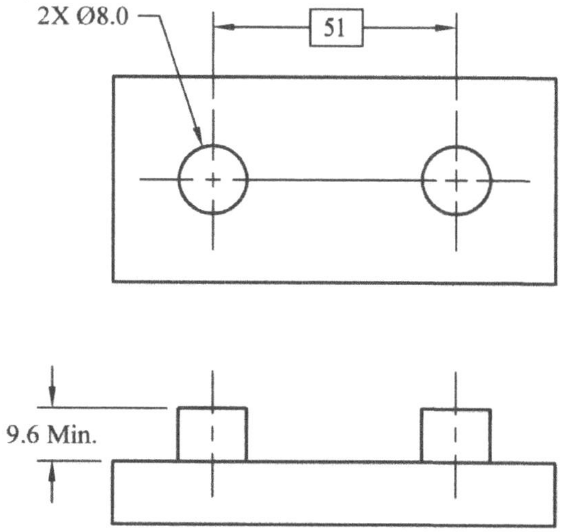

STEP 3:

There are two geometric controls that remain: a profile control and a position control for the 21 x 4.0-4.3 holes. Since both controls use exactly the same datum references, they would normally fall under the "**Simultaneous Requirement**" rule. **Simultaneous Requirement Rule: Simultaneous requirements apply to position and profile tolerances that are related to the same datum reference frame. The same datum reference frame is defined as using the same datum features referenced in the same order of precedence with same modifiers applied to the datum feature references. The default of simultaneous requirements requires that all applicable geometric tolerance zones are related to the same datum reference frame and are appropriately constrained as a single pattern to a single datum reference frame**. This would require them to be measured in the same set-up or with the same gage. It would also allow that if datum features B and C are produced at a size larger than their MMC and virtual condition (which are the same Ø8.0 in this case) the features (the profile and 21 holes) may shift as a group left or right, up or down (and any other direction) by the radial growth of B and C.

NOTE: Some, or all, of this growth may be negated by the datum feature B and C holes being out of perpendicularity to datum plane A.

But since they are considered one pattern of, in this case, 22 features (21 holes and a profile), they would have to shift in the same direction by the same amount (to a maximum of the difference between 8.0 and δ8.4, which is a diameter of 0.4 or a radius of 0.2). The "Simultaneous Requirement" rule was not considered desirable in this situation, so a local note was added beneath the profile control to negate it.

This local note of "SEP REQT" stands for "Separate Requirement" or, in the gaging community, a "Separate Gaging Requirement." It allows the profile to shift independently from the 21-hole pattern. For example, if the allowed shift was the maximum of Ø0.4 which is a radius of 0.2, the profile could shift to the left 0.2 and the 21 holes could shift to the right (or any direction) 0.2.

Since, beyond this explanation of pattern shift (sometimes called datum feature shift), it doesn't really matter which control we read next, let's read to position tolerance for the 21 holes.

Step 3 can be read:

⌖	=	Position
Ø	=	of the axes of these diameters
0	=	must be held to within zero tolerance
Ⓜ	=	if the holes are produced at their MMC size of Ø4.0 (to a maximum position tolerance of Ø0.3 if the holes are produced at Ø4.3, their least material condition). These holes mate.
A	=	to A for perpendicularity (this part seats on A in the assembly)
BⓂ	=	and B for location, if B is produced at its Maximum Material Boundary (and its MMC which are the same in this instance) (while B mates). All of the basic distances to the 21 holes (from B) are toleranced by this control.
C Ⓜ	=	and to C for angular orientation, if C is produced at its Maximum Material Boundary.

Again, B and C act together to control location and angular orientation equally as can be seen in the gage below. Per ASME Y14.5-2009 and Y14.5-2018, B and C's true geometric counterparts and therefore their datum feature simulators are located to each other at a basic dimension of 51 millimeters.

FIGURE 267 [Functional Gage for Inspection of the Position of the 21 Holes]

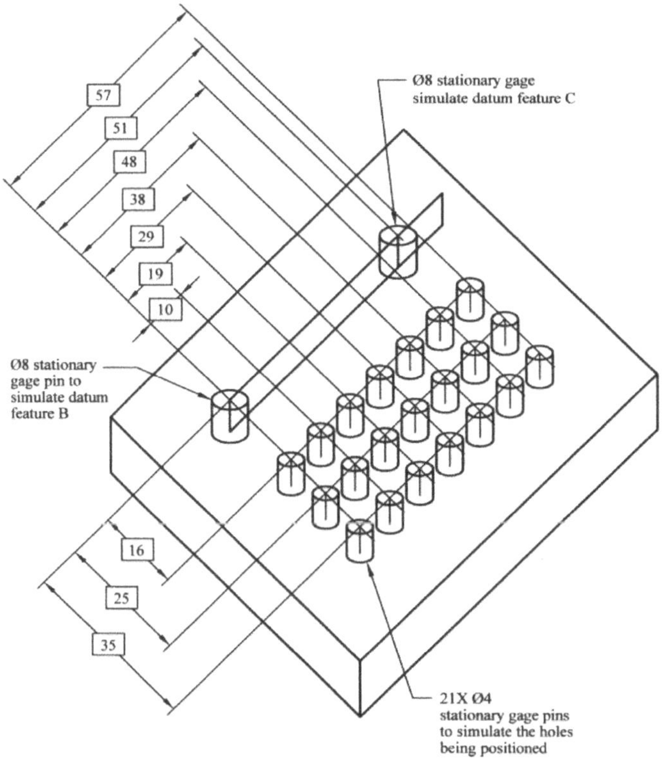

STEP 4:

The profile of the surface control is the last geometric characteristic to consider. Looking at the drawing, there is a basic size and shape to the part. There are also basic angles held to datum A (for perpendicularity). Distances from B are held while measuring the profile along the planes that cross at the middle (axis) of B. One of these planes also passes through the axis of C.

So, these are the relationships of size and shape that the profile of the surface control tolerances. This tolerance zone is equal-bilateral (since no phantom lines are shown) and the middle of this zone is depicted by the basic dimensions originating at datum axis B. This zone which looks vaguely like a picture frame must contain all elements of the surface all-around in this view. So like the 21 holes, the profile is perpendicular to datum plane A, located from the axis of B and oriented to the common plane between the axes of B and C.

FIGURE 268 [Illustration of a Profile Tolerance Zone and Pattern Shift]

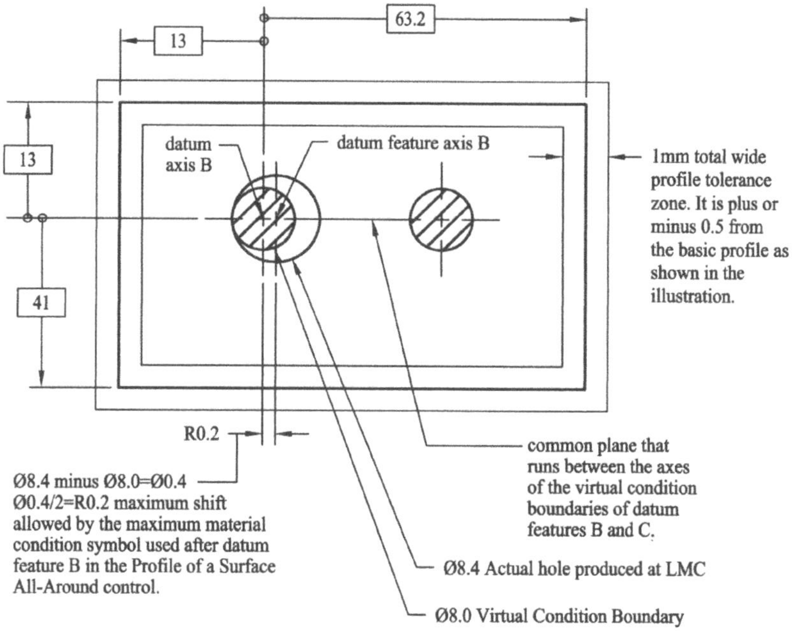

The Profile of a Surface control can be read:

⌓	=	Profile of this surface all-around in this view to control size and shape (3D form)
1	=	must be held to within 1mm (plus or minus 0.5mm) of the basic dimensions.
A	=	while the tolerance zone is held perpendicular to datum plane A
B	=	located the depicted basic dimensions from the axis of B
Ⓜ	=	allowed a shift of the profile (all-around) from the datum feature axis B as hole B departs from its MMB (Maximum Material Boundary) of ∅8.0 to a maximum of ∅0.4 if the B hole is produced at its LMC of ∅8.4,
C	=	while measured perpendicular and parallel to the datum plane common to the axes of datums B and C
Ⓜ	=	with an additional rotation of the profile (all-around) allowed by the growth of datum feature C from its MMB (Maximum Material Boundary) of ∅8.0.

Calculating Statistical Tolerancing

Statistical Tolerancing is applied to features on parts wherein it has been deemed acceptable to forego 100% interchangeability in assemblies. These tolerances are calculated **using a predetermined methodology** such as **Root Sum Square** or **Monte Carlo** or other formulas that attempt to predict how much tolerance is likely to be used during the manufacturing procedure. That "natural tolerance" likely to be consumed by the manufacturing is then compared to the arithmetically available tolerance that has been calculated using formulas such as the fixed and floating fastener assembly formulas discussed in this textbook.

The tolerance predicted to be consumed is put over the arithmetically available tolerance for the features in the assembly under consideration to be increased and converted to a percentage. This percentage of the arithmetically available tolerance that is predicted to be used during the manufacture of the features being considered is put under the number one. In other words, one divided by the percentage of the arithmetically calculated tolerance likely to be consumed during manufacturing becomes the percent the individual tolerances will be increased and redistributed back into the assembly.

For example, let's say we have an assembly tolerance for the considered features that adds up to 0.91. See Figure 269, 270 and especially Figure 271 in this section. Then we apply a formula such as the Root Sum Square formula that predicts, in a natural bell curve distribution of parts (a Gaussian Frequency Curve), that we are likely to consume (to within plus or minus 3 sigma) 0.71 of the 0.91.

We divide the 0.71 by the 0.91 and find that we have predicted that about 78% of the 0.91 will be consumed during the manufacturing procedure. We then divide the number 1 by 0.78 to get the factor by which we will increase the individual tolerances considered.

1 divided by 0.78 is approximately 1.28. Our individual tolerances will each be increased to 128% of their original value. If a Root Sum Square formula was run on the increased tolerances, it would predict (using the same plus or minus 3 sigma procedure as first applied) that of the sum of the increased tolerances, it is likely that only 0.91 (the original arithmetically available tolerance) would be consumed by the assembly.

This type of procedure, or one like it, can be used to increase tolerances and, in doing so, decrease the costs associated with production of features using the more tightly toleranced arithmetically calculated tolerances. These ST or Statistical Tolerances should only be allowed to manufacturers who have proven statistically capable in the judgment of the company or department vending the parts out to be produced. An ST tolerance should never be allowed for use by a supplier, who by contrast, is in statistical chaos. Such a supplier may produce entire batches of mating parts that have spanned their larger statistical tolerances and endanger the product's quality and its ability to function or assemble properly.

For an example of this methodology, we will use an illustration where, in a fixed fastener two-part assembly, we calculate the MIN GAP for the lower left. To do this, we have to choose the correct route to follow, convert the dimensions to equal bilateral with plus or minus tolerances, then calculate the gap. We used basic dimensions that have tolerances expressed as zero and slots and tabs whose tolerances included both size and position. Still, in the end, we came up with a series of plus and minus tolerances that we used to calculate the minimum gap.

6.095 is the radius of the mean dimension of the inner and outer slot boundary averaged. Likewise, the tolerance of plus and minus 0.055 is the tolerance on that slot radius derived from the inner and outer slot boundaries. The mean of the slot boundaries is 2x6.095 or 12.19 with a tolerance of plus and minus 0.11 (2x0.055). See the following calculations for the slot and the tab.

FIGURE 269

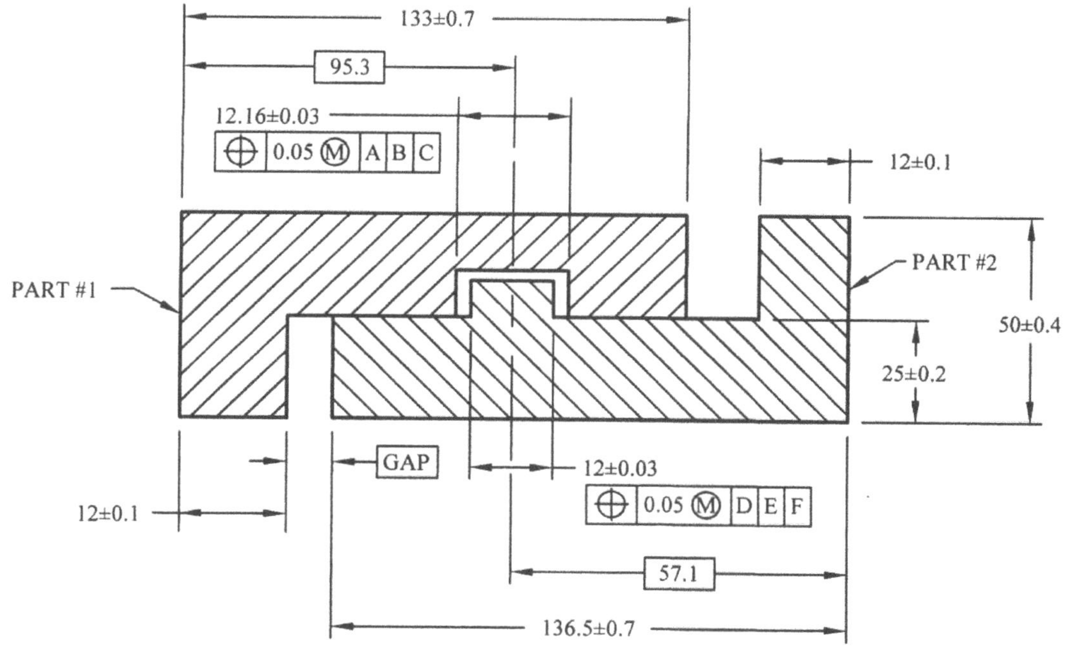

FIGURE 270 [Virtual and Resultant Condition Calculations for Tolerance Stack-Up Analysis for Slot and Tab – Summary]

Outer Boundary/Resultant Condition of Slot = 12.19+0.11 = 12.3

Inner Boundary/Virtual Condition of Slot = 12.13-0.05 = 12.08

Inner Boundary/Resultant Condition of Tab = 11.97-0.11 = 11.86

Outer Boundary/Virtual Condition of Tab = 12.03+0.05 = 12.08

Slot			Tab		
Resultant Condition Slot	=	12.30	Virtual Condition Tab	=	12.08
- Virtual Condition Slot	=	12.08	- Resultant Condition Tab	=	11.86
Difference	=	0.22	Difference	=	0.22
1/2 Difference Slot	=	0.11	1/2 Difference Tab	=	0.11
Resultant Condition Slot	=	12.30	Virtual Condition Tab	=	12.08
+ Virtual Condition Slot	=	12.08	+ Resultant Condition Tab	=	11.86
Sum	=	24.38	Sum	=	23.94
1/2 Sum of R.C. and V.C. of Slot =		12.19	1/2 Sum of V.C. and R.C. of Tab =		11.97
1/2 Sum ±1/2 Diff. of Slot =		12.19 ±0.11	1/2 Sum ±1/2 Diff. of Tab =		11.97 ±0.11
1/2 of 1/2 Sum ±1/2 of 1/2 Diff. Slot =		6.095 ±0.055	1/2 of 1/2 Sum ±1/2 of 1/2 Diff. Tab =		5.985 ±0.055

NOTE: The numbers used are for the slot and the tab in Figure 269. Figure 271 shows how to route a calculation of the minimum gap in the lower left-hand corner of this assembly.

Rules for Stack-Up Analysis

To do the stack-up analysis, follow these rules. 1. Determine the gap you wish to calculate. 2. Start the routing at the left of the gap in the left to right calculations (or the bottom of the gap, if the route is up and down). 3. Numbers that lead you to the left (or down) are negative numbers. Numbers that lead you to the right (or up) are positive numbers. 4. Always take the shortest route of dimensions that gets you from the left of the gap to the right of the gap (or from the bottom of the gap to the top of the gap). See Figures 270 and 271 for an example of a left to right gap calculation.

FIGURE 271 [Calculated Minimum Gap Lower Left]

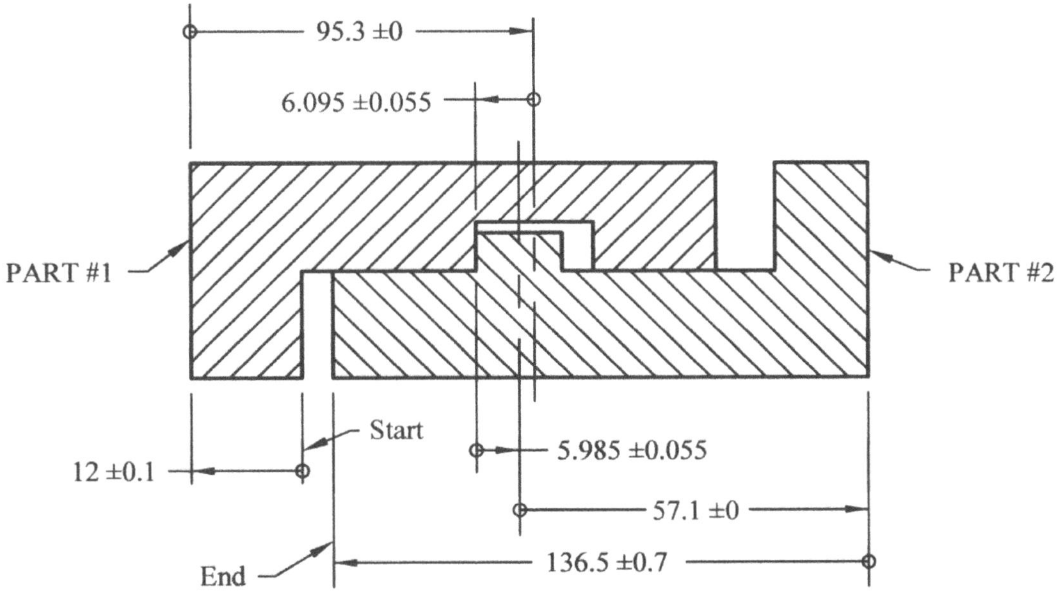

Right to Left	Left to Right		
-	+	±Tol	
12.000		0.100	Overall
	95.300	0.000	Basic Dim
6.095		0.055	Slot
	5.985	0.055	Tab
	57.100	0.000	Basic Dim
136.500		0.700	Wall
154.595	158.385	0.910	Totals
158.385 - 154.595 3.790		3.79 - 0.91 2.88	MIN GAP

Now we will use that illustration to show how we can convert to statistical tolerances with the same methods we used on the previous example that had only plus or minus tolerances.

FIGURE 272 [Calculating the Sum of the Squares]

–	+	±	± squared	Type
12.000		0.100	0.01	Wall
	95.300	0.000	0.00	Basic Dim
6.095		0.055	0.003025	Slot
	5.985	0.055	0.003025	Tab
	57.100	0.000	0.00	Basic Dim
136.500		0.700	0.49	Overall Dim
154.595	158.385	0.910	0.50605	Totals

158.385
- 154.595
+3.790 = difference between (technically the sum of) the positive and negative mean dimensions

$\sqrt{0.50605 \text{ Sum of Squares}}$ = 0.7113719 Square Root of Sum of Squares and tolerance likely consumed. So, about 78% of 0.91 tolerance likely to be consumed.

$$\frac{0.7113719 \text{ (tol. likely to be consumed)}}{0.91 \text{ (arithmetically calculated tol.)}} = 0.7817273$$

So, 0.7113719 is about **78%** of 0.91.

$\frac{1}{0.7817273}$ = 1.2792184 and 1.2792184 x 0.7113719 = 0.91 (the originally calculated arithmetic tolerance). So, tolerance can be increased to **128%** and still only use an amount likely to be 0.91.

The arithmetically calculated 100% tolerance allows a minimum gap that is 3.79 - 0.91 = 2.88.

The statistically calculated assembly tolerance allows a minimum gap that is 3.79 - 0.71 (rounded off to two decimal places). This 0.71 tolerance is the amount of tolerance likely to be consumed in a natural bell curved distribution of manufactured parts for this assembly. So, if we want to consume plus and minus 0.91 tolerance, the piece part tolerance should be increased to 128% (rounded off from 127.92184%).

So, the plus and minus 0.055 tolerance for the slot and the tab becomes 1.28 x 0.055 = 0.0704 or a tolerance for each that is plus and minus 0.070 (rounded off the three decimal places). The tolerance for the wall becomes 1.28 x 0.100 = ±0.128. The tolerance for the overall dimension becomes:

1.28 x 0.7 = plus and minus 0.896.

This is the answer to the problem of how much the tolerances can be increased. Increase all tolerances in the route of numbers to 128% of their original values.

Since, statistically, the tolerance originally assigned would not be fully consumed, the minimum gap calculation given originally as 3.79 - 0.91 = 2.88 MIN GAP becomes a statistical probability within ±3 sigma of consuming only 3.79 - 0.71 = 3.08 MIN GAP.

Given the newly assigned statistically calculated tolerances, with each piece part given a statistically calculated tolerance, we have a mathematical possibility of a MIN GAP that is 3.79 minus the sum of the statistical tolerances. They are: 0.070 (slot) + 0.070 (tab) + 0.128 (wall) + 0.896 (overall dimension) = ±1.164.

0.070 (slot)
+ 0.070 (tab)
+ 0.128 (wall)
+ 0.896 (overall dimension)
± **1.164**

And it leaves a MIN GAP of 3.790 - 1.164 = 2.626. So, arithmetically we might have a MIN GAP that is 2.626, but this is highly unlikely.

- The slot and tab statistical tolerances of 0.07 when squared become 0.0049 each.
- The wall statistical tolerance of 0.128 when squared becomes 0.016384.
- The overall dimension statistical tolerance of 0.898 when squared becomes 0.806404.
- When added, these tolerances are:

0.0049
+ 0.0049
+ 0.016384
+ 0.806404
0.832588 Sum of Squares of Tolerance

- The square root of 0.832588 = **0.91 (rounded off) Square Root of Sum of Squares**

So, we have again shown by the RSS (Root Sum Square) formula that even though the statistical assembly tolerance (plus and minus 1.164) is greater (to 128%) than originally calculated arithmetically as plus and minus 0.91, the amount the tolerance likely to be consumed is still only plus and minus 0.91.

This statistical approach assumes a zero mean shift for all the dimensions being used (and that the process capability equals 1). It is based on manufacturing processes that are in statistical control, not in statistical chaos. Those not employing Statistical Process Control in the manufacture of the workpieces should not use the RSS tolerancing methodology described here. Also, the RSS model assumes that parts produced for the assembly have been mixed and the components picked at random. The logic of the RSS Model is interesting. It basically allows more tolerances for those manufacturers that need it least—those using SPC controls. It calculates that the chances of producing a part that spans its larger Statistical Tolerance (ST) are so small that if it does happen, the randomly selected mating parts will make up for the potential problem by not spanning their tolerances. And, in fact, it presupposes that the mating part will be produced so much better than its tolerance extremes as to allow the parts to assemble well. If this is a false assumption, unacceptable functional conditions may arise, such as interference of material. In general, the RSS method is not used if there are less than four dimensions in the stack-up analysis.

The 100% tolerancing method scares many professionals when they see that a line fit possibly exists between mating features. This happens when the inner boundaries of holes (or slots) and the outer boundaries of shafts (or tabs) that are mating features are the same value. Professionals calculating the worst mating conditions of such features can see the line fit possibility and sometimes are uncomfortable with this. If that makes them uncomfortable, then allowing more tolerance using the RSS calculations, and consequently a greater possibility of interference, should make them even more unsettled.

Back in 1968, a man named A. Bender wrote a paper for SAE (Society of Automotive Engineers) entitled: *Statistical Tolerancing as it Relates to Quality Control and the Designer*. In this paper, he suggested a safety factor be added to the RSS formula. Instead of just taking the square root of the sum of the squares of the individual features' tolerances, he suggested a factor of 1.5 be multiplied by the answer of the standard RSS solution or, in other words, 1.5 times the square root of the sum of the squares of the individual features' tolerances. This means this factor estimates one is more likely to use more of the original tolerance than the regular RSS formula calculates is likely consumed. This was so that the additional tolerance given to the piece parts in the assembly was not quite so risky. It is known that, in most cases, producing features at their worst-case condition is unlikely, but it is also known that it happens. Some studies have shown that the RSS methodology does not accurately reflect the reality of what is produced, so to have a 'cushion' would be wise.

In years since, many authors and statisticians have suggested other 'safety' or 'correction' factors. They are often based on studies they have done that include probable or measured repeatability and/or accuracy rates of particular manufacturing processes used on specific products. These 'correction' factors often range from 1.4 to 1.8, although one of the most common remains the 1.5 suggested by Bender all those years ago. Before trying an arbitrary 'correction' factor, it is wise to look to your company to see if they have established which, if any, of these factors have been approved.

To practice any of the methods shown in this unit, pick examples from earlier units in this book and calculate the statistical tolerances with the RSS method with or without 'correctional' factors such as 1.5.

Of course, there are many methods that can be used to reintegrate the tolerances that distribute them differently. Some try to help the more difficult to manufacture features by drawing tolerance from other features in the assembly. This allows the difficult to manufacture features to get more of the tolerance. But, if that was a factor, it probably should have been thought of and handled when the tolerances were being arithmetically calculated and before the calculation of the Statistical Tolerances began.

The tolerances calculated through the methods shown in this unit are identified with the following ST symbol. When both the statistical tolerance and the smaller arithmetic tolerance are shown, only those facilities using SPC controls are to be allowed the larger ST tolerance.

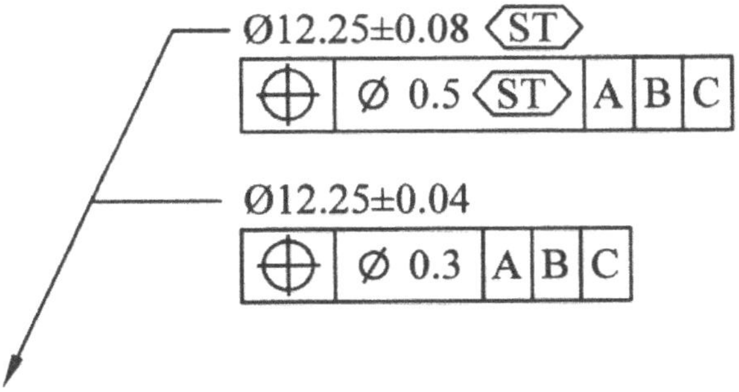

A note such as the following must be placed on the drawing: FEATURES IDENTIFIED AS STATISTICALLY TOLERANCED ⟨ST⟩ SHALL BE PRODUCED WITH STATISTICAL PROCESS CONTROLS, OR ELSE THEY SHALL BE HELD TO THE MORE RESTRICTIVE ARITHMETIC LIMITS.

Methods for calculating statistical tolerances vary from company to company. Some are best done by computer programs. **One such method is known as the Monte Carlo Method**. The term relates to methods that simulate manufacturing probabilities using random numbers. For example, we may use it to simulate the manufacture of dimensions on parts in an assembly. Given a knowledge of manufacturing capability, random numbers are generated to simulate possible process results. After extensive averaging, one arrives at the likely amount of tolerance that will be consumed. In fact, there are a wide variety of methods that are used that are called Monte Carlo Methods. Because of this, one method and its results may be quite different than another. These methods use inferential statistics. Inferential statistics use the principle that a random sample tends to exhibit the same properties as the entire population from which it comes. **If one takes too few samples, it is possible that the results will not reflect the entire population.**

These simulations can be done with simple and available software, such as might be found in a spreadsheet program like Microsoft's Excel. A database can be set up that will simulate uniform distributions of dimensions for each variable using a random number generator. The calculated mean and standard deviation of the sample can be used to determine if an opportunity exists to increase tolerances and still have parts that meet functional requirements from a statistical standpoint.

In other words, although in a worst-case analysis one would find functional difficulties with these increased tolerances (such as interferences), it is proven statistically unlikely that these difficulties will occur. In these programs, each variable can be treated differently. One is allowed to pick a distribution that is normal (given historical data about the variables) or uniform (given no historical data). Uniform distributions tend to give a more conservative answer since it is more of a guess rather than based on results given in past processing data.

Although it is never possible to achieve 100% accuracy through sampling without using the entire population, results are likely to be correct, given the statistics under the conditions we have put into place.

A Method for Reintegrating the Statistical Tolerancing into the Assembly Keeping the Same Size Mean Dimension instead of the Same Functional Boundary Mean

The Slot

A simple approach could be used if one wanted to keep the same mean size dimension on the slot of 12.16 instead of the same mean functional boundary of 12.19. That calculation would begin by looking at the original illustration in Figure 1. The tolerance on the size of 12.16 of the slot is plus and minus 0.03. If we multiplied 0.03 by 1.28, we would get the new size tolerance of plus and minus 0.0384. Since the original size mean was 12.16, the new size limits would be 12.16 plus and minus 0.0384. We would then multiply the position tolerance of 0.05 at MMC by 1.28. This would give us the new position tolerance of 0.064.

So, the new control would be:

Size: 12.16 plus and minus 0.0384 Statistical Tolerance
(12.1216 MMC and 12.1984 LMC)

Position: 0.064 at MMC Statistical Tolerance

In this way, both the size tolerance and the position tolerance would be increased to 128%. This would generate an inner boundary of 12.1216 minus 0.064 = 12.0576 (instead of 12.05) and an outer boundary of 12.1984 LMC plus 0.064 position tolerance at MMC plus the 0.0768 bonus tolerance = 12.3392 (instead of 12.328 shown as a rounded-off 12.33). The mean of these boundaries is 12.0576 plus 12.3392 = 24.3968 divided by 2 = 12.1984 (instead of 12.19).

Another, more accurate way of reintegrating the Statistical Tolerance into the assembly can be found in my book on Tolerance Stack-Up Analysis on my website www.geotolmeadows.com. However the method shown here is much simpler to implement and only skews the mean of 12.19 slightly (12.1984 vs. 12.19).

FINAL ANSWER

12.16±0.0384⟨ST⟩

⊕	0.064 Ⓜ ⟨ST⟩	A	B	C

A note such as the following must be placed on the drawing: FEATURES IDENTIFIED AS STATISTICALLY TOLERANCED ⟨ST⟩ SHALL BE PRODUCED WITH STATISTICAL PROCESS CONTROLS.

The Tab

The tab would be done the same way.

Chapter 18

Datums

Chapter Objectives

Readers will learn:

1. The six spatial degrees of freedom and how to assign datum features to stabilize a part.
2. How to fixture a part properly on its datum features.
3. How to designate the degrees of spatial freedom eliminated by each datum feature on a part.
4. How to apply angular orientation datum features and what affect they have on part stability.
5. The effects of differing datum precedence on part acceptance.
6. How to define and fixture a curved surface as a datum feature.
7. How to establish 3 mutually perpendicular datum planes from 3 surfaces that are not mutually perpendicular.
8. How to utilize complex and constant cross-section datum features.
9. How to specify degrees of freedom in the feature control frame.
10. How to identify multiple datum reference frames.
11. How to specify the desired datum material boundary with a number.
12. How to calculate the correct MMB and LMB.
13. How to use the translation modifier.
14. How to utilize the term 'basic' or its abbreviation in a feature control frame.
15. The meaning of planar datum features referenced at RMB and MMB.
16. How to specify and interpret irregular datum features of size at RMB and MMB.
17. How to specify and interpret repetitive patterns.

Specifying Datum Axes to be able to Designate Degrees of Freedom

FIGURE 273 [Six Degrees of Freedom—3 rotational and 3 linear] Part

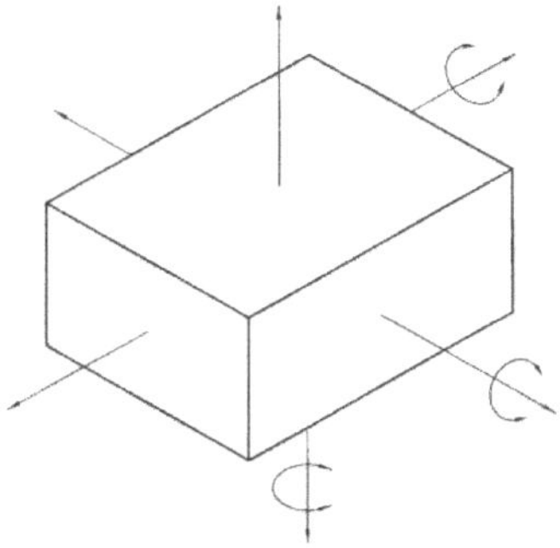

FIGURE 274 [Datum Reference Frame showing Six Degrees of Spacial Freedom to be Stabilized for any Part Configuration]

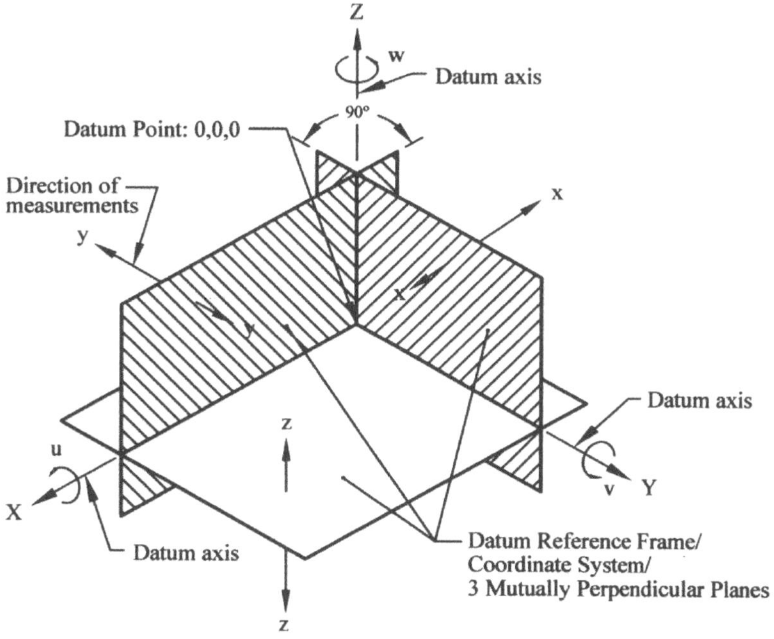

<u>Degrees of Spatial Freedom</u>

Linear	Rotation
x = translation along the X axis	u = about the X axis
y = translation along the Y axis	v = about the Y axis
z = translation along the Z axis	w = about the Z axis

NOTE:

Datum axis refers to an Axis of the Datum Reference Frame.

FIGURE 275 [Primary Datum Plane]
Simulated from Feature High Point Contact; 3 Point Contact

Note: Datum feature simulators are labeled as examples only.

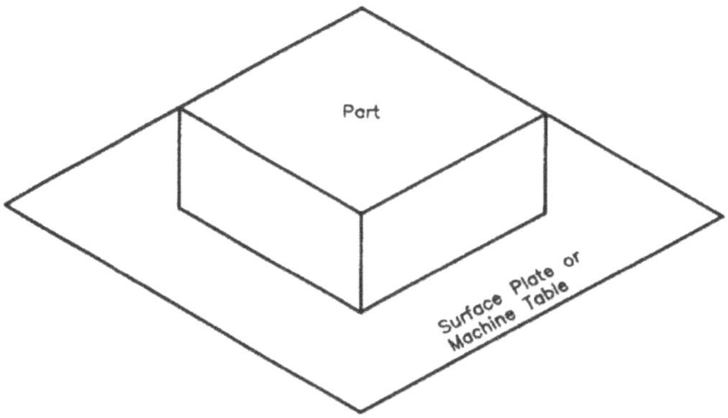

To simulate the primary datum plane, the produced part may be placed in contact with the surface plate (or machine table) and rests on its high points of contact. A primary datum plane is simulated by the imperfect surface plate (or machine table) from the datum feature (surface). A minimum of 3 high points on the surface are used to create the primary datum plane. If it happened that the surface did not rock, then the part would be stabilized in 3 of its 6 degrees of spatial freedom. Two rotational degrees of freedom, as well as 1 linear (up and down) degree of freedom, would be eliminated simply by placing the part surface in contact with the plate or table. If the part rocks on the plate, then it is to be adjusted to an optimum position to stop this rock. It may be rocked one way or another to stabilize the part to establish (simulate) the primary datum plane. It may, instead, be leveled while still touching the surface plate using the same concept as shimming.

This actually creates a candidate set of datum planes from which is chosen the optimum one plane that will serve as the primary datum. The chosen plane has some limitations in its selection. If one were to view the datum feature (surface) as being comprised of three thirds, the middle third may contain all points that create the valid datum plane. But neither of the outer thirds of the surface may individually contain all of the points that create the plane. Some of the high points may reside in both of the outer thirds of the surface simultaneously--but not all in either individually--in order to create a valid primary datum plane. This is a simplistic explanation of a more complicated concept on selection of a valid datum plane from a set of candidate planes. For more information, see the illustrations and explanations in this text on the creation of a valid datum plane.

To simulate a planar secondary datum feature, the part may be placed in contact with an angle plate (or rail) which is perpendicular to the primary datum feature simulator. It rests on the high points of contact (2 point high point contact minimum). A secondary datum plane is simulated by the imperfect angle plate (or rail) from the secondary feature (surface). If the surface does not rock on the datum feature simulator, the part is further stabilized by the elimination of 2 of the remaining 3 degrees of freedom (the other 3 degrees of freedom having already been eliminated/ stabilized by the primary plane).

FIGURE 276 [Primary and Secondary Datum Simulation]
3 Point Contact on Primary; 2 Point Contact on Secondary

Note: Datum feature simulators are labeled as examples only.

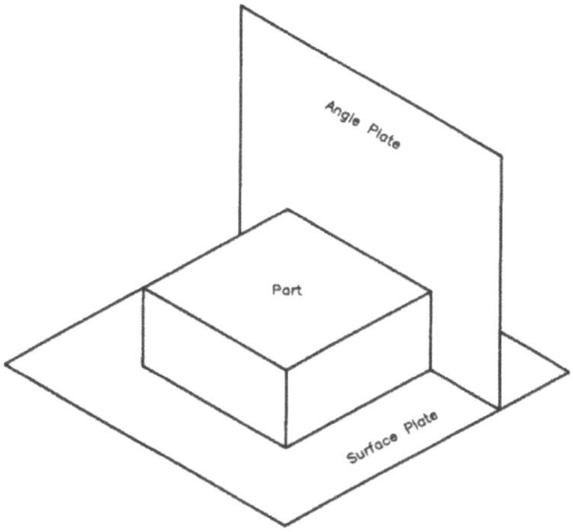

The secondary datum eliminates the remaining rotational degree of freedom and one of the remaining linear degrees of freedom. If the part does rock on the secondary datum feature simulator, the part must be adjusted optimally to stabilize the part into the datum reference frame to which, in this case, a third plane will be added. For guidance in adjusting the rocking surface to an optimum position, see the explanation for optimizing a primary rocking datum feature.

FIGURE 277 [Full Datum Reference Frame Simulation]
3 Point Primary; 2 Point Secondary; 1 Point Tertiary

Note: Datum feature simulators are labeled as examples only.

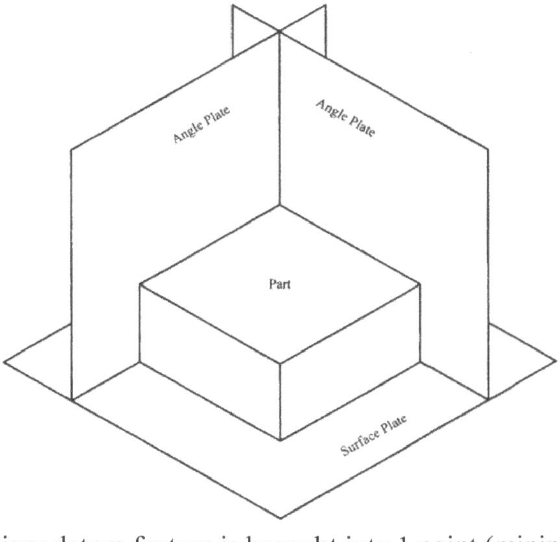

In FIGURE 277, the tertiary datum feature is brought into 1 point (minimum) of high point contact with the tertiary datum feature simulator (angle plate, rail, etc.) and eliminates the last degree of part freedom (linear). The last datum feature is brought into contact with the tertiary datum feature simulator while the part maintains its relationship with the primary and secondary datum feature simulator. The part is now stabilized in the datum reference frame (three mutually perpendicular planes).

All 6 spatial degrees of freedom (three rotational and three linear) are eliminated in this set-up. The part features that have been related to this datum reference frame may then be produced and/or inspected to the specified datums as represented by the manufacturing/inspection equipment.

At times, it may be desirable to designate the spacial degrees of freedom that are eliminated by each datum feature. This can be done by specifying X, Y and Z on the design drawing of the part. The letters x, y, and z designate linear degrees of freedom. The letters u, v and w designate rotational degrees of freedom and are not shown on the part drawing. See the "Means This" illustration for an explanation of what they control.

FIGURE 278 and 279 demonstrate this methodology. By simply putting an X, Y and Z on the part drawing (FIGURE 278), we are able to deduce which datum features stem all six degrees of spacial freedom (see FIGURES 279 and 280).

FIGURE 278 [Designating Datum Axes on the Part to Determine Degrees of Freedom]

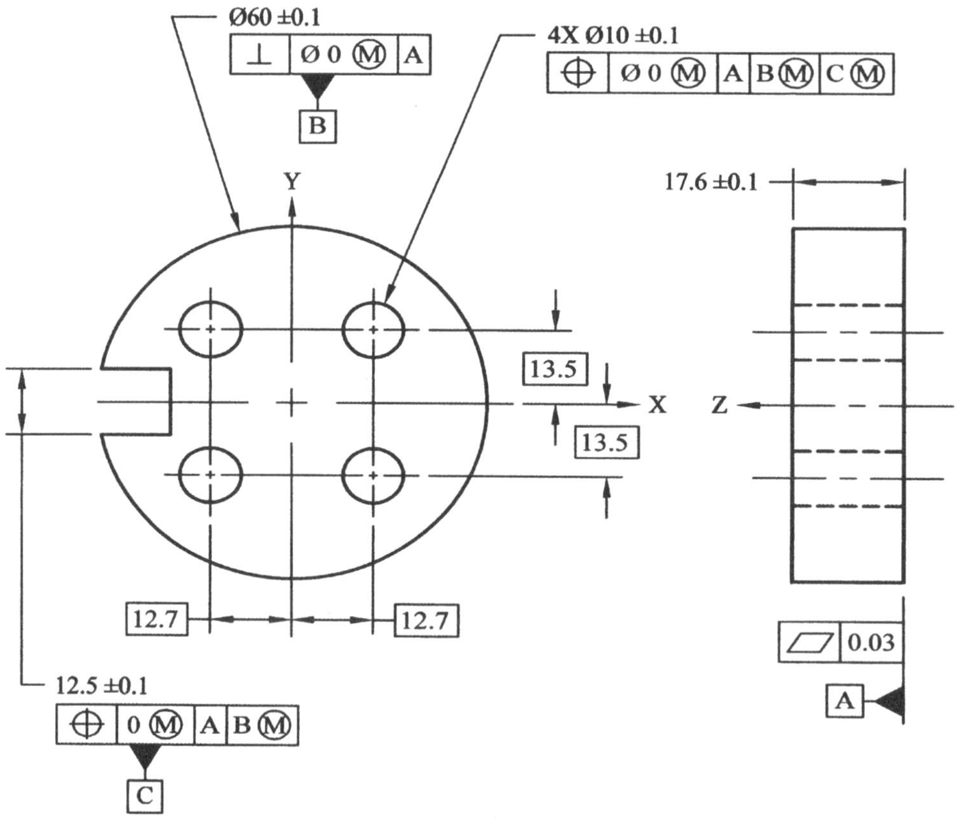

FIGURE 279 Means This: [Degrees of Spatial Freedom]

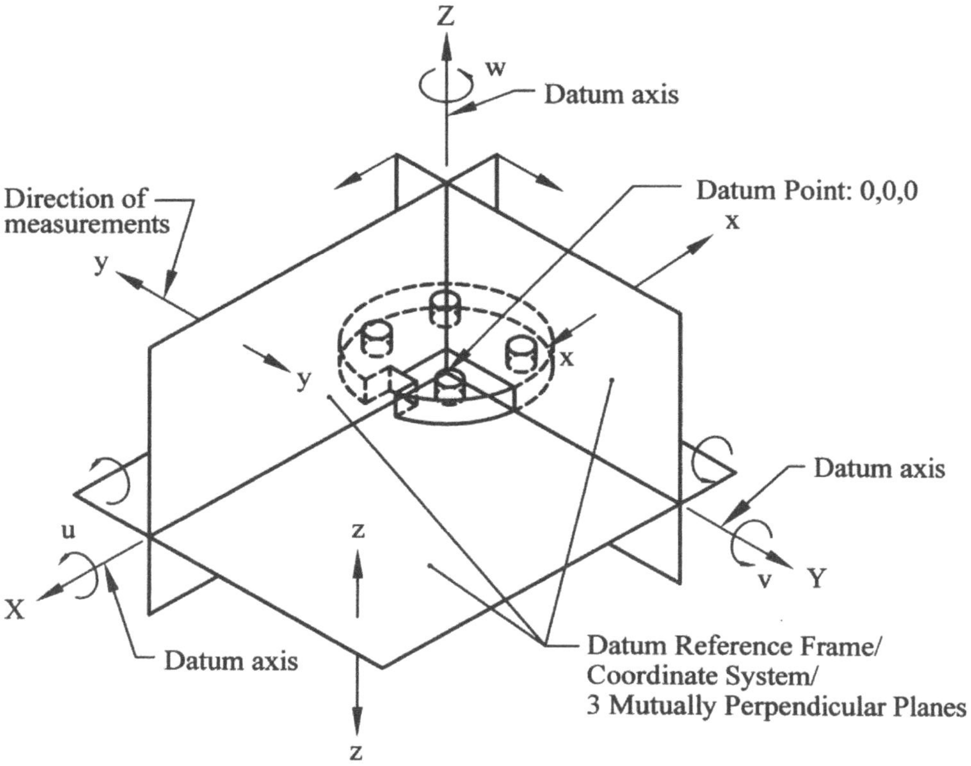

Figure 280-Explanation of Figure 279

<u>Degrees of Spacial Freedom</u>

Linear	Rotation
x = translation along the X axis	u = about the X axis
y = translation along the Y axis	v = about the Y axis
z = translation along the Z axis	w = about the Z axis

See Figures 293 and 314 for Examples of Releasing Degrees of Freedom on Parts from their Datum Structure.

Equalizing Datums

If a part has two opposing rounded features, they may be used to establish datums, such as a centerplane. In these instances, opposing pairs of datum targets, such as points or lines, are shown on the surfaces being used to establish the equalizing datum plane. If datum target points are used and coordinately dimensioned, equalizing pins with hemispherical tips may be used to establish the central datum. If the target points are designated by angles tangent to the feature, vee-type equalizers designed to touch the surface with opposing point contact may be used to establish the central datum.

If datum targets depicted by lines, but intended as point contacts, are used at a basic angle from one another or from the centerplane they hope to construct, then are designated for a height distance from another datum plane in another view, Vee block knife edge equalizers designed to contact only a point on each of these line elements may be used to establish the central datum. If a point on each line element at a specific height from a datum plane is not the desired contact, but rather an entire line contact for each line element is to be attempted, then Vee planes may be indicated by showing only the datum target lines at a basic angle from one another or from the central plane they hope to create, in the top view.

It should be noted that datum target points, lines or areas may be used to construct a central/equalizing plane on other part configurations as well as those with rounded ends. In the case of the use of equalizing datums in any functionally acceptable situation, it is all right to specify the datum feature on the centerplane or axis constructed by the equalizing datum targets. A note such as, "Datum axis A" may be used in place of the datum feature symbol. This, if necessary, for clarification, will help to identify the equalized theoretical datum axis or plane. One must remember that this technique is only to augment the use of the datum targets and should not be done without the targets, since this would make the drawing ambiguous as to how the datum axis or plane is constructed.

In order to assign basic dimensions to locate features, we must ask ourselves, "What have we established as the datums (not the datum features or datum targets)?" and assign the basic dimensions from those datums (planes, axes or points). For example, in the position of the three holes and slot in the part shown in FIGURE 282, datum targets are provided, and basic dimensions have been assigned as needed from datums to controlled features. **It should be noted that the practice shown in Figure 282 was essentially replaced by the introduction of the moveable datum target symbol (introduced in ASME Y14.5-2009) and the vector specification option. Please see pages 431-434 of this textbook for examples of those more current options.**

FIGURE 282 [Building a Datum Reference Frame with Equalizing Datums from Irregular Surfaces] Common Practice Prior to the Introduction Of the Moveable Datum Target Symbol in Y14.5-2009

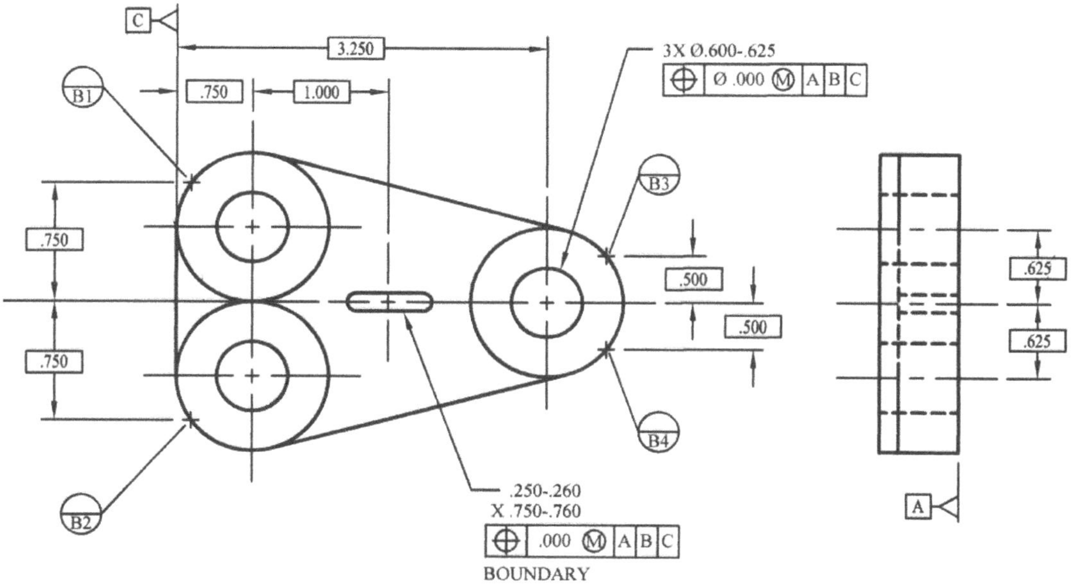

FIG. 283 [Datum Targets used to Create a Datum Axis]

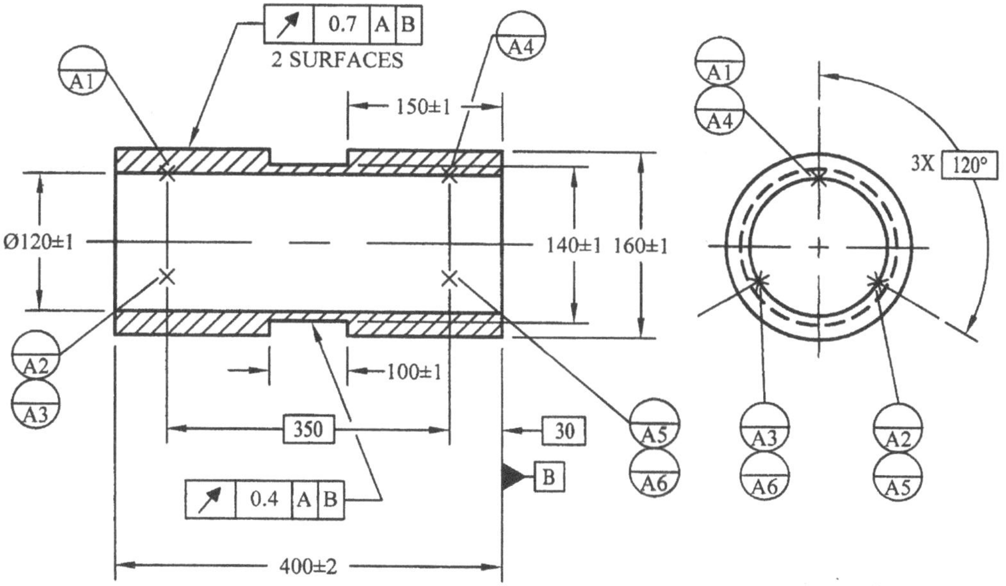

Another method of using equalizing datum targets is to create a datum axis. It is recommended that at least six different datum target points be used in situations similar to the example shown in FIGURE 283. In this example, three points are defined at one circular line element and three at another. These circular line elements should be separated by a distance sufficient to create maximum part stability without endangering or interfering with part functionality.

It is common to define the three datum target points on each circular line element as being 120° basic from one another, and the circular line elements a distance apart and a distance from a surface. These distances (shown in the example as 350mm apart and 30mm from datum B) may be specified with either basic or toleranced dimensions. If shown as basic (as in the example), they are toleranced by established tooling or gaging tolerances. If you are unaware that such tolerances have been established, you may wish to tolerance the dimensions on the field of the drawing or in the drawing notes. Since the 6 target points in this example establish a primary datum axis A, and therefore the location and angle at which the specified circular runout controls will be assessed, datum B is useful mainly as an origin of measurement from which to establish the datum target A point locations.

FIGURE 284 [Datum Targets to Control Pattern Rotation]

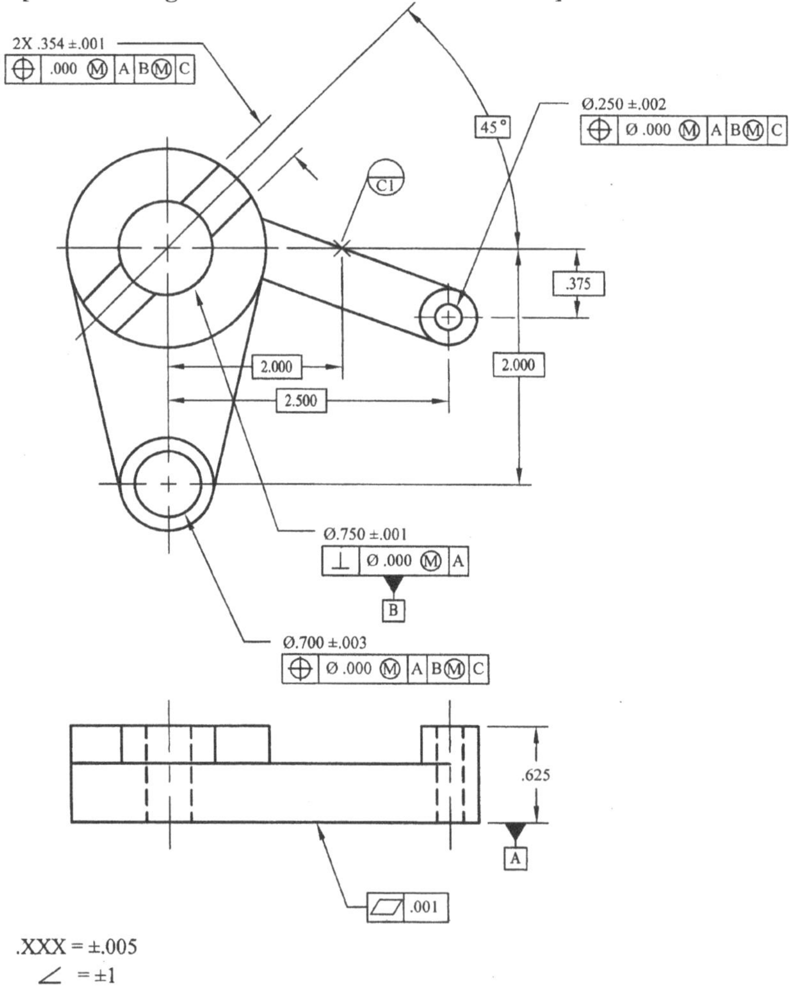

Establishing a Valid Datum Plane and New Options for Use as Datum Features

The problems faced in complying with datum establishment procedures are complicated by imperfect datum surfaces that rock or wobble when contacted at the high points. This creates a candidate set of planes or axes from which to choose and from which to measure. Where a nominally flat surface is specified as a datum feature, the corresponding datum is simulated by a plane contacting the high points of that surface. If irregularities on the surface of a primary or secondary datum feature are such that the part is unstable (that is, it wobbles) when brought into contact with the corresponding surface of a fixture, the part may be adjusted to an optimum position, if necessary, to simulate the datum.

Unfortunately, the statement "may be adjusted to an optimum position" is so ambiguous it is difficult to understand. Should one merely rock the surface one way or another and apply pressure until it stabilizes? Or should the surface be shimmed up until it no longer wobbles or rocks? If it is to be shimmed, is one to use shims of equal heights on each end, shim up only one end, or shim to create an equal angle (if the rock point is not in the middle of the part)?

If the irregularities of a datum feature surface are such that the part is unstable (that is, it rocks) when brought into contact with the corresponding true geometric counterpart, the default requirement is that the part be adjusted to a single solution that minimizes the separation between the feature and the true geometric counterpart per ASME Y14.5.1. If a different procedure is desired (candidate datum set, Chebychev, least squares, translational least squares, etc.), it must be specified.

The ASME Y14.5.1 committee (Mathematical Definition of Dimensioning and Tolerancing Principles) has considered this problem and issued clarifying statements. These statements guide us as to what we can do when a rocking datum feature (primary) or set of datum features (primary and secondary) could generate more than one possible datum plane or axis each, and/or multiple datum reference frames which may be established for a single feature control frame. Another way of looking at it is that, if we don't follow these rules and guidelines, the part may move in its datum reference frame. Without the specificity given in Y14.5 and Y14.5.1 and any additional clarification given by the design team in notes and measurement plans, the datum reference frame may by viewed as a candidate datum reference frame set from which one must choose to measure.

Due to these rules and formulas per Y14.5.1, Y14.5-2009 and Y14.5-2018, insight to compliance with "may be adjusted to an optimum position" is much better defined and, as such, much less ambiguous, and constructing a valid datum plane or axis is easier. In Y14.5.1-1994, one such recommendation is that (in simplified terms) the primary datum feature be divided into thirds in every possible viewing angle, and any candidate datum plane may not have all of the high-point contact points residing solely in one of the outer thirds of the surface. This type of clarification on constructing a valid datum plane makes for a more complete explanation.

FIGURE 285 [Simulating a Datum Feature]

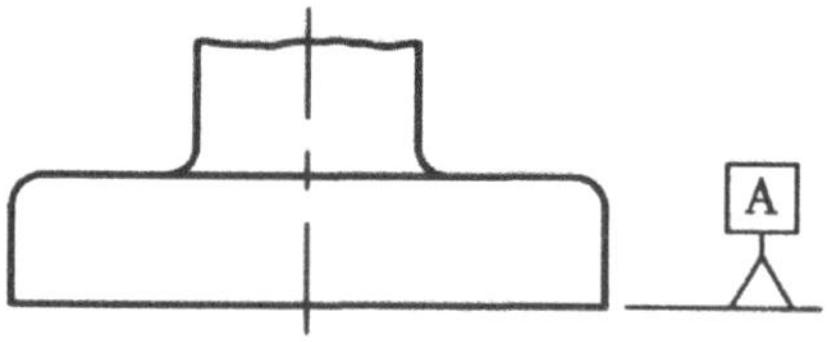

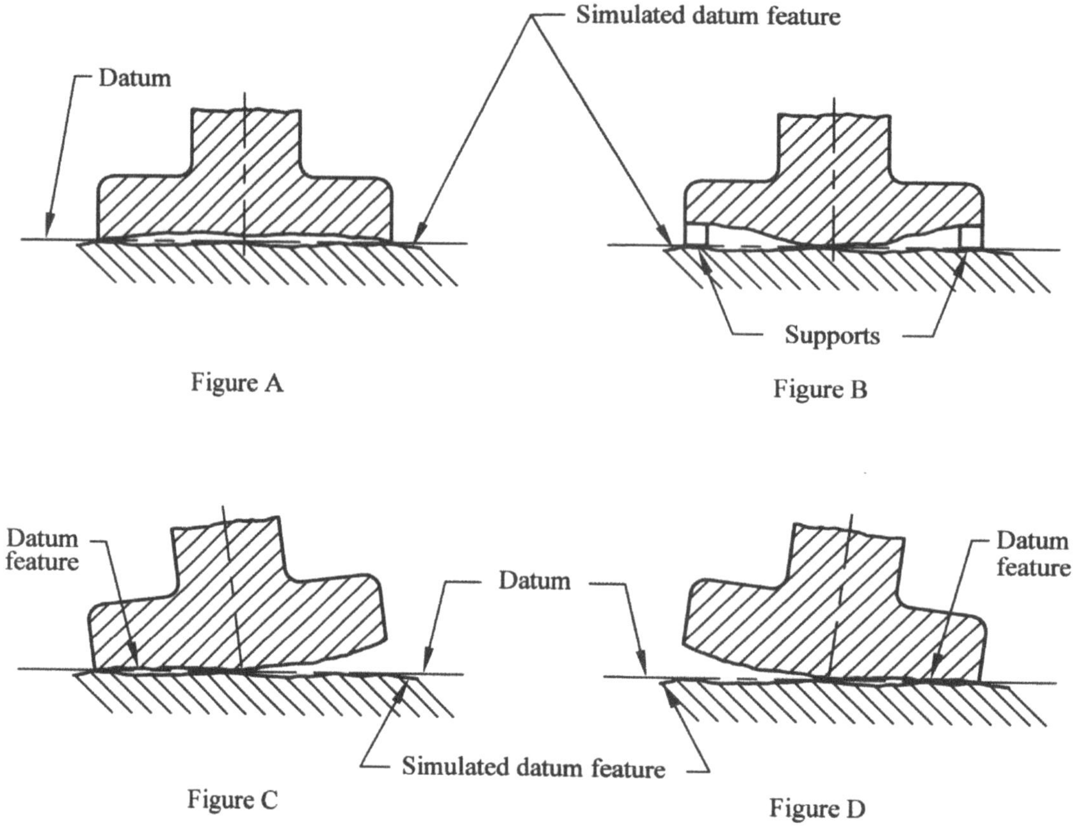

The Figure A portion of FIGURE 285 shows a datum feature that does not rock. If a primary datum feature, it merely rests on the 3 highest points of contact on, for example, a surface plate. This constitutes a valid simulated datum feature as long as all 3 high points of surface contact do not reside in any **one** of the outermost thirds of the datum feature. A more detailed explanation of this restriction is given in the text of this unit.

The Figure B portion of FIGURE 285 shows a datum feature with a rock point toward the middle of the surface. Adjusting the part to an optimum position, in this case, uses supports to stop the part from rocking. The supports may be shims. The inspector may decide the optimum position equalizes either the angle or the shim height or neither, so long as the part is stabilized and functionality of features measured from the simulated datum feature preserved.

The Figure C and D portions of FIGURE 285 show parts that, instead of using supports to stabilize the part (as in Figure B), are just rocked one side or the other. Again, provided this method uses high points on the surface that do not solely reside in any one of the outer thirds of the feature (as described in more detail in the text of this unit) and preserves part function, it is acceptable to simulate and stabilize the datum feature. **However, a new revision of how to establish the optimal datum using a math formula can be found in the revised Y14.5.1 (math) standard, which has not been released yet at the time of this writing. It is believed that it will be available sometime in the year 2020. Software will probably be required to do this.**

FIGURE 286 [Constructing a Valid Primary Datum Plane]

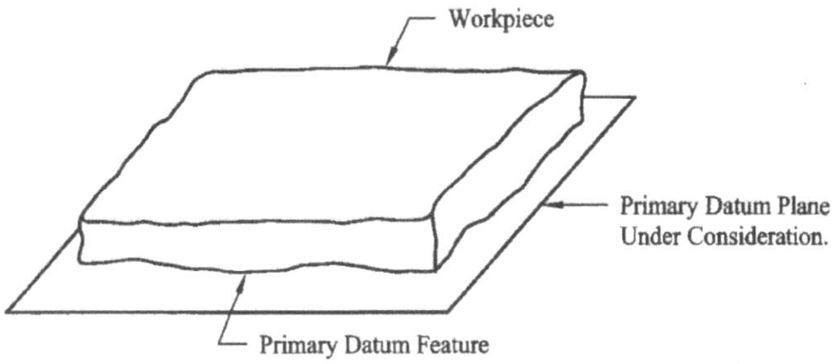

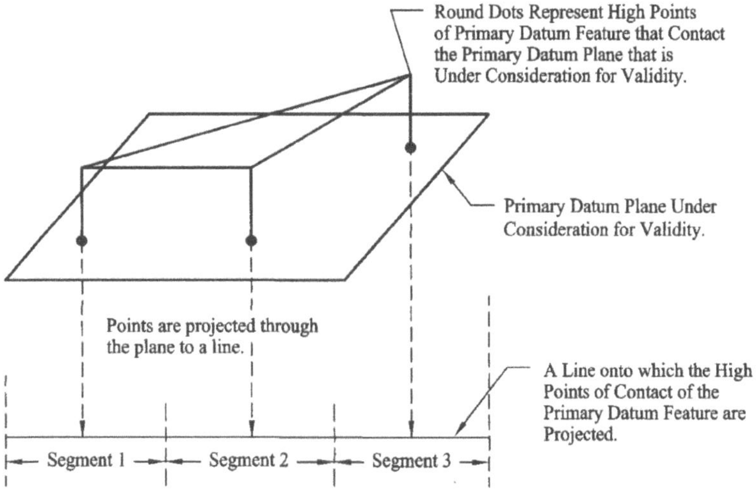

Sections 1, 2 and 3 are each equal to one-third of the line onto which the high points of the primary datum feature that contacts the datum plane are projected. The datum feature must be rotated to all possible orientations while these 3 high points continue to contact the datum plane. The points are projected onto the line in each orientation. All 3 points, once projected onto the line, may not lie solely in either section 1 or section 3 to qualify as the 3 high points to be used to construct a valid primary datum plane.

FIGURE 287 [Other Possible Orientations per Y14.5.1-1994]

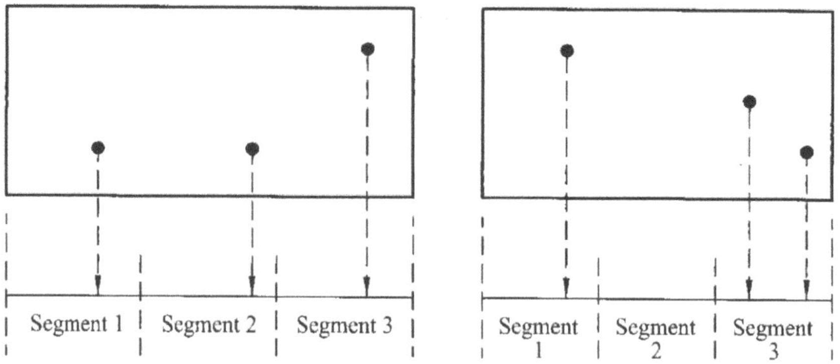

To understand the description given here, one must visualize the three chosen high points as resting on a perfectly flat plane. The surface of the part is examined from all aspects as though one is walking around the plane. The plane may be pictured as a surface plate seated parallel to the horizon at eye level.

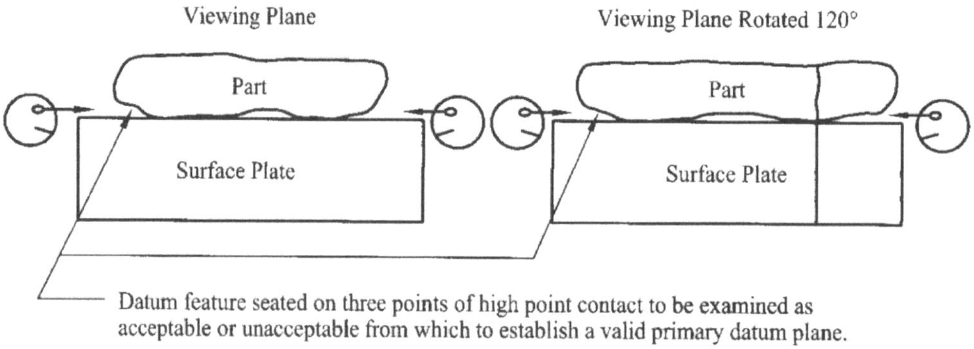

To be considered a valid primary datum plane, the surface high points that construct the plane are projected onto a line. The line which is the length of the surface in that viewing plane, is divided into three segments. The points that have been projected onto the line may not solely reside in either segment 1 or segment 3. The points may solely lie in segment 2. If all high points that construct the datum plane occupy two or more of the segments of the line they are projected onto, the datum plane is a valid datum plane candidate.

Once the high points under consideration are projected onto the projection line, their position on the line is examined. If all reside in section 1, the 3 points are rejected as being able to construct a valid primary datum plane. Likewise, should all points as projected reside only in section 3 of the projection line, these 3 points are rejected. Only if all points are projected into section 2, or if the points are spread out in such a way as they reside in more than one of the three segments of the projection line, will they be accepted as constructing a valid primary datum plane.

It is possible that a surface may rock to more than one orientation and create a situation wherein these different rocked orientations create more than one set of high points that fulfill the requirements of a valid datum plane (whether primary or even secondary). In these situations, one set of high points of contact is chosen as optimum from the candidate sets under consideration. Once this is done, the same set must be used for all set-ups that reference that datum and are part of a simultaneous requirement. Using the same set of points to construct the simulated datum feature and, consequently, the origin of measurement for all features that function in the same orientation is important. If this is done, one can be more assured that the features that pass inspection while measured simultaneously from the same datums (as simulated) will function as one pattern simultaneously.

As can be seen from the following illustrations, the order in which datum features are specified in a feature control frame can have considerable effect on part orientation in the datum reference frame. Consequently, the order of datum features has an effect on whether a part is acceptable or not to inspectors and, also, on how--or even if--the part will assemble.

FIGURE 288 [Effects of Differing Datum Precedence on Part Acceptance]

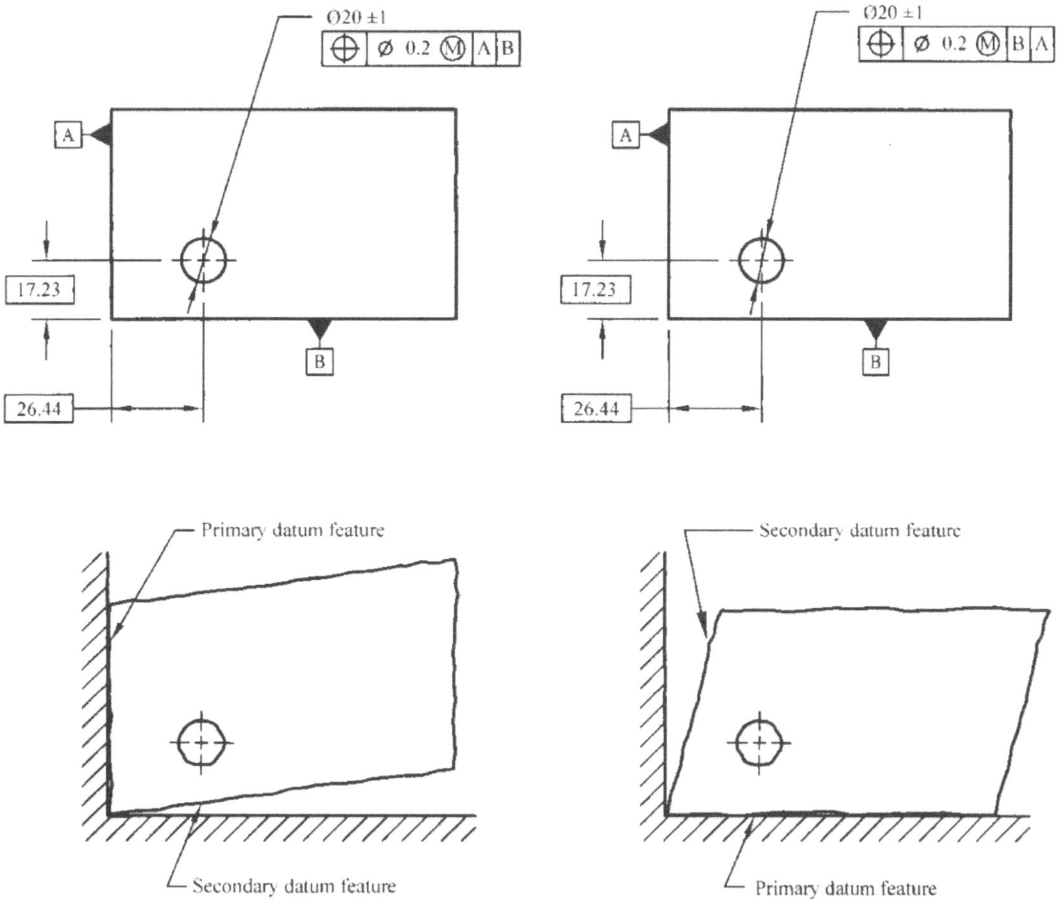

Although datum features are drawn at specific angles to one another, such as 90°, they are seldom, if ever, produced at exactly that angle. The primary datum feature will (generally speaking) have its surface elements in closer proximity to the simulated datum feature (angle plate or surface plate) in that the primary datum feature will be constructed by 3 points of high point contact and the secondary by only 2.

FIGURE 289 [Curved Surface as a Datum Feature]

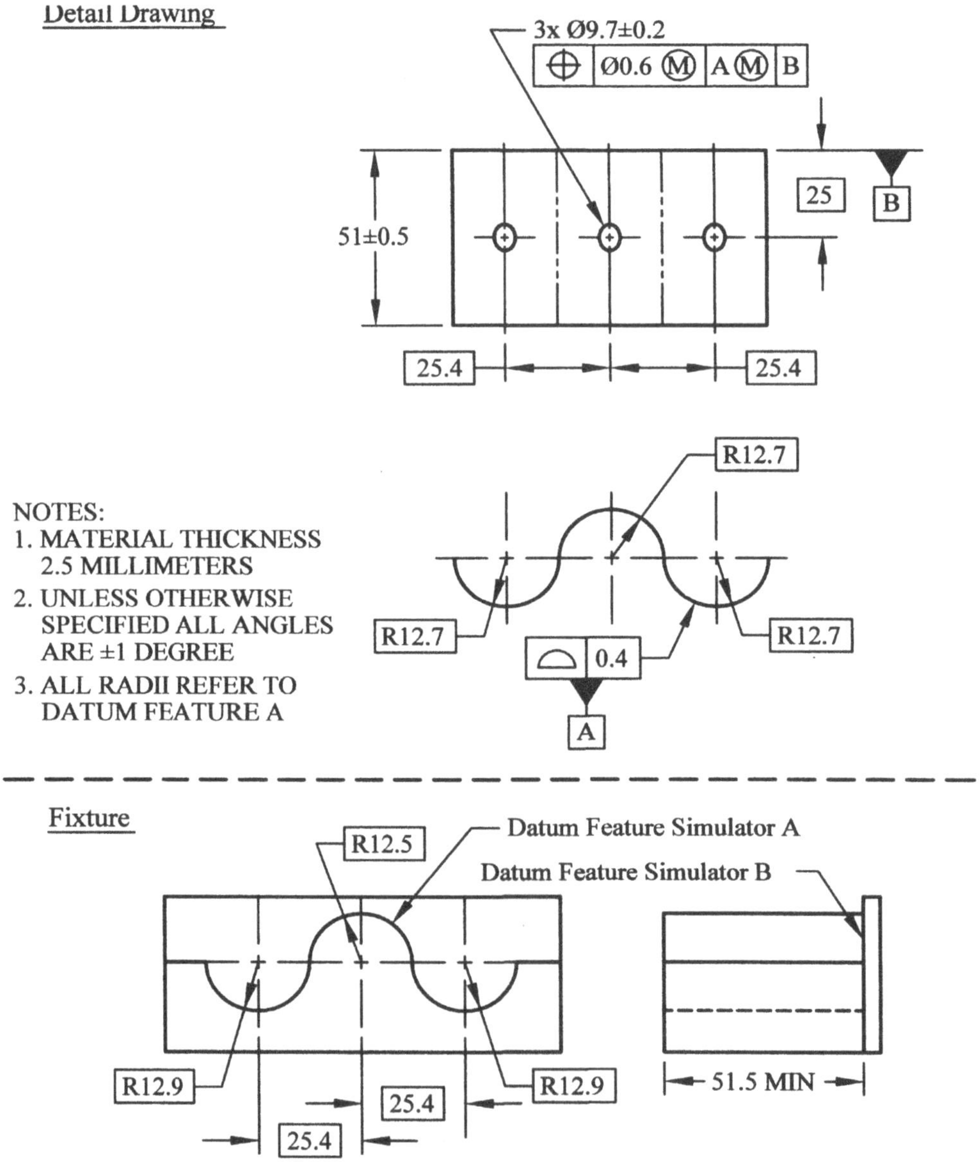

A curved surface may be used as a datum feature. These complex datum features stem more spacial degrees of freedom than either planar datum features or cylindrical datum features. For example, the curved surface in FIGURE 290 depicted as datum feature A stabilizes 5 of the 6 degrees of spacial freedom. Three rotational and two linear degrees of freedom are eliminated by datum A. This means that it generates 2 datum planes (the primary and secondary). B is the secondary datum feature but creates the tertiary datum plane. It, therefore, receives a minimum of 1 high point of contact from datum feature B. B is a location datum. For an example of a closed loop, oddly-configured hole as a datum feature, see the illustration in the section of this textbook on Functional Gages and Fixtures.

FIGURE 290 [Conical Datum Feature]

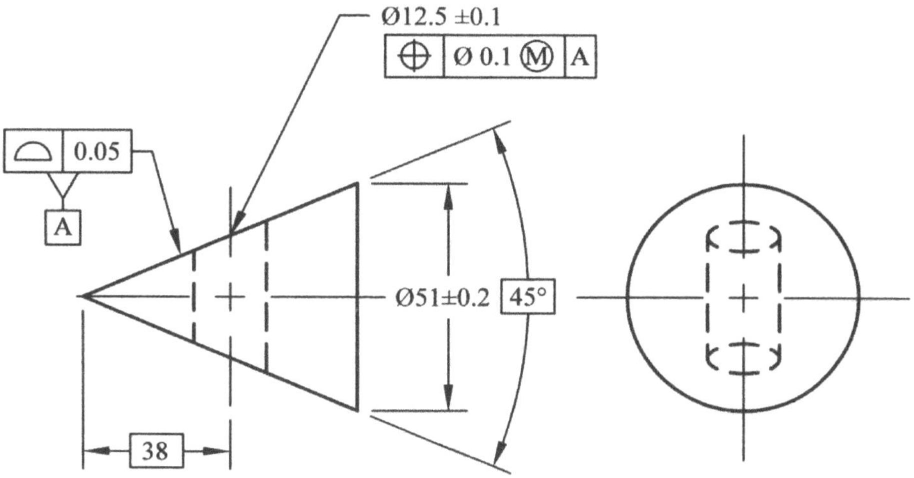

Means this: Datum feature A generates an axis (which consists of two planes intersecting at a 90 degree angle) and (from a point) a datum plane. This means that conical datum feature A generates three datum planes. The 12.5 hole's axis must intersect datum axis A and be 38mm from the tertiary datum plane also constructed by datum feature A to within the hole's position tolerance of a diameter of 0.1 at MMC.

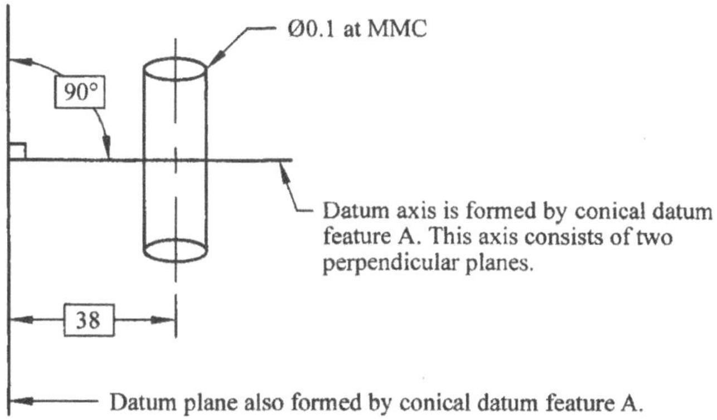

FIGURE 291 [Inclined Datum Features]

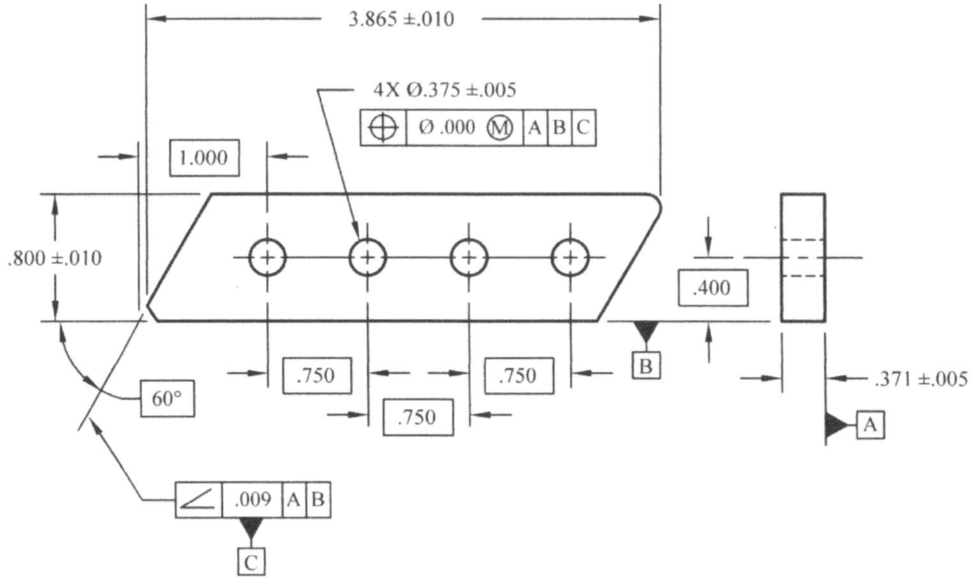

Since the tertiary datum feature is at an angle other than 90 degrees from the primary and secondary datum features, it is important that this angle is shown on the drawing as a basic dimension. That way, the tertiary datum plane can be simulated at 90 degrees to the primary and secondary datum planes through the use of an appropriately-angled wedge block. Measurements are taken from the datum planes once properly established.

MEANS THIS

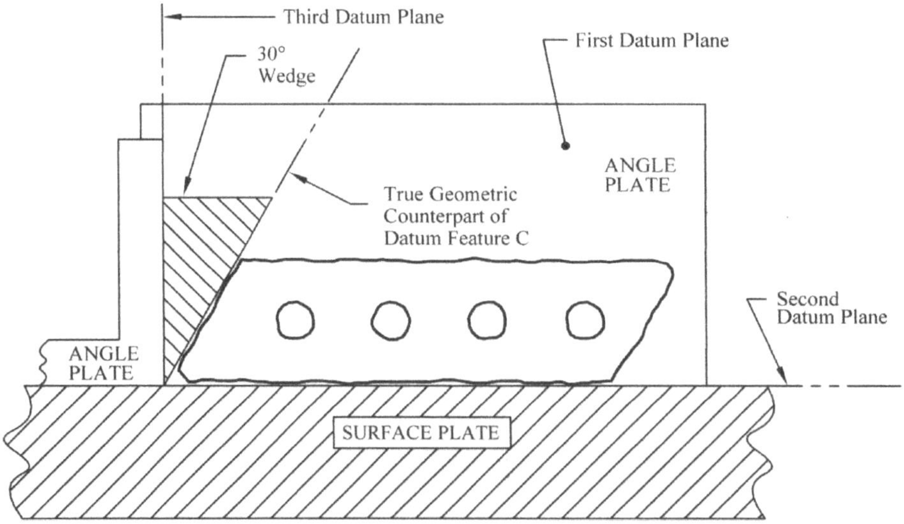

Although datum planes within a datum reference frame are, by definition, mutually perpendicular, the features from which these planes are established do not have to be mutually perpendicular.

Datum Features and Datum Feature Categories and What They Mean

In the ASME Y14.5-2018 standard, the following examples show a few of more complicated datum features and datum feature types that have been exhibited since ASME Y14.5-2009.

FIGURE 292 [Cones, Constant Cross-Sections (Linear Extruded Shapes) and Complex Datum Features]

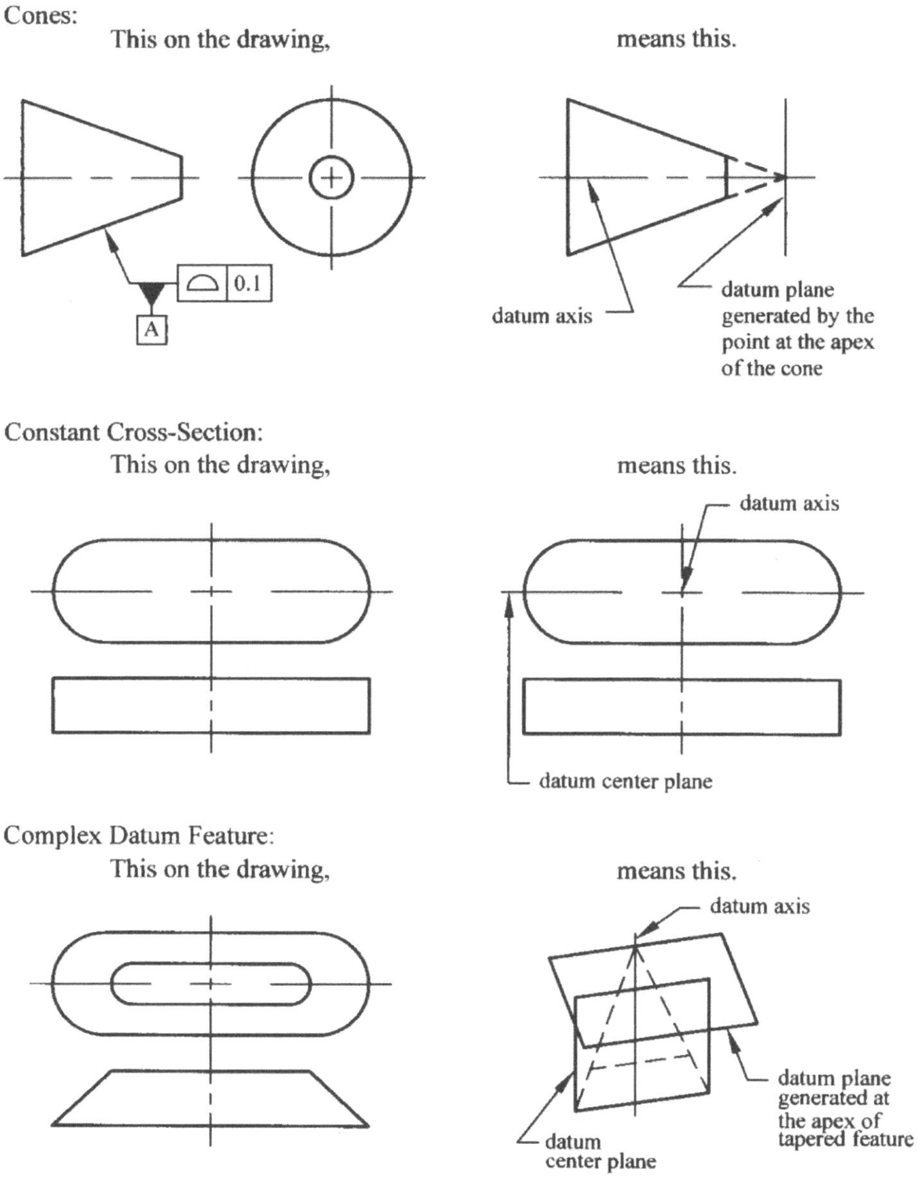

Datums Established from Complex or Irregular Surfaces

Datum targets may be used to establish a datum from a complex or irregular surface. Where a datum reference frame has been properly established but its planes are unclear, the datum reference frame coordinate axes may be labeled to appropriate extension or center lines as needed. The datum feature symbol should be attached only to identifiable datum features. Where datums are established by targets on complex or irregular surfaces, the datum may be identified by a note such as: DATUM AXIS A or DATUM PLANE A.

Datum Features Established from Datum Targets with Fewer Than Three Mutually Perpendicular Planes

When using datum features that are defined by datum targets in a feature control frame established by fewer than three mutually perpendicular planes, the datums that are the basis for the feature control frame shall be referenced. The targets that provide location for the datums referenced in the feature control frame shall be specified in a note, such as:
DATUM FEATURES B AND C ARE USED WHERE ONLY DATUM FEATURE A IS REFERENCED TO RELATE THE PART TO THE TARGETS THAT CREATE DATUM A

Specifying Degrees of Freedom in the Feature Control Frame

ASME Y14.5-2009, for the first time in any Y14.5 standard, allowed us to state which degrees of spatial freedom each datum feature is to control. This is done by putting the letters of the degrees of freedom to be stemmed by that datum feature in the feature control frame in brackets next to the letter representing the datum feature reference. On the part drawing, X,Y and Z are shown as upper case letters representing the X,Y and Z axes. The letters are shown in the feature control frame in lower case as x,y,z for linear degrees of freedom and u,v,w for rotational degrees of freedom. This allows the user to override the order of the datum feature references and the degrees of freedom they would normally control. Whereas, if the degrees of freedom to be constrained are not specifically designated for the datum features that are referenced in the feature control frame, the datum features arrest any degree of freedom that they are capable of arresting. The datum feature listed is assessed for the datum it creates and the degree or degrees of freedom it stems. If a datum feature is referenced as secondary or tertiary in the feature control frame, it may not constrain degrees of freedom already constrained by the datum feature references that precede it in the control. However, if it is desired to override the control of a datum feature that precedes it in the feature control frame, it may be done by listing the degree or degrees of freedom specifically constrained by each datum feature (and the datum it constructs), For an example of how to use this **option**, please see the following illustration.

FIGURE 293 [Specifying or Overriding the Degrees of Spatial Freedom]

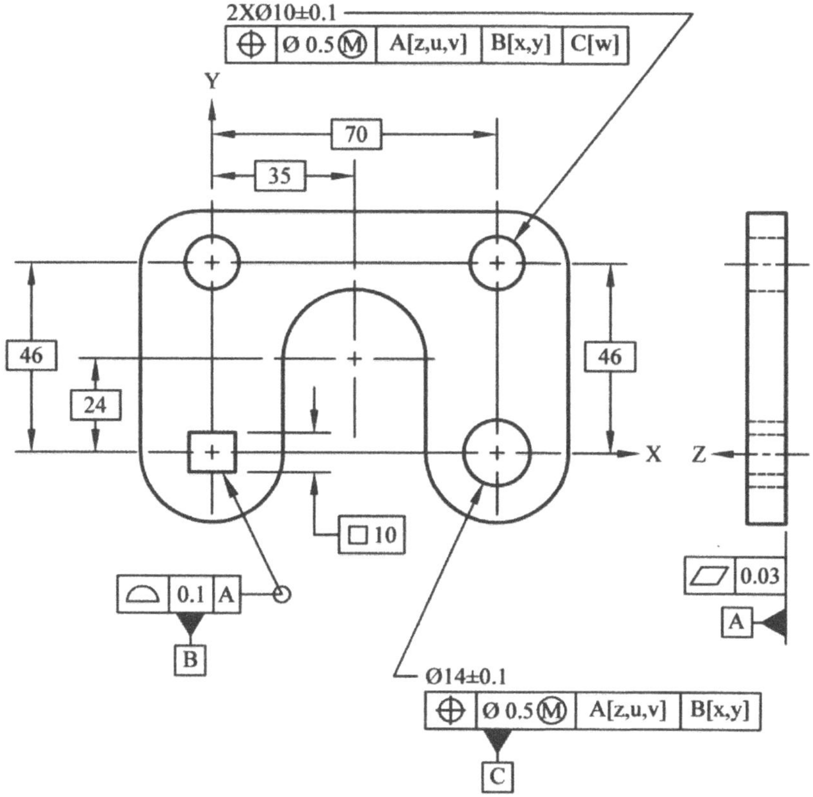

In FIGURE 293, the position control assigned to the two-hole pattern states that the holes must be:

1) Positioned to within a diameter of 0.5 at maximum material condition
2) to datum A for perpendicularity (with constraints in the linear z direction and rotational constraints u and v). Note: This is exactly what datum feature A and the plane that is constructed from it would commonly control. If the part is placed on a surface plate or machine table onto surface A and kept there, the part could not move up and down in the z direction or rotate in the u or v directions.
3) and to B for location in the x and y linear directions
4) and to C to stop rotation in the w direction.

Normally, because datum feature B is a rectangular hole, it would stop rotation in the direction of w. If we picture a rectangular gage pin being inserted into datum feature B, this w rotational degree of freedom would clearly be constrained. In this instance, however, the reader is asked to ignore that physical reality and treat datum feature B as generating a center, like an axis, instead of the two center planes that it actually generates. If datum feature B generates only a center, as though it was a round hole, instead of a rectangular hole, the rotational degree of freedom would not be constrained by B. This allows C to stop rotation around the center of B in the direction of w.

This option of "releasing degrees of spatial freedom" may be used at the discretion of the tolerancing specialist for unusual circumstances. It should not, however, be used in lieu of correct datum feature selection and assignment.

Note: Although the letters u, v and w are allowed in the feature control frames, they are not otherwise shown on the part design drawing. Just be aware that u is the rotational degree of spatial freedom about the X axis, v is the rotational degree of spatial freedom about the Y axis and w is the rotational degree of spatial freedom about the Z axis.

Multiple Datum Reference Frame Identification

Where multiple datum reference frames exist, and it is desirable to label the axes (X, Y and Z), any labeled axes shall include a reference to the associated datum reference frame. In the following figure the X, Y, and Z axes for the three datum reference frames are identified by the notation [A, B, C] , [A.B,D], and [A,B,E]. These labels represent the datum features (without modifiers) for each datum reference frame, and follow the X, Y, and Z identification letters.

FIGURE 294 [Multiple Datum Reference Frame Identification]

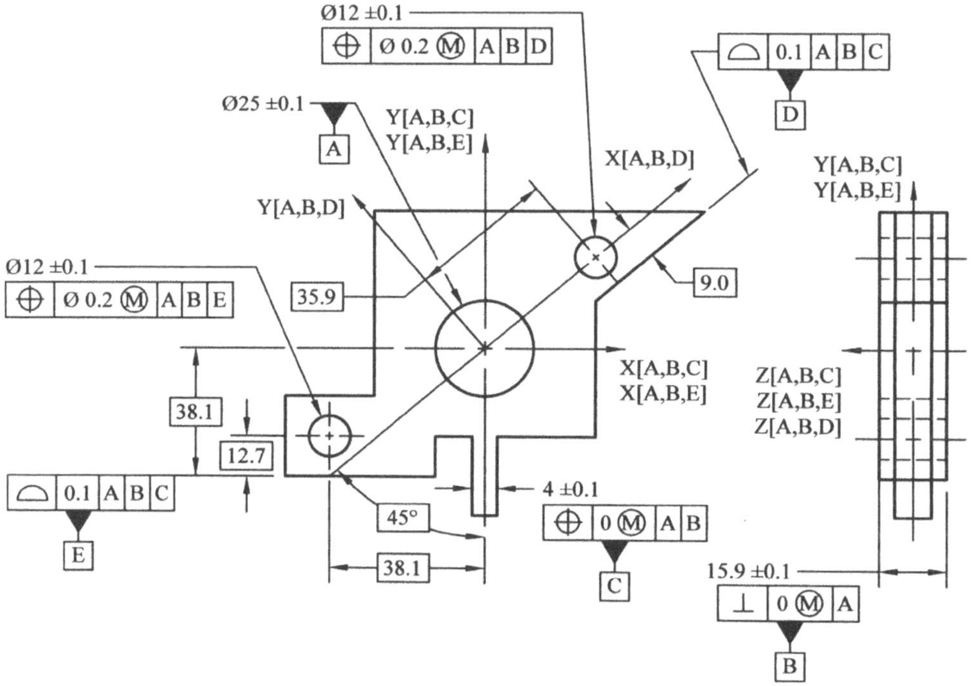

NOTE 1: UNLESS OTHERWISE SPECIFIED
ALL SURFACES [profile of a surface 0.5 A B C]

NOTE 2: FOR A COMPLETE DEFINITION
OF THIS PART SEE THE CAD MODEL

Specifying the Desired Datum Material Boundary with a Number

Where the boundary to be represented may be unclear to those interpreting a requirement, or a boundary is desired to be different than the correct drawing interpretation would give, a number representing the size of the desired boundary may be shown after the datum feature in the feature control frame. When this is done, the size of the boundary is displayed in brackets. See the figure below for an example of the boundary being specified where the correct maximum material boundary may be unclear. This option was first allowed by the ASME Y14.5-2009 standard.

FIGURE 295 [Correct Maximum Material Boundary Specified]

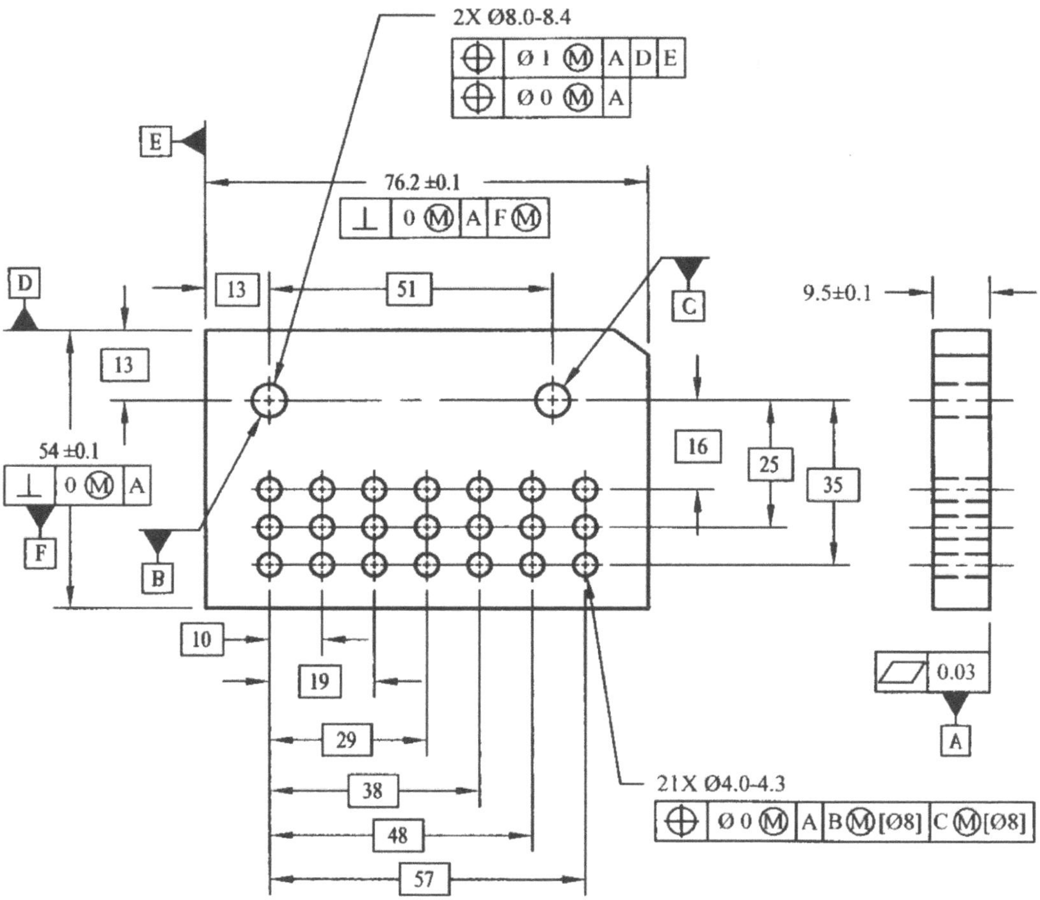

However, it is usually not necessary to specify the correct boundary if one knows how to calculate boundaries. The following illustrations show methods for calculating the correct datum feature boundary size.

FIGURE 296 [Correct Maximum Material Boundary Calculations]

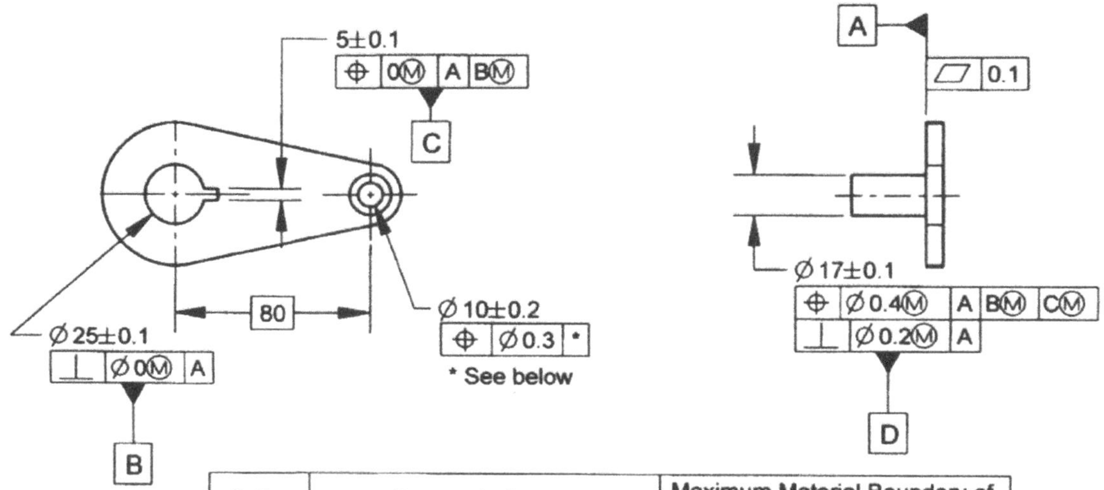

Option	Geometric Control	Maximum Material Boundary of datum feature D
(a)	⌖ Ø0.3 DⓂ	17.1
(b)	⌖ Ø0.3 A DⓂ	17.1+0.2 = 17.3
(c)	⌖ Ø0.3 A BⓂ DⓂ	17.1+0.4 = 17.5

FIGURE 297 [Correct Least Material Boundary Calculations]

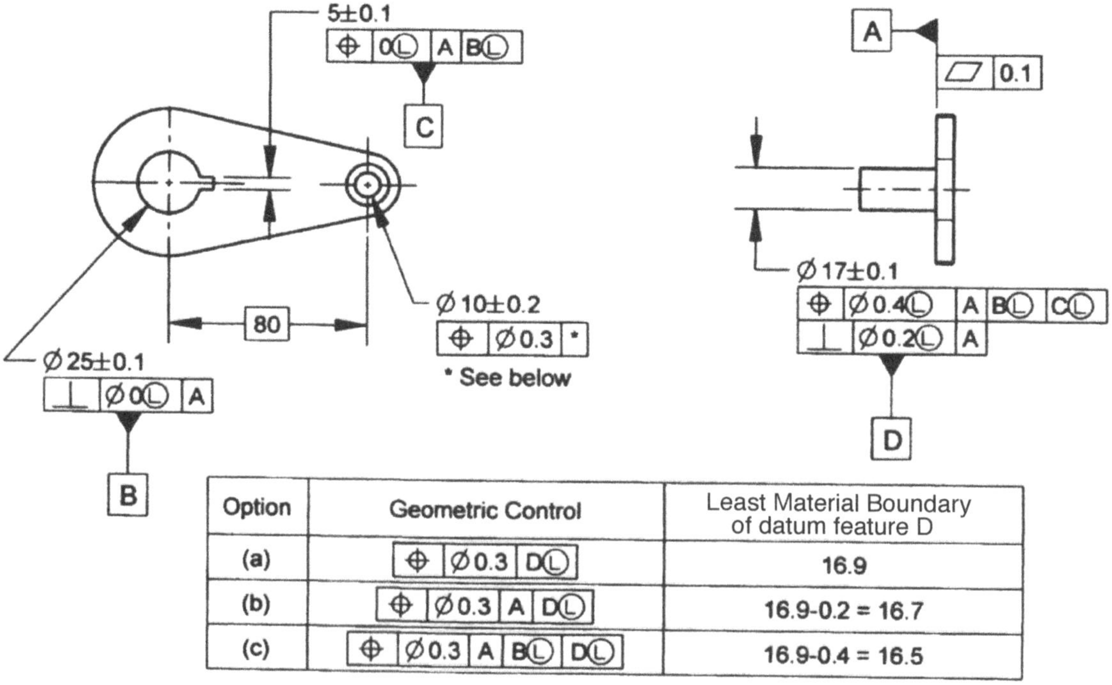

Option	Geometric Control	Least Material Boundary of datum feature D
(a)	⌖ Ø0.3 DⓁ	16.9
(b)	⌖ Ø0.3 A DⓁ	16.9-0.2 = 16.7
(c)	⌖ Ø0.3 A BⓁ DⓁ	16.9-0.4 = 16.5

Using the Translation Modifier

Prior to the publication of ASME Y14.5-2009, there was some disagreement as to whether angular orientation datum features were to be represented by simulators in gages, fixtures and software as stationary or moveable. A decision was made by the Y14.5 committee to make a new rule that unless otherwise specified, all datum feature simulators were stationary at their basic angle and location to the datum feature simulators that are referenced before them in the feature control frame. This rule had far reaching ramifications and even blurred the meaning of the datum feature sequence. For example, in FIGURE 298, if no translation modifier was used after datum feature C, the gage would be the same, whether B was used as secondary before C or tertiary after C. Also, without the translation modifier after C, if the primary datum feature is a planar surface, as is datum feature A in FIGURE 298, and the secondary datum features are holes, like B and C in that figure, let's consider the gage. B and C are simulated in a gage or fixture at a diameter of 8 millimeters and are located a basic 51 millimeters from one another. They are identical. If they are identical, which of them controls location (the traditional job of B in such a control), and which one controls angular orientation/pattern rotation (the traditional job of C)? In fact, in such an instance, there would be no difference in the gage to represent B and C as datum features if, instead of using B and C as secondary and tertiary datum features, B and C were used as a pattern datum (both holes called B and referenced as Bμ). Likewise, if the holes were used in compound as B at MMB dash C at MMB, they would be simulated in gages and fixtures the exact same way, with two 8 millimeter gage pins that were 51 millimeters apart.

Since this Y14.5-2009 rule blurs the importance of datum referencing, a way had to be found to make a distinction between the datum features, when a distinction was desired. So, as FIGURE 298 shows, the translation modifier used after datum feature C states that while B is simulated by a stationary pin, C is simulated by a moveable pin. That means that B controls location and C controls rotation of the pattern of 21 holes and the rotation of the profile. See gage in Figure 299.

FIGURE 298 [Translation Modifier]

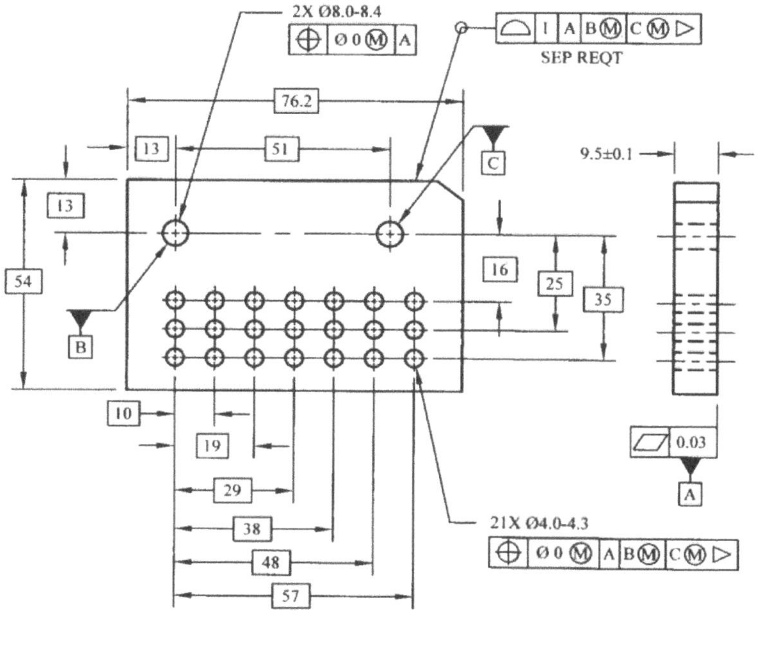

Figure 299: Gage with Translation Simulator.
The gage to simulate datum features B and C and to gage the 21-hole pattern's position tolerance is shown below.

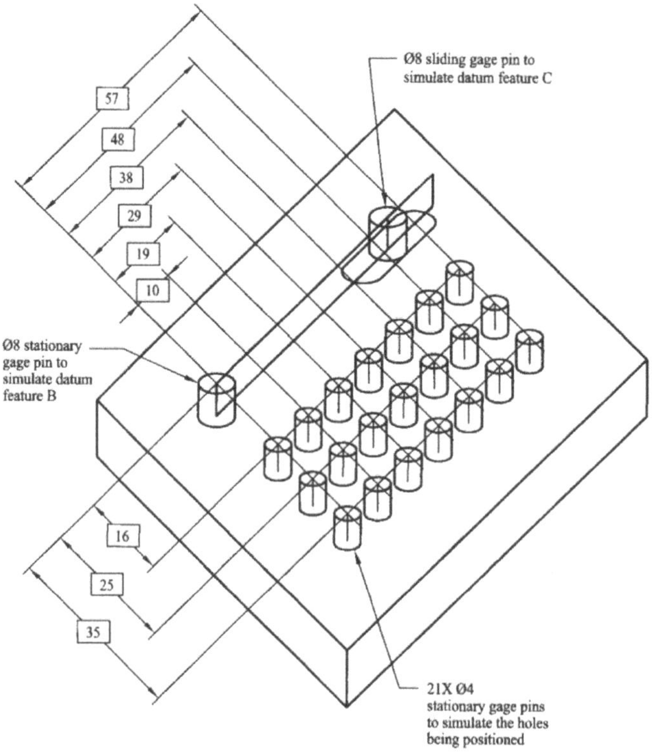

Using the Term "BASIC" or its Abbreviation "BSC" in a Feature Control Frame

Datum features are simulated at their basic locations, but at their appropriate boundary size. The Y14.5-2009 and the Y14.5-2018 standards allow datum features that have profile tolerances on them to be represented at their basic location from datums that are referenced before in the feature control frame. This distance can be viewed as a size to help us understand the options. In FIGURE 300, datum feature C is referenced in the position control for the 3 holes. If simulated as regardless of material boundary (RMB), the datum feature C simulator would grow from its basic dimension until it touches datum feature C on the part. If datum feature C had been referenced at MMB due to C being shown in the feature control frame as C with a circled M after it, the simulator would have been made at 6.4 from the axis of B. Instead, due to the term [BASIC] used after C in the feature control frame, datum feature simulator C is represented at a size/distance from datum axis B of 6.3 (the basic dimension).

FIGURE 300 [Using BASIC after a datum feature reference]

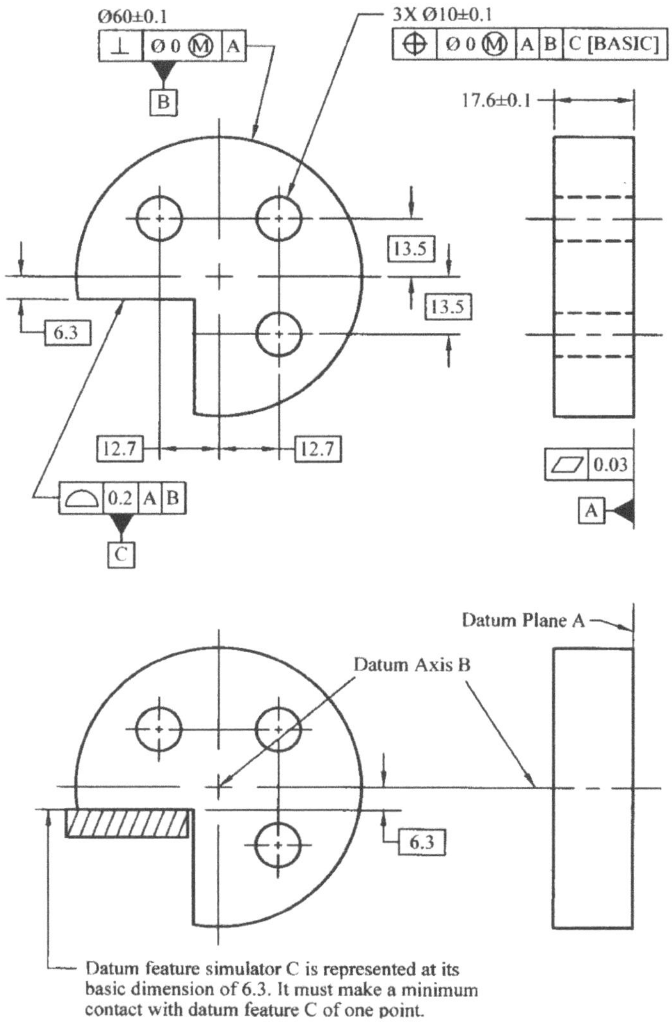

The abbreviation for BASIC may be used instead. The abbreviation is BSC. See FIGURE 301.

FIGURE 301 [BSC instead of BASIC]

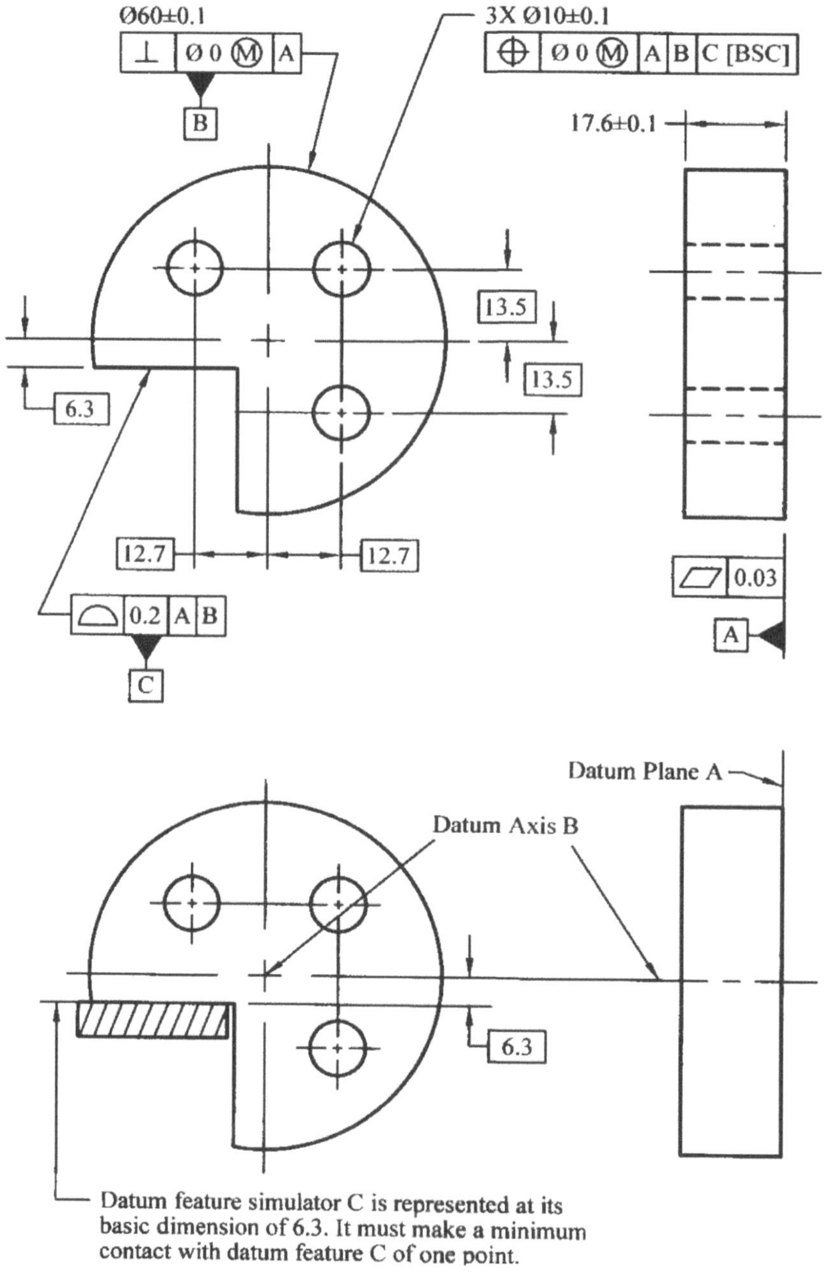

Datum feature referenced (implied) at regardless of material boundary (RMB)
When datum feature C in the previous illustration is referenced without a material boundary symbol, translation symbol or the term BASIC, the regardless of material boundary (RMB) concept is implied. Datum feature simulator would then be represented as shown in FIGURE 302.

FIGURE 302 [Datum feature simulated at RMB]

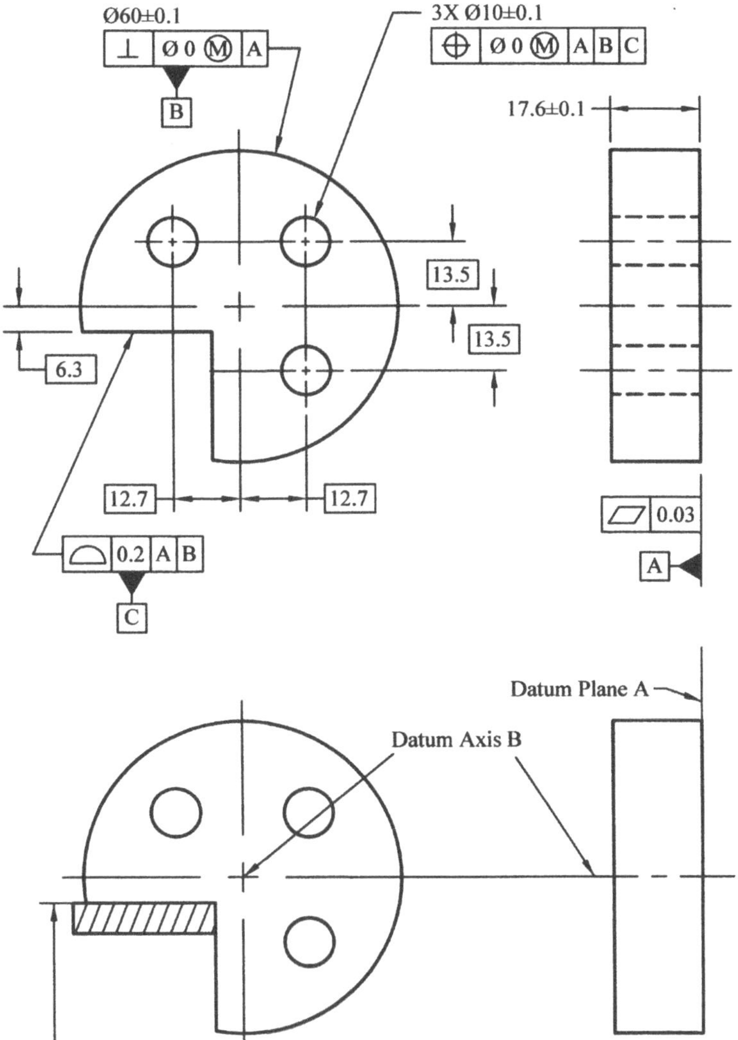

Datum feature simulator C progresses normal from the MMB until it makes maximum contact with datum feature C. No shift or rotation of the datum feature is allowed.

Datum features referenced at maximum material boundary (MMB)

When a datum feature is referenced in a feature control frame at MMB, such as C is shown in FIGURE 303, the datum feature simulator is represented at the location that is calculated by taking the basic dimension and either adding or subtracting the applicable geometric tolerance. In this case, the surface used as datum feature C has been controlled with an equal bilateral profile tolerance of 0.2 (±0.1), so C is simulated at the MMB that is calculated by adding the basic dimension of 6.3 and the growth allowed by the profile tolerance of 0.1. The MMB of surface C is 6.4. See FIGURE 303.

FIGURE 303 [Datum features referenced at MMB]

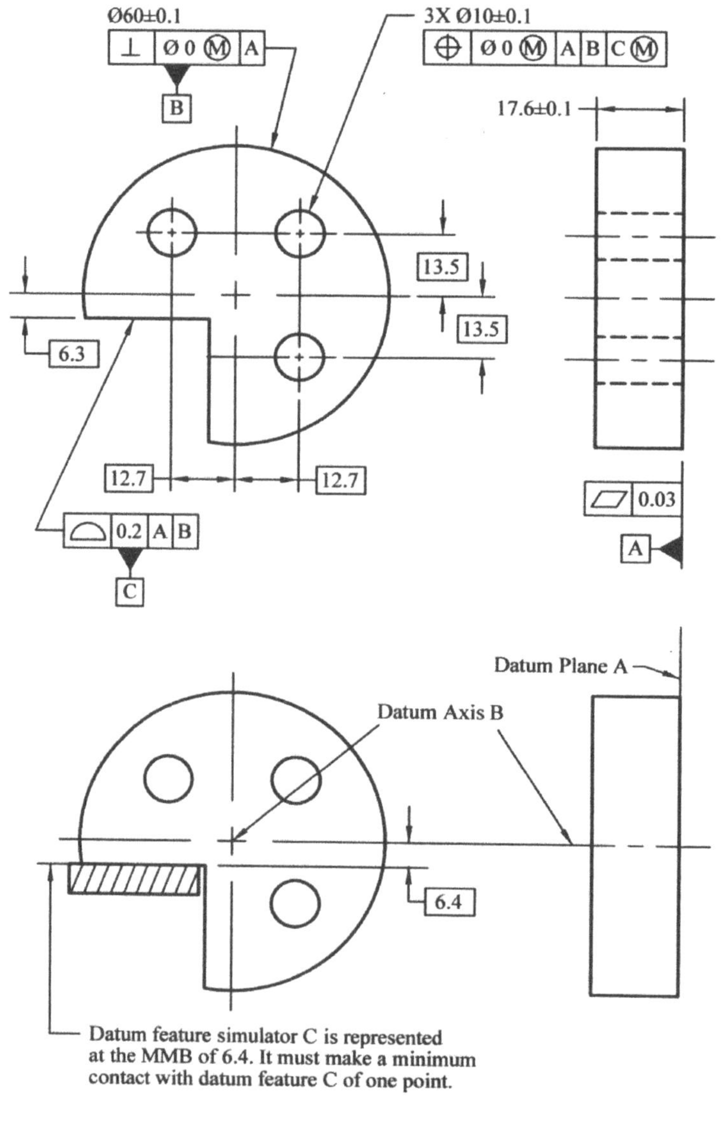

Moveable Targets And Finding the Datum Planes

The old adage in detective movies is, “Follow the money.” In FIGURE 304, to display movable datum targets, the phase could be changed to, “Follow the basic dimensions.” The two holes on the part are positioned to datums A, B and C.

Datum plane A is easy. It is formed by the three highest points of datum feature A. This planar surface creates a datum plane. Datum planes B and C are a little harder to ferret out. But to find datum plane B, “Follow the basic dimensions.” The holes are located in the left to right direction from the apex of the Vee (created by datum targets B1 and B2). A basic dimension of 50 millimeters leads us from here to the first hole. Then, another basic dimension leads us to the second hole. Both are traceable to the apex of the Vee. Datum plane B resides at this apex.

Datum plane C is created by the two movable datum targets C1 and C2 but must work in conjunction with datum targets B1 and B2 to equalize the part. Working together, they create a centerplane datum. The datum feature simulator pins for C1 and C2 must move in normal to the true profile and the moveable datum target symbol, though not required, may be used for clarity. In other cases, where the datum feature simulator is required to move and where the movement is not normal to the true profile, the moveable datum target symbol shall be used and the direction of movement shall be clearly defined.

Datum center plane C is the origin of measurement for the two holes in the last direction. The two holes are to be centered to this plane. The centerplane datum is between targets C1 and C2 helped along by B1 and B2. With the assistance of datum targets B1 and B2 (and the stability that datum plane A gives), datum target points C1 and C2 create the centerplane 18 millimeters from each C target point. It might appear as though the targets are located from the centerplane, but, in fact, the centerplane is halfway between the targets.

The movable datum target symbol was first established in the ASME Y14.8M-1989 Standard on Castings and Forgings, then again depicted in its 1996 revision. Later, it was shown in the ASME Y14.43 Standard on Dimensioning and Tolerancing Principles for Gages and Fixtures. Now it is displayed in ASME Y14.5-2009 and Y14.5-2018.

The moveable datum target symbol is used as an optional clarifying symbol to indicate movement of the datum target simulator. When datum targets are used to create a centerplane, axis or centerpoint and are referenced Regardless of Material Boundary (RMB), datum target simulators move normal to the basic profile. In some cases when the datum feature simulator is supposed to move, but its movement isn’t normal to the basic profile, the moveable datum target symbol must be used to show the direction of the movement.

Movable datum target symbol

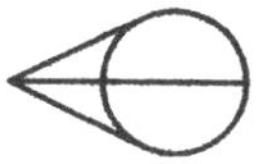

FIGURE 304 [Moveable Datum Target Symbol]

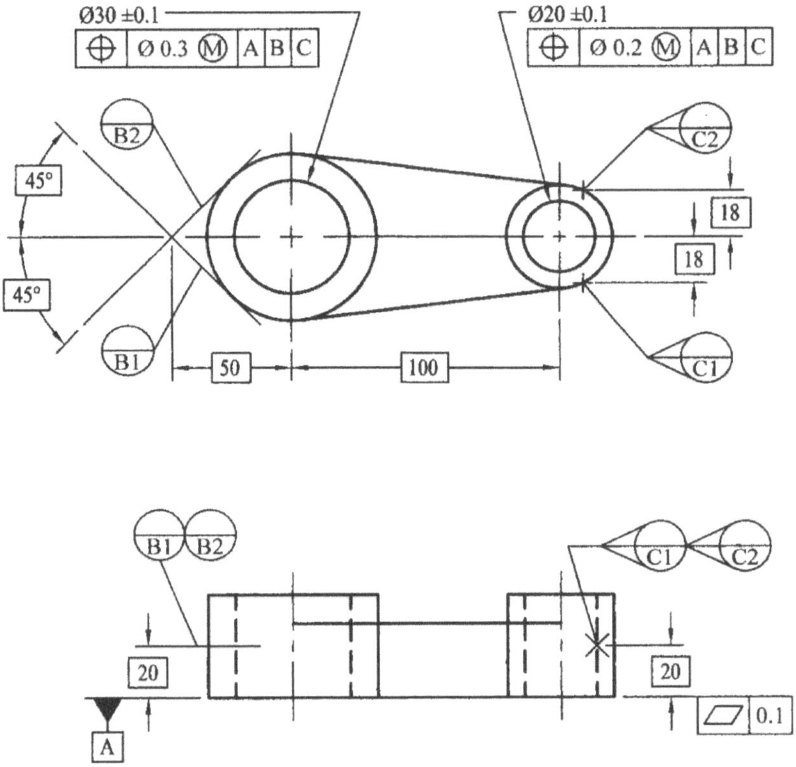

FIGURE 305 [Workpiece Mounted in Fixture]

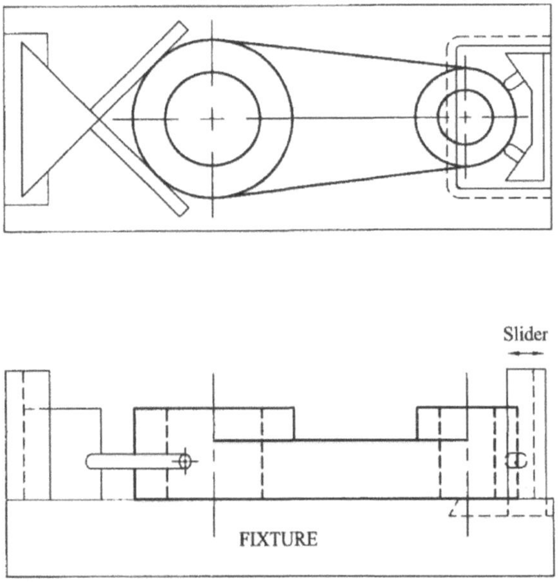

Moveable datum target simulators for C1 and C2 slide through tight fitting holes on the fixture and contact the part surface normal to the desired geometric configuration.
For orthographic views, a line may be added, showing the direction of movement. For a datum target point, this is done by placing the line element at the point of contact, along the line for a datum target line, or is shown with one or more basic angles. **For drawings that include X, Y, and Z axes to represent the datum reference frame(s), the direction of movement may be indicated using a unit vector designation consisting of i, j, k components (corresponding to the X, Y, and Z axes of the coordinate system), placed in brackets and adjacent to the "movable datum target" symbol.** The vector direction is toward the surface of each datum feature. The vector notation is shown in at least one view where the target is depicted. For drawings that include more than one datum reference frame, the particular datum reference frame to which the i, j, k components are related shall be specified by placing the applicable datum feature letters within square brackets following the closing bracket that contains the i, j, k components (such as "[i, j, k]" or [A, B, C]). See Figure 305.

Figure 306 [Vector Indicators]

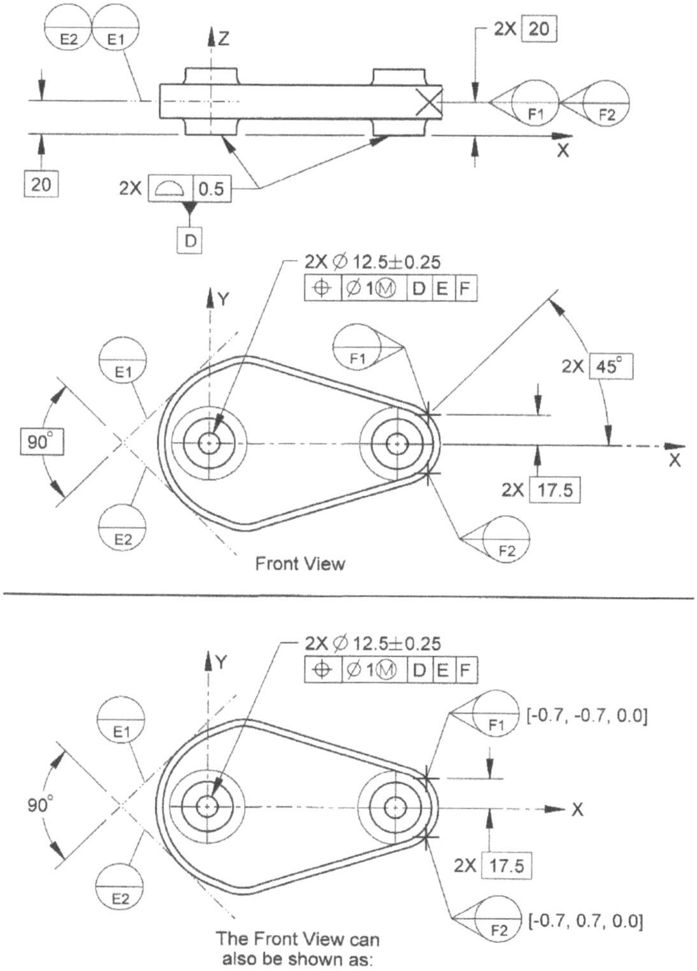

Movement in Figure 306 is shown in i, j, k vectors. As shown in brackets in the optional front view for datum target F2, movement for i is in the x direction (-0.7). Movement for j is in the y direction (0.7). Movement for k is in the z direction (0.0).
For a model, the direction of movement is indicated by the addition of a represented line element to show the direction of movement. The line element is placed on the outside of the material. One end of the line element terminates at the point of contact for a datum target point, at a point on the line for a datum target line, or at a point within the area for a datum target area.

FIGURE 307 [Datum Target Symbol for Spherical Diameters]

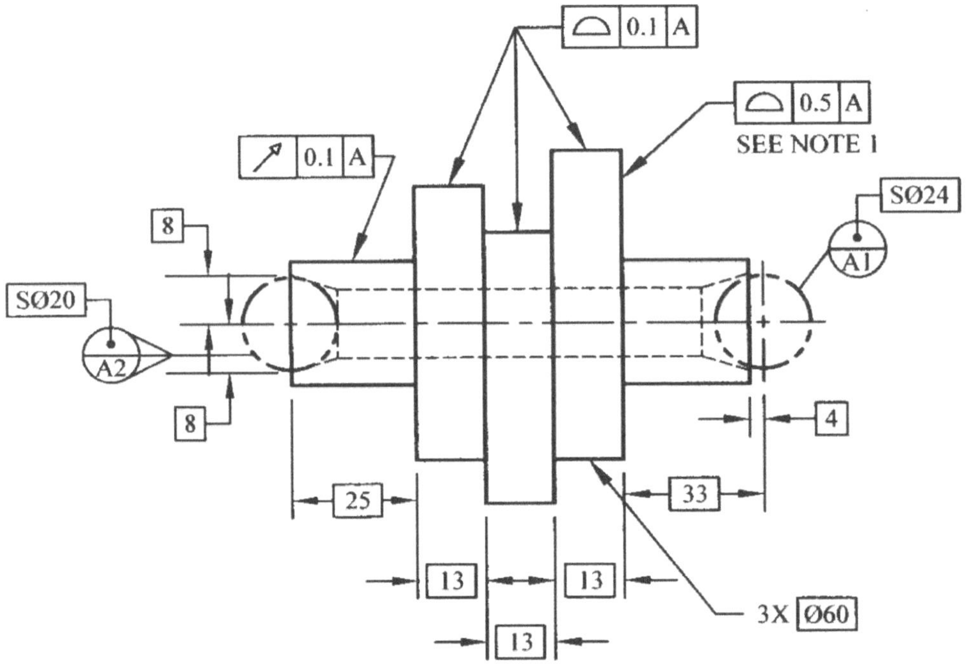

NOTE 1:

THIS CONTROL APPLIES TO ALL PLANER SURFACES LOCATED BY BASIC DIMENSIONS IN THIS VIEW.

Spherical datum target diameters A2 and A1 in FIGURE 307 create an axis that runs between the two sphere centers and a plane that exists at the center of A1. This 3 plane datum reference system (3 mutually perpendicular planes, 2 from the axis and one from the plane at the center of A1) is used to control runout and profile of the surfaces as shown.

FIGURE 308 [Datum Planes--Establishing Datum Centerplanes from Datum Features-Datum Reference Frame (Three Plane Concept)]

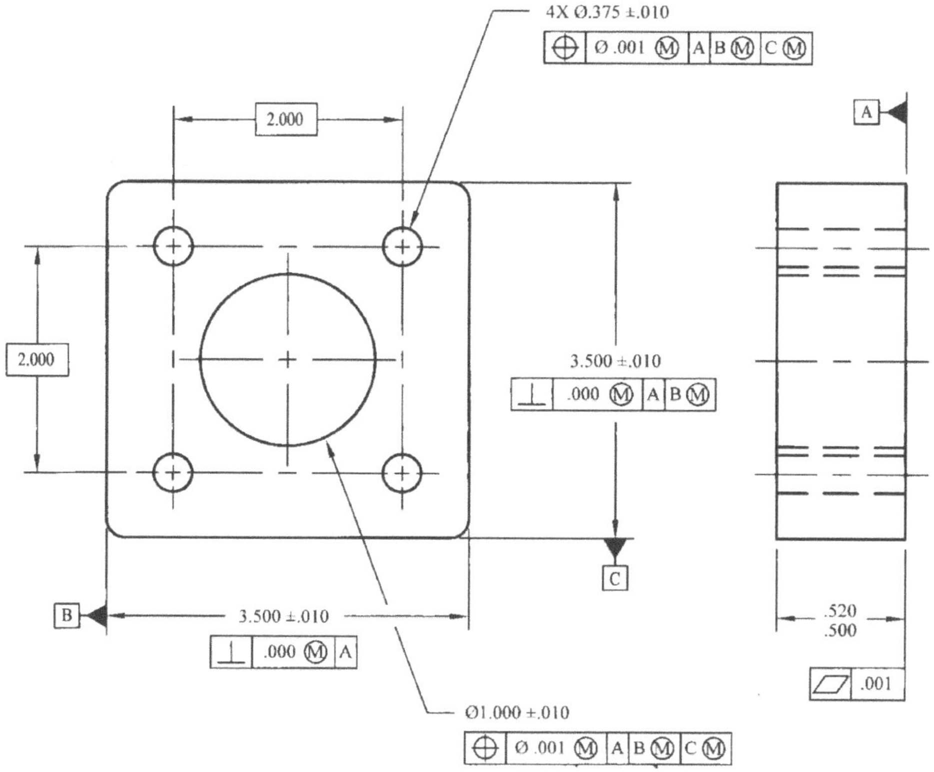

It is known that this part mates with a square cavity with five pins inside. This square cavity has an area available to receive this square part that measures 3.51 by 3.51 worst case. Datum feature A seats on the bottom surface of this cavity. In order to seat well, without rocking, it must be flat. Therefore, a flatness control of .001 has been added to refine the .020 of flatness given under Rule #1 (size limits control surface form). Although .500 to .520 is .020 of flatness, straightness, and parallelism control for datum feature A and its opposing surface, it must be still flatter because datum feature A must seat.

Secondary datum feature B is a width. It will generate a centerplane perpendicular to datum plane A (a plane taken from a minimum of the three highest points of datum feature A). This centerplane is formed at the center of two parallel planes at a 3.51 separation (while perpendicular to datum plane A). Datum feature of size B must be perfectly perpendicular to datum plane A if datum feature B is produced at maximum material condition.

Tertiary datum feature C is a width. It will generate a centerplane perpendicular to datum planes A and B. This centerplane is formed at the center of two parallel planes at a 3.51 separation (while perpendicular to datum planes A and B). Datum feature of size C must be perfectly perpendicular to datum planes A and B if datum features C and B are produced at MMC. The datum center planes B and C will be used as the origins of measurement for the location of all five holes on this part. All holes must be perpendicular to datum plane A to within their applicable position tolerance. The center hole must be centered to datum planes B and C to within its applicable position tolerance. The four hole pattern must be 1 inch in each applicable direction from datum planes B and C to within a positional tolerance of .001 if produced at MMC to .021 if produced at LMC.

Irregular Datum Features of Size

The ASME Y14.5-2009 and the ASME Y14.5-2018 standards allow the use of patterns of features such as holes, shafts, slots and tabs to be used to construct datums. Depending on whether they are referenced at MMB, RMB or LMB, these patterns are simulated by their geometric counterpart.

In some instances, it may be desired to use the minimum circumscribed cylinder that will fit around a pattern of pins or holes. In others it may be functional to use the maximum inscribed cylinder that will fit within a pattern of pins or holes. These concepts would be used Regardless of Material Boundary, so the gaging or fixturing element would expand or contract to meet the requirements of datum feature simulation.

In other cases, it may be functional to calculate the Maximum Material Boundary or Least Material Boundary of these patterns to simulate the irregular datum features of size that represent the outer or inner boundary of the pattern.

The following illustrations show just a few of the possibilities for this type of datum feature use.

FIGURE 309(a) [Outer boundary of a pattern of pins referenced at MMB]

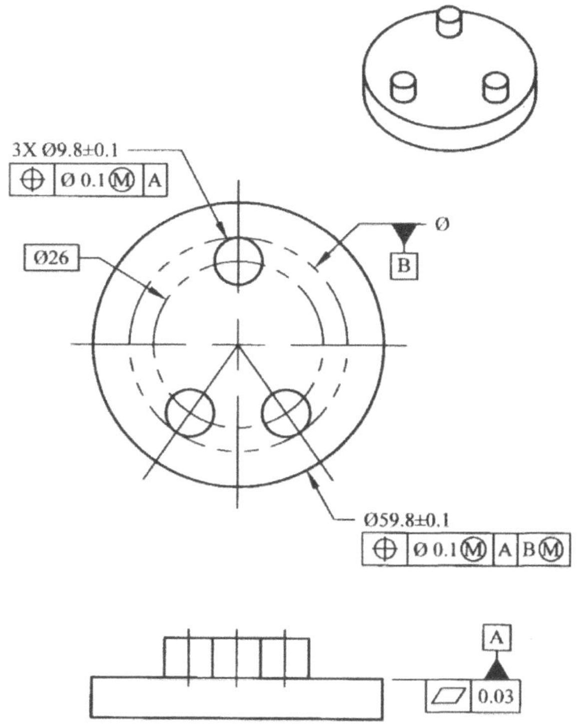

FIGURE 309(b) [A partial gage design for the part in Figure 36(a)]

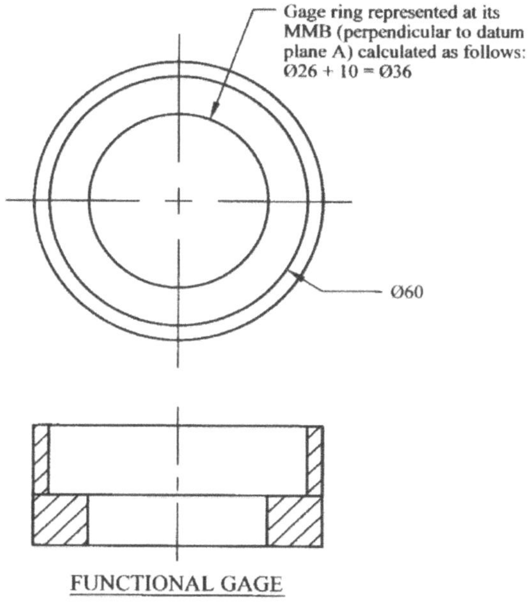

FIGURE 310(a) [Inner boundary of a pattern of pins referenced at RMB]

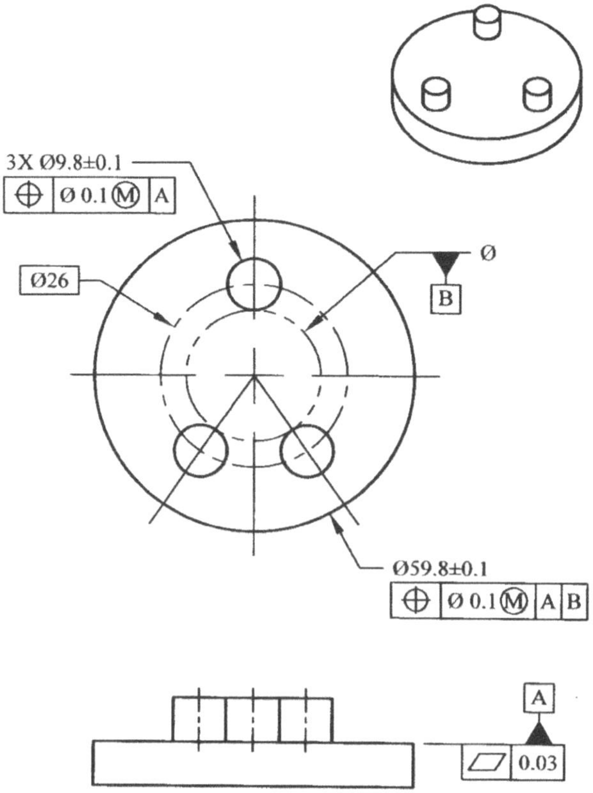

FIGURE 310(b) [A partial gage design for the part in Figure 37(a)]

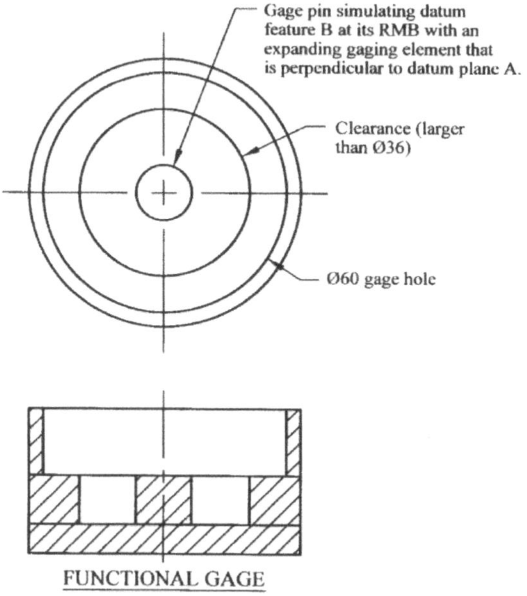

FIGURE 311(a) [Inner boundary of a pattern of pins referenced at MMB]

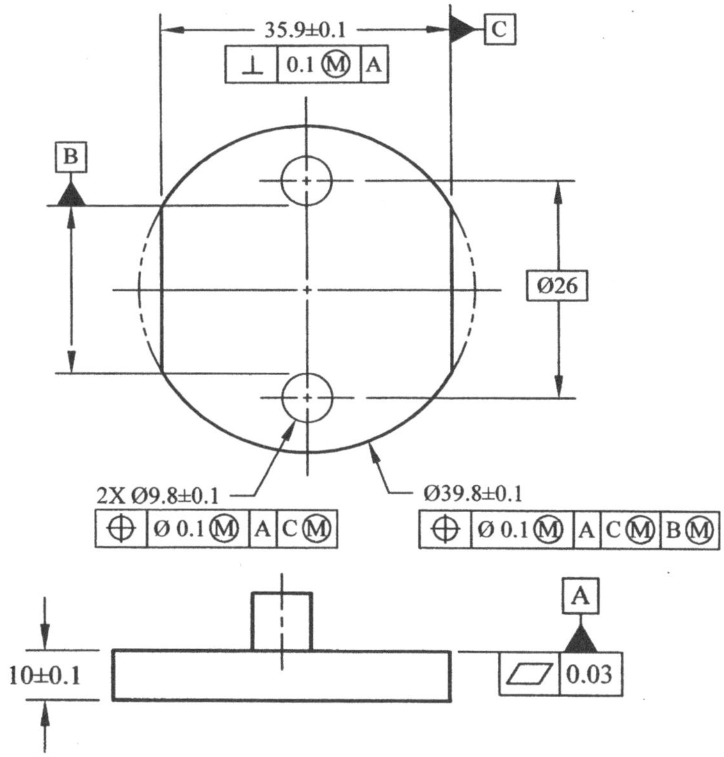

FIGURE 311(b) [Datum feature simulator B]

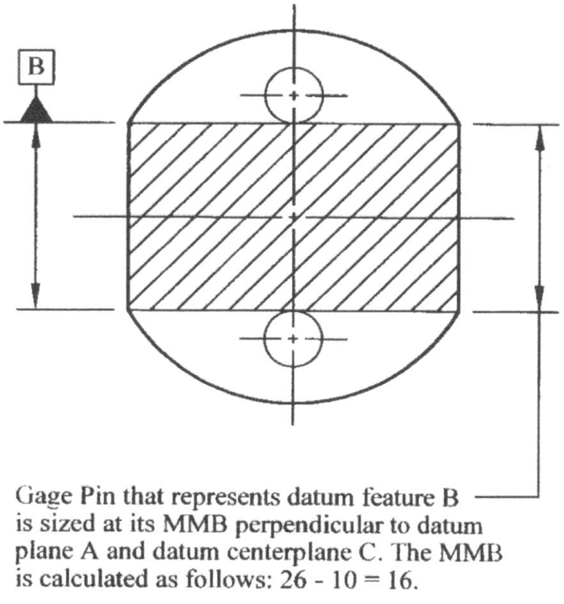

FIGURE 312 [Planar Datum Feature Referenced at MMB]

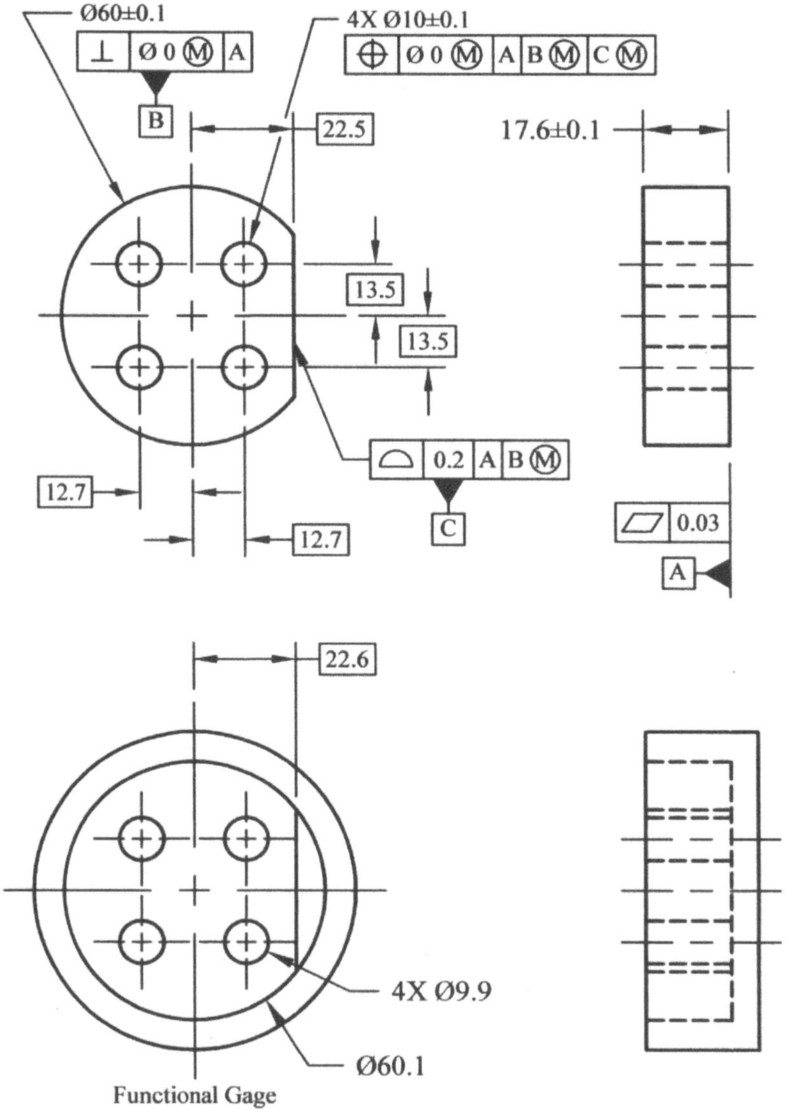

Repetitive Patterns

Sometimes patterns of features are repeated on a part. In these instances, where it is desired to relate a pattern to its own unique datum reference frame, a single datum feature symbol letter may be used to name many datum features. An analogy would be when counterbores are each related to their own pilot holes. The pilot holes are first positioned as a group to one datum reference frame, perhaps A, B and C. Then each pilot hole is named, perhaps D. Even though there may be many datum feature D's, since there are many pilot holes, each counterbore is positioned to its own datum D. This is done by using the local note "INDIVIDUALLY" next to D and stating how many D's there will be. Subsequently, each counterbore is positioned to its own D by stating how many times the pattern will repeat and again using the local note "INDIVIDUALLY". See FIGURE 313 for a similar example of repetitive patterns and how to control them.

FIGURE 313 [Repetitive Patterns]

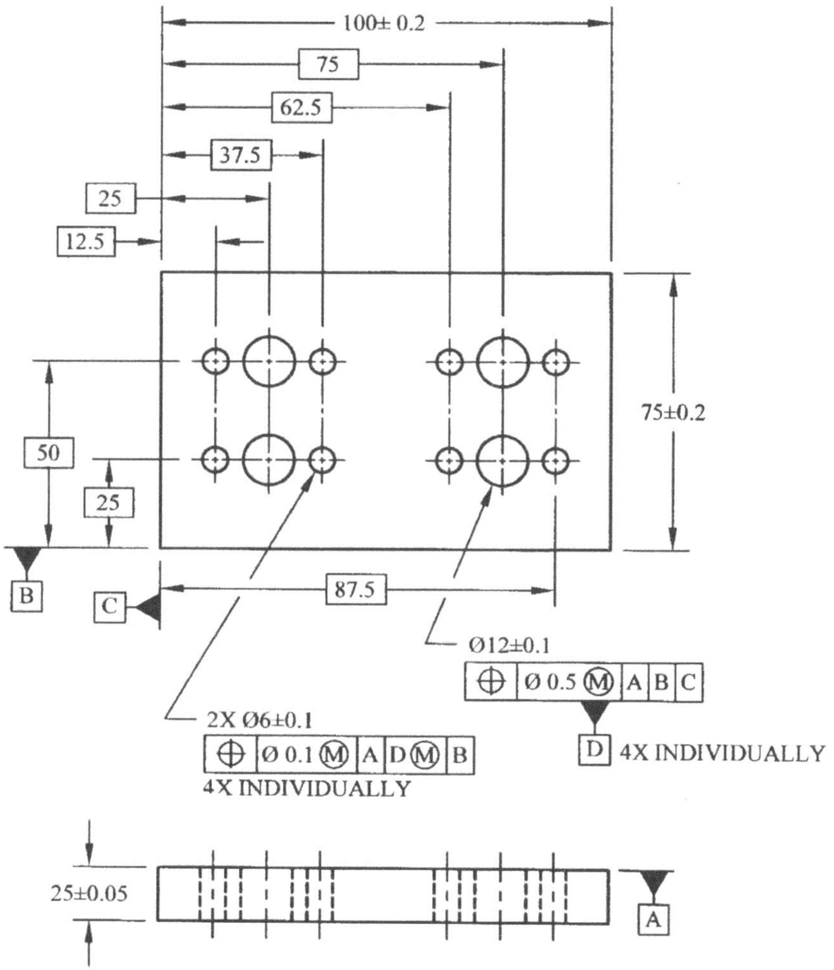

FIGURE 314 [Conicity - Profile Tolerance for a Conical Feature • Related Datums • Releasing Degrees of Freedom]

The following illustration shows a conical datum feature. In the position feature control frame, the spatial degrees of freedom that each datum reference constrains are listed in brackets after the datum reference. Datum feature A generates an axis (through the center of the cone) and a plane (at the apex of the cone) that would normally control the special degrees of freedom for the x, y and z linear movements, as well as the u and v rotations. But, in this feature control frame, since z does not appear within the brackets after datum reference A, the z location is no longer controlled from the apex of the cone. Therefore, datum B has been added to the feature control frame to control the distance in the z direction. Since w (the rotational degree of freedom around the Z axis) is unnecessary to stem, it is not included in any brackets.

Notice that when the degrees of freedom that each datum constrains are to be spelled out in the feature control frame, the datum axes are shown on the part representation. When this is done, datum axes X, Y and Z are shown using capital letters and the degrees of freedom in the feature control frame are shown using lower case letters. Degrees of freedom letters are only allowed to appear in the feature control frame and within the brackets, while the datum axes appear outside the feature control frame on the illustration of the part.

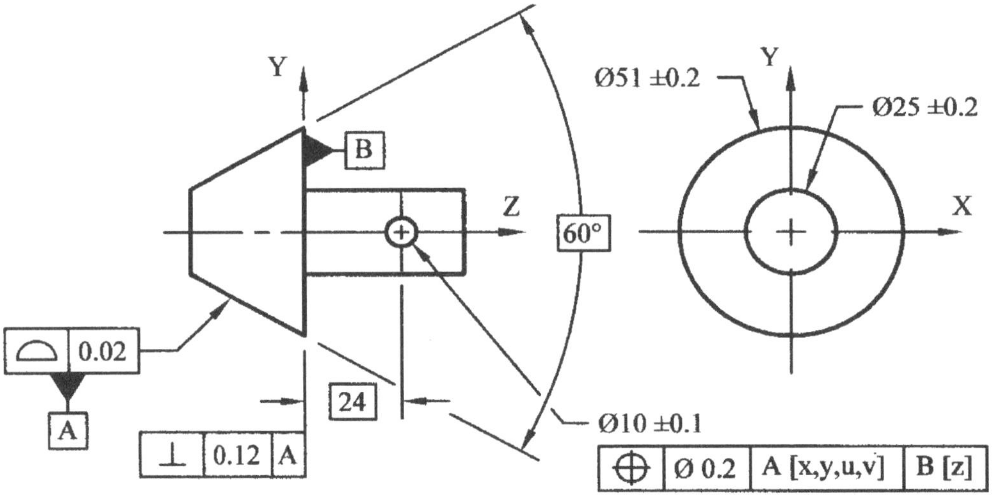

Chapter 19

Dimensioning and Tolerancing of Gages

[per the ASME Y14.43 Dimensioning and Tolerancing Principles for Gages and Fixtures Standard]

Steps for Writing a Dimensional Inspection Plan

Chapter Objectives

Readers will learn:

1. To design, dimension and tolerance GO gages for MMC, NOGO gages for LMC and Functional gages for geometric tolerances per ASME Y14.43.

2. How to calculate whether the gage is likely to accept borderline out-of-tolerance parts, reject borderline parts that are in-tolerance, or if the possibility exists that the gage might do either.

3. The ramifications of using different modifiers (MMC, LMC or RFS implied) on gage tolerances.

4. The differences between Absolute, Practical Absolute, Optimistic and Tolerant gages and which policies are preferred per ASME Y14.43.

5. The steps necessary in writing a Dimensional Inspection Plan

Dimensioning and Tolerancing of Gages
[per ASMEY14.43]

In 2003, an ASME standard was approved called *ASME Y14.43 Dimensioning and Tolerancing Principles for Gages and Fixtures*. It has since been revised as Y14.43-2011. But, in 2003, it marked the first time a nation had issued a standard (ANSI and Department of Defense approved) on the proper design, dimensioning and tolerancing of gages and fixtures for the inspection of geometric tolerances. This standard not only governs the principles for the appropriate procedures for the creation of gages for geometric tolerances (called functional gages) but also continues the practices to measure maximum material conditions with GO gages and least material conditions with NOGO gages that were originally shown in ANSI B4.4. B4.4 has been retired, but its principles were absorbed into Y14.43 and extended to apply to the more difficult Functional Gages (to inspect geometric tolerances).

The basic premise of all gaging is to reject all bad parts (those that violate the tolerances) and to accept all of the good parts (those parts that are in compliance with specified tolerances).

But since all gages need to be toleranced, it is understood that they will fail to achieve these lofty goals. They will either reject a small percentage of technically 'in-tolerance' parts or they will accept a small percentage of technically 'out-of-tolerance' parts. The parts that are on the borderline of exceeding their tolerances (whether just barely exceeding tolerances or just barely in tolerance) are the ones in question.

What is critical is that companies decide which side they would rather error on. Would you rather 'buy' a few bad parts or reject a few good ones? This is the question whose answer will determine whether gage pins will have a plus tolerance or a minus tolerance. It will also determine whether gage holes are toleranced on the plus or the minus side of their acceptable boundaries.

For example, if a GO gage pin designed to check a maximum material condition is dimensioned at the MMC of the hole to be gaged, but then toleranced with a plus only tolerance, the plus only tolerance will infringe on and, therefore, subtract from the tolerance assigned to the hole being gaged. Therefore, some of the borderline, but in-tolerance, holes being gaged could be rejected. This could have the effect of increasing manufacturing costs for the parts containing the holes, but increase the quality of the parts.

Conversely, if the GO gage pins are sized at MMC and then toleranced with a minus only tolerance, some of the borderline, but technically 'out-of-tolerance', holes being gaged could be accepted. This could have the effect of decreasing manufacturing costs but also decreasing the quality of the parts.

So, a company must choose which they will do--take the risk of rejecting a few borderline good parts or accepting a few borderline bad parts. Their decision will commonly set the course for all gages and fixtures the company designs (or has designed for them) in the future.

> **The ASME Y14.43 standard has taken as its preferred practice two policies on gage and fixture tolerancing. These policies are called Absolute (also called pessimistic) Gage Tolerancing and Practical Absolute Gage Tolerancing.**

For GO gages that inspect the maximum material condition sizes of features, the Absolute Gage Tolerancing policy is preferred. It sets as the goal never to accept an out-of-tolerance part. Therefore, all GO gage pins are designed at the MMC and toleranced to have only a plus tolerance on the size (no minus tolerance). All GO gage holes are dimensioned at the MMC of the pins being gaged and then toleranced so that the gage hole may only be produced at that size or smaller (all minus tolerance, no plus tolerance). This has the effect of never accepting features (holes, shafts, slots and tabs) that are outside of their tolerance range. It also has the effect of rejecting a small percentage of technically in-tolerance parts. For example:

FIGURE 315 [Part with Hole]

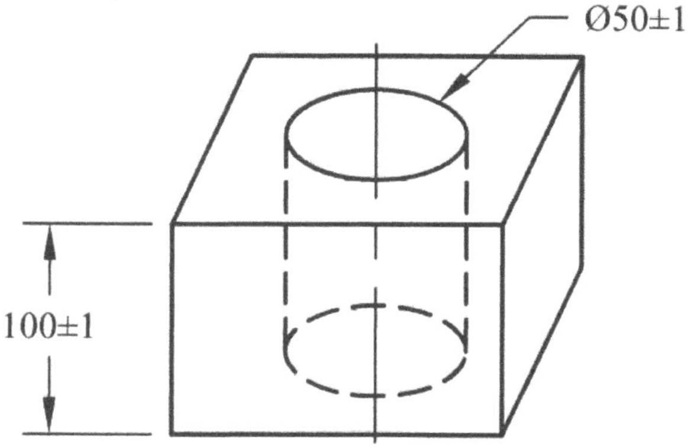

FIGURE 216 [GO Gage pin with Absolute Tolerancing]
This gage is shown using 10% of the part tolerance.

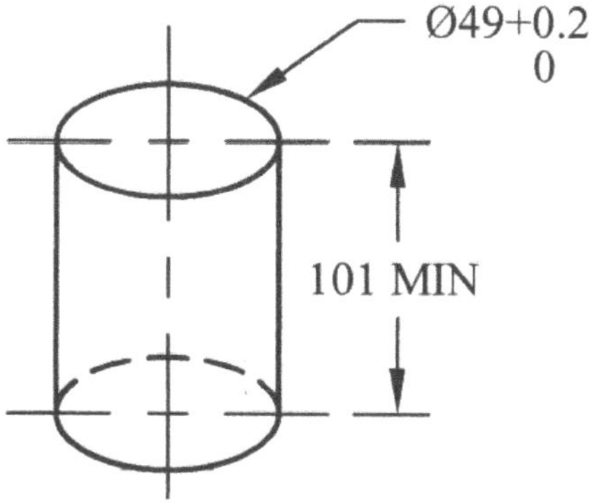

FIGURE 317 [NOGO Gage] This gage uses 10% of the part tolerance.

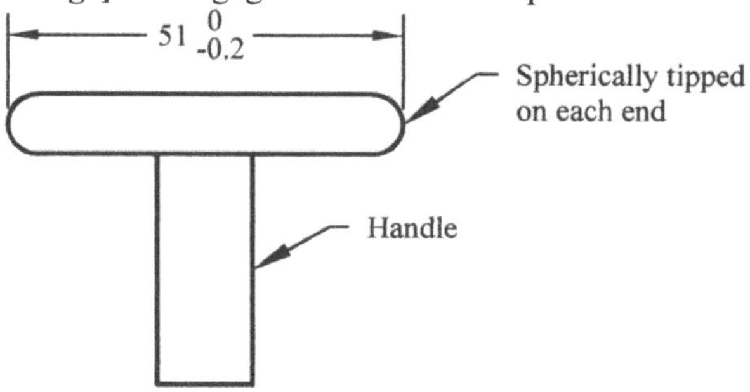

Both the GO gage and the NOGO gage have been toleranced so as to subtract tolerance from the hole being gaged. The GO gage pin is all plus tolerance, to accept no parts that are outside of the MMC size limit. This has the effect of also rejecting a very small percentage of in-tolerance holes. The NOGO gage also accepts no bad parts, but may reject a small percentage of borderline, but technically good, parts. **Remember, the job of the NOGO gage is to 'not go' into the hole.** By reducing the size of the gage (from the 51mm LMC) with a minus only tolerance, the gage is more likely to go into the hole and, therefore, reject the hole as being too large (in violation of the least material condition).

FIGURE 318 [Detail Drawing of a Workpiece to be Gaged]

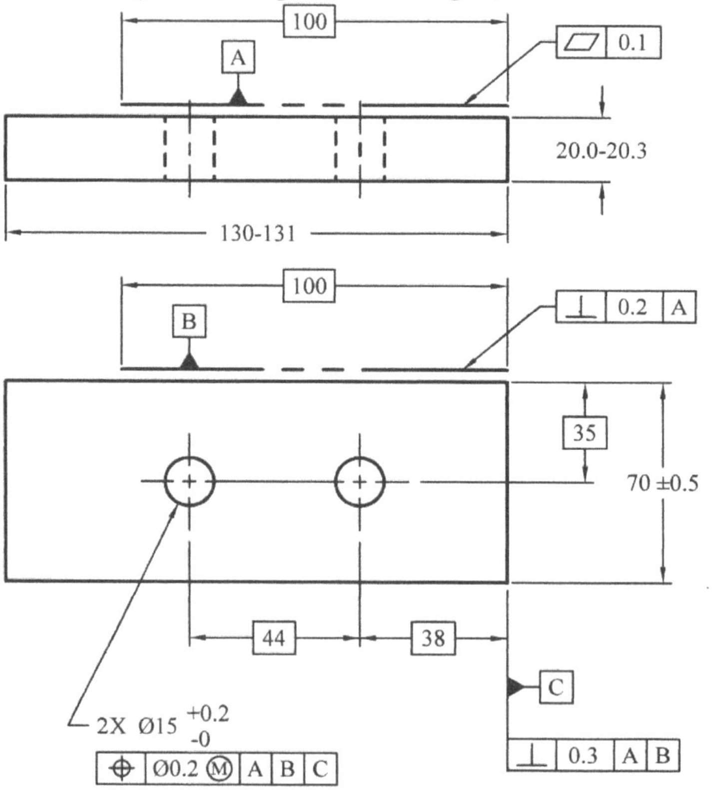

For the workpiece shown, a gage is constructed using 10% of the part tolerance for each element being represented on the gage--for example, 10% of the flatness tolerance, each perpendicularity tolerance, position tolerance and hole size tolerance. **This gage is called a Functional Gage and is toleranced with the Practical Absolute Gage Tolerancing methodology.**

FIGURE 319 [Functional Gage using the Practical Absolute Gage Tolerance]

GAGE

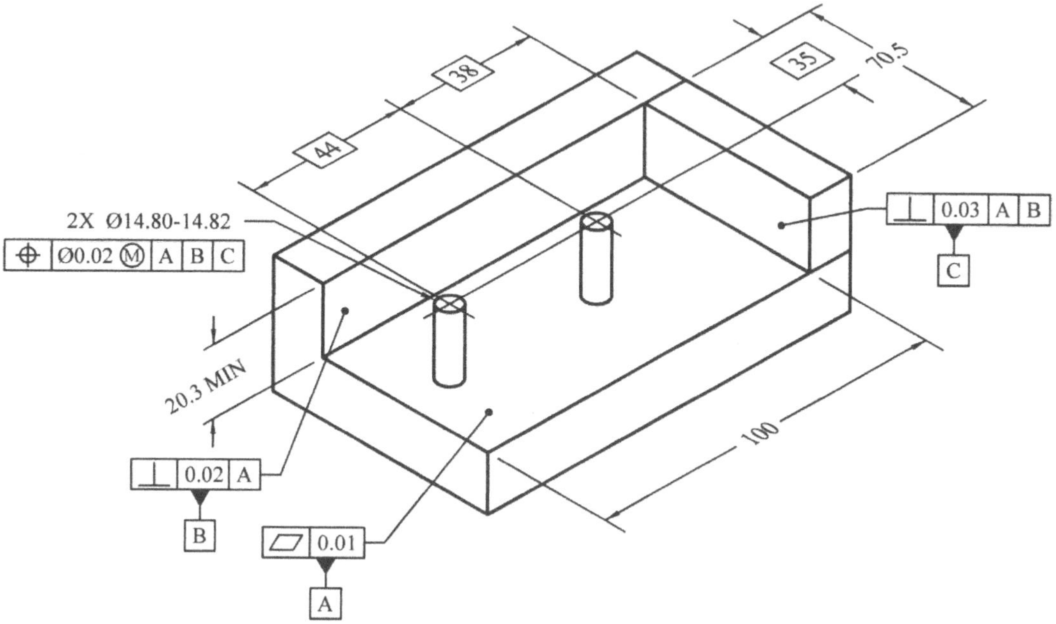

WORKPIECE APPLIED TO GAGE

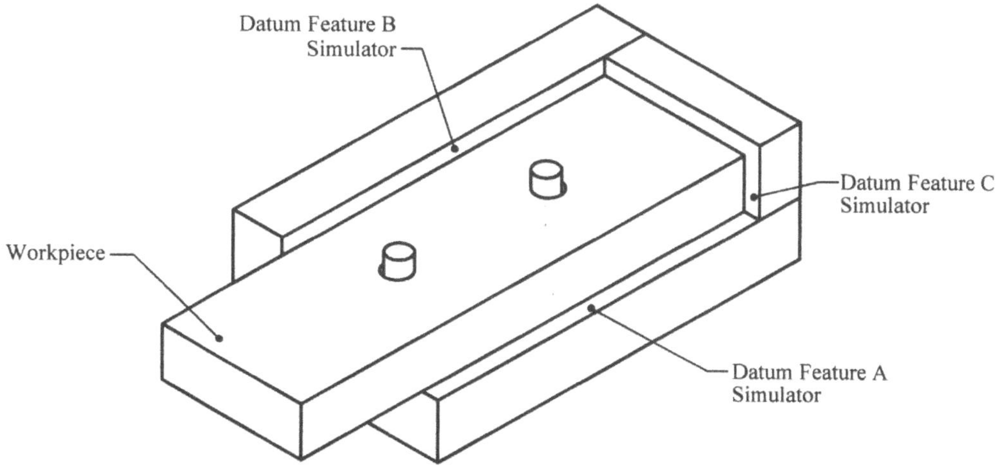

As you can see, datum feature simulators are constructed to represent datum features A, B and C. Datum feature A is a portion of the entire surface, so the datum feature simulator is that large (70.5 x 100). It is assigned a flatness tolerance of 0.01 (10% of the 0.1 flatness tolerance on the workpiece). The gagemaker's tolerance ideally ranges from 5% to 10% of the workpiece feature's tolerance that is being simulated.

Datum features B and C both have a tolerance on the workpiece of perpendicularity, so the datum feature simulators on the gage have been assigned perpendicularity tolerances of 10% of these tolerances. Datum feature simulator B references only datum A in its perpendicularity control, but datum feature simulator C references both datums A and B.

The holes on the workpiece are represented by gage pins on the gage. These gage pins are sized at the virtual condition of the holes on the workpiece that are being gaged.

15.0	=	MMC Holes
- 0.2	=	Geometric Tolerance at MMC
14.8	=	Virtual Condition of Holes

Functional gage pins are dimensioned to be the virtual condition of the holes being gaged. So, the two gage pins are sized at 14.8. With Absolute and Practical Absolute Gage Tolerancing methods, the gage pin tolerance is all on the plus side of the 14.8 virtual condition boundary size. Since the holes have a size tolerance of 0.2, the gage pins will have a plus only size tolerance of 10% of that, which is 0.02. The gage pins will be:

$Ø14.8 \begin{smallmatrix} +0.02 \\ 0 \end{smallmatrix}$

The gage pins are given a position tolerance. Since this gage is shown with fixed pins, the pins are given a position tolerance directly that is 10% of the position tolerance on the holes being gaged. The holes being gaged have a position tolerance of 0.2 at MMC, so the gage pins are given a position tolerance of 10% of 0.2 at MMC, which is 0.02 at MMC. Now the control reads:

$2X\ Ø14.8 \begin{smallmatrix} +0.02 \\ 0 \end{smallmatrix}$

⌖	Ø 0.02 Ⓜ	A	B	C

If this gage used push pins that are to be shown separate from the gage base, the gage pins would be dimensioned as

$2X\ Ø14.8 \begin{smallmatrix} +0.02 \\ 0 \end{smallmatrix}$

for the portion of the pin diameter doing the gaging. Then the holes in the base of the gage that the pins would be pushed in to (once the workpiece was mounted onto datum feature simulators A, B and C appropriately) would be given a position tolerance of:

⌖	Ø 0.02 Ⓜ	A	B	C

This position tolerance would govern the allowed movement of the holes in the base of the gage. The fit between the gage pins and these holes in the gage base is to be a 'Sliding Fit' as governed by *ANSI B4.2 on Preferred Metric Limits and Fits*. Had the workpiece and accompanying push pin gage been dimensioned and toleranced in inches, the fit between the gage pins and the holes in the base of the gage would have been a 'Sliding Fit' per ANSI B4.1. ANSI B4.1 and ANSI B4.2 have been combined into ASME Y14.39.

The way to determine the likelihood of a good part being rejected by this gage or a bad part being accepted is to construct a chart of the hole's (being gaged) virtual condition boundary and the gage pin's inner and outer boundaries. The virtual condition of the holes being gaged is 14.8. Any gage pin outer boundary larger than 14.8 runs the risk of rejecting good parts. Any gage pin inner boundary smaller than 14.8 runs the risk of accepting bad parts. The risk of rejecting good (but borderline) parts is very real. The risk of accepting bad parts is mostly theoretical in that the physical gage pin diameter is a minimum of 14.8. Any reduction of this number is caused by the position tolerance allowing the pin to move away from its perfect location (as shown by the basic dimensions on the gage drawing. But wherever the gage pin ends up in its location, it is still at least 14.8 in size.

Also, remember that for every action, there is an equal and opposite reaction. So, as the gage pin moves in on one side (acting smaller on that side), it moves out on the opposite side (acting larger on that side). This means that even though this movement may generate an inner boundary smaller than 14.8, it will (because of its movement) simultaneously generate an outer boundary larger than 14.8.

Think of yourself (as the gage pin) trying to walk through a door (the hole being gaged). If you center yourself to the middle of the door, you walk easily through it. But if you move a step to the right of center, your left shoulder easily clears the left side of the door. You are acting as though you are smaller on the left side of your body. But at the same time, your right shoulder bangs into the door frame and you don't fit through the door.

While your left side might be occupying less than its half of the door entrance, your right side is occupying more than its half of the door entrance (acting as though you have grown on your right side). So, you are really the same size as you always were, but because you have moved to the right, the left side of your body acts smaller and the right side of your body acts bigger.

Now the important part... You don't fit through the door. And, likewise, the gage pin doesn't fit into the hole being gaged. If the gage pin moves, it is more likely to reject a good part than accept a bad one.

When could it accept a bad one in this scenario? ...when the door (hole) moves in the same direction, by the same amount, as you (gage pin) move. This, in a practical sense, is most unlikely to happen. **That is why this type of gage tolerancing is called Practical Absolute Gage Tolerancing. It means that a gage toleranced in this manner will practically absolutely not accept a bad part.** Now that the practicality has been explained, we can look at the numbers and not panic when we see them wander down into the 'accepts bad parts' range.

Outer Boundary of the Gage Pins

14.82	=	MMC Pins
+0.02	=	Geometric Tolerance at MMC
δ14.84	=	Outer Boundary Gage Pins

Inner Boundary of the Gage Pins

14.80	=	LMC Pins
- 0.04	=	Geometric Tolerance at LMC
14.76	=	Inner Boundary Gage Pins

FIGURE 320 [Graph]

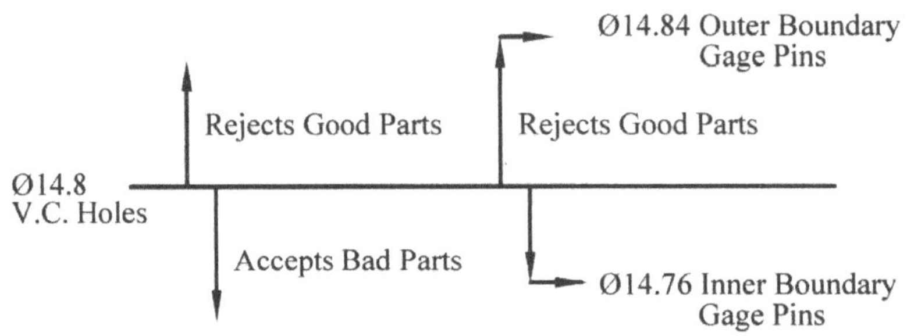

The graph would seem to imply that there was just as much of a possibility of accepting bad parts as rejecting good ones, until we remember that the physical diameter of the gage pins are a minimum of 14.8 and a maximum diameter of 14.82. The rest is movement. Moving the pin to the left or right is rarely going to allow a 14.80-14.82 gage pin to fit into a hole acting smaller than that. You will probably reject a very small percentage of technically good, but borderline, parts. You will absolutely, practically never accept any bad parts using this gage tolerancing policy.

The pin gage dimension and its tolerances can be manipulated to get any result you desire. For example, if I wanted an Absolute Gage (instead of Practical Absolute) where, even in theory, no bad parts would be accepted, we could increase both gage pin size limits by the difference between them and the position tolerance. Since the difference between the gage pin MMC of 14.82 and the LMC of 14.80 is 0.02, we would take that 0.02 and add it to the position tolerance, which is also 0.02 at MMC for a total of 0.04.

This 0.04 would then be added to the size limits as follows:

14.82	=	MMC Gage Pins
+ 0.04	=	Increase Factor
14.86	=	New Gage Pin MMC

and

14.80	=	LMC Gage Pins
+ 0.04	=	Increase Factor
14.84	=	New Gage Pin LMC

These new gage pins would be as follows:

LMC MMC
2X ∅14.84-14.86 Gage Pins

⌖	∅0.02 Ⓜ	A	B	C

This would generate new boundaries:

14.86	=	MMC
+ 0.02	=	Geometric Tolerance at MMC
14.88	=	Outer Boundary Gage Pins

and

14.84	=	LMC
- 0.04	=	Geometric Tolerance at LMC
14.80	=	Inner Boundary Gage Pins

So, our new graph would be as follows.

FIGURE 321 [Graph]

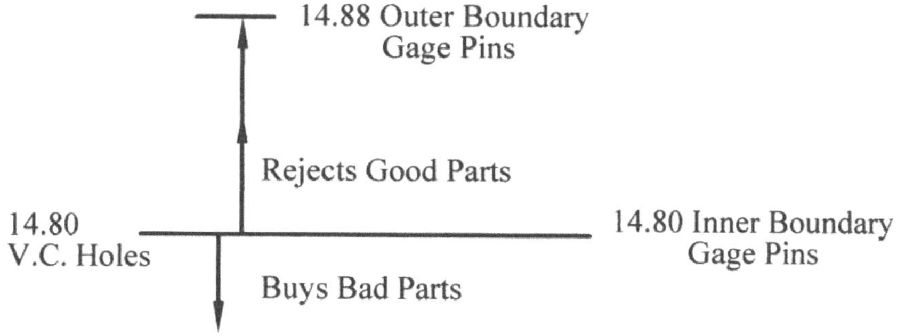

This graph shows that we can't buy a bad part with these new gage pins, even in theory. However, it also shows that the chance of rejecting good parts is much greater. With the original gage drawing, we had only wandered into the 'rejects good parts' range to δ14.84. Now, with the new gage pin dimensions, we have gone twice as deep into that range to δ14.88. This potentially raises the cost of the workpiece being gaged, with more technically good parts being rejected.

Another possibility that gives similar results is the use of the LMC modifier on the gage pins in the position control. For example:

LMC MMC

2X ⌀14.82-14.84 Gage Pins

⊕	⌀0.02Ⓛ	A	B	C

As you can see, the MMC and LMC of the gage pins have been increased by the 0.02 position tolerance. This is to keep us out of the 'accepts bad parts' (in theory) range. If the position tolerance had been zero, instead of 0.02, the gage pin MMC (14.82) and LMC (14.8) would have remained the same. But, with the MMC raised to 14.84 and the LMC raised to 14.82, the outer and inner boundaries are as follows:

14.84	=	MMC Gage Pins
+ 0.04	=	Geometric Tolerance at MMC (0.02 Geo. Tol.. plus 0.02. Bonus Tol.)
14.88	=	Outer Boundary Gage Pins

and

14.82	=	LMC Gage Pins
- 0.02	=	Geometric Tolerance at LMC
δ14.80	=	Inner Boundary Gage Pins

As calculated, it becomes apparent that these boundaries are the same for both possibilities that follow.

2X ⌀14.84-14.86 Gage Pins

⊕	⌀0.02Ⓜ	A	B	C

and

2X ⌀14.82-14.84 Gage Pins

⊕	⌀0.02Ⓛ	A	B	C

Both generate gages categorized as Absolute and will never, even in theory, accept bad parts. But both run the risk of rejecting more in-tolerance parts than the original ASME Y14.43 favored method of Practical Absolute Gages toleranced as:

2X ⌀14.80-14.82 Gage Pins

⊕	⌀0.02Ⓜ	A	B	C

The figures shown next use the Practical Absolute Gage Tolerancing method.

FIGURE 322 [The Workpiece]

Workpiece

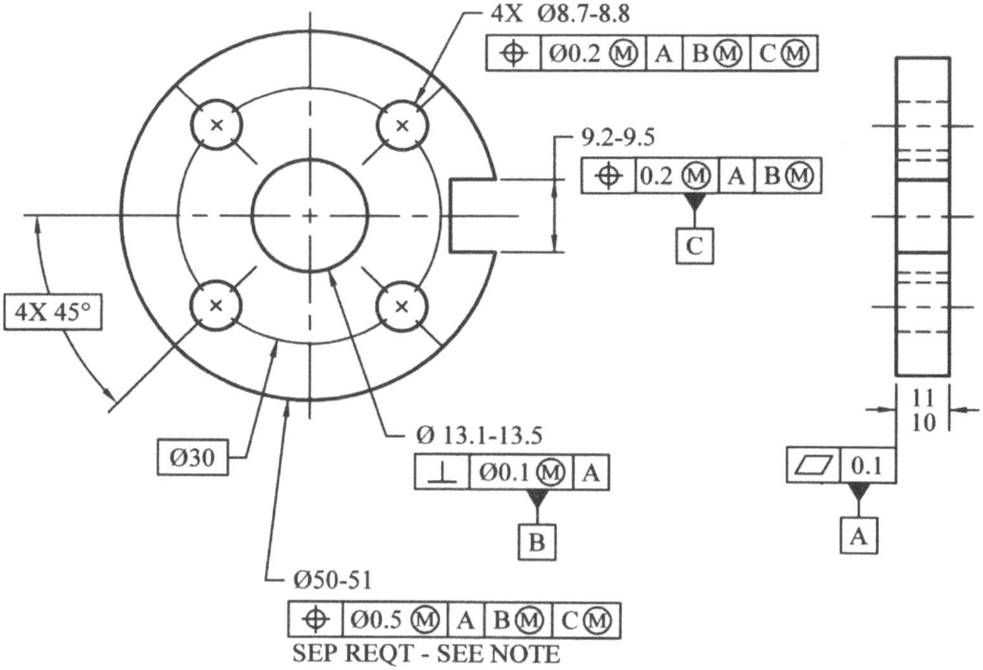

NOTE: THIS POSITION CONTROL IS A SEPARATE GAGING REQUIREMENT. IT WILL NOT BE GAGED WITH THE SAME GAGE AS THE 4 HOLE PATTERN.

FIGURE 323 [Functional Gage for the 4-Hole Pattern Position Tolerance]

GAGE

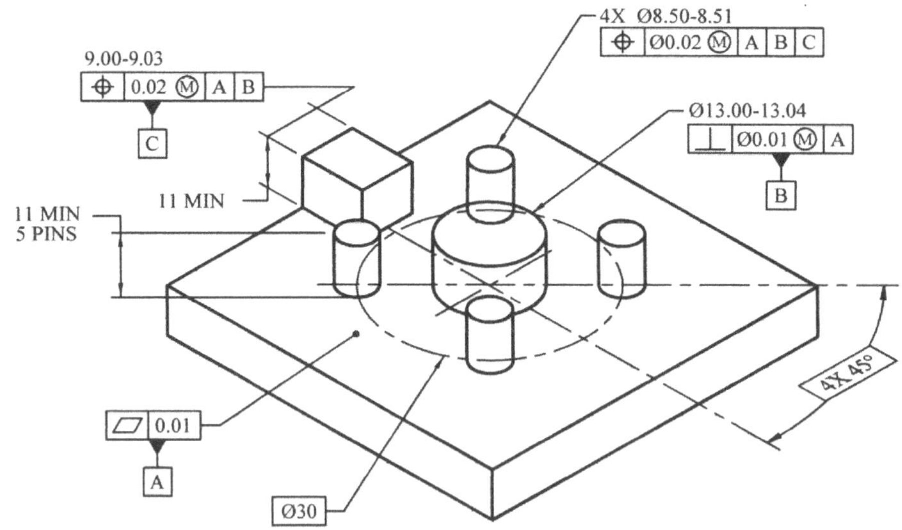

WORKPIECE APPLIED TO GAGE

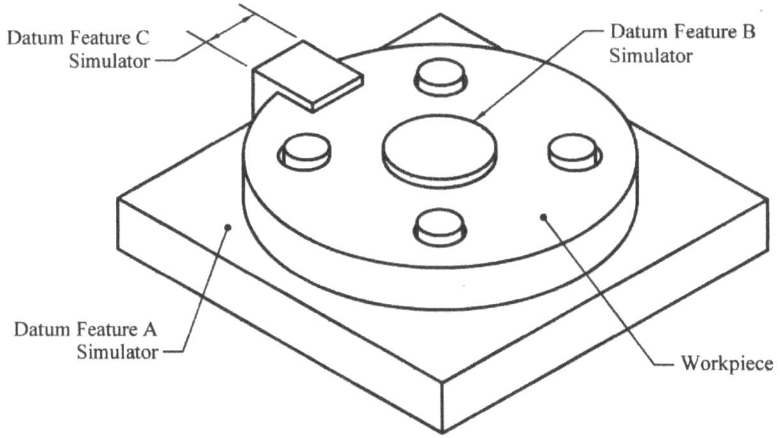

All gages in this section have used either the Absolute Tolerancing Method (shown on the GO and NOGO gages) or the Practical Absolute Tolerancing Methods (shown on both Functional Gages). These gaging practices use as the premise that all gage pins have plus tolerances and all gage holes have minus tolerances for all GO gages and Functional Gages. For NOGO gages, all gage pins have minus tolerances and all gage holes have plus tolerances. This is to achieve a gage that does not accept parts that are out of their tolerance ranges.

There are two other gage tolerancing practices that are NOT RECOMMENDED by the ASMEY14.43 standard on Dimensioning and Tolerancing Principles for Gages and Fixtures. One of these is called ***Optimistic Gage Tolerancing***. This policy tolerances gages in ways that are the opposite of those described in this unit. GO gage pins and functional gage pins would begin at the same sizes shown in this unit but would have no plus tolerances. These gage pins would be tolerancing entirely in the minus direction. The 5% to 10% policy would still apply, just in the opposite direction as shown for Absolute and Practical Absolute Gage Tolerancing.

For gage holes on Go gages and Functional Gages, the Optimistic Gage Tolerancing would be all plus and no minus. For NOGO gage pins, the Optimistic gage would have a plus tolerance and the Optimistic gage holes would have a minus tolerance. Optimistic Gage Tolerancing risks buying a small percentage of out-of-tolerance parts. Optimistic gages buy all parts within tolerance and also a few that are not. This is generally perceived as lowering production costs of parts but sacrificing a small portion of quality and the parts' ability to function or mate with other parts in the assemblies.

The third policy NOT RECOMMENDED by ASME Y14.43 is known as *Tolerant Gaging*. Tolerant Gaging sizes GO gages at MMC, NOGO gages at LMC and Functional Gages at virtual condition, just as do the Absolute, Practical Absolute and Optimistic methods. But instead of just tolerancing to either the plus or to the minus side only, the Tolerant Gaging policy gives gage pins and gage holes both a plus and a minus tolerance. The problem with this non-recommended practice is that it does not take a stance as to a company's policy. It does not decide to reject a few good parts and not buy any bad ones (Absolute), and it does not decide to buy all good parts and also to accept a few bad ones (Optimistic). They do not know whether their gages will buy a few bad parts or reject a few good ones.

The ASME Y14.43 standard suggests making the decision up front and, therefore, using one of the other gage tolerancing policies explained in this unit or in the ASME Y14.43 standard.

Steps in the Development of a Dimensional Inspection Plan

The concept of a dimensional inspection process designer creating a Dimensional Inspection Plan is recommended for the verification of most, if not all, product designs. Times when design needs for part functionality are abandoned because of inferior, ill-conceived inspection plans and a lack of knowledge of equipment potential belong in the past. This unit leads one through a process for the creation of a Dimensional Inspection Plan. It shows some of the items worthy of consideration when writing the step-by-step process, a part must go through to assess tolerance and cosmetic compliance and assure functionality. It explores the uncertainties common to the inspection processes and what uncertainties are and are not permissible. It raises the point of what to do with information gained from the inspection process to improve the manufacturing procedure and quality of parts produced from that time on. A step-by-step example is given for one possible Dimensional Inspection Plan for a part to be produced.

For each workpiece design, the measurement process designer may prepare a Dimensional Inspection Plan. The plan should include a list of measurements to be made, what gage to use for each measurement, the procedure for each measurement and the gaging limits for each measurement. It is recommended that the measurement process designer document logic for the plan. The Dimensional Inspection Plan may be developed in the following steps:

1) For every part to be inspected:
   a) Learn how the part functions.
   b) Decide which dimensions will be inspected and which dimensions will not be inspected.

c) Determine the ramifications of approving a dimension that is not within its tolerance and of rejecting a dimension that is within its tolerance.

2) For every dimension that is to be put under test:
   a) Determine if an inspection plan already exists that will suffice if minor modifications are introduced. If one exists, tailor it to the new part as needed, and go to Step 2 (e). Otherwise,

   b) Find out how the part is produced and what errors of geometric perfection are common to that manufacturing procedure.
   c) Determine the best inspection approach/scheme to follow.
      - Decide the most appropriate tools, gages and major equipment needed
   d) Discover which measurement uncertainties will be introduced.
      - Does your measurement approach/scheme contain inherent uncertainties?
      - Do the gages and/or inspection equipment and/or inspector know-ledge have uncertainties that need to be considered?
      - Does the environment within which the part will be inspected have the capability of introducing uncertainty?
   e) Decide on the acceptable probabilities of accepting a bad feature or rejecting a good feature.
   f) Analyze the most likely distribution of measurement data.
   g) Calculate/analyze the gaging limits.

3) Create a Dimensional Inspection Plan for every part design.

The Dimensional Inspection Plan Format

A Dimensional Inspection Plan should be created for each part design. Among the items to be considered are:

1) Which characteristics of which part features will be measured.
2) The tools needed for the measurement of each feature characteristic.
3) When in the best interest of the feature, the position and number of points on the surfaces where sample data will be taken may be specified.
4) The steps to be followed in the measurement procedure.
5) How collected data will be analyzed.
6) What to do with the collected data to improve the manufacturing process.

The inspection planner may consider it wise to document the reasoning behind each decision in the plan. This may explain to those scrutinizing the plan the logic behind the decisions made by the planner.

Plan Development

To determine the best possible plan for the inspection of any part, the designer of the measurement process should know:

1) How the part functions.
2) Which characteristics of which features must be inspected in order to insure the part's functionality.

He or she will also need to determine how lots will be measured and what type of sampling is necessary to insure optimization of inspection time and collection of the most valuable data. Whether the information is used merely to insure the functionality of that one part or is to be used as variable data for the continued production of parts within geometric tolerances, the inspection plan is a vital piece of the process. It can help insure functionality, interchangeability and a product produced at the lowest possible costs.

Each Measurement

The portion of the Dimensional Inspection Plan that deals with how each measurement should be taken should consider:

1) Has a previous plan been developed either for this part or for this type of part? If so, it may be used as a guide to rewrite the plan, or to create the plan.
2) Should the feature be measured? Sometimes measurement of a feature is not necessary. For example:
   a) If the dimension is controlled by a proven die or mold.
   b) If attribute can be verified by a means other than measurement. For example: strength vs. diameter.

c) If dimension is unimportant to workpiece function and is known to be not a factor worth consideration in the batch or part under test.
d) If a feature can be either accepted or rejected at a lower cost by simply seeing if it will fit into the assembly, it is sometimes better to do so (for example, if interchangeability is not a factor).

What Is Being Verified?

In the case of geometric controls applied per the ASME Y14.5 Standard, if the feature is to be inspected, verification of either **boundaries or tolerances zones** is required. When the designer specifies a positional MMC control, for example, the surface of the feature may not lie outside a boundary of virtual condition or else it often will infringe on space supposed to be occupied by the surface of the mating feature. Position of the virtual condition boundary may be fixed or movable (as in the case of features controlled to datum features of size at MMC).

In some controls, line elements are to reside within tolerance zones. In others, centerplanes, surfaces, axes or points must reside within the specified tolerance zone. In most instances where a centerplane or axis is being controlled, verification of either the tolerance zone or the virtual condition boundary it generates is acceptable. Verification of both is rarely necessary. Although it is generally understood that the verification of tolerance zones or boundaries is appropriate, and considered to be roughly equivalent, these two concepts are not always mathematically equivalent. Where there is considered to be a conflict between the two concepts for verification, the virtual condition boundary concept is given greater weight. It is considered that the MMC concept virtual condition boundary is usually more descriptive of the space needed for assembly in mating situations.

The intent of the designer should always be clear and interpretable per the ASME Y14.5 Standard. Where intent is not clear, the designer should be contacted for clarification whenever possible.

Hard vs. Soft Gages

In order to verify feature compliance with size or geometric tolerance, we may use either hard or soft gages. Hard gages are mechanical in nature, like for example Coordinate Measuring Machines, micrometers, vernier calipers, ring gages and snap gages. Depending on the situations, these gages are capable of taking either direct measurements or comparative (to a standard) measurements.

Hard gages are often used to determine coordinates of a set of points on a feature surface to estimate the true shape of the feature surface. This information is sometimes then compared to mathematical soft gages in computer software. The comparison can be either direct (measured values to the ideal values) or comparative (measured quantity determined to be either larger than, equal to, or less than the standard). If the information fed to the computers involved in this method is correct, soft gages can perform verification techniques for geometric controls quite well. Remember, however, that the verification is only as good as the procedures followed during the collection of data. Hard gage measurements can be compared directly with gaging limits to determine acceptance or rejection.

Soft gages can use knowledge of the manufacturing process for features to be able to augment data that is represented by the part configuration. For example, in terms of deviations which generate smooth profiles, these deviations from geometric perfection may be caused by thermal bending of the machine or workpiece during machining, clamping distortion or bad fixturing. Deviations which tend to form a trend like a periodic pattern over the workpiece may be caused by machine tool feed, tool form errors--as in rolling components, tool stiffness, cutter form errors, or cutter alignment problems.

Dividing these errors from one another and analyzing their causes can allow one to take action that will prevent them from reoccurring.

Choosing Gages

The choice of appropriate gages should be based on:

1) Capability
2) Availability
3) Cost Effectiveness

We would all like to check our parts in the most accurate manner available without spending more inspector time and machine time than the part can warrant. For this purpose, the Dimensional Inspection Plan designer should know:

1) How many parts are being made from this design?
   a) In this run.
   b) In future runs.
2) What gages do I have available from other inspection procedures that could be used?
3) How long will set-ups take using the various gages available?
4) How valuable is the machine time needed to inspect this part in relation to its importance?
5) What is the cost of investment of acquiring the use of gages not currently available?
6) Is training of personnel a factor in the implementation of this inspection plan?
7) Is the environment required available for use during inspection?

Determine Uncertainty

A) **Uncertainty of the Measurement Plan**

- Has every piece of important information concerning this part and its measurement been used to create the plan?

B) **Uncertainty of the Gages**

- What is the quality of the gages that have been chosen?
- What is the repeatability and the accuracy of the gages to be used?
- Software errors

With the advent of computer analysis of collected information about part features, we were faced with telling the computerized mechanisms how to collect data, how much data to collect and what to do with the data that was collected. Sometimes this was done by those with skills in one area of expertise, such as computer programming and/or mathematics, but that were not knowledgeable in the guidelines that existed in the standards for what was being inspected for (or what uncertainties were introduced by real world mechanisms and environments), not to mention operator error. Standards are now being written worldwide to combine the knowledge of those involved in all aspects of the work environment.

C) **Environmental Uncertainty**

1) **Temperature -** All part dimensions and tolerances apply at a temperature of 20° C (68° F). If both the gage and the workpiece are at 20° C there is no measurement uncertainty caused by temperature. For other conditions the effects of thermal expansion on the gage and the workpiece must be considered.

Should the gage and the workpiece be at the same temperature which is other than 20° C, both will expand by an amount that can be calculated by the formula: **KL (T - 20)**

K = coefficient of expansion
L = length
T = temperature in degrees Centigrade

With the same coefficients of expansion, no measurement uncertainty caused by temperature is introduced. With different coefficients of expansion, temperature related measurement uncertainty is introduced. If the gage or the workpiece is constructed of more than one component and these components have different coefficients of expansion, the structure should be examined to see if an additional uncertainty could occur because of twist or bend. Among the many other factors to consider are:

2) **Slowly changing temperature.** Should the air temperature slowly change over time, and thermal conductivities that are in the structures of the gage and the workpiece happen to be high, uncertainty caused by temperature can be scrutinized based on the premise that the temperatures of the gage and the workpiece are uniform, but not equal.

3) **Quickly changing temperature.** If minor, fast air temperature changes take place and if the gage and the workpiece are of large mass, the effect of the temperature changes could be small. In these situations, not enough heat flows in and out of the gage and the workpiece to change the temperature significantly. Big, fast air temperature fluctuations may impose temperature changes on the gage and the workpiece which can cause them to twist and bend. The uncertainty of this must be taken into consideration and, if possible, avoided.

4) **Other effects on temperature can be caused by:**

a) Radiant energy such as sunlight and artificial lighting. Sunlight should be avoided. Artificial lighting and radiant energy outside the visible spectrum should be brought to a minimum. Indirect lighting is often effective. Lighting should be as uniform as possible to prevent uneven heating of gage and workpiece. One of the most prevalent problems caused by radiant energy is in the **flatness** of large surface plates.
b) Workpieces not stabilized to the inspection environment.
c) Air from heating or cooling ducts.
d) Handling

D) **Other Environmental Uncertainty Effects**

1) Vibration - The characteristic of vibration which causes uncertainty is the relative motion between the gage sensing or contacting element and the measured point on the workpiece.
2) Magnetic Fields
3) Gravity
4) Acoustic Noise
5) Operator Skill
6) Electronic Drift
7) Humidity. The presence of excessive humidity can cause deterioration of gage elements due to corrosion of metal surfaces. It is important to have a measurement environment where humidity is maintained at a level that does not exceed 45%.

A DIMENSIONAL INSPECTION PLAN EXAMPLE

The set of measured values made on the workpiece constitutes the data up-on which calculations are carried out to determine position, size and departure from nominal (desired geometric) form. Data collection may be under manual or automatic control. The use of too few data points or data points inappropriately distributed may provide an unreliable, non-representative description of the actual feature being considered.

Also, to be given consideration is the knowledge and experience of the inspector. Whether operating a CMM or collecting data on a surface plate (or any other) situation, the inspector may know the likely deformations of a particular workpiece. The measurement procedure should take into account that existing user knowledge.

FIG. 324

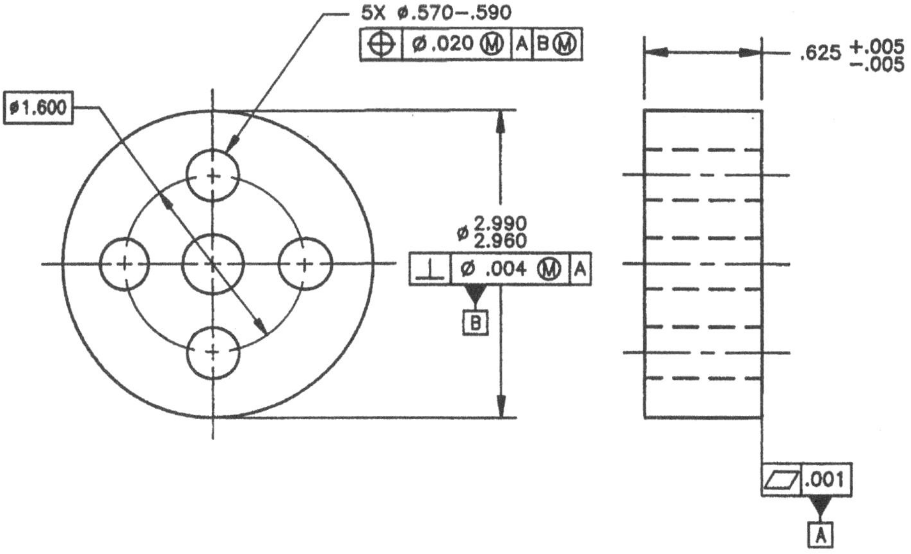

Dimensional Inspection Plan - An Example

Run of 20,000 Parts

- 100 parts per shift, 2 shifts per day, five days per week over 20 weeks.
- 10 parts per shift will be inspected (10% of production per shift) for variable data collection.
- One part per shift will be paper gaged.
- All other parts will be inspected for functionality by use of a positional functional gage for 4-hole pattern's feature control frame.

Brief Manufacturing Procedure Summary for Inspector Information

a) Part is Blanchard ground top and bottom for thickness.

b) Four holes are drilled on a four-spindle drill press.

c) Outside diameter is turned on a CNC lathe, while fixtured to hole pattern.

Brief Functional Description of This Part

This is Part 2 of a two-part assembly. The mating part consists of a cylindrical cavity with four pins inside. The parts assemble and do not spin. This sub-assembly is internal and will not be viewed on the finished product (a helicopter blade assembly).

Features to be Measured

a) For size:

Step 1 Ø 2.960 - 2.990

Step 2 4 x Ø .570 - .590

b) For geometric tolerance:

Step 3 Perpendicularity to within a diameter of .004 at MMC.

Step 4 Position to within a diameter of .020 at MMC to L and D at MMC.

Gages Needed

a) For size:

- One 2-3-inch external micrometer for LMC verification.
- One 2.990 diameter GO-GAGE for external envelope check.
- One 0-1-inch external micrometer for LMC verification with one set of telescoping gages for .570-.590 holes (small hole gages may be substituted).
- One .570 diameter GO-GAGE pin for internal envelope check

b) For geometric tolerance:
   - One surface plate.
   - One height gage.
   - One indexing head with inspection surface plate chuck, 90° indexing capability and magnetic face plate attachment.

Steps for the Development of a Dimensional Inspection Plan

- One gage pin set-.0005 increments between .5700 and .5900.
- One dial indicator (non-plunger type), one-inch long spindle, ball not to exceed Ø .100.
- One functional gage for positional control verification of the 4-hole pattern and the perpendicularity of D to L.
- Graph paper and vellum for paper gaging.

Measurement Procedural Steps

1) Use micrometer to verify 2.960. Record sizes and recommendations for pass, fail, rework.
2) Use GO GAGE to verify 2.990. Record pass/fail.
3) Use micrometer and telescoping gages to verify .590. Record sizes and recommendations for pass, fail, rework.
4) Use GO GAGE to verify .570. Record pass/fail.
5) Use indexing head with magnetic face plate, height gage and dial indicator to verify perpendicularity tolerance. Take differential measurements.
   a) Chart results and recommendations.
   b) Replace magnetic face plate with chucking device. Chuck up on datum feature D while perpendicular to datum L. Record inside diameter of chucking device for use as pattern shift tolerance if needed.
6) Use chucking device, pin gages, height gage and dial indicator to verify position of 4 hole pattern.
   a) Mount part in chucking device:
      1) Up against datum feature L, with a minimum of 3 high point contact.
      2) Grabbing datum feature D, minimum circumscribed cylinder perpendicular to L.
   b) Put gage pins in holes, maximum inscribed cylinders.
   c) Use one hole (gage pin) for angular orientation (if hole pattern does not check within tolerance in this orientation, other orientations may be tried).
   d) Mount dial indicator to Vernier height gage and set both, (along with chuck) onto surface plate.

e) Zero indicator out at height of chucking device's axis and record Vernier scale reading.
f) Go to top of gage pin from which orientation has been taken and find same indicator zero while placing indicator as close to the surface of the part as possible. Record Vernier scale reading.
g) Subtract Vernier scale readings, and from the difference also subtract one-half the diameter of the gage pin.
h) Record the difference between the .800 desired location and the actual location from center of chucking device to center of hole (this hole will only be out of position on the Y axis since it has been used for orientation).
i) While in this orientation, all other hole deviations should be inspected for deviations from the desired height of each from the chucking device's axis and deviations from perfect recorded.
j) The chucking device should then be rotated 90° and deviations of each hole's axis from the desired location recorded by following the same procedures as were used in the first orientation (again, the first hole will not be out at all in this second orientation unless the chucking device has created error while rotating).
k) Each hole may be validated as within tolerance by using the formula of 2 times the square root of the sum of the squares of the deviations on the x and y axes.
l) If the answer to step k is less than or equal to the .020 positional tolerance in each case, the holes have acceptable axial position at the top where inspected. Position must be verified in a similar manner along the axis of each hole. Access to the hole may be made through direct contact with the dial indicator if it is known that the hole is cylindrical in shape to such an extent as to not affect the validity of data collected. If not, differential measurements may be taken to give more valid data.
   - If holes check to be outside of the original .020 diameter positional tolerance, then each hole should be considered for individual bonus tolerance drawn one-for-one as much as that hole has grown from Ø.570 without exceeding its size limits. This bonus tolerance will be determined by the difference in size of the largest gage pin the hole will accept and the Ø.570 MMC.
   - If all holes are still not within their allowed positional tolerances, pattern shift may be considered. It is the deviation of the minimum circumscribed cylinder of datum feature D that is perpendicular to datum plane L (the difference between the inside diameter of the chucking device recorded in step 5b and the 2.994 virtual condition of datum feature D. In order to use this pattern shift for part acceptance, it is recommended that paper gaging be used.

SOURCES OF MEASUREMENT ERROR

WHAT WE KNOW IS OFTEN UNTRUE AND OTHER SOURCES OF MEASUREMENT ERROR

Consider the inspector who wants to do a good job, but also seeks a bit of comfort in his life. He oversees the construction of an environmentally controlled room in which to store his inspection equipment and to conduct his measurement procedures.

He makes certain the room temperature is set at a constant 68° F/20° C. The first time the thermostat is mounted, he stares at it proudly, then walks the thirty or forty feet to his CMM and places the first of a series of workpieces to be measured onto the machine table. He pauses for a moment to turn his face toward the window and bathe in the warm sunlight, then opens his eyes to admire the view of the nearby electrical power plant. Soon, he notices the air conditioning vent overhead which emits a soft blast of cold air that eventually reaches the thermostat and settles the room temperature over there at the appropriate number of degrees.

He feels the floor vibrant with the passing of a forklift just outside the confines of the room and smiles at his isolation from all that. The sounds of presses pounding out parts is muffled by the walls of his room and even the large man grinding burrs off of sheet metal is faint although the sparks are bouncing against the window of his door. The grinding dust is so thick outside this entrance that it engulfs his friend as he enters to admire the new set-up. His friend comments on the great set-up. The inspector beams with pride as he laughs and nods in agreement, and places his hand on the part to be measured, leaning his weight and warmth into it.

Eventually his friend leaves and he begins the measurement. He probes a minimum number of points on the part's surfaces to establish the datum reference frame, the three perpendicular planes the computer will create and that will act as his origins of measurement. He inadvertently probes some low as well as high points. The computer gives equal weight to each in establishing the datums. If, on the second part he measures, he decides to probe more points to establish his datum planes, the computer algorithms in the program he's using will establish an average "best fit" plane from all the points probes. These are not the high points of the surface most likely to seat in the assembly and orient the part. If he had taken the time to fixture the part and probe the fixture or parallels and angle plates that would naturally rest against the highest points on the surface, his results would be more representative of the actual feature locations as described in the ASME, ANSI approved Y14.5 standard on dimensioning and tolerancing, but the person who sold him the machine told him the computer negated the need for fixturing.

He finishes his measurements of this part in record time and decides to use the time he has saved to take a coffee break.

Maybe he should have taken the coffee break instead of measuring the part. Each act has the same amount of value added to the product. Or perhaps the coffee break, having less negative value, is the one that, in the future, should take the higher priority.

We all want to believe that we are hard workers and competent in what we do. Sometimes the reality of what we think we know just doesn't happen to be true or even remotely correct. But don't tell this inspector. He might take it as a personal affront. After all, to him, it seems as though you're trying to negate his knowledge and skills, a major part of his existence. His ignorance is blissful.

The sad truth is that messengers of what could be liberating knowledge are often killed. Study your history!

Still, the problem with doing nothing to improve these errors is that it eventually shows up in product quality. Let's face it, the job of the inspector is to validate whether or not a feature meets its requirements and contributes to the overall part functionality. And if the measurement errors are (as I have described here) legion, how can we place even the remotest confidence in the findings?

Chapter 20

Flexible Parts

Chapter Objectives

Readers will learn:

1. To tolerance flexible parts.
2. The difficulties in fixturing and inspecting flexible parts to collect repeatable measurement data.
3. About free state variation.
4. That part should be inspected in the way that they function.
5. To apply temporary datum features for in-process inspection.
6. Common misconceptions regarding the tolerancing and inspection of flexible parts.
7. The criteria for selecting datum features on flexible parts.
8. The "free state" symbol and what it means.
9. How to write a "restrained state" note.
10. That all parts are to be measured in the free state, unless restraint applied does not distort measurements, or unless a note is written to allow inspection in the restrained condition.
11. To dimension and tolerance a fixture for simulate datum features for inspection.
12. How to apply and fixture "step datums".
13. How to use multiple slot centerplanes to create a datum reference frame.
14. The most popular and effective datum scheme for tolerancing sheet metal panels.

Flexible Parts

Fixturing consistency problems are a constant for sheet metal workers. The soon-to-be-panel is often fixtured to be punched and drawn, with bends added and the panel trimmed. Moved from die to die or station to station and refixtured each time, it is often again fixtured in a completely different fashion to be welded into the assembly. Somewhere in the middle of all that, it may be inspected, sometimes while restrained, either in a similar or dissimilar way, to one of the aforementioned procedures. Hopefully, it is one that duplicates the way the panel fits into the assembly; but many times, it is not even close.

Often, the panel is inspected while it sits on the table of a Coordinate Measuring Machine. The panel is probed while it lays there in the "free state", sagging, varying greatly from what it is in the "restrained condition". The inspector is getting nowhere. Without proper fixturing, he or she could inspect the part until retirement and still not be able to tell us if it functions. Some of these procedures will continue, cannot be changed, and will not be made more repeatable. Others can be changed.

Parts should be inspected in the way they work!

They should be inspected in as close a re-creation of their operating orientation as possible, using the same type of restraint (or the lack thereof). The drawing should reflect restraints or free state inspection by the datums, datum targets, and augmenting notes it uses. Design drawings should reflect functionality. They are not exclusively in-process (process engineering) drawings. They can include in-process information if a company chooses, but that information should not conflict with or be confused with functional information. One thing is certain: anyone reading an engineering drawing should be able to distinguish between:

(a) information put there to help process the part through manufacturing, and
(b) information meant to impart functional requirements.

All too often this information becomes indistinguishable. Designing for producibility, a most desirable goal, is sometimes so sought after that it gradually pushes aside or confuses the information put on the drawing that would have enabled the part to work/assemble in the desired fashion.

For clarification, in-process information is sometimes conveyed on a separate process engineering drawing. If done, this extra drawing should be created in such a fashion as to allow the finished part to comply not only with this process engineering drawing but also with the functional information conveyed in the design drawing. Many process engineering drawings are unnecessary. If the manufacturing personnel receive a complete and correct drawing that relates functional requirements, they will use their experience and knowledge to devise a manufacturing procedure that produces such a part. This process is made easier by the early involvement of the manufacturing personnel during the design process (concurrent engineering).
At other times, it is decided that it is best to convey some manufacturing "principle locating points" (in-process datums/set-up points, lines and areas), along with the functional datums, right

on the field of the same (one-for-all) drawing. In these cases, it is especially important that one kind of information be distinguishable from the other. See FIGURE 325.

A good functional design drawing will have geometric controls on it that, if set up and inspected, include all of the critical interrelated features on a part. When inspected, if possible, all these interrelated features are inspected at one time or in one set-up, tying together all the important interrelationships between all controlled critical features and their functional datums.

FIGURE 325 [Temporary Datum Features]

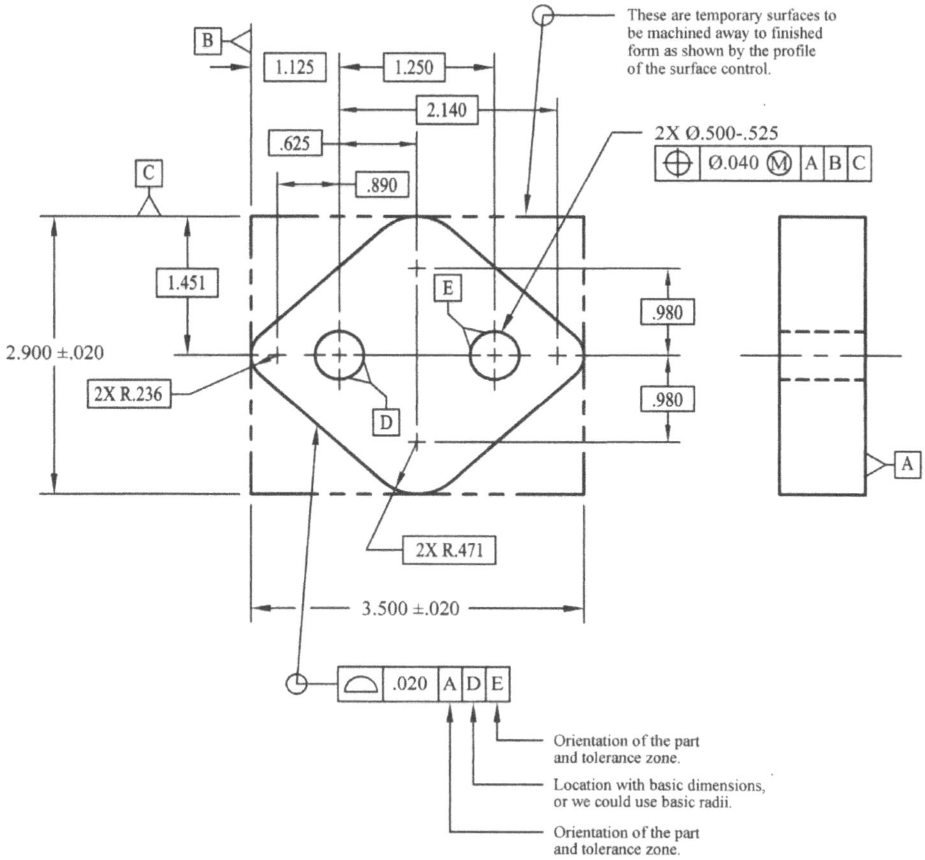

When a profile control is specified and uses a datum feature such as datum targets, the tolerance zone, if equal bilateral, is equally distributed about the part's perfect (basic) configuration. The datum plane formed by the datum target points helps to stabilize the part for measurement purposes. However, care must be taken if the datum targets are on surfaces that are part of what the profile control is applied to. If the profile tolerance zone is equal bilateral, the points that create the datum plane carry the datum plane with them. The profile tolerance zone, as it approaches the datum plane would split in two, with half on each side of the datum target points that have created the datum plane. The rest of the surface that contains the datum targets would be able to lie anywhere within the entire profile tolerance zone. A situation such as this is depicted in the GD&T sheet in FIGURE 333 where datum target areas A1, A2, A3 and A4 are shown on a surface profiled to datums A, B and C.

Common Misconceptions

- Datum targets are often used to establish a datum reference frame on molded plastic parts, sheet metal parts and cast parts. There have long been common misunderstandings about datum targets. For example, when datum target areas are used, it is often misunderstood that you can set up on points or a single point anywhere within the area. The principle is actually that contact should be attempted with the full area of the datum target (really, contact with the high points of the area).

- On cast or molded parts, many believe the area, by virtue of the fact that it simply is a datum target area, is machined. It is not required that the area be machined unless an additional control stipulates that it must.

- Another **misconception is that sheet metal parts are <u>required to be restrained</u>** (during inspection) to something like spherically-tipped pins (for datum target points), the side of a cylindrical pin (for datum target lines) and flat-topped pins or blocks (for datum target areas). Although the parts are to be placed on these datum target simulators, they are not restrained unless a note such as "to be inspected while restrained" is placed below the feature control frames that invoke the datums or such a note is specified in the general notes on the drawing. This is to prevent the part from being distorted arbitrarily to achieve compliant results.

- **Parts are to be inspected in the "free state" unless otherwise specified.** However, the ASME Y14.5 standard on dimensioning and tolerancing makes provision for clarification of intent (a clarifying redundancy). The symbol for "free state" (a circled F) is placed inside the feature control frame next to the geometric tolerance indicating that the feature will be inspected while unrestrained (in the free state). **If the feature and/or part is to be inspected while restrained, it must be noted on the drawing as a local or general note.** The note should be as clear and specific as necessary to ensure the feature and/or part is inspected in a way that very closely reflects the way it functions or fits into the assembly. **This note must specify the maximum clamping force allowed during inspection.**

The flexibility of the non-rigid parts often adds to confusion as to how to sufficiently stabilize them. Datum targets, when used, can number far beyond the three points of surface contact needed for construction of the primary datum plane recommended for rigid parts. In fact, some engineers will blend datum holes and targets (points, lines and/or areas) to stabilize a part and are often unable to distinguish the location of datum planes constructed by this conglomeration of features. Nor are they able to tell which of the large group of datum features and targets used creates the primary from the ones that create the secondary or tertiary datum planes.

When many datum features are used to create one datum reference frame, it is often the engineer's way of saying that the relationship between all controlled features and all datum features must be held as one simultaneous requirement. It is, however, not usually the most efficient way of saying so. As mentioned, the origin axes and planes can be blended in a way that is too confusing to determine the actual datum reference frame (three mutually perpendicular planes) from which to measure. At times, it is wiser to create a simultaneous requirement by relating many features to one simple datum reference frame. A clarifying note may even be added beneath each control that states that all are SIM REQT (Simultaneous Requirements). This is one reason the newest revision of ASME Y14.5 does not require the datum planes to be labeled when using datum targets. Basic dimensions will merely emanate from the targets or from planes traceable by basic dimensions from the datum targets.

Free State Variation in Sheet Metal Parts

Some concern has been expressed by designers of sheet metal parts about free state variation. Thin-walled materials such as sheet metal, as well as some plastics and rubber parts, are often prone to wide variations in form when released from the restraints used to machine or form them. Holding these features to specified tolerances of geometric control requires close examination of when and how the part function dictates the need for these controls.

For example, if a sheet metal part is to be used in a restrained state (bolted, riveted or otherwise clamped against several surfaces of the part), then perhaps the final shape or position (form, orientation, runout, profile or location) of features should be judged to be in or out of tolerance while in this restrained state.

A note added to the drawing, such as **"This feature shall be inspected while restrained (bolted, clamped, etc.) to the specified datum(s) with the following torque . . ."**, will allow the controlled features to be judged based on a simulation of the actual conditions the part will experience while in use. As such, any geometric tolerances assigned the features take on a realistic value and the datum features are chosen and utilized based on proper criteria:

- function
- representation of the mating situation
- repeatability
- accessibility.

Figure 326 states that all geometric controls are to be inspected in the restrained condition UNLESS OTHERWISE SPECIFIED. Two of the profile controls have been specified to apply in the free state. The first free state control is to qualify the datum features as complying with their profile tolerance. Once qualified, the datum features that comprise A are utilized to measure the surfaces for coplanarity and parallelism to datum plane A to within 2 millimeters.

The top of the surface plate in FIGURE 326 simulates datum plane A from the high points of both surfaces simultaneously while the part is in the free state. Distortion due to free state variation is not controlled. The part is not clamped onto the datum features to try to compensate for the warp experienced by the part during the inspection procedure.

The dial indicator traverses both surfaces controlled for coplanarity with the profile characteristic. This is measured while the part is in the free state to determine if the surfaces comply with the 2-millimeter profile tolerance. The surfaces must be coplanar (straight, flat, parallel and in the same plane) within a tolerance zone that consists of two parallel planes 2 apart. The tolerance zone must contain all elements of both surfaces simultaneously. Only then is the part to be restrained to datum feature A in order to measure the coplanarity of the top two surfaces to within 0.4 millimeters.

FIGURE 326 [Free State Inspection of Flexible Parts] The free state symbol may only be applied on feature tolerances on parts that include one or more restraint notes.

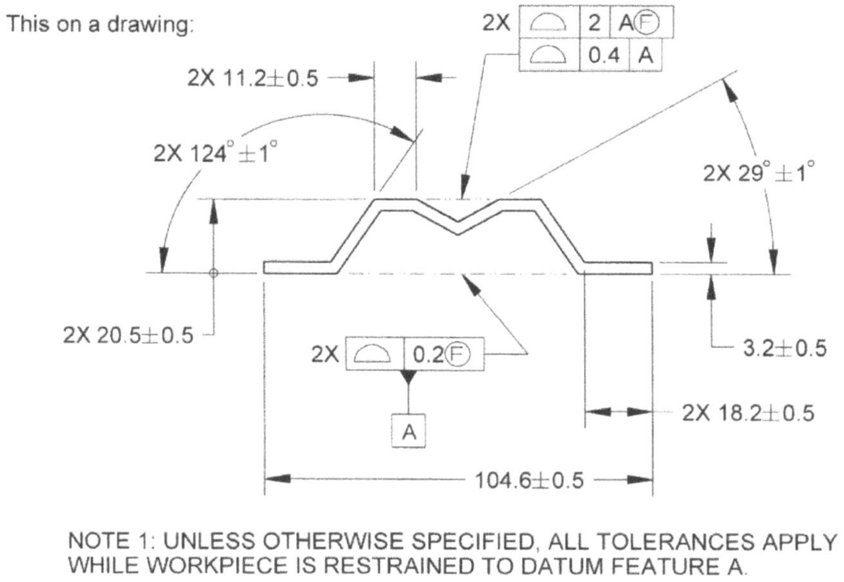

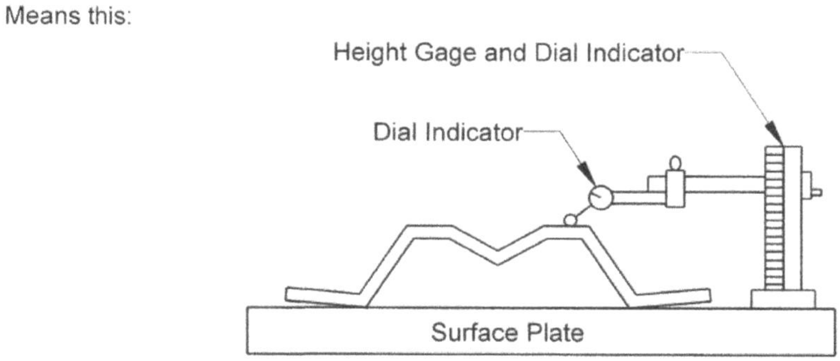

The tolerance zone is parallel to datum plane A simulated by the surface plate. Note that since a 0.2 coplanarity control has been applied to the datum features that form datum plane A from high point contact, the datum surfaces may be angled to and depart from the surface plate which simulates the primary datum plane A by that 0.2 tolerance.

Unless otherwise specified, all geometric controls are considered to apply in the free state, unless a restrained state condition is specified on the drawing.

As stated above, when it is specified by note that a geometric tolerance is applied to a restrained part or assembly, and that part or assembly requires one or more free state (unrestrained) datum references, unrestrained datum features are designated by the free state symbol applied to a datum feature referenced in the feature control frame following any other modifying symbols.

When some tolerances are specified as applying in the restrained condition (usually by a note) and additional tolerances are applied in the free state, the symbolic means of indicating that the toleranced features or datum features apply in their free state, it is done in the following ways:

1) When the free state symbol is applied to a tolerance in the feature control frame, it shall follow the stated tolerance and any modifier.
2) When the free state symbol is applied to a datum feature reference, it shall follow the datum feature reference and any modifier.
3) When the free state symbol is given to a directly toleranced dimension, it shall follow the dimension and its tolerance.
4)

If this free state inspection condition is deemed non-functional and/or inappropriate, then a restrained state inspection condition may be used instead. This is accomplished by a local or general note. See FIGURE 327.

FIGURE 327 [Restrained State Control of Flexible Parts]

This on a drawing:

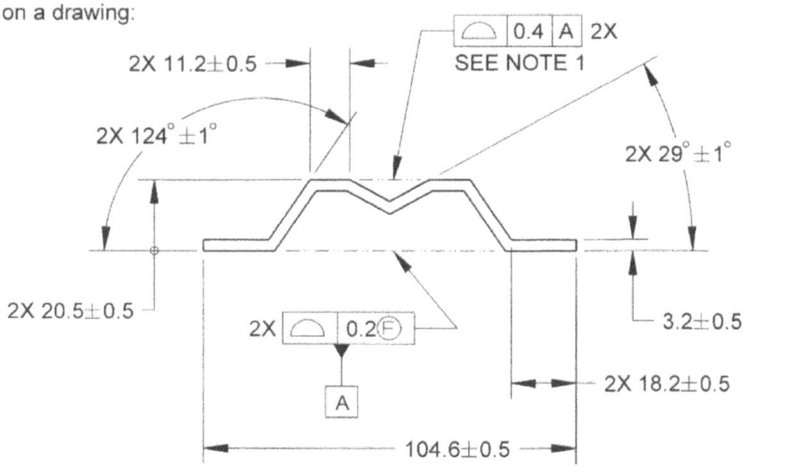

NOTE 1: THIS CONTROL TO BE INSPECTED IN A RESTRAINED STATE WHILE THE DATUM SURFACES, THAT FORM DATUM PLANE A, ARE MOUNTED AGAINST A FLAT RIGID SURFACE USING CLAMPS ON EACH SIDE OF THE TWO SURFACES. THE CLAMPS ARE TO BE TORQUED TO 9-15 N·m.

Means this:

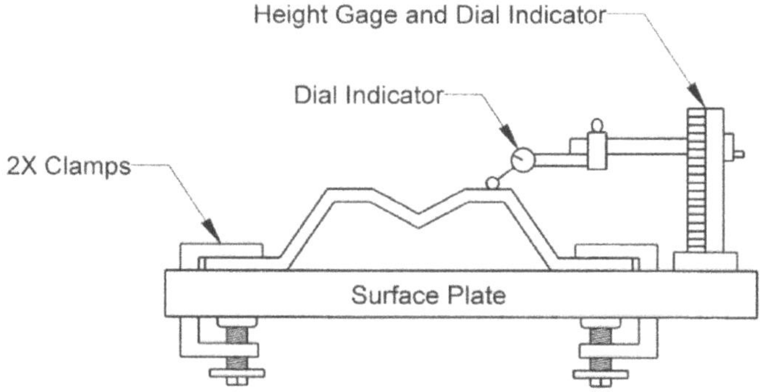

In Figure 327, the top of the surface plate simulates datum plane A from the high points of both surfaces simultaneously while the part is in the restrained state. Distortion due to free state variation is controlled. The part is clamped onto the datum features to try to compensate for the warp that may otherwise be experienced by the part during the inspection procedure.

The dial indicator traverses both surfaces controlled for coplanarity with the profile characteristic. This is accomplished while the part is in the restrained state to determine whether or not the surfaces comply with the 0.2 profile tolerance. The surfaces must be coplanar (straight, flat, parallel and in the same plane) within a tolerance zone that consists of two parallel planes 0.2 apart. The tolerance zone must contain all elements of both surfaces simultaneously while the part is restrained as prescribed in general note 1. Any free state variation of the datum features is negated by the restraint created by the clamping devices. This restraint should be, whenever possible, reflective of how the part is restrained during actual use (for example, assembly). The tolerance zone is parallel to datum plane A simulated by the surface plate.

For over-constrained datum reference frames, the datum features or datum targets referenced as free state are not required to contact the physical datum feature simulator when in the free state. For example, if the part has four datum target areas that are used to create a datum plane in the free state, the part might not (probably will not) rest on all of the areas that are used to establish the datum plane. **See Figure 328.**

FIGURE 328 Datum Features Referenced in the Free State

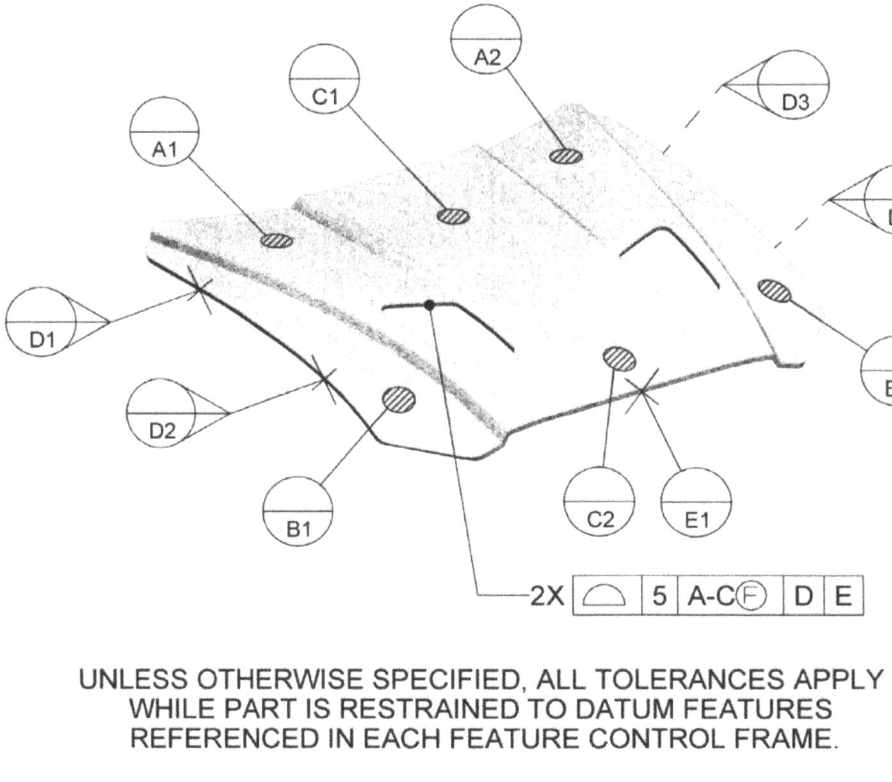

However, when restraint specified to be applied, all restrained non-size datum features shall contact the physical datum feature simulators, unless otherwise specified. Features of size referenced MMB are not required to make contact with the simulator.

FIGURE 329 [Fixturing Sheet Metal Parts]

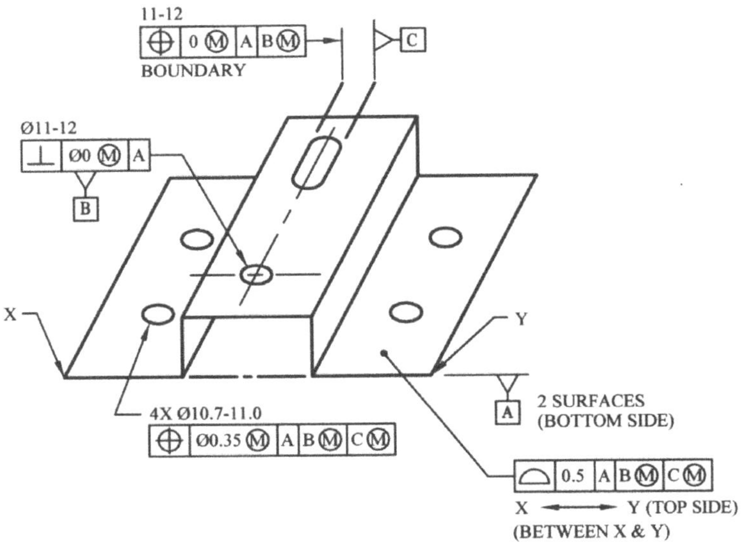

Means this:

The part in Figure 329 may be fixtured with a device similar to the fixture shown in the next illustration. This fixture may be used during inspection of the part to establish the appropriate datum reference frame. The part will seat on the surface that simulates datum feature A, while shafts that are sized at the virtual conditions are inserted into the holes that are datum features B and C.

Only the width of datum feature C will be used to establish the orientation of the measurements originating from the axis of datum feature B. The axis of datum feature B is simulated by the center of the cylindrical pin/shaft. The centerplane of the C width of the rectangular shaft that is aligned with B will serve as the angular orientation datum only. See FIGURE 330.

FIGURE 330 [Fixture]

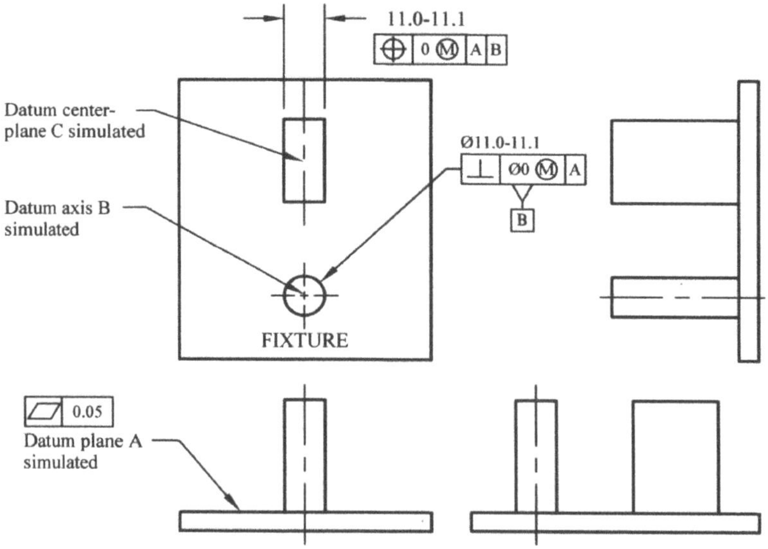

If the geometric controls used the datum features at regardless of feature size (RFS), the fixturing pins would be required to be expanding pins (sometimes simulated by tapered pins on very thin parts). This would locate the part and the holes on the part on the fixture without the shift allowed by the datum features controlled at MMC. This shift is equal to the difference in the actual size of datum features B and C and the fixture pin size (the virtual condition of the features).

FIGURE 331 [Fixturing Sheet Metal Parts with Step Datum Targets]

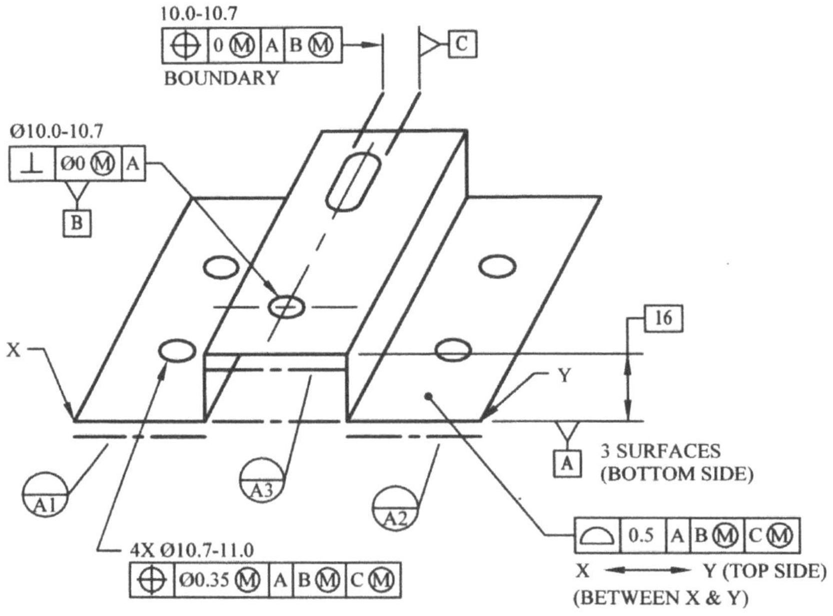

Means this:
The part in FIGURE 331 may be fixtured with a device similar to the one shown in the next illustration. This fixture may be used during inspection of the part to establish the appropriate datum reference frame. The part will seat on the surfaces that simulate datum feature A. This is termed a step datum.

One of the seating surfaces (A3) for the establishment of datum A will be 16mm taller than the other two (A1 and A2). The shafts that represent datum features B and C are inserted into the holes. Only the width of datum feature C will be used to establish the orientation of the measurements originating from the axis of datum B. The axis of datum B is simulated by the center of the cylindrical pin/shaft.

If the geometric controls used the datum features at regardless of feature size (RFS), the fixturing pins would be required to be expanding pins. This would locate the part and the holes on the part on the fixture without the shift allowed by the datum features controlled at MMC. The radial shift away from the datum axis allowed because of the datum features being referenced at MMC is equal to one-half the difference in the actual size of datum features B and C and the fixture pin size (the virtual condition of the datum features).

FIGURE 332

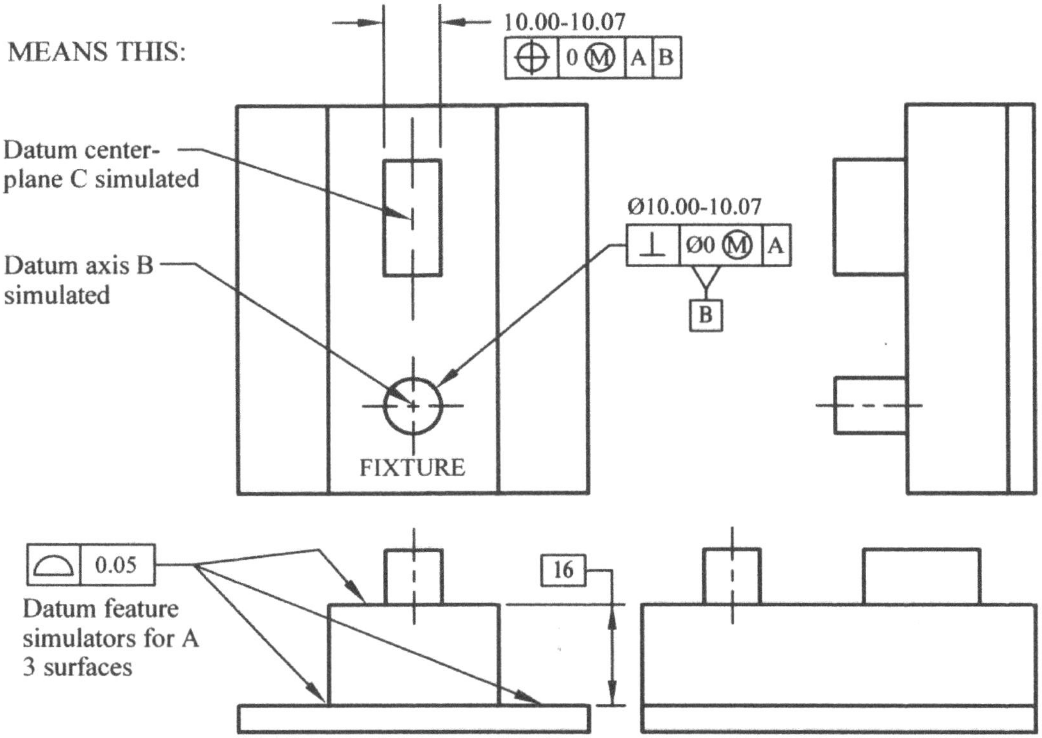

FIGURE 333:

FIGURE 333 is an illustration of the use of a geometric tolerance wherein the entire three-dimensional sheet metal panel is located using profile of the surfaces "ALL OVER" from three mutually perpendicular planes. In ASME Y14.5-2009, the note "ALL OVER" was replaced by a symbol consisting of two concentric circles placed on the elbow of the leader line from the profile feature control frame (See FIGURE 333 for an example). This symbol is also preferred practice in ASME Y14.5-2018. However, use of the note "ALL OVER" in place of this symbol is still acceptable.

The primary datum plane is created through the use of datum target areas. Eight areas are used because of the size and flexibility of the panel. If less were used, the panel would sag and not be stable enough to assess for compliance with functional requirements. Two slot widths are used to form a common compound centerplane B-C that acts as the secondary datum. Two more slot widths are used to form a common compound centerplane F-G as a tertiary datum that is mutually perpendicular to datums A and B-C.

FIGURE 333 [Datum Features and Target Areas for Sheet Metal Parts – Example 1]

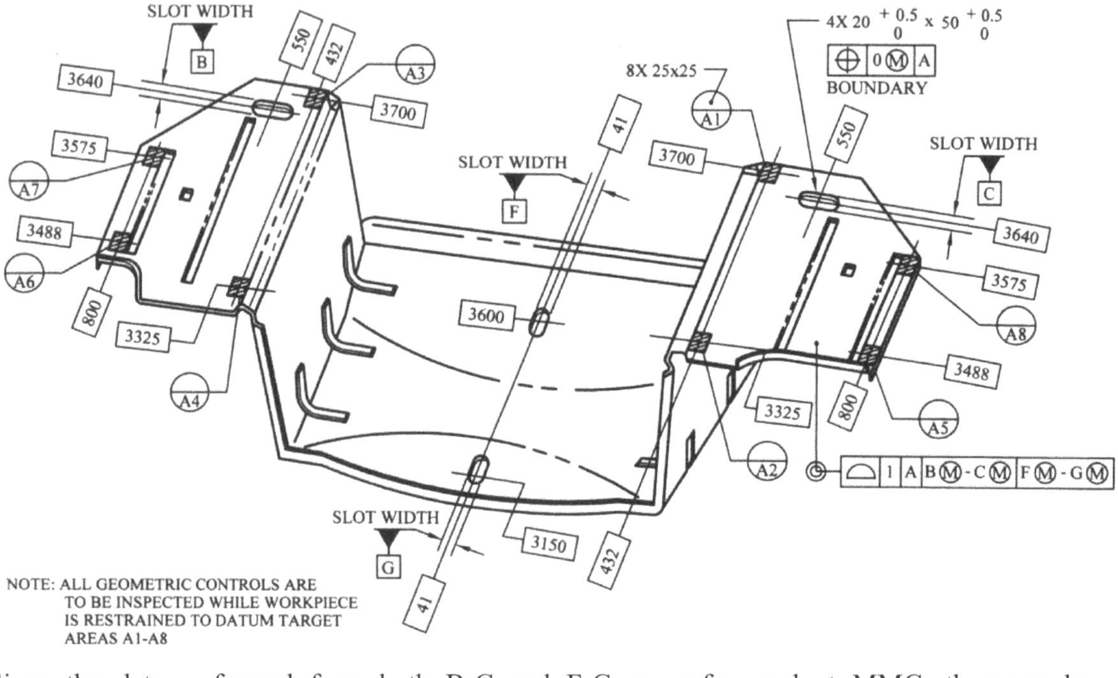

Since the datums formed from both B-C and F-G are referenced at MMC, they can be represented by gage or fixturing blocks sized at virtual condition and located at true position. Each fixturing block is represented only in terms of the virtual condition of the width of the elongated hole it represents. Minimal data as to the location of datums has been shown on this drawing. The remainder of the data regarding panel surface sizes, locations, angles and curvatures must be retrieved from the data base CAD model concerning this panel or subsequent drawing sheets not shown here.

As mentioned, the symbol for "ALL OVER" (two concentric circles) is used on the elbow of the leader line from the profile of a surface feature control frame. **This means that the control applies in all views simultaneously.** The produced surface of the part must be within plus or minus 0.5mm of the CAD model to meet the requirement. This part is from the automobile industry. Commonly, it is measured while fixtured to the datums on a CMM. The CAD model is downloaded to the software of the CMM and tolerance zones created within the software. Many points along the surface are then probed and their locations in space compared to the CAD model. If all points are within the 1mm total wide tolerance zone, the profile of the part is judged to be in compliance.

FIGURE 334:

FIGURE 334 on the following page is an illustration of the use of a datum reference frame wherein segments of the surface of a sheet metal panel are located using profile of the surface. Local notes are given beneath each feature control frame of profile of the surface to explain the extent of that control.

The primary datum plane is created through the use of four datum target areas. These four areas have been deemed appropriate to act as the primary datum features for part stabilization because of the size, material type, weight and configuration of the part. One hole has been used as the secondary datum feature. This hole will construct a secondary datum axis consisting of two intersecting planes which will be perpendicular to each other and to datum plane A. One of the planes crossing at the axis of datum feature B will also be the centerplane of tertiary datum feature C (the slot width).

Since the datum features are referenced at MMC, they are represented at their virtual condition. As shown in the section views, rectangular blocks may be used to simulate targets A1 - A4. These blocks will be the size and shape of the datum target areas. All missing information regarding datum features and the profile will be found on subsequent drawing sheets (not shown here) or in the computer data base CAD model for this part.

FIGURE 334 [Datum Features and Target Areas for Sheet Metal Parts – Example 2]

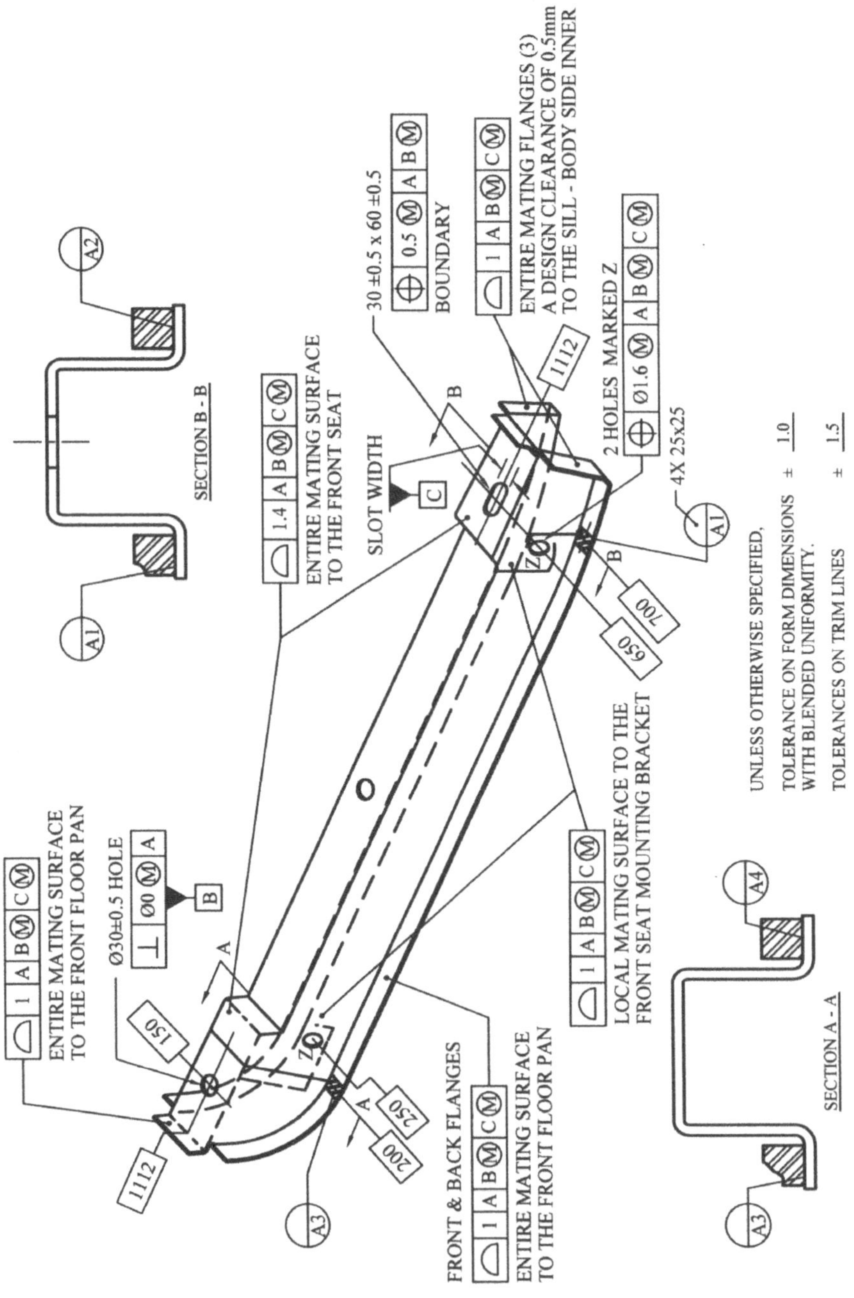

TROUBLED TIMES AND TECHNICAL LEAPS

Henry Miller, a famous dead author wrote in 1941 "Over and over again I have said that there is no way out of the present impasse. If we were wide awake we would be instantly struck by the horrors which surround us, we would drop our tools, quit our jobs, deny our obligations, pay no taxes, observe no laws, and so on. Could the man or woman who is thoroughly awakened possibly do the crazy things which are now expected of him or her every moment of the day?"

Well, Henry, where ever you are, I'd like you to know that here it is the 21st century and things aren't any better yet. The airline seats are too small and we have to use them too often and they don't serve peanuts because the guy in row 38 has an allergy, or maybe he just wants to ruin your day! We institute the best programs of design, manufacturing and quality into our systems by teaching our employees the buzz words but not making certain they are clear on the concepts. They walk around speaking in what must seem to them completely non-coherent tongues, saying things like TQM, SPC, Just-in-Time, GD&T and Team Building. They say these things because they are part of the industry jargon they know is needed to survive within their department/company/environment, but they lament over the poor souls (not themselves, of course), that "won't be able to deal with these complicated concepts" [translation - "Do we really have to do this stuff? I'm too busy doing the real work to try out your stupid newest Program-Of-The-Day."] But they wear a really concerned look when they say it, which they spent a great deal of time trying out in the mirror that morning, and consequently seem so sincere that another insecure soul who was thinking exactly the same thing thinks it's all right to agree with him and, therefore, not do the follow through necessary to make the program really work.

Well, I understand their real concern. There is a whole lot of new "stuff" we have to learn to survive in today's industry. It is difficult and frightening and wonderful and exciting. And contrary to dead Henry's suggestion, throwing up one's hands and running for the safety of hiding under the covers is only one of the options.

It's time to get on with the business of business, which I'm afraid in this day of chips (that are not of bovine origin), means the business of education. Cast aside your memories of mean old Mrs. Crookshank making you stand in front of the class holding two heavy books at arm's length until you suddenly, and somehow miraculously, were struck with the answer to the question, or soiled yourself in front of the entire class from the fear and injustice of it all (a memory that still haunts you to this day). These are new times and as the old song goes "You're not a kid anymore". Learn. Pack up your troubles in an old kit bag and study, study, study.

I've often said "Talk the Talk....We'll Work on the Walk Later." Well, it's later and maybe later than you think. It's time to put into action all the concepts behind the buzz words you have spent so much time using on people. Become a champion for the cause of better drawings, better manufacturing techniques, better communication between departments and a higher-quality product at a lower price to the consumer - even if you are the consumer.

Apply what you have learned, learn more, extend the research, personalize the techniques to your product line, and spread not only the word, but the zeal and the commitment. Make it work and teach others how it does. Become a mentor. Don't hoard the knowledge, thinking it gives you

power over others. Share the knowledge and confer. Confide that you don't know everything there is and allow others to add a missing piece to your puzzle.
Open up and admit to yourself and to others that you are human and, therefore, flawed. Admit that you haven't really believed you were perfect since you were about 16 years old and that you need the help of others for you to continue to learn and to extend what is currently known (hopefully to share with the rest of us).

We really do need one another, and until you truly believe that, "Teamwork" will just be another of those words that you mouth to get you through the next day of politics while you continue to do what you think is the "Real Work".

"I inherited it brick and left it marble". (Of the city of Rome) - Emperor Augustus 63 B.C.- A.D. 14

Chapter

Definitions

Definitions

Absolute Gage Tolerancing (pessimistic tolerancing). A policy of tolerancing gages that insures the complete random interchangeability of parts to assemble by applying the gagemaker's tolerances, the wear allowances, the measurement uncertainties and the form controls--all within the workpiece limits of size. This policy featured for GO and NOGO gages by the ASME Y14.43 standard on Dimensioning and Tolerancing Principles for Gages and Fixtures.

Actual Local Size. The value of any individual distance at any cross section of a feature.

Actual Mating Envelope. For an external feature of size, it is the perfect feature counterpart that is the smallest size that contacts the surface at the highest points. For an external cylindrical feature, it would be the minimum circumscribed cylinder. For two parallel planar external opposing surfaces, it would be two parallel planes at minimum separation. For an internal feature of size, it is the perfect feature counterpart that is the largest size that contacts the surface at the highest points. For an internal cylindrical feature, it would be the maximum inscribed cylinder. For two parallel planar internal opposing surfaces, it would be two parallel planes at maximum separation.

For features controlled with a geometric characteristic inside a feature control frame, such as perpendicularity or position, an actual mating envelope can be created at perfect orientation or position to specified datums, as appropriate.

An actual mating envelope that is not constrained by datums is known as an Unrelated Actual Mating Envelope. An actual mating envelope that is constrained by datums is known as a Related Actual Mating Envelope.

Actual Value. The smallest tolerance zone capable of containing the actual produced feature's surface, axis or centerplane as appropriate to the applicable geometric control assigned.

Algorithm. A step-by-step description in mathematical terms of a process for solving a particular problem--for example, the determination of the parameters describing a geometric feature.

All Checks in Free State. Workpiece measurements shall not be distorted to obtain compliant results. Per ASME Y14.5-2018, unless otherwise specified, all tolerances apply in the free state condition. Also see ASME Y14.43-2011 states, "Unless otherwise specified, all workpieces are to be inspected in the free state.

Angle Plates and Angle Blocks. A measurement accessory to give the inspector perpendicular surfaces to clamp or rest workpieces (which are under test) against. These plates and blocks act as a fixture or fixturing elements.

Angular Gage Blocks. Precisely ground (or otherwise manufactured) blocks of recorded angles to be used or stacked to create needed angles which a workpiece may be rested on or compared to.

Angularity. The condition of a surface, centerplane, or axis at a specified angle (other than 90°) from a datum plane or axis. Per ASME Y14.5-2009 and ASME Y14.5-2018, "As an alternative practice, the angularity symbol may be used to control parallel and perpendicular relationships."

Angularity Tolerance. An angularity tolerance specifies one of the following:

(a) a tolerance zone defined by two parallel planes at the specified basic angle from one or more datum planes, or axis, within which the surface of the considered feature must lie.

(b) a tolerance zone defined by two parallel planes at the specified basic angle from one or more datum planes, or axis, within which the axis or centerplane of the considered feature must lie.

(c) a cylindrical tolerance zone at the specified basic angle from one or more datum planes, or axes, within which the axis of the feature must lie.

(d) a tolerance zone defined by two parallel lines at the specified basic angle from a datum plane or axis, within which the line element of the surface must reside.

Angular Orientation Datum. A datum used to control the radial orientation (rotation) of a feature or a pattern of features.

Attributes data. Information obtained from an inspection process that indicates only whether a part is acceptable or not acceptable.

Attribute Gage. The family of receiver gages used to collect attributes (good vs. bad) data. For example, Go Gages, NOGO Gages and Functional Gages.

Average Diameter. The diameter derived from averaging a minimum of four diametral measurements across the axis of a circular cross section of a cylindrical or spherical feature.

Axis. The imaginary perfect center line of a cylindrical feature created by the intersection of two planes in space at 90° to one another.

Basic Dimension. The theoretically exact size, angle, profile, location or orientation desired by design. It can be shown by putting the untoleranced dimension inside a box (rectangle), or with a drawing note such as "Unless otherwise specified, all untoleranced dimensions shall be considered basic". Since all dimensions need a tolerance, the tolerance for these basic dimensions can be shown in feature control frames, drawing notes or other documents specified, such as established tooling or gaging tolerance documents of a facility.

Book Mold. A split mold hinged like a book.

Boundary, Inner. A worst-case envelope of the same configuration as the feature being considered. It is generated by the smallest feature (MMC for an internal feature and LMC for an external feature) minus the stated geometric tolerance and any additional geometric tolerance (if applicable) from the feature's departure from its specified material condition.

Boundary, Least Material (LMB). The limit defined by a tolerance or combination of tolerances which exists on or inside the material of a feature(s). For a datum feature hole referenced in a feature control frame at LMB, the LMB is equal to the Least Material Condition plus any applicable geometric tolerance. For a datum feature shaft referenced in a feature control frame at LMB, the LMB is equal to the Least Material Condition minus any applicable geometric tolerance.

Boundary Maximum Material, (MMB). The limit defined by a tolerance or combination of tolerances which exists on or outside the material of a feature. For a datum feature shaft referenced in a feature control frame at MMB, the MMB is equal to the Maximum Material Condition plus any applicable geometric tolerance. For a datum feature hole referenced in a feature control frame at MMB, the MMB is equal to the Maximum Material Condition minus any applicable geometric tolerance.

Boundary, Outer. A worst case envelope of the same configuration as the feature being considered. It is generated by the largest feature (LMC for an internal feature and MMC for an external feature) plus the geometric tolerance and any additional geometric tolerance (if applicable) from the feature's departure from its specified material condition.

Candidate Datum. A datum point, axis or plane established from a datum feature. It is most often one of many such imaginary datums. The candidate set is created when irregular surfaces rock or wobble when they come into contact with their perfect geometric counterparts or simulations of their perfect geometric counterparts.

Candidate Datum Reference Frame. A datum reference frame established from one set of candidate datums.

Candidate Datum Reference Frame Set. The set of all candidate datum reference frames established from the referenced datums of one feature control frame.

Candidate Datum Set. The set of all candidate datums that can be established from a datum feature.

Cartesian. Pertaining to Descartes or his philosophical or mathematical ideas.

Cartesian Coordinates.

a) A pair of numbers that locate a point by its distances from two intersecting, often perpendicular lines in the same plane. Each distance is measured as parallel to the other line.

b) Three numbers which locate a point by its distance from three fixed planes that intersect one another at right angles.

Casting. An object at or near finished shape obtained by solidification of a substance in a mold; pouring molten metal into a mold to produce an object of desired shape.

Centroid. The point described by coordinates which are the geometric means of the corresponding data points. The point or axis which exists at the center of a corresponding set of coordinate data points or axes.

Centroid of Revolution. The point or axis around which anything revolves.

Circularity (Roundness). A condition of a surface of revolution where: (a) for a cylinder or cone, all points of the surface intersected by any plane perpendicular to an axis are equidistant from that axis; (b) for a sphere, all points of the surface intersected by any plane passing through a common center are equidistant from that center.

Circularity Tolerance. Specifies a tolerance zone bounded by two concentric circles within which each circular element of the surface must lie, and applies independently at any plane (as defined in the definition of circularity). The circularity tolerance form control must be less than the form control given by the size tolerance, except for those parts subject to free state variation. A circularity tolerance specifies that all points of each circular element of the surface must lie in some zone bounded by two concentric circles whose radii differ by the specified tolerance. Circular elements are obtained by taking cross sections perpendicular to some spine. For a sphere, the spine is a point, and for a cylinder or cone the spine is a simple, non-self-intersecting, tangent-continuous curve. The concentric circles defining the circularity zone are centered on, and are in a plane perpendicular to the spine.

Coaxiality. The condition wherein two or more axes share the same place in space.

Comparator Gage. A gage that uses an indicator to compare differences between a master and a produced workpiece.

Complex Feature. A single surface of compound curvature or a collection of other features that constrains up to six degrees of freedom.

Composite Positional Tolerancing. This provides a composite application of positional tolerancing for the location of feature patterns as well as the interrelation (position and orientation) of features within these patterns. It involves two levels of positional control using one position geometric characteristic symbol. Datum features may be specified in the lower level control, but if specified must be one or more of the datums that appear in the upper level control, and used in the same order as used in the upper level control. Different datum features may not be specified in the lower level control than have been used in the upper level control. Since it is the main purpose of the lower level control to refine the location between the controlled features, any datum features used are in addition to that purpose, and are useful only as a refinement of the orientation of the feature pattern to the datums generated.

***Concentricity.** The condition where the median points of all diametrically- opposed elements of a figure surface of revolution are congruent with the axis (or centerpoint, as in the case of a spherical datum feature) of a datum feature. ***Concentricity Symbology was removed from ASME Y14.5 in the 2018 revision. This definition is valid per ASME Y14.5-2009.**

***Concentricity Tolerance Zone.** A cylindrical (or spherical) tolerance zone whose axis (or centerpoint) coincides with the axis (or centerpoint) of the datum feature, within which the median points of all diametrically-opposed elements of the feature being controlled, regardless of feature size, must lie. ***Concentricity Symbology was removed from ASME Y14.5 in the 2018 revision. This definition is valid per ASME Y14.5-2009.**

Conformance. That condition of a part wherein the feature does not violate the constraints of the tolerance. If datums are referenced, the feature conforms if it does not violate the tolerance zone constructed from at least one datum reference frame in the candidate datum reference frame set.

Conical Taper. The ratio of the difference in the diameters of two sections of a cone, to the distance between these sections.

Conicity (Profile Tolerance for a Conical Feature). The tolerance zone for profile when used to protect cones consisting of two concentric cones at the specified basic angle, the distance apart stated in the feature control frame, within which the surface of the controlled feature must lie. Datums may be used to orient or locate the tolerance zone, but are not always needed.

Continuous Feature of Size. The note CONTINUOUS FEATURE or the continuous feature symbol is used to identify a group of two or more features of size where there is a requirement that they be treated geometrically as a single feature. When using the continuous feature symbol (or note), extension lines between the features may be shown between the features or omitted. Extension lines by themselves do not constitute a continuous feature. The continuous feature note or symbol may be used on features of size such as coaxial diameters or features of size such as multiple planar widths.

Controlled Radius Tolerance. A controlled radius (CR) creates a tolerance zone defined by two arcs (the minimum and maximum radii) that are tangent to the adjacent surfaces. When the controlled radius is specified by the CR symbol (for example: CR4.8±0.2) the part contour must lie within the crescent-shaped tolerance zone, be a fair curve and have no reversals.

Constraint. A limit to one or more degrees of spatial freedom (for example linear x, y or z or rotational u, v or w degrees of freedom.

Coordinate Measuring Machine. A piece of equipment that probes, either through contact or non-contact mode, a workpiece to determine size, form, orientation, location and other geometric requirements.

Coplanarity (Profile Tolerance for the Preservation of Coplanar Surfaces). Two or more surfaces sharing the same plane in space. The profile of a surface tolerance may be used where it is desired to treat two or more surfaces as a single interrupted or non-continuous surface. The tolerance zone established is defined by two parallel planes within which the considered surfaces must lie. Datum references may be used to orient or locate the tolerance zone, but are not always needed.

Cutting Plane. A plane used to establish a planar curve in a feature. The curve is the intersection of the cutting plane with the feature.

Cutting Vector. A unit on the actual mating surface which, together with the normal to the actual mating surface, defines the direction of the cutting plane.

Cylindricity. A condition of a cylindrical surface of revolution in which all points of the surface are equidistant from a common axis.

Cylindricity Tolerance. Specifies a tolerance zone bounded by two concentric cylinders within which the entire surface must lie. Unlike circularity, the tolerance applies simultaneously to both the circular and longitudinal elements of the entire surface. It is a composite control of surface elements that includes roundness, straightness and taper. The tolerance states that all points that comprise the surface must reside within a zone consisting of two concentric coaxial cylinders whose radii differ by the tolerance specified in the feature control frame.

Datum. A theoretically exact point, axis, plane or combination thereof derived from the theoretical true geometric counterpart. A datum is the origin from which the location or geometric characteristics such as the orientation of features of a part are established. Datums act as the origin of measurement for distance and/or angle of controlled features and can be used to stabilize the part for manufacturing purposes.

Datum Axis. The axis of a theoretical true geometric counterpart established from the datum feature.

Datum Centerplane. The centerplane of a theoretical true geometric counterpart established from the datum feature.

Datums Established From Complex or Irregular Surfaces. Datum targets may be used to establish a datum from a complex or irregular surface. Where a datum reference frame has been properly established but its planes are unclear, the datum reference frame coordinate axes may be labeled to appropriate extension or center lines as needed. The datum feature symbol should be attached only to identifiable datum features. Where datums are established by targets on complex or irregular surfaces, the datum may be identified by a note such as: DATUM AXIS A or DATUM PLANE A.

Datum Features Established From Datum Targets With Fewer Than Three Mutually Perpendicular Planes. When using datum features that are defined by datum targets in a feature control frame established by fewer than three mutually perpendicular planes, the datums that are the basis for the feature control frame shall be referenced. The targets that provide location for the datums referenced in the feature control frame shall be specified in a note, such as:
DATUM FEATURES B AND C ARE USED WHERE ONLY DATUM FEATURE A IS REFERENCED TO RELATE THE PART TO THE TARGETS THAT CREATE DATUM A

Datum Feature. An actual feature of a part that is used to establish a datum. A datum feature is a feature that is identified with either a datum feature symbol or a datum target symbol(s).

Datum Feature, Simulator. For the purposes of the ASME Y14.5-2018 on Dimensioning and Tolerancing and the ASME Y14.43 standard on Dimensioning and Tolerancing Principles for Gages and Fixtures, a datum feature simulator is a real physical surface of adequately precise form (such as a surface plate, a bearing, or a mandrel, etc.) contacting the datum feature(s) and used to establish the simulated datum(s). Simulated datum features are used as the practical embodiment of the datums during manufacture and inspection. The Y14.5-2009 standard used the term "theoretical datum feature simulator", but now uses the term "true geometric counterpart" instead. True Geometric Counterpart is a term resurrected from ASME Y14.5-1994. For a detailed explanation, see True Geometric Counterpart.

Datum Feature Symbol. A box (rectangular frame) containing a datum letter designation, showing the intention to use that feature to establish a datum from which to orient or locate other features on the part. These symbols relate to real physical features on the part and are not assigned to imaginary items such as centerlines, centerplanes or axes (except as reference information for clarity usually in conjunction with datum targets).

Datum Reference Frame. Three mutually perpendicular planes established by contacting real features on the part for purposes of geometric definition, including: orientation and/or location of features on the part and stabilization of the part during function, manufacture and/or inspection.

Datum Reference Frame Customized. In order to override the degrees of freedom constrained by datum features referenced in an order of precedence, a customized datum reference frame may be used. When applying the customized datum reference frame, the following requirements govern the constraint on each datum feature reference. 1) The rectangular coordinate axes shall be labeled in at least two views on the drawing. 2) The degree(s) of freedom to be constrained by each datum feature referenced in the feature control frame shall be explicitly stated by placing the designated degree of freedom to be constrained in lower case letter(s) [x, y, z, u, v or w] in brackets following each datum feature reference and any applicable modifier.

Datum Reference Frame Identification. Where a datum reference frame has been properly established and it is considered necessary to illustrate the axes of a datum reference frame on the drawing, the axes or center planes may be labeled to determine the translational degrees of freedom x, y and z. Where multiple datum reference frames exist, and it is desirable to label the axes (X, Y and Z), any labeled axes shall include a reference to the associated datum reference frame. These labels represent the datum features [without modifiers] for each datum reference frame and follow the X, Y and Z identification letters.

Datum, Simulated. A point, axis or plane established by or derived from processing or inspection equipment such as the following simulators; machines tables, surface plates, collets, gage surfaces, a mathematical simulation, etc.
Datum Target. A specified point, line, or area on a part used to establish a datum.

Datum Target Location. Datum targets should be located as follows:
a) on features produced by one segment of a die or pattern
b) on features opposite machining cuts to establish subsequent machining datum reference frames
c) on features not subject to processing variables, such as parting lines, flash extensions etc.
d) on features not subsequently altered or removed
e) with optimum spacing considering function and producibility
f) on their respective datum planes, except when they establish an axis, a centerplane or a plane where area or location requires one or more targets offset from the datum features.

Derived Median Line. An imperfect (abstract) line formed by the center points of all cross-sections of the feature. These cross-sections are normal (perpendicular) to the axis of the unrelated actual mating envelope.

Derived Median Plane. An imperfect (abstract) plane formed by the center points of all line segments bounded by the feature. These line segments are normal (perpendicular) to the centerplane of the unrelated actual mating envelope.

Deviation. The straight (Euclidean) distance of a data point from the reference, measured normal to the geometric element.

Dimension. A numerical value expressed in appropriate units of measure and indicated on a drawing and in other documents along with lines, symbols, and notes to define the size or geometric characteristics, or both of a part or part feature.

Dimension, Basic. A numerical value used to describe the theoretically exact size, profile, orientation, or location of a feature or datum target. It is the basis from which permissible variations are established by tolerances in feature control frames, or in notes or in other documents such as those that establish tooling or gaging tolerances for datum targets.

Dimension, Reference. A dimension, usually without tolerance, used for information purposes only. It is considered auxiliary information and does not govern production or inspection operations. A reference dimension is a repeat of a dimension or is derived from other values shown on the drawing or on related drawings. It is designated by showing the dimension inside of parentheses.

Dimensional Inspection Plan. A plan for dimensional inspection of a workpiece.

Direction Vector. A unit vector. Conventionally, directions are associated with various geometries such as: The direction vector of a straight line, or pair of parallel lines is parallel to the line(s). The direction vector of a plane, or a pair of parallel planes, is normal to the plane(s). The direction vector of a cylinder is the direction vector of the cylinder's axis.

Draft Angle. Draft angle and tolerances shall be specified. Draft angle shall be accounted for in determining datum targets and datum positions. Draft may be shown for drawing clarity. Draft shall be additive to dimensional limits and may exceed the perfect form boundary at maximum material condition (MMC) unless it is otherwise specified. Draft is the taper applied to selected surfaces to aid in the removal of a part from a die or a pattern from a mold. Draft adds mass to the part unless otherwise specified. See ASME Y14.8M-1989 paragraph 3.5.

Drawing Note Items. Drawing notes should be included in the following as applicable:

a) reference to applicable standards
b) surface texture requirements
c) surface protrusion requirements, such as: fins, flash, gates, risers, knockout bosses
d) general features notes, such as: wall thickness, corner radii, fillet radii, requirements of surface intersections shown as sharp corners, draft allowances, tolerances, marking requirements, in-process weld requirements, surface treatment requirements, grain direction requirements and packaging requirements.

Effect of Draft and Parting Lines. The relationship of datums established by targets to the features of a part can be affected by draft and parting lines.

Envelope, Actual Mating. For an external feature of size, it is the perfect feature counterpart that is the smallest size that contacts the surface at the highest points. For an external cylindrical feature, it would be the minimum circumscribed cylinder. For two parallel planar external opposing surfaces, it would be two parallel planes at minimum separation. For an internal feature of size, it is the perfect feature counterpart that is the largest size that contacts the surface at the highest points. For an internal cylindrical feature, it would be the maximum inscribed cylinder. For two parallel planar internal opposing surfaces, it would be two parallel planes at maximum separation.

For features controlled with a geometric characteristic inside a feature control frame, such as perpendicularity or position, an actual mating envelope can be created at perfect orientation or position to specified datums, as appropriate.

An actual mating envelope that is not constrained by datums is known as an Unrelated Actual Mating Envelope. An actual mating envelope that is constrained by datums is known as a Related Actual Mating Envelope.

Envelope, Actual Minimum Material. This envelope is within the material.

a) For an external feature: A similar perfect feature counterpart of largest size which can be inscribed within the feature so that it just contacts the surface at the lowest points.
b) For an internal feature: A similar perfect feature counterpart of smallest size which can be circumscribed about the feature so that it just contacts the surface at the lowest points.

There are two types of actual minimum material envelopes.

1) Unrelated actual minimum material envelopes are a similar perfect feature(s) counterpart contracted about an internal feature(s) or expanded within an external feature(s), and not constrained to any datum reference frame.
2) Related actual minimum material envelopes are similar perfect feature(s) counterpart contracted about an internal feature(s) or expanded within an external feature(s) counterpart, and constrained in either orientation and/or location to a datum or datums.

Envelope Principle (also known as the Taylor Principle). This principle creates an interdependency between the size limits of a feature and its surface form. Unless otherwise specified, the limits of size control the allowed deviation from perfect geometric form. No envelope of perfect perpendicularity, coaxiality or symmetry is created by this principle.

Feature. The general term applied to a physical portion of a part, such as a surface, pin, tab, hole, or slot or its representation on drawings, models, or in digital data files.

Feature, Axis. The axis of the unrelated actual mating envelope of a feature.

Feature, Centerplane. The centerplane of the unrelated actual mating envelope of a feature.

Feature Control Frame. A single or composite frame that gives vital information about the form, orientation, profile, runout or location of a feature or pattern of features. It can contain the following components as appropriate: a geometric characteristic symbol, a diameter symbol, a geometric tolerance, a material condition symbol after the geometric tolerance and after any datum features of size, datum references.

Feature of Size. There are two types of Features of Size.

1) **Regular Feature of Size.** One cylindrical or spherical surface, a circular element, a set of two opposed parallel elements, or opposed parallel surfaces, each of which is associated with a directly toleranced dimension.
2) **Irregular Feature of Size.** There are two types of irregular features of size.
    a) a directly toleranced feature or collection of features that may contain or be contained by an actual mating envelope which is a sphere, cylinder or pair of parallel planes.
    b) a directly toleranced feature or collection of features that may contain or be contained by an actual mating envelope other than a sphere, cylinder or pair of parallel planes.

Fillet Radii. Fillet radii size and tolerance shall be specified on the drawing. A fillet radius is the concave radius on the surface of a part connecting intersecting surfaces.

Fitted Surface. An estimate of a real surface derived from coordinate data.

Fixed Fastener Situations. The condition wherein a part has an internal feature or features such as threaded holes or tight fitting holes meant to assemble with threaded fasteners such as screws or studs, or tight fitting members such as dowel pins. Also, situations with shafts at a stable orientation on a plate which will assemble with clearance holes on a mating plate.

Fixed Limit Gage. A device of defined geometric form and size used to assess the conformance of a feature or features of a workpiece to a dimensional specification. Also referred to as a limit gage.

Fixture. A device used to hold parts securely in the correct position in a tool or gage during manufacturing, assembly or inspection. A functional fixture has integral gage elements that represent the part's datum features.

Flash. Excess material (a fin of material) which results from leakage between mating surfaces of a mold or die.

Flash Extension. The limits of permissible flash extension shall be specified on the drawing. Flash extension shall be additive to dimensional limits and may exceed the perfect form boundary at maximum material condition (MMC) unless it is otherwise specified.

Flatness. Flatness is the condition of a surface or derived median plane having all elements in one plane.

Flatness Tolerance.

a) A flatness tolerance specifies a tolerance zone defined by two parallel planes within which the surface must lie.
b) A flatness of a derived median plane control uses a tolerance zone consisting of two parallel planes within which the derived median plane of the considered feature must lie.

[Note: When applied on an MMC basis for derived median plane, as each actual local size departs from MMC, an increase in the local size of the tolerance zone is allowed that is equal to the amount of such departure.]

Flexible Parts. Unless otherwise specified, all flexible parts are to be inspected in the free state. If restrained state inspection is desired, it shall be so noted on the workpiece drawing and accompanying inspection plan.

Floating Fastener Situations. The condition wherein two parts with clearance holes or hole patterns are to be assembled with rivets or through bolts and nuts. In these situations, the clearance holes do not lend definite fixed orientation to the members that pass through them.

Form. A straightness, flatness, circularity or cylindricity.

Form Tolerances. Tolerances that control straightness, flatness, circularity and cylindricity.

Free State Inspection. Unless otherwise specified, all tolerances apply in a free state condition. See ASME Y14.5-2018.

Free State Symbol. When tolerances are specified as applying in the restrained condition and an additional tolerance is applied in the free state, the symbolic means of indicating that the toleranced feature or datum feature applies in its free state is used. When the free state symbol is applied to a tolerance in the feature control frame, it shall follow the stated tolerance and any modifier. When the free state symbol is applied to a datum feature reference, it shall follow the datum feature reference and any modifier. When the free state symbol is given to a directly toleranced dimension, it shall follow the dimension and its tolerance.

Free State Variation. The variation in feature form, orientation or other geometric relationships that occurs once parts are released from any restraining forces (such as those used in manufacture). These distortions in the free, unrestrained state are often due to forces such as internal stresses, non-rigidity and/or weight.

Full Indicator Movement (FIM). The total movement of an indicator when appropriately applied to a surface to measure its variations.

Functional Gage. A gage which uses fixed elements such as plates, rails, pins and holes/bushings to simulate the worst mating condition of a part's geometric requirement. The part feature or features under test must be received by the functional gage. It assures interchangeability of accepted parts.

Fundamental Rules. Dimensioning and tolerancing clearly defines engineering intent and conforms to the following:

a) Each dimension has a tolerance, except for those dimensions specifically identified as reference, maximum, minimum, or stock (commercial stock size). The tolerance can be specified directly on the dimension (or indirectly in the case of basic dimensions), indicated by a note, or located in a block that is a portion of the drawing format.

b) Dimensioning and tolerancing needs to be complete so that there is a full understanding of the geometric characteristics of each feature. Drawings may not be scaled (measuring the size of a feature directly from the engineering drawing). Assumption of a distance or size is also not permitted, except for: Undimensioned drawings, such as loft, printed wiring, templates, master layouts or tooling layout (if they are prepared on stable material), provided the needed control dimensions are given.

c) All necessary dimensions of end products must be shown. No additional dimensions than those necessary for complete definition need be given. Reference dimensions on a drawing should be held to a minimum.

d) Dimensions are chosen and arranged to accommodate the function and mating relationship of a part and are not to be subject to more than one interpretation.

e) The drawing defines a part without usually specifying manufacturing methods. For example, only the diameter of a hole is stated without noting whether it will be drilled, reamed, punched, or made by any other operation. However, in situations where it is important to the engineering requirements, manufacturing methods may be specified on the drawing or in a document referenced by the drawing.

f) It is acceptable practice to identify as non-mandatory some processing dimensions that allow for finish allowance, shrink allowance, and other requirements, provided the final dimensions are stated on the drawing. Non-mandatory processing dimensions are identified by a note, such as NON-MANDATORY (MANUFACTURING DATA) or other unambiguous method.

g) Dimensions are arranged to provide required information for optimum clarity and readability. Dimensions are usually shown in true profile views and refer to visible object lines.

h) Wires, cables, sheets, rods and other items manufactured to gage or code numbers are specified by linear dimensions indicating diameter or thickness. Gage or code numbers can be shown in parentheses after the dimension.

i) A 90° angle is implied where centerlines and lines that show features are depicted on a drawing at right angles and no angle is stated.

j) A 90° basic angle exists where centerlines of features in a pattern or surfaces depicted at right angles on the drawing are located or defined by basic dimensions and no angle is stated.

k) Unless otherwise stated, all dimensions apply at a standard temperature of 20°C, which is 68°F. Compensation can be made for measurements taken at temperatures other than this standard.

l) Unless otherwise stated, dimensions and tolerances all apply in a free state condition.

m) Unless otherwise stated, geometric tolerances apply for full depth, length and width of the specified feature being controlled.

n) Unless otherwise specified, all dimensions and tolerances only apply at the drawing level where they are specified. Any dimension stated for a feature on one level of drawing, (for example, a detail drawing) is not applicable as mandatory for that feature at any other level of drawing (for example, an assembly drawing).

o) A zero basic dimension applies where axes, center planes or surfaces are shown coincident (in the same place) on a drawing and geometric tolerances establish the relationship between the features. [Although this concept is not new to previous Y14.5 standards, this statement explicitly states the concept for the Y14.5-2009 and the Y14.5-2018 standards.]

p) Where a coordinate system is shown on the drawing, it shall be right-handed unless otherwise specified. Each axis shall be labeled and the positive direction shown. [These rules were new to the Y14.5 2009 standard and restated in Y14.5-2018]. Where a model coordinate system is shown on the drawing, it shall be in compliance with ASME Y14.41M.

q) Unless otherwise specified by drawing or model notes or reference to a separate document, the as-designed dimension value does not establish a functional or manufacturing target.

r) Unless otherwise specified, elements of a surface include surface texture and flaws (for example, burrs and scratches). **All elements** of a surface shall be within the applicable specified tolerance zone boundaries. **Note: This rule is new to ASME Y14.5-2018 and very controversial, since it seems to include surface flaws down to the molecular structure.**

Gage. A real or simulated artifact used for determining dimensional data.
Hard Gage - a mechanical gage
Soft Gage - a mathematical simulation of a hard gage in a software program. For example, a simulation of a functional gage in a coordinate measuring machine software program.

Gagemakers' Tolerance. A manufacturing tolerance allowed a gage-maker. It is applied to fixed gages and comparator setting masters. ASME Y14.43 recommends the gage makers' tolerances to be 5% of the part tolerance being gage with an optional additional 5% of the part tolerance for wear allowance (for a total of 5-10%). Part tolerance is defined for Go and NOGO gages as the difference between the MMC and the LMC of the part being gaged. Part tolerance for a Functional Gage is defined as the difference between the Virtual Condition and the LMC of the part being gaged.

Gate. A channel in a mold used for introducing molten material to the mold cavity.

Geometric Characteristic Symbols. Symbols found in feature control frames that represent flatness, straightness, circularity, cylindricity, profile of a line, profile of a surface, angularity, perpendicularity, parallelism, circular runout, total runout and position. ***Concentricity and Symmetry Symbology were removed from ASME Y14.5 in the 2018 revision.**

Geometric Element. Line, plane, circle, cylinder, cone or sphere.

Geometric Feature. Part of an object nominally in the shape of one of the geometric elements. A real surface on an object.

Geometric Tolerances. All tolerances relating to allowed deviations from perfection in form, orientation, profile, runout and location. Geometric tolerance zones fall into three major types: surface controls, axial controls and centerplane controls.

Geometry. The branch of mathematics that deals with points, lines, surfaces and solids and examines their properties, measurement and mutual relation-ships in space.

Geometry, Solid. A branch of geometry dealing with solid or three-dimensional figures.

GO Gage. A fixed gage that checks a workpiece at the maximum material condition (MMC).

Implied 90° Angles. Where centerlines and surfaces of features of a part are shown on engineering drawings intersecting at right angles, no 90° angle is stated. A 90° angle is implied. The tolerance on these implied 90° angles is the same as for all other angular features shown on the field of the drawing governed by general tolerance notes or general tolerance block values. Where centerlines and surfaces of a part are shown on engineering drawings intersecting at right angles and basic dimensions or geometric controls have been stated, a 90° basic angle is implied. Tolerances on features associated with the implied 90° basic angles are provided by feature control frames governing the location, orientation, profile, or runout of these features.

Implied Zero Basic Dimensions. A zero basic dimension applies where axes, center planes or surfaces are shown coincident (in the same location) on a drawing and geometric tolerances establish the relationship between the features.

Independency Symbol. The circled I symbol stands for Independency. It is used to supersede the rule that states that size tolerance controls form tolerance (Rule #1). When used, the Independency symbol states that all size dimensions are taken at cross-sections. This means that the size tolerance and the form tolerance of a regular feature of size are independent of one another. It means that perfect form at MMC is no longer required.

Inspector. A person who accepts or rejects workpieces after examining compliance with part specifications.

Interferometer. A device using light interference phenomena to determine precise measurements of distance and angle. It is also used for optical component testing.

Invalid Data (Outliers). Coordinate data which, as a result of faulty measurement, lie far from a fitted surface.

Knurling. Knurling is given in terms of type, pitch, and diameter. This specification is complete enough to define the part before and after knurling. Where this much control is not required, the diameter after knurling is omitted. Where only a portion of a feature requires knurling, the location and length of the knurl shall be specified.

Knurling for Press Fit. Where knurling is used to provide a press fit between parts, a note that specifies the type of knurl, its pitch, the toleranced diameter of the feature before knurling, and the minimum acceptable diameter is required.

Laser Range Finder. A device used to measure distance to an object based on the time light takes to go to and from the object.

Least Material Boundary (LMB). The limit defined by a tolerance or combination of tolerances which exists on or inside the material of a feature. For a datum feature hole referenced in a feature control frame at LMB, the LMB is equal to the Least Material Condition plus any applicable geometric tolerance. For a datum feature shaft referenced in a feature control frame at LMB, the LMB is equal to the Least Material Condition minus any applicable geometric tolerance.

Least Material Condition (LMC). The condition in which a feature of size contains the least amount of material within the stated limits of size--for example, the maximum hole diameter, minimum shaft diameter. If a tolerance is specified as applying at LMC, the tolerance zone grows an amount equal to that feature's departure from LMC within size limits (for other than those self-centering situations such as threaded features).

Limit Dimensioning. In limit dimensioning, the limits of a dimension are specified instead of giving a dimension somewhere in the middle, then specifying a plus and minus tolerance. For example, instead of specifying .700 ± .005, a limit dimension would specify .695 -.705, or
.705
.695.

Location. The category of symbols that includes position. ***Concentricity and Symmetry Symbology were removed from ASME Y14.5 in the 2018 revision. These were Location Geometric Characteristic symbols per ASME Y14.5-2009.**

Master. The standard against which the desired dimension of the part is compared. The standard may be in the form of the wavelength of light, the length of a gage block, line standard, lead screw, etc.

Mastering. The action of nulling or setting a comparator with a master.

Match Draft. Additional draft allowance permitted on matching surfaces at parting lines when the normal draft allowance would result in an offset of the surfaces at the parting lines. Match draft shall be specified when applicable.

Material Condition Symbols. Sometimes called modifiers, there are three material condition symbols. They are the circled letters M, L and the S which is implied. They stand for: maximum material condition, abbreviated as MMC and uses a circled M as the symbol; least material condition, abbreviated as LMC and uses a circled L as the symbol; and regardless of feature size, abbreviated as RFS and uses a circled S as the symbol (although according to current ASME and ISO rules, the circled S is implied in feature control frames for all geometric characteristics, when no other modifier is specified). When used in feature control frames, these material condition symbols (or their implication, as appropriate) specify that a tolerance applies if a feature is made at that size.

Mating Surface Normal. For a given point on a part feature, the direction vector of a line passing through the point and normal to the actual mating surface at the point of intersection of the line with the actual mating surface.

Maximum Material Boundary (MMB). The limit defined by a tolerance or combination of tolerances which exists on or outside the material of a feature. For a datum feature shaft referenced in a feature control frame at MMB, the MMB is equal to the Maximum Material Condition plus any applicable geometric tolerance. For a datum feature hole referenced in a feature control frame at MMB, the MMB is equal to the Maximum Material Condition minus any applicable geometric tolerance.

Maximum Material Condition (MMC). The condition in which a feature of size contains the maximum amount of material within the stated limits of size--for example, minimum hole diameter, maximum shaft diameter.

Measured Value. A number representing in appropriate units the result of a direct measurement.

Measurement. The process of comparing a quantity with a standard, or the result of such a process.

Direct Measurement - Measurement where the result is a measured value or set of measured values. An example is determination of a length by comparison with a line scale.

Comparison Measurement - Measurement where the result is a determination that the measured quantity is larger than, equal to, or less than, the standard. An example is measurement of an outside diameter with a ring gage.

Measurement Cycle Time. The time between measuring and the previous mastering.

Measurement Procedure. A strategy for obtaining a representative set of points on a workpiece.

Measurement Process Designer. A person who prepares a dimensional inspection plan.

Measurement Uncertainty. The difference between the corrected measured size and the actual size. In cases where there is adequate information based on a statistical distribution, the estimate may be associated with a specific probability. In other cases, an alternative form of numerical expression of the degree of confidence to be attached to the estimate may be given.

Minimum Material Envelope, Actual. This envelope is within the material.

a) For an external feature: A similar perfect feature counterpart of largest size which can be inscribed within the feature so that it just contacts the surface at the lowest points.

b) For an internal feature: A similar perfect feature counterpart of smallest size which can be circumscribed about the feature so that it just contacts the surface at the lowest points.

There are two types of actual minimum material envelopes.

1) Unrelated actual minimum material envelopes are a similar perfect feature(s) counterpart contracted about an internal feature(s) or expanded within an external feature(s), and not constrained to any datum reference frame.
2) Related actual minimum material envelopes are similar perfect feature(s) counterpart contracted about an internal feature(s) or expanded within an external feature(s) counterpart, and constrained in either orientation and/or location to a datum or datums.

Mismatch. The offset of features on a part caused by the misalignment of opposing segments of a mold or die. Mismatch, when applicable, must be specified as a maximum value. Unless otherwise specified, mismatch is additive to dimensional limits and may exceed the perfect form boundary at maximum material condition (MMC).

Mold. A form made of sand, metal, or other material which contains the cavity into which molten material is poured to produce a part.

NOGO Gage. A fixed gage that checks a workpiece at the least material condition (LMC). Also referred to as a NOT GO gage.

Nominal Form. The ideal geometric shape of which the geometric feature under inspection is a machined or otherwise manufactured manifestation--for example, a cylinder or a sphere.

Non-Rigid Parts. When a part in free space distorts because of its own weight, release of its internal stresses and strains and its flexibility.

Normal. A line passing through a point on a curve or surface and perpendicular to the tangent line or plane at the point.

Not-To-Scale-Dimension. When a drawing shows dimensions that are not to scale and, consequently, out of proportion to the rest of the drawing, it is indicated by underlining the dimension with a thick straight line.

Open Set-Up. Standard measurement devices (such as a surface plate set-up) arranged to determine conformance of a produced part's geometric requirements as within specified tolerances.

Optical Comparator. A piece of equipment that projects an enlarged image of a workpiece onto a viewing screen to be compared to a comparator master chart or linear, angular, size or distance requirements from the design drawing.

Optical Flat. A very precise transparent surface capable of extremely accurate gaging using light wave interference. It is commonly used to inspect flatness of optically flat surfaces and gage blocks against a master block.

Optimistic Gage Tolerancing. The policy of tolerancing gages that assures all part features within tolerance that are gaged are accepted by the gage. This policy allows some borderline, but out of tolerance parts to be accepted by the gage. This is done by making the gagemaker's tolerances additive to the part tolerances. This policy is explained, but not featured by the ASME Y14.43 standard on Dimensioning and Tolerancing Principles for Gages and Fixtures.

Order Of Precedence. The order in which the datums are presented in the feature control frame is supposed to represent a physical description of how a part functions. Consequently, the order of precedence given, reading from left to right, is considered appropriate restraint during inspection, and, if possible, during manufacture. It represents the design requirements that must be met in order to assure part functionality.

Orientation. Orientation is the category of geometric characteristic symbols that contains perpendicularity, angularity and parallelism.

Outlier. A data point that is not regarded as a member of a set of data points representative of the geometric feature. An outlier may arise from a malfunction of the CMM, an error by its operator, or as a result of data taken from portions of the part not considered appropriate for assessment of the controlled feature.

Parallelism. Parallelism is the condition of a surface or centerplane, equidistant at all points from a datum plane; or an axis, equidistant along its length from one or more datum planes or a datum axis.

Parallelism Tolerance. A parallelism tolerance specifies one of the following:

a) a tolerance zone defined by two parallel planes parallel to a datum plane or axis, within which the surface or centerplane of the considered feature must lie.
b) a cylindrical tolerance zone parallel to one or more datum planes or a datum axis within which the axis of the considered feature must lie.
c) a tolerance zone defined by two parallel planes parallel to a datum plane or axis, within which the axis of the controlled feature must lie.
d) a tolerance zone defined by two parallel lines parallel to a datum plane or axis, within which a line element of the surface must lie.

First seen in ASME Y14.5-2009, ASME Y14.5-2018 also states that as an alternative practice, the angularity symbol may be used to control parallel and perpendicular relationships.

Parameters. Algebraic variables that represent the size and location of a geometric element--for example, the radius and center coordinates of a circle.

Parameter Values. Numerical values of parameters.

Parametrization. A choice of algebraic variables to represent a geometric element.

Partial Datum Surface. When function dictates only a portion of a feature be used to establish a datum, the area to be used is shown by means of a chain line (a line that looks like a very thick centerline) drawn either next to the surface or sometimes on the surface (in a face-on view) that establishes the extent of the partial datum surface.

Parting Lines. A line on a part that corresponds to the separation between the mold or die segments. Parting lines are usually not shown on casting or molding drawings.

Pattern of Features of Size at LMB. Where LMB is applied in a feature control frame to multiple datum features of size used to establish a single datum, the true geometric counterpart of each feature shall be fixed in location relative to one another. The true geometric counterparts shall be represented at their applicable Least Material Boundary size. This concept is theoretical and cannot be represented in a physical fixture or gage.

Pattern of Features of Size at MMB. Where MMB is applied in a feature control frame to multiple datum features of size used to establish a single datum, the true geometric counterpart of each feature shall be fixed in location relative to one another. The true geometric counterparts shall be represented at their applicable Maximum Material Boundary size. This concept is theoretical, but can usually easily be represented in a physical fixture or gage with datum feature simulators.

Pattern of Features of Size at RMB. Where RMB is applied in a feature control frame to multiple datum features of size used to establish a single datum, the true geometric counterpart of each feature shall be fixed in location relative to one another. The true geometric counterparts shall expand or contract simultaneously from their MMB to their LMB until the true geometric counterparts make maximum possible contact with the extremities of the datum feature(s). This concept is theoretical and difficult to represent in a physical fixture or gage. However, although usually more expensive than representing patterns at MMB in a physical fixture or gage, it is possible.

Perfect Form at MMC. The constraint the boundary of the maximum material condition envelope places on features of size. If a regular feature of size is produced at MMC, it must have perfect form, unless the feature:

a) is stock in the as purchased condition
b) is non-rigid
c) uses a drawing note stating perfect form at MMC is not required
d) is dimensioned with AVERAGE dimensions
e) has an applicable feature control frame that uses the least material condition symbol, or
f) specifies a straightness of the derived median line control
g) specifies a flatness of the derived median plane
h) associates the Independency symbol with the size dimension

Perfect form at MMC for two parallel planar opposing surfaces would include the surfaces being perfectly flat, straight and parallel. Cylindrical features would have to be perfectly circular and straight and not tapered at all. Spherical features would have to maintain a perfect spherical shape, which includes roundness. Only as the features of size depart from MMC may their form become imperfect.

Permanent Datum Feature. A datum feature that is not removed or altered during the manufacturing procedure.

Perpendicularity. The condition of a surface, centerplane, or axis at a right angle to a datum plane or axis.

Perpendicularity Tolerance. A tolerance that specifies one of the following:

a) a tolerance zone defined by two parallel planes perpendicular to a datum plane, or axis, within which the surface or centerplane of the controlled feature must lie.
b) a tolerance zone defined by two parallel planes perpendicular to a datum axis, within which the axis of the controlled feature must lie.
c) a cylindrical tolerance zone perpendicular to a datum plane within which the axis of the controlled feature must lie.
d) a tolerance zone defined by two parallel lines perpendicular to a datum plane, or axis, within which a line element of the surface must lie.

ASME Y14.5-2009 and ASME Y14.5-2018 both state that as an alternative practice, the angularity symbol may be used to control parallel and perpendicular relationships.

Plane. A theoretically perfect surface which has width and breadth without depth.

Plus-Or-Minus Tolerance. A tolerance applicable to a dimension stating an allowed variance from the dimension in both a negative and positive manner.

Point. The intersection of two axes in space, giving it position but not width, depth or height.

Polar Coordinate Method of Dimensioning. Polar coordinates use radial and angular dimensions to establish location relative to a datum reference frame.

Position. Establishes:

a) center distance between such features as holes, slots, bosses and tabs.

b) location of features of size as a group, from datum planes and/or axes formed by datum features such as planar and cylindrical surfaces.

c) coaxiality of features.

d) center distances of features located from datum(s).

e) location of feature surfaces on the basis of an inviolate boundary (MMC or LMC concept virtual condition).

Positional Tolerancing. A positional tolerance defines a) a zone within which the center point, axis, or centerplane of a regular feature of size is permitted to vary from true (theoretically exact) position. b) where specified on an MMC or LMC basis, a boundary of virtual condition located at the true position, that may not be violated by the surface or surfaces of the controlled feature. Basic dimensions establish the true position from specified datum features and between interrelated features. At times a basic dimension of zero is implied. A positional tolerance is indicated by the position symbol, a tolerance, specified (MMC or LMC) or implied (RFS) material condition symbols and appropriate datum references placed in a feature control frame.

Practical Absolute Gage Tolerancing. The policy of tolerancing gages that predicts most part features within tolerance will be accepted by the gage, some borderline part features within tolerance will not be accepted by the gage, and a very low probability that some borderline part features not within tolerance will be accepted by the gage. This policy is featured for Functional Gaging of geometric tolerances by the ASME Y14.43 standard on gages and fixtures.

Pre-Processing. Operations performed on measured data intended to render it more suitable for purposes of assessment.

Profile. A profile is the outline of an object in a given plane (two-dimensional figure). Profiles are formed by projecting a three-dimensional figure onto a plane or by taking cross sections through the figure. The elements of a profile are straight lines, arcs, and other curved lines.

Profile Tolerancing. The profile tolerance specifies a uniform boundary along the true profile within which the elements of the surface must lie. It is used to control form, or combinations of size, form, orientation and location. A profile tolerance may be applied to an entire three-dimensional surface (Profile of a Surface), or to individual two-dimensional profiles taken at various cross sections through the part (Profile of a Line). Datums may be used as required, but are not mandatory.

Projected Tolerance Zone. Where the variation in perpendicularity of threaded or tight-fitting holes could cause fasteners such as screws, studs, or pins to interfere with mating parts, Projected Tolerance Zone is recommended. The specified tolerance zone extends and exists only beyond the surface of a hole designed to receive a fixed internal fastener. The tolerance zone is coaxial to the desired location of the hole. The specified value for the projected tolerance zone is a minimum and represents the maximum permissible mating part thickness, or the maximum installed length or height of components such as studs or dowel pins. A projected tolerance zone is applicable where threaded or plain holes for studs or pins are located on a detail part drawing. In these cases, the specified projected height of projection should equal the maximum permissible height of the stud or pin after installation (which may not necessarily be the mating part thickness).

Quality Assurance Requirements (QAR) Notes. Generally, a description of inspection procedures that become part of the engineering drawing to ensure quality assurance requirements are understood and complied with.

Radius. A straight line extending from the center to the periphery of a circle, cylinder or sphere.

Radius Tolerance. A radius symbol R gives a tolerance zone defined by two arcs that are the minimum and maximum radii. The surface of the part may not violate this tolerance zone. The radius symbol R precedes the size of the radius.

Radius Tolerance, Controlled. A controlled radius symbol CR gives a toleranced zone defined by two arcs (the minimum and maximum radii) that are tangent to the adjacent surfaces. When specifying a controlled radius, the part contour within the crescent-shaped tolerance zone must be a fair curve without reversals. Additionally, radii taken at all points on the part contour shall neither be smaller than the specified minimum limit nor larger than the maximum limit. A CR precedes the size of the radius.

Range of Deviations. The difference between the largest and smallest signed deviations.

Raw Data. Measured coordinates of points on the boundary or surface of the geometric feature.

Rectangular or Cartesian Coordinates. The location of a point in space by specifying its perpendicular distance from three mutually perpendicular planes commonly known as a datum reference frame.
Rectangular Coordinate Dimensioning Without Dimension Lines. Dimensions may be shown on extension lines without the use of dimension lines or arrowheads.

Reference. A computed geometric element to be used as a basis for assessment. For example, a reference for assessment of circularity might be the center coordinates and radius of a circle which is the smallest circle enclosing the data points.

Reference Dimension. A dimension, usually without tolerance, used for information purposes only. It is considered auxiliary information and does not govern production or inspection operations. A reference dimension is a repeat of a dimension or is derived from other values shown on the drawing or on related drawings. It is designated by showing the dimension inside of parentheses.

Regardless of Feature Size (RFS). The term used to indicate that a geometric tolerance or datum reference applies at any increment of size of the feature within its size tolerance.

Regardless of Feature Size Tolerance. A geometric tolerance specified to apply RFS remains unchanged as the controlled feature varies in size within its size limits.

Regardless of Material Boundary (RMB). Indicates that the true geometric counterpart progresses from MMB toward LMB until it makes maximum contact with the extremities of a feature(s).

Relationship Between Individual Features. The limits of size do not control the orientation or location relationship between individual features. Features shown perpendicular, coaxial, or symmetrical to each other must be controlled with a tolerance for these interrelationships to avoid incomplete drawing requirements.

Repetitive Patterns. A way of defining multiple patterns of features, each with its own relationship within the pattern and each with its own relationship to a critical datum feature or features. The critical datum feature(s) for each pattern is a separate feature, but identical to the others, each is called the same datum letter but has how many there will be. For example,

D 4X INDIVIDUALLY

Then each pattern would be positioned to its own datum D by specifying under the control 4X INDIVIDUALLY.

Represented Line Element. A supplemental geometry line or curve segment indicating the orientation of a direction dependent tolerance

Representative Set of Data Points. A set of points that, for the purpose of assessment, represent the geometric feature.

Resolved Geometry. An all-encompassing term for the centerpoint of a sphere, the axis of a cylinder, or the centerplane of a width. Conceptually, it is the representation of a regular feature of size having perfect form and zero size.

Restrained Condition. Unless otherwise specified, all tolerances apply in a free state condition. In some cases, it may be desirable to restrain a part on its datum features to simulate function or interaction with other features or parts. To invoke a restrained condition, a note is specified or referenced on the drawing defining the specific requirements. See ASME Y14.5-2018.

Resultant Condition. The worst-case boundary generated by the collective effects of a size feature's specified MMC or LMC material condition, the geometric tolerance for that material condition and the size tolerance used as the additional geometric tolerance derived from the feature's departure from its specified material condition. For example: for a hole (MMC concept), the resultant condition is the hole's LMC, plus its geometric tolerance applicable at MMC, plus the difference between the MMC and LMC within the hole's limits of size.

Roundness. (See Circularity)

Rule #1. The Taylor Principle. Unless otherwise specified, the limits of size of a feature create the extent within which variations of geometric form, as well as size, are allowed. This control applies solely to individual features of size. Maximum material condition is inspected for violations of an envelope of perfect form at MMC, and least material condition is inspected for violations at cross-sectional opposing points (local size).

a) The surface or surfaces of a feature may not extend beyond a boundary (envelope) of perfect form at MMC. This boundary is the true geometric form represented by the drawing. No variation in form is permitted if the feature is produced at its MMC limit of size.

b) Where the actual local size of a feature has departed from MMC toward LMC, a variation in form is allowed equal to the amount of such departure.

c) There is no requirement for a boundary of perfect form at LMC. Thus, a feature produced at its LMC limit of size is permitted to vary from true form to the maximum variation allowed by the boundary of perfect form at MMC.

   Note: See Perfect Form at MMC for exceptions (and ways of specifying otherwise) to this rule. Also see Relationship Between Individual Features for a further explanation of the limitations of this rule.

Rule #2. For all applicable geometric tolerances RFS applies, with respect to the individual tolerance, datum reference, or both where no modifying symbol is specified. MMC or LMC must be specified on the drawing as required.

Runout. Runout is a composite surface geometric tolerance used to control the functional relationship of one or more features of a part to a datum axis.

Runout Tolerance.

a) Circular Runout. Circular runout provides control of circular elements of a surface. The tolerance is applied independently at any circular line element as the part is rotated 360°. Where applied to surfaces constructed around a datum axis, circular runout may be used to control the cumulative variations of circularity and coaxiality. Where applied to surfaces constructed at right angles to the datum axis, circular runout controls circular elements of a planar surface (wobble).

b) Total Runout. Total runout provides composite control of all surface elements. The tolerance is applied simultaneously to all circular and profile measuring positions as the part is rotated 360°. Where applied to nominally cylindrical surfaces constructed around a datum axis, total runout is used to control cumulative variations of circularity, straightness, taper and coaxiality of a surface. Where applied to surfaces at right angles to a datum axis, total runout controls variations of perpendicularity (to detect wobble) and flatness (to detect concavity or convexity).

Scale. The ratio between the size as depicted on an engineering drawing and the actual size of an object.

Simultaneous Requirement. Simultaneous requirements apply to position and profile tolerances that are related to the same datum reference frame. The same datum reference frame is defined as using the same datum features referenced in the same order of precedence with same modifiers applied to the datum feature references. The default of simultaneous requirements requires that all applicable geometric tolerance zones are related to the same datum reference frame and are appropriately constrained as a single pattern to a single datum reference frame.

They combine the patterns of features into one pattern and must be gaged with the same gage or measured in a single set up. Any pattern shift/datum feature shift experienced must be simultaneously experienced by all features within the combined patterns in the same direction by the same amount. This rule also applies to the Pattern Locating Tolerance Zone Framework (PLTZF) of multiple composite position or profile controls. This rule **does not apply** to the Feature Relating Tolerance Zone Framework (FRTZF) of multiple composite position or profile controls. Since datum features referenced in a Feature Relating Tolerance Zone Framework (FRTZF) for composite position or profile controls lose their ability to locate and become orientation-only datums, these FRTZF controls are not part of the simultaneous requirement rule.

Size. Dimensional extent of a feature determined, for example, by diameter or distance between opposite faces (of a slot, for example).

Size, Actual. The general term for the size of a produced feature. This term includes the actual mating size and the actual local sizes.

Size, Actual Local. The value of any individual distance at any cross section of a feature, measured between any two opposite points normal to the derived median line or the derived median plane.

Size, Actual Mating.

a) **External feature:** The dimension of the smallest similar perfect feature which can be circumscribed about the feature so that it just contacts the surface at the highest points. For example, the size of the smallest cylinder of perfect form or the smallest distance between two parallel planes of perfect form which just contact the highest point(s) of the surface(s).
b) **Internal feature:** The dimension of the largest similar perfect feature which can be inscribed within the feature so that it just contacts the surface at the highest points. For example, the size of the largest cylinder of perfect form or the largest distance between two parallel planes of perfect form which just contact the highest point(s) of the surface(s).

Size, Actual Minimum Material. The dimension of the actual minimum material envelope. This dimension may be a radius or diameter for a spherical or cylindrical envelope, or width or half-width for a parallel-plane envelope, depending on the context. (See Envelope, Actual Minimum Material)

Size, Limits of. The specified maximum and minimum sizes.

Size, Nominal. The designation used for purposes of general identification.

Size, True Position Mating. The size, optimized over the candidate datum reference frame set of the actual mating envelope constrained to be located and oriented at true position.

Size, True Position Minimum Material. The size, optimized over the candidate datum reference frame set of the actual minimum material envelope, constrained to be located and oriented at true position.

Size, Virtual Condition. The size of the boundary generated by the collective effect of MMC and any applicable geometric tolerance (when the MMC concept and modifier is used). The size of the boundary generated by the collective effect of LMC and any applicable geometric tolerance (when the LMC concept and modifier is used). These are constant boundaries that do not change with the size of the produced feature when the bonus tolerances are calculated into the equation.

Slope. The angle between the height and the length of a feature expressed as a ratio with one of the variables expressed as equal to 1.

Software (Software Implementation). A computer implementation of an algorithm.

Spine. A point, simple, non-self-intersecting curve, or simple surface. Spines are used in the definitions and inspection of size and circularity.

Spotface Symbol. When the symbol for spotface is used, either the remaining thickness of material may be specified or the depth of the spotface may be given. The spotface depth is the minimum needed to clean up the material of the surface to the diameter of the spotface. The spotface may be noted using the symbol and its diameter without showing it on the design drawing or in the CAD model. The surface to receive the spotface must be clearly indicated. If desired, a filet radius may be given on the drawing to relieve sharp corners for the spotface.

Statistical Tolerancing. The assignment of tolerances to a feature that allows a less restrictive tolerance than would be allowed if the features were to be produced without using statistical process control methods of manufacture. When statistical tolerances are assigned, a note shall be used on the drawing that states that tolerances identified with the statistical tolerancing symbol are to be produced using statistical process controls. If both the more liberal statistical tolerances and the more restrictive non-statistical tolerances are shown on the same features, a note must be added to the drawing that states the statistical tolerances are only allowed if features are produced with statistical tolerancing methods of manufacture. Otherwise, the more restrictive non-statistical tolerance must be held. **CAUTION: When using the "statistical tolerancing" symbol, the required statistical indices should be specified.**

Stock Size. The as furnished sizes of commercially produced stock materials such as bars, sheets, tubing and structural shapes. These items are produced to established industry or government standards other than ASME Y14.5, that prescribe limits for straightness, flatness and other geometric characteristics. Unless geometric tolerances are specified on the drawing of a part made from these items, those standards govern the surfaces that remain in the "as furnished" condition on the finished part, and these surfaces do not have to conform to the Rule #1 "Perfect Form at MMC" requirement.

Straightness. A condition where an element of a surface or a derived median line is a straight line. It also can be applied as a centerplane control.

Straightness Tolerance. A straightness tolerance specifies a tolerance zone within which the considered element or derived median line must lie.

a) A straightness of the surface control uses a tolerance zone consisting of two parallel lines between which the considered surface line element must lie.

b) A straightness of a derived median line control uses a tolerance zone consisting of a cylinder within which the derived median line of the considered feature must lie.

[Note: When applied on an MMC basis for derived median line, as each actual local size departs from MMC, an increase in the local diameter of the tolerance zone is allowed that is equal to the amount of such departure.]

Symmetry. Symmetry is that condition where the median points of all opposed or correspondingly located elements of two or more feature surfaces are congruent with the axis or centerplane of a datum feature.

Symmetry vs. Position Tolerance. Where it is required that a feature be equally disposed with respect to the centerplane of a datum feature, positional tolerancing may be used to achieve a control similar but not identical to that of what the symmetry symbol gives. Positional controls allow the MMC modifier to be used, whereas symmetry controls are always RFS. For example:

a) Position: When position is used to control the location of the centerplane of a slot, within the limits of size at either RFS, MMC or LMC (as specified), the centerplane of the slot's mating size must lie between two parallel planes equally disposed about the centerplane or axis of the location datum.

b) Symmetry: When symmetry is used to control the location of a slot, within the limits of size and regardless of feature size, all median points of opposed elements of the slot must lie between two parallel planes equally disposed about the centerplane or axis of the location datum.

Tabular Dimensions. Tabular dimensioning is a type of rectangular co-ordinate dimensioning in which dimensions from mutually perpendicular planes are listed in a table on the drawing rather than on the pictorial delineation. This method is used on drawings which require the location of a large number of similarly shaped features. Tables are prepared in any suitable manner that adequately locates the features.

Tabulated Tolerances. An entry in a feature control frame used instead of a tolerance, which directs the reader to a value found in a table of tolerances.

Tangent Plane. A plane contacting the high points of a surface.

Taylor Principle. The principle of GO gages. A shaft must be able to fit inside of a hole (at least as deep as the shaft is long) that has a size equal to the maximum allowable diameter of the shaft. A hole must be able to receive a shaft (at least as long as the hole is deep) that has a size equal to the minimum allowable diameter of the hole. (Shafts and holes in this definition have been used as general terminology to illustrate how this principle applies to all external and internal features of size). For further information see Rule # 1.

Temporary Datum Features. Selected datum features of in-process parts, such as sheet metal, castings, forgings, machinings, or fabrications, may be used temporarily for the establishment of machined or otherwise manufactured surfaces to serve as permanent datum features. Such temporary datum features may or may not be subsequently removed.

Tolerance. The total amount by which a specific dimension is permitted to vary. The tolerance is the difference between the maximum and minimum limits.

Tolerance Accumulation. The accumulation of error that occurs when the location of one feature is based on the produced location of another feature. When this occurs, subsequent features experience not only their own error, but the error experienced by the feature they are being measured from. These direct relationships between features are sometimes the most appropriate and cost effective. However, if a reduction of accumulated tolerance is desirable, then designation of the same origins of measurement (datums) for all features (or as many as possible) is most effective.

Tolerance, Bilateral. A tolerance in which variation is permitted in both directions from the specified dimension.

Tolerance, Geometric. The general term applied to the category of tolerances used to control form, profile, orientation, location, and runout.

Tolerance, Unilateral. A tolerance in which variation is permitted in one direction from the specified dimension.

Tolerance on 90° Angles. Where centerlines and surfaces of features of a part are shown on engineering drawings intersecting at right angles and no 90° angle is stated, a 90° angle is implied. The tolerance on these implied 90° angles is the same as for all other angular features shown on the field of the drawing governed by general angular tolerance notes or general tolerance block values. Where centerlines and surfaces of a part are shown on engineering drawings intersecting at right angles and BASIC dimensions or GEOMETRIC controls have been specified, a 90° BASIC angle is implied. The geometric tolerance on the feature associated with these implied 90° BASIC angles is provided by feature control frames which govern the location, orientation, profile, or runout of these controlled features.

Tolerant Gage Tolerancing. The policy of tolerancing gages with both a plus and a minus tolerance around the boundary that is being gaged. This assures that most part features within tolerance that are gaged are accepted and most part features not within tolerance are rejected. However, it is possible that some good parts will be rejected or some bad parts will be accepted, depending on which side of the boundary the gage is produced. This gage concept is explained, but not featured by the ASME Y14.43 standard on Dimensioning and Tolerancing Principles for Gages and Fixtures.

Translation Modifier. A symbol use after a datum feature that looks like a triangle on its side that means the datum feature simulator is moveable.

True Geometric Counterpart. For the purposes of the ASME Y14.5-2018 standard on Dimensioning and Tolerancing, a true geometric counterpart is a theoretically perfect boundary used to establish a datum from a specified datum feature. A theoretical true geometric counterpart may be; 1) a maximum material boundary (MMB), 2) a least material boundary (LMB), 3) an actual mating envelope, 4) a minimum material envelope, 5) a tangent plane, 6) a datum target or targets or 7) a mathematically defined contour.
Note: This term only applies to datums.

True geometric counterpart Requirements. Theoretical true geometric counterparts shall have the following requirements: 1) perfect form 2) basic orientation relative to one another for all the datum references in a feature control frame 3) basic location relative to other true geometric counterparts for all the datum references in a feature control frame, unless a translation modifier or moveable datum target symbol is specified 4) movable location when the translation modifier or moveable datum target symbol is specified 5) fixed at the designated size, when the MMB or LMB is specified 6) adjustable in size, when the datum feature applies at RMB.

True Position. The theoretically exact location of a feature established by basic dimensions (which may include implied basic dimensions such as a distance of zero, or an angle of 90°), most often in a particular datum reference frame.

True Profile. The profile defined by basic dimensions, such as basic radii, basic angles, basic coordinate dimensions, basic sizes, undimensioned drawings, formulas, or math data, including design models.

True Radius. A term used as a local note placed before a given radius that is dimensioned in a view that does not show the true shape of the radius.

Uncertainty. The extent to which an actual dimension may vary from the measured dimension.

Virtual Condition.

a) When the circled M is used in the feature control frame, the virtual condition is the boundary generated by the collective effects of the specified MMC limit of size of a feature and the applicable geometric tolerance at MMC. For an internal feature of size, the geometric tolerance is subtracted from the MMC limit. For an external feature of size, the geometric tolerance is added to the MMC limit.

b) When the circled L is used in the feature control frame, the virtual condition is the boundary generated by the collective effects of the specified LMC limit of size of a feature and the applicable geometric tolerance at LMC. For an internal feature of size, the geometric tolerance is added to the LMC limit. For an external feature of size, the geometric tolerance is subtracted from the LMC limit.

[Note: These boundaries are constant, and will not change as the features under control are produced at various sizes within their size limits (because of applicable bonus tolerances).]

Wear Allowance. An optional allowance which is normally 5% of the workpiece tolerance. It is used to add material to the gage so that the gage can be used for a longer period of time before wearing down beyond acceptable limits. Used in conjunction with the 5% of the workpiece tolerance that is gage tolerance, it is usually applied to GO gages and Functional Gages to comprise a total gage tolerance of 10% of the part tolerance.

Workpiece. The object or component under test or measurement, containing the geometric feature being assessed.

WORKBOOK SECTION

FOR

Geometric Dimensioning and Tolerancing

Textbook and Workbook

(per ASME Y14.5-2018)

Section 1-Word Problems

Section 2-Basic Applications

Section 3-Interpretation and Analysis

Section 4-Tolerancing Parts & Assemblies

Section 5-Analysis and Mating Parts

Section 6-Wall Thickness & Composite Tol.

Section 7-New in Y14.5-2009 & 2018

Section 8-Advanced Applications

SECTION I. WORD PROBLEMS

1. Terms and Definitions pertaining to Datums and Datum Features

Please circle all correct answers. Questions may have multiple correct answers.

1) Datums are:
   a. real
   b. simulated
   c. perfect
   d. irregular and flawed
   e. derived from datum features

2) Features are:
   a. real
   b. surfaces
   c. perfect
   d. always measurable for size
   e. sometimes measurable for size

3) Datum features are:
   a. never measurable
   b. real
   c. perfect
   d. sometimes size features
   e. surfaces

4) Surfaces are:
   a. features
   b. datums
   c. real
   d. perfect
   e. flawed
   f. sometimes chosen to serve as datum features

5) Datum features should be chosen based on:
   a. function
   b. roughness average, waviness and lay only
   c. repeatability
   d. accessibility
   e. representation of mating features
   f. the part itself with no consideration for mating features

6) List five common physical datum feature simulators.

7) A dimension with a box around it, like [100], is called?
   a. size dimension
   b. basic dimension
   c. plus and minus tolerance
   d. limit dimension

8) What is the [A] symbol called?
   a. datum feature symbol
   b. tolerance
   c. allowance
   d. geometric characteristic symbol

9) What kind of tolerance is expressed by 50 ± 1?
   a. limit dimensioning
   b. location tolerance
   c. bilateral plus and minus tolerancing
   d. unilateral plus and minus tolerancing

10) What is the Ⓜ called when used after the geometric tolerance?
   a. datum identification symbol
   b. Zone descriptor
   c. basic size
   d. maximum material condition symbol

11) What is third datum feature in the feature control frame called?
   a. primary
   b. tertiary
   c. geometric characteristic
   d. modifier

12) What is the Ⓜ called when used after a datum feature of size in **Y14.5-2009**?
   a. a geometric characteristic symbol
   b. datum feature symbol
   c. tolerance
   d. maximum material boundary symbol

13) If a Ø2 appears in a feature control frame, it is:
   a. the hole size.
   b. the size tolerance.
   c. basic size.
   d. the geometric tolerance

14) List at least two components (other than the perpendicularity symbol) that you feel are needed in a perpendicularity control.

15) A clearance fit implies the mating features, when produced within size limits,
    a. will always encounter material interference.
    b. will sometimes encounter material interference.
    c. will never encounter material interference.

16) An interference fit implies the mating features, when produced within size limits,
    a. will always encounter material interference.
    b. will sometimes encounter material interference.
    c. will never encounter material interference.

17) In a fixed fastener assembly condition, the geometric tolerance to be divided between the two parts containing the mating features is equal to,
    a. the difference between the Least Material Conditions of mating features.
    b. the difference between the Maximum Material Conditions of mating features.
    c. the difference between the shaft MMC and the hole LMC.
    d. the difference between the shaft LMC and the hole MMC.

2. MMC and LMC

Please identify the Ⓜ and Ⓛ of these.

HOLES:

1) 62.5 ± 0.1 Ⓜ ________________

   Ⓛ ________________

2) 59-62 Ⓜ ________________

   Ⓛ ________________

3) 83.5 ± 0.05 Ⓜ ________________

   Ⓛ ________________

4) 69.5-70.5 Ⓜ ________________

   Ⓛ ________________

5) 1.245-1.265 Ⓜ ________________

   Ⓛ ________________

SHAFTS:

6) 7.5-8.1 Ⓜ ________________

   Ⓛ ________________

7) 71 ± 0.2 Ⓜ ________________

   Ⓛ ________________

8) 12.45-12.65 Ⓜ ______________________

Ⓛ ______________________

9) 21.885-22.875 Ⓜ ______________________

Ⓛ ______________________

10) 0.85-0.90 Ⓜ ______________________

Ⓛ ______________________

3. Size Controls Form

1) Unless otherwise specified, individual features of size must:
   a. not violate size limits of MMC at any cross section.
   b. not violate an envelope of perfect form at LMC.
   c. not violate an envelope of perfect form at MMC.

2) The concept discussed in Question 1 is often inspected with a:
   a. NOGO gage
   b. GO gage
   c. segmented gage

3) Least material condition on most parts is checked
   a. at cross sections.
   b. for an envelope.
   c. never checked because it serves no purpose.

4) When checking for maximum material condition violations in a hole with size limits of Ø50 - 51, what size and type of gage would you use?
   a. a full form plug gage with a diameter of 50
   b. a full form plug gage with a diameter of 51
   c. measurement tool capable of checking size at any cross section

5) What effect on size measurements would adding the Independency symbol have?

6) Ø25 ± 0.1 AVG. This is a hole in metal. It is the I.D. of a thin walled tube.
   What is the AVG. MMC?
   What is the AVG. LMC?

7) Ø.500 ± .005 This is a flexible rubber shaft.
   What is the MMC?
   What is the LMC?

8) .625 - .635 This is a slot milled into a thick rigid plate.
What is the MMC?
What is the LMC?

9) .010 ± .002 This is a steel tab that is non-rigid.
What is the MMC?
What is the LMC?

10) Ø100 ± 1
| ⌖ | Ø0Ⓛ | A | B | C |
This control refers to a glass shaft outside diameter. A hole runs through its center making it look like a drinking straw. It is mounted on a plate.
What is the MMC?
What is the LMC?
How does the use of the LMC symbol change the way the size is measured?

11) Ø37.5 ± 0.2
| — | Ø0.5Ⓜ |
This is a metal rod (shaft) that is twenty-five feet long.
What is the MMC?
What is the LMC?
How does the use of straightness of the derived median line change the way size is measured?

12) 10.0 - 10.1. This forms two sides of a rectangular part. It is considered an external feature of size. It is rigid.
What is the MMC?
What is the LMC?

4. Rules, Concepts, Characteristics and Untoleranced Dimensions

1) What purpose does referencing a datum feature in a feature control frame serve?

2) Please categorize all the geometric characteristic symbols as:
INDIVIDUAL (not related to datums); RELATED (related to datums); INDIVIDUAL or RELATED:

3) When we check an individual feature for form, what are we comparing it to?
   a. a datum
   b. a perfect geometric form of itself
   c. a point, line or plane

4) Once on a drawing, how can we tell if a feature is individual or related?

5) Name three ways to eliminate the requirement for an envelope of perfect form at MMC for a feature of size.

6) If you do one of the three things stated in Question 5, can you exceed MMC at any cross sectional check?

7) If you assign the LMC concept and symbology to the geometric tolerance in the feature control frame, where is the envelope of perfect form size requirement?
   a. at MMC
   b. it does not exist
   c. at LMC

8) Which material condition symbol or symbols (MMC, LMC or RFS implied) commonly provides the most geometric tolerance to draw from when used?

9) Which parts are considered to be most economical to manufacture?
   a. those with features controlled at MMC
   b. those with features controlled RFS

10) What material condition symbol provides for the easiest use of functional (receiver-type) gaging?

11) Is it ever possible to use the Ⓜ symbol both for a feature of size and a datum feature of size in one Feature Control Frame? Yes No

    For example: | ⌖ | ⌀0.5Ⓜ | A | BⓂ | C |

12) When features are controlled RFS, they are given an original tolerance of form, orientation, profile, runout or location. Where can we draw extra geometric tolerance from?
    a. the size limits
    b. the title block
    c. we can't

13) As a minimum, the feature control frame must contain:
    a. a geometric characteristic symbol and datum feature reference.
    b. a geometric characteristic symbol and a geometric tolerance.
    c. a datum feature reference, tolerance and a material condition symbol.

14) Draw the symbols for the following geometric characteristic symbols and categorize them as form, orientation, profile, runout or position.
    a. Flatness
    b. Straightness
    c. Roundness
    d. Cylindricity
    e. Profile of a Line
    f. Profile of a Surface
    g. Parallelism

h. Perpendicularity
i. Angularity
J. Feature Control Frame
k. Position
l. Runout (Circular)
m. Concentricity
n. Runout (Total)

15) Identify the following symbols:
a. Ⓜ
b. ↧
c. ∅
d. Ⓟ
e. ⟨ST⟩
f. Ⓛ
g. ⌴

5. Datums, Datum Features and Datum Feature Simulators

1) What geometrically perfect elements are representative of datums?

2) How are datum features simulated in manufacturing or inspection?

3) What criteria is used to choose a datum feature when tolerancing?

4) What is a datum reference frame and what is it used for?

5) What six degrees of spatial freedom does a part have before it is restrained?

6) How many high points of contact are required to simulate:

-a primary datum plane from a planar datum feature?

-a secondary datum plane constructed from a planar surface after the primary is constructed as described above?

-a tertiary datum plane constructed from a planar surface after the primary and secondary are constructed as described above?

7) Are three datums always necessary for part orientation? If yes, why? If no, why not?

8) Is a datum feature perfect?

9) Is a datum feature ever given form control of its own? If not, why is it not necessary? If so, why it is necessary?

10) How is the order of precedence of datum features shown in a feature control frame (for example-per their order of alphabet, per their order read from left to right, etc.)?

11) Why does the order of precedence of datum features make such a difference in feature orientation?

12a) If a shaft is chosen as a datum feature RMB, how is the datum feature simulated?

12b) If a hole is chosen as a datum feature RMB, how is the datum feature simulated?

13) Of what use is an angular orientation datum in a datum reference frame when it is referenced after a cylindrical datum feature?

14) How many datum planes are derived from a cylindrical datum feature?

15) Why are datum targets necessary?

16) How are cylindrical or width datum features controlled Regardless of Material Boundary simulated?

17) At what size or condition are secondary datum features of size referenced at Ⓜ that have their own tolerance of orientation or position simulated as datum features?

18) At what size is a primary datum feature of size represented for the features controlled to it?

19) What does a datum feature such as AⓂ-BⓂ mean?

20) When a pattern of holes is controlled to a datum feature hole, as the datum feature hole departs from MMB within its limits, what happens to the controlled hole pattern and how is it calculated?

21) Do screw threads make good datum features?

22) If more than one datum reference frame is specified for a part, what effect will this have on manufacturing and inspection?

23) Why would a designer choose to use only a portion of a surface as a datum feature instead of the entire surface?

24) In what forms are datum targets specified?

25) If a designer wants to show a datum target as originating from the opposite side of a part than the view it has been shown in, how may it be done without another view?

26) What kind of contact is required from the physical datum feature simulation equipment, in a datum target area, on a non-cylindrical surface?

27) How is the shape and location of a datum target area defined?

28) What does the dimension origin symbol look like and of what use is it?

6. Virtual Condition and Resultant Condition Boundaries

1) When a feature has errors of orientation, location and certain tolerances of form, its effective mating boundary can be virtually altered at that orientation or at that location or by its form.
T F

2) In virtual condition (MMC concept), the effective mating boundary of a hole is virtually increased and the shaft size virtually decreased, even though they still measure within their limits of size.
T F

3) A virtual condition boundary (MMC concept) of the feature can represent the worst condition of assembly of the mating components at a specific orientation or location.
T F

4) If a feature is allowed a virtual condition, it may exceed its limits of size as an individual feature when measured.
T F

5) How does calculating the virtual condition of an external feature of size differ from calculating the virtual condition (MMC concept) of an internal feature of size?
T F

6) Functional gages are made to the virtual condition (MMC concept) of features they inspect.
T F

7) Functional gages are usually only made to check features controlled with MMC. Otherwise, for checking features controlled RFS, a functional gage would have to be made to take into account every size the feature could be produced at.
T F

7. Form Controls

FLATNESS

1) Describe the configuration of a flatness tolerance zone?

2) Is flatness considered individual or related to datums?

3) Is a datum feature reference allowable in the feature control frame?

4) May a diameter sign be included in the feature control frame?

5) Is flatness a surface or an axis control?

6) Is flatness a two-dimensional or a three-dimensional control?

7) To obtain a high level of certainty (for example: 90% certainty) of how flat a surface is, when measuring flatness will a few points from the surface suffice, or does this require hundreds of points per square inch?

8) If a surface is controlled with flatness, may the controlled surface in combination with its opposing side on the part violate an envelope of perfect form at MMC?

9) Must the flatness of a surface tolerance refine the size tolerance or may it be larger?

10) May a MMC symbol be included in the feature control frame if flatness is applied to a single surface?

11) How is the flatness control directed to the controlled surface?

12) May flatness be applied on a unit or a rate basis?

13) May flatness be applied to control a derived median plane per ASME Y14.5-2009?

STRAIGHTNESS

General:

1) Are datum feature references allowable/required in the feature control frame?

Surface Straightness (questions 2 through 8):

2) What view should the straightness of a surface tolerance be applied in?

3) In straightness of a surface, what is the shape of the tolerance zone, and what must lie within it if the feature is to be accepted?

4) What material condition symbols may be included in the feature control frame?

5) May a diameter sign appear in the feature control frame?

6) Without a straightness feature control frame, what controls the straightness of a part surface?

7) Is perfect form of the controlled surface and its opposing surface required at MMC?

8) Is bonus tolerance ever possible?

If so, when?

If not, why not?

Straightness of the derived median line (questions 9 through 15):

9) In straightness of a derived median line, what is the tolerance zone shape and what must be contained within it for the feature to be acceptable?

10) Is perfect form of the controlled feature required at MMC?

11) Is this control considered individual or related to datums?

12) How might straightness of a derived median line that is controlled with a MMC symbol be gaged?

13) How might straightness of a derived median line that is controlled with the RFS concept be measured?

14) Is a virtual condition allowed for both controls described in Questions 12 and 13?

15) Must the straightness tolerance be smaller than the feature's size tolerance?

CIRCULARITY

1) Is circularity a two- or a three-dimensional control?

2) What is the tolerance zone shape?

3) What material condition symbols are allowed in the feature control frame?

4) Are datum feature references allowable?

5) Is a feature controlled with circularity allowed to exceed its size limits?

6) Is any extra tolerance derived from the controlled feature's size limits as the feature departs from MMC?

7) How is circularity inspected if done with a dial indicator?

8) Does the tolerance in the feature control frame have to be smaller than the feature's size limits?

9) Is circularity considered individual or related to datums?

10) Can circularity controls be assigned to cones?

11) May a diameter sign be included in the circularity feature control frame?

12) How is the feature control frame connected to the controlled surface?

13) Circularity controls the surface
   a. between two concentric cylinders.
   b. at every cross section.
   c. but is mainly an axis control.

14) Without a circularity feature control frame, what would normally control a feature's roundness?

15) A circularity control is assumed to apply
   a. at LMC.
   b. RFS.
   c. at MMC.

CYLINDRICITY

1) Name at least two things cylindricity controls that circularity does not.

2) List three different ways cylindricity can be checked. Order them with the best way first and the least desirable method last.

3) What is the tolerance zone configuration a cylindricity control generates?

4) Is cylindricity a surface or an axis control?

5) May a least material condition symbol be used in the feature control frame?

6) Can cylindricity and circularity be applied to the same surface? If so, which tolerance must be smaller and why?

7) Does a cylindricity tolerance ever reference datum features inside the feature control frame?

8) Must the controlled feature, if produced at MMC, have perfect form?

9) Is the cylindricity tolerance additive to the controlled feature's MMC size limit?

10) Is bonus tolerance ever allowed?

11) A cylindricity tolerance is considered
   a. form.

b. orientation.
c. location.
d. profile.
e. runout.

12) A produced feature's cylindricity is compared to a perfect geometric counterpart of
a. a datum.
b. an axis.
c. itself.

13) List all items allowed in the cylindricity feature control frame.

14) A cylindricity tolerance controls roundness and
a. perpendicularity.
b. flatness.
c. straightness and taper.
d. angularity.

15) The tolerance applies simultaneously to circular and ____ elements of the controlled surface.
a. longitudinal
b. perpendicular
c. flat
d. positional

8. Orientation Controls

1) What is the configuration of the tolerance zone defined by a parallelism of a planar surface control?

2) What does that tolerance zone take its orientation from?

3) If we wanted the following orientation tolerance zone | ∠ | 0.5 | for a planar surface to be exactly 35° from datum plane R as well as perfectly perpendicular to datum plane S, what could we do to the given feature control frame per ASME Y14.5-2009 to reflect that?

4) If a part has a maximum height of 1.000. Is a .002 parallelism control additive to that height?

5) Is at least one datum feature reference always required in the feature control frame for angularity?

6) How does the tolerance configuration for the angularity in the control shown in Question 3 differ from the tolerance zone that would be generated if the 35° was ± 2° ?

7) What material condition symbols would be appropriate in feature control frames for orientation controls if they were used on a single planar surface?

8) What is the usual configuration and orientation of a controlled feature's perpendicularity tolerance zone if it is used on a planar surface?

9) What does the use of a datum feature reference add to the orientation geometric control?

10) ∅.250±.005 Hole

∠	∅.005Ⓜ	A

If the controlled hole was produced at a diameter of .250, what angularity tolerance would apply?

11) Does the controlled hole in Question 10 have a virtual condition? If so, what is it? If not, why not?

12) Would it be appropriate to include a maximum material boundary symbol in the feature control frame for datum features that are planar surfaces? Please explain.

13) How many contact points are required to establish datum A from a planar surface if A is primary?

14. Is an orientation geometric control related to datums or individual?

15) For parallelism used on a planar surface to a datum plane generated by an opposing planar surface, must the tolerance in the feature control frame be smaller than the size tolerance of the part's opposing sides that are included in the control?

16) ∅25±0.5 Hole

⊥	∅0.5Ⓜ	AⓂ	B

If the controlled hole departs from MMC to LMC, how much extra perpendicularity tolerance can be drawn from the size?

17) List the MMC and LMC of the datum feature in Question 16.

18) If the datum feature A referenced in Question 16 is a hole and has a size tolerance of ∅50±0.1, as it departs from MMB to LMC, how much additional shift of the controlled hole's tolerance zone is gained?

19) If both the controlled hole and datum feature A were made at LMC, how much could their axes be out of perpendicularity to each other?

20) What is the shape of the tolerance zone generated by this geometric control?

9. Profile

1) In a profile of a curved surface control, the profile must be dimensioned with
   a. plus and minus dimensions.
   b. limit dimensions.
   c. radial dimensions.
   d. basic dimensions.

2) What is the difference between profile of a line and profile of a surface?

3) For profile controls, datum feature references are
   a. allowable but not required.
   b. not allowed.
   c. required.

4) If a datum (or datums) was used for a profile control, what purpose would be served?

5) Are material condition symbols allowed for the feature being controlled? Why or why not?

6) If both profile of a surface and profile of a line are used on the same feature, which must have the smaller profile tolerance?

7) What is the tolerance zone shape of profile of a line?

8) What is the tolerance zone shape of profile of a surface?

9) How is profile tolerance designated as unequal-bilateral or unilateral?

10) How is the extent of where the profile control applies on a surface designated?

11) Suppose the entire outer periphery of the part profile must be controlled. How may it be done?

10. Runout

1) Circular runout checks are taken
   a. at cross sections.
   b. over the entire surface without lifting the indicator.

2) Runout checks rely on
   a. fine indicator movement.
   b. full indicator movement.
   c. full indicator motivation.

3) Runout checks are taken (choose the single most correct answer)
   a. while the feature is stationary.
   b. while the feature rotates 180°.
   c. while the feature rotates 90°.
   d. while the feature rotates 360°.

4) Runout checks are taken on the controlled feature as the feature rotates about
   a. the datum axis.
   b. its own axis.
   c. any axis on the part.

5) Total runout includes a variety of geometric controls. List them.

6) What does circular runout control?

7) Give an example of when circular runout would be a suitable control.

8) Runout is a
   a. datum axis to feature axis control.
   b. datum surface to feature axis control.
   c. surface to surface control.
   d. feature surface to datum axis control.

9) Is runout a control related or individual?

10) Are datums allowable or required?

11) Must the tolerance in the feature control frame be smaller than the controlled feature's size limits?

12) In its 360° rotation relative to the datum axis, must the controlled feature remain within an envelope of perfect location at MMC?

13) While stationary, must the controlled feature fit within a boundary of perfect form at MMC?

14) May a maximum material condition symbol be included in a runout feature control frame?

15) If circular runout and total runout are used on the same feature, which control must have the smaller runout tolerance?

11. Location

CONCENTRICITY

1) Concentricity is a _____ control.
   a. datum axis to entire feature surface
   b. datum axis to cross sectional feature surface
   c. feature median points to datum axis

2) Concentricity (per ASME Y14.5) is always assumed to apply at
   a. RFS.
   b. MMC.
   c. LMC.

3) Is concentricity a refinement of controls that already exist in the size limits?

4) Must a tolerance zone descriptor (such as a diameter sign) be included in all concentricity feature control frames?

5) What is the tolerance zone shape in a concentricity control?

6) Is concentricity a related or individual control?

7) Concentricity is a ______ control.
   a. orientation
   b. form
   c. location
   d. runout
   e. profile

8) If a feature is given a concentricity control,
   a. it probably rotates in assembly (or use).
   b. it probably does not rotate in assembly (or use).

9) Under what circumstances would a designer want to apply a concentricity control to a feature?
   a. When the part mates, so that the MMC principle can be applied.
   b. When surface form is just as important as balance.
   c. When balance for a rotating feature is important, surface form is not important.
   d. When orientation, but not location is important.

10) A feature is given a concentricity control. While that part is in a stationary state, is the feature allowed to violate an envelope of perfect form at MMC?

11) If a cylindrical feature is controlled with concentricity, when the feature rotates about the datum axis, must the rotating part stay within an envelope of perfect form at MMC that is centered to the datum axis?

SYMMETRY

12) Symmetry is a _______ control.
    a. centerplane to datum
    b. cloud of median points to datum plane
    c. datum centerplane to centerplane

13) Per ASME Y14.5, symmetry is always assumed to apply at
    a. RFS.
    b. MMC.
    c. LMC.

14) Is a symmetry a refinement of controls that already exist in the size limits?

15) May a tolerance zone descriptor, such as a diameter sign, appear in the feature control frame?

16) Describe the tolerance zone for a symmetry control used on a width.

17) Is symmetry required to use a datum?

18) Symmetry is a __________ control.
    a. form
    b. orientation
    c. profile
    d. runout
    e. location

19) If a feature is given a symmetry control,
    a. it probably mates in the assembly.
    b. it probably does not mate in the assembly; otherwise, position at MMC would be used.

20) If a rigid slot is controlled with symmetry, unless otherwise specified, must the feature have perfect form (flatness, straightness and parallelism) if produced at MMC?

POSITION

21) What is generally needed to dimension and tolerance the perfect position of a hole pattern?
    a. basic dimensions, datum features and a feature control frame.
    b. size dimensions, plus and minus location dimensions and zero lines.
    c. basic angles, plus and minus location tolerances and drawing notes.

22) What is the definition of a datum reference frame?

23) From what elements on the part are datum planes simulated?

24) What might cause a rocking effect when trying to establish a datum plane?

25) Why is tolerance accumulation not a problem when chain dimensioning is use to locate the true position of the feature?

26) What is the purpose of referencing a full datum reference frame?

27) What information may be contained within the feature control frame?

28) In addition to the original tolerance of position, when a feature is controlled at MMC, extra tolerance is sometimes available. Where is the extra positional tolerance drawn from?

FLOATING FASTENERS

29) Define what a floating fastener assembly is.

30) Show at least one floating fastener formula?

31) Calculate floating fastener position tolerance for the following assembly situations:

a. MMC fastener = .250
   MMC hole = .265

b. MMC hole = .520
   MMC fastener = .500

32) Calculate fastener MMC size from the following:
   a. MMC hole = .775
      Positional tolerance = .025

   b. Positional tolerance = .030
      MMC hole = 1.030

33) Calculate hole MMC size for the following situations.
   a. Fastener MMC - .3125
      Tolerance = .010

   b. Tolerance = .020
      Fastener MMC = 6.25

PROJECTED TOLERANCE ZONES

34) What determines the distance a tolerance zone is projected?

35) Why is the use of projected tolerance zones necessary?

36) On what types of features are projected tolerance zones used?

37) State at least one formula for calculating geometric tolerance for a fixed fastener assembly?

12. General Rules, Formulas and Measurement Questions

Complete the following questions by circling the appropriate answer.

1) Unless otherwise specified, limits of size are measured in the following manner:
   a) LMC at cross sections and MMC for violations of a full form envelope.
   b) LMC and MMC are inspected for violations at every two diametrically opposed sets of points.
   c) LMC and MMC are both inspected with a full form envelope gage.
   d) LMC is a full form envelope check and MMC is a cross sectional check.

2) When basic dimensions are given to locate a hole,
   a) no tolerance is allowed on the hole location.
   b) the location tolerance in the general title tolerance block on the drawing applies.
   c) the location tolerance is found in a position feature control frame.
   d) the tolerance will appear next to the basic dimension as a plus or minus tolerance.

3) The correct formula for 'bonus tolerance' for a hole controlled with the MMC concept is:
   a) LMC plus MMC.
   b) MMC minus virtual condition.
   c) actual size minus LMC.
   d) actual size minus MMC.

4) Total runout can be inspected the best with
   a) a micrometer.
   b) a vernier caliper.
   c) a dial indicator.
   d) a fixed size receiver gage made at virtual condition.

5) Position deviation from 'true position' can be calculated using the following formula:
   a) two times the square root of the sum of the squares of the x and y deviations.
   b) the difference between the sum of the square roots of the x and y deviations.
   c) the dividend of the squares of the x and y deviations.
   d) the square root of two times the difference in the x and y deviations.

6) A coordinate measurement machine can most accurately simulate a primary datum by
   a) probing three arbitrary points on the primary datum feature.
   b) probing a hundred arbitrary points on the primary datum feature.
   c) probing the surface plate on which the highest points of the datum feature rest.
   d) probing the surface of the mating feature.

7) The datum axis generated by an external cylindrical datum feature is the axis of the
   a) minimum circumscribed cylinder.
   b) the least squares center.
   c) the best fit center of the probed points.
   d) the minimum radial separation center of the two cylinders that are able to contain all points of the cylinder's surface.

8) An optical comparator is most often associated with the inspection of
   a) flatness.
   b) profile.
   c) perpendicularity.
   d) concentricity.

9) Rocking datum features may
   a) be rejected.
   b) bow beyond their maximum material condition boundary.
   c) be restrained with enough torque to allow controlled features to be accepted.
   d) be adjusted to an optimum position.

10) A cylindricity geometric characteristic symbol with a geometric tolerance of .003 allows:
   a) pits as deep as .006.
   b) bumps as tall as .006.
   c) the geometric tolerance to be additive to the size tolerance.
   d) size variations for the cylinder's diameter as large as .006.

11) Which of the following would be inappropriate to include in a feature control frame for a straightness control?
   a) a diameter symbol
   b) a maximum material condition symbol
   c) a datum reference
   d) a geometric tolerance

12) Runout is what type of control?
   a) surface to datum axis
   b) centerplane to datum axis
   c) an axis to datum axis
   d) derived median line to datum axis

13) Circular runout of a diameter does not control which of the following characteristics?
   a) coaxiality
   b) circularity
   c) straightness of the surface
   d) straightness of the axis

14) Position is a geometric control that is not capable of controlling
   a) perpendicularity
   b) distance
   c) coaxiality
   d) circularity

15) Of the following geometric characteristic symbols, which is the most restrictive in its control of surface characteristics?
   a) circular runout
   b) position
   c) total runout
   d) concentricity

16) Of the following geometric characteristic symbols, which is considered a 2-D line element control?
   a) circularity
   b) flatness
   c) concentricity
   d) cylindricity

17) Of the following geometric characteristic symbols, which is least capable of controlling flatness?
    a) perpendicularity
    b) angularity
    c) parallelism
    d) straightness

18) Limits of size on a rectangular part are not a control of which of the following?
    a) flatness
    b) perpendicularity
    c) parallelism
    d) straightness

19) Virtual condition (MMC concept) is calculated for a hole by the following formula:
    a) MMC of the hole minus the geometric tolerance applicable at MMC.
    b) MMC of the hole plus the geometric tolerance applicable at MMC.
    c) MMC of the hole divided by the geometric tolerance applicable at MMC.
    d) MMC of the hole multiplied by the geometric tolerance applicable at MMC.

20) Positional tolerance for all holes in a floating fastener assembly can be calculated in the following manner:
    a) MMC of the hole minus the LMC of the mating fastener.
    b) MMC of the hole minus the MMC of the mating fastener.
    c) LMC of the fastener minus the LMC of the hole.
    d) LMC of the hole minus the MMC of the hole.

21) What is the virtual condition of the following shaft?

    ∅8.3 – 8.5

⊕	∅0.3Ⓜ	A	B	C

    a) ∅8.0
    b) ∅2.49
    c) ∅8.8
    d) ∅27.67

22) When a pattern of holes is positioned to a datum feature of size such as a hole and this datum feature is referenced at MMB, as the datum feature (hole) grows,
    a) each hole in the pattern receives an individual growth in its tolerance zone.
    b) the pattern of holes as a group may shift.
    c) the tolerance zone for each hole shrinks.
    d) the pattern of holes must maintain its relative location to the datum feature.

23) In the concept of composite positional tolerancing,
   a) the features within a pattern hold a tighter tolerance in relation to one another than to any datums that are listed only in the upper level control.
   b) the features within a pattern hold a looser tolerance in relation to one another than to any datums that are listed only in the upper level control.
   c) the features within a pattern hold an equally tight tolerance in relation to one another and to any datums listed only in the upper level control.
   d) the features within a pattern are held in place to one another by a positional tolerance, but are located to datums by a plus and minus tolerance.

24) When a pattern of holes is used to establish a datum reference frame from which to locate other part features,
   a) only one hole may be referenced to form the datum reference frame.
   b) only two holes may be referenced to form the datum reference frame.
   c) only three holes may be referenced to form the datum reference frame.
   d) an infinite number of holes may be referenced to form the datum reference frame.

25) When multiple patterns of features are located from the same datum reference frame, they are considered to be
   a) a simultaneous requirement.
   b) a separate requirement.
   c) a composite tolerance control.
   d) a combination tolerance control.

26) If a pattern of holes is positioned and no datum features are used in their positional feature control frame,
   a) the control is incorrect.
   b) the holes are considered positioned to one another.
   c) implied datum features are considered as the origin of measurement.
   d) the part must be accepted as produced.

27) Two hole patterns are positioned to the same primary planar datum feature, and this datum feature, due to imperfections of flatness, tends to rock. If the patterns of holes are considered as separate requirements,
   a) the part may be rocked one way to accept one pattern of holes and another way to accept the other pattern of holes during inspection.
   b) the part may not be rocked one way to accept one pattern of holes and another way to accept the other of holes during inspection.
   c) the hole patterns must be inspected in two different (separate) orientations.
   d) the rocking datum must be brought under control by re-machining it (or another suitable re-manufacturing method) until it no longer rocks.

28) A four-hole pattern has size limits of ∅10.7 – 11.0 and a position tolerance to a datum reference frame of ∅0.3 at MMC. Its mating pin pattern has size limits of ∅9.7 – 10.0 and is positioned to a representative set of datums. How much position tolerance is available for the pins within the pin pattern at MMC?
a) ∅0.7
b) ∅1.0
c) ∅0.5
d) ∅0.4

29) For the circumstances given in Question 28, the position tolerance that would be allowed for a hole produced at ∅11 is
a) ∅0.4
b) ∅0.5
c) ∅0.6
d) ∅0.3

30) For the circumstances given in Question 28, what is the virtual condition (MMC concept) for the pins within the pin pattern?
a) ∅10.7
b) ∅11.0
c) ∅10.5
d. ∅10.4

Section 2-Basic Applications

1a. Calculate the MMC and LMC for the following holes and shafts.

Calculate the MMC and LMC for the following holes and shafts.

Holes:

1. Ø.625 ±.010 — Ⓜ ________ Ⓛ ________

2. Ø .620 / .590 — Ⓜ ________ Ⓛ ________

3. Ø.835 ±.005 — Ⓜ ________ Ⓛ ________

4. Ø .705 / .695 — Ⓜ ________ Ⓛ ________

5. Ø 1.265 / 1.245 — Ⓜ ________ Ⓛ ________

Shafts:

6. Ø .810 / .750 — Ⓜ ________ Ⓛ ________

7. Ø.710 ±.020 — Ⓜ ________ Ⓛ ________

8. Ø 1.265 / 1.245 — Ⓜ ________ Ⓛ ________

9. Ø 22.875 / 21.885 — Ⓜ ________ Ⓛ ________

10. Ø .090 / .085 — Ⓜ ________ Ⓛ ________

1b. Calculate the MMC and LMC for the following holes and shafts.

HOLES:

1) 62.5 ± 0.1 Ⓜ ____________

Ⓛ ____________

2) 59-62 Ⓜ ____________

Ⓛ ____________

3) 83.5 ± 0.05 Ⓜ ____________

Ⓛ ____________

4) 69.5-70.5 Ⓜ ____________

Ⓛ ____________

5) 1.245-1.265 Ⓜ ____________

Ⓛ ____________

SHAFTS:

6) 7.5-8.1 Ⓜ ____________

Ⓛ ____________

7) 71 ± 0.2 Ⓜ ____________

Ⓛ ____________

8) 12.45-12.65 Ⓜ ____________

Ⓛ ____________

9) 21.885-22.875 Ⓜ ____________

Ⓛ ____________

10) 0.85-0.90 Ⓜ ____________

Ⓛ ____________

2. Please convert the following specifications to equal bilateral toleranced dimensions generated by their inner and outer boundaries. RFS boundaries have been calculated to reflect holes and shafts that are axially straight, and tabs and slots have straight centerplanes.

	Outer Boundary	Inner Boundary
1) Ø20±2 Hole — ⌖ | Ø2 Ⓜ | A | B | C		
2) Ø20±2 Hole — ⌖ | Ø2 Ⓛ | A | B | C		
3) Ø20±2 Hole — ⌖ | Ø2 | A | B | C		
4) $Ø20^{+3}_{-2}$ Shaft — ⌖ | Ø0.5 Ⓜ | A | B | C		
5) Ø15-16 Shaft — ⌖ | Ø0.4 Ⓛ | A | B | C		
6) $Ø30^{+0.4}_{-0.2}$ Shaft — ⌖ | Ø0.3 | A | B | C		
7) $28.5^{+0.2}_{-0.1}$ Slot — ⌖ | 0.8 Ⓜ | X | Y | Z		
8) 23.08-24.04 Tab — ⌖ | 0.5 Ⓜ | T | E | C		
9) 75±2 Slot — ⌖ | 1 Ⓛ | C | A | B		
10) 70±0.5 Tab — ⌖ | 0.5 Ⓛ | D | E | F		
11) 12-13 Slot — ⌖ | 0.3 | G | A | F		
12) 9.5-10.6 Tab — ⌖ | 1.2 | H | J | K		

3. Please convert the following mating parts from plus and minus location tolerancing to position tolerancing. For Parts #1 and #2:
   Step 1) Assign a primary datum feature and give it a flatness tolerance of 0.01.
   Step 2) Assign a secondary datum feature and a perpendicularity tolerance of 0.03.
   Step 3) Assign a tertiary datum feature and a perpendicularity tolerance of 0.09.
   Step 4) Calculate and assign positional tolerances to both parts.

Part #1

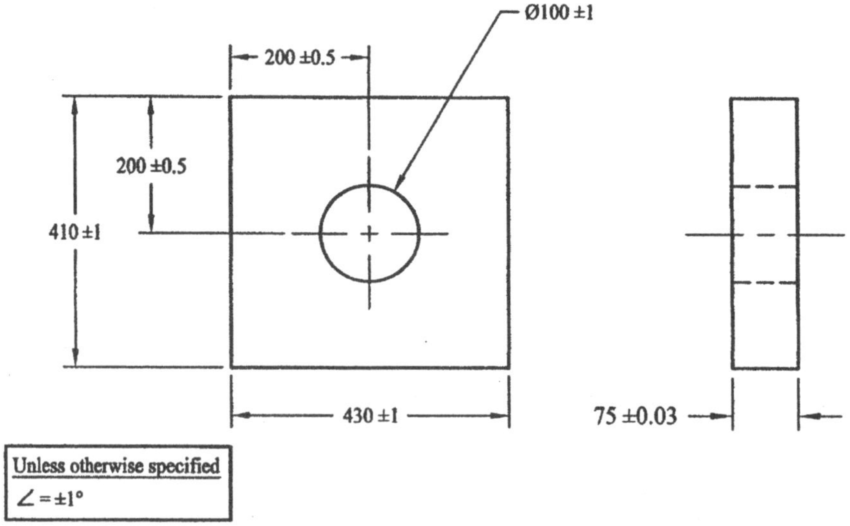

Part #2

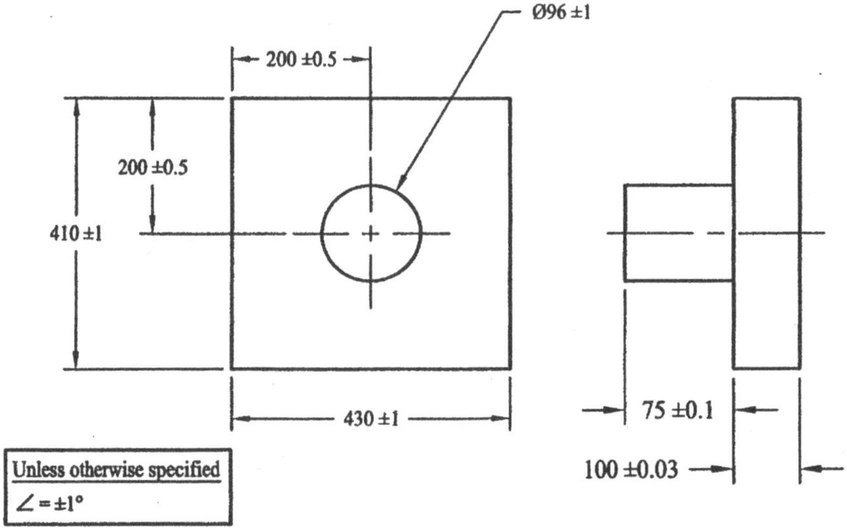

4. Using the illustration below, fill in the rest of the shaft's feature control frame to assure that it will mate with the hole.

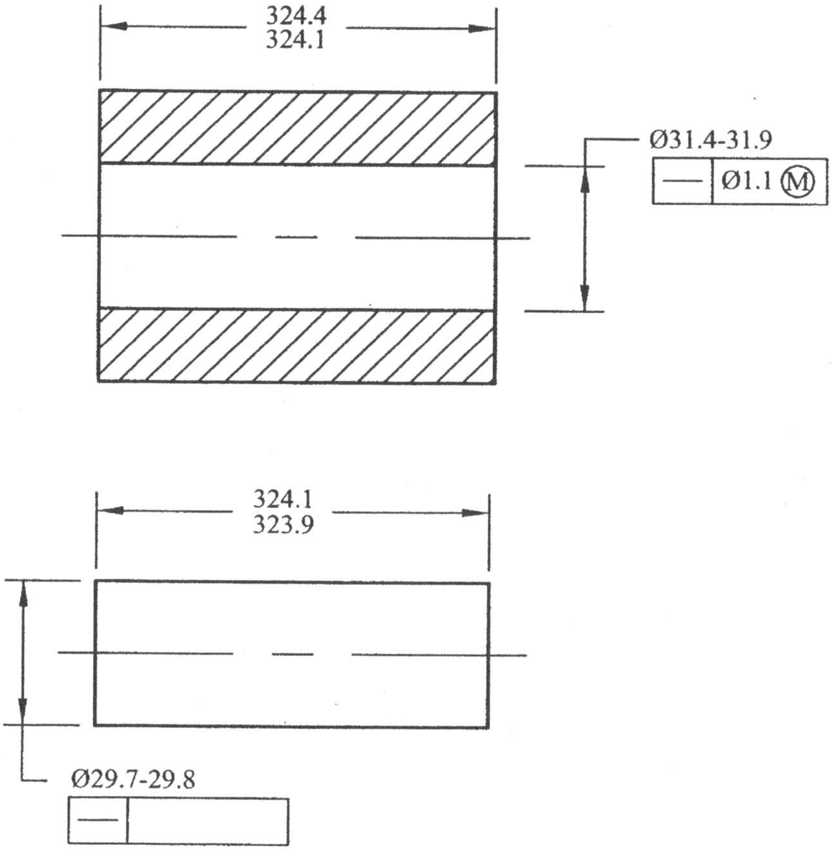

5. Using the illustration below, fill in the rest of the hole's feature control frame to assure that it will mate with the shaft.

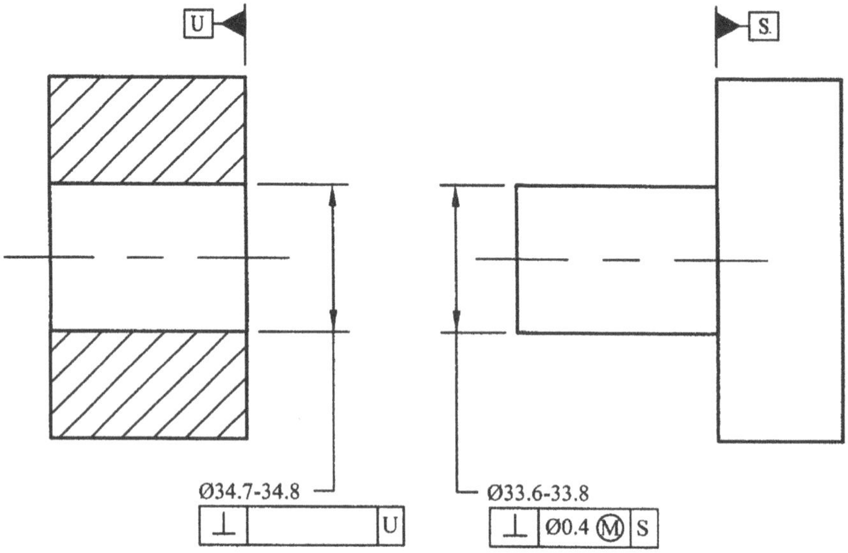

6. Fill in the rest of the shaft's feature control frame to assure that it will mate with the hole.

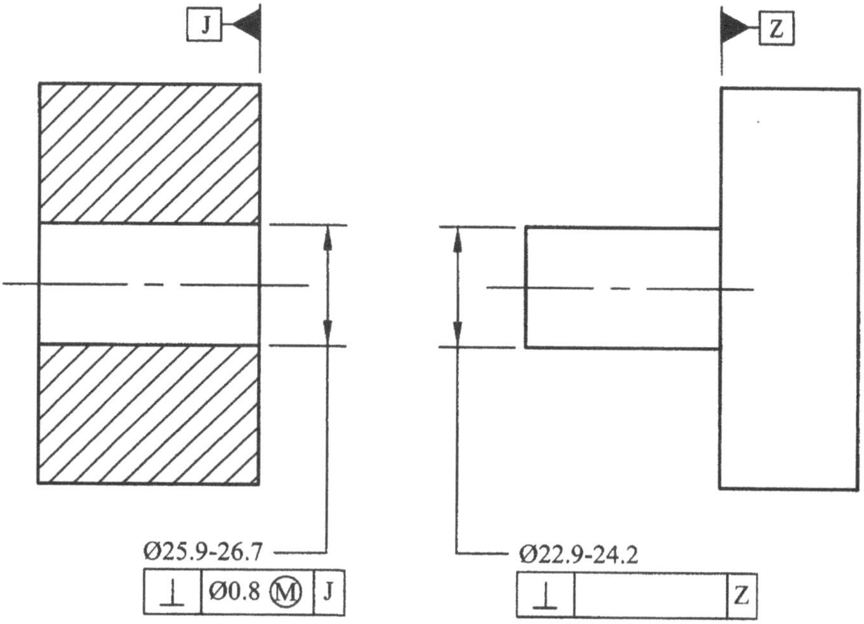

7. Using the illustration below, fill in the feature control frame so:
   a) the hockey puck (below the thick horizontal line) fits down inside of the cylindrical cavity (above the thick horizontal line).
   b) the holes in the hockey puck fit over the pins in the cavity.
   c) the load is not born by the pins but rather by L and T.

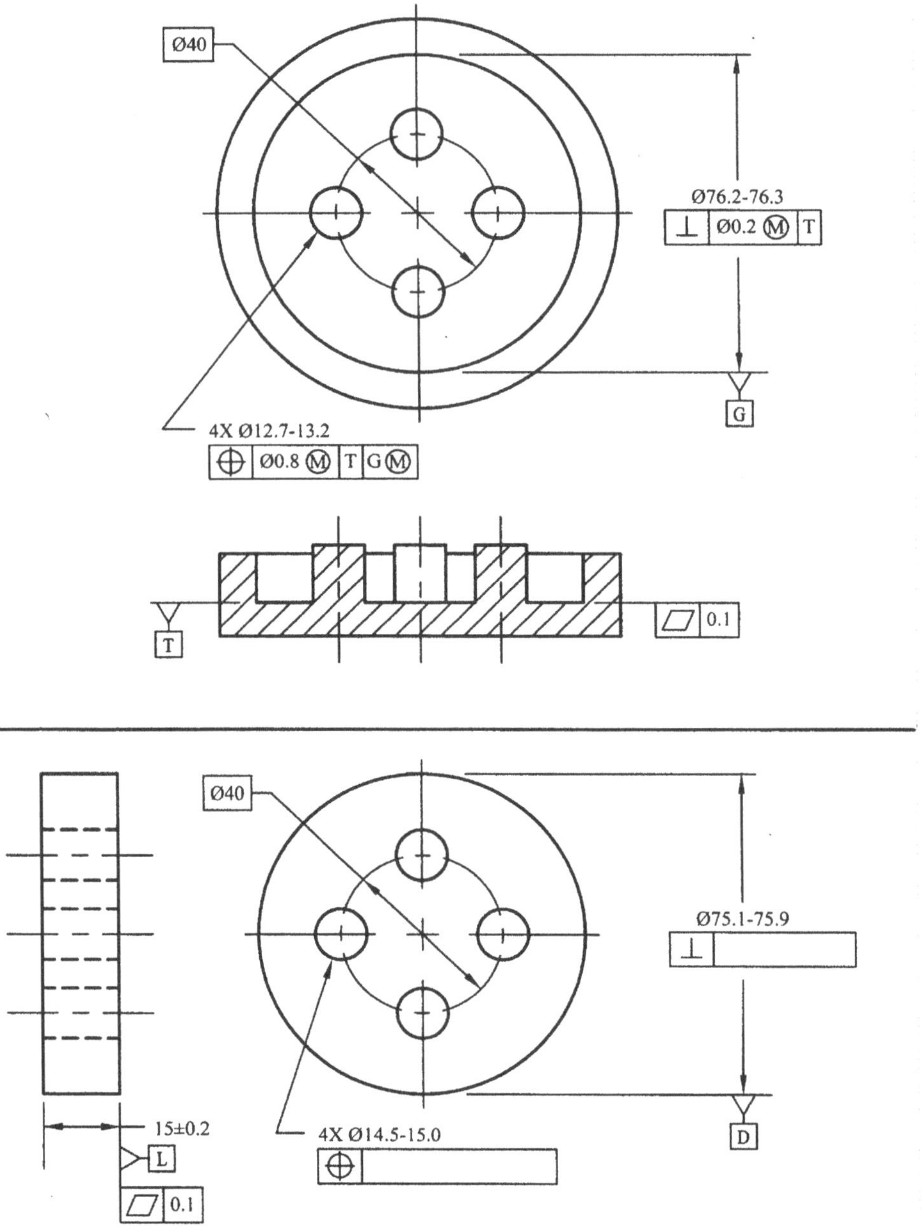

8.

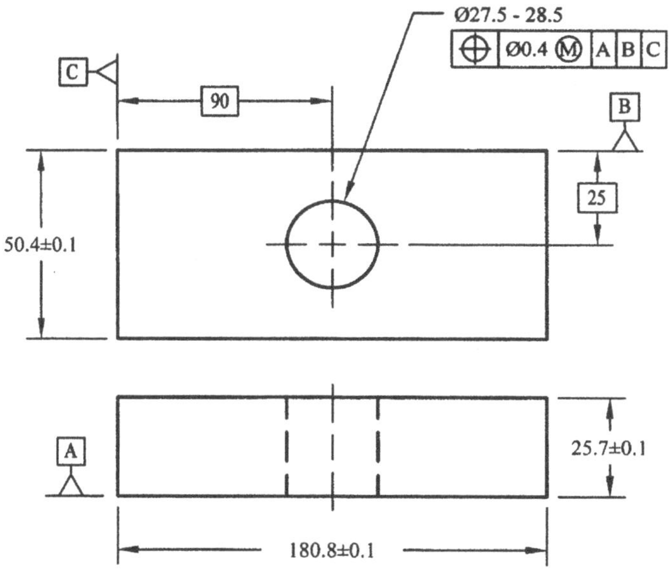

MMC
\- Geo. Tol.
① ⬭ Virt. Cond. to protect
② ⬭ protected
③

AND

① MIC
\- MMC
Bonus Geo. Tol.
\+ Orig. Geo. Tol.
Ø ⬭ Total Geo. Tol.

② MIC
\- MIC at ⊕
Ø ⬭ out of ⊕

③

Part 1 as produced being inspected:

Gage pin: Ø28.25 maximum inscribed cylinder

[The gage pin represents the actual mating envelope without regard to angle or position. Ø28.25 is the actual mating size without regard to angle or position.]

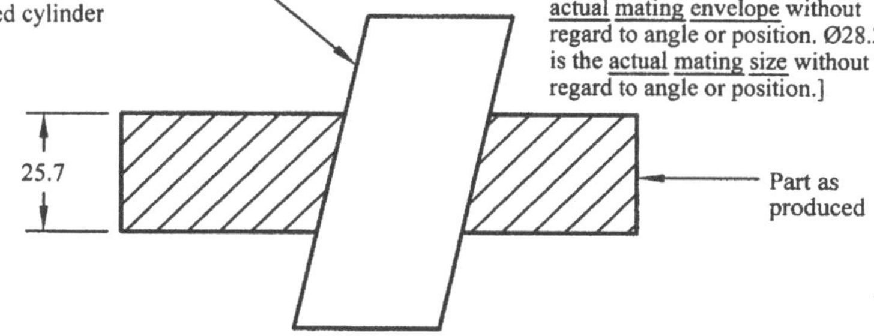

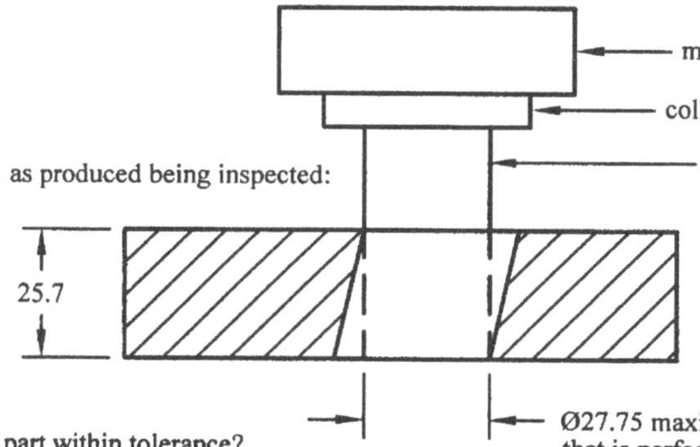

Part 1 as produced being inspected:

Ø27.75 gage pin (MIC at true position)

[The gage pin represents the actual mating envelope with regard to position.]

1. Is this part within tolerance?

Ø27.75 maximum inscribed cylinder that is perfectly perpendicular to primary datum A, 25mm from datum B and 90mm from datum C.

9. Using the drawing below, complete the following charts.

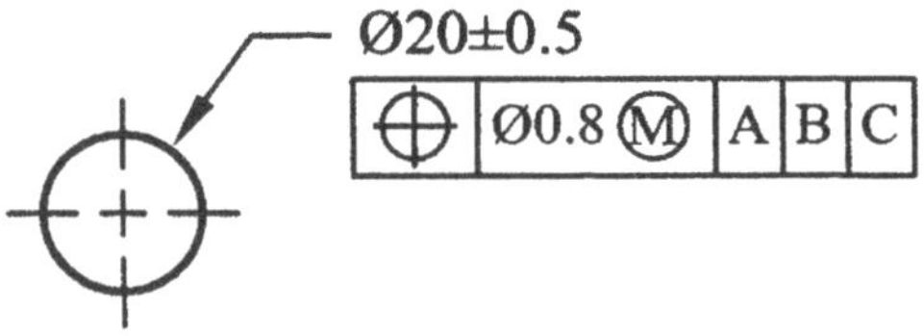

IF A HOLE

	Actual	Less MMC	Bonus Equals	Original Pos. Tol.	Total
A.	20.5				
B.	20.3				
C.	20.1				
D.	20				
E	19.8				
F.	19.6				
G.	19.5				

IF A SHAFT

	MMC	Less Actual	Bonus Equals	Original Pos. Tol.	Total
H.		19.5			
I.		19.6			
J.		19.8			
K.		20			
L.		20.1			
M.		20.3			
N.		20.5			

10. Complete chart sections A, B and C using the illustration below.
    a) What is the diameter of the allowed deviation from true position?
    b) What is the diameter of the deviation zone formed by the actual deviation from true position?
    c) Is the part good?

Dimension from Datum C	Dimension from Datum B	Actual Hole Size	Answers		
			A	B	C
(1) 1.901	.924	.500			
(2) 1.905	.929	.505			
(3) 1.902	.926	.510			
(4) 1.925	.932	.502			
(5) 1.895	.922	.500			
(6) 1.890	.925	.507			
(7) 1.900	.910	.509			
(8) 1.901	.926	.511			
(9) 1.870	.909	.509			

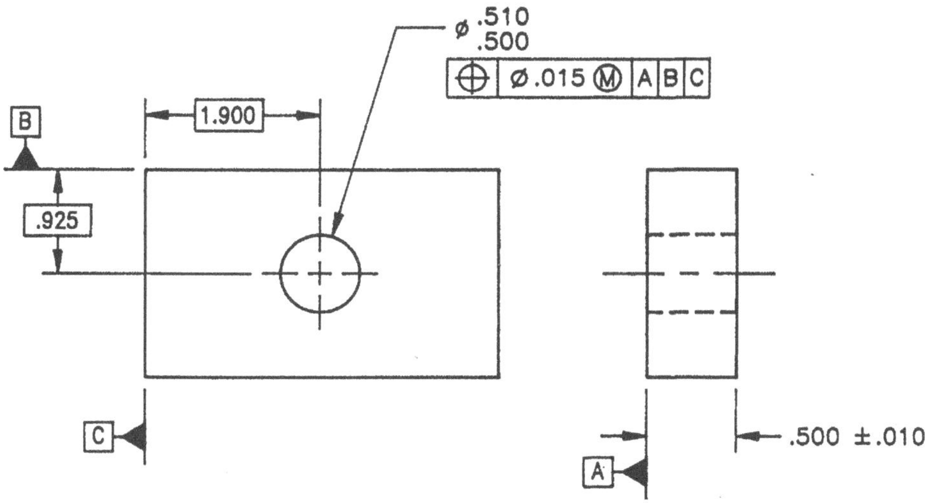

11. Size Tolerance: Using the drawing below, answer the following questions.

as drawn:

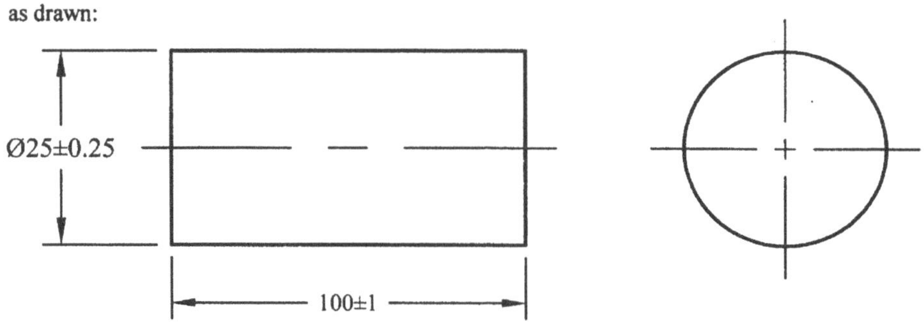

as produced:

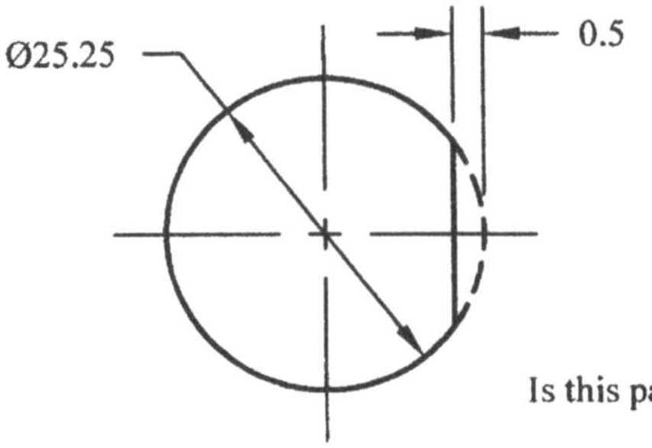

Is this part within tolerance ? _____

as produced:

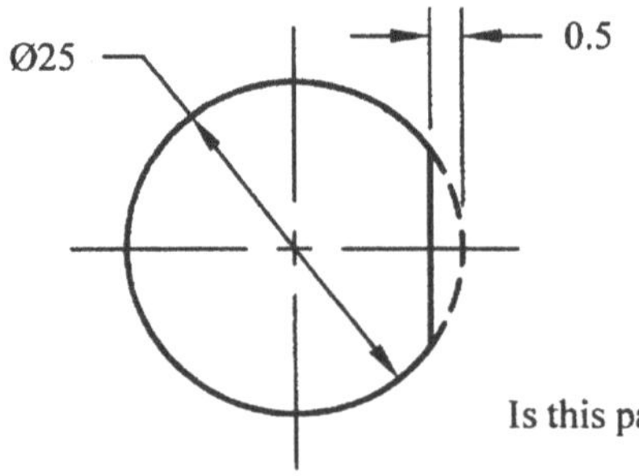

Is this part within tolerance ? _____

as produced:

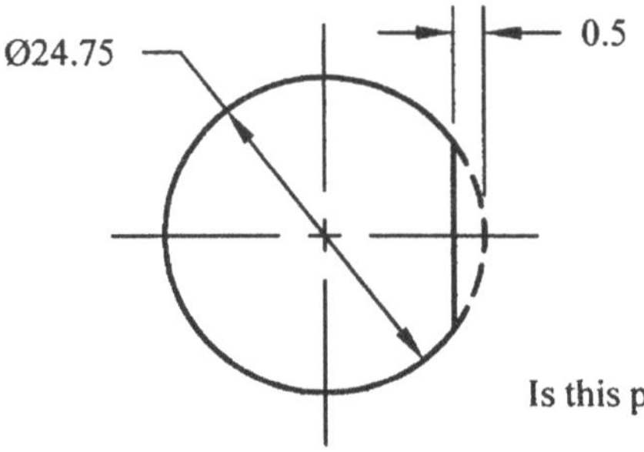

Is this part within tolerance ? _____

12. Size & Circularity Tolerance: Using the drawing below, answer the following questions.

as drawn:

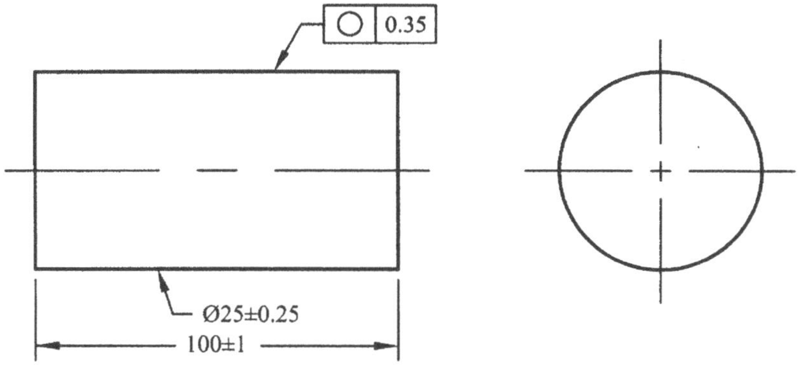

as produced:

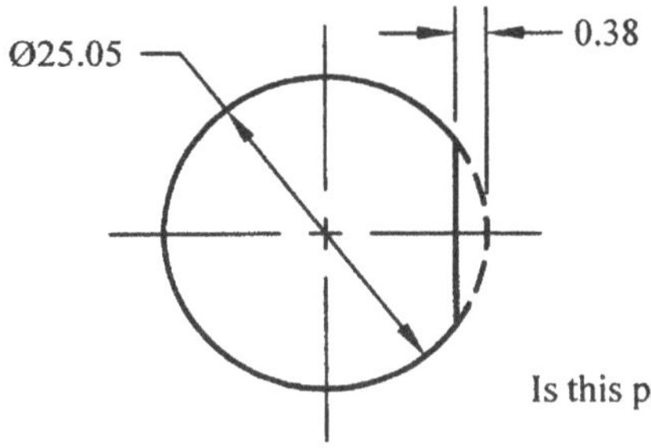

Is this part within tolerance ? _____

as produced:

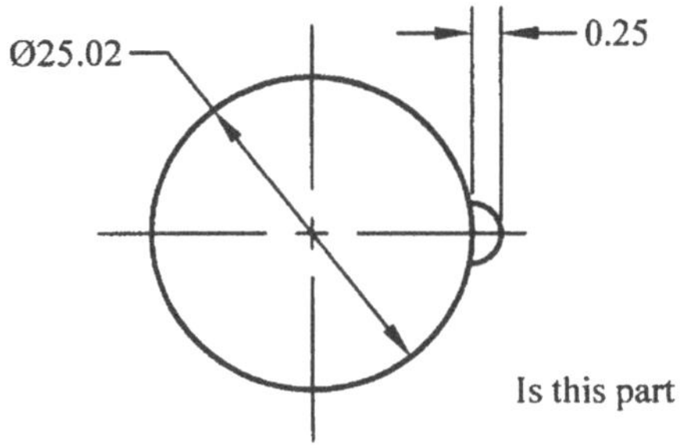

Is this part within tolerance ? _____

as produced:

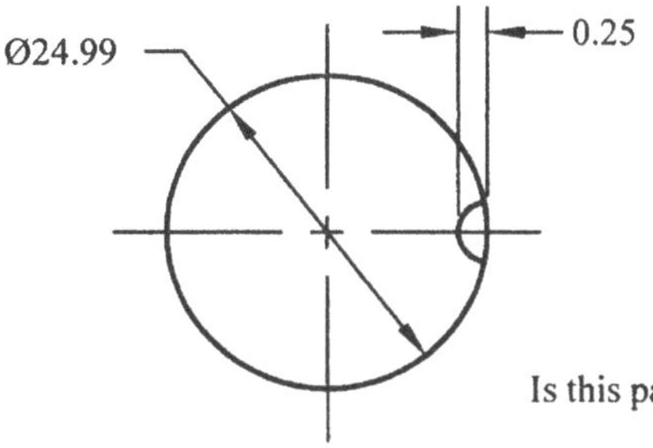

Is this part within tolerance ? _____

13. Cylindricity: Do any of the 'as produced' measurements shown indicate the part should be rejected? Why or why not?

Design Requirements:

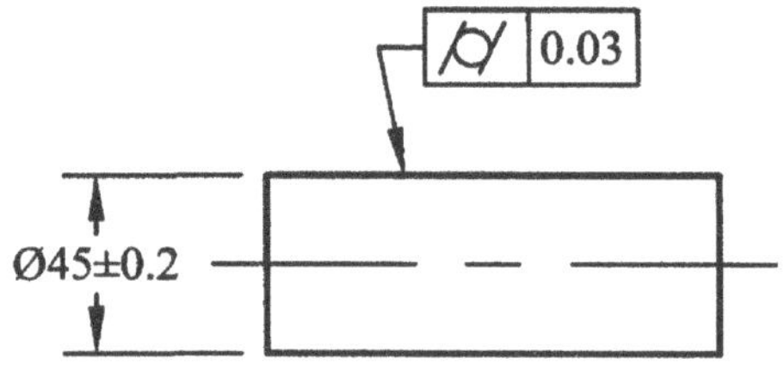

as produced:

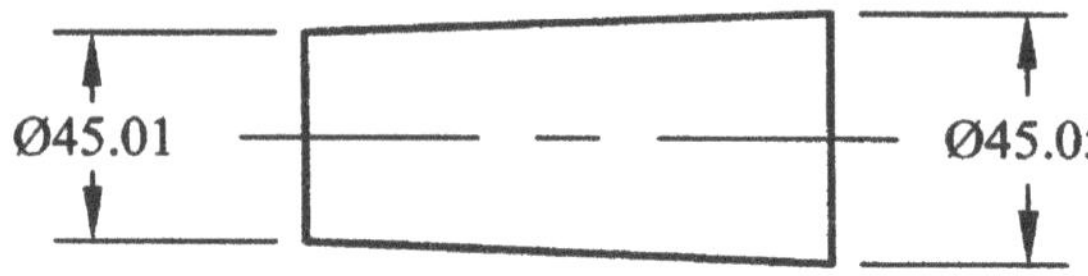

14. Circularity & Circular Runout: Answer the questions for each of the produced parts.

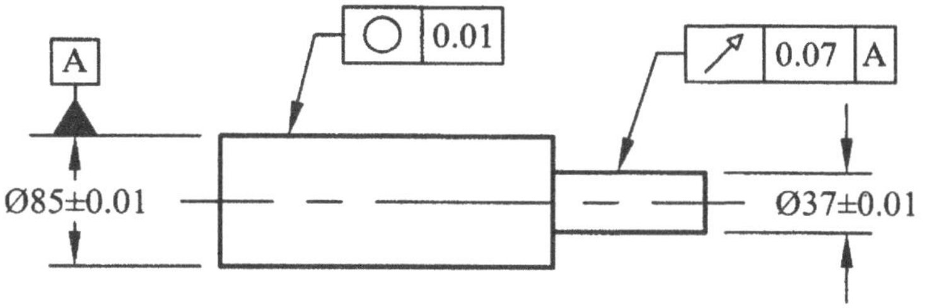

as produced:
1st part

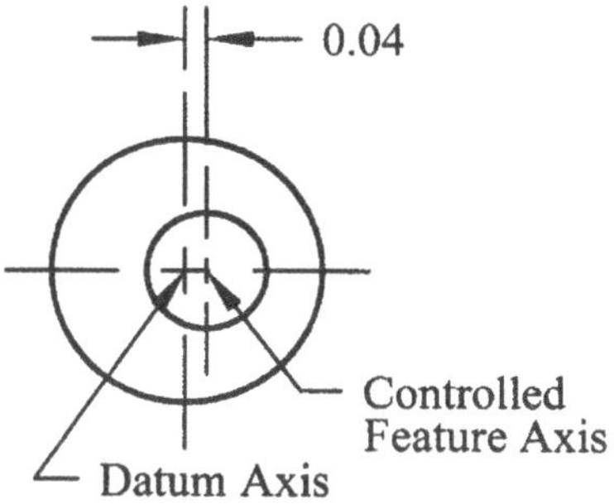

Is the part good?
Why, or why not?

2nd part

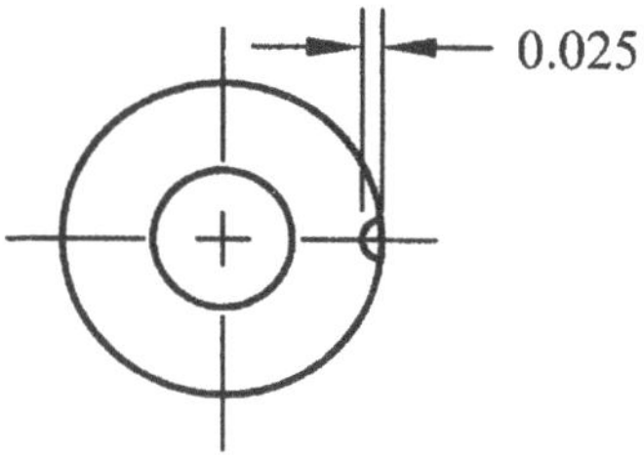

Is the part good?
Why, or why not?

3rd part

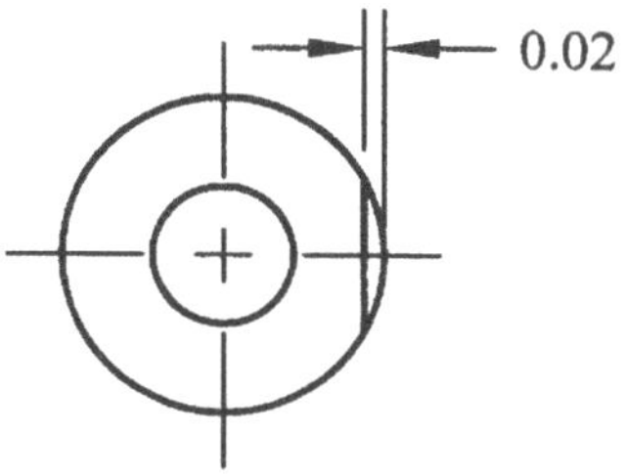

Is the part good?
Why, or why not?

15. Circularity, Straightness & Total Runout

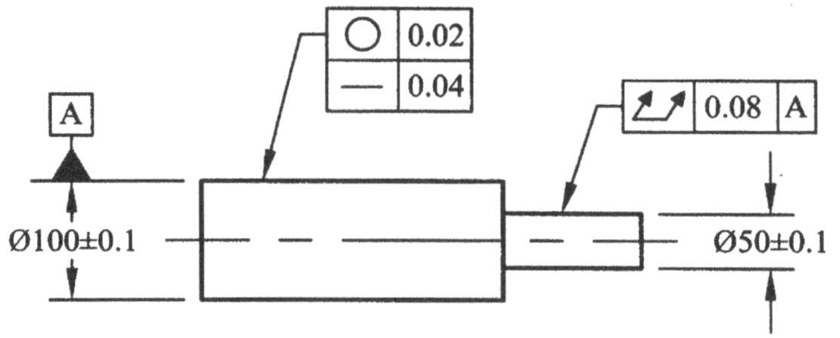

as produced:

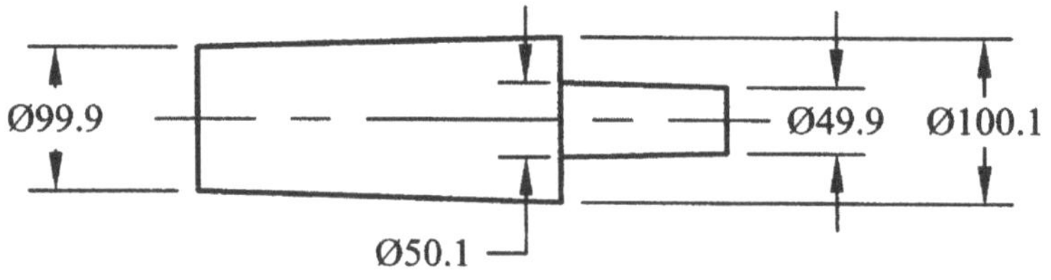

Do any of the "as produced" measurements shown indicate that the part should be rejected? Why, or why not?

16. Cylindricity & Circular Runout

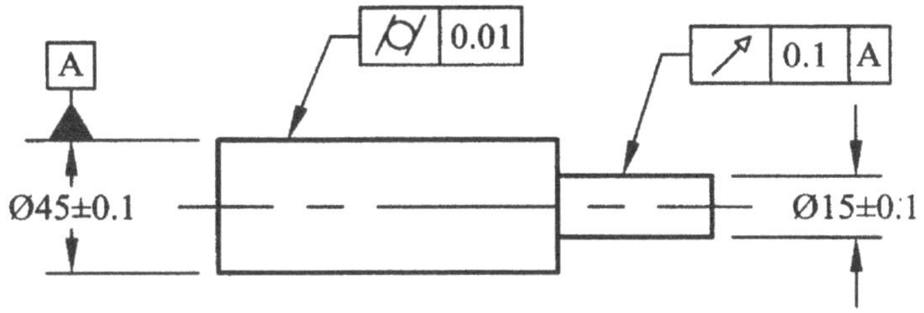

as produced:

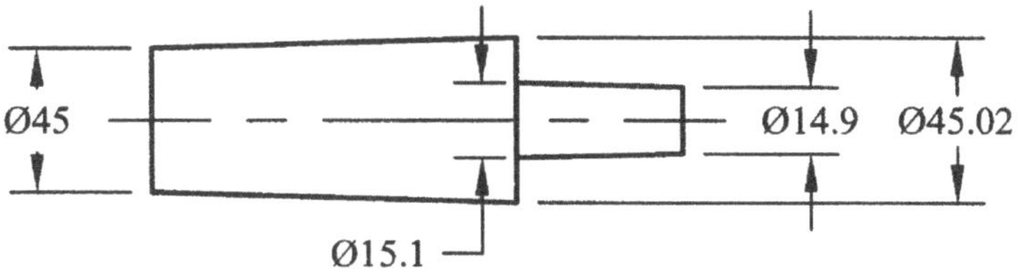

Do any of the "as produced" measurements shown indicate that the part should be rejected? Why, or why not?

17. Cylindricity & Total Runout

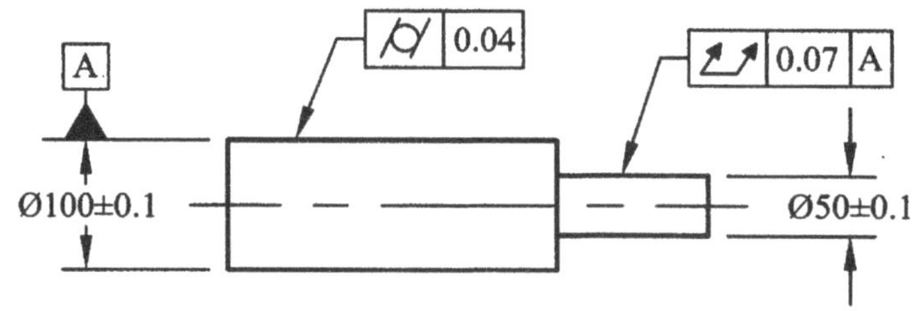

as produced:

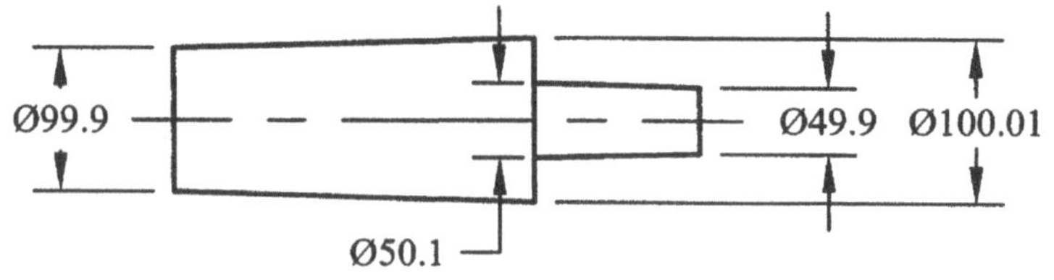

Do any of the "as produced" measurements shown indicate that the part should be rejected? Why, or why not?

18. Cylindricity & Total Runout: Answer the questions below.

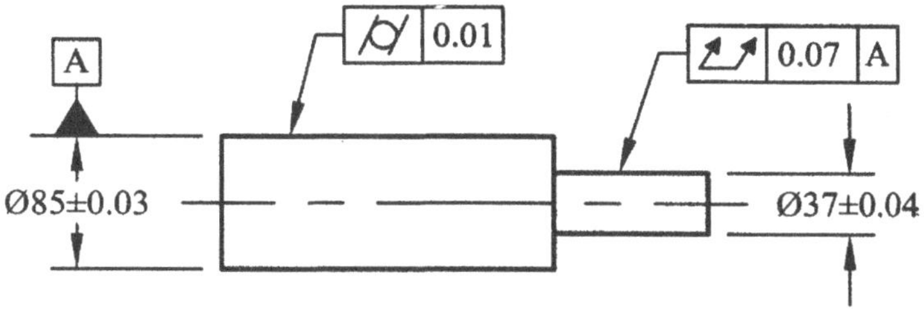

as produced:
1st part

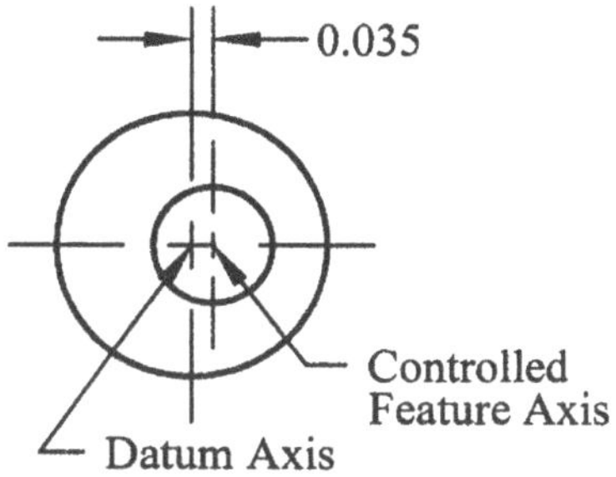

Is the part good?
Why, or why not?

2nd part

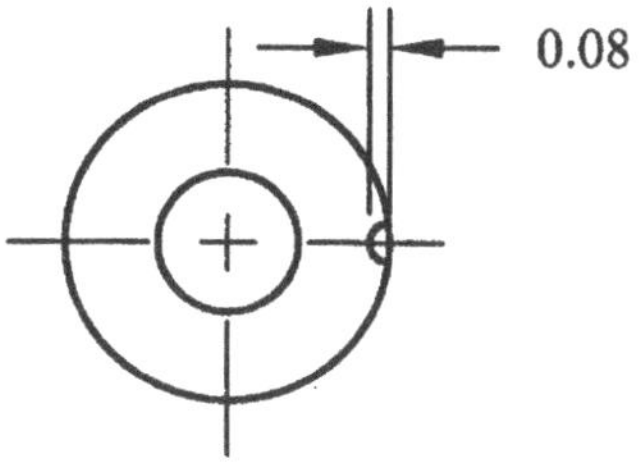

Is the part good?
Why, or why not?

3rd part

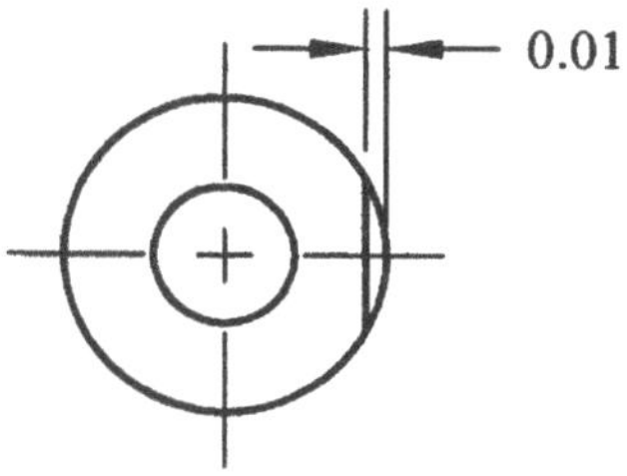

Is the part good?
Why, or why not?

19. Part 12135 – Bushing: Using the following variables data, assess the part on the following page and make assessments pertaining to its compliance to the tolerances depicted.

    Available collected measurement data:

    - Outside diameter measures .525 at its largest and .519 at its smallest.
    - Part outside diameter is round to within .0002.
    - FIM of I.D. while centered on and rotating about datum axis D is .002 at each cross section.
    - Inside diameter is .2550 at its largest and .2549 at its smallest.
    - Part thickness (shown in the left side view) varies between .196 and .195. It is rigid.
    - Part thickness (shown in the left side view) has a slight uniform bow. Its derived median plane is out of straightness .003.

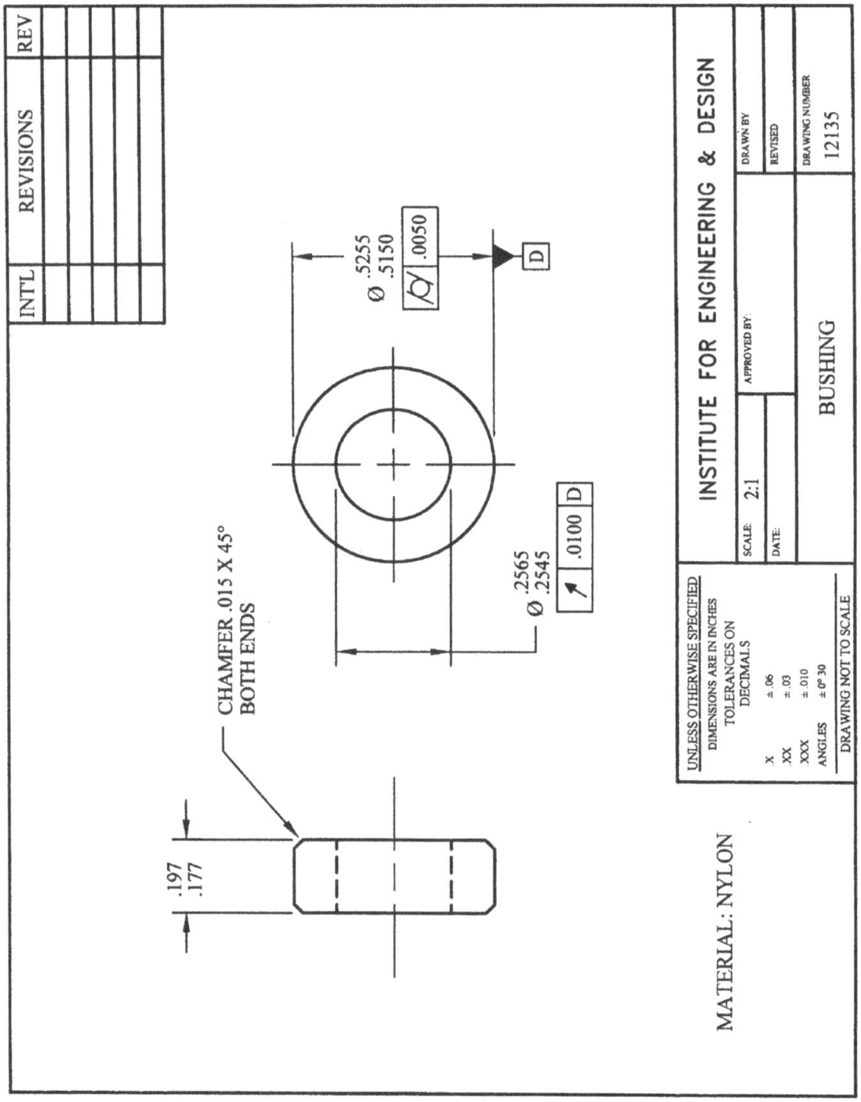

SECTION 3. Interpretation & Analysis Problems

Figure 1. Interpretation

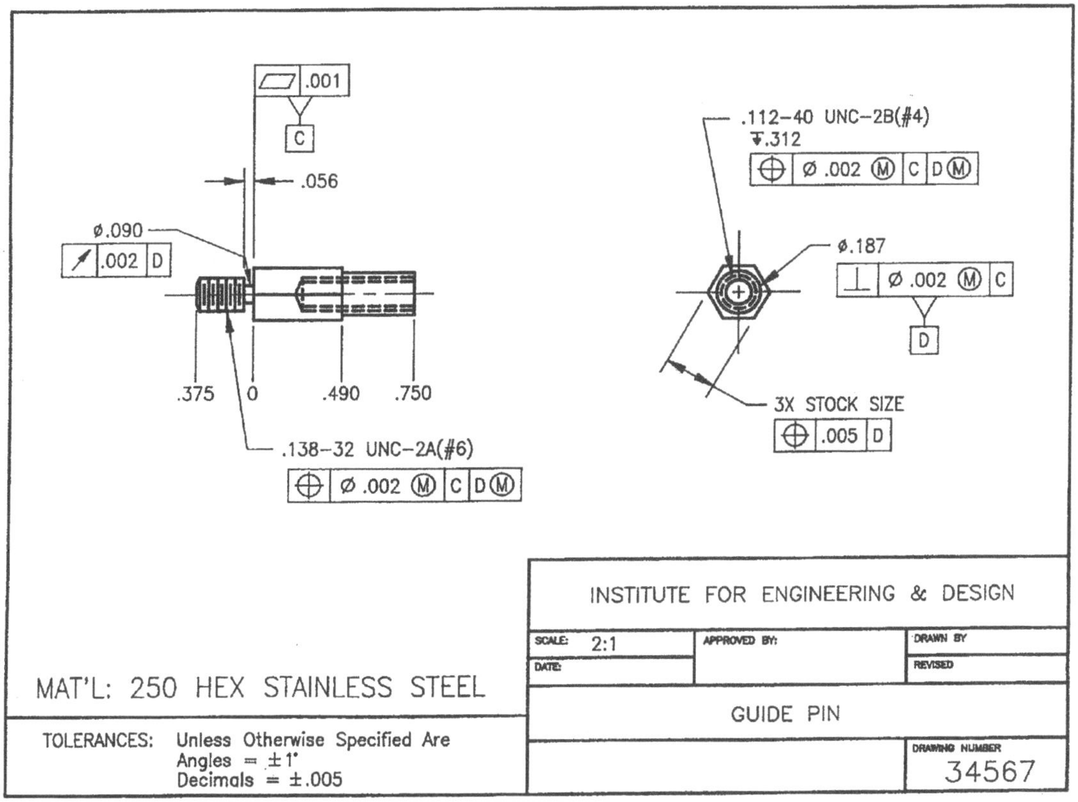

1) How deep a pit in datum feature C is acceptable?

2) How tall a bump is acceptable in the surface with the runout control?

3) How far off center radially may the axis of the diameter be that is controlled with circular runout?

4) How tapered may the diameter be that has a circular runout control on it?

5) What is the maximum bonus tolerance that could be allowed to the .112 threaded hole due to the MMC symbol after the positional tolerance of .002?

6) What is the maximum datum feature shift that could be experienced due to datum feature D being referenced at MMB for the threaded hole discussed in question 5?

7) What is the maximum **bonus** tolerance that datum feature D could experience?

8) What is the shape and size of the position tolerance zone for the six-sided figure?

Figure 2. **Interpretation**

Answer the following questions.

1) What are the largest acceptable produced dimensions for the outside profile of this part?
2) What is the maximum axial separation allowed between datum features B and C?
3) What is the minimum axial separation allowed between datum features B and C?
4) What is the maximum allowed distance between the axis of datum feature B and the edge of the part dimensioned .500 basic from it?
5) What is the minimum allowed distance between the axis of datum feature B and the edge of the part dimensioned .500 basic from it?

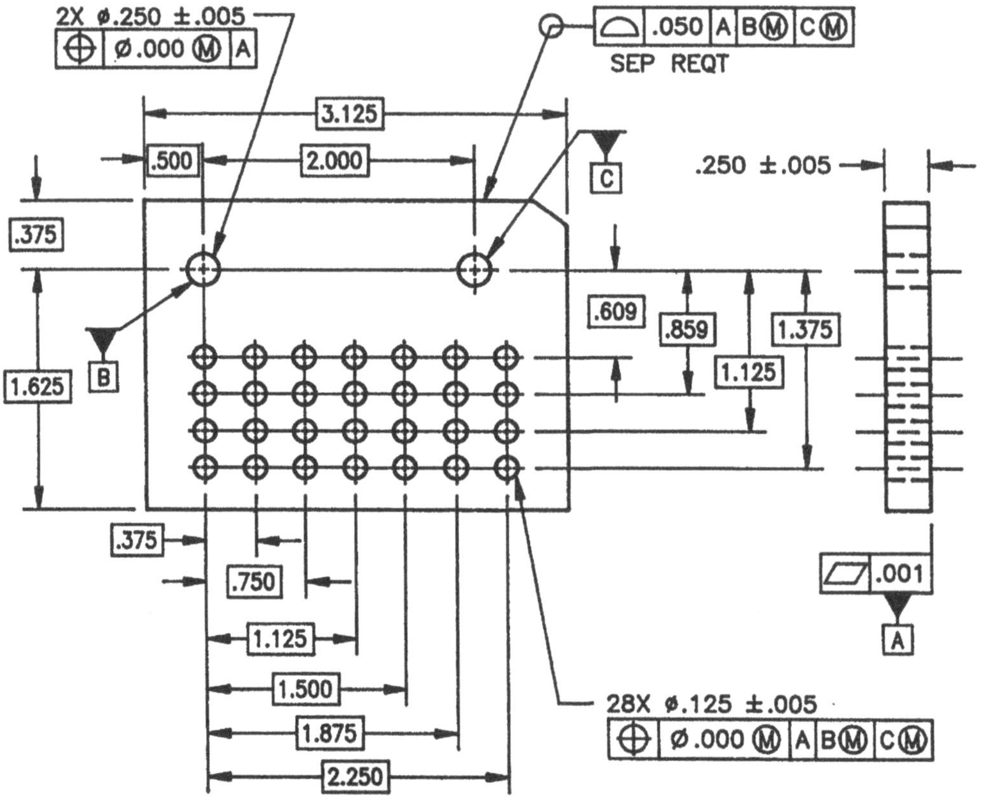

Figure 3. Interpretation

1) What are the largest acceptable produced dimensions for the outside profile of this part?
2) What is the maximum axial separation allowed between datum features B and C?
3) What is the minimum axial separation allowed between datum features B and C?
4) What is the maximum allowed distance between the axis of datum feature B and datum plane D using the .500 basic dimension?
5) What is the minimum allowed distance between the axis of datum feature B and datum plane D using the .500 basic dimension?

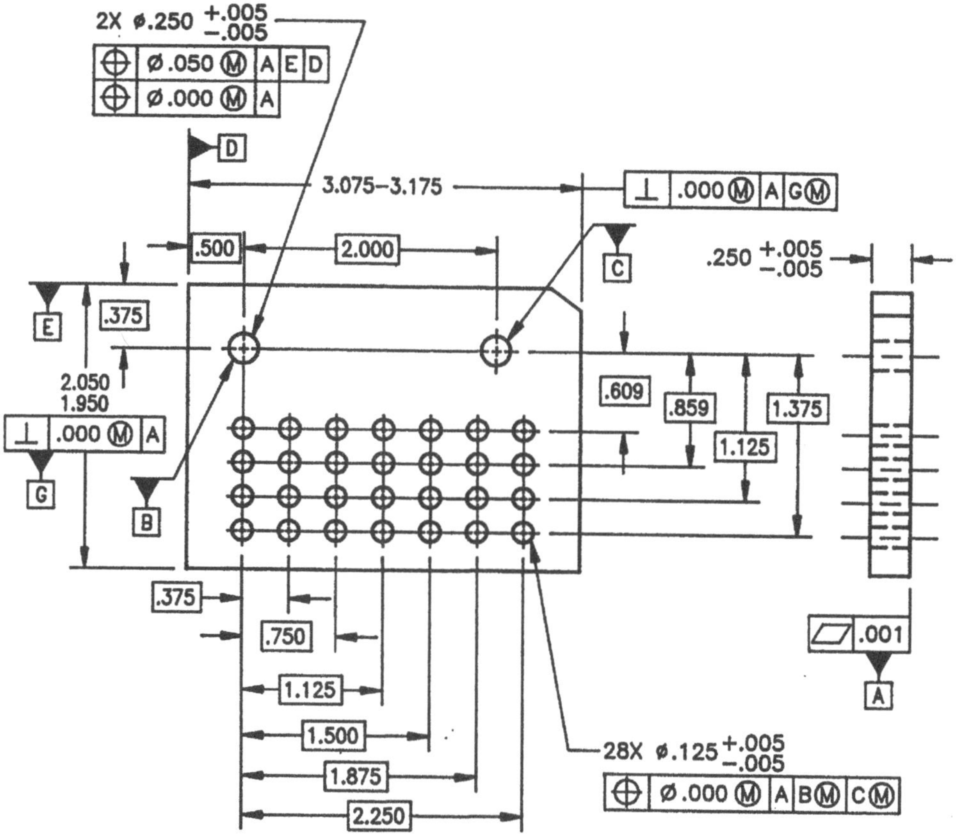

Figure 4. **Interpretation**

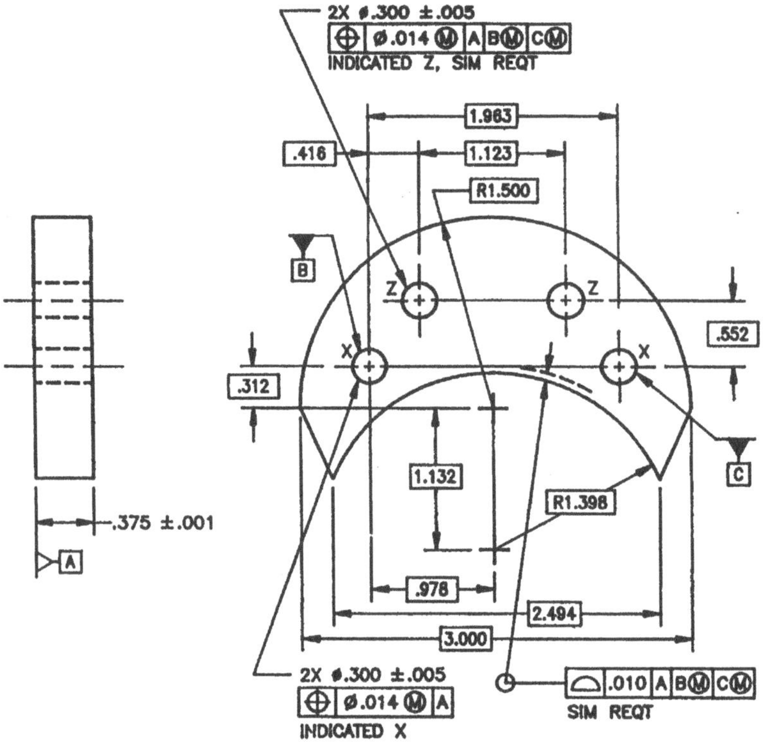

1) How flat must datum feature A be when produced?

2) For the two holes INDICATED X, what are they being positioned to besides datum feature A?

3) How many holes does the profile control tie the profile to?

4) What acts as the origin of measurement for distance for the two holes that are INDICATED Z?

5) What characteristic of geometry does datum A tolerance when used in each feature control frame?

6) Is the profile control a total wide tolerance zone of .010, or is it plus or minus .010?

7) What is the virtual condition of the X holes? ...the Z holes?

8) What acts as the angular orientation datum?

9) If the holes are produced at a diameter of .300, how much positional tolerance are they entitled to?
...at a diameter of .295? ...at a diameter of .305?

Figure 5. Interpretation

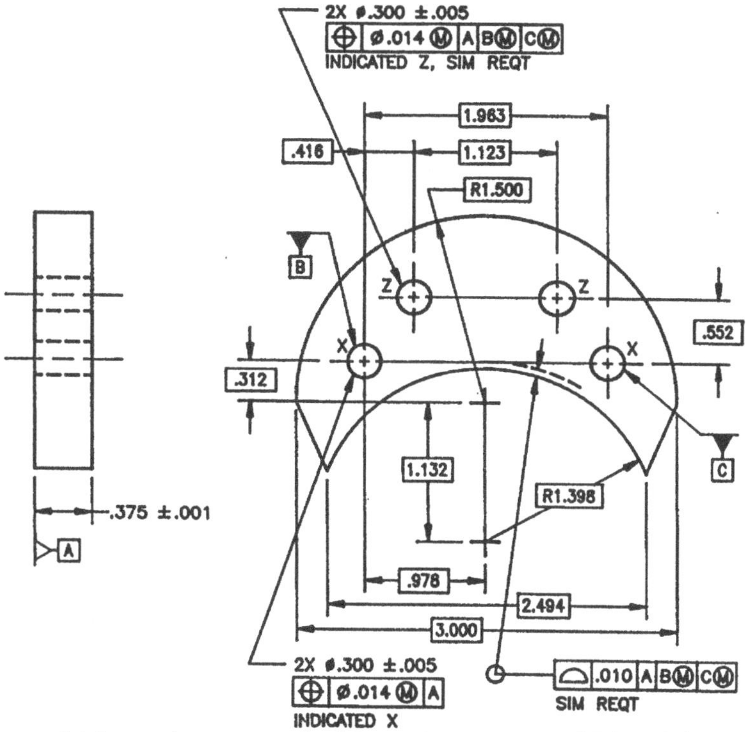

1) If the part thickness is measured with a micrometer at .374 and the median plane bows .003, is the part within its size tolerance?

2) If the two holes INDICATED X are produced at a size of .300 and their axes a distance apart of 1.982, are they acceptable?

3) If the profile size shown as a basic 3.000 is produced at 2.980, does it exceed its tolerance?

4) If the two Z holes are produced at a diameter of .305 and their axes at a distance of 1.099 from each other, are they within tolerance?

5) If the holes are produced at a diameter of .300 and out-of-perpendicularity to datum A a diameter of .015, are they out of tolerance?

6) If the surface profile has a bump on the top most radius that exceeds the basic radius shown by .005, is the surface to be rejected as out of tolerance?

7) If the holes measure with a micrometer .299 and can receive a gage pin (without regard to orientation or location) no larger than a .292, do they exceed their size tolerance?

8) If the largest gage pin that can be inserted into datum feature B measures .305, the largest gage pin that can be inserted into datum feature C measures a .295 and a Vernier caliper measurement from the outermost points on the gage pin diameters is a distance of 2.282, would that indicate that they exceed their positional tolerance?

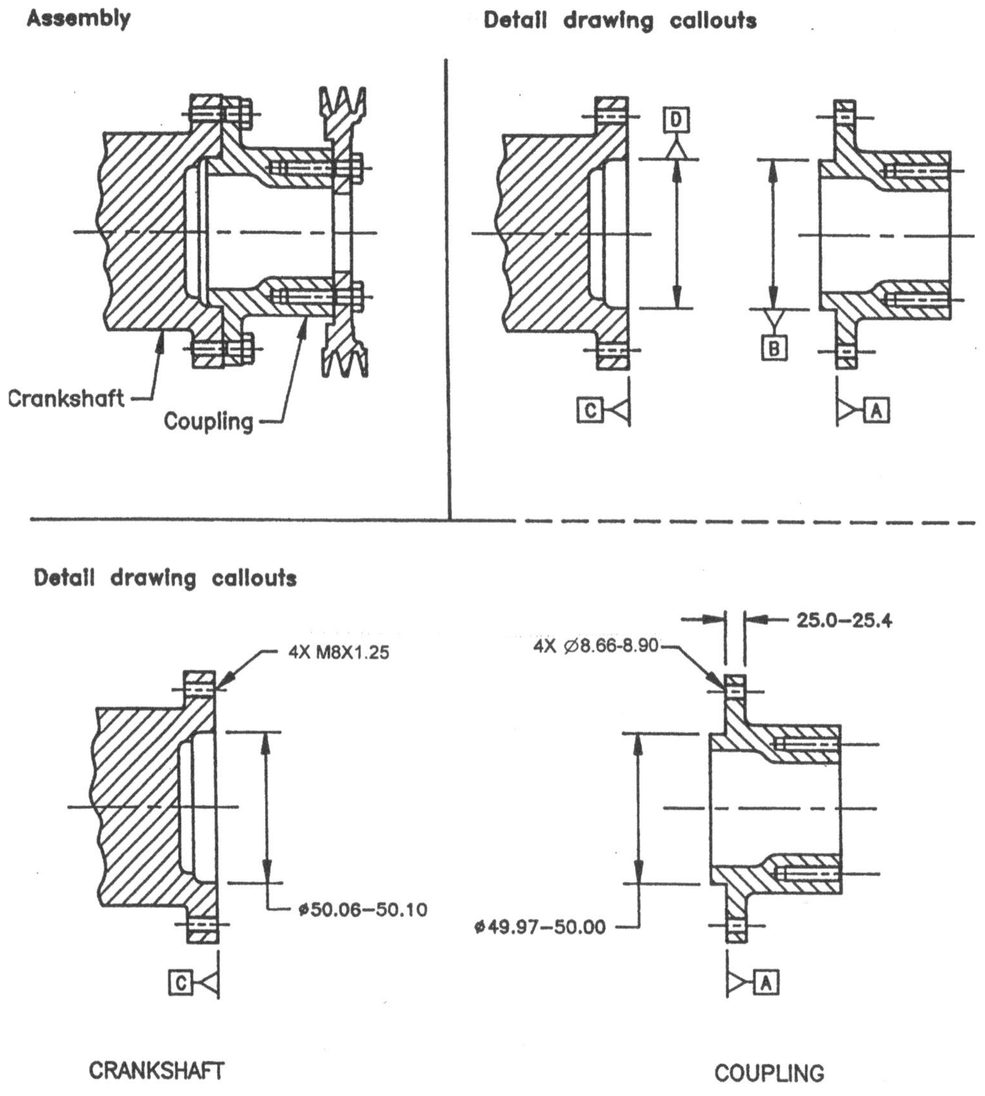

1. Center Datum Feature of Size Mating Part Design:
   1) Consider a flatness control for datum features A and C of 0.01.
   2) Give each hole pattern a basic hole to hole dimension of 200.
   3) Give each part's largest outside diameter a dimension of 301 plus and minus 0.5 with a circular runout tolerance of 0.05.
   4) Using the datum features shown, tolerance these parts to mate at assembly. Tolerance only those features showing size dimensions. Give the location datum features orientation controls only.

2. Consider these parts cylindrical. Tolerance them to mate at assembly. The assembly drawing is shown below. All work is to be done on the next page on the detail drawing callouts (Part 1 and Part 2).

1) Assign each part a primary datum feature and a flatness tolerance of 0.03.

2) Assign each part a secondary datum feature of size and a perpendicularity tolerance that you calculate.

3) Assign each hole pattern a positional tolerance with appropriate datum references.

4) Assign each outside diameter a positional tolerance of 0.8 at MMC

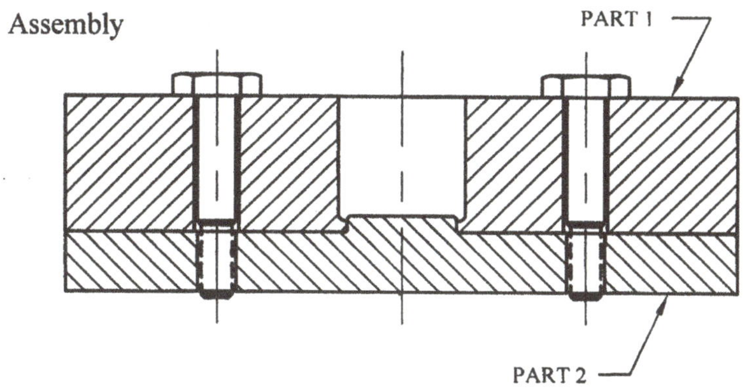

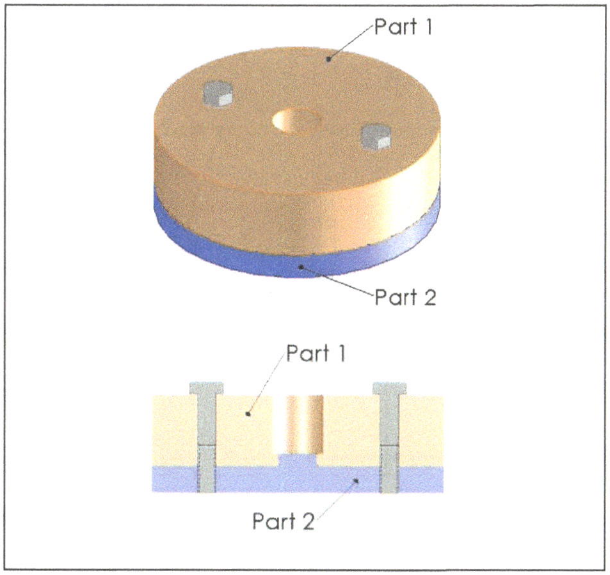

2. (Continued)

PART 1

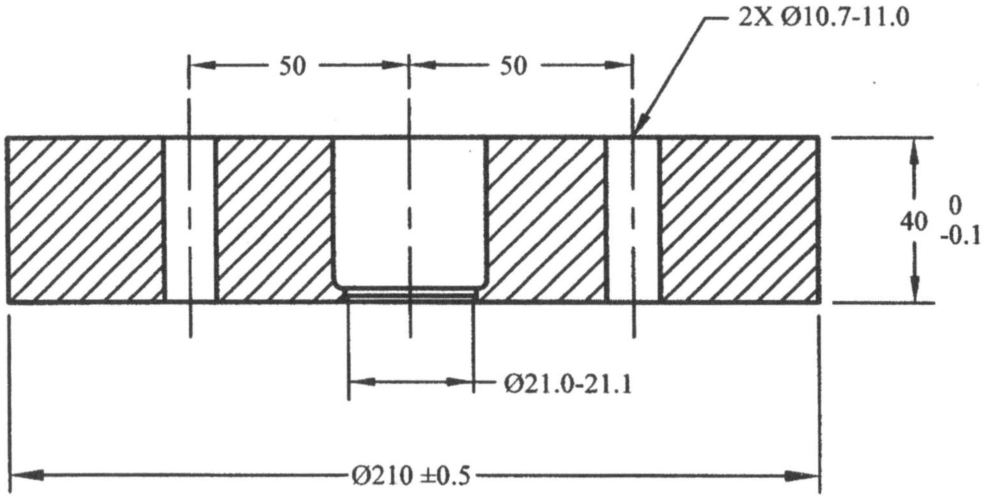

PART 2

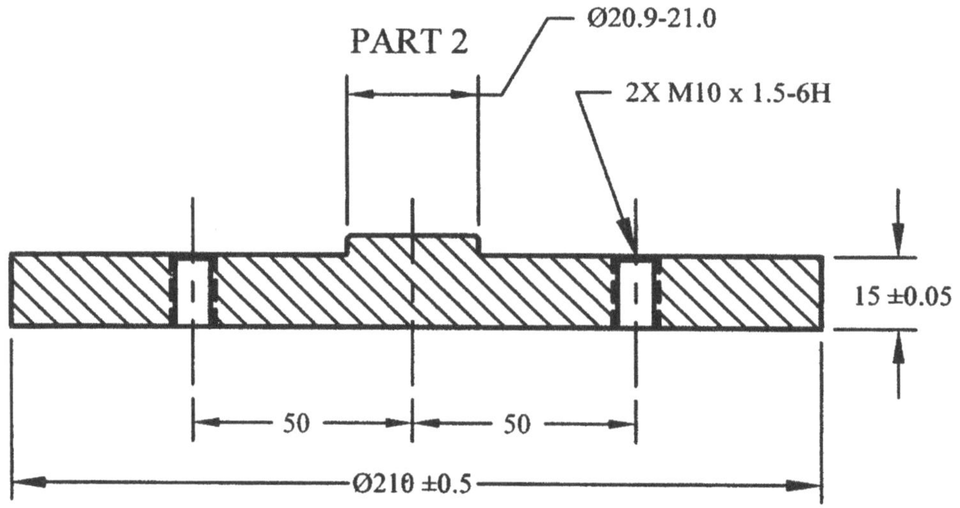

3.

Punch Pattern (Positioning Rectangular Holes): The virtual condition of the shaft that will mate with the ∅.250-.260 holes is ∅.250. The virtual condition of the rectangular shafts that mate with these rectangular holes is .690 x .890. Use exactly the same datum feature references for the .700-.710 widths as you use for the .900 -.910 widths. The illustration in this exercise appears on the following page.

1) Assign this part a primary, secondary and tertiary datum feature as the edges of the part. Use the surface marked 'mating surface' and the two edges from which the 1.00 dimensions originate.

2) Position the ∅.250-.260 hole within a positional tolerance that is the diameter of a circumscribed circle about a .120 x .120 square tolerance zone. To calculate this, use the formula that says the diameter of the circumscribed circle around a square is equal to one side of the square (.120) times the square root of 2 (the square root of 2 is approximately 1.4142, but for this calculation, use 1.4).

3) Assign the ∅.250-.260 hole a refinement of perpendicularity to the primary datum. Calculate this tolerance using the formula: MMC of the hole minus the virtual condition of the mating shaft equals the geometric tolerance for the hole. Make this hole a datum feature.

4) Assign the 6 x .700 -.710 widths a positional tolerance to
   a) the primary datum feature (for perpendicularity)
   b) the hole described in steps 2 and 3 above (for location)
   c) one side of the part (for angular orientation/anti-rotation).

5) Assign the 6 x .900 -.910 widths a positional tolerance to
   a) the primary datum feature (for perpendicularity)
   b) the hole described in steps 2 and 3 above (for location)
   c) one side of the part (for angular orientation/anti-rotation).

3. (continued)

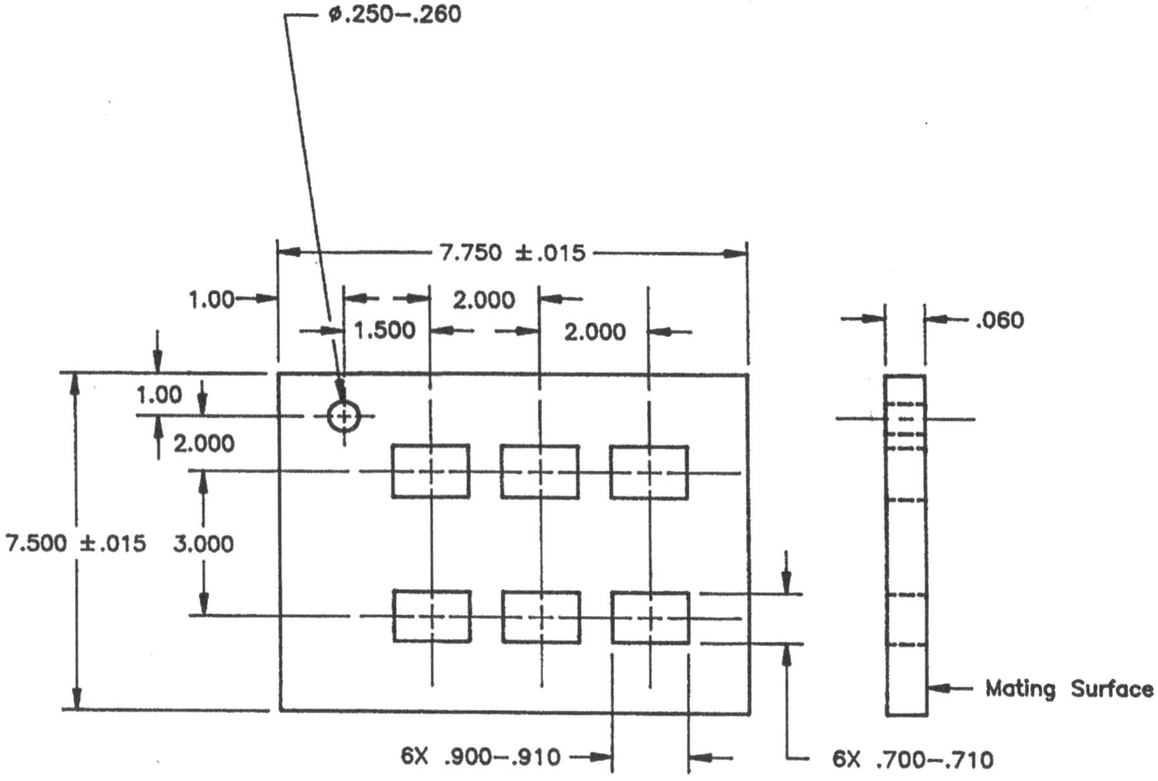

.XX = ±.06
.XXX = ±.005
∠ = ±1°

4. Calculating Minimum and Maximum Wall Thicknesses

Consider this object the shape of a hockey puck that has four holes drilled into it. Calculate the minimum and maximum wall thicknesses between the surface of one hole to the outside diameter of the hockey puck. In other words, calculate the MIN and MAX for X.

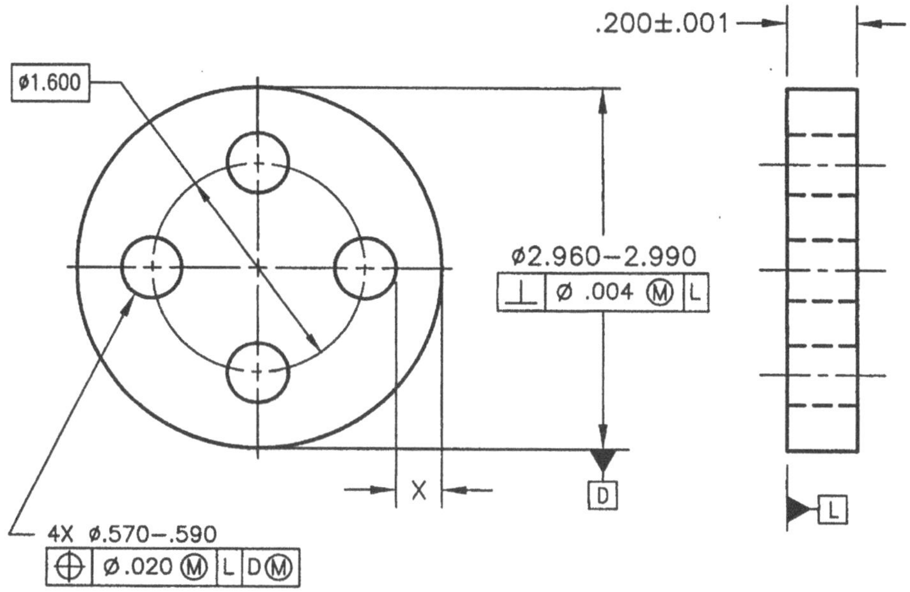

5. Using the sheet metal drawing below:

1) Convert this form from plus and minus tolerancing to position and profile. It is a fixed fastener situation in that these three clearance holes bolt to three threaded holes. The virtual condition of the threaded holes using a projected tolerance zone is a diameter of .128.
2) The workpiece is to be inspected while restrained.
3) Use the bottom of the part and two edges of the part as datum features.
4) Use Profile of a Surface within .020 for the curved surface.

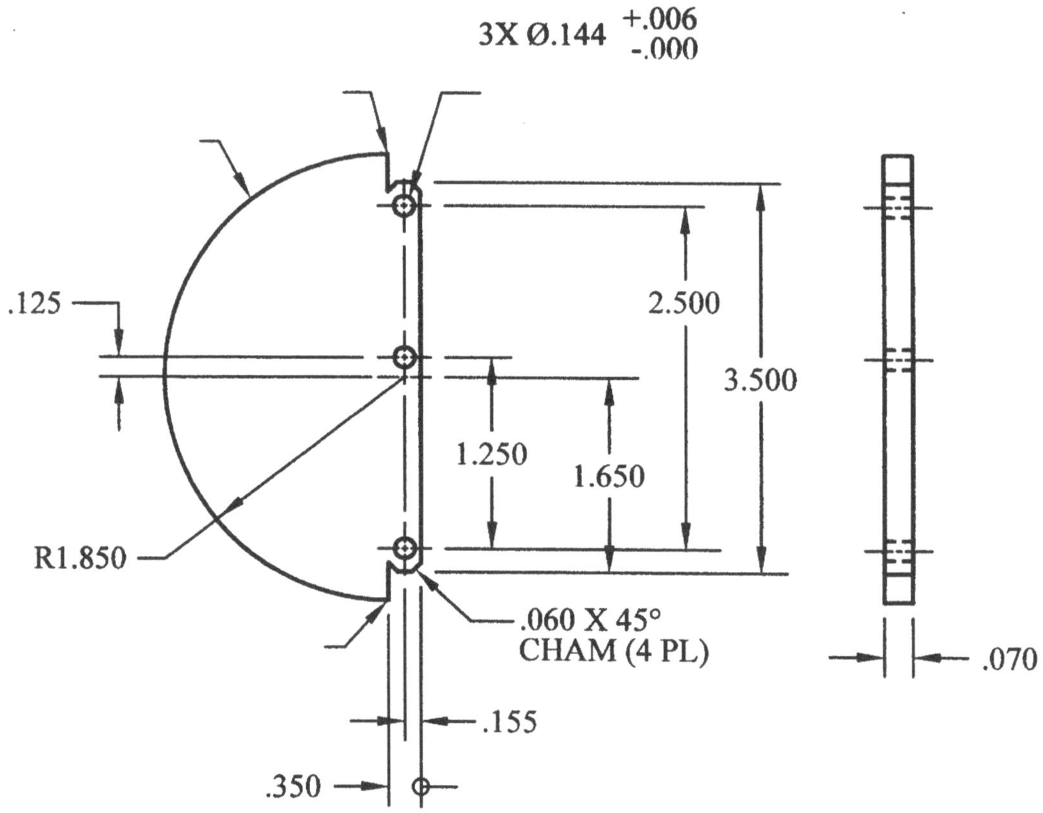

UNLESS OTHERWISE SPECIFIED:
.XXX = ±.010
∠ = ±1°

6. Using the sheet metal drawing below:
1) Convert this form from plus and minus tolerancing to position and profile. It is a fixed fastener situation in that these three clearance holes bolt to three threaded holes. The virtual condition of the threaded holes using a projected tolerance zone is a diameter of .128.
2) The workpiece is to be inspected while restrained.
3) Use the bottom of the part and two holes as datum features.
4) Use Profile of a Surface within .020 for the curved surface.

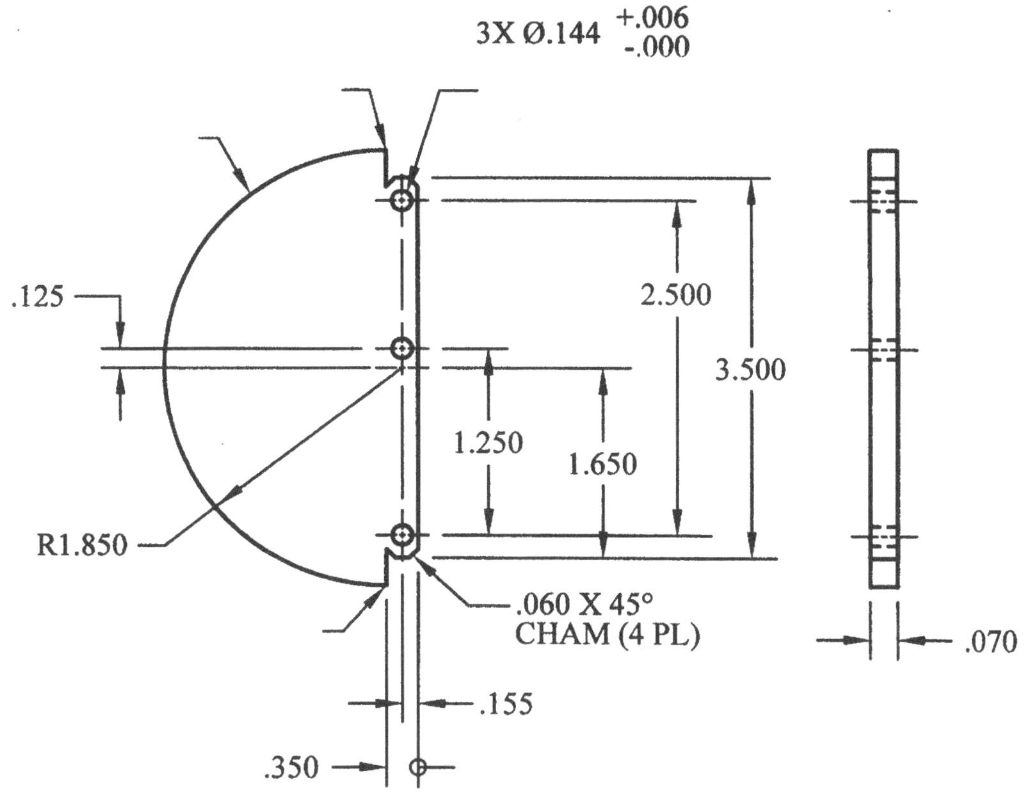

UNLESS OTHERWISE SPECIFIED:
.XXX = ±.010
∠ = ±1°

1. Answer the following questions.

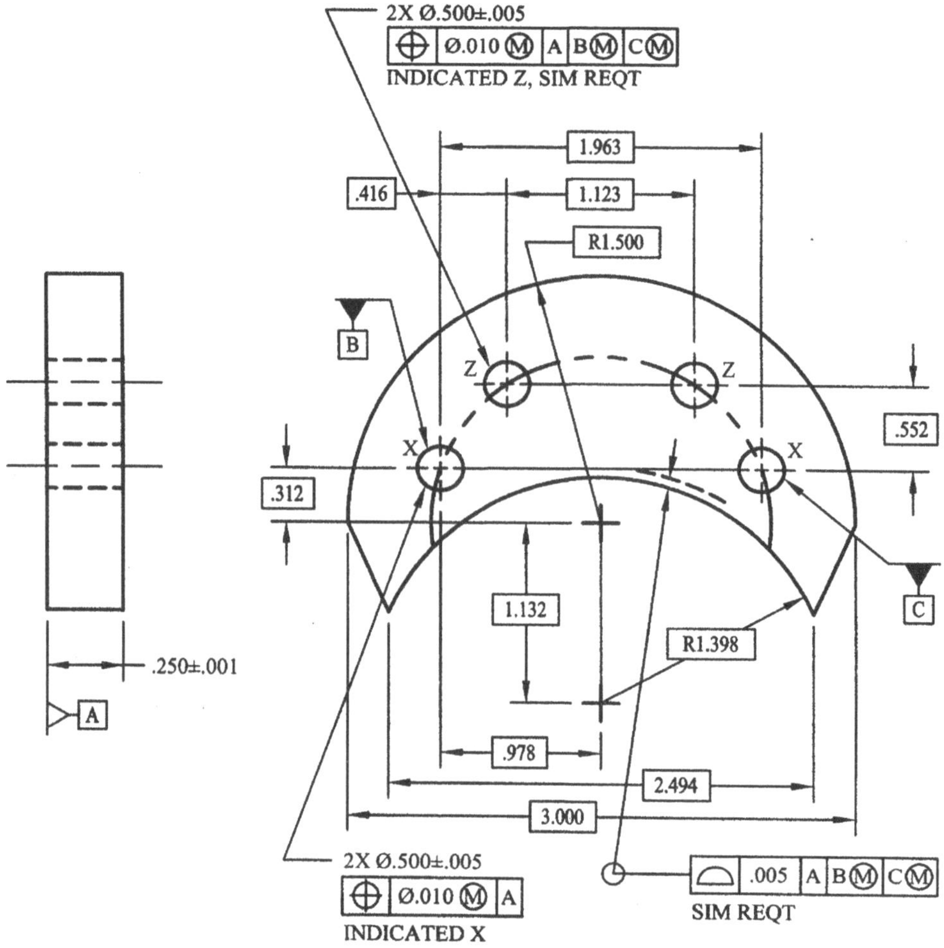

1) If the part thickness is measured with a micrometer at .249 and the median plane bows .002, is the part within its size tolerance?

2) If the two holes 'INDICATED X' are produced at a size of .500 and with their axes a distance apart of 1.982, are they acceptable?

3) If the profile size shown as a basic 3.000 is produced at 2.980, does it exceed its tolerance?

4) If the two Z holes are produced at a diameter of .505 and their axes at a distance of 1.099 from each other, are they within tolerance?

5) If the holes are produced at a diameter of .500 and out-of-perpendicularity to datum A a diameter of .015, are they out of tolerance?

6) If the surface profile has a bump on the upper left radius that exceeds the basic radius shown by .005, is the surface to be rejected as out of tolerance?

7) If the holes measure a diameter of .499 at each circular cross section and can receive a gage pin (without regard to orientation or location) no larger than a diameter of .492, do they exceed their size tolerance?

8) If the largest gage pin that can be inserted into datum feature B measures a diameter of .505, the largest gage pin that can be inserted into datum feature C measures a diameter of .495, and a vernier caliper measurement from the outermost points on the gage pin diameters is a distance of 2.458, would that indicate that they exceed their positional tolerance?

2. Answer the questions 1-11 below for the illustration shown on Page 575. Question 12 refers to the illustration on Page 576.

   1) How flat must datum feature A be when produced?

   2) For the four holes that become datum pattern D, what are they being positioned to besides datum feature A?

   3) How many holes does the profile control tie the profile to?

   4) What acts as the origin of measurement for location for the three-hole pattern?

   5) What characteristic of geometry does datum A tolerance when used in each feature control frame?

   6) Is the profile control a total wide zone of .01 or is it ±.01?

   7) What is the virtual condition of the holes in the three-hole pattern?

   8) How many gages would be necessary to gage the MMC of all seven holes, the two positional controls and the profile control (if only the outer tolerance zone boundary of the profile control was to be gaged)?

   9) If the two outermost .51-.52 holes are produced at a diameter of .514 and their axes a distance apart of 1.126, are they within tolerance?

   10) If the surface profile has a bump on the upper left radius that exceeds the basic radius shown by .005, is the surface in violation of the tolerance zone under all circumstances?

   11) If the largest gage pin that can be inserted into the middle .510-.520 hole measures a diameter of .518 and the hole just to its left can receive a gage pin no larger than .512 and a vernier caliper measurement from the outermost points on the gage pin diameters is a distance of 1.080, would that indicate that they exceed their positional tolerance?

   12) How does the gaging you visualized when answering Question 8 differ from the gaging that would be used for the part on Page 575?

2. (continued)

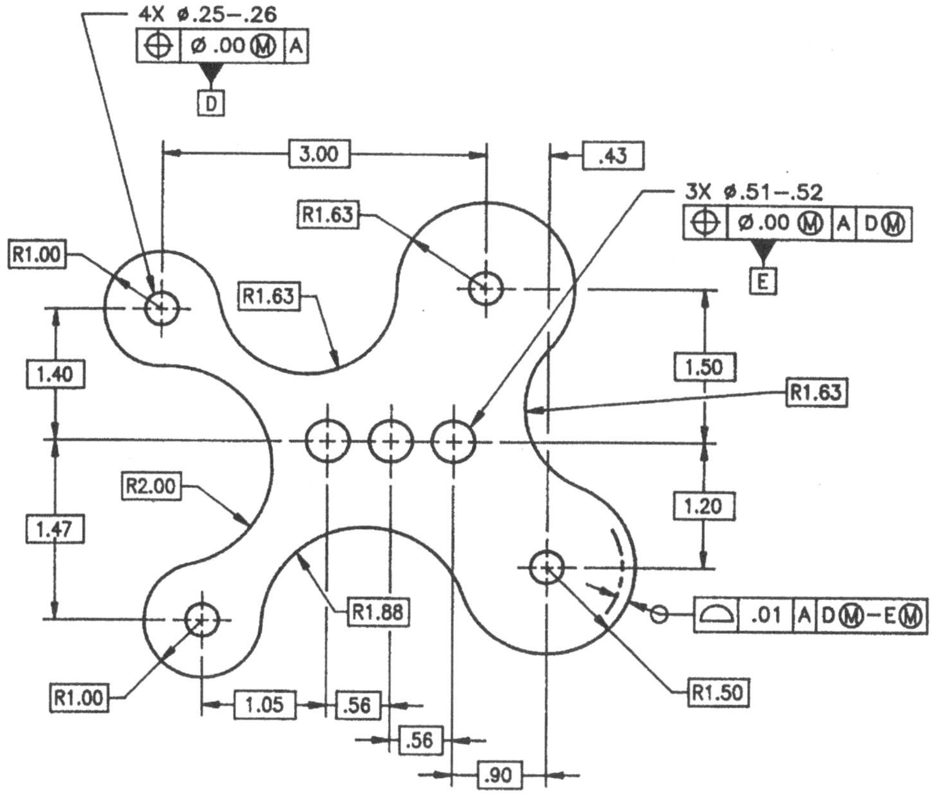
4X ⌀.25–.26
⌀.00 Ⓜ A
D
3.00
.43
3X ⌀.51–.52
⌀.00 Ⓜ A DⓂ
E
R1.63
R1.00
R1.63
1.50
R1.63
1.40
R2.00
1.20
1.47
R1.88
.01 A DⓂ–EⓂ
R1.00
R1.50
1.05
.56
.56
.90

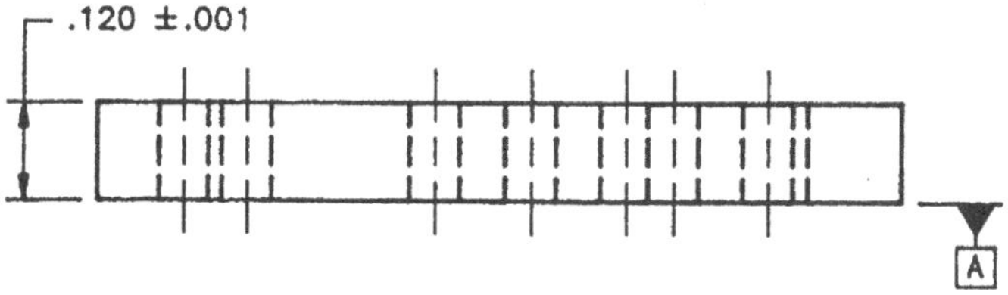
.120 ±.001
A

2. (continued)

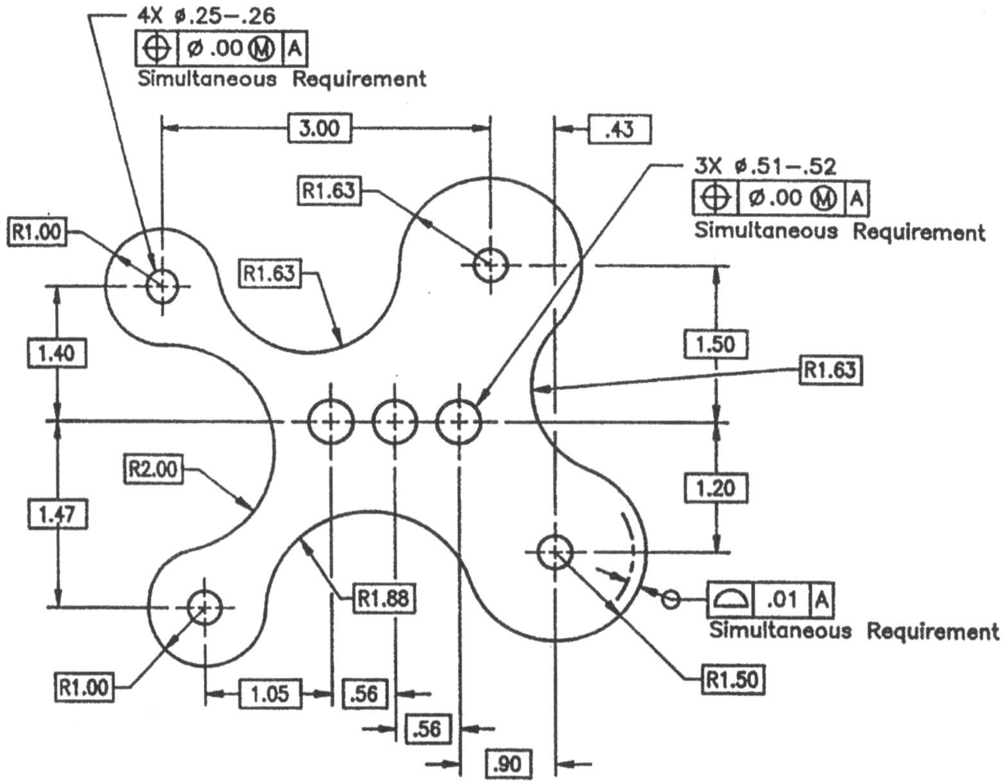
4X ⌀.25–.26
⌀.00 Ⓜ A
Simultaneous Requirement
3.00
.43
3X ⌀.51–.52
⌀.00 Ⓜ A
Simultaneous Requirement
R1.63
R1.00
R1.63
1.40
1.50
R1.63
R2.00
1.47
1.20
R1.88
.01 A
Simultaneous Requirement
R1.00
R1.50
1.05
.56
.56
.90

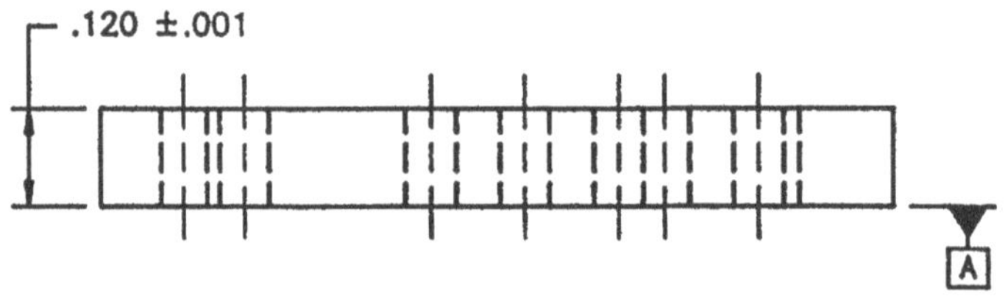
.120 ±.001
A

3. Please answer the following questions as they pertain to the illustration on Page 578.
   1) If the holes on the 1.600 diameter bolt circle have been produced at a distance from datum axis B of .820, under what conditions might the holes meet their positional tolerance?

   2) If the 2 holes opposite each other on the bolt circle are produced at a distance apart of 1.630 and the holes measure a diameter of .578, do they meet their positional tolerance?

   3) If the center hole is produced at a diameter of .586 and is off center of datum axis B a radius of .017, it is within its positional tolerance?

   4) If the part thickness measures .626 with a vernier caliper and curves like a banana .006 in the right side view, does it meet its size requirements?

   5) If datum feature B is produced at a diameter of 2.976 and is perfectly perpendicular to datum A, how much pattern shift would be allowed for the 5-hole pattern?

   6) If the outside diameter is produced at a size that will fit into a hole that is 2.980 and seat on datum feature A with three high points of contact, how much perpendicularity tolerance is the O.D. allowed?

   7) If none of the allowed perpendicularity tolerance in question #6 was used by the O.D., what could the tolerance be used for instead?

   8) If gage pins were inserted into two opposite holes on the 1.600 bolt circle and a measurement taken over the pins was 2.230, could the holes be within their positional tolerance? Under what circumstances?

   9) If two holes on the bolt circle were produced at a diameter of .580, what would their maximum allowed axial separation be?

   10) What is the maximum amount of pattern shift for the five-hole pattern that may ever occur on this part?

3. (continued)

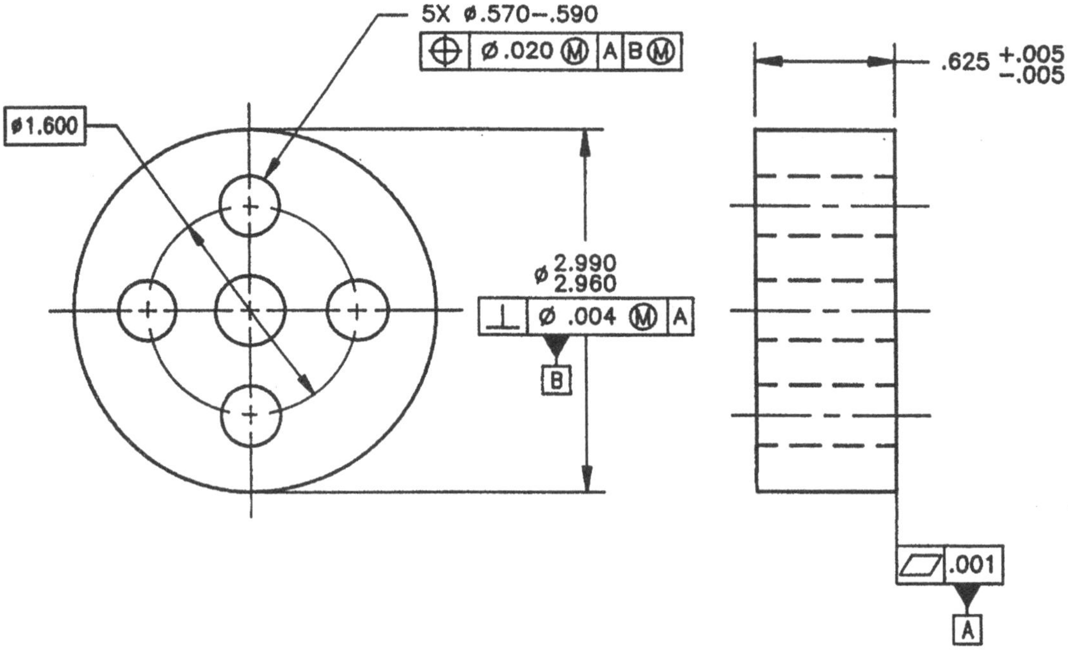

4. Fixed Fastener Assembly: Tolerance the adaptor depicted in the detail drawing callout, saving a portion of the positional tolerance for the next part in the assembly shown in the assembly drawing. To calculate the positional tolerance, use an 8mm screw.

Extra information: The bore that mates with the adaptor's 49.95-50.00 diameter has a virtual condition of 50.00.

Sequence reminder:

Step 1 – assign a primary datum feature and give it a flatness tolerance of 0.03.
Step 2 – assign a secondary datum feature and give it a perpendicularity control. Calculate the tolerance using the information given above.
Step 3 – position the 4-hole pattern to the datum reference frame created in Steps 1 and 2.

Assembly

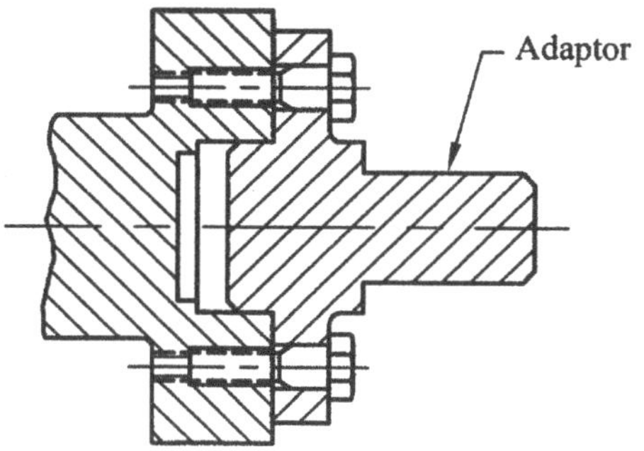

Detail drawing callout

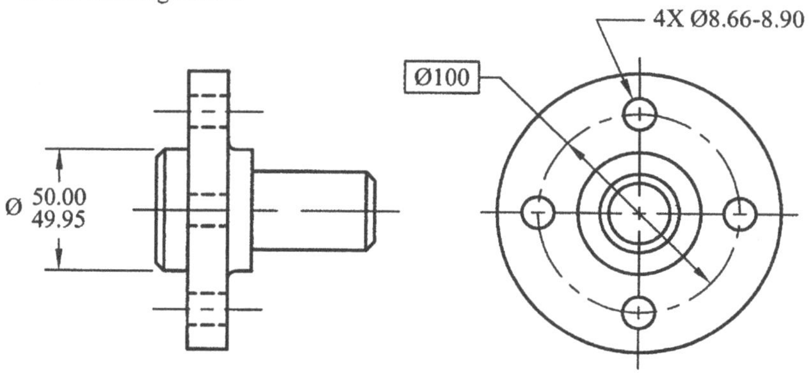

5. Multiple Patterns Made One for illustration on page 581: The following instructions are to be applied to the detail drawing callout on the next page. The purpose of this approach is to tie all ten holes together as one pattern so that they may be gaged with one functional gage to inspect their positional tolerances.

Multiple patterns of features located from the same datum references, using the same material condition symbols after those datum features referenced, are considered one pattern and, consequently, are gaged with the same gage. This is to be done on both hole patterns so all ten holes are considered one pattern.

1) Give the primary datum feature a flatness tolerance of 0.03.
2) Consider parallelism of 0.05 between surfaces into which holes are produced.
3) The virtual condition of the mating hole for the 99.95-100.00 shaft is 100.00. Assign this shaft a perpendicularity tolerance.
4) Locate the ten holes with geometric controls that dictate they are to be considered one pattern of features. Calculate appropriate positional tolerances where applicable. **Save a portion of the available tolerance for the mating parts depicted in the assembly drawing.** The threaded holes mate with holes of a diameter of 8.6-8.8 on a part where the thickness of that mating part is 50.0-51.0. Use your own judgment for missing screw sizes.

5. (continued)

Assembly

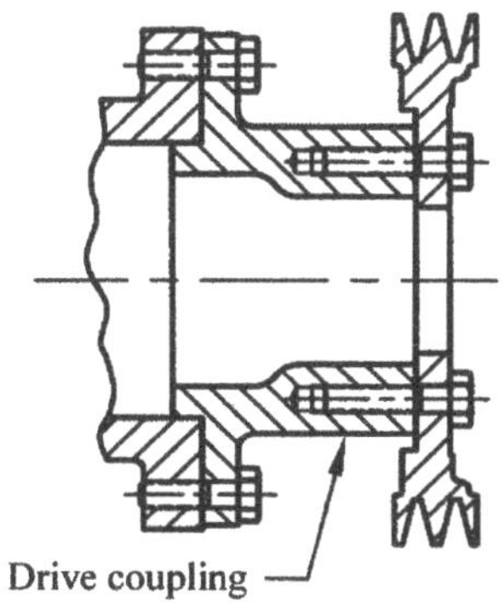

Detail part and gage for checking position of M8X1.25 tapped holes

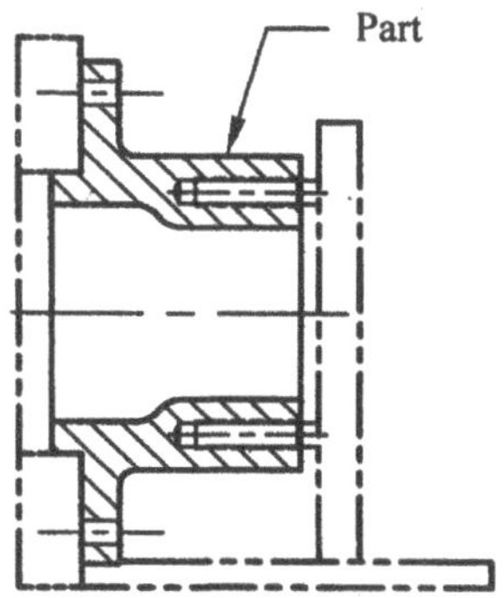

Detail drawing callouts

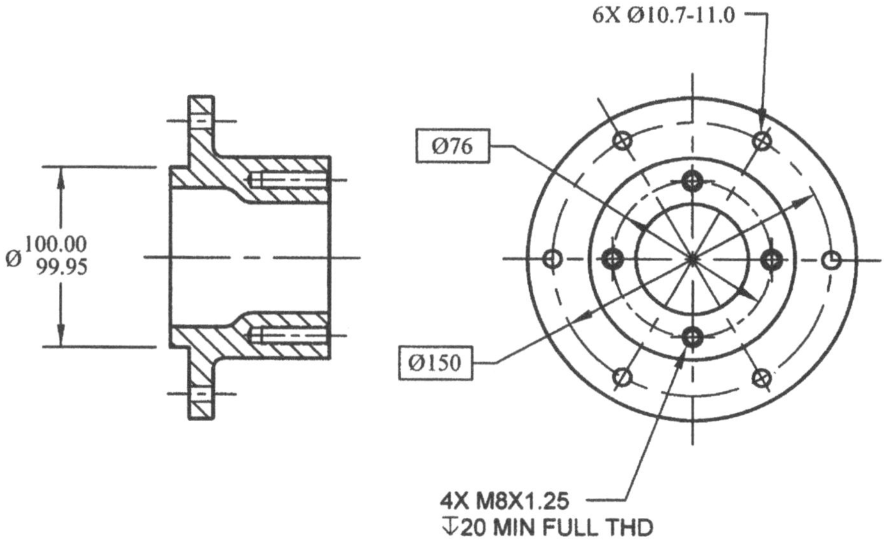

6. Centerplane Datums Fixed Fastener Assembly for illustrations on page 583: Tolerance the following rectangular parts to mate at assembly. The assembly drawing is shown below. All work is to be done on the next page on the detail drawing callouts (Part 1 and Part 2).
   1) Assign each part a primary datum feature and a flatness tolerance of 0.03.
   2) Assign each part a secondary datum feature and a perpendicularity of the centerplane tolerance that references the primary datum.
   3) Assign each part a tertiary datum feature and a perpendicularity of the centerplane tolerance that references the primary and secondary datums.
   4) Assign each part a positional tolerance for its hole pattern referencing all three datums. Use maximum material condition symbols everywhere it is appropriate. Use a projected tolerance zone on the threaded holes.
   5) Assign any remaining widths positional tolerances to within 1mm regardless of feature size. Reference only two datum features in each of these controls. Use no maximum material condition symbols in this step.

<u>Assembly Drawing</u>

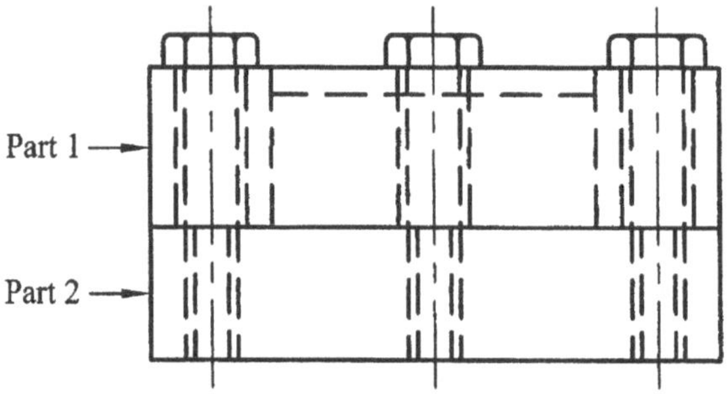

6. (continued)

Part 1

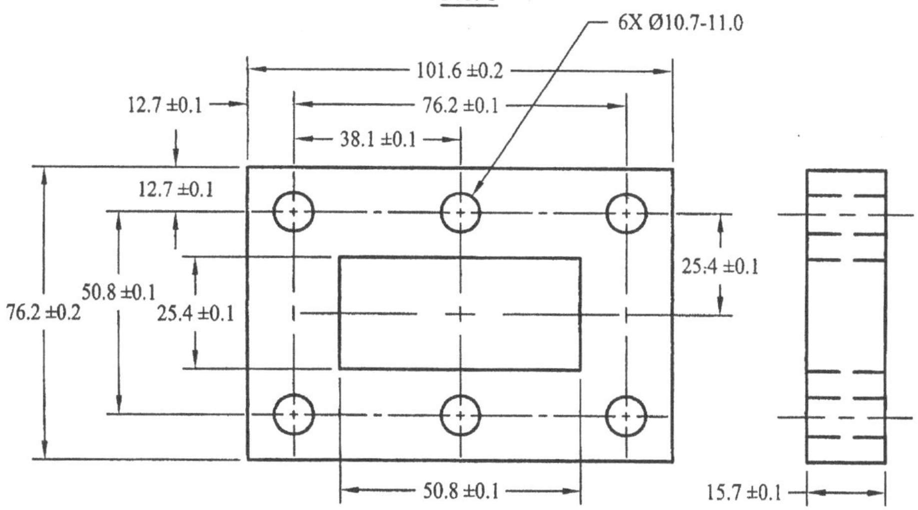

UNLESS OTHERWISE SPECIFIED ALL ANGLES ARE ±1°

Part 2

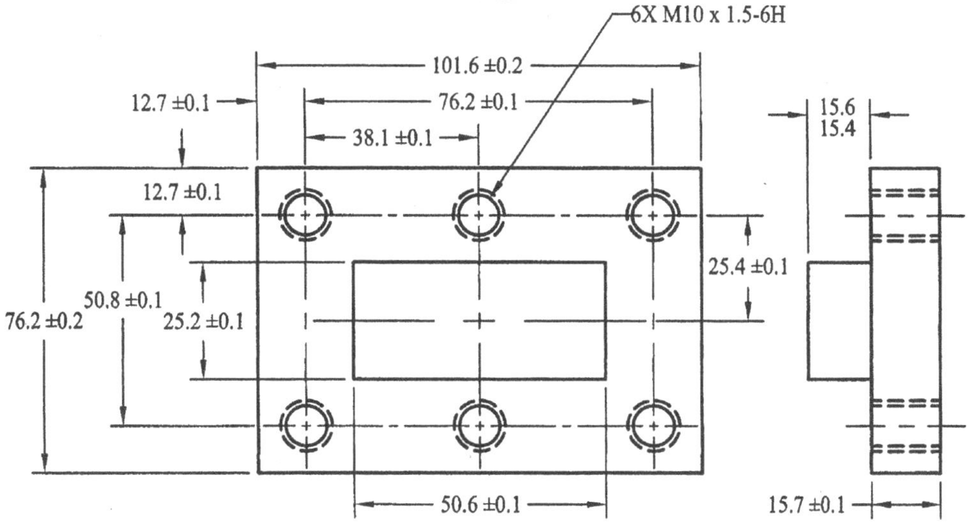

UNLESS OTHERWISE SPECIFIED ALL ANGLES ARE ±1°

7. Consider this object the shape of a hockey puck that has four holes drilled into it. Calculate the minimum wall thickness between the surface of one hole to the outside diameter of the hockey puck.

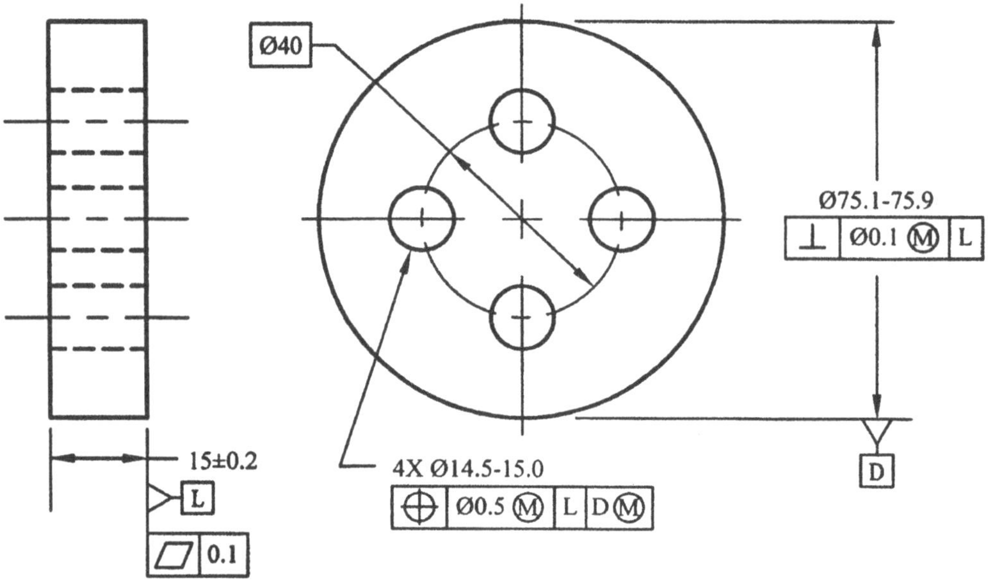

8. What is the minimum wall thickness of the hole pattern to datum feature C?

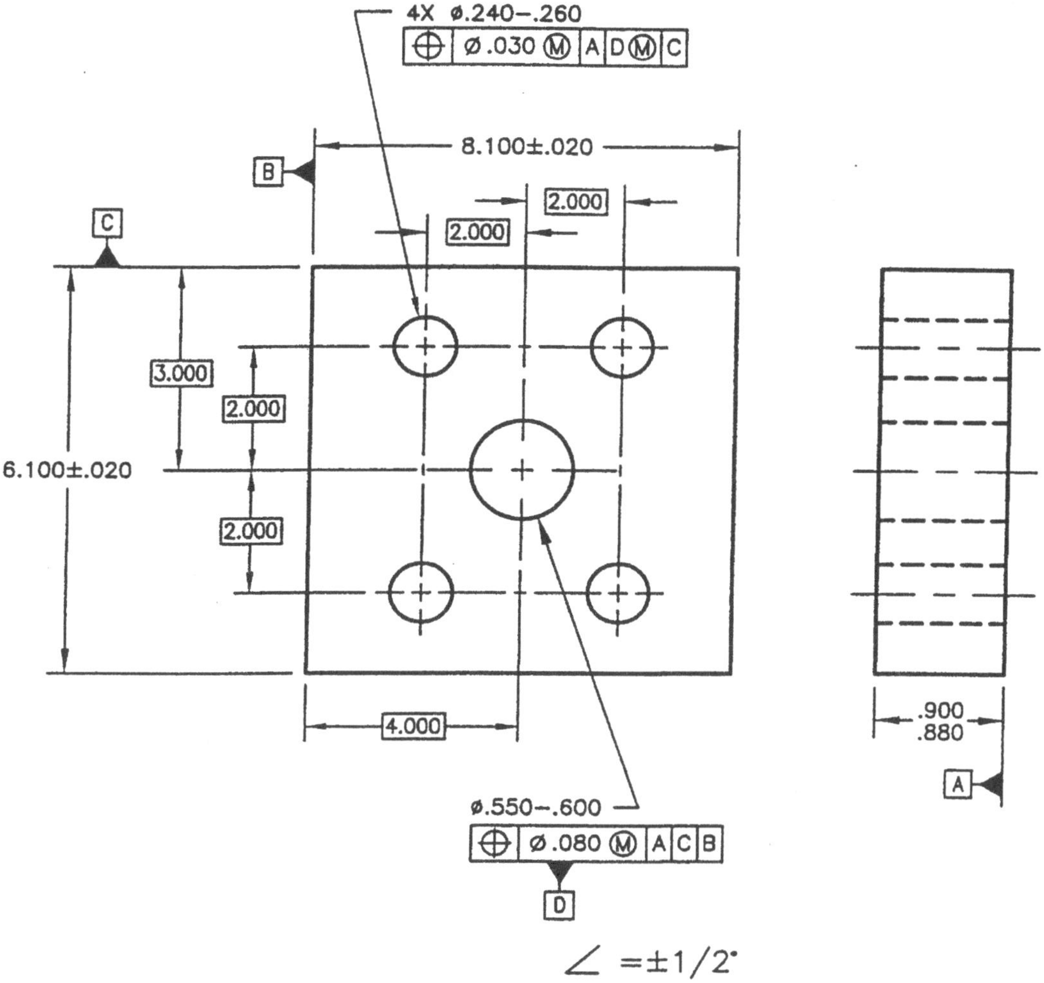

Note: 1/2° over .900=.008 (rounded off)

9. Please dimension and tolerance the sizes and positional controls for the gages depicted at the bottom of the page. Leave x's as is, but fill in all blanks. Make all positional tolerances zero at MMC. Use 10% of the difference between the LMC and the applicable virtual condition of the holes on the part as gage tolerance. Put it all into the plus tolerance on the gage pins.

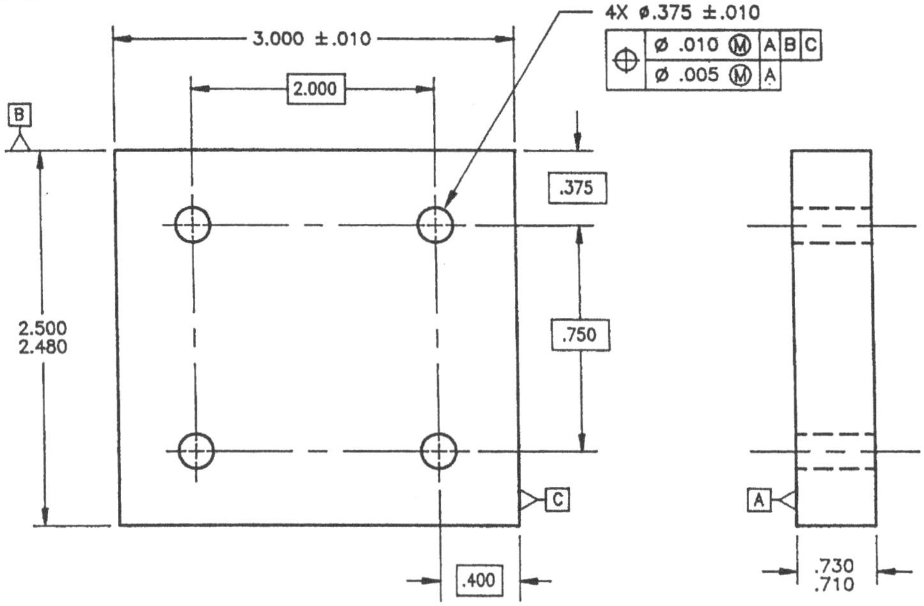

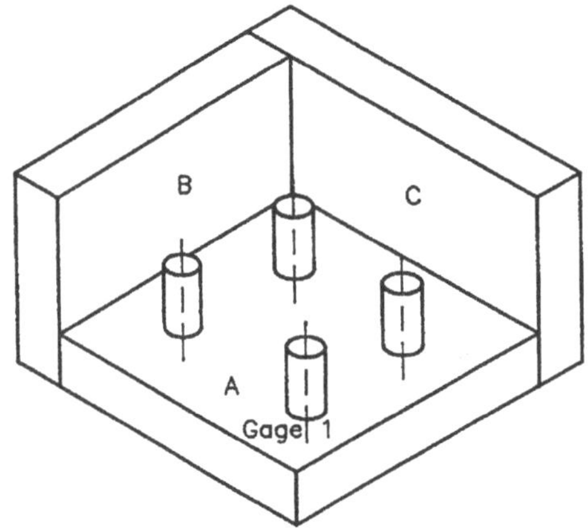

GAGE 1 – Datum to pattern
All four pins are located by the same basic dimensions (from the datum feature simulators) as used on the part drawing.

Pin sizes are ø ______ $^{+}_{-}$

with | ⌖ | ø ______ Ⓜ | A | B | C |

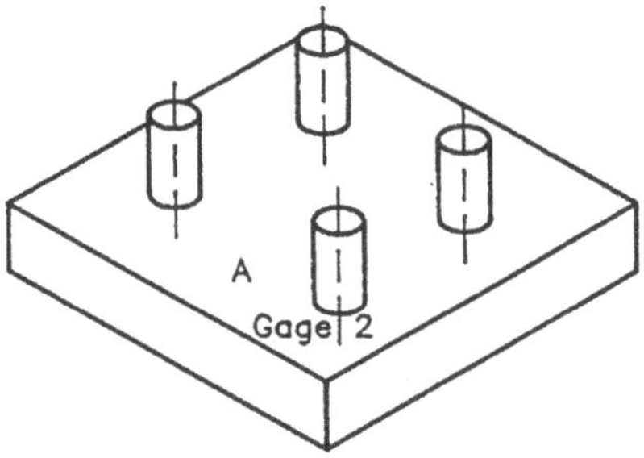

GAGE 2 – Feature to feature
pin axis to pin axis basic dimensions are the same as gage 1 (and part drawing).

Pin sizes are ø ______ $^{+}_{-}$

with ⌖ | ø .xxx Ⓜ | A | ___ | ___
⌖ | ø ______ Ⓜ | A

10. For the part shown at the top of this page, please answer the questions about each of the three gages shown (one on this page and two on the following page). These questions deal with the appropriate diameter of the gages' push pins and whether or not to project the tolerance zones on the holes that receive the push pins.

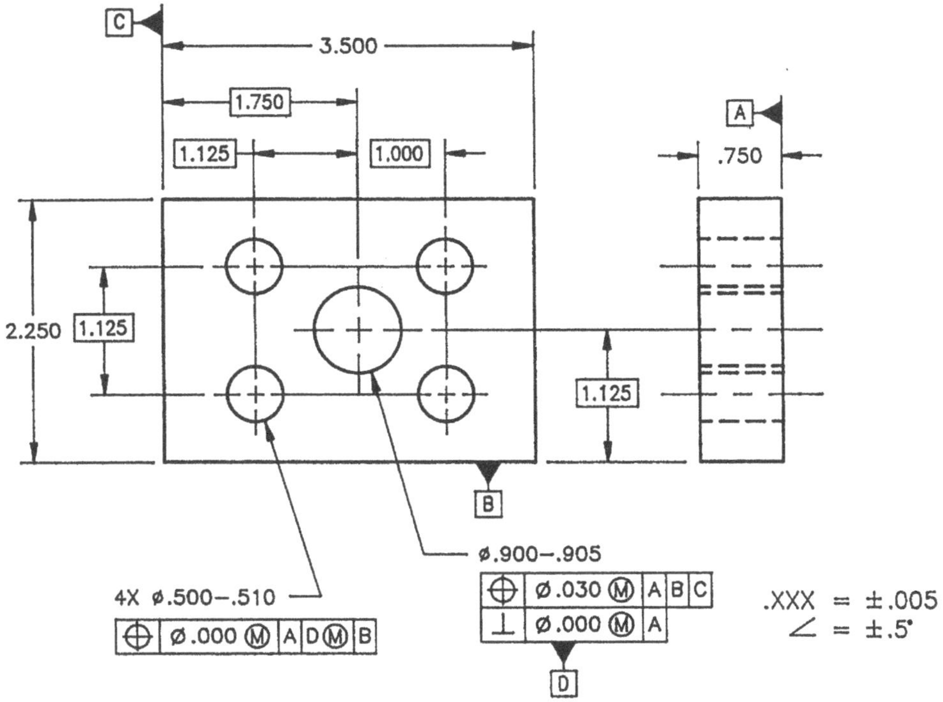

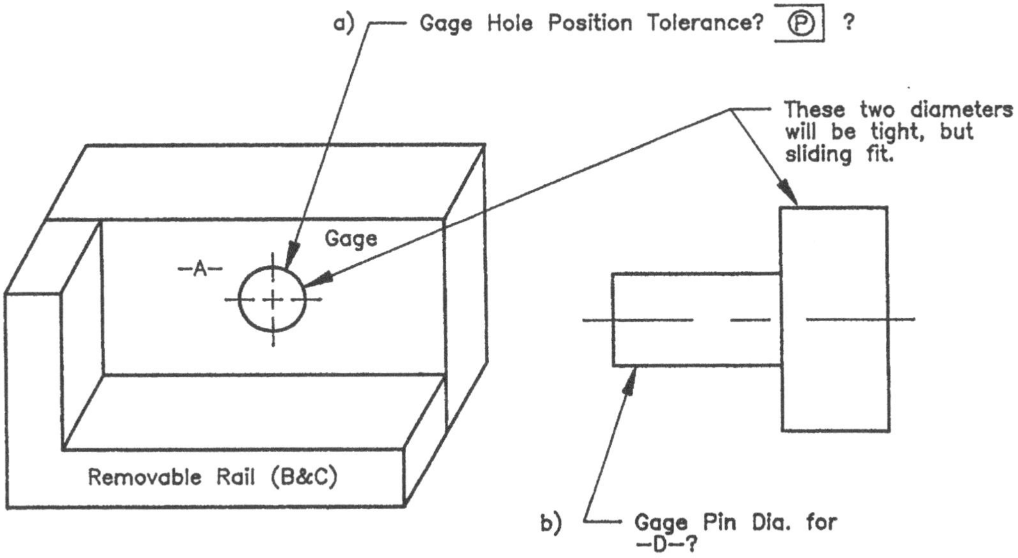

10. (continued)

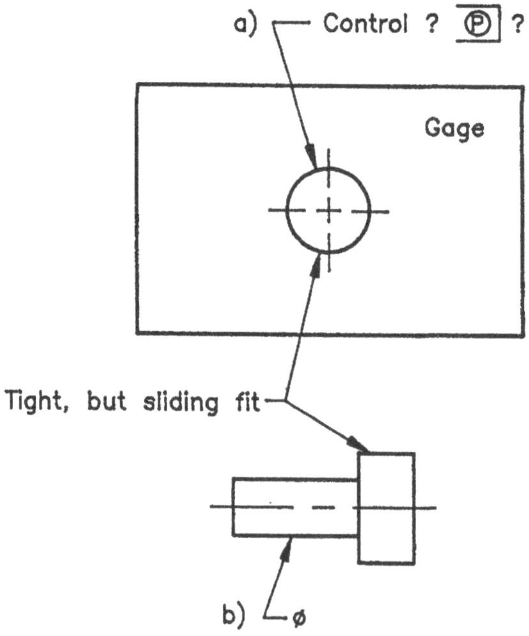

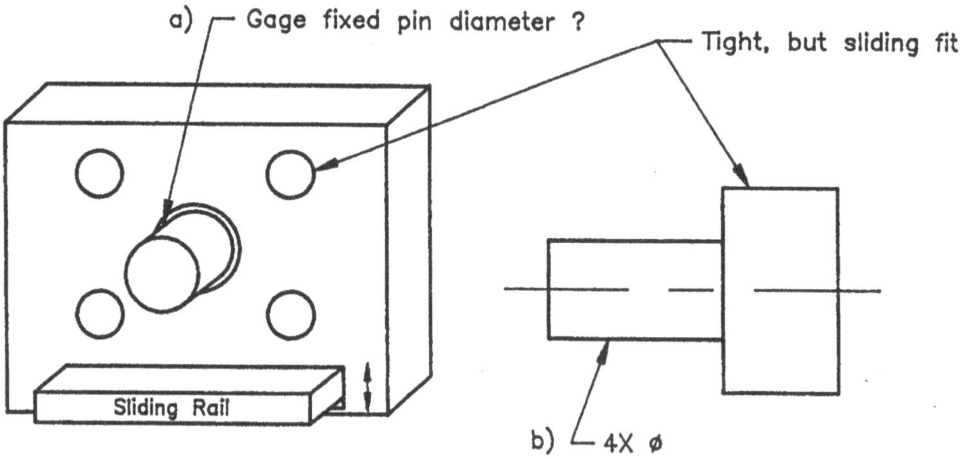

1. The following example demonstrates factors to determine minimum wall thicknesses. It is a good beginning for learning how to calculate any wall thickness problem. I call it the "Drinking Straw" problem. The premise is that most parts can be turned in drinking straws (or gun barrels) with a little thought.

   Please calculate the minimum wall thickness between the inside diameter and the outside diameter of this part.

Inner and Outer Boundaries for LMC

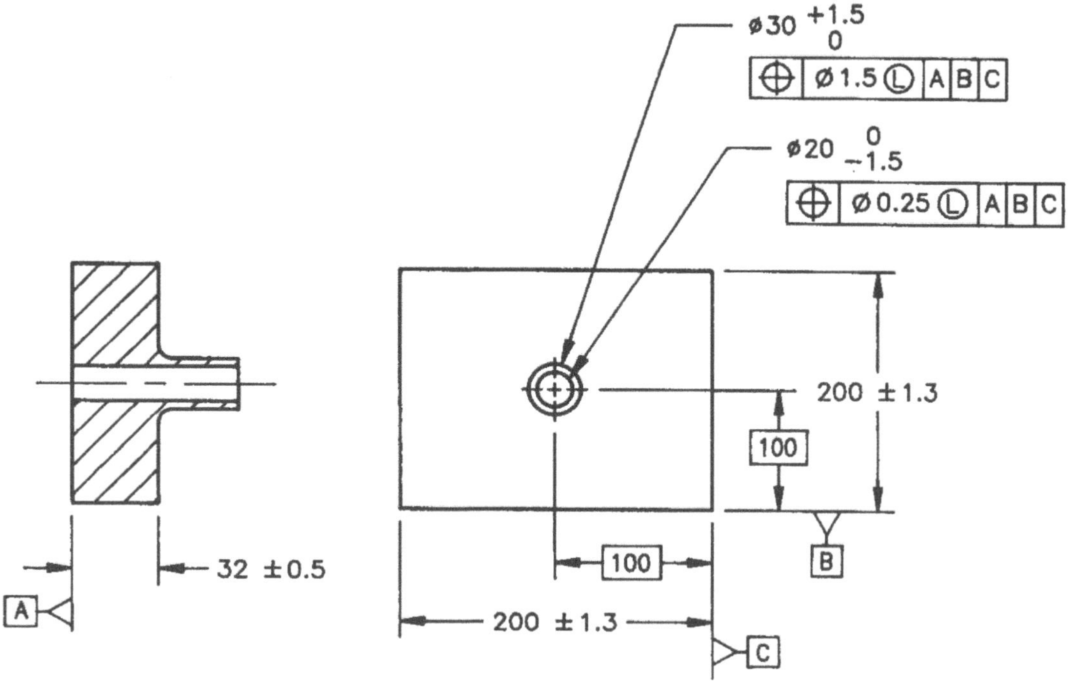

2. Displacement and a Simultaneous Requirement   Please calculate the minimum wall thickness between one of the four holes within the pattern and the outside diameter of the part.

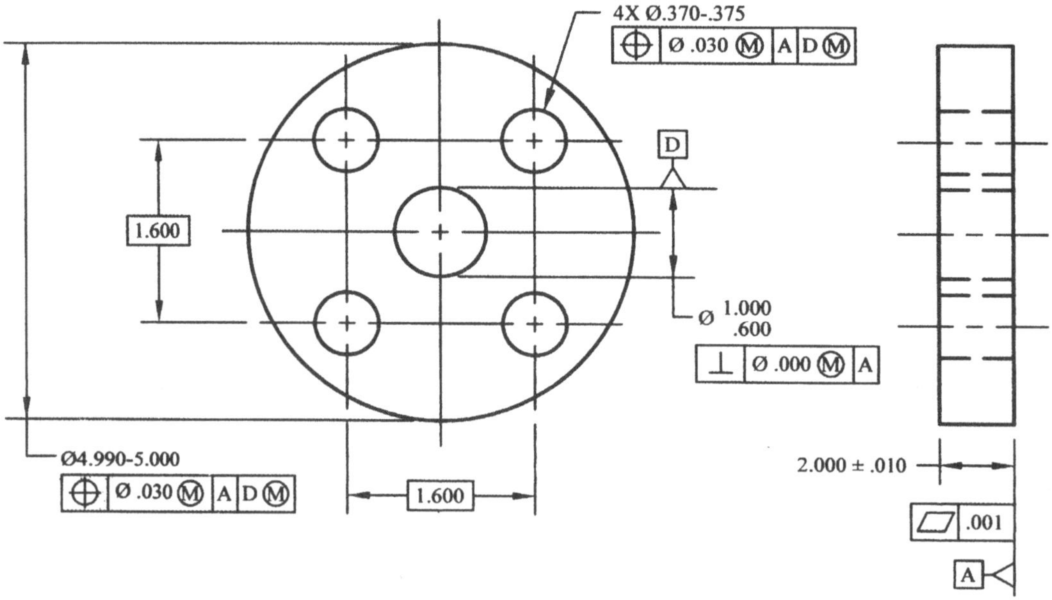

3. Displacement with a Separate Requirement Please calculate the minimum wall thickness between one of the four holes within the pattern and the outside diameter of the part.

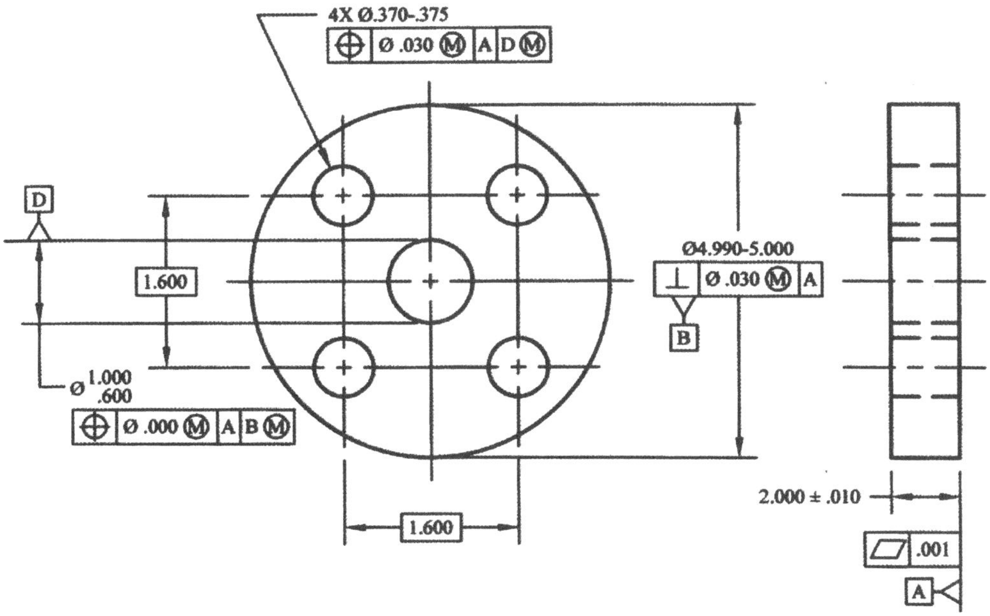

4. Convert this part to use composite positional tolerancing.
   a) Assign 3 datum features.
   b) Convert the general tolerance of ±.06 for the 1.62 and 1.50 dimensions by multiplying the square tolerance zone of .120 by 1.4 (the approximate square root of 2). This will circumscribe a cylindrical tolerance zone around the original square zone. This will serve as the Pattern Locating Tolerance Zone Framework.
   c) Calculate the Feature Relating Tolerance Zone Framework using the formula: MMC holes minus the virtual condition of the shafts equals the geometric tolerance for the holes. The virtual condition of the shafts is a diameter of .256.
   d) Assign the primary datum feature a flatness tolerance of .002.
   e) Assign the 4.0 and 5.0 sizes a perpendicularity tolerance of zero at MMC. Make one of these widths a fourth datum feature to orient the other width to.

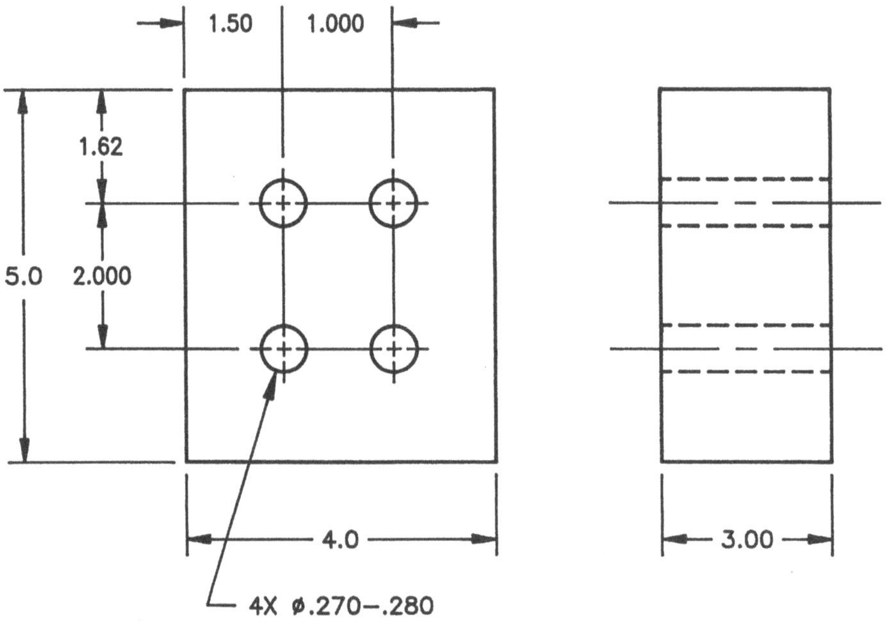

.X = ±.130
.XX = ±.06
.XXX = ±.010
∠ = ±1°

5. This is the mating part for the previous problem 4. The last part must fit down inside of this cavity and the holes must fit over the pins in the cavity.
   a) Assign all dimensions and tolerances.
   b) Square the part up as you did on the previous part. Assign the pins in the cavity only a single level of position tolerance.
   c) Assign the primary datum feature a flatness tolerance of .002.

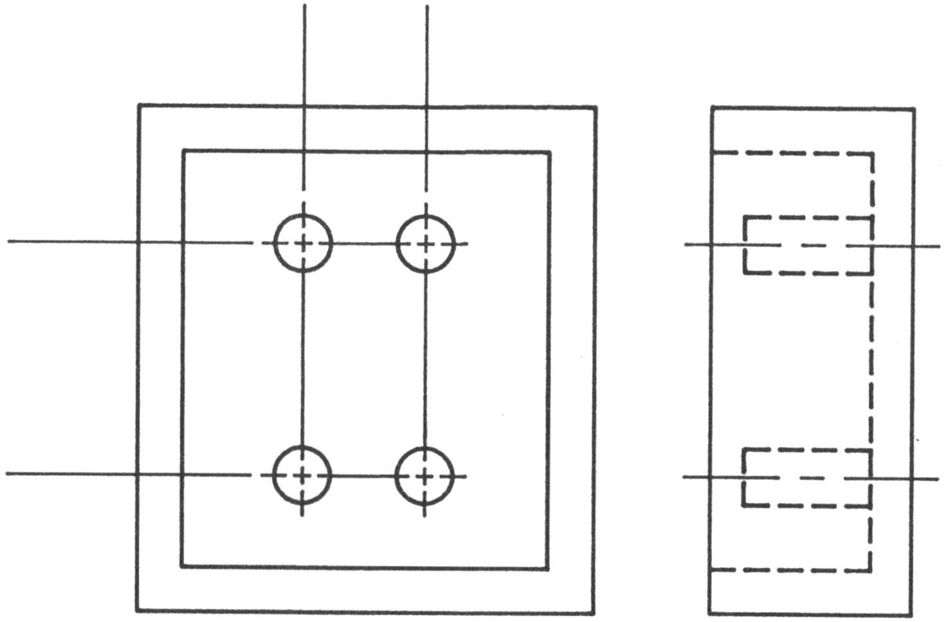

6. Please follow the directions on this page using the following two pages.

Core Board and Receiver Problem

Using the illustrations on the next two pages, complete the following questions. The .056 - .058 holes in the core board will each have the body of a nail with a .070 head (LMC) pressed into them. Nail is shown below. All nail heads must contact pogo pins on the receiver board.

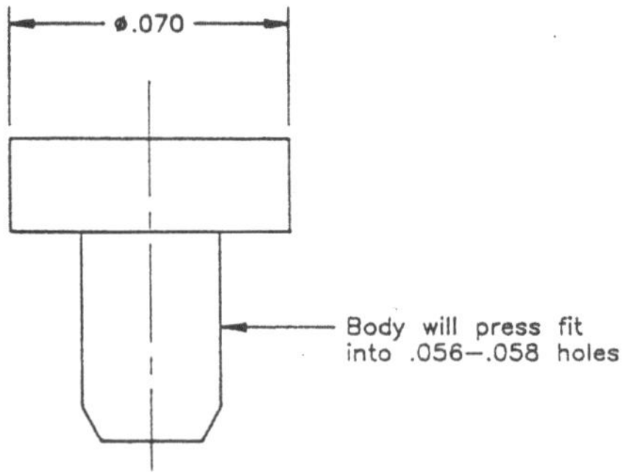

a) Calculate possible movement of heads to determine reduction of .070 contact area.

b) Determine the possible shift from the datum features by subtracting the LMC of the datum pins from the LMC of the datum holes.

c) Each pogo pin will be considered a point contact. Using the shift calculated in b) and the positional tolerance allowed each pogo pin, determine the amount of movement in terms of a diameter away from true position allowed each pogo pin.

d) Will the pogo pin points contact the nail head in all cases? If so, by how much in terms of a diameter? If not, by how much could it miss (in terms of a diameter)?

6. (continued)

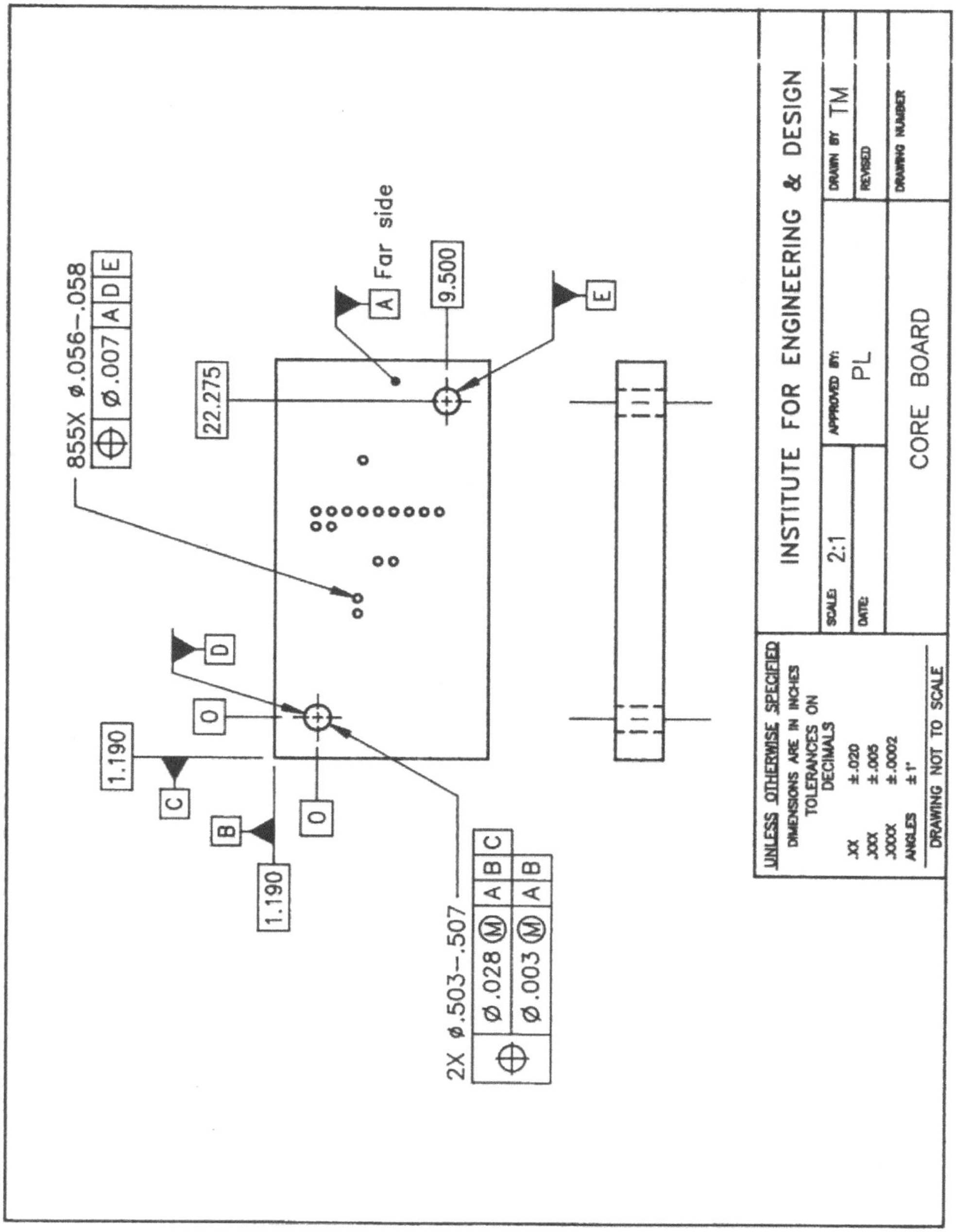

855X ⌀.056-.058
⌀.007 | A | D | E
22.275
A
Far side
9.500
E
D
0
0
1.190
C
B
1.190
2X ⌀.503-.507
⌀.028 Ⓜ | A | B | C
⌀.003 Ⓜ | A | B
INSTITUTE FOR ENGINEERING & DESIGN
DRAWN BY TM
REVISED
DRAWING NUMBER
APPROVED BY: PL
SCALE 2:1
DATE
CORE BOARD
UNLESS OTHERWISE SPECIFIED
DIMENSIONS ARE IN INCHES
TOLERANCES ON DECIMALS
.XX ±.020
.XXX ±.005
.XXXX ±.0002
ANGLES ±1°
DRAWING NOT TO SCALE

6. (continued)

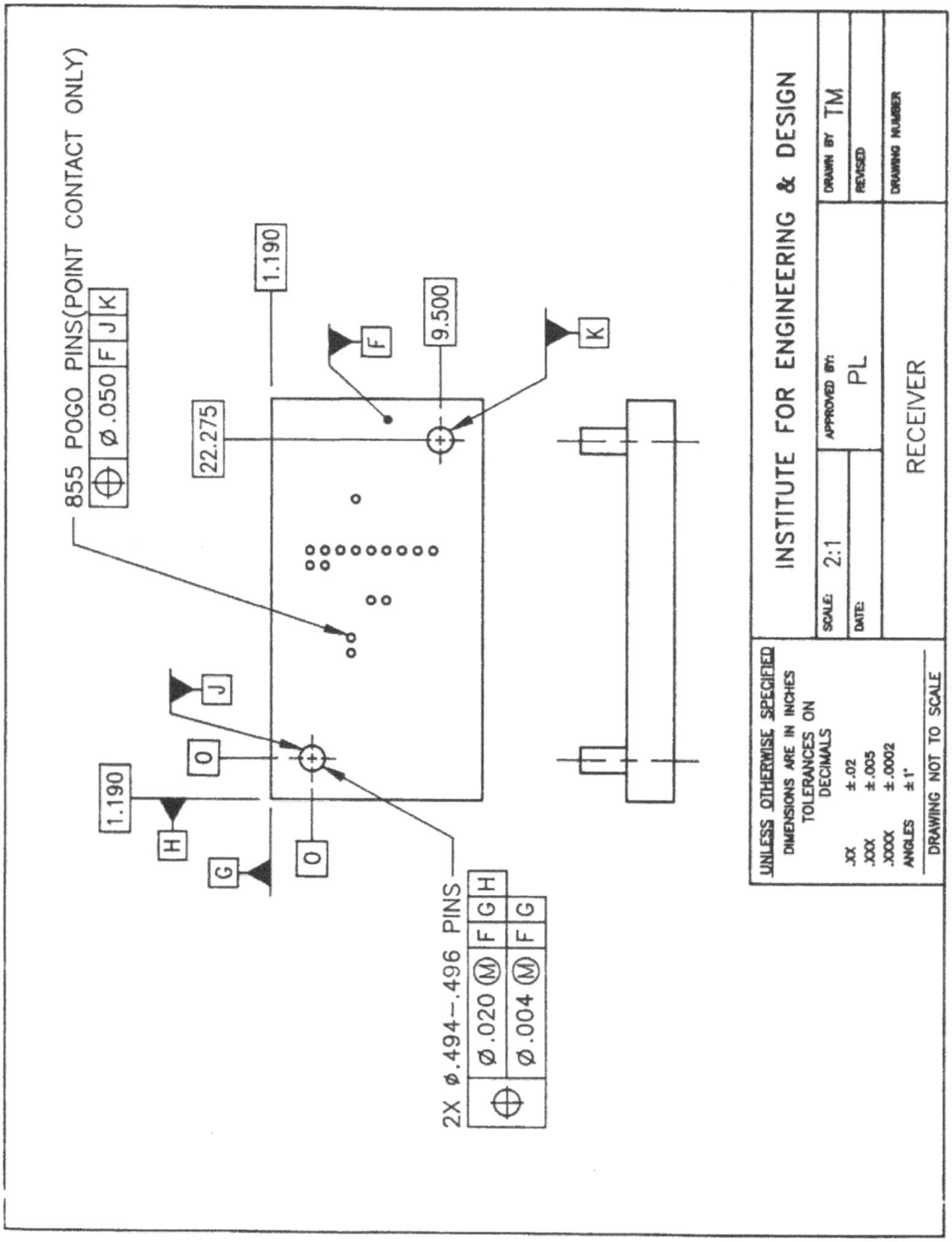
855 POGO PINS(POINT CONTACT ONLY)
⌖ Ø.050 F J K
1.190
9.500
22.275
F
K
J
O
O
1.190
H
G
2X Ø.494-.496 PINS
⌖ Ø.020 Ⓜ F G H
Ø.004 Ⓜ F G
INSTITUTE FOR ENGINEERING & DESIGN
DRAWN BY TM
REVISED
DRAWING NUMBER
APPROVED BY: PL
SCALE: 2:1
DATE:
RECEIVER
UNLESS OTHERWISE SPECIFIED
DIMENSIONS ARE IN INCHES
TOLERANCES ON DECIMALS
.XX ±.02
.XXX ±.005
.XXXX ±.0002
ANGLES ±1°
DRAWING NOT TO SCALE

SECTION 7. Concepts Introduced in Y14.5-2009 and 2018

1) Please complete the following feature control frame to indicate that the datum feature simulator for datum feature C translates (moves).

21X Ø4.0-4.3

2) Complete the following feature control frame to indicate a unilateral profile of a surface control that is 0.8 all inside of the basic profile.

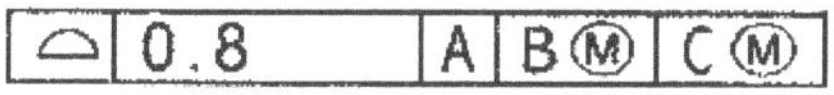

3) Eliminate the envelope of perfect form at maximum material condition by using the appropriate symbol after the size limits of the following hole.

Ø10.5-10.7

4) Use the correct symbology (a new symbol in Y14.5-2009) to make the following two diameters treated as though they are one diameter.

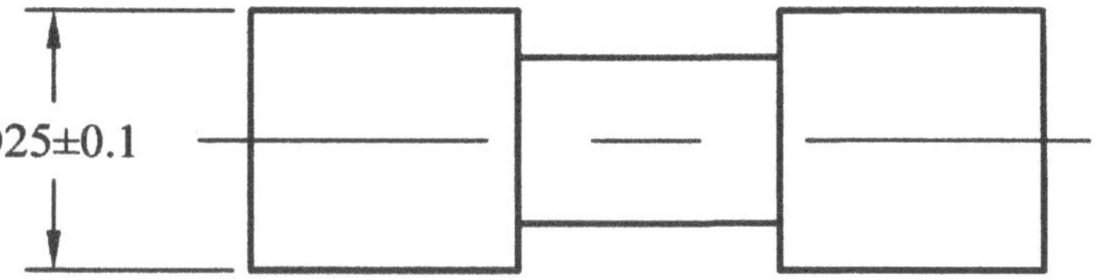

5) When the symbol Ⓜ is used after a datum feature of size referenced in a feature control frame, it stands for: (Circle correct answer)
   a. Minimum Material Condition
   b. Maximum Material Condition
   c. Maximum Material Boundary
   d. Minimum Material Boundary

6) When the symbol Ⓛ is used after a geometric tolerance in a feature control frame, it stands for: (Circle correct answer)
   e. Least Material Boundary
   f. Least Material Condition
   g. Least Squares Center
   h. Least Boundary Condition

7) Put a symbol in the blank area on the elbow of the leader line from the feature control frame that means ALL OVER, APPLIES IN ALL VIEWS SIMULTANEOUSLY.

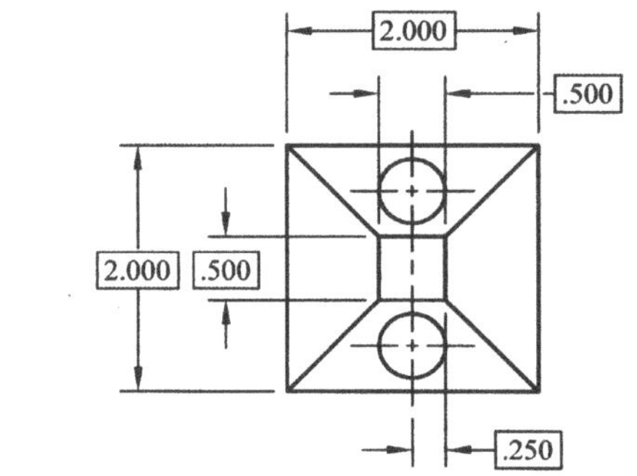

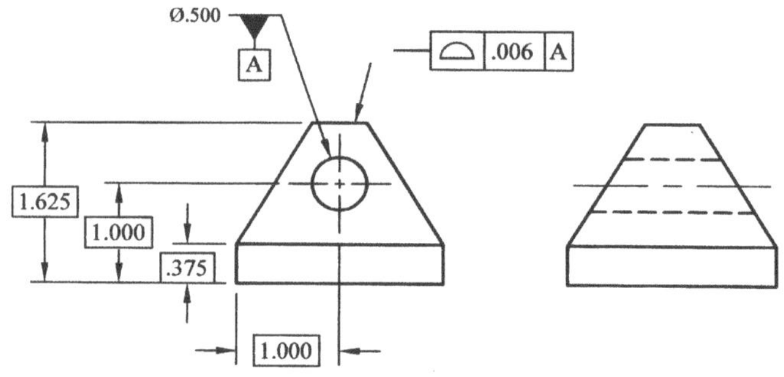

8) When no material condition symbol is used inside a feature control frame after any datum features of size referenced, what is the abbreviation for the symbol that is implied? (Circle correct answer)

a. MMC
b. RMB
c. LMC
d. MMB
e. RFS
f. LMB

9) What does each of the abbreviations in question 8 stand for?

g. MMC stands for ______________________________.
h. RMB stands for ______________________________.
i. LMC stands for ______________________________.
j. MMB stands for ______________________________.
k. RFS stands for ______________________________.
l. LMB stands for ______________________________.

10) What does the symbol that includes the SF letters in the following drawing stand for?

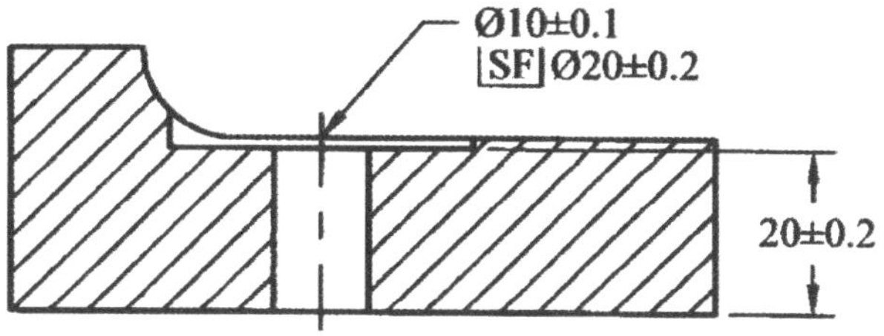

11.) According to the rules in ASME Y14.5-2009 and ASME Y14.5-2018, may the Ⓢ symbol be used in a feature control frame?

12) Label each of the envelopes to which arrows point in the following illustrations as either the "actual mating envelope" or the "actual minimum material envelope".

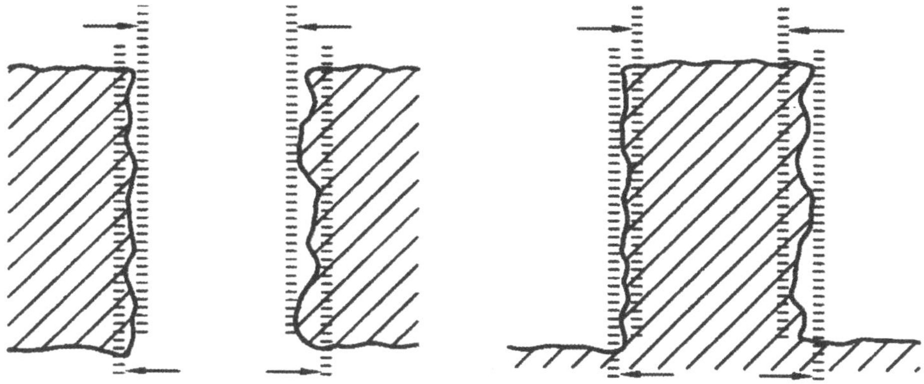

13) Per ASME Y14.5-2009 and ASME Y14.5-2018, what geometric characteristic symbol (if placed below the 25±0.5) would be capable of creating the following tolerance zone for the part shown? (Circle the correct answer below the illustrations)

Part Specification

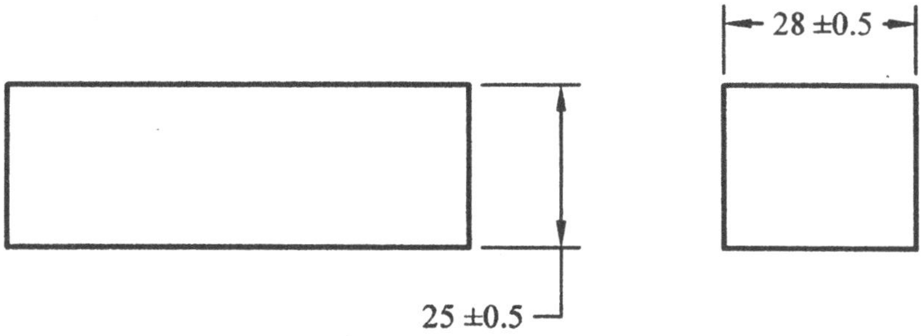

Tolerance Zone

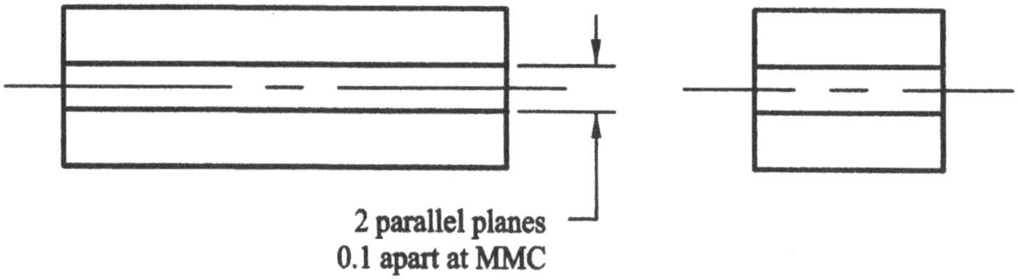

a. | ⏤ | 0.1 Ⓜ |
b. | ○ | 0.1 |
c. | ⌒ | 0.1 |
d. | ▱ | 0.1 Ⓜ |

14) An "Actual Mating Envelope" is a similar perfect feature(s) counterpart of smallest size that can be contracted about an external feature(s) or of largest size that can be expanded within an internal feature(s) so that it coincides with the surface(s) at the highest points. What is the difference between an "Unrelated Actual Mating Envelope" and a "Related Actual Mating Envelope"? (Circle one)
   a. Related is not tied to the feature control frame, but unrelated is.
   b. Related is oriented or located to a datum(s), while unrelated is not.
   c. Unrelated is not a "Simultaneous Requirement" with other feature patterns, but Related is related to other feature patterns.
   d. Related is a combined control using one feature control frame stacked on another.

15) Per ASME Y14.5-2009 and ASME Y14.5-2018, which geometric characteristic symbol that is categorized as "Orientation" may be used to tolerance any angle no matter how many degrees it is specified as?
   m. Angularity
   n. Perpendicularity
   o. Parallelism
   p. Profile of a Surface

16) Where axes, center planes or surfaces are shown coincident on a drawing and geometric tolerances establish the relationship between the features, what applies?
   a. a plus and minus tolerance
   b. a zero basic dimension
   c. a general tolerance note
   d. a reference dimension

17) Where a coordinate system is shown on the drawing, unless otherwise specified, it shall be:
    q. left handed
    r. labeled in the negative direction
    s. neither labeled in the positive or negative directions
    t. right handed

18) Per ASME Y14.5-2018, unless otherwise specified, a theoretical datum feature simulator is:
    u. called a true geometric counterpart
    v. called a false geometric counterpart
    w. called a real physical surface
    x. rails, machine table, surface planes, collet chucks and parallels used to fixture parts.

19) Match the following symbols with what they represent by listing the correct letter of the definition after each symbol.
    y. Continuous Feature of Size (treats multiple features of size as one)
    z. Translation symbol (datum feature simulator translates/moves)
    aa. All Over (control applies in all views-commonly used with profile of a surface)
    bb. Unequally Disposed or Unilateral Tolerance
    cc. Independency (size tolerance does not control form)

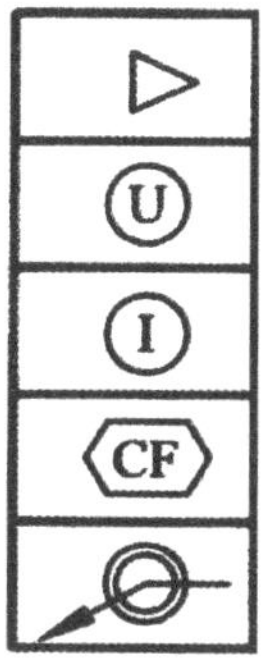

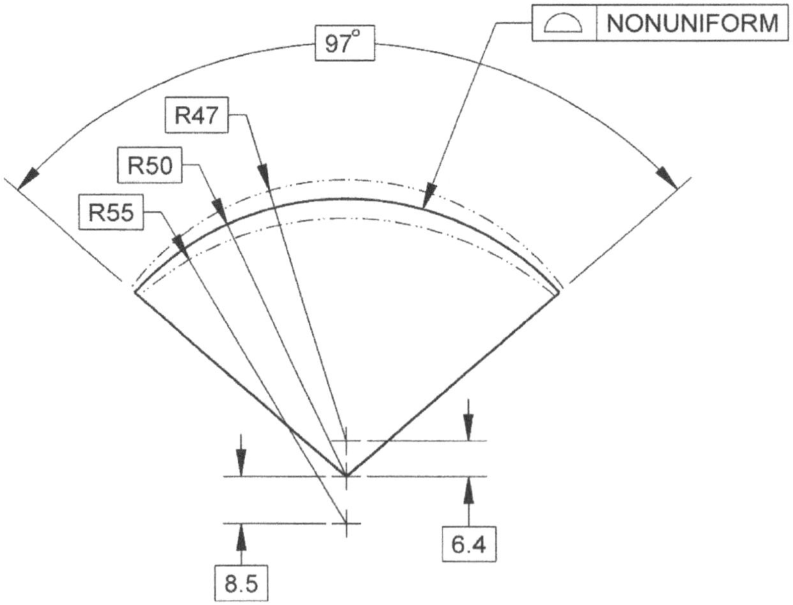

20) The illustration above is of a NONUNIFORM profile of a surface control. The surface being toleranced by this control must reside within the tolerance zone depicted by the phantom lines. The purpose of such a tolerance zone can be to: (Circle all correct answers)

a. smooth the transition between profile segments
b. allow greater variations in the surface form at the ends of the zone
c. allow abrupt surface variations at the ends of the profile zone
d. allow smaller surface variations at the ends of the profile zone
e. allow larger surface variations in form in the middle of the zone

21) The following illustration compares Composite Profile Tolerancing (one profile symbol) with Two Single Segment Profile Tolerancing (two profile symbols).

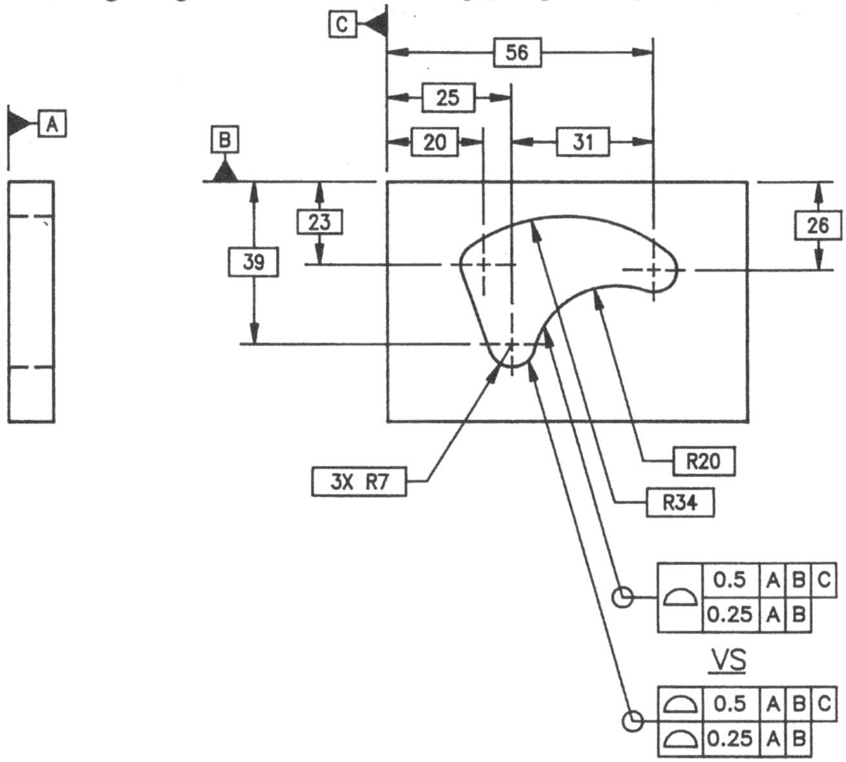

In the composite profile control:

a. What is the tolerance on the size and form of the hole?
b. What is the tolerance on the perpendicularity of the hole to datum plane A?
c. What is the tolerance on the angular orientation of the hole to datum plane B?
d. What is the tolerance on the location of the hole to datum plane B?

In the two single segment profile control:

e. What is the tolerance on the size and form of the hole?
f. What is the tolerance on the perpendicularity of the hole to datum plane A?
g. What is the tolerance on the angular orientation of the hole to datum plane B?
h. What is the tolerance on the location of the hole to datum plane B?

22) The following illustration a composite profile control with another profile control that applies INDIVIDUALLY to each hole.

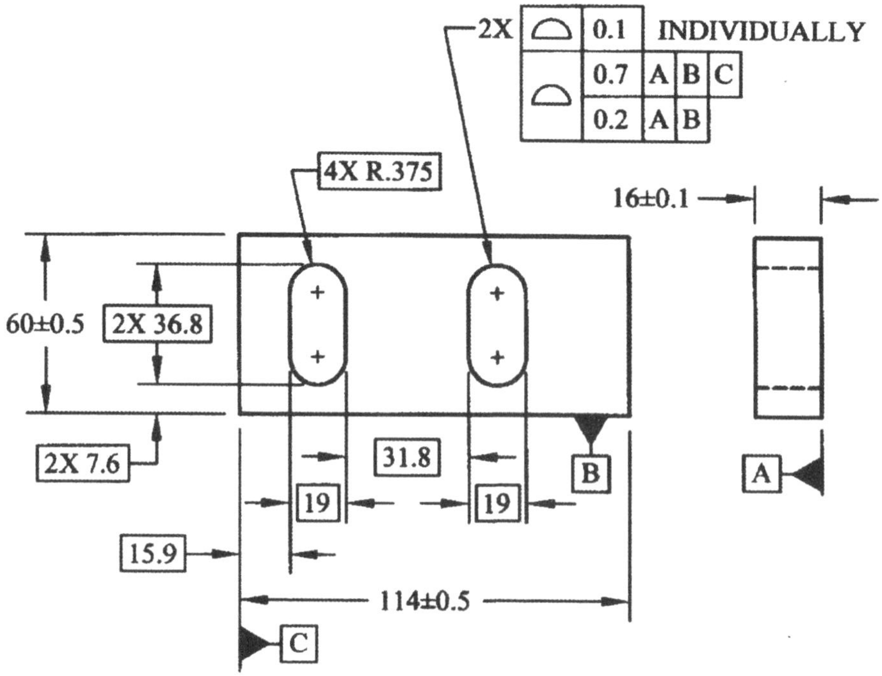

a. What is the tolerance on the size and form of the holes?
b. What is the tolerance on the location of the holes to datum B?
c. What is the tolerance on the location of the holes to datum C?
d. What is the tolerance on the location of the holes to each other?
e. What is the tolerance on the perpendicularity of the holes to datum plane A?
f. What is the tolerance on the angular orientation (rotation) of the 2-hole pattern to secondary datum plane B?
g. What is the tolerance on the angular orientation (rotation) of the 2-hole pattern to tertiary datum plane C?

23) The following illustration shows a conical feature used as a datum feature.

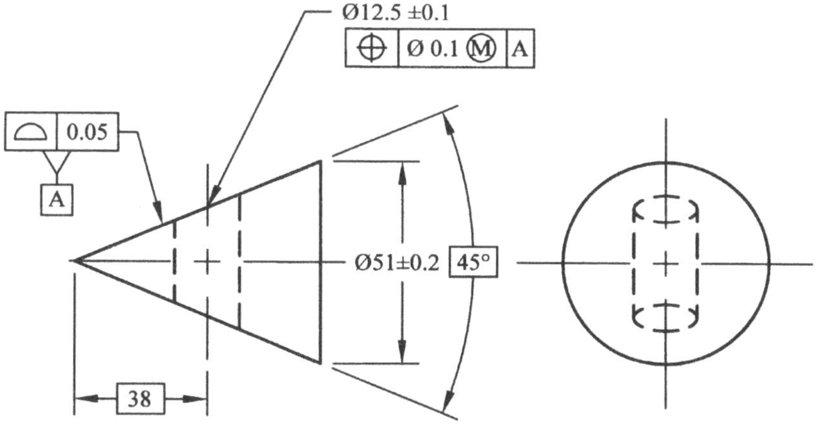

a. How many spatial degrees of freedom are controlled by referencing datum feature A in the position control on the hole?
b. Does datum feature A generate:
   1. an axis only
   2. an axis and a plane (from the point at the apex of the cone)
   3. an axis and a plane from the right edge in the view on the left
   4. a plane only

24) The following illustration shows a conical datum feature. In the position feature control frame the spatial degrees of freedom that each datum reference controls are listed in brackets after the datum reference.

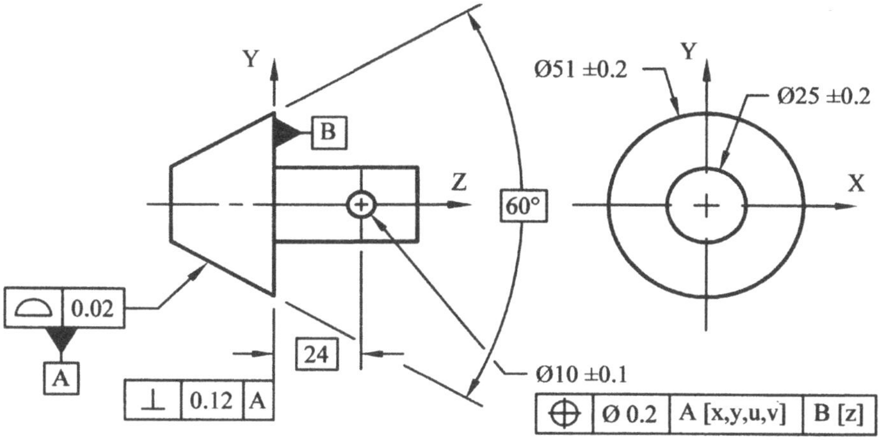

a. What controls the rotational spatial degree of freedom known as w?
b. If datum feature B was referenced as primary instead of secondary, would it be necessary to specify the bracketed z after B in the feature control frame?
c. If datum feature B was not referenced in the position control, how would the dimensions on the part drawing change?

25) Detail Drawing with Degrees of Spatial Freedom Indicated by Showing X, Y and Z Axes on the Drawing.

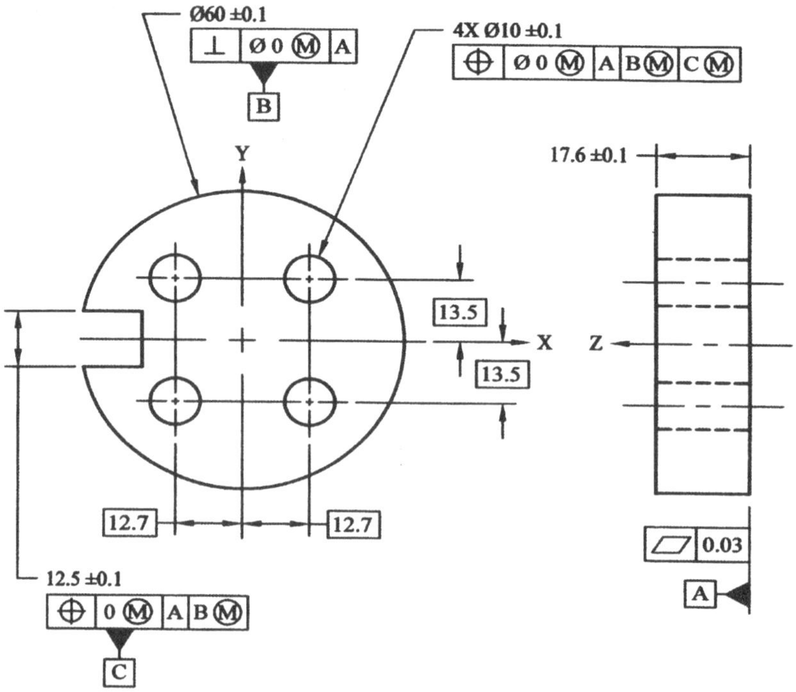

There are 6 spatial degrees of freedom. 3 are linear (x, y and z) and 3 are rotational (u, v and w). Which datum features control each of these? Specify the datum feature that controls each degree of spatial freedom listed below.

a. Movement in the x direction is controlled by datum________.
b. Movement in the y direction is controlled by datum________.
c. Movement in the z direction is controlled by datum________.
d. Rotation in the u direction is controlled by datum_________.
e. Rotation in the v direction is controlled by datum_________.
f. Rotation in the w direction is controlled by datum_________.

26) Overriding the Degrees of Spatial Freedom Controlled by Datums.

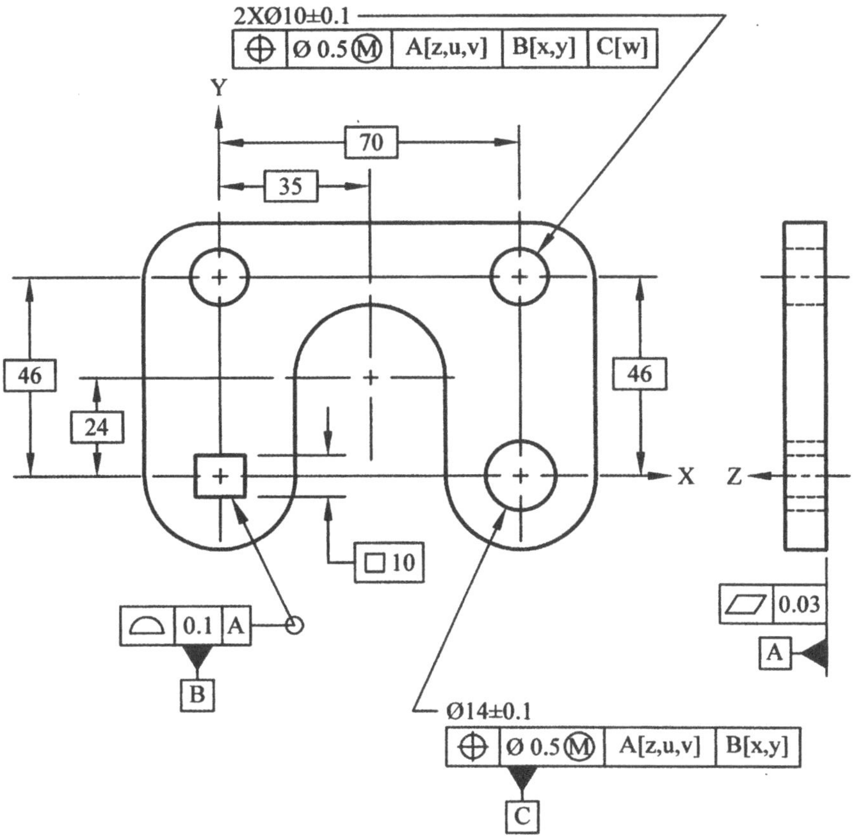

The positional controls shown in the illustration above designate which spatial degrees of freedom are controlled by the datum references. In the 4-hole positional control, what degrees of spatial freedom would each datum reference have controlled if not overridden by the bracketed designations?

a. Datum A would have controlled the following degrees of spatial freedom:

__________________________________.

b. Datum B would have controlled the following degrees of spatial freedom:

__________________________________.

c. Datum C would have controlled the following degrees of spatial freedom:

__________________________________.

27) Please fill in the correct MMB or LMB in the blank spaces on the following illustrations.

Correct Maximum Material Boundary Calculations

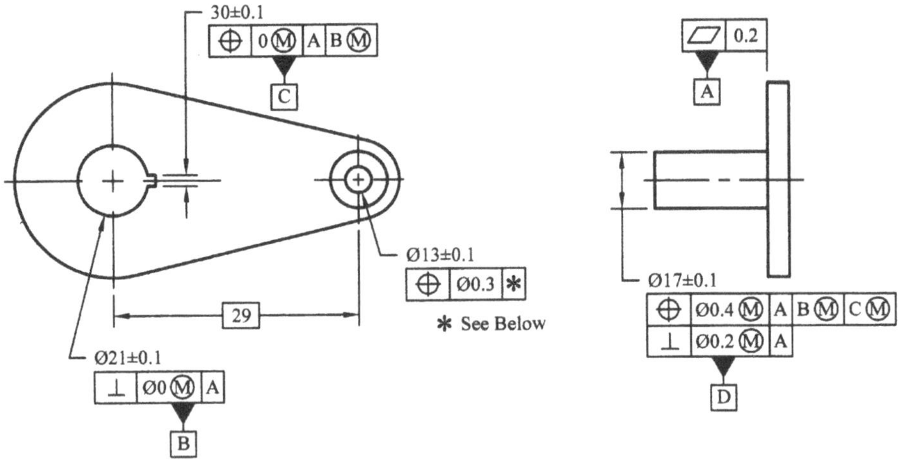

Option	Geometric Control	Maximum Material Boundary of datum feature D
(a)	⊕ Ø 0.3 DⓂ	
(b)	⊕ Ø 0.3 A DⓂ	
(c)	⊕ Ø 0.3 A B DⓂ	

Correct Least Material Boundary Calculations

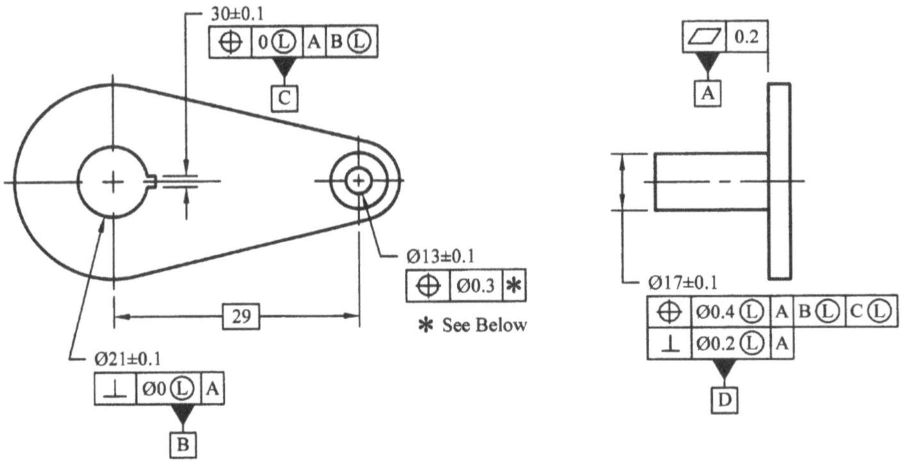

Option	Geometric Control	Least Material Boundary of datum feature D
(a)	⊕ Ø 0.3 DⓁ	
(b)	⊕ Ø 0.3 A DⓁ	
(c)	⊕ Ø 0.3 A B DⓁ	

28) Answer the questions for the following illustration.

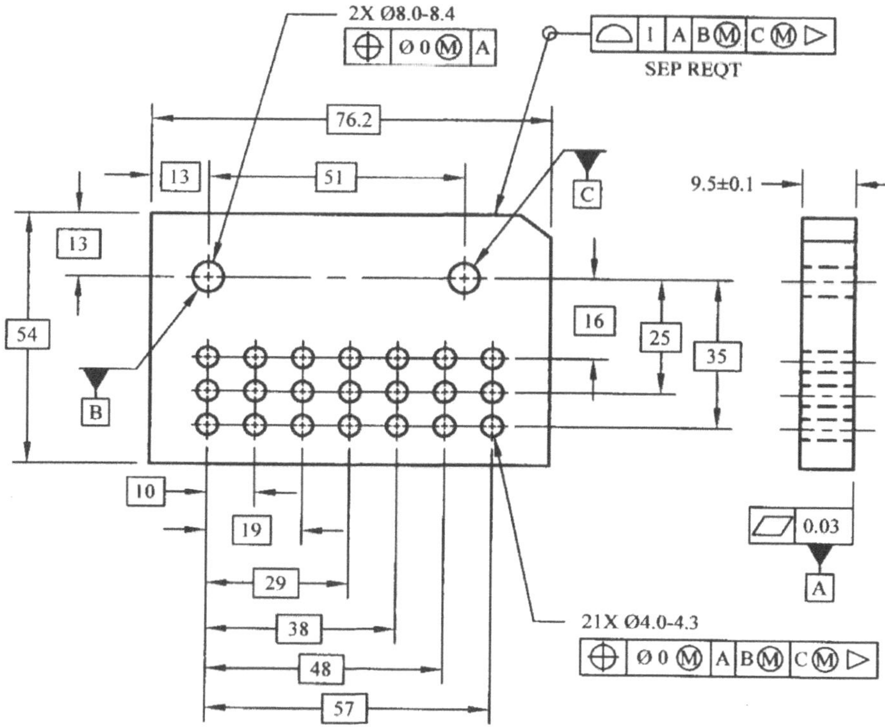

For the 21-hole pattern:

a. What does the symbol Ⓜ used after the zero tolerance in the feature control frame (FCF) stand for?

b. Would the datum feature simulator for datum feature B be stationary or moveable?

c. What does the symbol Ⓜ used after datum feature B in the FCF stand for?

d. What does the symbol used after the µafter datum feature C stand for?

e. What size gage pin would be used to represent datum feature B?

f.. What size gage pin would be used to represent datum feature C?

g. How would the datum feature simulator used to represent datum feature C differ from the one used for datum feature B?

For the profile of a surface control:

h. What does the SEP REQT note stand for?

i. How would datum feature shift due to datum feature B's departure from its virtual condition differ if the SEP REQT note was not on the drawing?

29) Using the correct datum feature designations, feature control frames and local notes:

a. Designate each of the 12 millimeter holes as a datum feature using one datum feature symbol (the letter D).

b. Locate the smaller 6 millimeter holes to the left and right of each large hole to that 12 millimeter hole only, using datum plane A for perpendicularity, datum axis D for location and datum plane B for angular orientation.

c. Use the local note 4X INDIVIUALLY twice.

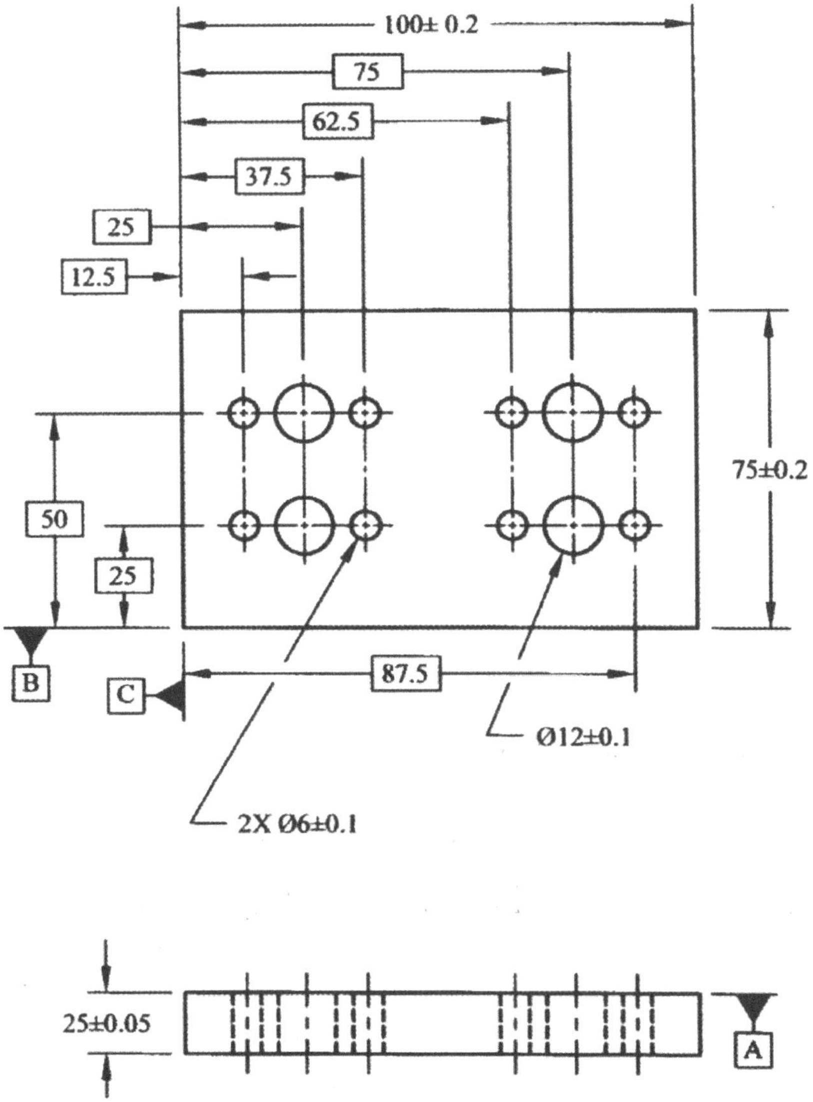

30) Please apply a three-level composite positional tolerancing control to the three-hole pattern to mean the following:

1. The top level control locates the three holes to datums B and C. Regardless of the material boundary of datum feature B, the three holes must be centered to datum centerplane B and hold the 19 millimeter basic dimension from datum plane C to within a diameter of 3 millimeters if the holes are produced at their maximum material condition.
2. The middle level control is ultimately only holding the orientation of the three holes (as a pattern) to datum centerplane B and datum plane C to within a diameter of 1.5 millimeters if the holes are produced at their maximum material condition.
3. The third level control is aligning the three holes' axes to one another for coaxiality to within a diameter of 0.3 millimeters if the holes are produced at their maximum material condition. It is worth noting that no datum feature references are used here, in that, none are necessary to position the three holes to each other. This tolerance is the one that allows the 8 millimeter screw to pass through the three holes. It generates a virtual condition (inner boundary) for the three holes that is calculated by subtracting the 0.3 millimeter diameter position tolerance from the 8.3 millimeter maximum material condition of the holes. 8.3 minus 0.3 equals an 8 millimeter diameter virtual condition. This is the worst mating condition and is compatible with the 8 millimeter screw that must pass through. Since 8 millimeter screws are actually a little smaller than 8 millimeters, there is even a little clearance built into this assembly condition.

Detail Drawing

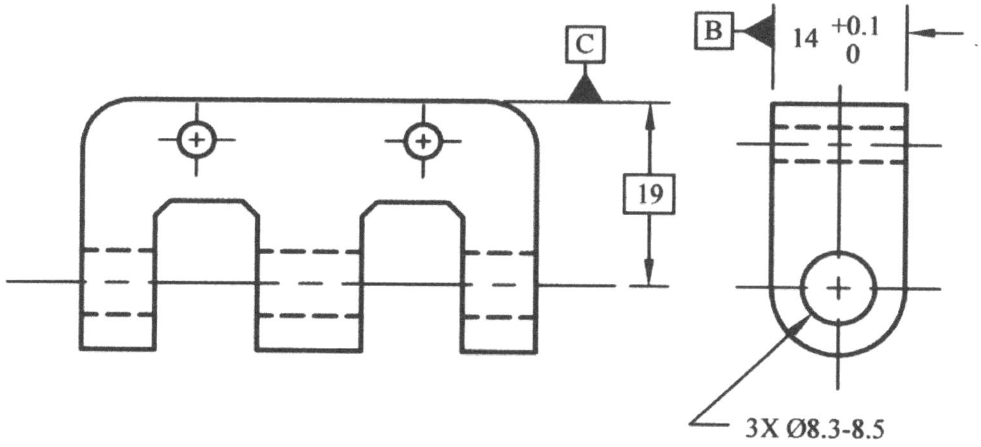

SECTION 8. Assemblies-Fixed and Floating Fasteners

1. Combined Positional and Perpendicularity vs. Composite Positional Tolerancing
   1) What would be the difference between the composite positional control given to datum feature D and a control that, instead, combined position to within a diameter of .060 at MMC to A, B and C on the top and perpendicular to A to within .000 at MMC on the bottom? Which would be most appropriate?
   2) In controls such as the perpendicularity of C to A, the word 'Total' was used in the distant past inside of some feature control frames. What was meant by the word 'Total'?
   3) What purpose does datum feature C serve in the 3-hole pattern's positioning?
   4) What would be the result of leaving datum feature C out of the control?

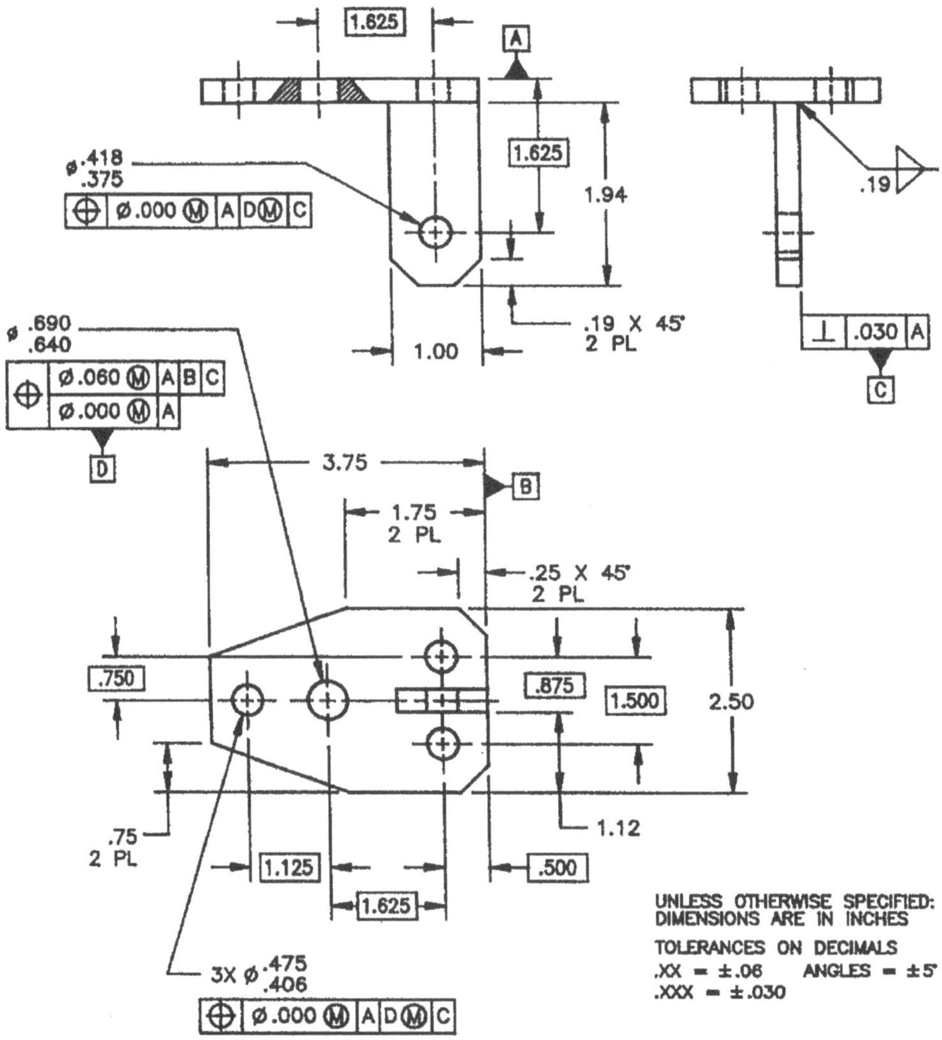

UNLESS OTHERWISE SPECIFIED:
DIMENSIONS ARE IN INCHES
TOLERANCES ON DECIMALS
.XX = ±.06 ANGLES = ±5°
.XXX = ±.030

2. Problem #2 Mating Part Design; Holes Perpendicular to Part Axis
   1) Assign the primary datum feature on each part as the large mating diameters.
   2) Assign a secondary datum feature on each part as the end of the part.
   3) Assign a two-single segment position tolerance on the 6.2-6.3 holes on each part. In the upper level segment, use a 1mm position tolerance to both datums. In the lower level segment, calculate the position tolerance using the floating fastener formula and reference only the functional diameter datum feature at MMC.

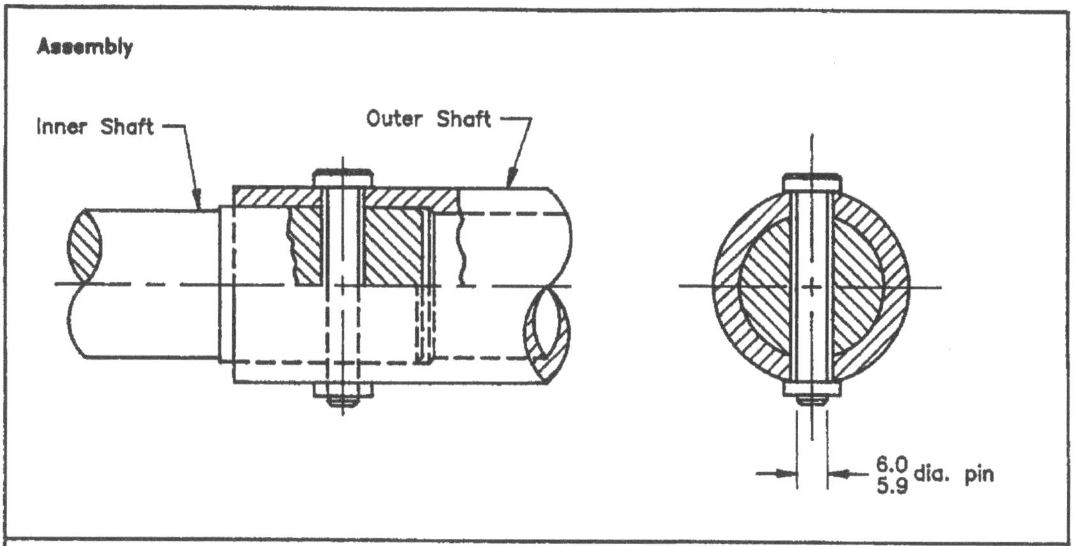

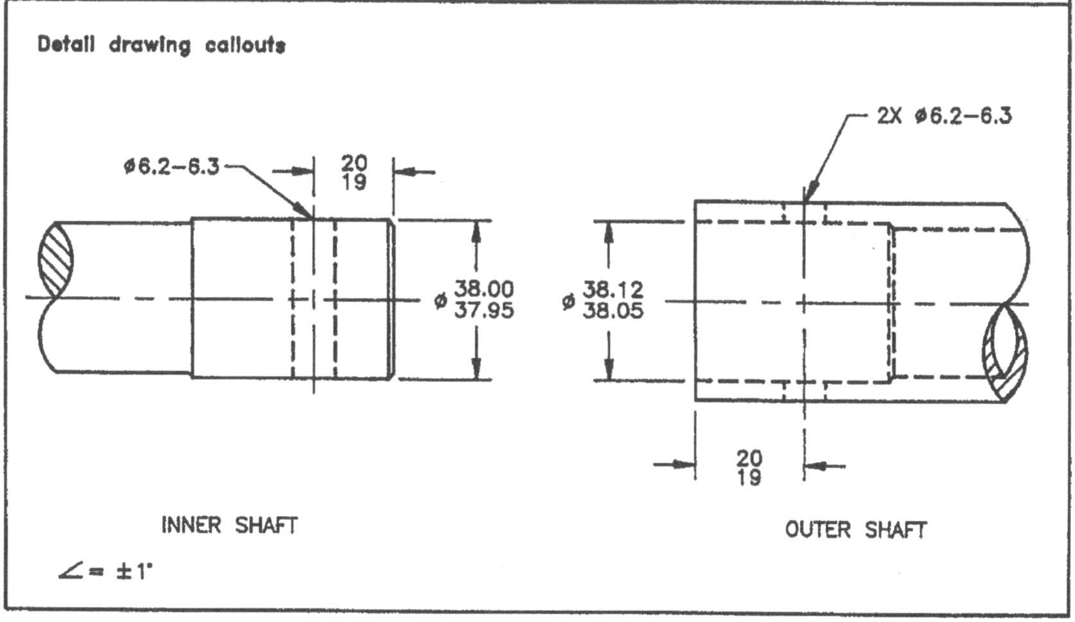

3. Multiple Datum Reference Frame Complete the detail drawings on the next page using the following information:
   1) Give a flatness tolerance to datum feature A of 0.03mm. Give a perpendicularity tolerance of 0.03mm to datum feature C. Give datum feature D a combined control of position to within 1mm at MMC followed by a perpendicularity tolerance requirement to be calculated. The mating shaft has a virtual condition of a diameter of 50mm.
   2) Assign a two-single segment position tolerance for the 8.04-8.07 holes. Use 1mm of position tolerance to three datums in the upper level control. In the lower level control, calculate the position tolerance between the holes and to the primary datum using the fixed fastener formula. These holes use 8mm screws.
   3) Give the three threaded holes a two-single segment position tolerance. In the upper level control, reference three datum features and use a 1mm at MMC position tolerance. In the lower level control, reference two datum features and calculate the position tolerance using the fixed fastener formula. Assume 8mm screws are to be used. Save some of the tolerance for the mating 8.04-8.09 holes. The screws must pass through 25.4mm of material before reaching the threaded holes, so use a projected tolerance zone in both levels of control.

[**NOTE:** The assembly drawing appears below and the detail drawings appear on the next page.]

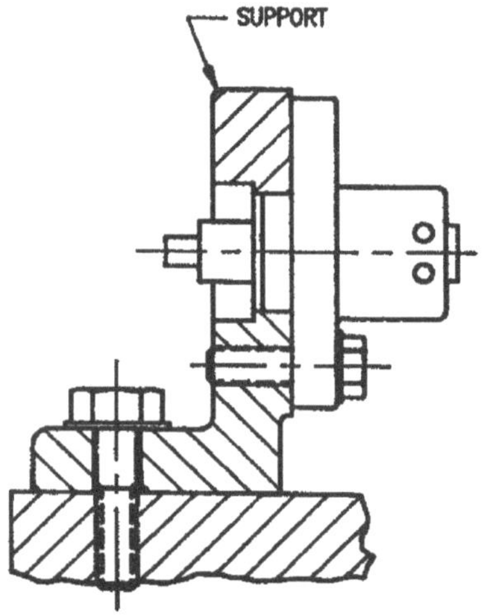

3. (continued)

Detail drawing callout

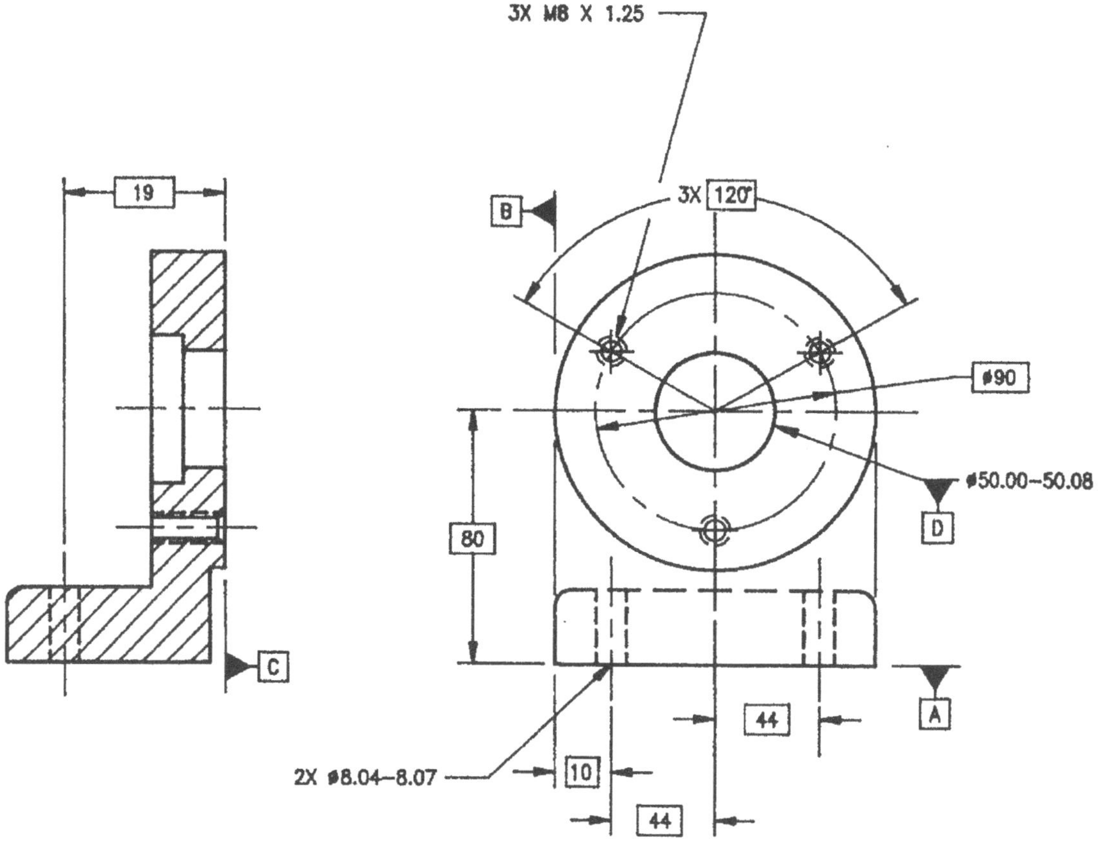

4. Rigid Sheet Metal Part Using the sheet metal drawing on the next page:
   1) Assign a primary datum feature. Control its form to within .010.
   2) Assign a secondary and a tertiary datum feature as two of the six holes. Relate them to each other and the primary datum by giving the secondary datum feature a perpendicularity control and the tertiary datum feature a position control. [**NOTE:** These six holes bolt the part to six threaded holes which have yet to be toleranced. The assembly will use #10-24 screws (MMC of diameter of .190).]
   3) Locate the four holes that remain to the datum reference frame. Use the MMC concept where possible.
   4) Locate the .06 x 5.22 slot to the datum reference frame. Make it part of the same inspection set-up used for the features in item 3 above. Give it a (MMC concept) virtual condition of .03 x 5.19. Use a local note stating BOUNDARY beneath the position control.
   5) Form, orient, locate and size the profile in the front view to within .02 per surface. Use the datum reference frame already established. Make it a simultaneous requirement with the features in items 3 and 4 above.

4. (continued)

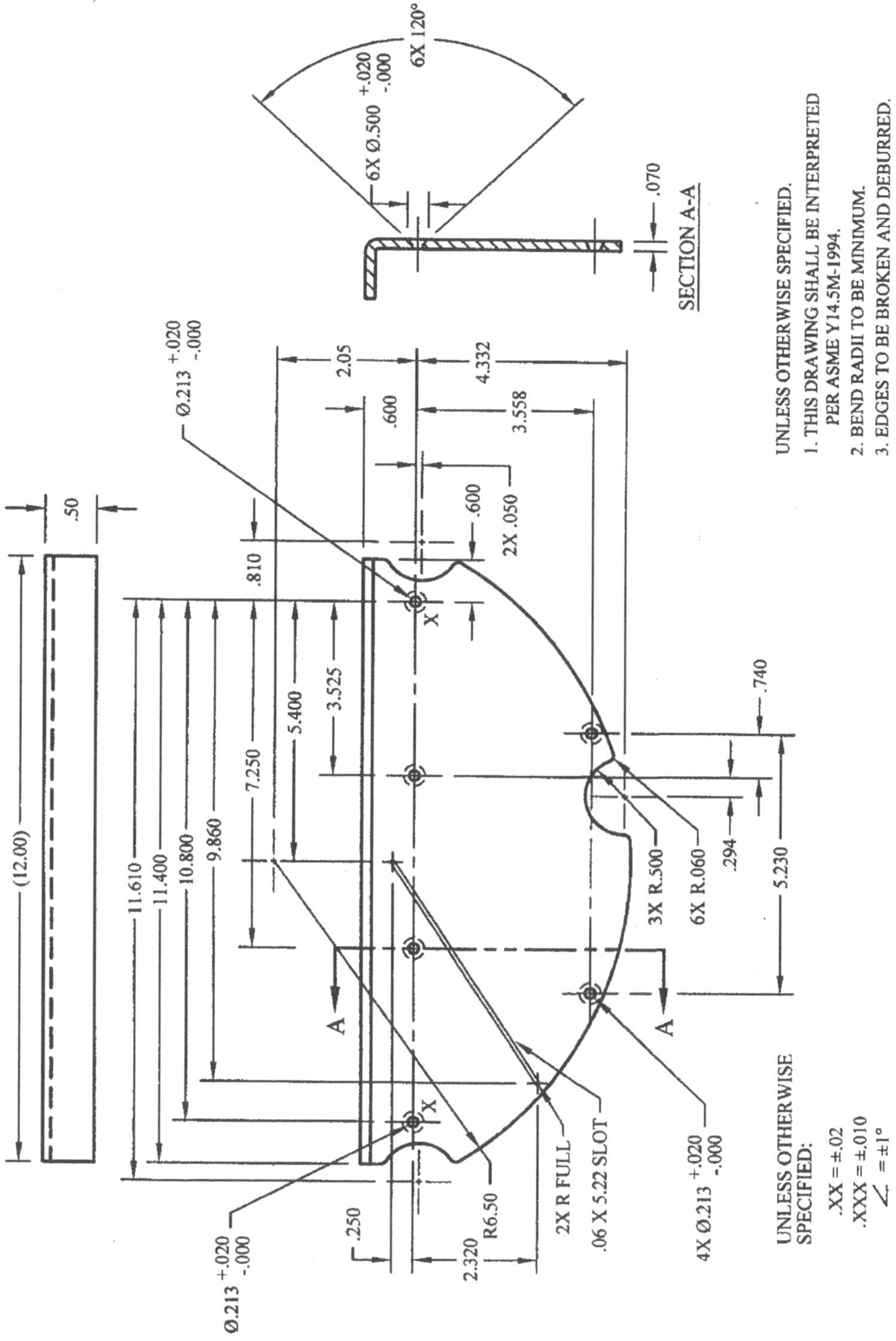
6X 120°
6X Ø.500 +.020 -.000
.070
SECTION A-A
UNLESS OTHERWISE SPECIFIED.
1. THIS DRAWING SHALL BE INTERPRETED PER ASME Y14.5M-1994.
2. BEND RADII TO BE MINIMUM.
3. EDGES TO BE BROKEN AND DEBURRED.
Ø.213 +.020 -.000
2.05
4.332
.600
3.558
.50
.600
2X .050
.810
(12.00)
11.610
11.400
10.800
9.860
7.250
5.400
3.525
.740
.294
5.230
3X R.500
6X R.060
A
A
X
X
.250
2.320
R6.50
2X R FULL
.06 X 5.22 SLOT
4X Ø.213 +.020 -.000
Ø.213 +.020 -.000
UNLESS OTHERWISE SPECIFIED:
.XX = ±.02
.XXX = ±.010
∠ = ±1°

5. Converting from Plus and Minus to Composite Position Tolerancing: Converting one part from plus and minus tolerancing to composite position tolerancing, then redesigning its mating part(s) around its virtual condition boundaries, is sometimes desirable. However, it is recommended parts be designed initially using the appropriate geometric controls. Convert this drawing from plus and minus to composite positional tolerancing. Consider the following information:
   1) Assign the part three datum features.
   2) To calculate the pattern locating tolerance to all three datums, circumscribe a circle around the existing square tolerance zone using the tolerances on both 1.000 inch dimensions.
   3) To calculate the feature-relating tolerance (from hole to hole), consider this part as one piece in a floating fastener assembly, using a screw that has a diameter of .212 MMC. Reference only the primary datum in this control.

[**NOTE:** Don't forget to convert all location dimensions to basic dimensions.]

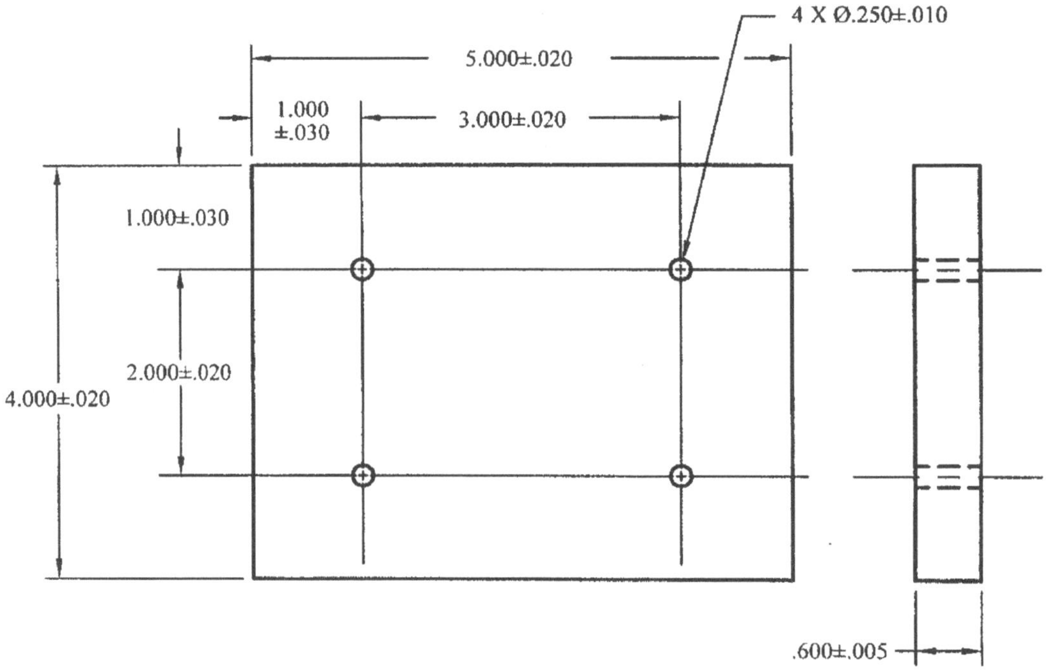

6. Convert this part from plus and minus tolerancing to composite position tolerancing.
   1) Assign the part three datum features.
   2) To calculate the pattern locating tolerance to all three datums, circumscribe a circle around the existing square tolerance zone using the tolerances on both 25mm dimensions.

   3) To calculate the feature-relating tolerance (from hole to hole), consider this part as one piece in a floating fastener assembly, using a screw that has a diameter of 10mm MMC. Reference only the primary datum in this control.

[**NOTE:** Don't forget to convert all location dimensions to basic dimensions.]

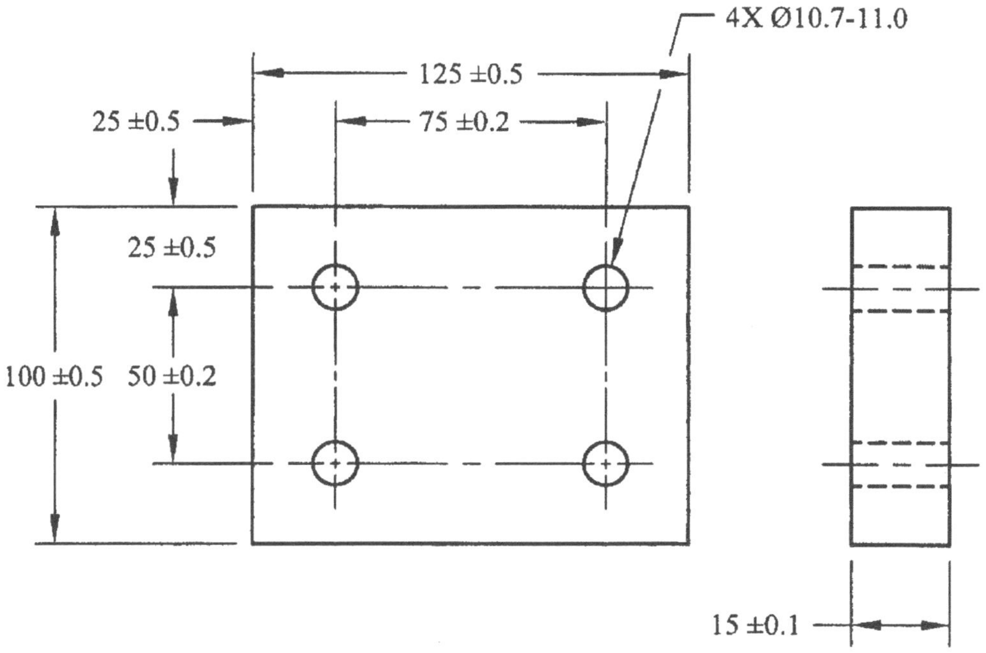

7. The three 8.3-8.5 holes in the drawing on the next page must be held coaxial to within the proper geometric tolerance to receive an 8mm bolt. However, other than that, their placement on the part can be loosely controlled.
   1) Tolerance the part using a composite position control. Use only the datums shown.
   2) To calculate the hole-to-hole coaxiality tolerance, use the floating fastener formula.
   3) To calculate the pattern-locating tolerance, assume a minimum wall thickness at X of 1.1mm. The following formula may be used to calculate the position tolerance (Unknown Position Tol. at MMC listed below):

      LMC of Holes =
      \+ Unknown Position Tol. at MMC =
      <u>+ Bonus Tol. =</u>
      Outer Boundary of Holes

      MMC of Datum Feature B =
      <u>- Outer Boundary of the Holes =</u>
      Material on Both Sides of Hole

      Material on Both Sides of Hole ÷ 2 = Wall Thickness without Pits

      Wall Thickness without Pits =
      <u>- Size/Flatness Tolerance on Datum Feature B (Pits) =</u>
      Minimum Wall Thickness = 1.1
   4) Could this design be changed (while retaining the same virtual conditions) so as to allow inspection to accept more parts that would mate at assembly? If so, make the change. **Hint:** Use a zero at MMC tolerance to hold coaxiality of the three holes to each other.

7. (continued)

Assembly

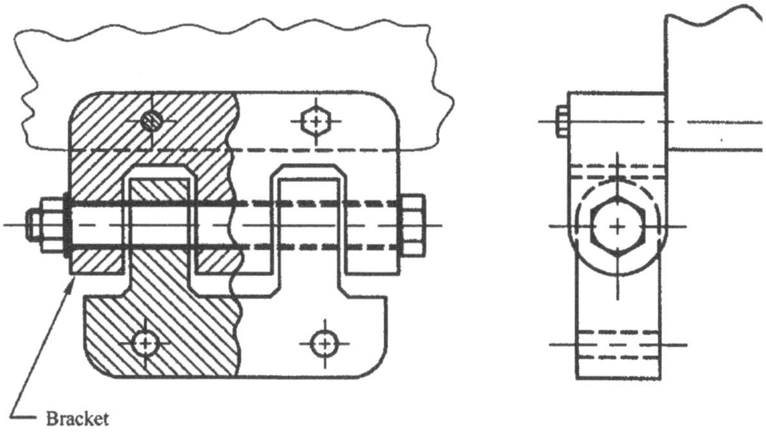

Detail Drawing Callout

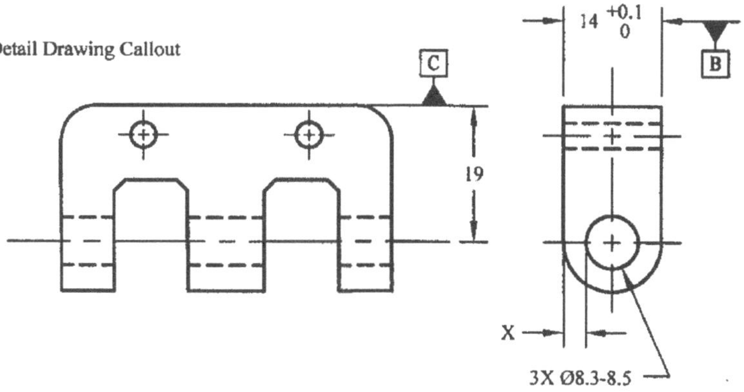

INDEX